QUALITY PLANNING AND ANALYSIS

McGraw-Hill Series in Industrial Engineering and Management Science

CONSULTING EDITORS

Kenneth E. Case, *Department of Industrial Engineering and Management, Oklahoma State University*
Philip M. Wolfe, *Department of Industrial and Management Systems Engineering, Arizona State University*

Barnes
Statistical Analysis for Engineers and Scientists: A Computer-Based Approach

Bedworth, Henderson, and Wolfe
Computer-Integrated Design and Manufacturing

Blank and Tarquin
Engineering Economy

Ebeling
Reliability and Maintainability Engineering

Grant and Leavenworth
Statistical Quality Control

Gryna
Quality Planning and Analysis: From Product Development through Use

Harrell, Ghosh, and Bowden
Simulation Using PROMODEL

Hillier and Lieberman
Introduction to Operations Research

Kelton, Sadowski, and Sadowski
Simulation with ARENA

Khalil
Management of Technology

Kolarik
Creating Quality: Concepts, Systems, Strategies, and Tools
Creating Quality: Process Design for Results

Law and Kelton
Simulation Modeling and Analysis

Nash and Sofer
Linear and Nonlinear Programming

Nelson
Stochastic Modeling: Analysis and Simulation

Niebel and Frievalds
Methods, Standards, and Work Design

Pegden
Introduction to Simulation Using SIMAN

Riggs, Bedworth, and Randhawa
Engineering Economics

Sipper and Bulfin
Production: Planning, Control, and Integration

Steiner
Engineering Economics Principles

QUALITY PLANNING AND ANALYSIS

From Product Development through Use

FOURTH EDITION

Frank M. Gryna

Distinguished Professor of Industrial Engineering Emeritus
Bradley University

Earlier editions of this book were written by
J. M. Juran, Chairman Emeritus of the Juran Institute, Inc., and Frank M. Gryna

Boston Burr Ridge, IL Dubuque, IA Madison, WI New York
San Francisco St. Louis Bangkok Bogotá Caracas Kuala Lumpur
Lisbon London Madrid Mexico City Milan Montreal New Delhi
Santiago Seoul Singapore Sydney Taipei Toronto

McGraw-Hill Higher Education

A Division of The McGraw-Hill Companies

QUALITY PLANNING AND ANALYSIS

Published by McGraw-Hill, an imprint of The McGraw-Hill Companies, Inc. 1221 Avenue of the Americas, New York, NY, 10020.

This book is printed on acid-free paper.

1 2 3 4 5 6 7 8 9 9 FGR/FGR 0 9 8 7 6 5 4 3 2 1 0

ISBN 0-07-039368-0

Publisher: *Thomas Casson*
Executive editor: *Eric Munson*
Developmental editor: *Maja Lorkovic*
Marketing manager: *John Wannemacher*
Project manager: *Christina Thornton-Villagomez*
Production supervisor: *Rose Hepburn*
Coordinator freelance design: *Gino Cieslik*
Supplement coordinator: *Mark Sienicki*
New media: *Phillip Meek*
Cover design: *Gino Cieslik*
Cover image: ©PhotoDisc
Compositor: *Lachina Publishing Services*
Typeface: *10.5/12 Times Roman*
Printer: *Quebecor Printing Book Group/Fairfield*

Library of Congress Cataloging-in-Publication Data

Gryna, Frank M.
Quality planning and analysis / Frank M. Gryna.—4th ed.
p. cm.
Earlier ed. by J.M. Juran.
ISBN 0-07-039368-0 (alk. paper)
1. Quality control. I. Juran, J. M. (Joseph M.), 1904–. Quality planning and analysis. II. Title.

TS156.G78 2001
658.5'62—dc21

00-056621

www.mhhe.com

This book is dedicated to J. M. Juran:

. . . a pioneer

. . . a practitioner

. . . a professional

who discovers and learns for the benefit of society.

ABOUT THE AUTHOR

FRANK M. GRYNA has degrees in industrial engineering and more than 50 years' experience in the managerial, technological, and statistical aspects of quality activities.

From 1991 to 1999 he served first as director of the Center for Quality and then as Distinguished University Professor of Management at the University of Tampa. From 1982 to 1991 he was with the Juran Institute as senior vice president. Prior to 1982, Dr. Gryna was based at Bradley University, where he taught industrial engineering and served as acting dean of the College of Engineering and Technology. He is now Distinguished Professor of Industrial Engineering Emeritus. In addition, he has been a consultant for many companies on all aspects of quality and reliability programs from initial design through field use.

Dr. Gryna also served in the U.S. Army Signal Corps Engineering Labs and the Esso Research and Engineering Company. At the Space Systems Division of the Martin Company, he was manager of reliability and quality assurance.

He coauthored *Quality Planning and Analysis* with J. M. Juran and was associate editor of the second, third, and fourth editions of *Juran's Quality Handbook*. His research project, *Quality Circles,* received the Book of the Year Award sponsored by various publishers and the Institute of Industrial Engineers. He has received recognitions as a Fellow of the American Society for Quality, a Fellow of the Institute of Industrial Engineers, a Certified Quality Engineer, a Certified Reliability Engineer, and a Professional Engineer (Quality Engineering). He has also received various awards, including the E. L. Grant Award of the American Society for Quality, Engineer of the Year Award of the Peoria Engineering Council, teaching and professional excellence awards, and the Award of Excellence of the Quality Control and Reliability Engineering Division of the Institute of Industrial Engineers. Dr. Gryna is also the recipient of the Ott Foundation Award, presented by the Metropolitan Section of the American Society for Quality.

He currently does research and writing in quality management.

PREFACE

This is a textbook about achieving customer satisfaction and loyalty. Not just about meeting specifications, not just about statistical process control, but a textbook about—to use an old-fashioned word—quality.

Meeting the quality needs of society requires an active role by all the major activities of an organization. Market research must discover the quality needs of the users; product development must create designs responsive to these needs; manufacturing and operations planning must devise processes capable of executing the product designs; production and operations must regulate these processes to achieve the desired qualities; purchasing must obtain adequate materials; inspection and test must prove the adequacy of the product through simulated use; marketing must sell the product for the proper application; and customer service must observe the usage, remedy failures, and report on opportunities for improvement. In addition, the administrative and support activities must meet the needs of their customers, both internal and external.

The quality subactivities present in these major activities are collectively a major activity that has become known as the "quality function." It may be defined as that collection of activities, no matter where performed, through which the company achieves customer satisfaction and loyalty. The outlines of this quality function have been emerging clearly. It has now become evident that, in common with other major company functions, successful conduct of the quality function demands much specialized knowledge and many specialized tools, as well as trained specialists to use these tools and apply this knowledge. This book rejects the concept that the control of quality is primarily a matter of statistical techniques. Instead, it develops the viewpoint that product and service quality requires managerial, technological, statistical, and behavioral concepts throughout all the major functions of an organization. This action will seem like strong medicine; it is.

Since the publication of the third edition in 1993, we have seen the emergence of new concepts. The new fourth edition explains concepts such as six sigma, customer satisfaction versus customer loyalty, the balanced scorecard, quality function deployment, process management, project management, the learning organization, supply chain management, electronic commerce, outsourcing, customer relationship management, and preferred practices in developing a quality information system. Also, this edition provides additional examples from the service industries.

With material on a broad subject such as product quality, topics can be presented in several sequences. The sequence selected for this book takes the viewpoint that the practitioner is often faced with the urgent task of solving the quality problem of current products. An equally important task is proper planning for quality of future products.

This latter task must sometimes wait until the most urgent problems are solved. The foregoing priorities have guided the sequence of topics in this book.

A brief Chapter 1 has important definitions and concepts, including distinctions between manufacturing and service industries; Chapter 2 presents a four-step assessment process to help an organization determine where it stands on quality today; Chapters 3–5 explain some structure in the form of three quality processes; Chapter 6 covers process management; Chapters 7–9 cover strategy, organization, and culture. The remainder of the book presents the elements of a broad process covering the collection of activities through which we achieve customer satisfaction and loyalty. For many of these activities, the operative phrase is *prevention of quality problems*.

Throughout the book, chapters on statistical concepts have been included, as needed, to supplement the managerial concepts. There is no attempt to provide a state of advanced knowledge in statistical methodology. Many excellent books cover the statistical aspects in greater depth.

Readers may wish to rearrange the sequence of chapters to meet their specific needs.

All chapters include real-world problems. Such problems require the student to face the realities that confront managers, designers, engineers, marketers, operations personnel, users, and others involved in the quality function. Students must make assumptions, estimate economics, reach conclusions from incomplete facts, and otherwise adapt themselves to the imperfect world of the practitioner. An instructor's manual provides solutions, additional questions, case examples, and enlargements of selected figures for preparing transparencies.

Appendix I contains study questions from the brochures describing five certification examinations of the American Society for Quality.

We also draw attention to the relationship between *Quality Planning and Analysis* and *Juran's Quality Handbook,* fifth edition (J. M. Juran and A. B. Godfrey, coeditors, McGraw-Hill, New York, 1999). The handbook is a reference compendium that, through broad sale in the English language plus translation into other languages, has become the standard international reference work on the subject. This edition of *Quality Planning and Analysis* includes frequent references to the handbook (as well as to other sources).

Many organizations have kindly given us permission to reprint material from their publications. We particularly note the helping hands of the American Society for Quality, the Juran Institute, Inc., and McGraw-Hill, Inc. We are grateful to the literary executor of the late Sir Ronald A. Fisher, F.R.S.; to Dr. Frank Yates, F.R.S.; and to Longman Group Ltd., London, for permission to reprint Table III from their book *Statistical Tables for Biological, Agricultural and Medical Research* (4th ed., 1974).

Leonard A. Seder deserves special mention for his contribution of a case problem on quality costs. In addition, his many contributions to quality methodology have been reflected in this book.

The content of the fourth edition draws upon the concepts of previous editions. Many of these concepts were originally proposed by J. M. Juran as part of his pioneering work in the field of quality management. Of course, I accept full responsibility for my application of those concepts and for the additions of new ideas. But we all owe a debt of gratitude to Joe Juran for his crystal clear thinking in developing concepts for the benefit of society.

ACKNOWLEDGMENTS

Many people deserve to be recognized. I particularly want to thank the reviewers of the manuscript. Their efforts—a labor of love of quality under tight deadlines—provided me with extremely useful comments. These reviewers are Charles Aubrey of American Express; Lawrence Aft of Southern Polytechnic Institute; Roger Berger of Iowa State University; John Early of Empire Blue Cross, Blue Shield; James Freeland of the University of Virginia; Sheila Lawrence of Rutgers University; John Ramberg of the University of Arizona; Thomas Robinson of Gateway Consulting Group and California State University at Long Beach; Diane Schaub of the University of Florida; and Joseph Tsiakals of Amgen Corp.

A particular thank you is in order to Al Endres of the University of Tampa. His review of the manuscript and our many discussions have contributed greatly to this edition.

Over the years, my colleagues at Juran Institute, Bradley University, the University of Tampa, and in the business world were a continuing source of ideas, many of which are in this book. Among the practitioners who provided examples were Roger Eggena, Maureen Esposito, Vonda Hargrove, Steve Maple, Kevin Shuker, and Joseph Tsiakals.

As in the third edition, my son, Derek S. Gryna, provided input on many topics. As a former quality practitioner now functioning in the operations front line of the financial services industry, he kept me focused on the real world.

Where would authors be without the help of someone on the personal computer? That burden was carried with perseverance, competence, and cheer by my wife, Dee.

Any author knows how creating a book interferes with family life. In my case the impact was not only on my wife but also on our children Wendy Esslinger and her husband, Perry; Derek Gryna and his wife, Barbara; and Gary Gryna and his wife, Dina. Then the grandchildren: Jason, Sarah, and Elizabeth Esslinger; Wesley, William, and Whitney Gryna; and Emily and Samuel Gryna. Now maybe I can join their parade.

Throughout my career my wife, Dee, has set the standard for patience and understanding. She could write the book on those qualities.

CONTENTS

1

BASIC CONCEPTS

1.1
QUALITY—A LOOK AT HISTORY

Two dramatic examples illustrate how quality has an impact on both sales revenue and costs.

First, an international organization that claimed to have strong customer service refused to accept sales orders for delivery times less than 48 hours—even though competitors honored delivery requests of 24 hours. Imagine the millions of dollars of sales revenue lost each year because this company did not recognize the quality need of its customers.

Second, listen to the president of a specialty casting manufacturing company: "Our scrap and rework costs this year were five times our profit. Because of those costs, we have had to increase our selling price and we subsequently lost market share. Quality is no longer a technical issue; it is a business issue."

Does this company have a marginal quality reputation in the marketplace? No. Customers rate it as having the best quality available. But the old approach of inspection has failed, and the company has embarked on a new approach.

Our forefathers knew—as we know—that quality is important. Metrology, specifications, inspection—all go back many centuries before the Christian era.

Then came the 20th century. The pace quickened with a lengthy procession of "new" activities and ideas launched under a bewildering array of names: quality control, quality planning, continuous quality improvement, defect prevention, statistical process control, reliability engineering, quality cost analysis, zero defects, total quality control, supplier certification, quality circles, quality audit, quality assurance, quality function deployment, Taguchi methods, competitive benchmarking, six sigma. This book discusses all these concepts.

Following World War II, two major forces emerged that have had a profound impact on quality.

The first force was the Japanese revolution in quality. Prior to World War II, many Japanese products were perceived, throughout the world, to be poor in quality. To help sell their products in international markets, the Japanese took some revolutionary steps to improve quality:

1. Upper-level managers personally took charge of leading the revolution.
2. All levels and functions received training in the quality disciplines.
3. Quality improvement projects were undertaken on a continuing basis—at a revolutionary pace.

The Japanese success has been almost legendary.

The second major force to affect quality was the prominence of product quality in the public mind. Several trends converged to highlight this prominence: product liability cases; concern about the environment; some major disasters and near disasters; pressure by consumer organizations; and the awareness of the role of quality in trade, weapons, and other areas of international competition. This emphasis on quality has been further accented by the emergence of national awards such as the Baldrige and European Quality Awards (see Section 2.8, "Assessment of Current Quality Activities").

During the 20th century a significant body of knowledge emerged on achieving superior quality. Many individuals contributed to this knowledge, and five names deserve particular mention: Juran, Deming, Feigenbaum, Crosby, and Ishikawa. This chapter's Supplementary Reading furnishes basic references for each of these experts.

J. M. Juran emphasizes the importance of a balanced approach using managerial, statistical, and technological concepts of quality. He recommends an operational framework of three quality processes, i.e., quality planning, quality control, and quality improvement. The foundation for this book is the Juran approach. This book makes frequent references to Juran's *Quality Handbook,* fifth edition, which is denoted as *JQH5.*

W. Edwards Deming also focused on a broad view of quality, which he initially summarized in 14 points aimed at the management of an organization. These 14 points rest on a system of "profound knowledge" that has four parts: the systems approach, understanding of statistical variation, nature and scope of knowledge, and psychology to understand human behavior.

A. V. Feigenbaum emphasizes the concept of total quality control throughout all functions of an organization. Total quality control really means both planning and control. He urges that a quality system be created to provide technical and managerial procedures to assure customer satisfaction and an economical cost of quality.

Philip Crosby defines quality strictly as "conformance to requirements" and stresses that the only performance standard is zero defects. His activities demonstrated that all levels of employees can be motivated to pursue improvement but that motivation will not succeed unless tools are provided to show people how to improve.

Kaoru Ishikawa showed the Japanese how to integrate the many tools of quality improvement, particularly the simpler tools of analysis and problem solving.

The approaches of these masters have similarities as well as differences—particularly in the relative emphasis on managerial, statistical, technological, and behavioral elements. This book draws upon the contributions of these and other experts.

Juran (1995) provides a comprehensive history of managing for quality for different time spans (ancient, medieval, modern), geographical areas, products, and political systems.

The major forces affecting managing for quality led to a changing set of business conditions.

1.2 QUALITY—THE CHANGING BUSINESS CONDITIONS

The prominence of product quality in the public mind has resulted in quality becoming a cardinal priority for most organizations. The identification of quality as a core concern has evolved through a number of changing business conditions. These include

1. *Competition.* In the past higher quality usually meant higher price. Today customers can obtain high quality and low prices simultaneously. Quality is now a "given"; i.e., customers assume that they will receive adequate quality. In the service sector, deregulation in areas such as airlines and utilities created competition that did not exist before, and quality is a key dimension of that competition.
2. *The customer-focused organization.* The impact of quality as a tool of competition has lead to viewing quality as customer satisfaction and loyalty rather than conformance to specifications. In the United States the growth of the service sector (which now employs more than 70 percent of all workers) with so much customer direct contact has been a driver of customer focus. Also, the concept of "customer" now includes both external and internal customers (see Section 1.3 below).
3. *Higher levels of customer expectation.* Higher expectations, spawned by competition, take many forms. One example is lower variability around a target value on a product characteristic, even though all product meets the specification limits. Another form of higher expectation is improved quality of service both before and after the sale.
4. *Performance improvement.* Quality, cycle time, cost, and profitability have become interdependent. Many organizations now talk of "performance improvement" or "business excellence" rather than quality.
5. *Changes in organization forms.* Most organizations no longer try to be completely self-sufficient with layers of management for various functional activities. What has emerged are concepts like partnering with other organizations, outsourcing of complete functions, process management, and various types of permanent and temporary teams—and all this with fewer layers of management.
6. *Changing workforce.* These changes include a higher level of education for some parts of the workforce, a multilingual workforce, and downsizing.
7. *Information revolution.* The key operating parameters are now labor, material, equipment, capital—and information. The relative ease with which information can

be collected and disseminated throughout an organization now makes it feasible to do planning and control activities that were unthinkable a few decades ago. Automated real-time access to business data is now common. Internets and intranets have become a way of life.

8. *Electronic commerce.* Organizations now use the Internet for many activities—to provide customers with a wealth of information, enable customers to order products, collect information on customer needs and purchasing behavior, customize products based on customer needs, link suppliers, and link dealers. Electronic commerce applies not only to companies and their consumer customers but also to commerce between companies. The impact on functional activities will be pervasive—marketing, purchasing, product development, operations, and customer service. Evolving issues concerning quality include assuring product quality of e-commerce transactions, measuring the quality of information provided to customers, and gathering and analyzing information on customer needs and problems.
9. *Role of a "quality department."* In previous decades many organizations (particularly in the manufacturing sector) had a quality department that performed various roles involving formal evaluation of products and assistance to line departments on planning for quality. The recent emphasis on quality and the enormous training provided within an organization have resulted in transferring some activities from the staff quality department to the line departments. The integration of quality into line departments has decreased the size of (or even eliminated) the quality department. A reality is that in some organizations the integration was wishful thinking, and the reduction of the department was done prematurely—but the process has forced quality departments to review the services they provide to internal and external customers.

Each of these changing business conditions must be thoroughly understood if organizations are to survive in competitive world markets. This book addresses these changing conditions and describes the approaches needed for future quality leadership.

Before examining these approaches, we need to define some terms.

1.3
QUALITY DEFINED

The dictionary has many definitions for the word *quality.* One short definition of *quality* is "customer satisfaction and loyalty." "Fitness for use" is an alternative short definition. Although such a brief definition has a focus, it must be developed further to provide a basis for action.

Unfolding the meaning starts with defining the word *customer.* A *customer* is "anyone who is affected by the product or process."

1. *External customers* include not only ultimate users (current and potential) but also intermediate processors, as well as merchants. Other customers are not purchasers but have some connection to the product, e.g., government regulatory bodies, share-

holders, suppliers, partners, investors, the media, the general public. External customers clearly are of primary importance.

2. *Internal customers* include not only other divisions of a company that are provided with components for an assembly but also departments or persons that supply products to each other. Thus when a purchasing department receives a specification from an engineering department for a procurement, purchasing is an internal customer of engineering; when the procurement is provided, then engineering is the internal customer of purchasing. Similarly, at a bank the payroll department and operations department are internal customers of each other.

These external and internal customers are sometimes called "stakeholders."

A product is the output of any process. Three categories can be identified:

1. *Goods:* e.g., automobiles, circuit boards, reagent chemicals.
2. *Software:* e.g., a computer program, a report, an instruction.
3. *Service:* e.g., banking, insurance, transportation. Service also includes support activities within companies, e.g., employee benefits, plant maintenance, secretarial support.

Throughout this book *product* means "goods, software, or services."

Customer satisfaction and loyalty are achieved through two components: product features and freedom from deficiencies. Examples of the main categories of these components are shown in Table 1.1 for manufacturing and service industries. We see

TABLE 1.1
Two components of quality

Manufacturing industries	Service industries
Product features	
Performance	Accuracy
Reliability	Timeliness
Durability	Completeness
Ease of use	Friendliness and courtesy
Serviceability	Anticipating customer needs
Esthetics	Knowledge of server
Availability of options and expandability	Appearance of facilities and personnel
Reputation	Reputation
Freedom from deficiencies	
Product free of defects and errors at delivery, during use, and during servicing	Service free of errors during original and future service transactions
All processes free of rework loops, redundancy, and other waste	All processes free of rework loops, redundancy, and other waste

dramatic difference within manufacturing industries (assembly versus chemicals) and within services (restaurants versus banking). Each organization must identify the dimensions of quality that are important to its customers.

A closer examination of the two components reveals further insights:

1. *Product features* have a major effect on *sales income* (through market share, premium prices, etc.). In many industries the total external customer population can be segmented by the level or "grade" of quality desired. Thus the spectrum of customers leads to a demand for luxury hotels and budget hotels; to a demand for refrigerators with many special features as well as for those with basic cooling capabilities. Product features refer to the *quality of design*. Increasing the quality of the design generally leads to higher costs.
2. *Freedom from deficiencies* has a major effect on *costs* through reductions in scrap, rework, complaints, and other results of deficiencies. "Deficiencies" are stated in different units, e.g., errors, defects, failures, off-specification. Freedom from deficiencies refers to *quality of conformance*. Increasing the quality of conformance usually results in lower costs. In addition, higher conformance means fewer complaints and therefore decreased customer dissatisfaction.

How product features and freedom from deficiencies interrelate and lead to higher profits is shown in Figure 1.1.

To summarize, quality means external and internal customer satisfaction. Product features and freedom from deficiencies are the main determinants of satisfaction. For example, an external customer of an automobile desires certain performance features along with a record of few defects and breakdowns. The manufacturing department, as an internal customer of the product development department, wants an engineering specification that is producible in the shop and is free of errors or omissions. Both of these customers want "the right product right."

Finally, we note that quality experts offer different shorthand definitions of quality—"fitness for use" (Juran), "conformance to specifications" (Crosby), "loss to society" (Taguchi), "predictable degree of uniformity" (Deming). These definitions are

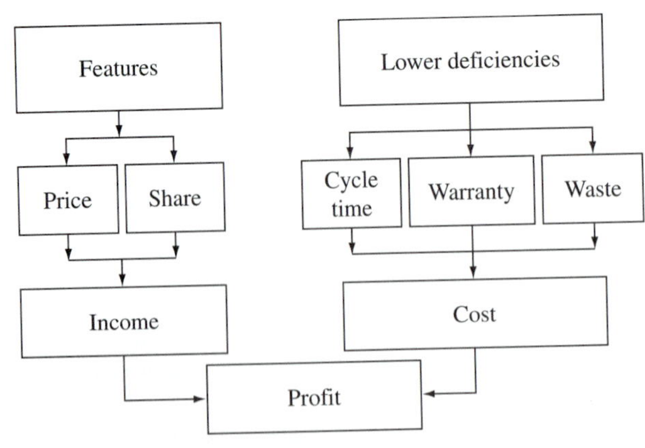

FIGURE 1.1
Quality, market share, and return on investment. (*From Juran Institute, Inc., 1990.*)

complementary and provide operational meaning at different phases of quality activities. For the record the International Organization for Standardization (ISO) defines quality as the "totality of characteristics of an entity that bear on its ability to satisfy stated and implied needs" (ANSI/ISO/ASQ 8402-1994).

1.4 THE QUALITY FUNCTION

Attainment of quality requires the performance of a wide variety of identifiable activities or quality tasks. Obvious examples are the study of customers' quality needs, design review, product tests, and field-complaint analysis. In a tiny enterprise all these tasks (sometimes called work elements) may be performed by a few persons. As the enterprise grows, however, specific tasks may become so time-consuming that we must create specialized departments to perform them. Corporations have created departments such as product design, operations, and customer service, which are essential to launching any new or changed product. These functions follow a relatively unvarying sequence of events (see the spiral in Figure 1.2). In addition to the main "line" activities in the spiral, we need many administrative and support activities such as finance, human resources, and information technology.

The spiral shows that many activities and tasks must be performed to attain customer satisfaction and loyalty. Some of these are performed within manufacturing or service companies. Others are performed elsewhere—by suppliers, merchants, regulators, etc. A convenient shorthand name for this collection of activities is "quality function." The quality function is the entire collection of activities through which we achieve customer satisfaction and loyalty, no matter where these activities are performed.

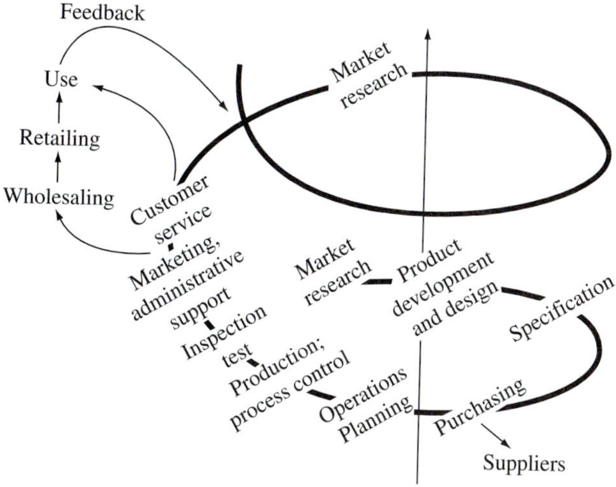

FIGURE 1.2
Spiral of progress in quality.

TABLE 1.2
Little Q and big Q

Topic	Content of little Q	Content of big Q
Products	Manufactured goods	All products, goods, and services, whether for sale or not
Processes	Processes directly related to manufacture of goods	All processes; manufacturing support; business, etc.
Industries	Manufacturing	All industries; manufacturing; service, government, etc., whether for profit or not

Some practitioners look upon the spiral or the quality function as a system, i.e., a network of activities or subsystems. Some of these subsystems correspond to segments of the spiral. Others, although not shown on the spiral, are nevertheless present and active, e.g., data processing, standardization. These subsystems, when well designed and coordinated, become a unified system that carries out the intended quality objectives.

The traditional scope of quality activities is undergoing a radical and exciting change from the historical emphasis on quality of physical products in manufacturing industries ("little Q") to what is now emerging as the application of quality concepts to all products, all functional activities, and all industries ("big Q"). Table 1.2 summarizes this change in scope.

Thus quality concepts are now being applied to order entry, inventory management, human resources, and new product development. Two books have even been written that apply quality concepts to personal activities—both on and off the job (see Roberts and Sergesketter, 1993, and Forsha, 1992).

Under this enlarged concept, all jobs encompass three roles for the jobholder: the customer who receives inputs of information and physical goods, the processor who converts these inputs into products (outputs), and the supplier who delivers the resulting products to customers. This concept is called the triple-role concept. Thus a product development department has three roles: as an internal customer of the marketing department, the product development department receives information on customer needs; as a processor, product development creates designs for new products; and as a supplier, it furnishes specifications to the operations department to create the new products. Figure 1.3 shows how the Paradyne organization identifies the actions to be taken in each role to pursue continuous improvement.

1.5 RELATIONSHIPS: QUALITY, PRODUCTIVITY, COSTS, CYCLE TIME, AND VALUE

Does an emphasis on quality have a positive or negative impact on productivity, costs, cycle time, and value? The easy answer is "positive," but the reality is that, although

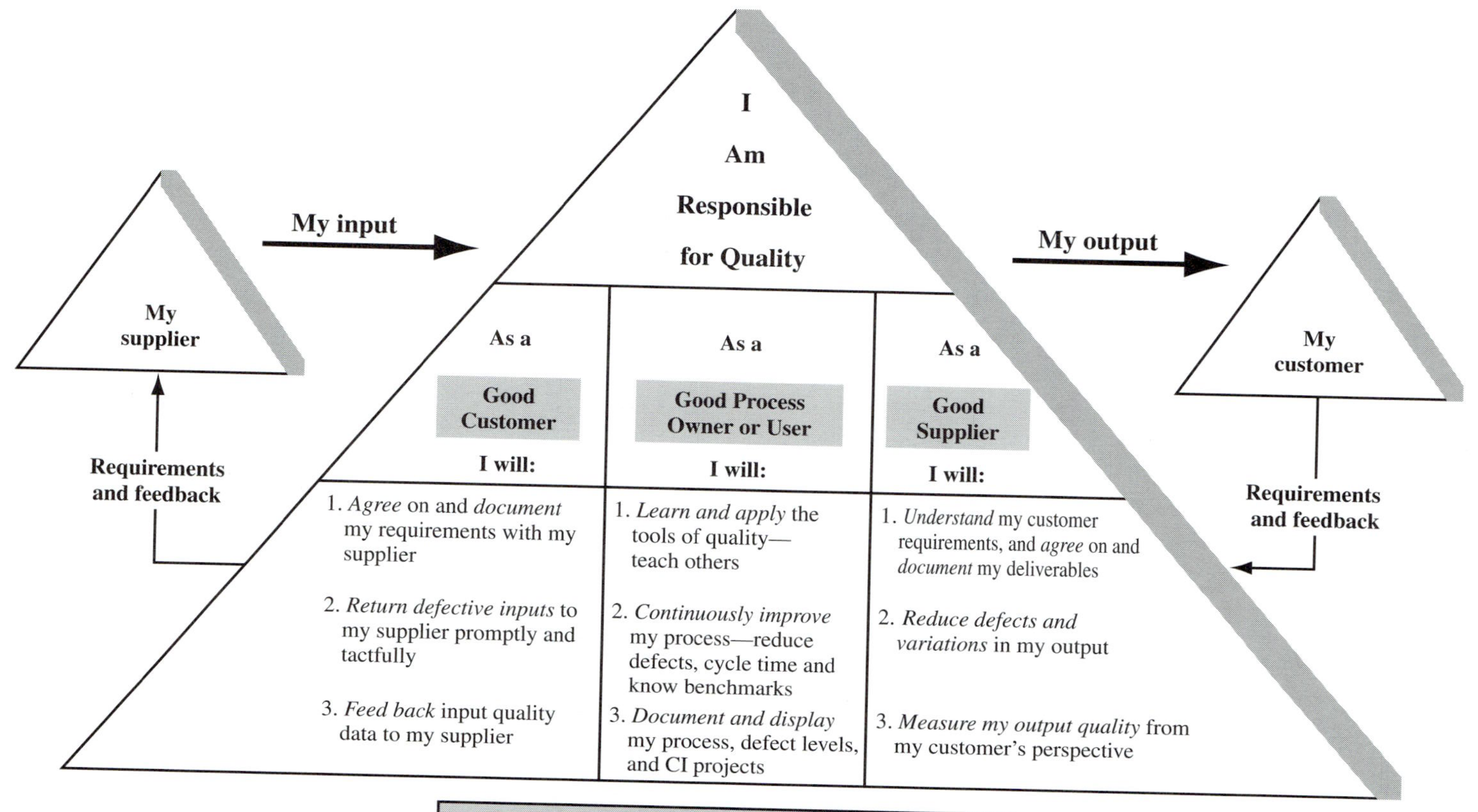

FIGURE 1.3

Triple-role concept. Continuous improvement practices. (*From Paradyne Corporation Continuous Improvement Leadership Team.*)

these parameters should be mutually compatible, they may be either supportive or opposing forces.

Quality and Productivity

Productivity is the ratio of salable output divided by the resources used. The resources include labor, raw materials, and capital. Any one of these (or the total) can be the denominator in the productivity ratio. McCracken and Kaynak (1996) discuss alternative definitions of productivity and their relationships to quality (a key conclusion: as quality increases, productivity increases).

A common measure of productivity is labor productivity, e.g., number of salable units per hour of direct labor. When quality is improved by identifying and eliminating the causes of errors and rework, more usable output is available for the same amount of labor input. Thus the improvement in quality directly results in an increase in productivity. Productivity is now a key issue in national economic policy decisions, and thus the definition and measurement of productivity is important to any country. For a discussion of five links between national economics and product quality, see Brust and Gryna (1997).

Quality and Costs

As the quality of design (product features) increases, costs typically increase. As the quality of conformance increases, the reduction in rework, complaints, scrap, and other deficiencies results in a significant decrease in costs. An ideal strategy calls for using the savings from reduced deficiencies to pay for any increase in product features without increasing the selling price, thus resulting in higher customer satisfaction and increased sales revenue. Many exemplary firms are quietly following this strategy.

Quality and Cycle Time

In both the manufacturing and service sectors, the cycle time to complete the activities required for customers is now a key parameter. In service industries customers view cycle time to provide a transaction as a quality parameter. Customers simply demand fast response time. Thus when a quality improvement effort reduces rework, redundant operations, and other deficiencies, a simultaneous reduction in cycle time occurs.

Quality and Value

Value is quality divided by price. The reality is that customers do not separate quality from price; they consider both parameters simultaneously. Improvements in quality that can be provided to customers without an increase in price result in better "value." As customers compare sources of supply, organizations must evaluate the value they provide relative to the competition. Advertising has trivialized the word *value,* but the concept itself is sound because value is what customers demand.

Quality, productivity, costs, cycle time, and value are interrelated. Quality activities must try to detect quality problems early enough to permit action to be taken without requiring a compromise in cost, schedule, or quality. The emphasis must be on prevention rather than on just correction of quality problems. Improving quality can be the driving force to improve results in the other parameters. In an end-of-chapter problem, readers are asked to calculate the benefits that a quality improvement program contributes to cost savings, delivery schedules, and productivity.

Companies that have high product quality typically achieve superior financial performance. For example, recipients of the Baldrige Quality Award (see Section 2.9) outperformed the Standard and Poor's 500 stock index by 2.5 to 1.0 (Hertz, 1998). Hardie (1998) summarizes the findings of 43 studies on quality and business performance.

1.6 MANAGING FOR QUALITY

Quality management is the process of identifying and administering the activities needed to achieve the quality objectives of an organization. A useful way to introduce the concept of quality management to upper and middle management is to relate it to another well-known management concept, i.e., financial management.

Financial management is accomplished by the use of three managerial processes: planning, control, and improvement. Some key elements of these three processes are shown in Table 1.3. The same three processes apply to quality. The three financial processes provide a methodical approach to addressing finance; the three quality processes provide a methodical approach to addressing quality. Of particular importance is that all three quality processes can be further defined in a universal sequence of activities. Table 1.4 summarizes these sequences. Later chapters develop further detail for these "road maps."

The three processes of Juran's quality trilogy are interrelated. Figure 1.4 shows the interrelationship as applied to one of the two components of the quality definition, freedom from deficiencies. This figure is of uncommon importance. For example, note the graphic distinction between the noisy *sporadic* quality problem and the muted *chronic* waste. The sporadic problem is detected and acted upon by the process of

TABLE 1.3
Financial processes

Process	Some elements
Financial planning	Budgeting
Financial control	Expense measurement
Financial improvement	Cost reduction

quality control. The chronic problem requires a different process, namely, *quality improvement.* Such chronic problems are traceable to an inadequate *quality planning* process. Later chapters develop these concepts in more detail.

The experience gained from quality improvement activities provides feedback ("lessons learned") for replanning activities.

The quality planning process is really operational quality planning directed at product and process planning. This activity is in contrast to strategic quality planning, which establishes long-term goals and the approach to meet those goals.

For the trilogy of quality processes to be a successful framework for achieving quality objectives, the processes must occur in an environment of inspirational leadership where the practices are strongly supportive of quality. Without such a quality "cul-

TABLE 1.4
Universal processes for managing quality

Quality planning	Quality control	Quality improvement
Establish the project	Choose control subjects	Prove the need
Identify customers	Establish measurement	Identify projects
Discover customer needs	Establish standards of performance	Organize project teams
Develop product	Measure actual performance	Diagnose the causes
Develop process	Compare to standards	Provide remedies, prove that the remedies are effective
Develop process controls, transfer to operations	Take action on the difference	Deal with resistance to change
		Control to hold the gains

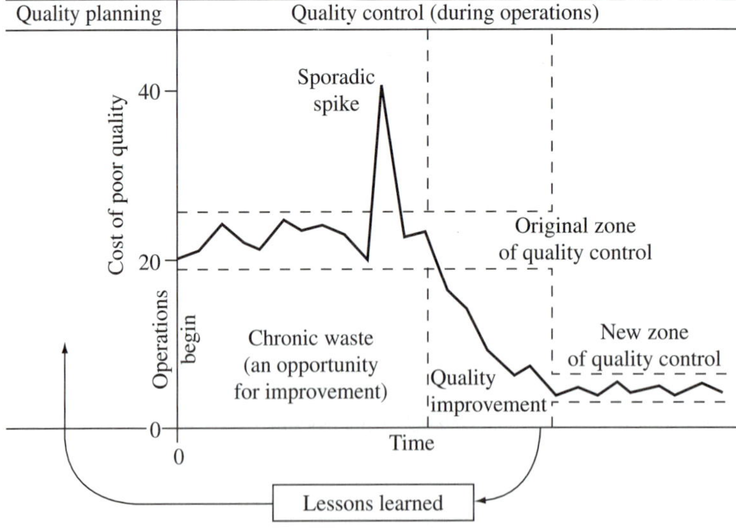

FIGURE 1.4
The Juran Trilogy diagram. (*From Juran, 1986.*)

ture," the trilogy of quality processes cannot be fully effective. These elements have an impact on people at all levels. Again, later chapters provide additional explanation.

1.7 QUALITY DISCIPLINES AND OTHER DISCIPLINES

This book discusses the concepts, techniques, and tools essential to modern competition in quality. *Quality disciplines* is the term used to denote the body of quality-related knowledge. Originally this knowledge was applied to the quality of manufacturing processes for physical goods (little Q). As the scope of quality activities has expanded to all processes—and to both external and internal customers—knowledge from other disciplines has become useful. The contributions of other disciplines are sometimes unique and sometimes overlap with quality disciplines. A summary of the contributions of other disciplines is shown in Table 1.5. People concerned with quality management must draw upon the contributions of all disciplines. Potential applications are identified throughout the book.

1.8 PERSPECTIVE ON QUALITY—INTERNAL VERSUS EXTERNAL

As a preview of the rest of this book, Table 1.6 shows two views of quality: the conventional internal view still followed by many organizations and the modern external view that many other organizations have found to be imperative for survival. Stay tuned for some obvious and some more subtle aspects of the entries of this table.

TABLE 1.5
Contributions of various disciplines

Discipline	Example of contribution
Finance	Measuring the cost of poor quality
Industrial engineering	Design of integrated systems, measurement, problem solving, work analysis
Information technology	Measurement, analysis, and reporting on quality
Marketing research	Competitive standing on quality; understand customer desires
Operations management	Management of integrated systems
Operations research	Analyzing product design alternatives for optimization
Organizational behavior	Understanding quality culture; making teams effective
Organizational effectiveness	Satisfying the needs of both internal and external customers
Strategic planning	Quality as a means of achieving a unique competitive advantage
Systems engineering	Translating customer needs into product features and process features
Value engineering	Analysis of essential functions needed by customer

TABLE 1.6
Two views of quality

Internal view	Customer-focused view
Compare product to specification	Compare product to competition and to the best
Get product accepted at inspection	Provide satisfaction over product life
Prevent plant and field defects	Meet customer needs on goods and services
Concentrate on manufacturing	Cover all functions
Use internal quality measures	Use customer-based quality measures
View quality as a technical issue	View quality as a business issue
Efforts coordinated by quality manager	Efforts directed by upper management

Source: Juran Institute, Inc. (1990).

The recent emphasis on customer satisfaction and loyalty, application of quality concepts beyond manufacturing to the service and public sectors, and participation of all employees has given rise to a new title—total quality management (TQM). TQM is the system of managerial, statistical, and technological concepts and techniques to achieve quality objectives throughout an organization. That system will unfold throughout this book.

A new approach to quality with a special name (e.g., TQM) initially commands attention. Inevitably, the enthusiasts for the new approach promise more benefits than are actually delivered, resulting in "quality" not living up to expectations. Further, other approaches emerge as competitors, e.g., reengineering, performance improvement, leadership, six sigma.

Quality-related problems will always be present—products and services don't have the features to meet customer needs; products and services have errors and defects. These quality issues must be addressed.

The reader is urged to learn about all approaches to quality, choose the best ideas, and then customize the framework for your own organization.

SUMMARY

- Quality is customer satisfaction and loyalty.
- Quality has two components: product features and freedom from deficiencies.
- Product features affect sales income.
- Freedom from deficiencies affects costs.
- Attainment of quality requires activities in all functions of an organization.
- Traditional quality activities have concentrated on manufacturing ("little Q"); modern quality activities encompass all activities ("big Q").
- All jobs have three roles: customer, processor, supplier.
- We can identify three quality processes: quality planning, quality control, quality improvement. Each process has a defined list of steps.

- Sporadic and chronic quality problems require different approaches.
- Quality, costs, and schedules can be mutually compatible.
- Quality management draws upon the knowledge of many other disciplines.
- Both internal and external views of quality are essential.
- Total quality management (TQM) is the system of managerial, statistical, and technological concepts and techniques to achieve quality objectives throughout an organization.

PROBLEMS

Note: An instructor's manual is available for instructors using this book as an assigned text.

1.1. For customer satisfaction, computer software must not only be free of errors or "bugs" but also have the necessary product features. List some important features of a word processing software package.

1.2. *Quality* has been defined as "customer satisfaction and loyalty" (or "fitness for use"). *Quality* may also be defined as "conformance to specification." In theory, creating the proper specifications and then manufacturing product that conforms to those specifications should lead to customer satisfaction. Alas, life isn't that simple. Consider these four situations for a product:

- Conforms to specifications; is competitive on fitness.
- Conforms to specifications; is not competitive on fitness.
- Does not conform to specifications; is competitive on fitness.
- Does not conform to specifications; is not competitive on fitness.

Which two of these four situations are theoretically not plausible, but in practice do occur (and cause much confusion and concern)? Can you cite any examples?

1.3. Select one functional department in a manufacturing or service organization, such as the product development department.

(*a*) For the selected department, list two internal customers. For example, the manufacturing department is a likely internal customer of product development.

(*b*) For each internal customer, describe in one sentence a likely requirement of that customer that the supplier department must meet.

(*c*) Propose a measurement that might be used to quantify how well the requirement is being met.

1.4. For each 100 units of product manufactured, a certain process yields 85 conforming units, five that must be scrapped, and 10 that must be reprocessed. Each unit scrapped results in a \$50 loss; each reprocessed unit requires an extra half hour of processing time. The resource time of producing the original 100 units is 20 hours.

(*a*) Calculate the scrap cost, reprocessing time, and productivity per hour. Productivity should be calculated in terms of conforming units per hour of resource input.

(*b*) A quality improvement program has recently been successfully introduced. For each 100 units manufactured, the process now yields 95 conforming units, one to be scrapped, and four for reprocessing. Repeat the calculations of paragraph (*a*). What are the quantitative benefits of the quality effort to costs, to delivery schedules, and to productivity?

1.5. Based on your experience, identify three examples of chronic or sporadic deficiencies that were the result of either poor work effort or poor quality planning.

REFERENCES

Brust, P. J. and F. M. Gryna (1997). *Product Quality and Macroeconomics—Five Links*. Report No. 904, College of Business Research Paper Series, University of Tampa, Tampa, FL.

Forsha, H. I. (1992). *The Pursuit of Quality through Change,* Quality Press, American Society for Quality, Milwaukee.

Hardie, N. (1998). "The Effects of Quality on Business Performance," *Quality Management Journal*, vol. 5, no. 3, pp. 65–83.

Hertz, H. (1998). "Lessons Learned from the First Decade of the Baldrige Award Program," Florida Sterling Quality Conference, Orlando, FL.

Juran, J. M., ed. (1995). *A History of Managing for Quality,* Quality Press, ASQ, Milwaukee.

McCracken, M. J. and H. Kaynak (1996). "An Empirical Investigation of the Relationship between Quality and Productivity," *Quality Management Journal,* vol. 3, no. 2, pp. 36–51.

Roberts, H. V. and B. F. Sergesketter (1993). *Quality Is Personal,* Free Press, New York.

SUPPLEMENTARY READING

Approaches of various experts: Lowe, T. A. and J. M. Mazzeo (1988). "Lessons Learned from the Masters—Experiences in Applying the Principles of Deming, Juran, and Crosby," *ASQC Quality Congress Transactions,* Milwaukee, pp. 397–402.

Juran, J. M. and A. B. Godfrey, eds. (1999). *Juran's Handbook on Quality,* 5th ed., McGraw-Hill, New York.

Deming, W. E. (1986). *Out of the Crisis,* MIT Center for Advanced Engineering Study, Cambridge, MA.

Feigenbaum, A. V. (1991). *Total Quality Control,* 3rd ed., revised, McGraw-Hill, New York.

Crosby, P. (1979). *Quality Is Free,* McGraw-Hill, New York.

Ishikawa, K. (1985). *What Is Total Quality Control?* Prentice-Hall, Upper Saddle River, NJ.

Future of quality: "21 Voices for the 21st. Century," *Quality Progress,* January 2000, pp. 31–39. "Charting A Course for the New Century," *Quality Progress*, December 1999, pp. 27–47, three papers discussing the future.

Quality and other disciplines: Brickley, J. A., C. W. Smith Jr., and J. L. Zimmerman (1997). "Managerial Economics and Organizational Architecture," in *Total Quality Management and Reengineering,* McGraw-Hill, New York.

Golomski, W. A. (1992). "Social Science Aspects of Quality," *Proceedings of Conference on Quality in the Year 2000,* Rochester Institute of Technology, June 10, Rochester, NY.

Quality and financial performance: Dusseau, S. P. (1996). *An Analysis of the Relationship between Financial Performance and Total Quality Management Implementation,* Ph.D. dissertation, UMI Dissertation Services, Ann Arbor, MI.

Sandholm, L. (2000). *Total Quality Management,* 2nd ed., Student litteratur, Sweden.

WEBSITES

American Society for Quality: www.asq.org
Deming Electronic Network: http://deming.eng.clemson.edu/pub/den/deming_map/htm
Juran Institute: www.juran.com

2

COMPANYWIDE ASSESSMENT OF QUALITY

2.1
WHY ASSESSMENT?

In recent decades the quality movement has been bombarded by a never-ending procession of new techniques with promises of near miracles. Some techniques have been successes; some have been failures. To help develop a quality strategy a formal assessment of quality provides a guide for improvement by telling us (1) the size of the quality issue and (2) the areas demanding attention.

The assessment approach described in this chapter provides the facts that are essential to stimulate upper management to take action on quality (that's why assessment is presented early in this book). This factual information not only inspires immediate action on quality but also provides a solid basis for developing long-range strategies for quality (see Chapter 7, "Strategic Quality Management").

This book uses the term *quality assessment* to describe a companywide review of the status of quality.

Assessment of quality comprises four elements:

1. Cost of poor quality.
2. Standing in the marketplace.
3. Quality culture in the organization.
4. Operation of the company quality system.

These elements constitute an analysis of a firm's quality status in terms of strengths, weaknesses, opportunities, and threats (SWOT) and become a formal part of the SWOT analysis for overall operations of an organization (see Chapter 7 under "Elements of Strategic Quality Management").

Other elements may be added as circumstances require. An annual or biannual assessment is usually warranted. Assessment can be conducted for an entire organization or for a division, a plant, or even a department. For a discussion of the four elements of quality assessment in banking, see D. Gryna and F. Gryna (1999).

We will start this discussion by assessing the cost of poor quality.

2.2 COST OF POOR QUALITY

During the 1950s the concept of "quality costs" emerged. Different people assigned different meanings to the term. Some people equated quality costs with the costs of *attaining* quality; some people equated the term with the extra costs incurred because of poor quality. This book focuses on the cost of poor quality because this component of assessment is important in reducing costs and customer dissatisfaction.

The *cost of poor quality* is the annual monetary loss of products and processes that are not achieving their quality objectives.

Companies estimate the cost of poor quality for several reasons:

1. Quantifying the size of the quality problem in the language of money improves communication between middle managers and upper managers. In some companies the need to improve communication on quality-related matters has become a major objective for embarking on a study of the costs of poor quality. Some managers say, "We don't need to spend time to translate the defects into dollars. We realize that quality is important, and we already know what the major problems are." But typically, when a study is done, these managers are surprised by two results. First, the quality costs turn out to be much higher than the managers had thought—in many industries in excess of 20 percent of sales. Second, while the distribution of the quality costs confirms some of the known problem areas, it also reveals other problem areas that were not previously recognized.
2. Major opportunities for cost reduction can be identified. Costs of poor quality do not exist as a homogeneous mass. Instead, they are the total of specific segments, each traceable to some specific cause. These segments are unequal in size, and relatively few of the segments account for the bulk of the costs. A major by-product of this evaluation is the identification of these vital few segments. We cover this issue in Chapter 3 under "The Pareto Principle."
3. Opportunities for reducing customer dissatisfaction and associated threats to product salability can be identified. Some costs of poor quality are the result of product failures that take place after a sale. In part, these costs are paid by the manufacturer in the form of warranty charges, claims, etc. But whether the costs are paid by the manufacturer or not, the failures add to customers' costs because of downtime and other forms of disturbance. Analysis of the manufacturer's costs, supplemented by marketing research into customers' costs of poor quality, can identify the vital few areas of high costs. These areas then lead to problem identification.
4. Measuring this cost provides a means of evaluating the progress of quality improvement activities and spotlighting obstacles to improvements.

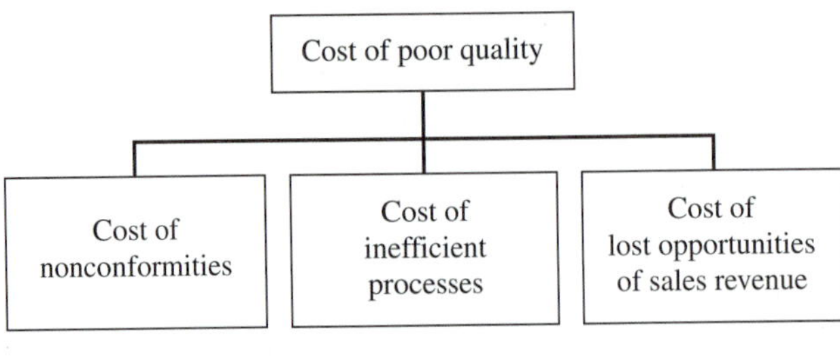

FIGURE 2.1
Cost of poor quality.

5. Knowing the cost of poor quality (and the three other assessment elements) leads to the development of a strategic quality plan that is consistent with overall organization goals.

The main components of the cost of poor quality (which apply to both manufacturing and service organizations) are shown in Figure 2.1. Note that this framework reflects not only the cost of nonconformities (sometimes called "quality costs") but also process inefficiencies and the impact of quality on sales revenue. These three categories are incorporated below in the categories of quality costs. Each organization must decide which cost elements to include in its cost of poor quality.

2.3 CATEGORIES OF QUALITY COSTS

Many organizations summarize the costs associated with quality into four categories: internal failures, external failures, appraisal, and prevention. These categories are discussed below. A useful reference on definitions, categories, and many other aspects is Campanella (1999). For an exhaustive listing of elements within the four categories see Atkinson, Hamburg, and Ittner (1994).

Collectively the four categories are often called the "cost of quality." The cost of poor quality includes the internal and external failure categories, whereas the appraisal and prevention categories are viewed as investments to achieve quality objectives.

Internal Failure Costs

Internal failure costs are the cost of deficiencies discovered before delivery that are associated with the failure to meet explicit requirements or implicit needs of customers. Also included are avoidable process losses and inefficiencies that occur even when requirements and needs are met. These costs would disappear if no deficiencies existed. Internal failure costs consist of (1) cost of failure to meet customer requirements and needs and (2) cost of inefficient processes.

Failure to meet customer requirements and needs

Examples of subcategories are costs associated with the following:

Scrap. The labor, material, and (usually) overhead on defective product that cannot economically be repaired. The titles are numerous—scrap, spoilage, defectives, etc.

Rework. Correcting defectives in physical products or errors in service products.

Lost or missing information. Retrieving information that should have been supplied.

Failure analysis. Analyzing nonconforming goods or services to determine causes.

Scrap and rework—supplier. Scrap and rework because of nonconforming product received from suppliers. This area also includes the costs to the buyer of resolving supplier quality problems.

One hundred percent sorting inspection. Finding defective units in product lots that contain unacceptably high levels of defectives.

Reinspection, retest. Reinspection and retest of products that have undergone rework or other revision.

Changing processes. Modifying manufacturing or service processes to correct deficiencies.

Redesign of hardware. Changing designs of hardware to correct deficiencies.

Redesign of software. Changing designs of software to correct deficiencies.

Scrapping of obsolete product. Disposing of products that have been superseded.

Scrap in support operations. Defective items in indirect operations.

Rework in internal support operations. Correcting defective items in indirect operations.

Downgrading. The difference between the normal selling price and the reduced price because of poor quality.

Cost of inefficient processes

Examples of subcategories are

Variability of product characteristics. Losses that occur even with conforming product (e.g., overfill of packages due to variability of filling and measuring equipment).

Unplanned downtime of equipment. Loss of capacity of equipment due to failures.

Inventory shrinkage. Loss due to the difference between actual and recorded inventory amounts.

Variation of process characteristics from "best practice." Losses due to cycle time and costs of process as compared to best practices in providing the same output. The best-practice process may be internal or external to the organization.

Non-value-added activities. Redundant operations, sorting inspections, and other non-value-added activities. A value-added activity increases the usefulness of a product to the customer; a non-value-added activity does not. The concept is similar to the 1950s idea of value engineering and value analysis.

External Failure Costs

External failure costs are associated with deficiencies that are found after the customer receives the product. Also included are lost opportunities for sales revenue. These costs also disappear if there were no deficiencies.

Failure to meet customer requirements and needs

Examples of subcategories are

Warranty charges. The costs involved in replacing or making repairs to products that are still within the warranty period.

Complaint adjustments. The costs of investigation and adjustment of justified complaints attributable to defective product or installation.

Returned material. The costs associated with receipt and replacement of defective product received from the field.

Allowances. The costs of concessions made to customers due to substandard products accepted by the customer as is or to conforming product that does not meet customer needs.

Penalties due to poor quality. This category applies to goods or services delivered or to internal processes such as late payment of an invoice resulting in a lost discount for paying on time.

Rework on support operations. Correcting errors on billing and other external processes.

Revenue losses in support operations. An example is the failure to collect on receivables from some customers.

Lost opportunities for sales revenue

Examples are

Customer defections. Profit on potential customers lost because of poor quality.

New customers lost because of lack of capability to meet customer needs. Profit on potential revenue lost because of inadequate processes to meet customer needs.

Appraisal Costs

Appraisal costs are incurred to determine the degree of conformance to quality requirements. Examples are

Incoming inspection and test. Determining the quality of purchased product, whether by inspection on receipt, by inspection at the source, or by surveillance.

In-process inspection and test. In-process evaluation of conformance to requirements.

Final inspection and test. Evaluation of conformance to requirements for product acceptance.

Document review. Examination of paperwork to be sent to customer.

Balancing. Examination of various accounts to assure internal consistency.

Product quality audits. Performing quality audits on in-process or finished products.

Maintaining accuracy of test equipment. Keeping measuring instruments and equipment in calibration.

Inspection and test materials and services. Materials and supplies in inspection and test work (e.g., X-ray film) and services (e.g., electric power) where significant.

Evaluation of stocks. Testing products in field storage or in stock to evaluate degradation.

In collecting appraisal costs, the decisive factor is the kind of work done and not the department name (the work may be done by chemists in the laboratory, by sorters in the operations department, by testers in the inspection department, or by an external firm engaged for the purpose of testing). Also note that industries use various terms for "appraisal," e.g., *checking, balancing, reconciliation, review.*

Prevention Costs

Prevention costs are incurred to keep failure and appraisal costs to a minimum. Examples are

Quality planning. This category includes the broad array of activities that collectively create the overall quality plan and the numerous specialized plans. It includes also the preparation of procedures needed to communicate these plans to all concerned.

New-products review. Reliability engineering and other quality-related activities associated with the launching of new design.

Process planning. Process capability studies, inspection planning, and other activities associated with the manufacturing and service processes.

Process control. In-process inspection and test to determine the status of the process (rather than for product acceptance).

Quality audits. Evaluating the execution of activities in the overall quality plan.

Supplier quality evaluation. Evaluating supplier quality activities prior to supplier selection, auditing the activities during the contract, and performing associated effort with suppliers.

Training. Preparing and conducting quality-related training programs. As in the case of appraisal costs, some of this work may be done by personnel who are not on the payroll of the quality department. The decisive criterion is again the type of work, not the name of the department performing the work.

Note that prevention costs are costs of special planning, review, and analysis activities for quality. Prevention costs do not include basic activities such as product design, process design, process maintenance, and customer service.

The compilation of prevention costs is initially important because it highlights the small investment currently made (usually) in prevention activities and suggests the potential for an increase in prevention costs with the aim of reducing failure costs. Upper management immediately grasps this point. Continuing measurement of prevention costs can usually be excluded, however, to focus on the major opportunity, i.e., failure costs.

One of the issues in calculating the cost of poor quality is how to handle overhead costs. Three approaches are common: include total overhead using direct labor or some other base, include variable overhead only (the usual approach), or do not include overhead at all. The allocation of overhead has an impact on the total cost of poor quality and on the distribution over various departments. Activity-based costing (ABC) can help to provide a realistic allocation of overhead costs. ABC is an accounting method that allocates overhead based on the activities that cause overhead cost elements to be incurred. For an explanation of the concept, see *JQH5,* page 8.14. Cokins (1999) discusses the impact of traditional accounting and ABC on quality.

An example for one plant of a tire manufacturer is shown in Table 2.1. This example resulted in some conclusions that are typical of these studies.

- The total of almost $900,000 per year is large.
- Most (79.1 percent) of the total is concentrated in failure costs, specifically in waste scrap and consumer adjustments.
- Failure costs are about five times the appraisal costs.

TABLE 2.1

Annual quality cost—tire manufacturer

Cost of quality failures—losses		
Defective stock	$ 3,276	0.37%
Repairs to product	73,229	8.31
Collect scrap	2,288	0.26
Waste—scrap	187,428	21.26
Consumer adjustments	408,200	46.31
Downgrading products	22,838	2.59
Customer ill will	Not counted	
Customer policy adjustment	Not counted	
Total	$697,259	79.10%
Cost of appraisal		
Incoming inspection	$ 32,655	2.68
Inspection 1	32,582	3.70
Inspection 2	25,200	2.86
Spot-check inspection	65,910	7.37
Total	$147,347	16.61%
Cost of prevention		
Local plant quality		
Control engineering	7,848	0.89
Corporate quality		
Control engineering	30,000	3.40
Total	$ 37,848	4.29%
Grand total	$882,454	100.00%

TABLE 2.2
Revenue lost through poor quality

$10,000,000	Annual customer service revenue
1,000	Number of customers
× 25%	Percent dissatisfied
250	Number of dissatisfied
× 75%	Percent of switchers (60–90% of dissatisfied)
188	Number of switchers
× $10,000	Average revenue per customer
$1,880,000	Revenue lost through poor quality

Source: The University of Tampa (1990).

- A small amount (4.3 percent) is spent on prevention.
- Some consequences of poor quality cannot be conveniently quantified, e.g., customer ill will and customer policy adjustment. However, these factors are listed as a reminder of their existence.

As a result of this study, management decided to increase the budget for prevention activities. Three engineers were assigned to identify and pursue specific quality improvement projects.

Strictly defined, the cost of poor quality is the sum of the internal and external failure costs categories. This definition assumes that the elements of appraisal costs—e.g., 100 percent sorting inspection—necessitated by inadequate processes are classified under internal failures. Some practitioners use the term *cost of quality* for the four broad categories.

Although many organizations use the categories of internal failure, external failure, appraisal, and prevention, the structure may not apply in all cases. A different approach employed by a bank is shown in Table 2.2.

Here the cost of poor service is calculated based on customer satisfaction data and customer loyalty/retention data. In this case the cost of poor quality is the cost of lost opportunities of sales revenue (see Figure 2.1).

In still another approach some organizations define cost of poor quality with a focus on the cost of key activities or processes, i.e., the difference between actual costs and cost in an organization that has the best practice for that activity. This approach addresses the cost of inefficient processes (see Figure 2.1). For a discussion of the traditional quality cost approach (four categories), the process cost approach, and the "quality loss approach," see Schottmiller (1996).

Additional examples of cost-of-poor-quality studies in both manufacturing and service industries are provided by Campanella (1999) and Atkinson et al. (1994).

Hidden Quality Costs

The cost of poor quality may be understated because of costs that are difficult to estimate. "Hidden" costs occur in both manufacturing and service industries and include the following:

1. Potential lost sales.
2. Costs of redesign of products due to poor quality.
3. Costs of changing processes due to inability to meet quality requirements for products.
4. Costs of software changes due to quality reasons.
5. Costs of downtime of equipment and systems including computer information systems.
6. Costs included in standards because history shows that a certain level of defects is inevitable and allowances should be included in standards.
   a. *Extra material purchased.* The purchasing buyer orders 6 percent more than the production quantity needed.
   b. *Allowances for scrap and rework during production.* History shows that 3 percent is "normal," and accountants have built this allowance into the cost standards. One accountant said, "Our scrap cost is zero. The production departments are able to stay within the 3 percent that has been added in the standard cost and therefore the scrap cost is zero." Ah, for the make-believe numbers game.
   c. *Allowances in time standards for scrap and rework.* One manufacturer allows 9.6 percent in the time standard for certain operations to cover scrap and rework.
   d. *Extra process equipment capacity.* One manufacturer plans for 5 percent unscheduled downtime of equipment and provides extra equipment to cover the downtime. In such cases the alarm signals ring only when the standard value is exceeded. Even when operating within those standards, however, the costs should be a part of the cost of poor quality. They represent opportunities for improvement.
7. Extra indirect costs due to defects and errors. Examples are space charges and inventory charges.
8. Scrap and errors not reported. Scrap may never be reported because employees fear reprisals, or scrap may be charged to a general ledger account without being identified as scrap.
9. Extra process costs due to excessive product variability (even though within specification limits). For example, a process for filling packages with a dry soap mix meets requirements for label weight on the contents. The process aim, however, is set above label weight to account for variability in the filling process. See "Cost of Inefficient Processes" under "Internal Failure Costs."
10. Cost of errors made in support operations, e.g., order filling, shipping, customer service, billing.
11. Cost of poor quality within a supplier's company. Such costs are included in the purchase price.

Hidden costs can accumulate to a large amount, three or four times the reported failure cost. Where agreement can be reached to include some of these costs, and where credible data or estimates are available, then they should be included in the study. Otherwise, they should be left for future exploration.

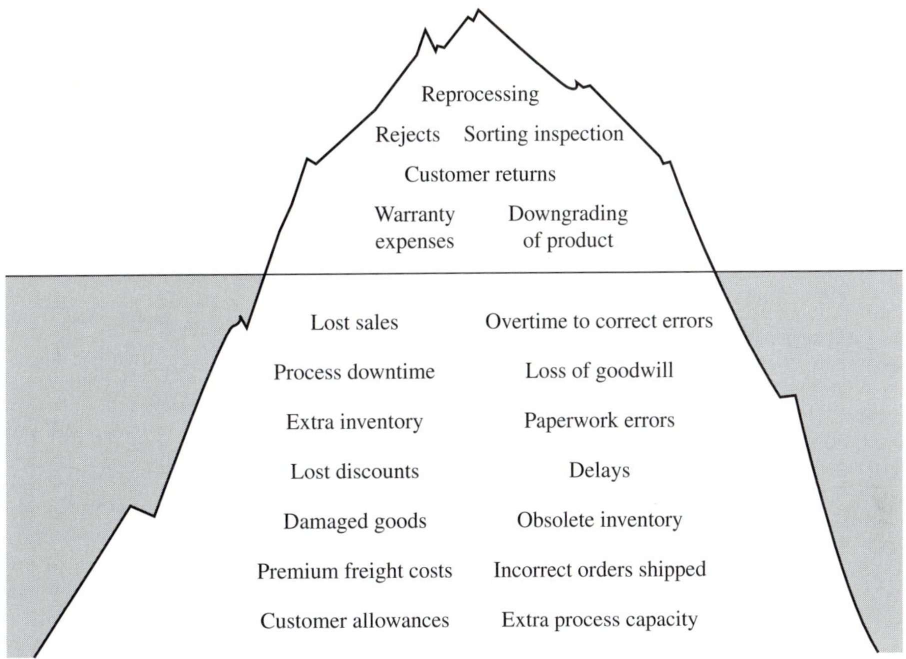

FIGURE 2.2
Hidden costs of poor quality.

Progress has been made in quantifying certain hidden costs, and therefore some of them have been included in the four categories discussed above. Obvious costs of poor quality are the tip of the iceberg (Figure 2.2).

2.4 RELATING THE COST OF POOR QUALITY TO BUSINESS MEASURES

Interpretation of the cost of poor quality is aided by relating this cost to other figures with which managers are familiar. Table 2.3 shows actual examples of the annual cost of poor quality related to various business measures.

Reducing the cost of poor quality has a dramatic impact on company financial performance, as illustrated by the Dupont Financial Model (Werner and Stoner, 1995). This model states:

$$\text{Return on assets} = \text{Profit margin} \times \text{Asset turnover}$$

Suppose the cost of poor quality (COPQ) was 10 percent of sales revenue, the profit margin was 7 percent, and the asset turnover was 3.0. The return on assets is then 7.0

TABLE 2.3
Languages of management

Money (annual cost of poor quality)
24% of sales revenue
15% of manufacturing cost
13 cents per share of common stock
\$7.5 million per year for scrap and rework compared to a profit of \$1.5 million per year
\$176 million per year
40% of the operating cost of a department
Other languages
The equivalent of one plant in the company making 100% defective work all year
32% of engineering resources spent in finding and correcting design weaknesses
25% of manufacturing capacity devoted to correcting quality problems
13% of sales orders canceled
70% of inventory carried attributed to poor quality levels
25% of manufacturing personnel assigned to correcting quality problems

× 3.0 or 21 percent. Further suppose that a quality improvement effort reduced the COPQ from 10 percent to 6 percent and the asset turnover remained at 3.0. The profit margin would then be 7.0 + 4.0 or 11 percent, and the return on assets would be 11.0 × 3.0 or 33 percent. Note the impact of asset turnover.

How do we convince upper management to institute a quality improvement program to reduce the cost of poor quality? The steps are presented in Section 3.4.

2.5 OPTIMUM COST OF QUALITY

When cost summaries on quality are first presented to managers, one typical question is, What are the right costs? The managers are looking for a standard ("par") against which to compare their actual costs so that they can make a judgment on whether there is a need for action.

Unfortunately, few credible data are available because (1) companies almost never publish such data and (2) the definition of cost of poor quality varies by company. But we can cite some numbers. For manufacturing organizations the annual cost of poor quality is about 15 percent of sales income, varying from about 5 percent to 35 percent depending on product complexity. For service organizations the average is about 30 percent of operating expenses, varying from 25 percent to 40 percent depending on service complexity. But three conclusions on cost data do stand out: The total costs are highest for complex industries, failure costs are the largest percentage of the total, and prevention costs are a small percentage of the total.

The study of the distribution of quality costs over the major categories can be further explored by using the model shown in Figure 2.3. The model shows three curves:

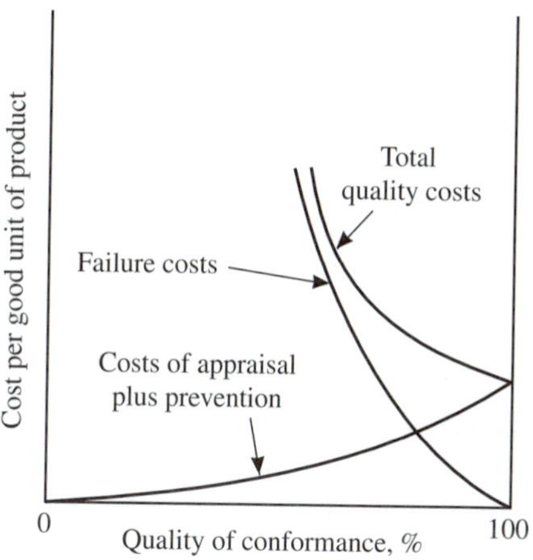

FIGURE 2.3
Model for optimum costs (*From JQH4, p. 4.19.*)

1. *The failure costs.* These costs equal zero when the product is 100 percent good and rise to infinity when the product is 100 percent defective. (Note that the vertical scale is cost per good unit of product. At 100 percent defective, the number of good units is zero, and hence the cost per good unit is infinity.)
2. *The costs of appraisal plus prevention.* These costs are zero at 100 percent defective and rise as perfection is approached.
3. *The sum of curves 1 and 2.* The third curve is marked "total quality costs" and represents the total cost of quality per good unit of product.

Figure 2.3 suggests that the minimum level of total quality costs occurs when the quality of conformance is 100 percent, i.e., perfection. This result has not always been the case. During most of the 20th century, the predominant role of (fallible) human beings limited the efforts to attain perfection at finite costs. Also, the inability to quantify the impact of quality failures on sales revenue resulted in underestimating the failure costs. The result was to view the optimum value of quality of conformance as less than 100 percent.

Although perfection is obviously the goal for the long run, perfection is not necessarily the most economic goal for the short run, or for every situation. Industries, however, are facing increasing pressure to reach for perfection. The prospect is that the trend to 100 percent conformance will extend to more and more goods and services of greater and greater complexity.

To evaluate whether quality improvement has reached the economic limit, we need to compare the benefits possible from specific projects with the costs involved in achieving these benefits. When no justifiable projects can be found, the optimum has been reached.

Let's identify a few key points about the cost of poor quality. This cost is usually high (sometimes larger than the annual profit), quantifying this cost can be the key to gaining approval from management to assign resources to quality improvement, and the main uses of the cost-of-poor-quality study are to identify opportunities for improvement projects and to provide supporting data to assist in improvement.

For elaboration on finalizing the definitions of cost categories, steps in making the initial study, and data collection methods, see *JQH5,* Section 8.

Next, we move to another element of assessment—standing in the marketplace.

2.6 STANDING IN THE MARKETPLACE

Estimating the cost of poor quality is an essential part of the assessment. But it is not enough. We also need to understand where the company stands on quality in the marketplace, relative to the competition. This component of assessment will prove to be important in increasing sales income.

Similar to the assessment of the cost of poor quality, the market study (1) gives a snapshot of standing relative to competition and (2) identifies opportunities and threats.

The approach must be based on a marketing research study. Such studies should be planned not by any one department but by a team involving members from marketing, product development, quality, operations, and other areas as needed. This team must agree beforehand on what questions need to be answered by the field study. Three types of questions should be considered: (1) What is the relative importance of various product qualities as seen by the user? (2) For each of the key qualities, how does our product compare with competitors' products, as seem by the users? (3) How likely is the customer to purchase from us again or recommend us to others?

Chapter 12, "Understanding Customer Needs," includes other questions to help in developing new or modified products.

Answers to such questions must be based on input from current customers, lost customers, and noncustomers. Opinions of company personnel, no matter how extensive the experience base, cannot and should not substitute for the voice of the customer.

Examples of Field Studies

The first example comes from a manufacturer of health products. In a multiattribute study customers were asked to consider several product attributes and indicate both their relative importance and a competitive rating. The results for one product are shown in Table 2.4. Note that an overall score is obtained for each manufacturer by multiplying the relative importance by the score for that attribute and then adding these results.

In another example a manufacturer of equipment was experiencing a decline in market share. Complaints about quality led to a proposal to "beef up the inspection." Discussions within the company revealed uncertainty about the nature of the complaints, so management decided to conduct a field study to learn more about customer views. A team was formed to plan and conduct the study. Visits were made to about 50 customers.

In one part of the study, six attributes were identified, and customers were asked to rate the company as superior, competitive, or inferior to the competition on each attribute (see Table 2.5). The results were a surprise. The equipment problems were confirmed, but the study revealed the presence of both design and manufacturing causes. Also, documentation and field repair service were identified as weak areas; these were surprises to the company involved. The company took dramatic action. This manufacturer created a broad approach to quality, starting with the initial design and continuing throughout the spiral of all activities affecting fitness for use. This outcome was in stark contrast to the original proposal of adding inspectors. The study required about seven worker-months of effort, including planning, customer visits, analysis of results, and preparation of a report—a small price to pay to develop a proper strategy.

Many organizations in service industries have extensive experience in marketing research. For example, one bank periodically conducts marketing research as part of a quality system. This research probes 20 attributes of banking service by asking

TABLE 2.4
Multiattribute study

		Company X		Competitor A		Competitor B	
Attribute	**Relative importance, %**	**Rating**	**Weighted rating**	**Rating**	**Weighted rating**	**Rating**	**Weighted rating**
Safety	28	6	168	5	140	4.5	126
Performance	20	6	120	7	140	6.5	130
Quality	20	6	120	7	140	4	80
Field service	12	4	48	8	96	5	60
Ease of use	8	4	32	6	48	5	40
Company image	8	8	64	4	32	4	32
Plant service	4	7.5	30	7.5	30	5	20
Total			582		626		488

TABLE 2.5
Heavy equipment case

	Comparison to competition, %		
Attribute	**Superior**	**Competitive**	**Inferior**
Analysis of customer needs			
Preparation of quality requirements and purchase order			
Preparation of specifications and technical documentation			
Quality of equipment			
Quality and availability of spare parts			
Quality of field repair service			

Source: Private communication.

consumers about the relative importance of attributes and consumer degree of satisfaction. Table 2.6 shows the format of the summarized results.

In an example from the GTE Corporation, customers are asked to rate GTE on overall quality and on five attributes of telephone service. One of the summary statistics is the percentage of excellent ratings. Other national local carriers serve as benchmarks. Table 2.7 shows the results for one franchise area.

Note that the performance is mixed: Scores are generally in the middle of the range, never at the top, but one (for installation) is at the bottom. For each attribute a goal can be set and a schedule set for achieving the goal based on knowledge of proposed process improvements and customer satisfaction results. After reviewing separate data on the relative importance of the attributes, GTE set the need to narrow the ratings gap for local dial service as its first priority.

In an example from the public sector, the United States Postal Service has used a Customer Satisfaction Index survey that asks customers for the relative importance and satisfaction rating on 10 attributes (e.g., responsiveness, carrier services, complaint handling). A separate question asks about "customer willingness to switch to another mail service."

Graphing the market research results can be helpful. Rust et al. (1994) show an example of how the mapping of satisfaction and importance ratings can relate cus-

TABLE 2.6
Marketing research at a bank

Satisfaction with . . .	Very satisfied, %	Sample size	Important and low in satisfaction, %	Important and high in satisfaction, %
1. Greeting you with a smile				
8. Processing transactions without error				
14. Easy-to-read and understand bank statements				
20. Prompt follow-up to questions and problems				

TABLE 2.7
GTE benchmark data for local telephone service

Attribute	Company score	Competitor scores
Overall quality	38.6	36.6–46.3
Local dial quality	40.6	36.7–47.9
Billing quality	34.5	28.7–37.2
Installation quality	41.2	43.2–53.3
Long distance quality	47.5	40.9–55.3
Operator quality	41.5	35.0–47.1

Source: Drew and Castrogiovanni (1995). Courtesy of Marcell Dekker, Inc.

tomer views and potential action (Figure 2.4). In this approach attributes in which importance is high and satisfaction is poor represent the greatest potential for gain.

In Figure 2.4 the four quadrants are roughly defined by the averages on the two axes. Interpretation of the quadrants is typically as follows:

Upper left (satisfaction strong, importance low). Maintain the status quo.

Upper right (satisfaction strong, importance high). Leverage this competitive strength through advertising and personal selling.

Lower left (satisfaction weak, importance low). Assign little or no priority on action.

Lower right (satisfaction weak, importance high). Add resources to achieve an improvement.

For additional examples of maps for customer retention modeling, see Lowenstein (1995, Chapter 9).

Collecting data on customer loyalty and retention goes beyond customer satisfaction research. Lauter (1997) reports on customer retention results at a bank. The actual customer retention percentage was calculated for each quarter in 1996—the percentage retained was typically more than 95 percent. Further analysis on the customers lost revealed a surprising result (Figure 2.5). For example, in the first quarter 120,000 households (HHs) were lost, and this loss translated into $26 million of profit.

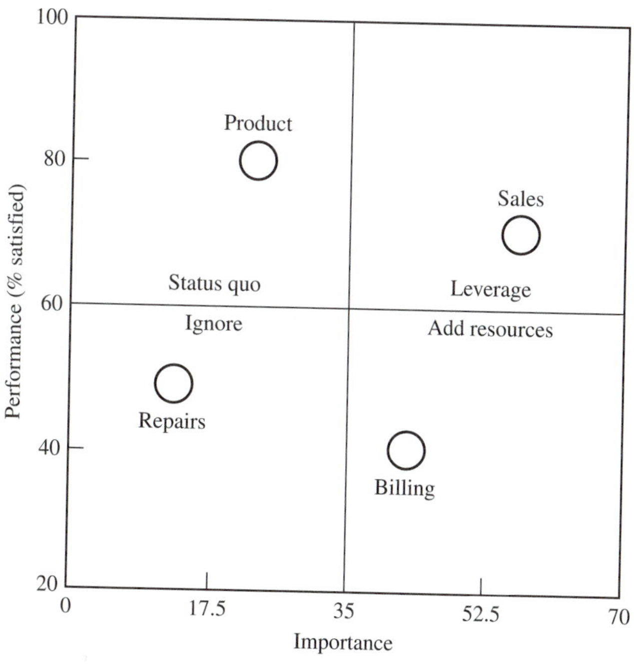

FIGURE 2.4
Performance versus importance in driving satisfaction: quadrant map. (*Rust et al., 1994.*)

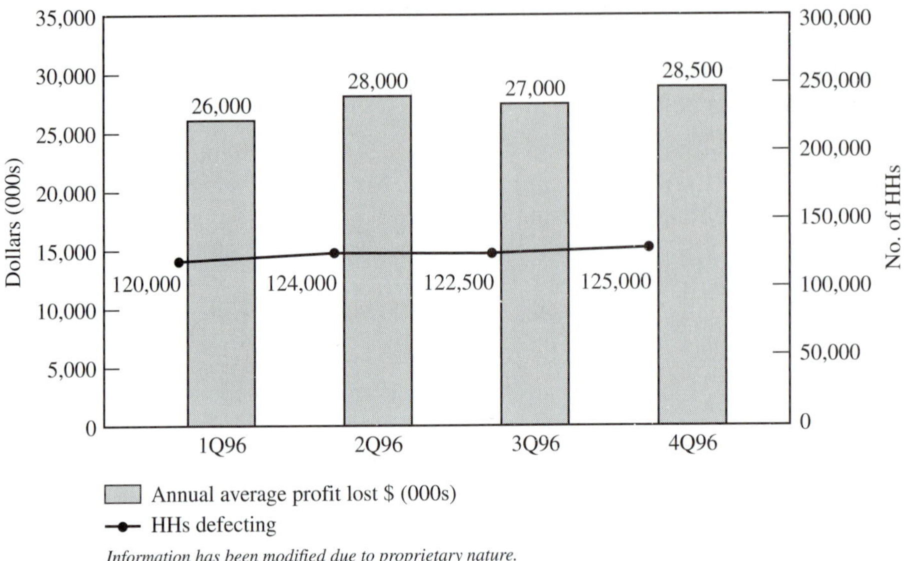

FIGURE 2.5
Average "opportunity cost of attrition." (*Lauter, 1997.*)

Although the information has been modified because of the proprietary nature, a small percentage of customers lost from a high volume of customers can result in a significant absolute dollar amount of revenue.

Generalizing from the examples

First, the customers in these studies were experienced in using the product. Also, the studies identified specific attributes leading to customer satisfaction, quantified their relative importance, determined ratings for each attribute, and compared the ratings to the competition. In some cases the organizations also collected data on customer loyalty and retention.

Terms like *customer satisfaction* and *quality* are too foggy to be meaningful. Instead we must identify the attributes or features of the product that collectively define satisfaction. Examples are the accuracy of a transaction at a bank or the workmanship in a residential home. A focus group (see Section 2.7) or a sample of managers will have ideas on which attributes to include. In addition, a sample of customers should be asked (in preliminary research) which attributes constitute high quality. Also, a review of trade journals in an industry may reveal standard measures for that industry. With these inputs the list of attributes should be finalized for the research with the larger sample of customers. Note that the list of attributes should not be limited to attributes of the product provided to customers but can span the entire cycle of customer contact from initial contact with a salesperson through use and servicing of the product.

It is possible to use a generic template of attributes. For example, in the area of customer service the SERVQUAL model (Zeithaml et al., 1990) identifies five dimensions: tangibles, reliability, responsiveness, assurance, and empathy. Such broad terms need to be defined. Research on these five areas is explored by presenting customers with 22 statements that examine the difference between customer expectations and perceptions. These statements can be customized for each product or company.

The relative importance of the product attributes can be determined by several methods. In one approach customers are asked to allocate 100 points over the various attributes. Another approach presents combinations of product attributes to customers and asks them to indicate their preferences. The importance ratings can then be calculated. Churchill (1991), Appendix 9B, describes this method ("conjoint measurement").

Obtaining data on competitor quality involves a variety of methods. For physical products, laboratory testing is one method. Other methods include directly asking customers for ratings on competitors or using mystery shoppers, focus groups, or some of the other market research methods discussed in Chapter 12. When a noncustomer is purchasing from a competitor, it is useful to ask an open-ended question, e.g., What are the chief reasons that you buy from company X? Another source for comparison to competition is the American Customer Satisfaction Index (ACSI). The index, calculated from about 50,000 telephone survey responses, covers 200 firms in seven sectors of the economy (both manufacturing and service). The index uses a scale of 0 to 100. One use of this index is to compare individual firms with the industry average. To benefit from this information, companies need to understand the assumptions and other elements of methodology. Additional information on the ACSI (including a methodologies report) can be acquired from the American Society for Quality. Still another source of information is the benchmarking program of the American Productivity and Quality Center.

Many of the concepts presented above for external customers can be adapted and applied to internal customers. Thus a quality assurance department can conduct internal market research for its internal customers, e.g., operations, product development, purchasing. The manager of a large maintenance department at Carolina Power and Light started internal research by visiting with an operations department after some routine maintenance had been performed. The results of these discussions were so useful that he instituted a more formal approach using service attributes, importance ratings, and performance scores. The United Parcel Service uses the Internet to conduct research on internal customers.

For additional examples and elaboration on the methods of measuring customer satisfaction, see *JQH5,* Section 18. Collecting information on customer satisfaction and retention is an essential part of a companywide assessment. But, as with the cost-of-poor-quality data, follow-up action is even more essential than data collection. The steps to achieve action are covered in Chapter 3, "Quality Improvement and Cost Reduction," and Chapter 4, "Operational Quality Planning and Sales Income."

2.7
ORGANIZATION CULTURE ON QUALITY

Employees in an organization have opinions, beliefs, traditions, and practices concerning quality. We will call this set of characteristics the "company quality culture." Gaining an understanding of this culture should be part of a company assessment of quality for two reasons: (1) The culture clearly has a major impact on quality results, and (2) knowing the present culture can identify barriers for developing a strategy and implementing an action plan based on the companywide assessment of quality.

Assessing the quality culture is usually performed by using questionnaires to survey employees, by conducting focused discussions with groups of employees, or by doing both.

An insurance company studied its quality culture by asking employees to respond to seven questions. Possible answers were greatly, moderately, minimally, or not at all. Listed below are the seven questions and the responses with the highest percentage of employees:

1. Do you feel that you understand what good quality is? Response with highest percentage: moderately (52 percent).
2. To what degree are you familiar with the company's emphasis on quality? Response with highest percentage: moderately (36 percent).
3. To what degree do you agree with the statement "My manager's actions and attitude convince me that quality is important"? Response with highest percentage: minimally (43 percent).
4. To what degree do you understand the quality measurements in your department? Response with highest percentage: minimally (48 percent).
5. Everything considered, how do you rate your department on providing high-quality service and outputs? Response with highest percentage: greatly (57 percent).
6. To what degree do you think your achievement of quality standards affects your performance evaluation? Response with highest percentage: greatly (35 percent).
7. Have you worked in a quality circle, quality improvement project, or quality improvement team in the last 12 months? Response with highest percentage: not at all (64 percent).

Overall, the conclusions were mixed, but several areas needing improvement were identified.

Some organizations employ an extensive array of questions. One organization uses 56 questions to explore eight areas: overall product quality, conformance to requirements, equipment, supplier quality, management commitment, work group performance, employee participation, and training. Another organization poses 82 questions to evaluate 18 factors that collectively address the quality culture.

One federal agency uses 70 questions to address job factors such as communication practices, participation in decision making, recognition for a job well done, and organizational trust. For each job factor the employee is asked to (1) indicate the degree of importance and (2) rate the job. If a job factor is considered absolutely essential or very important and the rating is less than excellent or very good, then the organization is vulnerable to reduced performance due to that job factor. The survey results are summarized in vulnerability percentages (VPs). Low percentages are desirable; high VPs pinpoint job factors of highest concern among employees. Two examples of major concern were

> "Having management that listens to, and is willing to act on, your suggestions" (VP = 75 percent).

"Having the necessary equipment, tools, and supplies to do your job" (VP = 72 percent).

If the VP is 60 or above, the job factor is classified as a major concern; a percentage from 40 through 59 is considered a developing concern; a percentage less than 40 means a job factor is a major strength.

A frequent response in both manufacturing and service industries concerns the time to carry out quality initiatives: "Management wants us to add on these quality activities to our regular duties without giving us additional time—it can't be done."

In determining the quality culture, we must choose the topical areas, create questions to meet the purposes of the study, and decide how to collect information on employee perceptions.

One list of topical areas is the framework of five elements of quality culture discussed in Chapter 9, "Developing a Quality Culture." These elements are goals and measurements, evidence of management leadership on quality, self-development and empowerment of employees, participation, and recognition and rewards. Specific questions can be created around these five elements. Other topical areas might be quality delivered by internal departments, work group performance, and adequacy of specifications and procedures. Each organization should develop its own framework of topical areas and specific questions. Management and operations personnel should be asked for ideas on topical areas. This input not only assures that the content of a questionnaire meets employee needs but also increases the likelihood that the assessment will result in action.

Questions can be scaled, open ended, or forced choice. Scaled questions have a limited range of responses as illustrated in the preceding examples. Open-ended questions provide for responses of wider scope, e.g., How would you describe management's attitude toward quality? Open-ended questions can also be posed to a small sample of employees in an interview that supplements the questionnaire. Forced-choice questions often ask for a ranking, e.g., Please rank cost, delivery schedules, and quality in their order of importance in our organization with a rank of 1 being most important, 2 the next important, and 3 the least in importance.

A useful source of specific questions is Getty and Getty (1993) who provide 30 questions for exploring management commitment to continuous improvement. Tabladillo and Canfield (1994) provide 25 questions for determining an employee's viewpoint on hospital quality.

Finally, some organizations incorporate questions on quality culture in a broader corporate culture survey. Thus one bank asks for employee perceptions on numerous characteristics, values, and attitudes. Of 116 questions about one-third are related to quality.

A second method of collecting employee perceptions is the focus group. This method can be used as a supplement or in place of a questionnaire. A focus group consists of 8 to 14 typical employees who would meet to discuss quality culture. Such a caucus can generate a melting pot of ideas. Supervisory and managerial personnel are excluded from a workforce focus group. If desired, several focus groups can be used and the results summarized. The focus group meets for a maximum of two hours.

Focus groups collect information in depth by asking employees to express their views on various issues. To be effective, the group must have a moderator who not only is skilled in group dynamics but also has a clear understanding of the information needed and a plan for guiding the discussion. A local university professor can often provide an objective role as a moderator.

A focus group provides a means of collecting a great deal of information at a reasonable cost. Also, people in a focus group are candid because they often receive reinforcement of their views from their peers in the group.

In summary, the quality culture has a major impact on quality results. Collecting employee perceptions on the quality culture is essential to the success of an action plan for quality improvement.

The discussion here covers the determination of the present quality culture. The steps needed to improve the quality culture are discussed in Chapter 9, "Developing a Quality Culture."

Sears Roebuck assembles information quarterly and conducts an annual assessment that relates employee satisfaction and customer satisfaction to measures of investor satisfaction (Rucci et al., 1998).

The elements of assessment apply to both large and small organizations. For application to small organizations, see the following reports available from the College of Business, University of Tampa, Tampa, Florida 33626: Report No. 902 "Assessing the Cost of Poor Quality in a Small Business" (by Nat R. Briscoe and Frank M. Gryna); Report No. 905 "Assessment of Quality Culture in Small Business" (by Mary Anne Watson and Frank M. Gryna); and Report No. 906 "Market Research for Quality in Small Business" (by William L. Rhey and F. M. Gryna).

2.8 ASSESSMENT OF CURRENT QUALITY ACTIVITIES

The fourth element of assessment is the evaluation of current quality-related activities in the organization. Such an evaluation could cover a wide range of both scope (global to a few activities) and examination (cursory to detailed). Our discussion concentrates on a companywide examination in sufficient depth to meet the needs of the overall quality assessment. Chapter 24 addresses the audit of individual activities for conformance to procedures.

Assessment of current quality activities can take two forms: assessments that focus on customer satisfaction results but include an evaluation of the current system of quality-related activities and (2) assessments that focus on evaluation of the current quality system, with little emphasis on customer satisfaction results. In either case the assessment can be performed by the organization itself (self-assessment) or by an external organization. Also, the assessment can be conducted with or without the use of explicitly defined criteria. Conti (1997) provides a comprehensive discussion on assessment with an emphasis on self-assessment.

Both types of assessments can identify improvement opportunities by comparing current quality activities with an accepted system model such as those described in this section (e.g., the Baldrige Award or ISO 9000).

Hannukainen and Salminen (1998) describe how Nokia Mobile Phones developed a self-assessment that local management teams apply to establish priorities for improvement and action plans.

Often an organization starts by having an outsider conduct a three- to five-day assessment of the current status on quality. This study includes assessment questions that address companywide issues of quality management. Questions on this category include

- Have studies on the cost of poor quality, standing in the marketplace, and quality culture been completed?
- Have quality objectives been established?
- Have quality objectives been linked to business objectives?
- Do policies and plans ensure that quality will be competitive in the marketplace?
- Is the approach to quality led by line management rather than by staff?
- Does the plan cover all processes (big Q)?
- Is the organizational machinery in place to identify and pursue opportunities for increasing sales income and for reducing costs of poor quality?
- Is there an effective system for providing early warning of potential quality problems?

Additional questions relate to functional areas. The following are common to *all* functional areas:

- What measures are used to judge output quality?
- What is the company's performance, as reflected in these measures?
- Have the resources spent in detecting and correcting quality-related problems been estimated?
- To what extent do personnel understand quality-related responsibilities?
- To what extent have personnel been trained in the quality disciplines?
- To what extent has the capability of key processes been quantified?
- Does the data system meet the needs of the personnel?

2.9 MALCOLM BALDRIGE NATIONAL QUALITY AWARD

A more detailed companywide assessment relies on established criteria, e.g., the criteria used in connection with the Malcolm Baldrige National Quality Award. The Baldrige Award is managed by the National Institute of Standards and Technology (NIST), a nonregulatory federal agency, and is based on comparative studies of company stock made by NIST. Baldrige Award recipients as a group have outperformed the Standard and Poor's 500 stock index by about 2.5 to 1.0.

This award is presented annually in the United States. A maximum of three organizations in each of five categories (manufacturing, service, small business, education, and health care) may be selected each year. An exhaustive evaluation is based on submitted written materials followed by site visits to qualifying organizations.

Since the inception of the award in 1988, the emphasis of the criteria has changed from quality to overall company performance. This book follows that theme of integrating quality into the business plan of an organization. The examination categories for the year 2000 and points assigned (out of 1000 points) are listed in Table 2.8.

Note the heavy emphasis on quality results and customer satisfaction. The Baldrige Award recognizes organizations that have achieved the highest levels of quality.

The seven Baldrige categories can be viewed as a system (Figure 2.6). Categories 1, 2, and 3 represent a leadership triad; categories 5, 6, and 7 represent a results triad; category 4 provides the foundation of fact-based information.

Many organizations use the Baldrige criteria to conduct a self-assessment without submitting an application to compete for the award. Winning the Baldrige or similar award is extremely difficult, and realistically most organizations start with a self-assessment. After making the assessment, they realize that the knowledge gained from that process is the real benefit. The *Quality Progress* journal, June through December 1997, contains papers that explain the meaning of each Baldrige category in making a self-assessment.

Understanding the Baldrige criteria requires a knowledge of the concepts of alignment and linkages. *Alignment* is the translation of organization goals into goals, subgoals, and standards at all levels—organization, key processes, and work unit levels.

TABLE 2.8
Baldrige Quality Award, 2000 categories/items

1. Leadership (125 points)
   - 1.1 Organizational leadership (85)
   - 1.2 Public responsibility and citizenship (40)
2. Strategic planning (85)
   - 2.1 Strategy development (40)
   - 2.2 Strategy deployment (45)
3. Customer and market focus (85)
   - 3.1 Customer and market knowledge (40)
   - 3.2 Customer satisfaction and relationships (45)
4. Information and analysis (85)
   - 4.1 Measurement of organizational performance (40)
   - 4.2 Analysis of organizational performance (45)
5. Human resource focus (85)
   - 5.1 Work systems (35)
   - 5.2 Employee education, training, and development (25)
   - 5.3 Employee well-being and satisfaction (25)
6. Process management (85)
   - 6.1 Product and service processes (55)
   - 6.2 Support processes (15)
   - 6.3 Supplier and partnering processes (15)
7. Business results (450)
   - 7.1 Customer-focused results (115)
   - 7.2 Financial and market results (115)
   - 7.3 Human resource results (80)
   - 7.4 Supplier and partner results (25)
   - 7.5 Organizational effectiveness results (115)

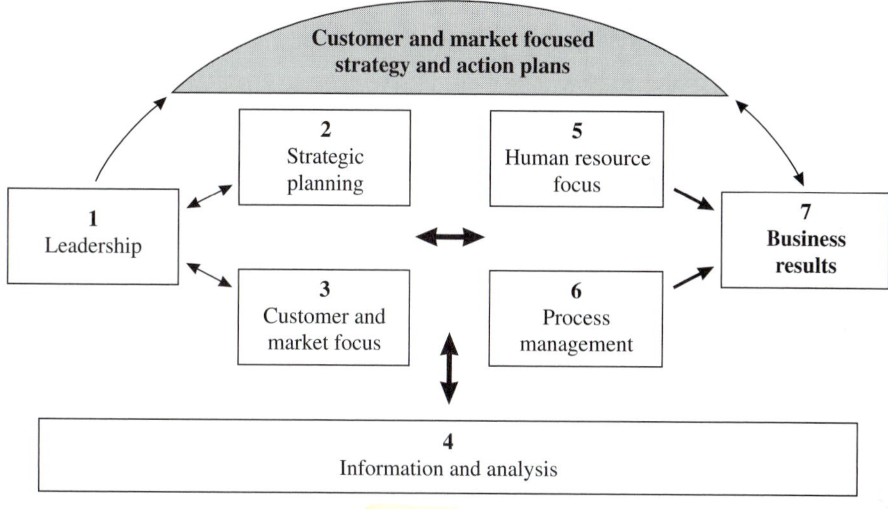

FIGURE 2.6
Baldrige criteria for performance excellence framework: a systems perspective.

Linkages are the interrelationships—the connections—between specific quality-related management activities so that the activities are mutually reinforcing to produce the desired results. These linkages can be as important as the activities themselves. The Baldrige criteria provide direction for excellence, but alignment and linkages are driving forces to achieve the exceptional results. In later chapters we show where alignment and linkages are important. Blazey (2000) provides an extensive discussion on linkages among the Baldrige criteria.

Eleven core values and concepts are embodied in the award criteria. These 11 are: visionary leadership, customer driven, organizational and personal learning, valuing employees and partners, agility, focus on the future, managing for innovation, management by fact, public responsibility and citizenship, focus on results and creating value, and systems perspective. For a discussion of the relationships between the core values and key management processes in the Baldrige criteria see Evans and Ford (1997).

Other countries have also created quality awards. (For a thorough discussion of international and national quality awards, see Hromi, 1995.) For example, the European Foundation for Quality Management has created the European Quality Award (see Figure 2.7). For a discussion of the similarities and differences between the European Quality Award and the Baldrige Award, see Conti (1997). Puay et al. (1998) provide a comparative study of nine national quality awards. In the United States, most states sponsor an award, and many of these awards are based on the Baldrige criteria. Some states recognize that most organizations are not ready to apply for the state or federal award but do need help to get started on assessment. For example, in Florida the Baldrige-based Sterling Award process helps organizations by providing a self-assessment process at either the beginning or intermediate level of quality assessment.

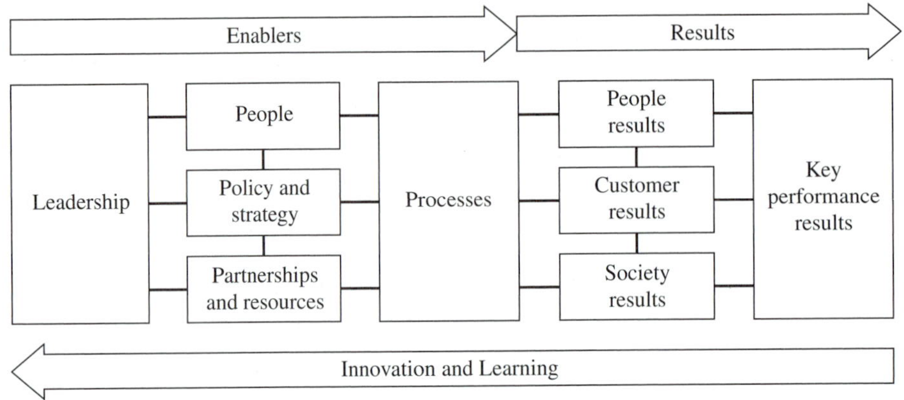

FIGURE 2.7
The enhanced EFQM Excellence Model. (*Bacivarof, 2000.*)
Courtesy of Marcell Dekker, Inc.

TABLE 2.9
ISO 9000 quality standards (1994)

ISO	Scope
9000-1	Selection and use of documents
9001	Design, development, production, installation, and servicing
9002	Production, installation, and servicing
9003	Final inspection and test
9004	Quality management and system elements

2.10 ISO 9000 QUALITY STANDARDS

An international effort to identify the key elements of a quality system for manufacturing and service organizations has resulted in a series of quality standards. The purpose is to facilitate international trade by establishing a common set of standards.

These standards have been developed by the International Organization for Standardization (ISO) and are known as the ISO 9000 series quality standards. Table 2.9 lists the main documents involved in the 1994 version of ISO 9000.

Figure 2.8 shows the main elements of each standard. (ANSI is the American National Standards Institute; ASQC is the American Society for Quality Control, now shortened to ASQ, the American Society for Quality.)

In more than 100 countries these standards are being used to certify that an organization meets the minimum criteria for a quality system as defined by the standards. Purchasers of products can require potential suppliers to be certified to the appropriate ISO criteria as a prerequisite to receiving a contract. Organizations find that they must achieve ISO 9000 certification to meet both domestic and international competition. Evaluation for such “quality systems registration” is made by an independent organization. Often the process includes surveillance visits once or twice a year to

External quality assurance				Clause title in ANSI/ASQC Q9001-1994	QM guidance ANSI/ASQC Q9004-1-1994	Road map ANSI/ASQC Q9000-1-1994
ANSI/ASQC Q9001-1994	Requirements ANSI/ASQC Q9002-1994	ANSI/ASQC Q9003-1994	Application guide ISO 9000-2			
4.1 ■	■	○	4.1	Management responsibility	4	4.1; 4.2; 4.3
4.2 ■	■	○	4.2	Quality system	5	4.4; 4.5; 4.8
4.3 ■	■	■	4.3	Contract review	X	8
4.4 ■	X	X	4.4	Design control	8	
4.5 ■	■	■	4.5	Document and data control	5.3; 11.5	
4.6 ■	■	X	4.6	Purchasing	9	
4.7 ■	■	■	4.7	Customer-supplied product	X	
4.8 ■	■	○	4.8	Product identification and traceability	11.2	5
4.9 ■	■	X	4.9	Process control	10; 11	4.6; 4.7
4.10 ■	■	○	4.10	Inspection and testing	12	
4.11 ■	■	■	4.11	Control of inspection, measuring, and test equipment	13	
4.12 ■	■	■	4.12	Inspection and test status	11.7	
4.13 ■	■	○	4.13	Control of nonconforming product	14	
4.14 ■	■	○	4.14	Corrective and preventive action	15	
4.15 ■	■	■	4.15	Handling, storage, packaging, preservation, and delivery	10.4; 16.1; 16.2	
4.16 ■	■	○	4.16	Control of quality records	5.3; 17.2; 17.3	
4.17 ■	■	○	4.17	Internal quality audits	5.4	4.9
4.18 ■	■	○	4.18	Training	18.1	5.4
4.19 ■	■	X	4.19	Servicing	16.4	
4.20 ■	■	○	4.20	Statistical techniques	20	
				Quality economics	6	
				Product safety	19	
				Marketing	7	

■ Comprehensive requirement
○ Less-comprehensive requirement than ANSI/ASQC Q9001-1994 and ANSI/ASQC Q9002-1994
X Element not present

FIGURE 2.8
ISO 9000 cross-reference list of clause numbers for corresponding topics.

ensure that a demonstrated quality system remains in place. The automotive industry has created its own definition of elements of a quality system, i.e., QS 9000. QS 9000 includes ISO 9000 requirements but incorporates additional quality system elements required by the automotive industry, e.g., process capability and process performance requirements.

The ISO 9000 series (1994 version) should be viewed as identifying the minimum elements of a quality system. The emphasis in ISO is not on results, but on the existence and conformance to the elements of a quality system. ISO 9000 covers only a portion of the scope of the Baldrige criteria. The Baldrige criteria focus on customer satisfaction results; ISO 9000 criteria focus on elements of a quality system. Thus Baldrige and ISO 9000 should be viewed as complementary. Fletcher (1999) describes how an organization can move beyond ISO 9000 to become a world-class firm by using the Baldrige criteria.

A major revision of the ISO 9000 standards was in the final stages in late 2000. The expected changes are explained in six issues of *Quality Progress,* October 1999 through March 2000.

It is expected that ISO 9001, 9002, and 9003 will be combined into a new ISO 9001 standard entitled Quality Management Systems—Requirements. Cianfrani, Tsiakals, and West (2000) explain the proposed document. The new ISO 9000 document is entitled Quality Management Systems—Fundamentals and Vocabulary. The new ISO 9004 is entitled Quality Management Systems—Guidelines for Improvements.

The structure of the proposed ISO 9001 has the clauses shown in Table 2.10.

ISO 9004 follows the same structure of clauses as ISO 9001. Also, ISO 9001 has been developed to be compatible with other standards such as ISO 14001: 1996 Environmental Management Systems—Specification with Guidance for Use.

ISO 9001 and 9004 have been developed as a consistent pair of quality management system standards. This pair of standards is based on eight quality management principles:

- Customer focus
- Leadership
- Involvement of people
- Process approach
- System approach to management
- Continual improvement
- Factual approach to decision making
- Mutually beneficial supplier relationships

To facilitate a convenient transition to the ISO 9000: 2000 series, it is expected that organizations will be permitted, if they desire, to apply for certification under the ISO 9001, 9002, 9003 (1994) requirements for a period of three years while the ISO 9000: 2000 series is being implemented.

This chapter discusses four components of a periodic assessment of quality. On a broader scale some organizations combine measurements from financial, customer, internal processes, and learning and growth areas into a "balanced scorecard." For elaboration, see Section 7.8.

TABLE 2.10
Contents of proposed ISO 9001

0	Introduction
0.1	General
0.2	Process approach
0.3	Relationship with ISO 9004
0.4	Compatibility with other management systems
1	Scope
1.1	General
1.2	Application
2	Normative reference
3	Terms and definitions
4	Quality management system
4.1	General requirements
4.2	Documentation requirements
5	Management responsibility
5.1	Management commitment
5.2	Customer focus
5.3	Quality policy
5.4	Planning
5.5	Responsibility, authority, and communication
5.6	Management review
6	Resource management
6.1	Provision of resources
6.2	Human resources
6.3	Infrastructure
6.4	Work environment
7	Product realization
7.1	Planning of realization process
7.2	Customer-related processes
7.3	Design and development
7.4	Purchasing
7.5	Production and service provision
7.6	Control of monitoring devices and measuring
8	Measurement, analysis, and improvement
8.1	General
8.2	Monitoring and measurement
8.3	Control of nonconforming product
8.4	Analysis of data
8.5	Improvement
Annex A (informative) Correspondence between ISO/DIS 9001:2000 and ISO 14001:1996	
Annex B (informative) Correspondence between ISO/DIS9001:2000 and ISO 9001:1994	

SUMMARY

- All organizations periodically need a companywide assessment of quality.
- Quality assessment comprises four elements:

  Cost of poor quality.
  Standing in the marketplace.

Quality culture.
Operation of the quality system.

- The Malcolm Baldrige National Quality Award provides criteria to identify organizations that have achieved the highest levels of quality. This award recognizes superior achievement.
- The ISO 9000 standards provide minimum criteria for a quality system. These documents provide some assurance to potential customers that an organization certified as meeting the standard has an adequate quality system.

PROBLEMS

2.1. The Federated Screw Company manufactures a wide variety of made-to-order screws for industrial companies. The designs are usually supplied by customers. Total manufacturing payroll is 260 people with sales of about $28 million. Operations are relatively simple but geared to high-volume production. Wire in rolls is fed at high speeds to heading machines, where the contour of the screw is formed. Pointers and slotters perform secondary operations. The thread-rolling operation completes the screw configuration. Heat treatment, plating, and sometimes baking are the final steps and are performed by an outside contractor located nearby.

You have been asked to prepare a quality cost summary for the company and have made the following notes:

- The quality control department is primarily a final inspection department (eight inspectors), which also inspects the incoming wire. Patrol inspection (one inspector) is performed in the Heading Room by checking the first and last pieces of each run. The quality control department also checks and sets all gages used by that department and by production personnel. An inspector's salary is approximately $24,000 a year.
- Quality during manufacture is the responsibility of the operator setup teams assigned to batteries of about four machines each. It is difficult to estimate how much of their time is spent checking setups or checking the running of the machines, so you have not tried to do this as yet. Production has two sorting inspectors, each earning $18,000, who sort lots rejected by final inspection.
- The engineering department prepares quotations, designs tools, plans the routing of jobs, and establishes quality requirements, working from customers' prints. The engineers also do troubleshooting, at a cost of about $20,000 a year. Another $16,000 is spent in previewing customers' prints to identify critical dimensions, trying to get such items changed by the customer, and interpreting customers' quality requirements into specifications for use by Federated inspectors and manufacturing personnel.
- Records of scrap, rework, and customer returns are meager, but you have been able to piece together a certain amount of information from records and estimates:

  Scrap from final inspection rejections and customer returns amounted to 438,000 and 667,000 pieces, respectively, for the last two months.
  Customer returns requiring rework average about 1 million pieces per month.
  Scrap generated during production is believed to be about half of the total floor scrap (the rest not being quality related) of 30,000 pounds per month.

Final inspection rejects an average of 400,000 reworkable pieces per month. These items can then be flat rolled or rerolled.

- Rough cost figures have been obtained from the accountants, who say that scrap items can be figured at $12.00 per thousand pieces, floor scrap at $800 per thousand pounds, reworking of customer returns at $4.00 per thousand pieces, and flat rolling or rerolling at $1.20 per thousand pieces. These figures are supposed to include factory overhead.

Prepare a quality cost summary on an annual basis. (This example was adapted from one originally prepared by L. A. Seder.)

2.2. Review the explanation of the marketing research study depicted in Table 2.5. What additional question would have been useful to ask during the study?

2.3. Information on the present quality culture in an organization is an important input to assessment. Asking employees their opinions about the quality culture incurs some risks. State two such risks.

2.4. In a survey used to learn about the quality culture in an organization, employees are asked questions that relate to three levels—upper management, their own manager, and the people in their work group. For each of these levels, state two questions that would identify employee perceptions about the quality culture.

2.5. When consumers eat at a fine restaurant, they expect excellent food and service. Research has identified seven features of restaurant service. What are these service features?

REFERENCES

Atkinson, H., J. Hamburg, and C. Ittner (1994). *Linking Quality to Profits,* ASQ Quality Press, Milwaukee, and Institute of Management Accountants, Montvale, NJ.

Bacivarof, I. C. (2000). "Current Events in Europe," *Quality Engineering,* vol. 12, no. 3, p. 478.

Blazey, M. L. (2000). *Insights to Performance Excellence,* Quality Press, ASQ, Milwaukee.

Campanella, J. ed. (1999). *Principles of Quality Costs,* 3rd ed., ASQ, Milwaukee.

Churchill, G. A. Jr. (1991). *Marketing Research Methodological Foundations,* Dryden Press, Chicago, pp. 399–400.

Cianfrani, C. A., J. J. Tsiakals, and J. E. West (2000). *ISO 9001: 2000 Explained,* ASQ Quality Press, Milwaukee.

Cokins, G. (1999). "Why Is Traditional Accounting Failing Quality Managers? Activity Based Costing Is the Solution," *Annual Quality Congress Proceedings,* ASQ, Milwaukee.

Conti, T. (1997). *Organizational Self-Assessment,* Chapman & Hall, London.

Drew, J. H. and C. A. Castrogiovanni (1995). "Quality Management for Services: Issues in Using Customer Input," *Quality Engineering,* vol. 7, no. 3, pp. 551–566.

Evans, J. R. and M. W. Ford (1997). "Value-Driven Quality," *Quality Management Journal,* vol. 4, no. 4, pp. 19–31.

Fletcher, A. C. (1999). "Moving beyond ISO 9000: Becoming a World-Class Organization Using the MBNQA Criteria," *Annual Quality Congress Proceedings,* ASQ, Milwaukee, pp. 306–312.

Getty, R. L. and J. M. Getty (1993). "Organizations Hold the Initiative for Improvement," *Annual Quality Congress Transactions,* ASQ, Milwaukee, pp. 710–716.

Gryna, D. S. and F. M. Gryna (1999). "Quality in Banking Starts with Four Assessments," *Quality Progress,* August, pp. 27–34.

Hannukainen, T. and S. Salminen (1998). "Setting the Course for Quality: A Case Study in Applied Self-Assessment," *Annual Quality Congress Proceedings,* ASQ, Milwaukee, pp. 316–323.

Hromi, J., ed. (1995). *The Best on Quality,* IAQ Book Series, vol. 5, ASQ Quality Press, Milwaukee.

Lauter, B. E. (1997). "Determining the State of Your Customers," Sterling Quality Conference, Orlando, FL.

Lowenstein, M. W. L. (1995). *Customer Retention,* ASQ Quality Press, Milwaukee.

Puay, S. H., K. C. Tab, M. Xie, and T. N. Goh (1998). "A Comparative Study of Nine National Quality Awards," *TQM,* vol. 10, no. 1, pp. 30–39.

Rucci, A. J., S. P. Kirn, and R. T. Quinn (1998). "The Employee—Customer—Profit Chain at Sears," *Harvard Business Review,* January–February, pp. 82–97.

Rust, R. T., A. J. Zahorik, and T. L. Keiningham (1994). *Return on Quality,* Probus, Chicago.

Schottmiller, J. C. (1996). "ISO 9000 and Quality Costs," *Annual Quality Congress Proceedings,* ASQ, Milwaukee, pp. 194–199.

Tabladillo, M. Z. and S. Canfield (1994). "Creation of Management Performance Measures from Employee Surveys," *Quality Management Journal,* July, pp. 52–66.

Werner, F. M. and J. A. F. Stoner (1995). *Modern Financial Managing,* Harper Collins College Publishers, New York, pp. 143–144.

Zeithaml, V. A., A. Parasuraman, and L. A. Berry (1990). *Delivering Service Quality: Balancing Customer Perceptions and Expectations,* The Free Press, New York.

SUPPLEMENTARY READING

Quality costs: *JQH5,* Section 8.

- Campanella, J., ed. (1999). *Principles of Quality Costs,* 3rd ed., ASQ, Milwaukee.
- Miller, J. R. and J. S. Morris (2000). "Is Quality Free or Profitable?" *Quality Progress,* January, pp. 50–52.

Standing in the marketplace: *JQH5,* Section 18.

Quality culture: *JQH5,* pp. 22.65–22.66.

Assessment of current quality activities: *JQH5,* Section 11 and pp. 14.17–14.28.

- ISO 9000-9004 Quality Standards.
- Malcolm Baldrige National Quality Award Criteria for Performance Excellence, National Institute of Standards and Technology, Gaithersburg, MD 20899-1020.
- Wilson, D. D. (1997). "An Empirical Study to Test the Causal Linkages Implied in the Malcolm Baldrige National Quality Award." Ph.D. dissertation, UMI Dissertation Services, Ann Arbor, MI.

WEBSITES

Baldrige Updates: www.quality.nist.gov
Directory of State Quality Awards: www.apqc.org/statequa.htm
European Foundation for Quality Management: www.efqm.org/
International Organization for Standardization (ISO): www.iso.ch/welcome.html
Quality standards: http://e-standards.asq.org/perl/catalog.cgi

3

QUALITY IMPROVEMENT AND COST REDUCTION

3.1 SPORADIC AND CHRONIC QUALITY PROBLEMS

The CEO of a large American automobile manufacturer said this: "Of the final cost from start to finish of delivering a vehicle and selling it, probably 1/3 is waste now" (*Business Week,* January 12, 1998). Of the trilogy of quality processes (see Section 1.6, "Managing for Quality"), the process of quality improvement plays a dominant role in reducing the costs of waste.

The costs associated with poor quality are due to both *sporadic* and *chronic* quality problems (see Figure 3.1). A sporadic problem is a sudden, adverse change in the status quo, which requires remedy through *restoring* the status quo (e.g., changing a depleted reagent chemical). A chronic problem is a long-standing adverse situation, which requires remedy through *changing* the status quo (e.g., revising an unrealistic specification).

"Continuous improvement" (called *kaizen* by the Japanese) has acquired a broad meaning, i.e., enduring efforts to act upon both chronic and sporadic problems and to make refinements to processes. For chronic problems, it means achieving better and better levels of performance each year; for sporadic problems, it means taking corrective action on periodic problems; for process refinements, it means taking such action as reducing variation around a target value.

The distinction between chronic and sporadic problems is important for two reasons:

1. The approach to solving sporadic problems differs from that to solving chronic problems. Sporadic problems are attacked by the control process defined and developed in Chapter 5. Chronic problems use the improvement process discussed in this chapter.
2. Sporadic problems are dramatic (e.g., an irate customer reacting to a shipment of bad parts) and must receive immediate attention. Chronic problems are not dramatic

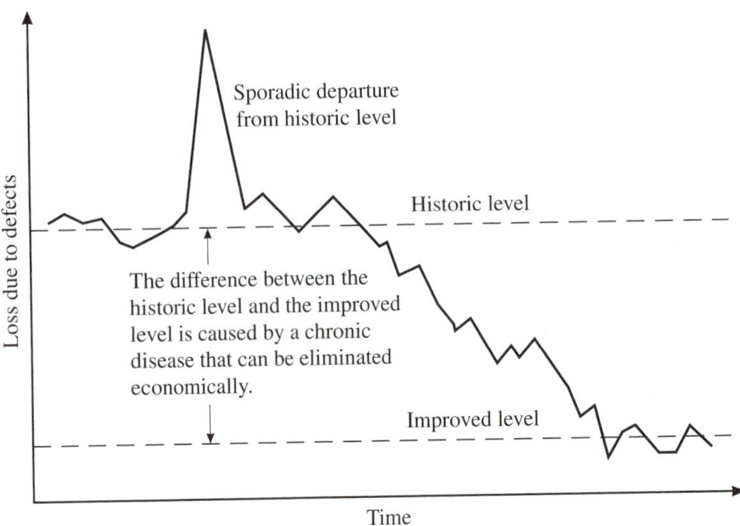

FIGURE 3.1
Sporadic and chronic quality problems.

because they occur for a long time (e.g., 2 percent scrap has been typical for the past five years), are often difficult to solve, and are accepted as inevitable. The danger is that the fire fighting on sporadic problems may take continuing priority over efforts to achieve the larger savings that are possible, i.e., on chronic problems.

3. A key reason for the chronic waste present in organizations is the lack of a structured approach to identify and reduce the waste. This chapter provides that structure.

We need to explain the sequence of this book's quality chapters. Logic would have it that first we plan (Chapter 4), then we measure and control (Chapter 5), and then we improve (Chapter 3). In the real world, logic does not always prevail. We will first cover the improvement process because most organizations have serious quality problems that practical people want to correct. People are more motivated to correct current problems and less motivated to do better planning. (Later you will see that the cause of many current problems is poor planning, but this relationship needs to be personally discovered before there is a burning desire to do good planning.)

A structured approach to improvement applies not only to quality but also to other parameters, e.g., productivity, cycle time, and safety. Addressing chronic problems achieves breakthrough to an improved level (Figure 3.1). Acting on chronic problems is best achieved by the "project by project" approach.

3.2 PROJECT-BY-PROJECT APPROACH

The most effective approach to improvement is project by project. Here a project is a chronic quality-related problem that has been chosen for solution. The project approach can apply to all three processes in the quality trilogy and thus be the basis of a total quality initiative. A quality improvement project addresses the deficiencies dimension of quality; a quality planning project focuses on planning (or replanning) product and process features; a quality control project analyzes a collection of sporadic problems. In practice, organizations benefit most by starting with quality improvement projects.

Setting up the approach for quality improvement, quality planning, or quality control projects comprises three main steps:

- Proving the need.
- Identifying projects.
- Organizing project teams.

Carrying out a quality improvement project involves these tasks:

- Verifying the project need and mission.
- Diagnosing the causes.
- Providing a remedy and proving its effectiveness.
- Dealing with resistance to change.
- Instituting controls to hold the gains.

This approach was originally proposed by Juran in 1964 as the "breakthrough sequence" for improvement (Juran, 1964). Other approaches to improvement include plan, do, study, act; reengineering, theory of constraints, and six sigma. Each approach brings fresh ideas, and organizations have learned to continuously integrate these new ideas with the older successful methods.

The phases of the six-sigma approach are define, measure, analyze, improve, and control. This chapter presents an integrated view of Juran's breakthrough sequence and the six-sigma approach. I am indebted to two former colleagues at Juran Institute for their help. Richard Chua supplied the raw materials on six sigma, and both he and William Barnard provided many useful comments on the two approaches.

To place individual projects in perspective, we first provide a summarized example. Next we discuss how to prove the need for a major improvement initiative consisting of many projects. Experiences with the project approach follow. Then we are ready to describe the steps for setting up and carrying out each project.

3.3 EXAMPLE OF A PROJECT

The problem (Betker, 1983) concerns the soldering process used at the GTE Corporation in the manufacture of printed circuit boards (PCBs). A typical PCB has 1700 connections. Any solder connection can cause testing problems or performance and reliability problems for the customer. In the following discussion we trace the steps of the breakthrough sequence for improvement (see above, "Project-by-Project Approach"). Each step in the breakthrough sequence is related to the corresponding step in the six-sigma approach.

Verify the Project Need and Mission (the Define Step in Six Sigma)

The process was statistically out of control and numerous solder connections required touch-up. (Statistical control is discussed in Chapter 18, "Statistical Process Control.") The project team's mission: reduce the number of defective solder connections.

Diagnose the Causes (the Measure and Analyze Steps in Six Sigma)

A team of people, not from one department but from several cross-functional departments, was set up to guide the project and do the diagnosis. Figure 3.2 (a Pareto diagram; for elaboration, see Section 3.7) depicts the distribution of symptoms by type of solder defect. Data on the defects were analyzed and theories were offered on the causes of the defects. Figure 3.3 is a cause-and-effect diagram summarizing the theories. These theories were grouped into three categories, thereby allowing a checklist to be developed that supervisors and the control inspector used to evaluate the theories. After additional data collection and analysis, low solder temperature was found to be the main cause of defects. Figure 3.4 shows part of the analysis.

Provide a Remedy and Prove Its Effectiveness (the Improve Step in Six Sigma)

Data and further analysis revealed that, for ideal soldering conditions, either the temperature of the solder should be raised or the conveyor speed of the wave soldering

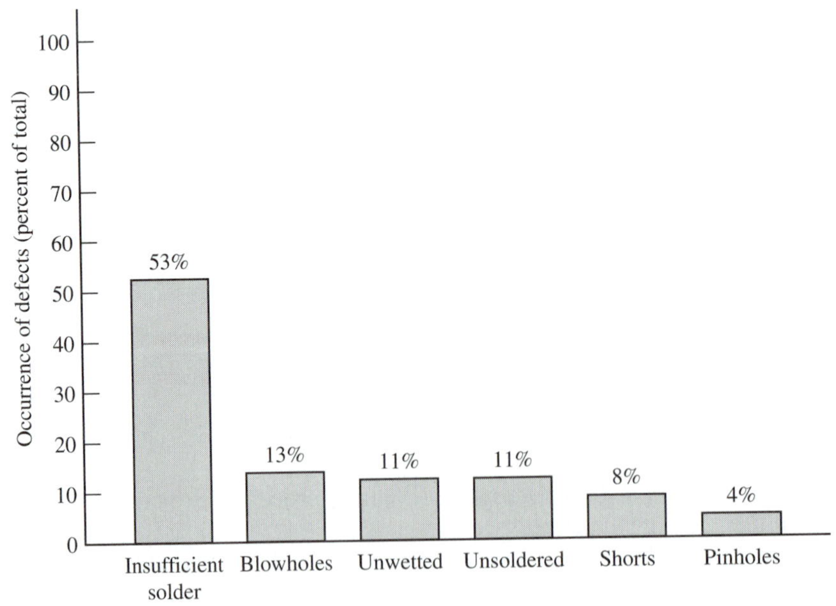

FIGURE 3.2
Solder defect types, Pareto analysis. (*From Betker, 1983.*)

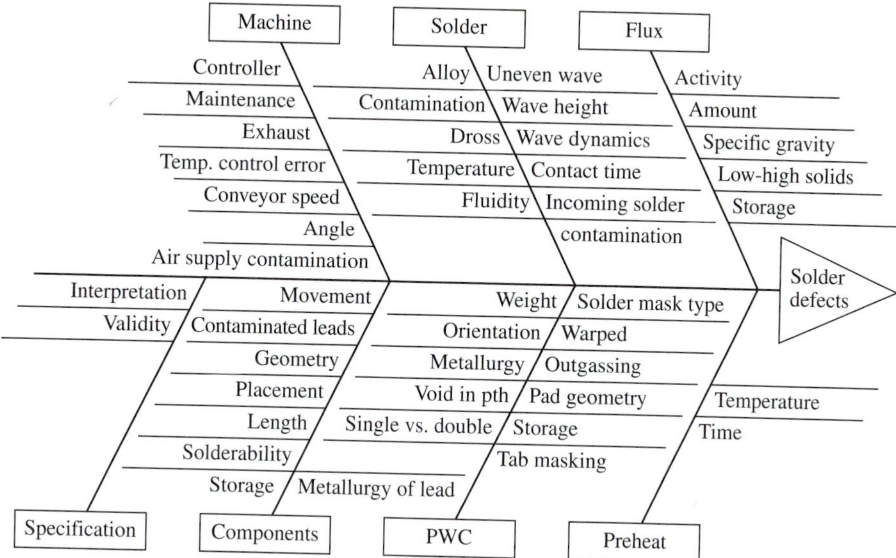

FIGURE 3.3
Ishikawa cause-and-effect diagram. (*From Betker, 1983.*)

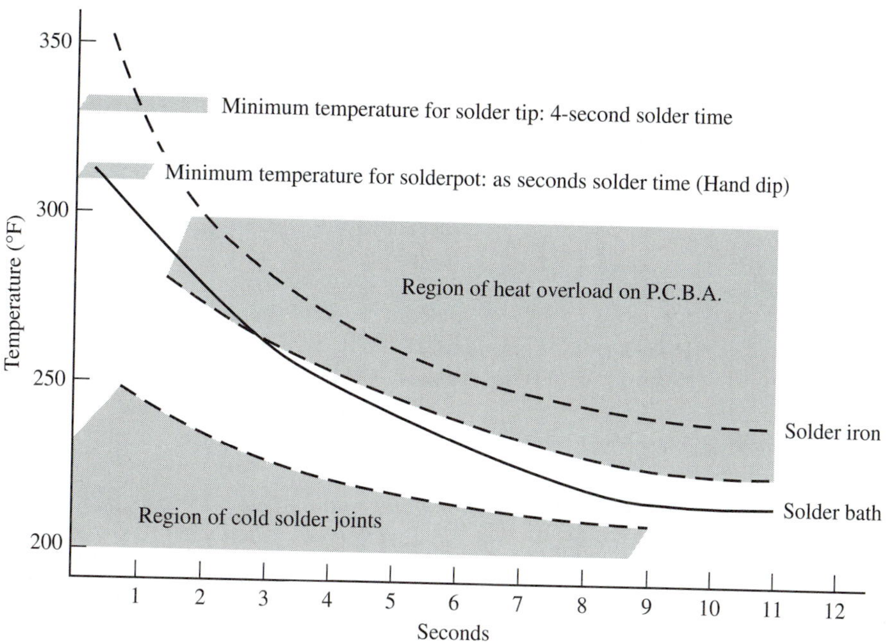

FIGURE 3.4
Temperature/time relationship when flow or dip soldering is applied. (*From Betker, 1983.*)

machine should be reduced. These were remedies to remove the cause. A trial was conducted using a higher temperature. This resulted in an improvement in solder defects without any adverse effects.

Deal with Resistance to Change (the Improve Step in Six Sigma)

From the start of the project, a manufacturing engineer on the team argued that the cause was outside the control of the machine. The diagnosis explained above convinced him otherwise. But he felt that raising the temperature would result in "reflow of tin under the solder mask, thereby causing solder shorts and peeling of the solder mask." This belief, based on a trial conducted 10 years earlier on other equipment, had been expressed so often that it was no longer questioned. The trial of the remedy overcame this resistance.

Institute Controls to Hold the Gains (the Control Step in Six Sigma)

The defect level was reduced by 62 percent, and the out-of-control points on statistical control charts were eliminated. To assure that the improved level was maintained, the process was monitored. Not only was the improved level maintained but the elimination of the dominant cause (low temperature) unmasked other causes. Performance has been improved to the point that the hand solder touch-up operation may be eliminated.

This example provides an overview of the breakthrough sequence and the six-sigma approach as applied to individual quality projects. Carrying out quality improvement projects on chronic problems takes resources, and so we next address how to prove the need for such resources.

3.4 PROVING THE NEED FOR A MAJOR QUALITY IMPROVEMENT INITIATIVE

To gain management approval for a major quality initiative requires several steps:

1. *Estimate the size of the chronic waste or other quality-related losses.* The key data for this step comes from studies on the cost of poor quality and on competitive standing in the marketplace as described in Chapter 2, "Companywide Assessment of Quality."

   Usually, managers are stunned by the size of the cost of waste and/or lost sales income due to poor quality. To establish proof of the need, different "languages" may be required for different levels of management. For upper management, the language of money works best; for lower levels, other languages are effective. Table 2.3 shows examples of statements in money and other languages.

2. *Estimate the savings and other benefits:*
   a. If the organization has never before undertaken a program to reduce quality-related costs, then a reasonable goal is to cut these costs in half, within five years.
   b. Don't imply that the quality costs can be reduced to zero.
   c. For any benefits that cannot be quantified as part of the return on quality, present these benefits as intangible factors to help justify the improvement program. Some benefits can be related to problems of high priority to upper management, such as meeting delivery schedules, controlling capital expenditures, or reducing a delivery backlog. In a chemical company a key factor in justifying an improvement program was the ability to reduce significantly a major capital expenditure to expand plant capacity. A large part of the cost of poor quality was due to having to rework 40 percent of the batches every year. The improvement effort was expected to reduce the rework from 40 percent to 10 percent, thus making available production capacity that was no longer needed for rework. What convinced management was not the savings in rework costs, but the reduction of capital expenditures.
3. *Calculate the return on investment resulting from improvement.* This return should reflect savings in the traditional cost of poor quality, savings in process capability improvement, and increases in sales revenue due to a reduction in customer defections and an increase in new customers. For elaboration, see Section 2.4 (including the Dupont model) and *JQH5,* Section 8 under "Return on Quality."
4. *Use a successful case history (a bellwether project) in the organization to justify a broader program.* One or more of these pilot projects not only shows the tangible benefits that can result but also demonstrates that a successful step-by-step approach has been developed in the organization. These short-term wins must be genuine, visible, and related to the overall improvement effort. The following scenario illustrates how the ABC electronics company used this approach.

The estimated cost of poor quality was $200 million per year, and a notorious quality problem was scrap for a certain major electronic component. This scrap ran to about $9 million per year. The principal defect type was defect X, and it was costing about $3 million per year ("proof of the need" for eliminating defect X).

The company took on a project to reduce the incidence of defect X. The project was a stunning success. The cost of defect X was cut from $3 million to $1 million—an annual profit improvement of $2 million. An investment of about $250,000 was needed.

Then followed an exciting extrapolation and contrast. Engineers estimated that extending the improvement to the entire $200-million cost of poor quality could cut the total in half—thus creating a profit improvement of $100 million annually.

The defect X project proved to ABC managers that they could get a big return on investment by improving quality. This in-house project was more convincing to results-oriented managers than any number of lectures, books, or success stories about other companies.

Next we summarize experiences with the project-by-project-approach and then examine the specific steps in more detail.

3.5
EXPERIENCES WITH THE PROJECT-BY-PROJECT APPROACH

Experiences in both manufacturing and service industries have led to encouraging conclusions:

- Large cost reductions and improved quality to customers have been achieved. For each dollar invested in improvement activity, the return is between $5 and $10.
- Investment required for improvement has been modest and *not* capital intensive. Most of the investment is in the time of people doing diagnosis for the projects.
- Most projects can be completed in six months if the scope in the mission statement is carefully defined.
- The key chronic quality-related problems cut across departments and thereby require cross-functional project teams.
- Some organizations consist of multiple autonomous units that have similar activities, e.g., manufacturing plants, hospitals, hotels. An improvement project that is successful in one unit may apply to another unit without the need for duplicating the diagnosis work. For elaboration on this "cloning," see *JQH5,* pages 5.29–5.30.
- The improvement approach should include suppliers—both by encouraging or requiring suppliers to have their own improvement program and by executing joint quality improvement projects with suppliers (see Chapter 15 under "Supplier Quality Improvement").
- Like any major activity (e.g., new-product development), improvement must include goals in the annual business plan that are aligned with other business goals and strategies, organizational infrastructure to execute improvement (process for selecting projects and the formation, training, and support of improvement teams), training, review of progress by upper management, and reward and recognition. All of these matters should be integrated into a road map for implementation (see Chapter 7 under "Implementing Total Quality").

Of particular concern is the need for sufficient time and resources for people to carry out their improvement responsibilities. In many companies employees believe that they are asked to work on improvement as an "add on" to their regular responsibilities and are not given sufficient time to do the improvement work.

An increasing number of companies have reported the completion of more than 1000 projects in about four years. Using the six-sigma approach, General Electric reports an increase in projects from 200 in 1995 to 47,000 in 1999 (Slater, 1999). Today's competitive business conditions dictate a revolutionary rate of improvement to replace the evolutionary rate of the past. Such improvement means replication of projects using the project-by-project approach.

Blakeslee (1999) provides a useful discussion of seven principles to ensure success in applying the six-sigma approach to project-by-project improvement.

3.6 INTRODUCTION TO SIX SIGMA

The six-sigma approach is a collection of managerial and statistical concepts and techniques that focus on reducing variation in processes and preventing deficiencies in product.

Variation in a process is denoted by sigma—the standard deviation of measurements around the process mean (see Chapter 9 under "The Concept of Variation"). In a process that has achieved six-sigma capability, the variation is small compared to the range of the specification limits; i.e., there are six standard deviations between the process mean and either specification limit. (In the earlier days of the quality movement, a process was considered adequate if there were three standard deviations between the process mean and either specification limit.) Even if the process mean shifts (by 1.5 sigma), no more that 3.4 units per million fall outside the specification limits. ("Units" may be parts, lines of code, transactions, or other forms of output.) Thus the higher the number of sigmas the better. Most processes are at about 3 to 4 sigma. For elaboration on the statistical concept, see Chapter 18 under the "Six-Sigma Concept of Process Capability."

A key focus is the relationship between the input variables and the outputs of a process, expressed as

$$Y = f(X_1 \ldots X_n)$$

Conceptually, product results (Y) are a function (f) of many process variables $X_1 \ldots X_n$. Thus Y is an output, an effect, a dependent variable; X are inputs, causes, dependent variables. The six-sigma approach identifies the process variables that cause variation in product results. Some of these process variables are critical to quality and are set at a certain value and maintained within a specified range (i.e., "controllable variables"). Other variables cannot be easily maintained around a certain value and are considered uncontrollable or "noise." For elaboration, see Sanders, Ross, and Coleman (1999).

Six sigma makes use of five phases:

1. *Define.* This step identifies potential projects, selects and defines a project, and sets up the project team.
2. *Measure.* This step documents the process and measures the current process capability.
3. *Analyze.* This step collects and analyzes data to determine the critical process variables.
4. *Improve.* This step conducts formal experiments, if necessary, to focus on the most important process variables and determine the process settings to optimize product results.

5. *Control.* This step measures the new process capability, documents the improved process, and institutes controls to maintain the gains.

The six-sigma approach and the breakthrough sequence are identical in objective and similar in the steps to achieve the objective. Both have an objective of finding the causes of deficiencies and developing remedies to prevent future deficiencies. Both employ the same tools and techniques, most of which were developed in the 1950s. Many of the quantitative tools were underutilized in the past because of their relative complexity (for some users), but recent advances in information technology have simplified the data collection and analysis, thereby facilitating the use of these tools. For a discussion of six sigma in product development, see Chapter 13 under "Designing for Six Sigma."

Harry and Schroeder (2000), Hahn et al. (2000), and Henson (1999) in the Supplementary Reading present overviews of six sigma, including examples from General Electric and other organizations. Bajaria (1999) in the Supplementary Reading identifies certain elements ("points") of six sigma but then raises some issues ("counterpoints") to provide perspective.

The remainder of this chapter examines the five phases of six sigma.

3.7 DEFINE PHASE

This phase identifies potential projects, selects and defines a project, and sets up the project team. The steps are

- Identify potential projects.
- Evaluate projects.
- Select project.
- Prepare problem and mission statement for project.
- Select and launch project team.

Identify Potential Projects

Project identification consists of nominating, screening, and selecting projects. *The focus must be on the vital few opportunities that will increase customer satisfaction and reduce the cost of poor quality.* All projects should be aligned with the organization's mission and quality goals. The six-sigma approach particularly uses financial measurements to select and measure the success of projects.

Nomination for projects

Nominations come from several sources:

- Analysis of data on the cost of poor quality, quality standing in the marketplace, employee satisfaction studies, or other forms of assessment (see Chapter 2).

- Analysis of other field intelligence, e.g., input from sales, customer service, and other personnel to retain current customers and attract new customers.
- Gaps between goals and actual performance.
- Inputs from all levels of management and the workforce.
- Benchmarking studies of other organizations (see Chapter 7 under "Competitive Benchmarking").
- Developments arising from the impact of product quality on society, e.g., government regulation, growth of product liability lawsuits.

Project nominations address both improvement of effectiveness (meeting customer needs) and efficiency (meeting those needs at minimum cost). Note that the six-sigma and the breakthrough-sequence approaches are strategic because projects should be based on gaps between actual performance and goals.

A data analysis tool for generating project nominations is the Pareto principle.

The Pareto principle

As applied to the cost of poor quality, the Pareto principle states that a few contributors to the cost are responsible for the bulk of the cost. These vital few contributors need to be identified so that quality improvement resources can be concentrated in those areas.

A study of quality-related costs at a paper mill showed a total of $9.07 million (Table 3.1a). The category called "broke" (paper mill jargon for paper so defective that it is returned to the beater for reprocessing) amounts to $5.56 million, or 61 percent of the quality costs. Clearly, no major reduction in these costs will occur without a successful attack on broke—which is where the loss of money is concentrated.

In that paper mill, 53 types of paper are made. When broke is allocated among the various types of paper, the Pareto principle is again in evidence (Table 3.1b). Six of the product types account for $4.48 million, which is 80 percent of the $5.56 million. No major improvement will occur in broke without a successful attack on these six types of paper. Studying 12 percent (6 types out of 53) of the problem results in attacking 80 percent of broke.

Finally, it is helpful to look at what kinds of defects are being encountered in these six types of paper, and the associated costs for broke. The matrix of Table 3.1c shows this analysis. There are numerous defect types, but five dominate. In addition, the cost figures in the table also follow the Pareto principle. The largest number is $612,000 for tear on paper type B, then comes $430,000 for porosity on A, and so on. Such analysis would be helpful in nominating projects for cost reduction.

Evaluate potential projects

Project nominations are usually reviewed by middle management, and recommendations are then made to upper management for final approval.

The review varies from an analysis of the project scope and potential benefit to a formal examination of several factors to help set priorities. For example, an insurance company screens potential projects by asking six questions: Can we impact? Can we analyze? Are data available? Are they measurable? What areas are affected? What is the level of control?

TABLE 3.1A
Pareto analysis by accounts—quality losses in a paper mill

Accounting category	Annual quality loss,* $ thousands	Total quality loss, %	
		This category	Cumulative
"Broke"	5560	61	61
Customer claim	1220	14	75
Odd lot	780	9	84
High material cost	670	7	91
Downtime	370	4	95
Excess inspection	280	3	98
High testing cost	190	2	100
Total	9070		

*Adjusted for estimated inflation since time of original study.

TABLE 3.1B
Pareto analysis by products—"broke" losses in a paper mill

Product type	Annual broke loss,* $ thousands	Broke loss, %	Broke loss, cumulative %
A	1320	24	24
B	960	17	41
C	720	13	54
D	680	12	66
E	470	8	74
F	330 (4480)	6	80
47 other types	1080	20	100
Total 53 types	5560	100	

*Adjusted for estimated inflation since time of original study.

TABLE 3.1C
Matric of quality costs*

Type	Trim, $ thousands	Visual defects,† $ thousands	Caliper, $ thousands	Tear, $ thousands	Porosity, $ thousands	All other causes, $ thousands	Total, $ thousands
A	270	94	None‡	162	430	364	1320
B	120	33	None‡	612	58	137	960
C	95	78	380	31	74	62	720
D	82	103	None‡	90	297	108	680
E	54	108	None‡	246	None‡	62	470
F	51	49	39	16	33	142	330
Total	672	465	419	1157	892	875	4480

*Adjusted for estimated inflation since time of original study.

†Slime spots, holes, wrinkles, etc.

‡Not a specified requirement for this type.

TABLE 3.2
Ranking by use of Pareto Priority Index (PPI)

Project	Savings, $ thousands	Probability	Cost, $ thousands	Time, years	PPI
A	100	0.7	10.0	2.0	3.5
B	50	0.7	2.0	1.0	17.5
C	30	0.8	1.6	0.25	60.0
D	10	0.9	0.5	0.50	36.0
E	1.5	0.6	1.0	0.10	9.0

Hartman (1983) describes an approach at AT&T that makes use of a Pareto priority index (PPI) to evaluate each project.

$$\text{PPI} = \frac{\text{Savings} \times \text{probability of success}}{\text{Cost} \times \text{time to completion (years)}}$$

Table 3.2 shows the application of this index to five potential projects. High PPI values suggest high priority. Note how the ranking of projects A and C is affected when the criterion is changed from cost savings alone to the index covering the four factors.

The result of the review by middle management is a recommended list of projects. Typically, one responsibility of an upper management quality council is reviewing the recommendations or creating the organizational machinery for review and final approval.

Pareto analysis is an example of *data mining,* i.e., a process for analyzing data to extract information that is not offered by the raw data alone. Koch (1998) in the Supplementary Reading presents a wide-ranging discussion of the Pareto principle that covers applications to organizations and individuals for both business and personal effectiveness. The Pareto concept is called "key-factor analysis" in disciplines such as sociology, animal behavior, ecology, and toxicology.

The Pareto principle identifies the "vital few" projects for improvement. These projects are the major contributors to quality leadership in terms of sales revenue and lower costs. Such projects usually go across functions and require cross-functional quality improvement teams. This chapter addresses the vital few projects. Beyond the vital few projects are the "useful many" projects. Typically the useful many projects are not cross functional and provide an opportunity for employee participation through workforce teams within a department; see Chapter 8 under "Workforce Teams." The total improvement effort must include both the vital few and the useful many projects.

Selection of Initial Projects

"The first project should be a winner." A successful project is a form of evidence to the project team members that the improvement process does lead to useful results. Ideally:

- The project should deal with a chronic problem—one which has been awaiting solution for a long time.

- The project should be feasible, i.e., have a good likelihood of being brought to a successful conclusion within about six months.
- The project should be significant. The results should be sufficiently useful to merit attention and recognition.
- The results should be measurable in money as well as in technological terms.
- The project should serve as a learning experience for the process of problem solving.

Problem and Mission Statement for Project

A problem statement identifies a visible deficiency in a planned outcome; e.g., "During the past year, 7 percent of invoices sent to customers included errors." Note that this statement is specific and manageable (names a specific and limited process) and observable and measurable (describes the size of the problem). A problem statement should never imply a cause or a solution.

A mission statement is based on the problem statement but provides direction to the project team. A goal or other measure of project completion and a target date should be defined. For example, the team is asked to reduce the error rate in invoices to 2 percent or less within the next three months. A mission statement should not imply cause or solution—that is the role of later steps based on data.

Select and Launch the Project Team

A project team usually consists of about six to eight persons drawn from multiple departments and assigned to address the selected chronic problem. Suppliers and customers may also be part of the team.

The team meets periodically, and members usually serve part time in addition to performing their regular functional responsibilities. When the project is finished, the team disbands.

A project team typically has a sponsor ("champion"), a leader, a recorder, team members, and a facilitator. Their roles on the project team are critical. Other improvement teams include "blitz teams" (for accelerated improvement) and "virtual teams" (involving different geographical locations). These matters are discussed in Chapter 8 under "Quality Project Team."

To help launch a team, some organizations develop a charter that defines *what* the team will do (e.g., mission) and *how* the team will function (e.g., principles used to make decisions). Van Aken and Sink (1996) describe 14 elements of a charter. In addition, their research concluded that teams having a charter ("team reviewed" or "team developed") had both higher team performance and higher team satisfaction than teams with no charter.

Early and Godfrey (1995) summarize a multi-industry study of 20 improvement projects that were identified as taking too long. Of the wasted time, 39 percent was related to management not providing preparation and support (e.g., vague mission, limited time dedicated to the project, management delay in confronting resistance) and

24 percent related to the teams not using best practices (e.g., not focusing on the vital few causes, too much flow diagramming too soon, prematurely jumping to remedy).

In practice, conducting effective team meetings requires certain skills that, frankly, many people do not possess. These skills include planning for team meetings and conducting the meetings. Planning involves matters such as logistics (time, location, use of electronic meeting software), setting meeting objectives, and preparing and distributing an agenda and other documentation. Conducting the meeting requires skills in developing participation, listening, and trust; handling problems (overly talkative members, quiet members, side conversations, absent members); resolving disagreements; and guiding the team to decisions. Some organizations have found it necessary to provide special training for team leaders.

3.8 MEASURE PHASE

This phase identifies key product parameters and process characteristics and measures the current process capability. The steps are

- Measure the baseline performance and verify the project need.
- Document the process.
- Plan for data collection.
- Validate the measurement system.
- Measure the process capability.

Measure Baseline Performance and Verify the Project Need

Presumably, a project has been selected because it is important. It is useful, however, to verify the size of the problem *in numbers*. This process serves two purposes: (1) assure that the time to be spent by the project team is justified and (2) help to overcome resistance to accepting and implementing a remedy. Verifying the need for an individual project makes use of the same type of information discussed earlier under "Prove the Need." In addition, the *scope* of the project must be reviewed after the team has met once or twice, to be sure that the mission assigned to the team can be accomplished within, say, about six months. Otherwise, the project should be divided into several projects. Failure is likely if a project stretches out like a freight train.

Document the Process

This step records the activities under study and information relating to actual or potential problems.

A useful tool is the process flow diagram (or "process map"). An example for the process of admitting patients to a hospital emergency room is given in Figure 3.5

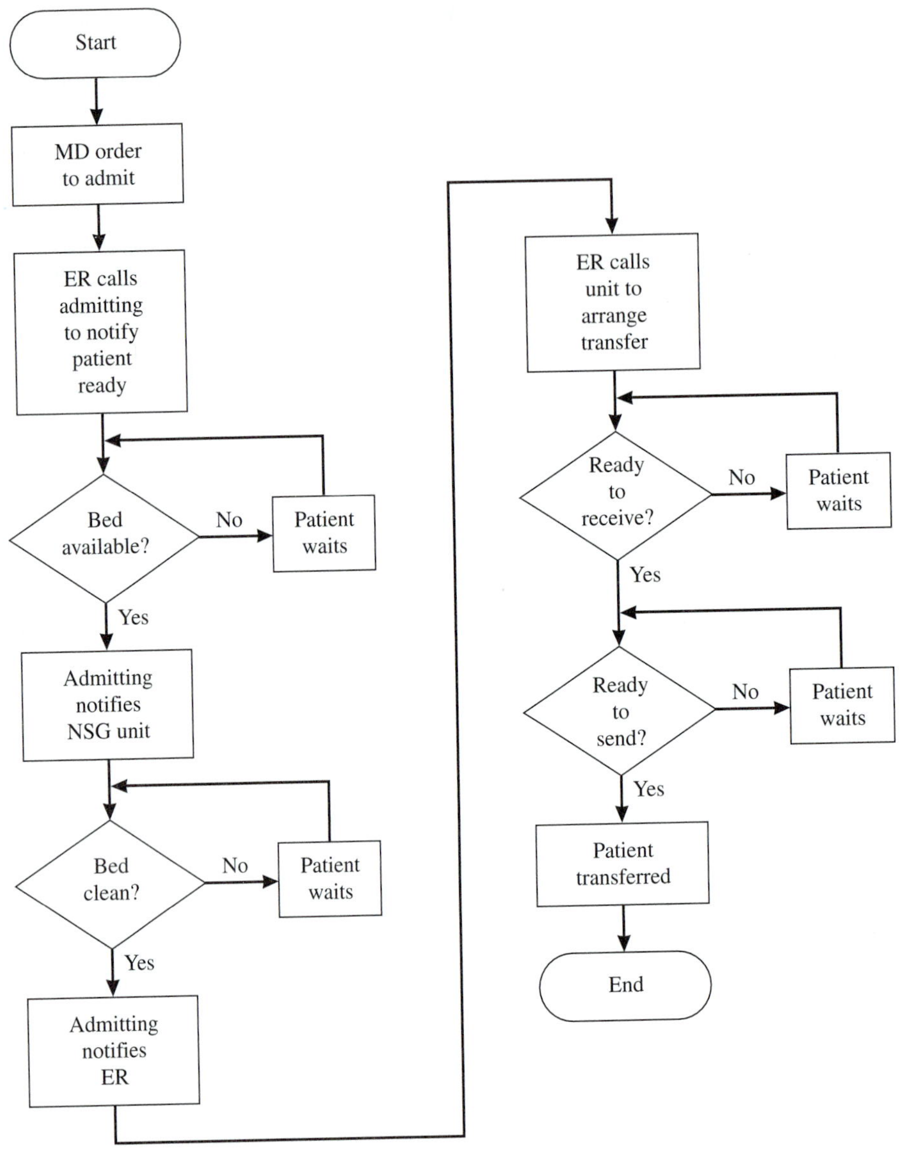

FIGURE 3.5
High-level flow diagram to admit ER patient.

(Lenhardt, 1993). The diagram shows the sequence of steps and their relationships in the process. Study of the flow diagram, in conjunction with other information (see below), enables us to develop a list of product characteristics (or key process output variables [KPOVs]) and key process parameters (or key process input variables [KPIVs]). The KPOVs and KPIVs provide the detail for the conceptual framework of $Y = f(X)$, i.e., relating process variables to process results.

Other information is useful in documenting the process, but first we need to define some terms:

A *defect* (or disconnect) is any nonfulfillment of intended usage requirements, e.g., oversize, low mean time between failures, illegible invoice. A defect can also go by other names, e.g., error, discrepancy, nonconformance.

A *symptom* is an observable phenomenon arising from and accompanying a defect. Sometimes, but not always, the same word is used both as a defect description and as a symptom description, e.g., open circuit. More usually, a defect will have multiple symptoms; e.g., "insufficient torque" may include the symptoms of vibration, overheating, erratic function, etc.

A *theory* is an unproved assertion as to reasons for the existence of defects and symptoms. Usually, several theories are advanced to explain the presence of the observed phenomena.

A *cause* is a proven reason for the existence of the defect. Multiple causes are common, in which case they follow the Pareto principle; i.e., the vital few causes will dominate all the rest.

A *remedy* is a change that can successfully eliminate or neutralize a cause of defects.

Two journeys are required for quality improvement: the diagnostic journey from symptom to cause and the remedial journey from cause to remedy. This distinction is critical. To illustrate, three supervisors were faced with a problem of burrs on screws at the final assembly of kitchen stoves. In their haste to act, they skipped the diagnostic journey and concluded that better screws were needed (a remedy). Fortunately, a diagnostician interceded. He pointed out that three separate assembly lines were feeding product into one inspection station and suggested that the data be segregated by assembly line. The data revealed that the burrs occurred only on line 3. Further diagnosis based on data led to agreement that the true cause was an improperly trained assembler. Then the remedy came easily.

The diagnostic journey has three steps:

- Study the symptoms surrounding the defects to serve as a basis for theorizing about causes.
- Theorize on the causes of these symptoms.
- Collect and analyze data to test the theories and thereby determine the causes.

Many analysis techniques are available to assist in these three steps. Some are illustrated in Section 3.9 below.

Plan for Data Collection

Chronic problems are usually not easy to solve and require careful planning and collection of data to confirm and analyze the input and output variables. This "management by facts" concept is basic to all problem-solving approaches. The time and effort involved may be considerable but is a necessary investment for analysis and improvement. Klenz (1999) discusses the role of a "quality data warehouse" in quality improvement.

Planning for data collection involves matters such as where in the process data will be collected, who will provide the data and how often, data collection forms, data accuracy, separation of data into categories ("stratification"), and whether the data is sufficient in content and quantity for the data analysis tools. Sometimes an apparently simple matter of describing the symptoms of a problem requires careful planning.

Description of symptoms

Understanding symptoms is often hindered because some key word or phrase has multiple meanings.

In one example a Pareto analysis of inspection data in a wire mill indicated a high percentage of defects due to "contamination." Various remedies were tried to prevent the contamination. All were unsuccessful. The desperate investigators finally spoke with the inspectors to learn more about the contamination. The inspectors explained that the inspection form contained 12 defect categories. If the observed defect did not fit any of the categories, the inspectors would report the defect as "contamination."

Imprecise wording also occurs because of generic terminology. For example, a software problem is described in a discrepancy report as a "coding error." Such a description is useless for analysis because there are many types of coding errors, e.g., undefined variables, violation of language rules, violation of programming standards.

A way out of such semantic tangles is to think through the meanings of the words used, reach an agreement, and record the agreement in the form of a glossary. The published glossary simplifies subsequent analysis.

Quantification of symptoms

The frequency and intensity of symptoms are of great significance in pointing to directions for analysis. The Pareto principle, when applied to records of past performance, can help to quantify the symptom pattern. Figure 3.6 displays a Pareto diagram for errors in sales order forms sent from a field sales office to the home office for order processing. The four vital few categories account for about 86 percent of the total errors. Note that the diagram includes three elements: (1) the contributors to the total effect, ranked by the magnitude of the contribution, (2) the magnitude of the contribution expressed numerically, and (3) the cumulative-percentage-of-total effect of the ranked contributors.

The Pareto principle applies to several levels of diagnosis: finding the vital few defects, finding the vital few symptoms of a defect, and finding the vital few causes of one symptom.

Formulation of theories

This process has three steps: generation of theories, arrangement of theories, and choice of theories to be tested.

Generation of theories. The best sources of theories are the line managers, the technologists, the line supervisors, and the workforce. A systematic way to generate theories is the brainstorming technique. Potential contributors are assembled for the purpose of generating theories. Creative thinking is encouraged by asking each person, in turn, to propose a theory. No criticism or discussion of ideas is allowed, and all

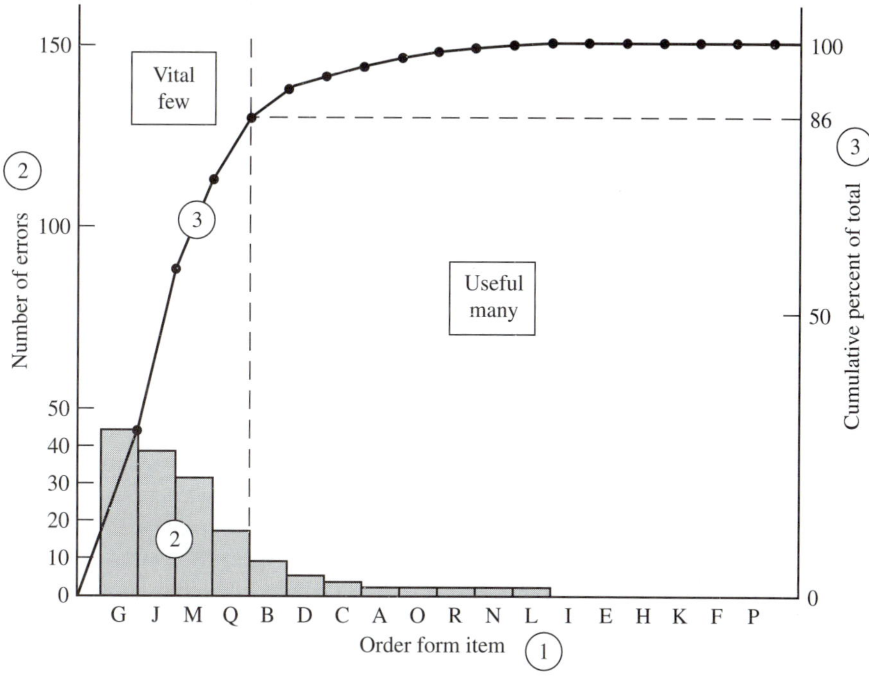

FIGURE 3.6
Pareto diagram of errors on order forms.

ideas are recorded. Following the brainstorming session, the resulting list of theories is critically reviewed.

A useful supplement to the brainstorming technique is "storyboarding." Each theory proposed is recorded on an index card or adhesive note. The cards are arranged on a board to form a visual display of the theories. Storyboarding provides a visual system for organizing theories and planning subsequent evaluation of these theories.

Arrangement of theories. Normally, the list of theories should be extensive, 20 or more. As a list grows, it is essential to create an orderly arrangement. Such order helps us to understand the interrelationships among theories and to plan for testing of the theories. Table 3.3 shows a tabular arrangement of theories contributing to low yield of a process making fine-powder chemicals. The theories consist of major variables and contributing subvariables. A second method, which is highly effective, is a graphical arrangement called the Ishikawa cause-and-effect (or "fishbone") diagram. Figure 3.7 shows such a diagram, which presents the same information as is listed in Table 3.3.

Writing and arranging theories on adhesive notes (storyboarding) can be a useful tool. Forsha (1995) describes storyboarding.

Sanders, Ross, and Coleman (1999) describe how combining the flow diagram and the cause-effect diagram can help to classify process parameters as controllable or noise.

TABLE 3.3
Orderly arrangement of theories

- Raw material
  - Shortage of weight
  - Method of discharge
- Catalyzer
  - Types
  - Quantity
  - Quality
- Reaction
  - Solution and concentration
  - B solution temperature
  - Solution and pouring speed
  - pH
  - Stirrer, RPM
  - Time
- Crystallization
  - Temperature
  - Time
  - Concentration
  - Mother crystal
    - Weight
    - Size
- Moisture content
  - Charging speed of wet powder
  - Dryer, RPM
  - Temperature
    - Steam pressure
    - Steam flow
- Overweight of package
  - Type of balance
  - Accuracy of balance
  - Maintenance of balance
  - Method of weighing
  - Operator
- Transportation
  - Road
  - Cover
  - Spill
  - Container

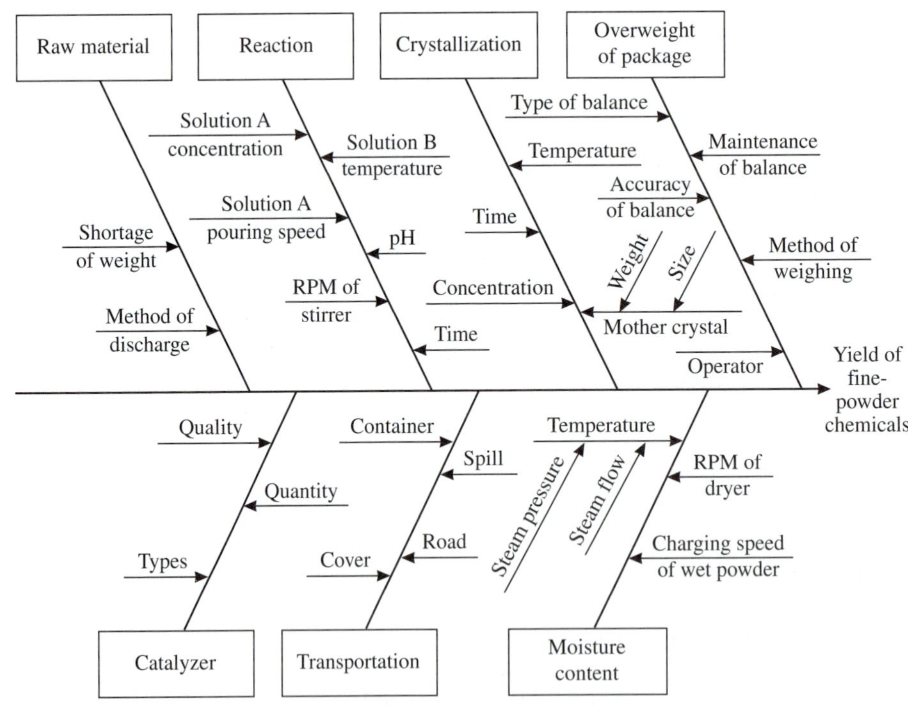

FIGURE 3.7
Ishikawa cause-and-effect diagram.

Thus the cause-and-effect diagram provides information for identifying the input and output variables. Additional tools include the quality function deployment matrix (see Chapter 13 under "Designing for Basic Functional Requirements") and the failure mode, effects, and criticality analysis (see Chapter 13 under "Designing for Time-Oriented Performance"). Thoroughly documenting the current ("as is") process is an essential step in developing an improved ("should") process.

Choice of theories to be tested. After the theories are arranged in an orderly fashion, priorities must be established for testing the theories. In practice the improvement team reaches a consensus on the most likely theory for testing. Whether to test one theory at a time, one group of interrelated theories at a time, or all theories simultaneously requires a judgment based on the experience and creativity of the team.

In determining the most likely cause to investigate, we can use the concept of "belief networks" (graphical models that represent the dependence between variables, using conditional probabilities). For a discussion of belief networks, see Jenzarli (1996). His dissertation applies belief networks to project management, but the concept can also help to find the most likely cause of a quality problem.

Validate the Measurement System

The variation in observed measurements from a process is due to the variation of the process itself and variation due to the measurement system. Often the variation due to the measurement system is assumed to be zero or at least small compared to process variation. This assumption, although it simplifies matters, is typically made without any data. Thus we often do not know how close the observed measurements are to their true (or "master") values.

With the current emphasis in defining quality in units of parts per million rather than percentage, the capability of the measurement system must be recognized as important and evaluated before measuring the capability of the process. Measurement capability involves both the ability of the people making the measurements and the capability of the measuring instruments. When necessary, a complete measurement capability study can involve matters such as reproducibility, repeatability, accuracy, stability, and linearity. These matters are discussed in Section 19.8. Fortunately, software tools such as Minitab® are of great help in analyzing the measurement study data. A measurement study provides assurance that the measurement system data can be trusted. With such assurance, then data collection for product and process analysis and improvement can start.

Measure the Process Capability

Process capability refers to the inherent ability of a process to meet the specification limits for a product. In the measure phase the initial process capability is established by obtaining measurements and observing how the process variability compares to the specification limits. For a static process to be capable at the six-sigma level, the specification limits must be at least six sigma above and below the process mean (see

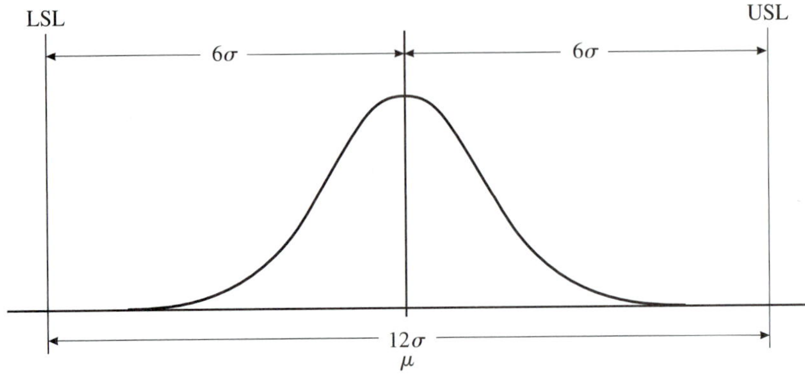

FIGURE 3.8
Process capability and six sigma.

Figure 3.8). In practice the capability study must recognize both short-term variation, in which the process average and spread are stable, and long-term variation, in which the process average may shift up or down. Knowing the initial process capability helps to define the work to be done in the analysis and improve phases to achieve a capability at the six-sigma level. The methodology for measuring process capability is discussed in Section 18.10.

3.9 ANALYZE PHASE

This phase analyzes past and current performance data to identify the causes of variation and process performance. The steps are

- Collect and analyze data.
- Develop and test theories (hypotheses) on sources of variation and cause-effect relationships (i.e., identify the determinants of process performance).

In this phase we do diagnosis—the process of studying symptoms of a problem and determining their cause(s). The beginning of diagnosis is collecting data on the symptoms; the end is agreement on the causes.

Collect and Analyze Data

Many managers harbor deep-seated beliefs that most defects are caused during operations and specifically are due to worker errors, i.e., that defects are mainly worker-controllable. The facts seldom bear this out, but the belief persists. To deal with such deep-seated beliefs, it can be useful to conduct studies to separate defects into broad categories of responsibility. Such studies include:

1. A study to determine the origin of the defects in the design, manufacture, etc. Such a study to determine the distribution of causes over functional areas often has some surprising results. In a classic study, Greenidge (1953) examined 850 failures of electronic products supplied by a number of companies. The results showed that 43 percent of the failures were caused by the product design, 30 percent by field operation conditions, 20 percent by manufacturing, and the remaining 7 percent by miscellaneous causes. For products of moderately high technology, it is not unusual for about 40 percent of field problems to be traceable to the product design.
2. A study to determine whether defects are primarily management-controllable or worker-controllable ("management" here includes not only people in supervisory positions but also others who influence quality, e.g., design engineers, process engineers, buyers, etc.). In general, defects are more than 80 percent management-controllable and less than 20 percent worker-controllable. Some authors use the term "system-controllable" for "management-controllable."

Such broad studies provide important direction for improvement, but individual projects require their own unique data collection and analysis as discussed below.

We proceed next to the test of theories—first management controllable, then worker controllable.

Test Theories of Management-Controllable Problems

Basic to the concept of diagnosis is the factual approach—the use of facts, rather than opinions, to reach conclusions about the causes of a quality problem. Obvious? In practice the urgency to take action on a dramatic problem often results in premature (and incorrect) decisions. Some problems become chronic because the true causes have not been determined, even though a flurry of action takes place. The factual approach not only determines the true cause but also helps to *gain agreement* on the true cause by all of the parties involved. The eloquence of facts carries the day.

Numerous diagnostic methods have been created to test theories. These "incisions" into product and process data can be fascinating and some of the basic methods are discussed below.

Flow diagram

Preparing a flow diagram (see Section 3.8) helps us understand the progression of steps in a process. Detailed steps on using a flow diagram in process analysis are given in Chapter 6 under "The Planning Phase of Process Management." The flow diagram provides an important tool for initiating process capability and process dissection.

Process-capability analysis

One of the theories most widely encountered is "the process can't meet the specifications." To test this theory, measurements from the process must be taken and analyzed to determine the amount of variability in the process. This variability is then compared to the specification limits. Figure 3.9a shows an example of the weight of glass rings from a glass-tube-cutting process (Payne, 1984). Note that the process is currently not meeting the specifications.

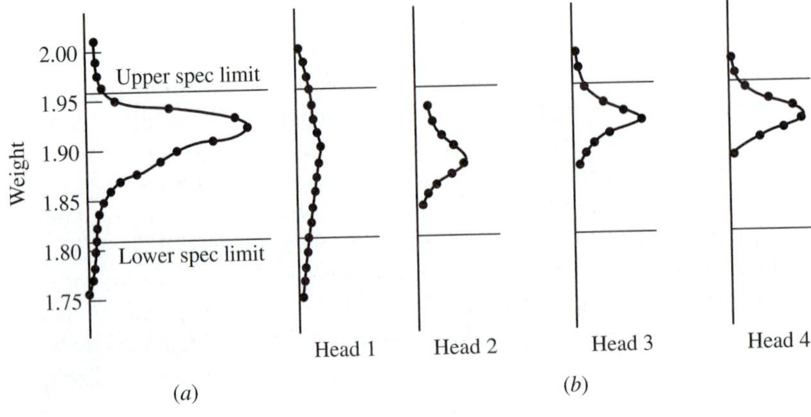

FIGURE 3.9
Distribution of glass bead weights: *(a)* sum of four heads, *(b)* weight distribution on each of the four heads. (*From QCH4, p. 22.42.*)

The weight of the rings was the critical element in determining the properties of the finished glass product. A sample of data (Figure 3.9a) apparently confirmed a theory that the machine was not functioning correctly. However, the machine contained four heads. Data collected separately from each head (stream) revealed that nothing was wrong with heads 2, 3, or 4—except for a need to recenter heads 3 and 4 (Figure 3.9b). However, something was wrong with head 1. Ultimately, the remedy was proper maintenance of the machine rather than a redesign of the machine, as had originally been contemplated based on Figure 3.9a.

Initially, operations people believed that the process was not capable, but the capability study revealed that, with proper maintenance, the process was capable of meeting the specifications.

The concept of process capability applies not only to attributes on physical products but also to attributes of service activities, e.g., the time taken to respond to a customer request at a call center or the time taken to conduct an automatic teller machine (ATM) transaction at a bank. When data shows that a process is not meeting a standard, then we must use the tools of product and process dissection to determine the cause.

Product and process dissection

Some products are produced by a "procession" type of process, i.e., a series of sequential operations. At the end of the series, the product is found to be defective, but which operation did the damage is not known. In some of these cases, it is feasible to dissect the process, i.e., make measurements at intermediate steps in the process to discover at which step the defect appears. Such a discovery can drastically reduce the subsequent effort in testing theories. Dissection of a process is aided by constructing a flow diagram showing the various steps in the process (see Section 3.8 above). Several other forms of dissection are discussed below.

Test at intermediate stages. When defects or errors are found at the end of a procession of steps, we must determine which operational step did the damage. Measuring or testing the product at intermediate steps to discover where the defect first appears can greatly reduce the effort of testing theories.

Stream-to-stream analysis

To meet production volume requirements, several sources of production ("streams") are often necessary. Streams take the form of different machines, machine operators, call center operator shifts, suppliers, etc. Although the streams may seem to be identical, the resulting products may not be. Stream-to-stream analysis consists of recording and examining data separately for each stream.

In the glass-ring example, the four heads on the machine were really four streams. Another example involves invoice errors (Juran Institute, Inc., 1989). A quality improvement team studied 60 incorrect invoices. The data showed the day of the week the invoice was issued, the week of the month, and which of four employees prepared the invoice. First, the data was separated into five "streams" representing Monday through Friday to test the theory that day of the week was the cause. No real difference was found. Next, to test the theory of employees being the cause, the data was separated into streams by employee. Again, no difference. Finally, the data were separated into streams by week of the month (see Figure 3.10). It was clear that at the end of the month the number and percentage of errors were about double what they were during the first three weeks. Further study revealed that during the last week of the month the workload increased dramatically, resulting in a higher error rate. At the start of the study, researchers believed that some employees were the most likely cause of errors. But the study concluded that the excessive workload of the fourth week was the primary factor for errors.

Stream-to-stream analysis uses the concept of "stratification." Stratification separates data into categories and helps to identify which categories (strata) are the main contributors to the problem.

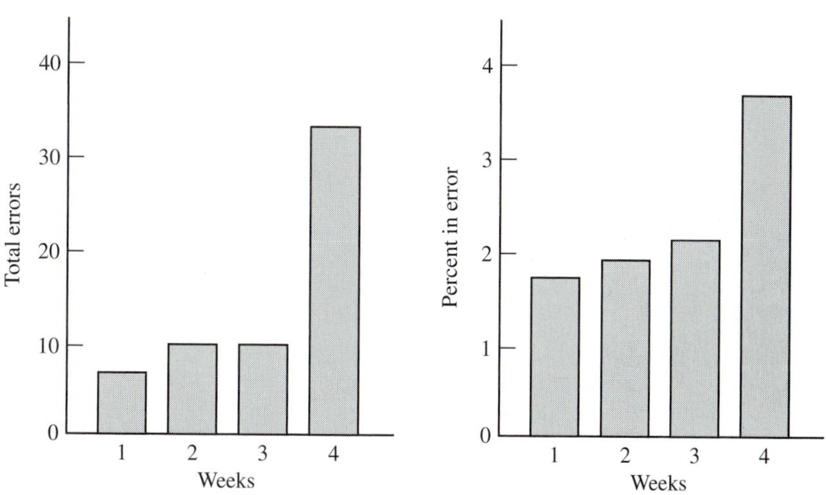

FIGURE 3.10
Stratification of invoice errors by week.

Time-to-time analysis

Time-to-time analyses include (1) a simple plot of data on a time scale; (2) analysis of the time between abnormalities or problems; (3) analysis of the rate of change, or "drift," of a characteristic; and (4) the use of cumulative data techniques with respect to time. Examples are given below.

In one example field failures of oil coolers were assumed to be due to manufacturing. A parade of remedies (skipping the journey from symptom to cause) resulted in zero improvement. An engineer decided to plot the frequency of failures by month of the year, which led to an important discovery. Of 70 failures over a nine-month period, 44 occurred during January, February, and March. These facts shifted the search to other causes such as winter climatic conditions. Subsequent diagnosis revealed the cause to be in design rather than manufacturing.

In analyzing time-to-time variations, the length of time between abnormalities can be a major clue to the cause. A textile-carding operation experienced a cyclic rise and fall in yarn weights, the cycle being about 12 minutes. The reaction of the production superintendent was immediate: "The only thing we do every 12 minutes or so is to stuff that feed box."

Within many streams, there is a time-to-time "drift"; e.g., the processing solution gradually becomes more dilute, the tools gradually wear, the worker becomes fatigued. Such drifts can often be quantified to determine the magnitude of the effect.

Cumulative data plots can help to discover differences that are hidden when the data are in noncumulative form. Figure 3.11 compares histograms (noncumulative) and cumulative plots for data from two years. A difference in adjustments for year 1 versus year 2 is apparent from the cumulative plot but is hidden in the histogram.

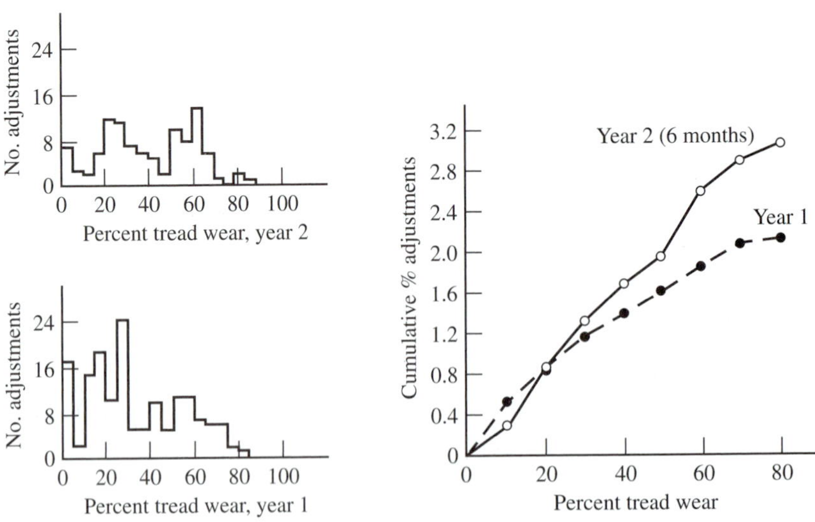

FIGURE 3.11
Comparison of histograms and cumulative plots. (*From QCH4, p. 22.44.*)

Control charts are a powerful diagnostic tool. Data are plotted chronologically, and the chart then shows whether the variability from sample to sample is due to chance or assignable causes of variation. Detection of assignable causes of variation can be the link to discovering the cause of a problem. Chapter 18, "Statistical Process Control," explains the concept.

Other helpful statistical tools include box plots and probability paper (see Chapter 10, "Basic Concepts of Statistics and Probability").

Simultaneous dissection

Some products exhibit several types of variation, e.g., piece to piece, within piece, and time to time. The multivari chart is a clever tool for analyzing such variation. In this chart a vertical line depicts the range of variation within a single piece of product. Figure 3.12 depicts three examples of the relationship of product variation to tolerance limits. In case 1 the within-piece variation alone is too great in relation to the tolerance. Hence no solution is possible unless within-piece variation is reduced. In case 2 within-piece variation is comfortable, occupying only about 20 percent of the tolerance. The problem, then, is piece-to-piece variation. The problem in case 3 is excess time-to-time variability. For elaboration on the multivari chart, see Bhote (1991), Chapter 6.

Defect-concentration analysis

A different form of piece-to-piece variation is the defect-concentration study used for attribute types of defects. The purpose is to discover whether defects are located in the same physical area. The technique has long been used by shop personnel when they observe that all pieces are defective and in precisely the same way. However,

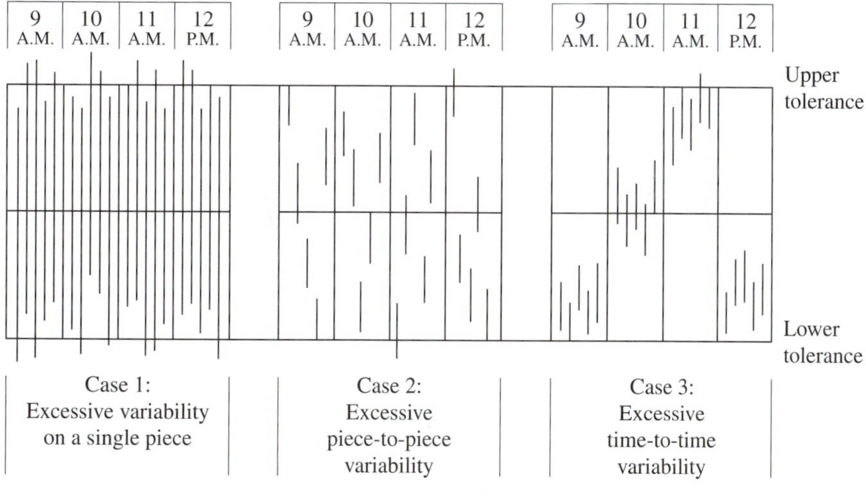

FIGURE 3.12
Multivari chart.

when the defects are intermittent or become evident only in later departments, the analysis can be beyond the unaided memory of the shop personnel.

For example, a problem of pitted castings was analyzed by dividing the castings into 12 zones and tallying up the number of pits in each zone over many units of product. The concentration at the gates (through which the metal flows) became evident, as did areas that were free of pits.

A quality improvement team at the Aid Association for Lutherans used the defect-concentration concept to analyze errors in changing beneficiaries on insurance policies. The team recorded data on the type of error, frequency, the form on which the error occurred, and the location of the error on the form.

Association searches

Sometimes diagnosis can be advanced by analyzing data relating symptoms to some theory of causation, pinpointing the process, piece of equipment, employee(s), or other factor(s). Possible relationships can be examined by using tools such as correlation and ranking.

Correlation analysis. This approach plots data that relate the incidence of symptoms of the problem to values of a potential causal variable.

In one case (Juran Institute, Inc., 1991), a previously successful drug was losing its reputation as being effective, and the medical community was complaining to the manufacturer. One theory suggested that the active ingredient in the compound could be breaking down more quickly than expected while the medication sat on the shelf. Researchers decided to investigate this shelf-life theory.

Samples of the drug were pulled from shelves and returned to the lab to determine the milligrams of active ingredient (the minimum for the drug to be effective was 120 milligrams). Data on time since production and the weight of the ingredient were examined in several ways. A key analysis tool was the scatter diagram shown in Figure 3.13. Note that the data were stratified by two suppliers who provided a certain "filler" ingredient. Lots produced with filler from supplier A showed slow degradation of the active ingredient; lots associated with supplier B showed rapid degradation. Further investigation revealed a substance in B's formulation that led to accelerated breakdown of the active ingredient.

Ranking. In this approach past or current data are collected on two or more variables of a problem and summarized in a table to see whether any pattern exists. In a study involving 23 types of torque tubes, the symptom was dynamic unbalance. One theory was that a swaging operation was a dominant cause. Table 3.4 tabulates the percentage of defective pieces (dynamic unbalance) and shows whether swaging was part of the process. The result was dramatic—the worst seven types of tubes were all swaged; the best seven were all unswaged. This pattern partially confirmed that swaging was a dominant cause. Later analysis revealed an inadequate specification on an important coaxial dimension.

Later in this chapter, a matrix-ranking technique for analyzing problems in insurance contracts is illustrated.

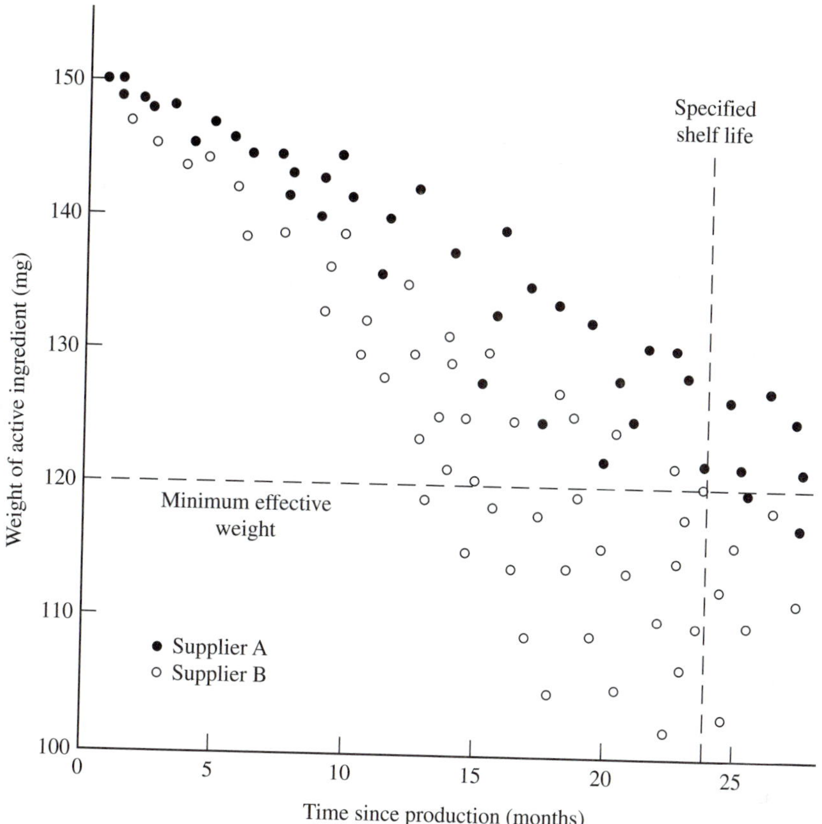

FIGURE 3.13
Stratified scatter diagram of shelf-life data.

TABLE 3.4
Test of theories by ranking

Type	Defective, %	Swaged (marked X)	Type	Defective, %	Swaged (marked X)
A	52.3	X	M	19.2	X
B	36.7	X	N	18.0	X
C	30.8	X	O	17.3	
D	29.9	X	P	16.9	X
E	25.3	X	Q	15.8	
F	23.3	X	R	15.3	
G	23.1	X	S	14.9	
H	22.5		T	14.7	
I	21.8	X	U	14.2	
J	21.7	X	V	13.5	
K	20.7	X	W	12.3	
L	20.3				

Test of theories by collection of new data

In some cases discovery of causes requires careful examination of additional stages in the process. This "cutting of new windows" can take several forms:

1. *Measurement at intermediate stages of a single operation.* An example concerned a defect known as "voids" in welded joints of pressure vessels. Initial diagnosis established six sources of variation: operator, time to time, joint to joint, layer to layer, within layers, and within one welding "bead." Available past data permitted analysis of the first two sources as possible causes of voids. The remaining three could not be analyzed, since the critical X-ray test was performed only when a joint was completely finished. The answer was to "cut a new window" by making an X ray after each of the several beads needed to make a joint. The data established that the main variable was within-bead variation and that the problem was concentrated at the start of the bead.

   An example from a human resources process concerns the time required for hiring new engineers. Measurements taken at six steps in the hiring process formed the basis for a diagnosis of excessive time taken to hire engineers.
2. *Measurement following noncontrolled operations.* This type of diagnosis includes the collection of additional information at individual steps in a process. One example involved the diagnosis of excessive shutdown time for removing a hot mold from a molding machine. The process of shutdown was divided into 11 steps, and measurements were taken to estimate the time required for each step. Two of the 11 steps accounted for 62 percent of the shutdown time. This Pareto effect was important in further diagnosis.
3. *Measurement of additional or related properties of the product or process.* Diagnosis sometimes requires measurement of characteristics other than those for which the specification is not being met. In the manufacture of disks, the symptom was a high percentage of disks with surface defects. Automatic timers controlled the processing cycle. Diagnosis revealed that the times for various steps should not be fixed, but should be determined based on additional measurements taken periodically. The remedy was to monitor pressure, temperature, viscosity, and other factors. An on-line computer evaluates these data for each disk and decides on the optimum molding conditions. Only then, and no sooner, does the process create the product.
4. *Study of worker methods.* In some situations, there are consistent differences in the defect levels coming from various workers. Month after month some workers produce more "good" product than others do. This consistent difference in observed performance must have a cause. Diagnosis of problems related to human performance is discussed later in this section.

Sometimes formally designed experiments are needed to test theories (see Section 3.10 in this chapter).

Most quality problems are management/systems controllable—although many managers find this statement impossible to believe. Consider the situation at a bank. A middle manager faced the problem of an employee whose work output was poor. The manager sincerely believed that most problems were due to the employee. When the situation was examined more closely (as part of a classroom assignment), the employee and her boss concluded that the process was much of the problem. In her words: "If you put a good guy up against a bad process, the bad process will win every time." Exactly. With that reminder, we next discuss human errors.

Test of theories involving human error

Diagnosis of human errors reveals "multiple species" of errors. To illustrate these species, Table 3.5 shows the distribution of 80 errors made by six office workers engaged in preparing insurance policy contracts.

The 29 types of errors follow the Pareto principle. Notice the data for error type 3. There were 19 of these, and worker B made 16 of the 19. The table also shows the rest of the work done by worker B. Except for error type 3, B made few errors. There is nothing basically wrong with the job specification or the method, since the other five workers had little or no trouble with error type 3. There is nothing basically wrong with worker B except for defect type 3. It follows that worker B and no one else is misinterpreting some instruction, resulting in that cluster of 16 errors of type 3.

Error type 5 is of a different species. There is a cluster of 13 of these and all the workers made this error, more or less uniformly. This pattern suggests some difference in approach between all the workers on the one hand the inspector on the other. Such a difference is usually of management-controllable origin, but the reality can soon be established by interviews with the respective employees.

Notice also the column of numbers associated with worker E. The total is 36 errors, the largest cluster in the table. Worker E made nearly half the errors for the entire team, and that worker made them in virtually all error categories. Why did worker E make so many errors? It might be any of a variety of reasons, e.g., inadequate training, lack of capacity to do such exacting work. Further study is needed, but it might be easier to go from symptom directly to remedy—find a less demanding job for that worker.

TABLE 3.5
Matrix of errors by insurance policy writers

Error type	Policy writer						Total
	A	B	C	D	E	F	
1	0	0	1	0	2	1	4
2	1	0	0	0	1	0	2
3	0	(16)	1	0	2	0	(19)
4	0	0	0	0	1	0	1
5	2	1	3	1	4	2	(13)
6	0	0	0	0	3	0	3
.							
.							
.							
27							
28							
29							
Total	6	(20)	8	3	(36)	7	80

TABLE 3.6
Interrelationship among error pattern, likely subspecies of worker error, and likely solution

Pattern disclosed by analysis of worker error	Likely subspecies of error causing this pattern	Likely solution
On certain defects no one is error prone; defect pattern is random.	Errors are due to inadvertence.	Error-proof the process.
On certain defects some workers are consistently error prone, and others are consistently "good."	Errors are due to lack of technique (ability, know-how, etc.). Lack of technique may take the form of secret ignorance. Technique may consist of known knack or of secret knowledge.	Discovery and propagation of knack. Discovery and elimination of secret ignorance.
Some workers are consistently error prone over a wide range of defects.	There are several potential causes: Conscious failure to comply with standards Inherent incapacity to perform this task Lack of training	Solution follows the cause: Increase motivation Transfer worker Supply training
On certain defects all workers are error prone.	Errors are management controllable.	Meet the criteria for self-control.

Thus this one table shows the presence of multiple species of worker errors. The remedy is not as simplistic as "motivate the worker." Understanding these species through diagnosis is important in identifying causes. The great majority of worker errors fall into one of four categories: inadvertent, technique, conscious, and communication. Table 3.6 shows the interrelationship among the error pattern, the likely subcategory, and the likely remedies. The four categories are examined below.

Inadvertent errors. Workers are unable to avoid inadvertent errors because of human inability to maintain attention. Centuries of experience have demonstrated that human beings are simply unable to maintain constant attention.

The usual examples involve a component omitted from an assembly or a process adjustment that is set incorrectly.

Diagnosis to identify errors as inadvertent is aided by understanding their distinguishing features. They are

- *Unintentional.* The worker does not want to make errors.
- *Unwitting.* At the time of making an error, the worker is unaware of having made it.
- *Unpredictable.* There is nothing systematic as to when an error will be made, what type of error will be made, or which worker will make the error. As a consequence of this unpredictability, the error pattern exhibits randomness. *A set of data that shows a random pattern of worker error suggests that the errors are inadvertent.* The randomness of the data may apply to the types of errors, to the persons who make errors, and to the times when the errors are made.

Remedies for inadvertent errors involve two approaches:

1. *Reducing the extent of dependence on human attention.* The tools used here are all of the error-proofing type: fail-safe designs, validation of processes, countdowns, redundant verifications, cutoffs, interlocks, alarm signals, automation, robots. Large reductions in errors can result from the use of bar codes to help identify items.
2. *Helping workers remain attentive.* Examples of remedies are reorganization of work to reduce fatigue and monotony; job rotation; and use of sense multipliers, templates, masks, and overlays.

Technique errors. These errors arise because the worker lacks some essential technique, skill, or knowledge needed to prevent the error from happening. Diagnosis to identify errors due to technique is aided by understanding their features. They are

- *Unintentional.* The worker does not want to make errors.
- *Specific.* Technique errors are unique to certain defect types—those types for which the missing technique is essential.
- *Consistent.* Workers who lack the essential technique consistently make more defects than workers who possess the technique. This consistency is readily evident from data on worker errors.
- *Unavoidable.* The inferior workers are unable to match the performance of the superior workers because the former do not know "what to do differently."

Discovery of the existence of technique errors makes use of the diagnostic tools for worker errors, as illustrated here in the assembly of shotguns.

THE GUN ASSEMBLY CASE. Guns were assembled by 22 skilled craft workers, each of whom assembled a complete gun from bits and pieces. After a safety test about 10 percent of the guns could not be opened to remove the spent cartridge—a defect known as "open hard after fire." For such defects, it was necessary to disassemble the gun and then reassemble it, which required about two hours per defective gun—a significant waste.

After an agony of fruitless discussion, it was clear that the missing element was factual information. Data already in the files by assembler and time were collected and arranged in a matrix (Table 3.7). Some helpful information became evident:

1. There was a wide month-to-month *departmental* variation in the defect rate, ranging from a low of 1.8 percent in January to a high of 22.6 percent in February. Since all workers seemed to be affected, this variation must have had its cause outside of the department. (Subsequent analysis confirmed this theory.)
2. The ratio of the five best performances to the five worst showed *a stunning consistency.* In each of the six months, the five worst performances add up to an error rate that is at least 10 times greater than the sum of the five best performances. There must be a reason for such a consistent difference, and it can be found by studying work methods—the techniques used by the respective workers.

THE KNACK. The study of work methods showed that the superior performers used a file to cut down one of the dimensions on a complex component; the inferior performers did not file that component. This filing constituted a "knack"—a small

TABLE 3.7
Matrix analysis

Assembly operator rank	Nov.	Dec.	Jan.	Feb.	Mar.	Apr.	Total
1	4	1	0	0	0	0	5
2	1	2	0	5	1	0	9
3	3	1	0	3	0	3	10
4	1	1	0	2	2	4	10
5	0	1	0	10	2	1	14
6	2	1	0	2	2	15	22
.							
.							
.							
17	18	8	3	37	9	23	98
18	16	17	0	22	36	11	102
19	27	13	4	62	4	14	124
20	6	5	2	61	22	29	125
21	39	10	2	45	20	14	130
22	26	17	4	75	31	35	188
Total	234	146	34	496	239	241	1390
% defective	10.6	6.6	1.8	22.6	10.9	11.0	10.5
Five best	9	6	0	20	5	8	48
Five worst	114	62	12	265	113	103	669
Ratio	13	10	∞	13	23	13	14

difference in method that accounts for a large difference in results. (Until the diagnosis was made, the superior assemblers had not realized that filing greatly reduced the incidence of defects.)

Usually the difference in worker performance is traceable to some superior knack used by the successful performers to benefit the product. In the case of the gun assemblers, the knack consisted of filing one component. In some cases, however, the difference in worker performance is due to unwitting *damage* done to the product by the inferior performers.

A useful rule for predicting whether the difference in worker performance is due to a beneficial knack or to a negative knack considers which workers are in the minority. If the superior performers are in the minority, the difference is probably due to the beneficial knack. If the inferior performers are in the minority, the difference in performance is probably due to a negative knack.

SUMMARY OF TECHNIQUE ERRORS. The sequence of events to identify, analyze, and remedy technique errors follows.

1. For the defect types under study, create and collect data that can disclose any significant worker-to-worker differences.
2. Analyze the data on a time-to-time basis to discover whether consistency is present.
3. Identify the consistently best and consistently worst performers.

4. Study the work methods used by the best and worst performers to identify their differences in technique.
5. Study these differences further so as to discover the beneficial knack that produces superior results or the negative knack that is damaging the product.
6. Bring everyone up to the level of the best through appropriate remedial action such as
   a. Training inferior performers in use of the knack or in avoidance of damage.
   b. Changing the technology so that the process embodies the knack.
   c. Error-proofing the process in ways that either require use of the knack or prohibit the technique that is damaging to the product.

Conscious errors. Diagnosis to identify errors as conscious is aided by understanding their features. They are

- *Witting.* At the time of making an error, the worker is aware of it.
- *Intentional.* The error is the result of a deliberate intention on the part of the worker.
- *Persistent.* The worker who makes the error usually intends to keep it up.

The outward evidence of conscious errors is likewise unique. Whereas inadvertent errors exhibit randomness, conscious errors exhibit consistency; i.e., some workers consistently make more errors than others. However, whereas technique errors are typically restricted to defect types that require some special knack, conscious errors tend to cover a wider spectrum of defect types. Knowing these types is helpful in diagnosing errors as conscious.

MANAGEMENT-INITIATED CONSCIOUS ERRORS. Many conscious errors are management initiated. The most common examples arise from the multiple standards that all managers must meet—cost, delivery, and productivity, as well as quality. Because of changes in the marketplace, managers keep shifting their priorities; e.g., in a seller's market, delivery schedules will prevail over some quality standards. The pressures on the managers are then transmitted to the workforce and can result in conscious violation of one standard in order to meet another.

WORKER-INITIATED CONSCIOUS ERRORS. Some conscious errors are worker initiated. Workers may have real or fancied grievances against the boss or the company. They get their revenge by not meeting standards. A few become rebels against the whole social system, and they use sabotage to show their resentment. Some instances are so obviously antisocial that neither fellow employees nor the union will defend the actions.

REMEDIES FOR CONSCIOUS ERRORS. Generally, the remedies listed here emphasize securing changes in behavior without making any special effort to secure a change in attitude. Either way, the approach is oriented primarily to the persons rather than to the "system"—the managerial or technological aspects of the job. Possible remedies include

- Explaining the impact of the error on internal or external customers.
- Establishing individual accountability.
- Providing a balance between productivity and quality.

- Conducting periodic audits.
- Providing reminders to workers about specific defects.
- Improving communication between management and workers on quality issues.
- Creating competition and incentives.
- Error-proofing the operation.
- Reassigning the work.

For elaboration on these remedies, see *JQH5,* pages 5.62–5.63.

Communication errors. These errors arise because of a failure in communication provided to the employee. Communication errors may occur along with inadvertent, technique, or conscious errors. Several forms of communication errors exist.

COMMUNICATION OMITTED. Some conscious errors seem to be worker initiated but have their origin in inadequate communication by management. For example, three product batches fail to conform to quality characteristic X. In each case, the inspector places a hold on the batch. In each case, the material review board concludes that the batch is fit for use and releases it for delivery. However, neither the production worker nor the inspector is told why. Not knowing the reason, these workers may conclude that characteristic X is unimportant. That sets the stage for unauthorized actions.

COMMUNICATION INHIBITED. Historically, upper management placed little emphasis on encouraging communication from the bottom up. This behavior resulted in a failure to capture both general ideas and specific work-design suggestions from those closest to the action, i.e., the workforce. Fortunately, management has taken significant steps to make greater use of this underemployed asset. Most managers now recognize the importance of workforce contributions to employee involvement teams, self-directed worker teams, and other forms of employee participation. Notwithstanding this progress, individual cases of management by fear and intimidation still occur and inhibit communication from the workforce.

TRANSMISSION ERRORS. Human communication is imperfect. Errors arise from lack of precise wording, confusion about similar words or acronyms, background noise that inhibits oral communication, and many other reasons. Some stock brokerage companies maintain a special account to cover expenses in connection with errors made in trading stock, e.g., purchase of the wrong stock because a broker confused the Copytel name with Datacopy. In the athletic arena a football game is lost because, on a key play near the end of the game, a player mistakenly hears a play called as "green" instead of "three." He misses his defensive assignment, and the opposing team scores a touchdown.

REMEDIES FOR COMMUNICATION ERRORS. Although "poor communication" is often given as the vague cause of a wide variety of problems, steps that can be taken to prevent these errors include the following:

- An alertness by upper and middle management to identify and carefully explain key decisions that affect the workforce. These include decisions that affect employee welfare and product and process decisions that affect quality. Such communication will take time, but the investment will reap rewards.

- An emphasis on improving the clarity and precision of specifications, standards, policies, and work instructions. This task can be accomplished by means such as preparing a written glossary of terms and definitions, standardizing key terms, preparing physical samples, and using numbers instead of adjectives to describe product features.

For elaboration on human error, see *JQH5,* Section 5.

As illustrated by the examples above, causes can sometimes be determined by a simple analysis of available data (current and past) from a process. Surprising agreement on causes can occur when people are confronted with data—particularly if data had never been collected before.

In other cases, however, the data analyses require the use of statistical techniques, i.e., setting up quantitative hypotheses and using statistical tests to accept or reject the hypotheses. The concepts and specific tests are described in Chapter 10, "Basic Concepts of Statistics and Probability," and in Chapter 11, "Statistical Tools for Analyzing Data." As explained in Chapter 10, a failure to understand the need for a statistical analysis of data can easily lead to incorrect conclusions.

Many techniques are available to assist in the analysis phase. The Japanese concluded that most quality-related problems could be solved with seven basic tools: cause-effect diagram, stratification analysis, check sheet, histogram, scatter diagram, Pareto analysis, and control charts. Later seven new tools were recommended (Table 3.8): affinity diagram, tree diagram (systematic diagram), process decision program chart, matrix diagram, interrelationship digraph (relations diagram), prioritization matrix (matrix data analysis), and activities network diagram (arrow diagram).

TABLE 3.8
Seven new Japanese tools

Tool	Concept
Affinity diagram	Organize facts, opinions, and issues into natural groupings as an aid to diagnosis
Tree diagram (systematic diagram)	Dissect a problem into subproblems and causes
Process decision program chart	Assess alternative processes to help select best process
Matrix diagram	Show presence or absence of relationships among collected pairs of elements in a problem situation
Interrelationship diagraph (relations diagram)	Help identify key factors for inclusion in a tree diagram
Prioritization matrix (matrix data analysis)	Evaluate options through a systematic approach of identifying, weighing, and applying criteria to the options
Activities network diagram (arrow diagram)	Analyze the sequence of tasks necessary to complete a project and determine the critical tasks to monitor in order to execute the project efficiently

These tools are drawn from the industrial engineering, operations research, statistics, and management disciplines and focus on understanding relationships among quality-related activities in a broad system. In contrast, the seven basic tools focus on problem solving for a specific product or process. For elaboration in the seven new tools, see Mizuno (1988).

Another broad approach to improvement, the theory of constraints (TOC), deserves mention. A *constraint* is the weakest link in a system (process) and therefore should be the focus for improvement. Constraints are mostly policy (procedures, past practice) but may also be physical (machines, people, other resources). TOC identifies the constraint(s) and analyzes the system to assure that all parts are aligned and adjusted to support the maximum effectiveness of the constraint. A presentation of the concept in the form of a novel is provided by Goldratt (1997). Dettmer (1995) explains the steps as they would apply to quality.

Some recent and classic books that provide extensive discussion on analysis include Wadsworth, Stephens, and Godfrey (2001); Smith (1998); Kolarik (1995); Mears (1995); Wilson, Dell, and Anderson (1993); and Juran Institute, Inc. (1989). Plsek (1997) applies the tools of creative thinking and innovation to various aspects of quality management including process design.

3.10 IMPROVE PHASE

This phase designs a remedy, proves its effectiveness, and prepares an implementation plan. The steps are

- Evaluate alternative remedies.
- If necessary, design formal experiments to optimize process performance.
- Design a remedy.
- Prove effectiveness of the remedy.
- Deal with resistance to change.
- Transfer the remedy to operations.

Evaluate Alternative Remedies

Remedial action responds to the findings of diagnosis. Usually several alternative remedies (solutions) are proposed to remove the cause found in diagnosis. The remedy selected should make a significant improvement on the original problem and should optimize both company costs and customer costs. In evaluating remedies, a useful tool for the improvement team is the remedy selection matrix (Table 3.9).

The team first discusses and agrees on the criteria it will use. If desired, weights can be assigned to the criteria. Then the team evaluates each remedy for each criterion. The rating can be quantitative or stated simply as high, medium, or low desirability.

TABLE 3.9
Remedy selection matrix

Criterion	Remedy 1	Remedy 2	Remedy 3
Remedy name			
Total cost			
Impact on the problem			
Benefit/cost relationship			
Cultural impact/resistance to change			
Implementation time			
Uncertainty about effectiveness			
Health & safety			
Environment			
Summary (1 for best, 2 for next, and so on.)			

Source: Juran Institute, Inc., 1993.

TABLE 3.10
Types of diagnostic experiments

Type of experiment	Purpose and approach
Evaluating suspected dominant variables	Evaluate changes in values of a variable by dividing a lot into several ports and processing each portion at some different value, e.g., temperature.
Exploratory experiments to determine dominant variables	Statistically plan an experiment in which a number of characteristics are carefully varied to yield data for quantifying each dominant variable and the interactions among variables.
Production experiments (evolutionary operation)	Make small changes in selected variables of a process and evaluate the effect to find the optimum combination of variables.
Response surface experiments	Make changes in selected variables of a process to generate an empirical map and mathematical model of the response.
Simulation	Use the computer to study the variability of several dependent variables that interact to yield a final result.

Design of Experiments

Experiments in the laboratory or outside world may be necessary to determine and analyze the dominant causes of a quality problem and to design a remedy. Five types of experiments are summarized in Table 3.10.

Exploratory (screening) experiments are often conducted to verify the vital few causes and dominant variables, and laboratory and production experiments are conducted to generate a mathematical model of the process and optimize process performance. These experiments require definition of the objective, response (output) variables, independent (input) variables, test levels for the variables, and the selection of the design of the experiment. Chapter 11 discusses several types of experimental designs, some tools for sound experimentation, and the classical (one factor at a time) versus modern methods of experimentation. *JQH5,* Section 47, provides a comprehensive coverage of both the design and analysis of experiments as part of the learning process in the scientific method.

The application of formal design of experiments (DOE) has been hindered by both the difficulties of selecting an experimental design and the relatively complex statistical calculations and data analysis required. Now the personal computer and associated software reduce these burdens and facilitate the use of powerful tools like DOE. Software such as Minitab® can help to select designs, analyze data (including computer-constructed graphical displays of data), and help to create a mathematical model of process performance.

A well-organized exploratory experiment has a high probability of identifying the dominant causes of variability. However, there is a risk of overloading the experimental plan with too much detail. A check on overextension of the experiment is to require that the analyst prepare a written plan for review. This written plan must define

1. The characteristics of material, process, environment, and product to be observed.
2. The control of these characteristics during the experiment; a characteristic may be
   a. Allowed to vary as it will and measured as is.
   b. Held at a standard value.
   c. Deliberately randomized.
   d. Deliberately varied, in several classes or treatments.
3. The means of measurement to be used (if different from standard practice).

If the plan shows that the experiment may be overloaded, a dry run in the form of a small-scale experiment is in order. A review of the dry-run experiment can then help decide the final plan.

Production experiments

Experimentation is often regarded as an activity that can be performed only under laboratory conditions. To achieve maximum performance from some manufacturing processes, however, the effect of key process variables on process yield or product properties must be demonstrated under shop conditions. Laboratory experimentation to evaluate these variables does not always yield conclusions that are completely applicable to shop conditions. When justified, a "pilot plant" may be set up to evaluate process variables. However, the final determination of the effect of process variables must often be done during the regular production run by informally observing results and making any changes that are deemed necessary. Thus informal experimentation *does* take place on the manufacturing floor.

To systematize informal experimentation and provide a methodical approach for process improvement, G. E. P. Box developed a technique known as "evolutionary operations" (EVOP). EVOP is based on the concept that every manufactured lot has information to contribute about the effects of process variables on a quality characteristic. Although such variables could be analyzed by an experimental design, EVOP introduces *small* changes into these variables according to a planned pattern of changes.

These changes are small enough to avoid nonconformance but large enough to gradually establish (1) which variables are important and (2) the optimum process values for these variables. Although this approach is slower than a formal experimental design, results are achieved in a production environment without the additional costs of a special experiment.

The steps are

1. Select two or three independent process variables that are likely to influence quality. For example, time and temperature were selected as variables affecting the yield of a chemical process.
2. Change these steps according to a plan (see Figure 3.14). This diagram shows the *plan*, not any data. For example, a reference run was made with the production process set to run at 130°C for 3.5 hours. The next batch (point 1 in Figure 3.14) was run at 120°C for three hours. The first *cycle* contains five runs, one at each condition. Samples were taken from each batch, and analyses were made.
3. After the second repetition of the plan (cycle 2) and each succeeding cycle, calculate the effects (see *JQH5,* pp. 47.46–47.51).
4. When one or more of the effects is significant, change the midpoints of the variables and perhaps their ranges.
5. After eight cycles, if no variable has been shown to be effective, change the ranges or select new variables.
6. Continue moving the midpoint of the EVOP plan and adjust the ranges as necessary.
7. When a maximum has been obtained or the rate of gain is too slow, drop the current variables from the plan and run a new plan with different variables.

EVOP is a highly structured form of production experimentation. Ott and Schilling (1990) present various practical design and analysis techniques for solving production quality problems. An associated technique that is applied in research and development laboratories is response surface methodology (RSM). RSM provides empirical maps (contour diagrams) illustrative of how factors under the experimenter's

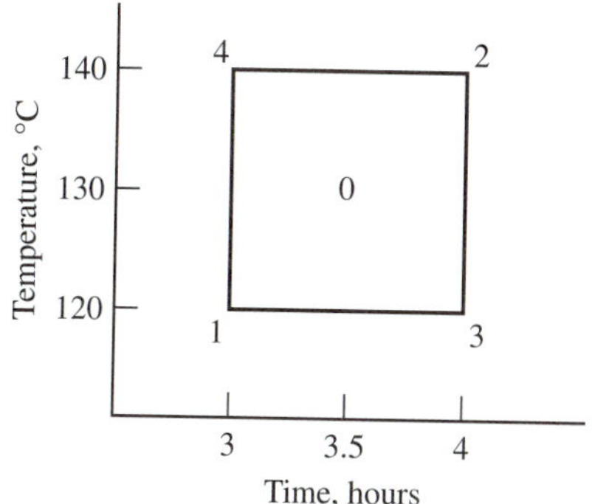

FIGURE 3.14
An EVOP plan. Numbers are in run order. 0 is the reference run.

control influence the response. Factor settings are viewed as defining points in the factor space (may be multidimensional) at which the response will be recorded. Thus the maps illustrate the nature of the response surface. For elaboration, see *JQH5*, Section 47.

Simulation experiments

From the field of operations research comes a technique called "simulation" that can be useful in analyzing quality problems. Simulation provides a method of studying the effect of a number of variables on a final quality characteristic—but all of this is done on paper without conducting experiments! A simulation study requires the following inputs:

1. Definition of the output variable(s)
2. Definition of the input variable(s).
3. Description of the complete system relating the input and output variables.
4. Data on the distribution of each input variable; thus variability is accepted as inherent in the process.

In simulation a system model is developed and translated into a computer program. This program not only defines the relationship between input and output variables but also makes provision for storing the distribution of each input variable. The computer then selects values at random from each input distribution and combines these values, using the relationship defined, to generate a simulated value of the output variable. Each repetition of this process results in a simulated output result. These can then be formed into a frequency distribution. The payoff is to make *changes* in the input variables or the relationships, run another simulation, and observe the effect of the change. Thus the significance of variables can be evaluated on paper, providing one more way of evaluating theories on causes of problems.

Simulation has been applied to many quality problems, including interacting tolerances, circuit design, and reliability.

Simulation has also been used in the service industry. Anderson, Abetti, and Savage (1995) present an application at Aetna Health Plans. The improvement project involved reducing the cycle time of patient cases that required additional medical review. The original average turnaround time (TAT) was 18.1 days and was a major source of customer complaints. The goals of the improvement project were to (1) reduce TAT to 50 percent of files resolved within one week, (2) improve overall quality, and (3) encourage a sense of ownership and accountability. The people involved included customer service representatives (CSRs), nurses, mailroom personnel, medical directors, and various supervisors. A simulation model for the computer was developed by using the process model (how work flows through the area), the resource model (who is available to do the work), and the input model (what kind of work is done). For example, detailed process flow diagrams, including times for each task, were developed for the simulation model. The model was then tested (and refined) by using the computer simulation to generate three months of TAT data and comparing the simulation results to three months of actual data. The result was a computer simulation model that could be used to evaluate proposed process changes.

Three sets of process changes were tested by using the simulation model. The simulation results identified the key drivers of TAT. The remedy in the form of process changes included assigning a dedicated CSR resource, scheduling meetings among resources to replace interoffice mail routing, using preprinted form letters, undertaking aggressive action to obtain outside information, and assigning work not requiring medical expertise to clerical resources. The improved process reduced TAT from 18.1 days to 6.6 days.

Aetna concluded that the simulation model provides an experimental test bed (in the computer) to try ideas, encourages the need for data, and expedites implementation of the improved process because "to delay implementation in the face of such compelling data would have been unthinkable." The model was a powerful force in obtaining upper-management focus on the problem.

Batson and Williams (1998) present brief descriptions of seven case studies in quality simulation in both manufacturing and service industries. The *New York Times* (January 11, 2000) describes a fascinating application of simulation to study the movement of bubbles in a glass of Guiness stout. Using computational fluid dynamics software, researchers ran simulations to alter various parameters, including the temperature of the stout, the size and concentration of the bubbles, and even the shape of the glass. Of course, the computer simulation was supplemented by field trials in local pubs—sample size not reported.

For a complete discussion of the methodology and application of simulation, see Banks (1998). Also, an emerging development is the combining of simulation with "fuzzy logic" concepts. *Fuzzy logic* is the application of mathematics to represent and manipulate data that possess nonstatistical uncertainty and uses imprecise information, e.g., cool, warm, or hot, rather than specific temperatures (for a brief explanation, see *JQH5,* page 10.8).

Design a Remedy

The remedy must fulfill the original project mission, particularly with respect to meeting customer needs. This step identifies the customers, defines their needs, and proves the effectiveness of the remedy. In some cases designing a remedy may require a major replanning of the product or process and involve a structured approach to the replanning (see Chapter 4, "Operational Quality Planning and Sales Income").

Prove Effectiveness of the Remedy

Before a remedy is finally adopted, it must be proven effective. Two steps are involved:

1. *Preliminary evaluation of the remedy under conditions that simulate the real world.* Such evaluation can make use of a "paper" reliability prediction, a dry run in a pilot plant, or the testing of a prototype unit. But these preliminary evaluations have assumptions that are never fully met, e.g., the prototype unit is assumed to be made under typical manufacturing conditions, when actually it is made in the engineering model shop.

2. *Final evaluation under real-world conditions.* There is no substitute for testing the remedies in the real world. If the remedy is a design change on a component, the final evaluation must be a test of the redesigned component operating in the complete system under field conditions; if the remedy is a change in a manufacturing procedure, the new procedure must be tried under typical (not ideal) factory conditions; if the remedy is a change in a maintenance procedure, the effectiveness must be demonstrated in the field environment by personnel with representative skill levels.

Finally, after a remedy is proven effective, an issue of communication remains. A remedy on one project may also apply to similar problems elsewhere in an organization. It is useful, therefore, to communicate the remedy to (1) others who may face similar problems and (2) those responsible for planning future products and processes. In one approach the remedy is entered into a database that can easily be examined by means of key words.

Deal with Resistance to Change

Various objections to the remedy may be voiced by different parties, e.g., through delaying tactics or outright rejection of the remedy by a manager, the workforce, or the union. "Resistance to change" is the usual name. Change consists of two parts: (1) a technological change and (2) a social consequence of the technological change.

People often voice objections to technological change, although the true reason for their objection is the social effect. Thus those proposing the change can be misled by the objections stated. For example, an industrial engineer once proposed a change in work method that involved moving the storage of finished parts from a specific machine to a central storage area. The engineer was confused by the affected worker's resistance to the new method. The method seemed to benefit all parties concerned, but the worker argued that it "would not work." The supervisor was perceptive enough to know the true reason for resistance—the worker's production was superb, and many people stopped at his machine to admire and compliment him. Who would want to give up that pleasure? To cite another example, some design engineers resist the use of computer-aided design (CAD), claiming that the technology is not as effective as design analysis by a human being. The real reason, for some older designers, may include the fear of having difficulty adapting to CAD. To achieve change, we must:

- Be aware that we are dealing with a pattern of human habits, beliefs, and traditions (culture) that may differ from our own.
- Discover the exact social effects of the proposed technological changes.

Based on the scars of experience, some rules can be identified for introducing change.

Rules of the road for introducing change

Important among these are

- *Provide for participation.* The most important rule for introducing change is to provide for participation. Those who are likely to be affected by the change should be members of the project team and participate in both diagnosis and remedy. Lack of participation leads to resentment, which can harden into a rock of resistance.

- *Establish the need for the change.* The change should be explained in terms that are important to the people involved rather than on the basis of the logic of the change.
- *Provide enough time.* How long does it take the members of a culture to accept a change? They must take enough time to evaluate the impact of the change and find an accommodation with the advocates of the change. Providing enough time takes various forms:
  a. *Starting small.* Conducting a small-scale tryout before going "all out" reduces the risks for the advocates as well as for the members of the culture.
  b. *Avoiding surprises.* A major benefit of the cultural pattern is its predictability. A surprise is a shock to this predictability and disturber of the peace.
  c. *Choosing the right year.* There are right and wrong years—even decades—for a change.
- *Keep the proposals free of excess baggage.* Avoid cluttering the proposals with extraneous matters not closely concerned with getting results. The risk is that debate will get off the main subject and onto side issues.
- *Work with the recognized leadership of the culture.* The culture is best understood by its members. They have their own leadership, which is often informal. Convincing the leadership is a significant step in getting the change accepted.
- *Treat people with dignity.* The classic example is that of the relay assemblers in the "Hawthorne experiments." Their productivity kept rising, under good illumination or poor, because in the laboratory they were being treated with dignity.
- *Reverse the positions.* Ask the question: What position would I take if I were a member of the culture? It is even useful to get into role playing to stimulate understanding of the other person's position.
- *Deal directly with the resistance.* There are many ways of dealing directly with resistance to change:
  a. Trying a program of persuasion.
  b. Offering a quid pro quo—something for something.
  c. Changing the proposals to meet specific objections.
  d. Changing the social climate to make the change more acceptable.
  e. Forgetting it; sometimes the correct alternative is to drop the proposal.

Dealing with resistance to change is an art. However, some approaches provide a methodical way of (1) understanding the impact of change and (2) resolving differences among the parties involved. One approach to understanding the impact is to identify the restraining forces and the driving forces for change ("force field analysis"). Another approach to resolving differences focuses on having the parties clearly state their positions to identify the exact areas of disagreement (see *JQH5,* page 5.66, for elaboration).

Transfer the Remedy to Operations

Transfer to operations may include revisions in operating standards and procedures; changes in staffing and responsibilities; additional equipment, materials, and supplies;

and extensive training on the why and how of the changes. A detailed schedule of tasks and dates can help to plan for the implementation. We move now to the final phase of six sigma, control.

3.11 CONTROL PHASE

In this phase we design and implement certain activities to hold the gains of improvement. The steps are

- Design controls and document the improved process.
- Validate the measurement system.
- Determine the final process capability.
- Implement and monitor the process controls.

Design Controls and Document the Improved Process

In this step we provide a systematic means for holding the gains—the process of *control*. Control during operations is done through use of a feedback loop—a measurement of actual performance, comparison with the standard of performance, and action on the difference. Elaboration of the concept of control is discussed in Chapter 5, "Quality Control"; Chapter 23 describes process audits as a means of verifying the presence of the required process conditions and other remedial steps; Chapter 18, "Statistical Process Control," explains some statistical process control techniques that are useful in detecting out-of-control conditions. The controls should also include mistake-proofing the process to prevent errors—see Chapter 16 for elaboration. Documentation of the improved process should include the steps taken during the define, measure, analyze, and improve phases.

Validate the Measurement System

The measurement system for the improved process must be evaluated and made capable. This step may involve new measurement devices, the collection of new data, and additional training for process personnel. Chapter 19 provides further discussion of the measurement system.

Determine the Final Process Capability

This step means providing the operating forces with a process capable of holding the gains under operating conditions. To the extent that is economically feasible, the process changes should be designed to be *irreversible*. For example, changing from hand insertion of components for printed circuit boards to automatic insertion by programmed tape rolls is an irreversible remedy. In wave soldering a remedy that requires a different specific gravity for a flux could be reversible because contamination or other factors may result in a flux having a specific gravity that was previously unacceptable.

Implement and Monitor the Improved Process

In this step the improved process is placed into operation, and the control steps described above are used to monitor process conditions and product performance. The team should provide for measuring the cost of poor quality to confirm that the remedies have worked.

Implementing and monitoring the improved process is the final step in a quality improvement project. We conclude the chapter with a discussion of how to maintain a focus on continuous improvement.

3.12 MAINTAINING A FOCUS ON CONTINUOUS IMPROVEMENT

Competitive pressures require many improvement projects—at a revolutionary rate. We must replicate the results of current projects, nominate new projects, and take other actions for improvement.

To replicate the results of a current project means to apply a proven remedy to similar problems in an organization. As part of completing a project, the improvement team documents all the steps taken in diagnosing causes and implementing remedies. A computer database of improvement projects (completed and in progress) should be maintained and made known to all managers. In addition, managers should be asked to identify problems in the organization that are similar to those solved by completed improvement projects. When a similar problem is identified, a "replication team" can be appointed to validate the problem in the new situation and determine whether the causes and remedy of the original problem apply to the new problem.

The foundation for maintaining a focus on improvement is the infrastructure for improvement. The infrastructure includes formalizing a process for nominating and selecting projects, forming improvement teams, and providing the training and support for the teams. Additional actions can help to maintain the focus:

1. Use information from assessment studies (see Chapter 2) to identify improvement opportunities. These assessments include cost of poor quality, market research on customer satisfaction and loyalty, quality culture, and the broad system. Note that these areas go far beyond nonconformance to specifications.
2. Address process improvement in terms of effectiveness, efficiency, adaptability, and cycle time (see Chapter 6, "Process Management").
3. Pursue radical forms of improvement (see Chapter 6).
4. Apply all the tools of improvement—technical and behavioral, simple and sophisticated. Increasingly, savings from improvement projects have "skimmed the cream off the top." The next round requires deeper analysis.

5. Build up a database of improvement information. We can start by simply recording completed and in-progress improvement projects, but other steps can also be taken. For example, a fast-food firm is creating an "intellectual network" of computer bulletin boards of information to include best-practices information. Rethmeier (1995) explains how an alliance of 300 hospitals uses a "learning center" to create and transfer knowledge for continuous improvement. The learning center uses lessons derived from the experience of prior events. Juran (*JQH5,* Section 5) calls this retrospective analysis a "Santayana review" in honor of George Santayana who observed, "Those who cannot remember the past are condemned to repeat it." Taking an example from early Native American tribes, organizations will require a "wisdomkeeper."

SUMMARY

- The quality improvement process addresses *chronic* quality problems.
- The six-sigma sequence and the breakthrough sequence are strategic approaches to improvement—strategic because the projects selected are based on gaps between actual performance and goals. The two approaches are complementary in both objective and content.
- The six-sigma steps are define, measure, analyze, improve, and control.
- The breakthrough steps are proving the need, identifying projects, organizing project teams, verifying the project need and mission, diagnosing the causes, providing a remedy and proving its effectiveness, dealing with resistance to change, and instituting control to hold the gains.

PROBLEMS

3.1. Selwitchka (1980) presents data on 10 types of errors using two measures—frequency and cost:

Type of error	Frequency	Cost ($, DM)
A	960	20,000
B	870	28,460
C	420	375,000
D	210	42,000
E	180	124,300
F	180	9,000
G	60	77,800
H	60	12,125
I	30	9,000
J	30	9,125

Note that the first two columns present a Pareto analysis based on frequency of occurrence. Prepare a second Pareto table based on cost. Comment on the ranking of errors using frequency versus the ranking based on cost.

3.2. Unplanned shutdowns of reactors have been a chronic problem. After much discussion, the consensus—called the "wisdom"—identified "cylinder changes" and "human error" as the primary causes. A diagnostic approach based on facts was instituted. Here are data on the causes of previous shutdowns:

Cause	Frequency
Cylinder changes	21
Human error	16
Hot melt system	65
Initiator system	25
Interlock malfunction	19
Other	23
	169

(a) Convert the above data into a Pareto table having three columns: cause, frequency, and percentage of total frequency.
(b) Calculate the cumulative frequencies for the table in part *(a)*.
(c) Calculate the percentage cumulative frequencies. Make a Pareto diagram by plotting percentage cumulative frequency versus causes.
(d) Comment about the wisdom versus the fact.

3.3. In diagnosing causes it is helpful to ask, Why? several times. Imai (1986) attributes the following example on a machine stoppage to Taiichi Ohno:

Question 1: Why did the machine stop?
Answer 1: Because the fuse blew due to an overload.

Question 2: Why was there an overload?
Answer 2: Because the bearing lubrication was inadequate.

Question 3: Why was the lubrication inadequate?
Answer 3: Because the lubrication pump was not functioning right.

Question 4: Why wasn't the lubricating pump working right?
Answer 4: Because the pump axle was worn out.

Question 5: Why was it worn out?
Answer 5: Because sludge got in.

What is the benefit of repeating Why? Can you cite a similar example from your own experience?

3.4. Draw an Ishikawa diagram for one of the following: *(a)* the quality of a specific activity at a university, bank, or automobile repair shop; *(b)* the quality of *one* important characteristic of a product at a local plant. Base the diagram on discussions with the organization involved.

3.5. The following data summarize the total number of defects for each worker at a company over the past six months:

Worker	Number of defects	Worker	Number of defects
A	46	H	9
B	22	I	130
C	64	J	10
D	5	K	125
E	65	L	39
F	79	M	26
G	188	N	94

A quality cost study indicates that the cost of these defects is excessive. There is much discussion about the type of quality improvement program. Analysis indicates that the manufacturing equipment is adequate, the specifications are clear, and workers are given periodic information about their quality record. What do you suggest as the next step?

3.6. An engineer in a research organization has twice proposed that the department be authorized to conduct a research project. The project involves a redesign of a component to reduce the frequency of failures. The research approach has been meticulously defined by the engineer and verified as valid by an outside expert. Management has not authorized the project because "other projects seem more important." What further action should the engineer consider?

3.7. A company that manufactures small household appliances has experienced high scrap and rework for several years. The total cost of the scrap and rework was recently estimated on an annual basis. The number shocked top management. Discussions by the management team degenerated into arguments. Finally, top management proposed that all departments reduce scrap and rework costs by 20 percent in the coming year. Comment on this proposal.

3.8. A small steel manufacturing firm has had a chronic problem of scrap and rework in the wire mill. The costs involved have reached a level where they are a major factor in the profits of the division. All levels of personnel in the wire mill are aware of the problem, and there is agreement on the vital few product lines that account for most of the problem. However, no reduction in scrap and rework costs has been achieved. What do you propose as a next step?

3.9. The medical profession makes the journey from symptom to cause and cause to remedy for medical problems of human beings. Compare this process with the task of diagnosing quality problems of physical products. If possible, speak with a physician to learn the diagnostic approach used in medicine.

3.10. From your experience, recall a chronic quality-related problem that was acted upon by an organization. Critique the approach for handling the problem in terms of the use—or nonuse—of the steps in the breakthrough sequence.

3.11. Select one chronic quality-related problem in your organization.
(a) Write a brief problem statement.
(b) Write a mission statement for a quality improvement team.
(c) What data could be collected to prove the need to address the problem?
(d) What departments should be represented on the team?

(e) State one or more symptoms of the problem.
(f) State at least three theories on the cause(s).
(g) Select one theory. What data or other information is needed to test the theory?
(h) Assume the data show that your selected theory is the true cause. State a remedy to remove the cause.
(i) What forms of resistance to the proposed remedy are you likely to encounter, and how will you deal with this resistance to change?
(j) What methods should be instituted to hold the gains?

3.12. You are trying to convince upper management of an insurance company to embark on a new approach to quality. Studies on the cost of poor quality, market standing, quality culture, and the quality system have shown a need for action. You have proposed a course of action but management has not acted on your proposal. Which additional step would be most useful to convince management?

3.13. Your hospital has embarked on the project-by-project approach to quality improvement. Projects have been selected, a team formed for each project, and diagnostic and other forms of training provided to the teams. Many of the teams have been working on their project for more than one year, and some teams have become discouraged. What is the most likely reason for the lengthy project times?

3.14. Describe how the principles of quality improvement could help to improve your game of golf. Consider matters such as setting goals, defining measurements, collecting and analyzing data, identifying weaknesses, implementing improvement actions, and training. For the views of a quality professional and a golf professional, see Karlin and Hanewinckel (1998).

REFERENCES

Anderson, D., F. Abetti, and P. Savage (1995). "Process Improvement Utilizing Computer Simulation: Case Study," *Annual Quality Congress Proceedings,* ASQ, Milwaukee, pp. 713–724.

Banks, J., ed. (1998). *Handbook of Simulation*, Engineering and Management Press, Institute of Industrial Engineers, Norcross, GA.

Batson, R. G. and T. K. Williams (1998). "Process Simulation in Quality and BPR Teams," *Annual Quality Congress Proceedings,* ASQ, Milwaukee, pp. 368–374.

Betker, H. A. (1983). "Quality Improvement Program: Reducing Solder Defects on Printed Circuit Board Assemblies," *Juran Report Number Two,* Juran Institute, Inc., Wilton, CT, pp. 53–58.

Bhote, K. R. (1991). *World Class Quality,* AMACOM, New York.

Blakeslee, J. A. Jr. (1999). "Achieving Quantum Leaps in Quality and Competitiveness," *Annual Quality Congress Proceedings,* ASQ, Milwaukee, pp. 486–496.

Dettmer, H. W. (1995). "Quality and the Theory of Constraints," *Quality Progress,* April, pp. 77–81.

Forsha, H. I. (1995). *Show Me, The Complete Guide to Storyboarding and Problem Solving,* Quality Press, ASQ, Milwaukee.

Goldratt, E. M. (1997). *Critical Chain,* North River Press, Great Barrington, MA.

Hartman, B. (1983). "Implementing Quality Improvement," *Juran Report Number Two,* Juran Institute, Inc., pp.124–131.

Juran Institute, Inc. (1989). *Quality Improvement Tools,* Wilton, CT.

Juran, J. M. (1995). *Managerial Breakthrough,* rev. ed., McGraw-Hill, New York.

Karlin, E. W. and E. Hanewinckel (1998). "A Personal Quality Improvement Program for Golfers," *Quality Progress,* July, pp. 71–78.

Klenz, B. W. (1999). "The Quality Data Warehouse: Serving the Analytical Needs of the Manufacturing Enterprise," *Annual Quality Congress Proceedings,* ASQ, Milwaukee, pp. 521–529.

Kolarik, W. J. (1995). *Creating Quality,* McGraw-Hill, New York.

Lenhardt, L. (1993). "Quality Improvement in the Emergency Department Admission Process," *Impro Proceedings,* Juran Institute, Inc., Wilton, CT, pp. 3A.3-1 to 3A.3-9.

Marquette, D. K. (1996). "Achieving Breakthrough Results in Six Months. . . . Really!: Rework Reduction in Insurance Service Processing," *Impro Proceedings,* Juran Institute, Inc., Wilton, CT, pp. 3C.-11 to 3C.-29.

Mears, P. (1995). *Quality Improvement Tools and Techniques.* McGraw-Hill, New York.

Mizuno, S., ed. (1988). *Management for Quality Improvement: The 7 New QC Tools*, Productivity Press, Portland, OR.

Plsek, P. E. (1997). *Creativity, Innovation, and Quality*, Quality Press, ASQ, Milwaukee.

Rethmeier, K. A. (1995). "Creating the Learning Organization: Toward Excellence in Knowledge Transfer," *Impro Proceedings*, Juran Institute, Inc., Wilton, CT, pp. 3C.1-1 to 3C.1-11.

Sanders, D., B. Ross, and J. Coleman (1999). "The Process Map," *Quality Engineering*, vol. 11, no. 4, pp. 555–561.

Selwitchka, R. (1980). "The Priority List on Measures for Reducing Quality Related Costs," *EOQC Quality,* vol. 24, no. 5, pp. 3–7.

Slater, R. (1999). *Jack Welch and the GE Way*, McGraw-Hill, New York.

Smith, G. F. (1998). *Quality Problem Solving*, Quality Press, ASQ, Milwaukee.

Wadsworth, H. M., K. S. Stephens, and A. B. Godfrey (2001). *Modern Methods for Quality Control and Improvement,* John Wiley and Sons, New York.

Wilson, P. F., L. D. Dell, and G. F. Anderson (1993). *Root Cause Analysis*, Quality Press, ASQ, Milwaukee.

SUPPLEMENTARY READING

Quality improvement, general: *JQH5,* Section 5, and applications in Sections 27–33.

Breakthrough and control: Juran, J. M. (1995). *Managerial Breakthrough*, 2nd ed., McGraw-Hill, New York.

Nadler, G. and S. Hibino (1990). *Breakthrough Thinking,* Prima Publishing and Communications, Rocklin, CA.

Data analysis: Tukey, J. W. (1977). *Exploratory Data Analysis*, Addison-Wesley Publishing Company, Reading, MA.

Medical diagnosis and quality diagnosis: White, G. L. (1999). "GM Takes Tips from CDC to Debug Its Fleet of Cars," *Wall Street Journal,* April 24.

Pareto concept: Koch, R. (1998). *The 80/20 Principle,* Currency Doubleday, New York.

Six-sigma approach: Harry, M. and R. Schroeder (2000). *Six Sigma,* Doubleday, New York.

Hahn, G. J., N. Doganaksoy, and R. Hoerl (2000). "The Evolution of Six Sigma," *Quality Engineering,* vol. 12, no. 3, pp. 317–326.

Bajaria, H. J. (1999). "Six Sigma—Moving beyond the Hype," *Annual Quality Congress Proceedings,* ASQ, Milwaukee, pp. 1–3.

Henson, D. (1999). "Quality Driving Customer Impact," *Proceedings of the 8th Annual Service Quality Conference,"* ASQ, Milwaukee, keynote address.

WEBSITES

Theory of constraints: www.goldratt.com
European Continuous Improvement Network: www.dipoli.hut.fi/org/TechNet/org/eurocinet/

4

OPERATIONAL QUALITY PLANNING AND SALES INCOME

4.1 CONTRIBUTION OF QUALITY TO SALES INCOME

Planning for quality always takes time and effort. Unfortunately, people simply won't spend the time on planning unless they are first convinced that thorough quality planning is essential for business success. To cite an old cliche—they don't have time to do thorough quality planning, but they do have the time to correct the mistakes of poor planning. Fortunately, we know how to do good quality planning—it's one of the three processes in the trilogy (see Chapter 1 under "Managing for Quality"). But before we discuss the steps in quality planning, we will first discuss why the time spent on quality planning is an investment that yields a strong measurable return.

In defining quality for goods and services, two components have been identified: product features and freedom from deficiencies (Figure 1.1). Although both components are essential to generating sales, in general the product-features component is more dominant. This chapter discusses the implication of quality on sales income and then provides an overview of an approach that provides the product features necessary to meet sales goals. Chapter 12, "Understanding Customer Needs," Chapter 13, "Designing for Quality," Chapter 16, "Operations—Manufacturing Sector," and Chapter 17, "Operations—Service Sector," provide further elaboration. Also, note that this chapter discusses quality planning for products and processes (operational quality planning) in contrast to strategic quality planning, which is discussed in Chapter 7, "Strategic Quality Management."

For profit-making organizations, the contribution of quality to sales income occurs through several means:

- Increasing market share.
- Securing premium prices.

- Achieving economics of scale through increased production.
- Achieving unique competitive advantages that cement brand loyalties.

For both profit and nonprofit organizations, quality is synonymous with providing satisfaction to customers, both internal and external. For most organizations, satisfaction must be viewed relative to the competition and thus goes far beyond the document called a "specification."

4.2 QUALITY AND FINANCIAL PERFORMANCE

Understanding the effect of quality on sales and financial performance can be aided by looking at some valuable research efforts. This research is based on a database known as Profit Impact of Market Strategies (PIMS). The PIMS program aims at determining how key dimensions of strategy affect profitability and growth. More than 450 companies from numerous manufacturing and service industries have contributed data.

Analysis of the PIMS data has led to important conclusions that center on quality relative to the competition. Assessment of relative quality involves the use of multiattribute comparisons (see Section 2.6, "Standing in the Marketplace"). A key conclusion is this: In the long run the most important factor affecting business performance is quality relative to the competition. Market share and profitability measure business performance. Table 4.1 shows return on investment (ROI) data. Note that businesses having both a larger market share and better quality earn returns much higher than businesses with a smaller market share and inferior quality. Further, although quality and market share are correlated, each has a strong separate relationship to profitability. Figure 4.1 shows the relationship of quality to profitability [return on sales (ROS) or ROI].

Of course, the greater profitability could be due to either higher prices or lower costs. But analysis of the PIMS data sheds some additional light. Quality affects relative price, but separate from quality, market share has little effect on price. The quality/price relationship is shown in Table 4.2. Superiority in quality commands a premium price.

TABLE 4.1
Quality and market share both drive profitability (ROI)

Relative quality percentile	Relative market share		
	Below 25%	25-59%	60% and above
Below 33%	7	14	21
33–66%	13	20	27
67% and above	20	29	38 (ROI)

Source: Buzzell and Gale (1987).

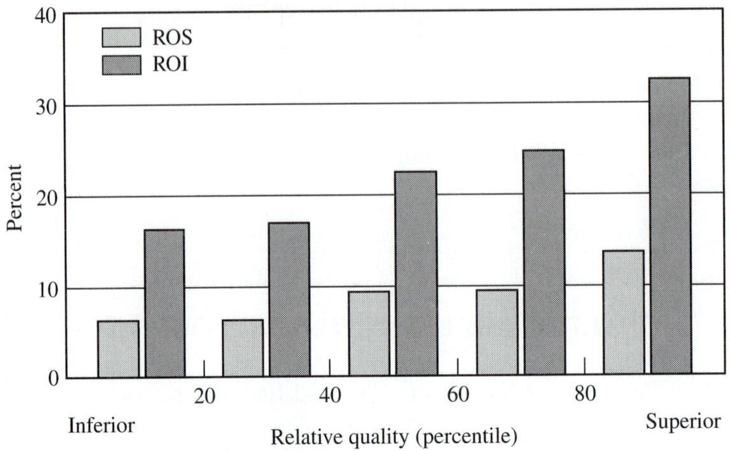

FIGURE 4.1
Relative quality boosts rates of return. (*From Buzzell and Gale, 1987.*)

TABLE 4.2
Quality and price

Relative quality	Relative price
Lowest	100
Commodity	102
Average	104
Better	106
Best	108

Source: Gale and Klavans (1985).

According to other PIMS data, relative quality has little effect on cost. Apparently, the savings from efforts to reduce scrap and rework (deficiencies) are offset by the increased costs for product attributes (features) that sell the product.

4.3 ACHIEVING A QUALITY SUPERIORITY

To achieve quality leadership for a product or service, an organization must focus on one or more parameters of quality such as performance features, long life, ease of use, freedom from deficiencies, personal service, or fast service.

In some cases superior quality may be obvious, e.g., a hotel room has outstanding features or an insurance claim adjuster arrives within two hours after a car accident is reported.

In other cases the superiority can be translated into users' economics, e.g., an automobile achieves higher mileage per gallon than a competitor. Sometimes the superiority is minor but can be demonstrated, e.g., one chemical has slightly less variation in a property than a competitor's chemical has, even though both meet the same specification. In still other cases clever marketing can present convincing illustrations of superiority in performance that are accepted by buyers. Many organizations try to establish a superiority in value (the combination of quality and price). Value is difficult to quantify, but because of the importance to buyers, value is often chosen as the focus for establishing superiority and quality leadership. Chapters 12 and 13 provide some examples of using value to achieve leadership. Also, Rust (1994) elaborates on the concepts of value, quality, and utility. Gale (1994) discusses how customer value analyses can help to plan improvement actions.

Whatever focus management chooses to establish quality leadership, subsequent actions must view quality relative to competition. In this regard the concept of competitive benchmarking can provide direction.

Competitive Benchmarking

A *benchmark* is a point of reference by which performance is judged or measured. For quality, possible benchmarks go from the traditional to the unusual:

- The specification.
- Customer desires.
- Competition (see Chapter 2 under "Standing in the Marketplace").
- Best in our industry.
- Best in any industry.

For survival in the marketplace, the traditional benchmark (the product specification) must be supplemented by measuring quality relative to competition; for quality leadership the benchmark must be the "best." Xerox, for example, defines competitive benchmarking as "the continuous process of measuring our products, services and practices against our toughest competitors or those companies renowned as leaders." Thus a benchmark for a photocopier product would be the best competitor in the photocopier industry; a benchmark for the Xerox system of packing warehouse orders might be the performance of a company in any industry, e.g., a mail-order firm selling consumer products. The initial benchmarking steps are

1. Determining the characteristics to be benchmarked.
2. Determining the organizations from which data will be collected.
3. Collecting and analyzing the data.
4. Determining the "best in class."

Strategic plans are then prepared to develop or adapt the "best practices." Such strategies are, of course, directed at both retaining present customers and generating new ones. This approach is illustrated later in this chapter.

The full process of competitive benchmarking as an approach to achieving quality leadership is discussed in Chapter 7.

TABLE 4.3
Customer satisfaction versus customer loyalty

Customer satisfaction	Customer loyalty
What customers say—opinions about a product	What customers do—buying decisions
Customer expects to buy from several suppliers in the future	Customer expects to buy primarily from one or two suppliers in the future
Company aims to satisfy a broad spectrum of customers	Company identifies key customers and "delights" them
Company measures satisfaction primarily with the product for a spectrum of customers	Company measures satisfaction with all aspects of interaction with key customers and also their intention to repurchase
Company measures satisfaction primarily for current customers	Company also analyzes and learns the reasons for lost customers (defections)
Company emphasizes staying competitive on quality for a spectrum of customers	Company continuously adds value by creating new products based on evolving needs of key customers

4.4 CUSTOMER SATISFACTION VERSUS CUSTOMER LOYALTY

It is useful to make a distinction between customer satisfaction and customer loyalty (see Table 4.3).

In brief, a satisfied customer will buy from our company but also from our competitors; a loyal customer will buy primarily (or exclusively) from our company. A dissatisfied customer is unlikely to be loyal, but surprisingly, a satisfied customer is not necessarily loyal (see Section 4.6).

The sales revenue lost due to customers with problems can be calculated—see Section 20.11.

4.5 ECONOMIC WORTH OF A LOYAL CUSTOMER

The sales revenue from a loyal customer measured over the period of repeat purchases can be dramatic, e.g., $5000 for a pizza eater; $25,000 for a business person staying at a favorite hotel chain; $300,000 for a loyalist to a brand of automobile. The economic worth is calculated as the net present value (NPV) of the net cash flow (profits) over the expected lifetime of repeat purchases. The NPV is the value in today's dollars of the profits over time. Profits generally rise as the customer defection rate (customers who do not repurchase) decreases; i.e., retention rate increases. Reichheld (1996) shows, for a variety of service industries, the increase in NPV of the customer profit stream in the future resulting from a decrease of five percentage points in defection rate (Figure 4.2). He concludes that a five percentage point decrease in defection rate can increase profits by 35 to 95 percent.

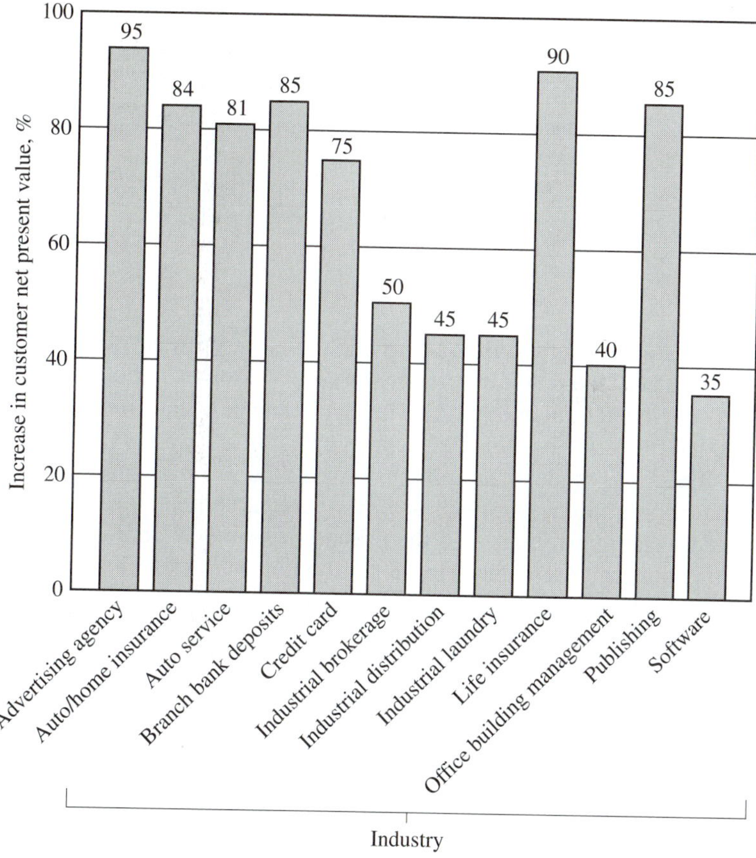

FIGURE 4.2
Profit impact of 5 percent increase in retention rate. (*Reichheld, 1996.*) Reprinted by permission of Harvard Business School Press.

4.6 IMPACT OF QUALITY ON LOST SALES

Sometimes the evidence of the importance of quality to retaining present customers is dramatic. Manufacturer B of household appliances had a leadership position for two of the four models of a product (see Table 4.4). Over a four-year period, leadership was lost, even though the company was competitive on product features, price, and delivery dates. It was not competitive, however, on field failures and warranty costs. The president had to personally lead the development and execution of a strategy to reduce the failures and warranty costs. He was successful.

Sometimes the sales lost because of poor quality can be quantified. Two manufacturers of washing machines were compared. One measure was the percentage of present customers who would not purchase the same brand again. For brand A, only 1.1 percent said they would not purchase it again; for brand B, 10.5 percent declined

TABLE 4.4
Change in suppliers of product

Model	1979	1980	1981	1982
High price	A	C	C	C
Middle price	B	B	C	C
Low price	C	C	C	C
Special model	B	B	B	C

Note: A, B, and *C* indicate suppliers.

to purchase it again. For brand B, the overwhelming reason was "poor quality." When the 10.5 percent was translated to the total present customer population for brand B, researchers concluded that an additional sales income of $5 million would be needed to make up the profit in the lost replacement sales. In a case involving an industrial product, a survey was made of present customers who were buying the product made by one manufacturer. Some of those customers had purchased a different brand. One-third of these former customers said the primary reason was poor quality. These lost sales due to poor quality amounted to $1.3 billion in sales income (and 2000 jobs).

For an example of sales revenue lost at a bank, see Chapter 2 under "Categories of Quality Costs."

For elaboration on quantifying the impact of quality on sales see Rust et al. (1994). See also Campanella (1999) and *JQH5* (Sections 8 and 18) in the Supplementary Reading.

4.7 LEVEL OF SATISFACTION TO RETAIN PRESENT CUSTOMERS

Sometimes acceptable levels of customer satisfaction with the product still result in a significant loss of new sales. Table 4.5 presents two examples from the service industry. Note that even when the customer view of quality is good, a quarter or more of the present customers may not return. A similar story exists in the manufacturing sector (Burns and Smith, 1991). At the Harris Corporation satisfaction with specific attributes of the product (goods and service) is measured on a scale of 1 to 10; loyalty is measured as the percentage of customers who will continue to purchase from Harris. The data showed that to achieve a 90 percent loyalty rate required a satisfaction level of at least 8.5. In some industries more than 90 percent of customers report that they are satisfied or very satisfied with the product—but repurchase rates are only about 35 percent.

Another dimension of this phenomenon is the level of customer satisfaction with the handling of their complaints. Table 4.6 shows the percentage of customers at a financial services firm who intend to purchase a product or service again, based on their level of satisfaction with the resolution of their complaints (*JQH5,* page 33.22).

Customers who have a problem but are unsatisfied with the resolution ("recovery") are unlikely to repurchase (30 percent). Customers who are very satisfied with

TABLE 4.5
Satisfaction and sales

Customer view of quality	Willing to recommend supplier, % (GTE)	Very willing to repurchase, % (AT&T)
Excellent	96	92
Good	76	63
Fair	35	18
Poor	3	0

Source: Bultmann (1989); Scanlan (1989).

TABLE 4.6
Repurchase intent

Type of customer	Percentage who intend to repurchase	Percentage who will recommend purchase to others
Typical satisfied customer with no problem	51	65
Customer with problem; unsatisfied with recovery	30	46
Customer with problem; very satisfied with recovery	79	88

the handling of a complaint have a much higher intention to repurchase (79 percent) and recommend purchase to others (88 percent). Some companies seize upon a complaint as a special opportunity to generate additional sales revenue by providing dramatic, and memorable, recovery action. Finally, note that some satisfied customers, with no problem, will not repurchase.

Electronic Commerce

Electronic commerce (or e-commerce) is the buying and selling of products across telecommunications networks. Three dimensions are emerging. First, e-commerce can contact large numbers of customers and present many products and options ("reach"). Second, e-commerce provides great depth of information to customers and can collect much information about customers ("richness"). Third, the depth of information provided by e-commerce (e.g., alternative products) helps to promote loyalty by providing customers with objective information for making decisions ("affiliation"). For elaboration of the reach, richness, and affiliation dimensions, see Evans and Wurster (1999).

4.8 LIFE CYCLE COSTS

For simple consumable products like food or transportation, the purchase price is also the cost of using the product. As products grow in complexity and as the length of use increases, the purchase price must also increase to include operational, maintenance, and other special costs. For some products, the after-sale costs can easily exceed the original purchase price.

A *life cycle cost* can be defined as the "total cost to the user of purchasing, using, and maintaining a product over its life." A study of all of the cost elements can lead to a redesign of a product that could result in a significantly lower life cycle cost, perhaps at the expense of a small increase in the original price. This situation presents a marketing opportunity to provide a product that will result in savings over the life of the product to potential customers. These customers are urged to make purchasing decisions by comparing life cycle costs for competing products. But the initial purchase price may be higher, and marketing people may find it more difficult to sell the product to potential customers whose first priority is initial price. Table 4.7 shows the ratio of life cycle costs to original price for various consumer products.

An associated concept, user failure cost, calculates the cost to the user of failures over the product life. Table 4.8 shows an example of annual failure-related costs (Gryna, 1977).

The life cycle cost concept is fundamentally sound in logic, but implementation has been slow. Two reasons for the slow pace of adoption predominate. First, estimat-

TABLE 4.7
Life cycle costs: consumer products

Product	Ratio, life cycle cost to original price
Room air conditioner	3.3
Dishwasher	2.5
Freezer	4.8
Range, electric	4.4
Range, gas	1.9
Refrigerator	3.5
TV (black and white)	2.5
TV (color)	1.9
Washing machine	3.6

TABLE 4.8
Annual quality cost of foundry equipment

	11 shell core makers	12 induction furnaces
Repairs	$16,000	$ 70,000
Effectiveness loss	25,000	90,000
Lost income	8,000	50,000
Extra capacity	2,500	8,000
Preventive maintenance	8,000	20,000
Total	$59,500	$238,000
Initial investment	$70,000	$700,000

ing the future costs of operation and maintenance is difficult. A greater obstacle, however, is the cultural resistance of purchasing managers, marketing people, and product designers. The skills, habits, and practices of these people have long been built around the concept of the original purchase price being of primary importance (see *JQH5*, pages 7.22–7.27, for further elaboration).

4.9 SPECTRUM OF CUSTOMERS

For both consumer and industrial products and for both physical goods and services, the variety of customers forms a wide spectrum. Some companies choose to address one part of the spectrum, whereas other companies pursue several types of customers. For purposes of planning for quality, we will identify three types of customers:

1. Those who emphasize initial purchase price as equal to or more important than quality.
2. Those who evaluate alternative products on both initial price and quality simultaneously.
3. Those who place emphasis on obtaining "the best."

Table 4.9 links these three categories with a translation into desires on product features and freedom from deficiencies.

All three categories need to be satisfied in the marketplace. In consumer products particularly, some customers change categories over a lifetime; e.g., some young couples with growing children change from "economy minded" to "value minded."

TABLE 4.9
Categories of customer emphasis as related to quality

Emphasis	Product features	Freedom from deficiencies
Initial economy	Willing to forego some features Like do-it-yourself features and add options later	Will tolerate some product deficiencies at delivery and during use
	Will tolerate a relatively short product life	Will tolerate some deficiencies in service before and after purchase
Value	Willing to make trade-offs between quality and price	Warranty provisions can be important
	Features must be justified by benefits and related price	Concerned about operating and repair costs
The "best"	Desire many convenience features	Greatly annoyed at deficiencies and associated inconveniences
	Emphasis on luxury, esthetics, brand image	Demand complete and timely response to all problems
	Desire high level of performance from product and from all personnel	

4.10 PLANNING FOR PRODUCT QUALITY TO GENERATE SALES INCOME

A starting point for planning quality for individual products is the definition of quality provided in Chapter 1. The components of the definition—product features and freedom from deficiencies—can be expanded into the major areas that must be addressed to achieve customer satisfaction. Table 1.1 lists some major subcategories of the two components. Note that these subcategories go far beyond the specification of a physical product.

In designing a specific product or service, the detailed characteristics supporting the subcategories must be identified, planned for, and executed. For example, Ford identified 429 characteristics for the Taurus model car; GTE identified 31 characteristics for a telecommunications service. Clearly, customer input must be taken into account when defining specific characteristics. That input takes the form of not only identifying the characteristics but also clarifying the importance of each, and finally providing a status in the current marketplace relative to competition. (This concept was introduced in Section 2.6, "Standing in the Marketplace." Chapter 12, "Understanding Customer Needs," elaborates on the concept.)

This emphasis on understanding customer needs as a prerequisite to meeting sales goals addresses the reality that warehouses are filled with products that meet specifications and are competitively priced but *don't satisfy customers' needs* as well as a competitor's product does. Examples from a spectrum of industries provide reasons that sales are lost to competition (Table 4.10). Note that some of the examples involve increased satisfaction for the ultimate user (e.g., the driver of an automobile); other examples involve increased satisfaction for an intermediate user (e.g., the processor of photographic film). In recent years lower variability around a target value within a set

TABLE 4.10
Hidden customer needs

Product	Hidden customer need satisfied by a competitor
Abrasive cloth	Lower internal cost of polishing parts due to better durability of cloth
Automobile	Less effort in closing door; better "sound" when door closes
Dishwasher	Sense of greater durability due to heavier parts making up the appliance
Software	Understandable owner's manual
Fibers	Lower number of breaks in processing fibers
Tire valve	Higher productivity when tire manufacturer uses valve in a vulcanizing operation
Photographic film	Fewer process adjustments when processing film due to lower variability
Commodity product	Delivery of orders within 24 hours rather than the 48-hour standard requirement
Home mortgage application	Decision in shorter time than that of competition

of specification limits has become increasingly important to customers who do further processing of a product.

Product features must satisfy customer needs, but features that delight customers today typically become the expected features of tomorrow.

4.11 A QUALITY PLANNING ROAD MAP FOR ENSURING PRODUCT SALABILITY

The quality planning road map (see Table 1.4) presents a framework for the planning (or replanning) of new products (or product revisions). This road map applies to both the manufacturing and service sectors and to products for both external and internal customers. Developing a remedial solution for a quality improvement project (see Chapter 3) may require the use of one or more steps of this quality planning process. Early and Colletti (1999) and Juran (1992) provide an extensive discussion of the steps. These quality planning steps must be incorporated with the technological tools for the product being developed. Designing an automobile requires automotive engineering disciplines; designing a path for treating diabetes requires medical disciplines. But both need the tools of quality planning to ensure that customer needs are met.

The road map is presented in more detail in Figure 4.3, and the steps are discussed in later chapters. It is useful, however, to present an overview now to briefly explain the

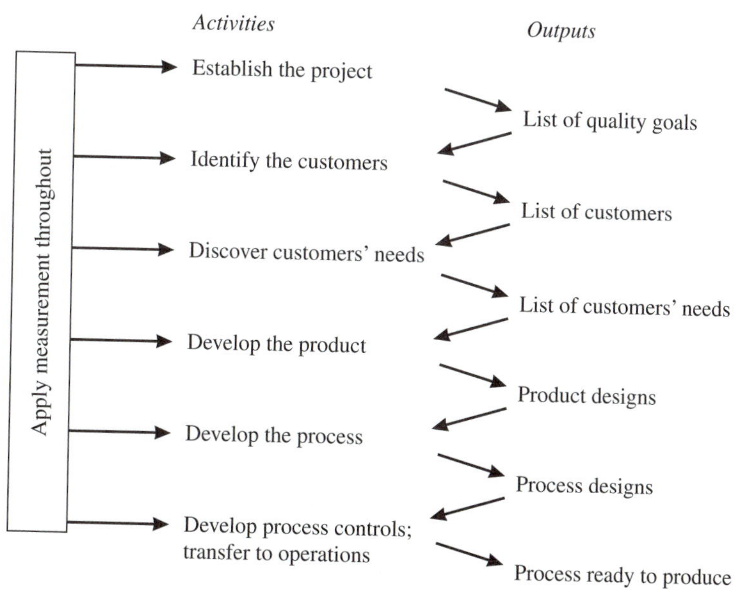

FIGURE 4.3
Quality planning road map. (*From Juran, 1990.*)

steps (see Early and Colletti, 1999). Our example is based on the planning work done for the Ford Motor Company's original Taurus and the 1994 Mustang automobiles.

In the early 1980s, Ford began the initial planning for a new front-wheel-drive midsize car. The business environment included some ominous elements: strong foreign competition, decreasing market share, and a projection of greatly increased fuel prices. Ford concluded that a new approach to designing the model was essential. Basic to the new approach was "customer satisfaction" with the objective being that the Taurus would be the best car in its class. This "best in class" focus spawned some unusual approaches to planning. One of the breaks with tradition was the organization of the planning for the Taurus. Historically, new-car design used the traditional organizational structure (Figure 4.4*a*). With such a structure, the main activities were executed *sequentially,* e.g., the planning department studied customer desires and then presented the results to design; design performed its tasks and handed the results to engineering; engineering created the detailed specifications; and the results were then given to manufacturing. Unfortunately, the sequential approach results in a minimum of communication between the departments as the planning proceeds—each department hands its output "over the wall" to the next department. This lack of communication often leads to problems for the next internal customer department. Under Taurus, activities were organized as a team (Figure 4.4*b*) from the beginning of the project. Thus, for example, manufacturing worked *simultaneously* with design and engineering before the detailed specifications were finalized. This approach allowed the team to address producibility issues during the preparation of the specifications.

We will now refer to the steps in Figure 4.3 to explain how the Taurus and Mustang projects proceeded. As a homework problem, readers are asked to provide further illustrations.

1. *Establish the project.* This step has three substeps:
   a. Create a mission statement—the purpose, scope, and goals of the project.
   b. Establish a team to do the planning.
   c. Plan the execution of the project—responsibilities, schedules, resources, and follow-up.

   Some tools used to establish the project are competitive analysis, benchmarking, and deployment of goals.

   For the original Ford Taurus (in the early 1980s), the quality goal was to be best in class. For the 1994 Ford Mustang, four numerical quality goals were set, e.g., "things gone wrong/1000."

   The Mustang team was cross functional and included representatives from various sections of design engineering, manufacturing and assembly, sales and marketing, purchasing, and finance. This team functioned as a quality planning project team. In addition, the design and development process benefited from the strong involvement of inputs from customers (external and internal) and suppliers—before the release of the design to production. Chapter 8 explains several types of teams and the roles of a team leader, team facilitator, and team members.
2. *Identify the customers.* The customer is anyone who is affected by the product—not just the purchaser. This step includes identifying both the external and internal customers (see Chapter 1 under "*Quality* Defined"). Examples of tools employed are flow charts, Pareto analysis, and spreadsheets.

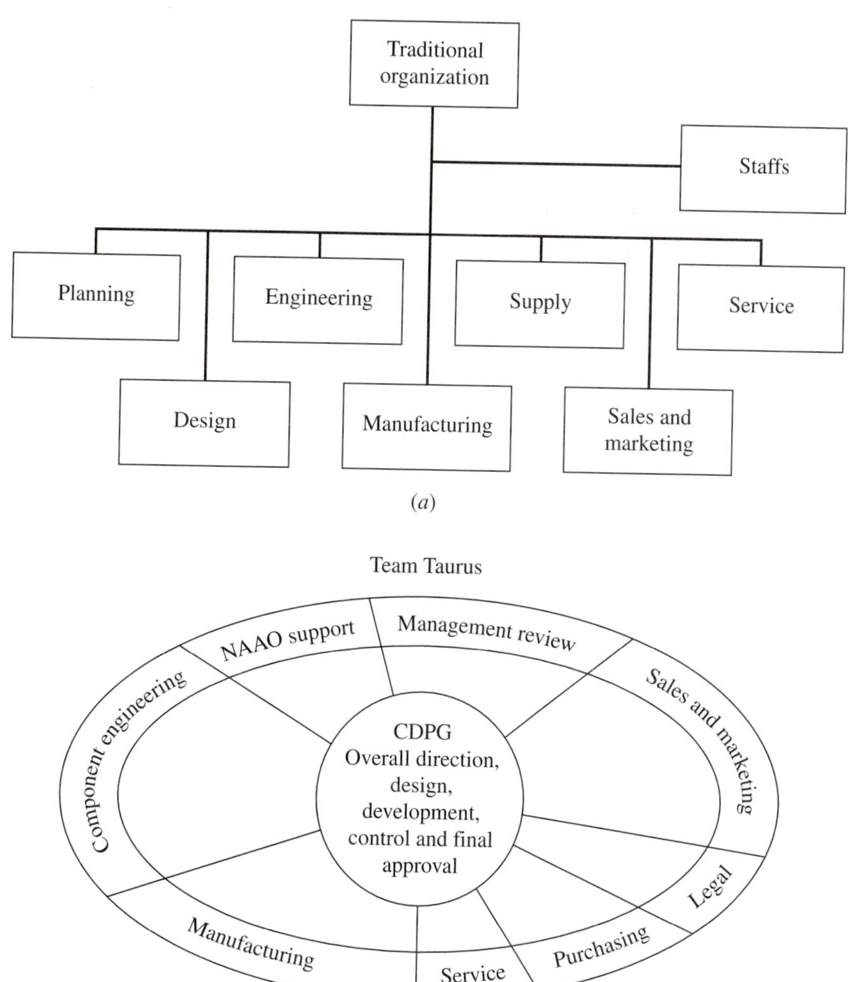

FIGURE 4.4
(a) Traditional organization. *(b)* Organization for Taurus. (*From Juran, 1990.*)

Some customers are obvious; others are not. Thus, for an automobile:

Company function	Customer
Sales	Consumer
Legal	U.S. Dept. of Transportation
Design engineering	Manufacturing
Parts manufacturing	Assembly plant
Advertising	Media

(Note that the customers are both external and internal.)

This step is discussed further in Chapter 12, "Understanding Customer Needs."

3. *Discover customers' needs.* This step has three substeps:
   a. Plan how to discover customer needs—e.g., market research, customer complaints, dealer input, competitive evaluations.
   b. Collect information on customer needs.
   c. Analyze and prioritize customer needs.

   Some tools for discovering customers' needs are multiattribute studies, focus groups, questionnaires, site visits, and spreadsheets (see Chapter 12, "Understanding Customer Needs").

   For the Taurus and the Mustang, extensive market research was conducted. For the Taurus, research was conducted on 429 potential product features. For example:

Customer	Need
Consumer	Effective heater
U.S. Dept. of Transportation	High-mount brake light
Assembly plant	Ease of assembly

   We develop this step in more detail in Chapter 12.

4. *Develop the product.* This step has four substeps:
   a. Group together related customer needs.
   b. Identify alternative product features.
   c. Develop detailed product features and goals.
   d. Finalize the product design.

   Some tools used to develop the product are competitive analysis, reliability, safety and value analysis, prototype tests, and spreadsheets (see Chapter 13, "Designing for Quality").

   The Taurus team used marketing research to provide product development with detailed guidelines for 429 product features that were important to achieving high product salability. These guidelines then became the basis of specific design projects. Two examples of these features were the amount of effort required to raise the hood of the car and the level of wind noise. For each feature, a means of measurement was defined, competitive data were obtained, and a numerical goal was set. For example, the effort required to raise the hood was measured in pounds by a spring scale. Competitive data showed that the best competitor had a design requiring 9 pounds of effort. For Taurus, a goal of 8 pounds was set; the final design exceeded the goal by requiring only 7 pounds. Taurus achieved best in class for 80 percent of the product features. Certain other product features were also incorporated, even though they were not directly related to product salability, e.g., the high mounting of the brake light, a feature that was desired by the Department of Transportation.

   This step is discussed further in Chapter 13.

5. *Develop the process.* This step has four substeps:
   a. Identify alternative process features.
   b. Develop detailed process features and goals.
   c. Establish the initial process capability.
   d. Finalize the process design.

Some tools used to develop process features are flow charts, process capability studies, pilot runs, and spreadsheets (see Chapter 16, "Operations—Manufacturing Sector").

Because the team for developing the new automobile was organized at the beginning of the project, manufacturing worked simultaneously with design and engineering before the detailed specifications were finalized. This approach enabled the team to address producibility issues during the preparation of the specifications; e.g., the assembly plan identified specific manufacturing issues to be addressed during design and manufacturing planning. The assembly plant listed 1400 issues, ranging from a desire to automate body-side assembly to having an annual assembly-plant shutdown for vacation. Much effort to achieve process capability ("process qualification") and to optimize the processes was also part of the planning. As a result, a set of process plans was ready at the start of production.

Elaboration on this step is provided in Chapter 16.

6. *Develop process controls and transfer to operations.* This step has five substeps:
   - *a.* Identify controls needed and design feedback loop.
   - *b.* Optimize self-control and self-inspection (see Chapter 5).
   - *c.* Establish audit of process.
   - *d.* Verify process capability in operations.
   - *e.* Transfer plans to operations.

   At Ford, as these plans were put into production, the coordination among all functions continued and resulted in final refinements to the product and process design.

For elaboration, see Chapter 16, "Operations—Manufacturing Sector," and Chapter 18, "Statistical Process Control."

An Example of Quality Planning for Services

In an example from the service sector, the quality planning process was applied to replanning the process of acquiring corporate and commercial credit customers for a major affiliate of a large banking corporation. Here is a summary of the steps in the quality planning process.

1. *Establish the project.* A goal of $43 million of sales revenue from credit customers was set for the year.
2. *Identify the customers.* This step identified 10 internal customer departments and 14 external customer organizations.
3. *Discover customers' needs.* Internal customers had 27 needs; external customers had 34 needs.
4. *Develop the product.* The product had nine product features to meet customers' needs.
5. *Develop the process.* To produce the product features, 13 processes were developed.
6. *Develop process controls and transfer to operations.* Checks and controls were defined for the processes, and the plans were placed in operation.

The revised process achieved the goal on revenue. Also, the cost of acquiring the customers was only one quarter of the average of other affiliates in the bank.

Quality planning generates a large amount of information that must be organized and analyzed in a systematic way. The alignment and linkages of this information are essential for effective quality planning for a product. A useful tool is the quality planning spreadsheet or matrix (basically, a table). Figure 4.5 shows five spreadsheets corresponding to steps in the quality planning process. Note how the spreadsheets interact and build on one another; they cover both quality planning for the product and quality planning for the process that creates the product. The approach is often called "quality function deployment" (QFD). Thus QFD is a technique for documenting the logic of translating customer needs into product and process characteristics. The use of spreadsheets in the quality planning process unfolds in later chapters (see particularly Section 13.4 under "Quality Function Deployment").

These six quality planning steps apply to a new or modified product (goods or service) or process in any industry. In the service sector the "product" could be a credit card approval, a mortgage approval, a response system for call centers, or hospital care. Also, the product may be a service provided to internal customers; e.g., Chapter 22 explains how the six steps are employed to develop a quality information system for use within an organization. Endres (2000) describes the application of the six quality planning steps at the Aid Association for Lutherans insurance company and the Stanford University Hospital.

SUMMARY

- Businesses having larger market share and better quality earn returns much higher than their competitors do. Quality and market share have strong separate relationships to profitability.
- Competitive benchmarking is the continuous process of measuring products, services, and practices against those of the toughest competitors or leading companies.
- Quality can be the decisive factor in lost sales, and sometimes its impact can be quantified.
- Customer complaints resolved with less-than-complete customer satisfaction will result in significant lost sales.
- Planning for product quality must be based on meeting customer needs, not just meeting product specifications.
- In-depth marketing research can identify suddenly arising customer needs.
- Planning for quality must recognize a spectrum of customers with different needs.
- For some products, we need to plan for perfection; for other products, we need to plan for value.
- Life cycle cost is the total cost to the user of purchasing, using, and maintaining a product over its life.
- Quality superiority can be translated into a higher market share or a premium price.

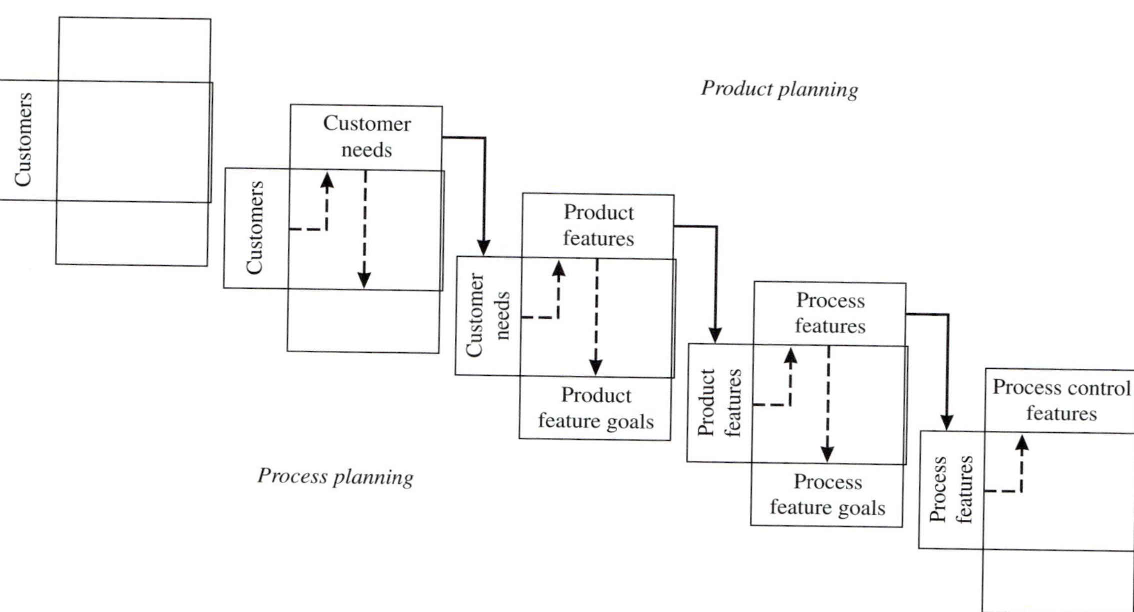

FIGURE 4.5
Spreadsheets in quality planning. (*Juran Institute, Inc. Copyright 1994. Used by permission.*)

- Quality planning for a new or modified product follows these steps: establish the project, identify the customers, discover customers' needs, develop the product, develop process features, develop process controls, and transfer the plans to operations. The measurement process must be applied during all of the steps.

PROBLEMS

4.1. In the explanation of the Taurus case, examples were given to illustrate each step in the planning road map; e.g., five customers were listed.

For each step in the planning road map, give two new examples. The examples may apply to any product or service.

4.2. For a consumer or industrial physical product, make a quality comparison of three brands at the low, medium, and high price levels. The comparison should list differences in product features and freedom from deficiencies.

4.3. For a consumer or business service, make a quality comparison as described in problem 4.2.

4.4. Select two physical products for which *Consumer Reports* magazine has conducted an analysis, provided ratings on quality, and reported the purchase price. For each product, summarize the relation between high quality and price, i.e., to what extent is high quality associated with a higher price.

4.5. Form a discussion group of five people. Select a physical product that you are likely to purchase and discuss how you might define quality for this product. How does the definition relate to customer satisfaction and loyalty, product features, and freedom from deficiencies?

4.6. Form a discussion group of five people. For a consumer or business service, conduct a discussion as described in problem 4.5.

REFERENCES

Bultmann, C. (1989). "How to Define Customer Needs and Expectations: An Overview," *Customer Satisfaction Measurement Conference Notes,* American Marketing Association and ASQ, February 26, 28. The written paper was "New Ways of Understanding Customers' Service Needs" by T. F. Gillett.

Burns, R. K. and W. Smith (1991). "Customer Satisfaction—Assessing Its Economic Value," *Annual Quality Congress Proceedings,* ASQ, Milwaukee, pp. 316–321.

Buzzell, R. D. and B. T. Gale (1987). *The PIMS Principles: Linking Strategy to Performance,* The Free Press, Macmillan, New York, reprinted with permission.

Early, J. F. and O. J. Coletti (1999). "The Quality Planning Process," in *Juran's Quality Handbook,* 5th ed., McGraw-Hill, New York, Section 3.

Endres, A. (2000). *Implementing Juran's Roadmap for Quality Leadership,* John Wiley & Sons, New York.

Evans, P. and T. S. Wurster (1999). "Getting Real about Virtual Commerce," *Harvard Business Review,* November–December, pp. 85–94.

Gale, B. T. (1994). *Managing Customer Value,* The Free Press, New York.

Gale, B. T. and R. Klavans (1985). "Formulating a Quality Improvement Strategy," *Journal of Business Strategy*, Winter, pp. 21–32, Warren, Gorham, and Lamont, used with permission.

Gryna, F. M. (1977). "Quality Costs: User vs Manufacturer," *Quality Progress*, June, pp. 10–15.

Juran, J. M. (1988). *Juran on Planning for Quality*, The Free Press, New York.

Reichheld, F. F. (1996). *The Loyalty Effect*, Harvard Business School Press, Boston.

Rust, R. T., A. J. Zahorik, and T. L. Keiningham (1994). *Return on Quality*, Probus, Chicago.

Scanlan, P. M. (1989). "Integrating Quality and Customer Satisfaction Measurement," *Customer Satisfaction Measurement Conference Notes*, American Management Association and ASQC, February, pp. 26–28.

SUPPLEMENTARY READING

Quality planning process: *JQH5,* Section 3.

Quality and income: *JQH5,* Section 7.

Quantifying the impact of quality on sales:

JQH5, Sections 8 and 18.

Rust, R. T., A. J. Zahorik, and T. L. Keiningham (1994). *Return on Quality,* Probus, Chicago.

Campanella, J., ed. (1999). *Principles of Quality Costs,* 3rd ed., ASQ, Milwaukee.

WEBSITE

Benchmarking studies of American Productivity and Quality Center: www.apqc.org

5

QUALITY CONTROL

5.1 DEFINITION OF CONTROL

As used in this book, *control* refers to the process employed to consistently meet standards. The control process involves observing actual performance, comparing it with some standard, and then taking action if the observed performance is significantly different from the standard.

The control process is in the nature of a feedback loop (Figure 5.1). Control involves a universal sequence of steps as follows:

1. Choose the control subject: i.e., choose what we intend to regulate.
2. Establish measurement.
3. Establish standards of performance: product goals and process goals.
4. Measure actual performance.
5. Compare actual measured performance to standards.
6. Take action on the difference.

This universal sequence applies to individuals at all levels from the chief executive officer to members of the workforce. The sequence can be applied as a framework for helping supervisors and work teams to understand and run everyday work processes. Such a framework becomes increasingly important as the team concept—particularly self-directed teams—emerges as an important form of business life. Chapter 8 explains several types of teams and the roles of a team leader, team facilitator, and team members.

When the natural work team in a department puts into practice the control process, three purposes are served:

- Maintain the gains from improvement projects.

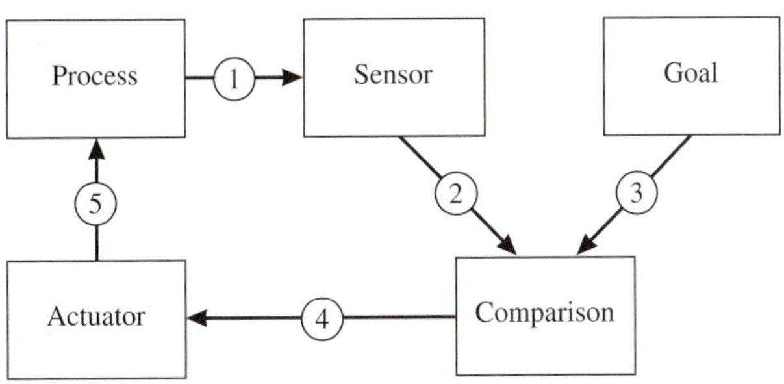

FIGURE 5.1
The feedback loop.

- Promote analysis of process variation, based on data, to identify improvement opportunities.
- Allow team members to clarify their responsibilities and work toward achieving a state of self-control.

The first three steps in the control process (choose the subject, establish measurement, and establish standards) require the *participation* of the department work team. The last three steps (measure, compare to standards, and take action) can be the *responsibility* of the department work team.

Control, one of the trilogy of quality processes, is largely directed at meeting goals and preventing adverse change, i.e., holding the status quo. In contrast, improvement focuses on creating change, i.e., changing the status quo. The control process addresses sporadic quality problems; the improvement process addresses chronic problems.

5.2 MEASUREMENT

Central to the process of quality control is the act of quality measurement: "What gets measured, gets done." Measurement is basic for all three operational quality processes and for strategic management: For quality control, measurement provides feedback and early warnings of problems; for operational quality planning, measurement quantifies customer needs and product and process capabilities; for quality improvement, measurement can motivate people, prioritize improvement opportunities, and help in diagnosing causes; for strategic quality management, measurement provides input for setting goals and later supplies the data for performance review.

Figure 5.2 shows the far-reaching impact of measurement in quality management. Note how measurement provides both alignment and linkages at several levels from daily work to strategic quality planning. These elements, in turn, become drivers to encourage the use of measurements for quality. This chapter presents concepts

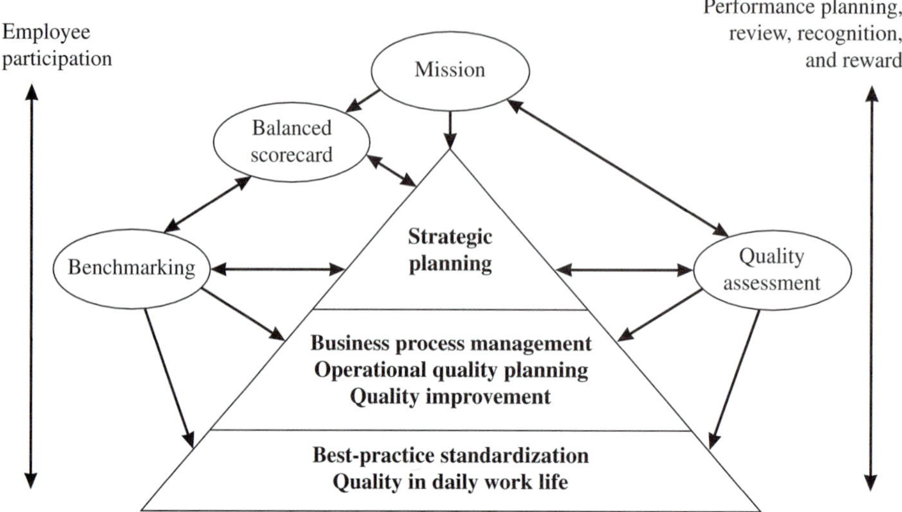

FIGURE 5.2
Measurement drivers. (*Reprinted by permission of John Wiley & Sons, Inc.*)

underlying measurement; later chapters present examples of quality measurement at both the operational and strategic levels.

The following principles can help to develop effective measurements for quality:

1. Define the purpose and use that will be made of the measurement. An example of particular importance is the application of measurements in quality improvement. Final measurements must be supplemented with intermediate measurements needed for diagnosis.
2. Emphasize customer-related measurements; be sure to include both external and internal customers.
3. Focus on measurements that are useful—not just easy to collect. When quantification is difficult, surrogate measures can at least provide a partial understanding of an output.
4. Provide for participation from all levels in both the planning and implementation of measurements. Measurements that are not put to use will eventually be ignored.
5. Provide for making measurements as close as possible to the activities they impact. This timing facilitates diagnosis and decision making.
6. Provide not only concurrent indicators but also leading and lagging indicators. Current and historical measurements are necessary but leading indicators help to look into the future.
7. Define, in advance, plans for data collection and storage, analysis, and presentation of the measurements. Plans are incomplete unless the expected use of the measurements is carefully examined.
8. Seek simplicity in data recording, analysis, and presentation. Simple check sheets, coding of data, and automatic gaging are useful. Graphical presentations can be especially effective.

9. Provide for periodic evaluations of the accuracy, integrity, and usefulness of the measurements. Usefulness includes relevance, comprehensiveness, level of detail, readability, and interpretability. For elaboration, see Section 22.7.
10. Realize that measurements alone cannot achieve improvement in products and processes. Measurements must be supplemented with the resources and training to enable people to achieve improvement. For elaboration on these and other principles of measurement systems, see *JQH5,* Section 9, and Zairi (1994).

This chapter presents concepts underlying measurement, but measurement is spread throughout the book. For example, Chapter 2 discusses measurements for broad quality assessment; Chapter 7 addresses strategic measurement, including the balanced scoreboard; Chapters 13, 15, 16, 17, 20, and 21 present examples of functional measurements; and Chapter 22 describes the quality information system. Thus measurements are for both product process control and management control.

5.3 SELF-CONTROL

Ideally, quality planning for any task should put the employee into a state of self-control. When work is organized in a way that enables a person to have full mastery over the attainment of planned results, that person is said to be in a state of self-control and can therefore be held responsible for the results. Self-control is a universal concept, applicable to a general manager responsible for running a company division at a profit, a plant manager responsible for meeting the various goals set for that plant, a technician running a chemical reactor, or a bank teller serving customers.

To be in a state of self-control, people must be provided with:

1. Knowledge of what they are supposed to do, e.g., the budgeted profit, the schedule, and the specification.
2. Knowledge of their performance, e.g., the actual profit, the delivery rate, the extent of conformance to specification (this is quality measurement).
3. Means of regulating performance if they fail to meet the goals. These means must always include both the authority to regulate and the ability to regulate by varying either (a) the process under the person's authority or (b) the person's own conduct.

If all the foregoing parameters have been met, the person is said to be in a state of self-control and can properly be held responsible for any deficiencies in performance. If any parameter has not been met, the person is not in a state of self-control and, to the extent of the deficiency, cannot properly be held responsible.

In practice, these three criteria are not fully met. For example, some specifications may be vague or disregarded (the first criterion); feedback of data may be insufficient, often vague, or too late (the second criterion); people may not be provided with the knowledge and process adjustment mechanisms to correct a process (the third criterion). Thus if we have a quality problem and we fail to meet any of the three criteria, the problem is "management controllable" (or "system controllable"); if we have a quality problem and if all three criteria are fully met, the problem is "worker controllable." Chapters 16 and 17 apply the concept of self-control to manufacturing and service industries.

TABLE 5.1
Classical control and self-control

Classical control	Self-control
Standard or goal	Knowledge of what people are supposed to do
Measurement	Knowledge of performance
Action on the difference	Means of regulating a process
Primary emphasis during execution	Primary emphasis before execution

Classical control and self-control are complementary (Table 5.1). An important difference, however, involves timing. Classical control takes place *during* the execution of a task; self-control provides useful criteria for evaluating plans *before* a task is executed.

Kondo and Kano (1999, p. 41.3) submit that there is a relationship among the control process; the "plan, do, check, act" cycle; and the concept of self-control. Figure 5.3*a* depicts the plan, do, check, act cycle, which corresponds to the main elements of the feedback loop (Figure 5.1) of the control process. They observe that individual worker performance during the "do" step comprises a plan, do, check, act cycle (Figure 5.3*b*). The extent to which the task of the worker is adequately planned reflects the degree to which the worker is placed in a state of self-control. The plan, do, check, act cycle is often called the "Deming cycle."

Some authors refer to the cycle as plan, do, study, act. Gitlow et al. (1995) emphasize that the cycle repeats over and over and provides a means of never-ending improvement.

For both self-control and the Deming cycle, the concept of standardization of work practices is important. Here employees apply a standardize, do, study, act (SDSA) cycle. Employees analyze the process to develop best-practice methods, use the best-practice methods on a trial basis, evaluate the effectiveness of the best practices, and document the standardized process. This standard process helps to stabilize the process and reduce variation. For elaboration, see Gitlow et al. (1995) and Imai (1986). One tool of standardization is the "5S method" of achieving an organized workplace. This method is discussed in Section 16.2 under "Plan for Neat and Clean Workplace."

Schonberger (1999) describes the concept of a self-adjusting system where front-line personnel employ simple, direct methods continuously. He proposes four elements: (1) process capacity management to minimize queues ("kanban"), (2) operator plotting of process data ("statistical process control"), (3) prevention of errors ("fail-safing"), and (4) quality checks before passing work output to the next worker ("source inspection"). As with self-control, Schonberger's concept aims to provide personnel with all that is needed for them to directly control their work output.

Be aware that another concept, self-inspection, is *not* the same as self-control. Self-inspection addresses the examination of the product; self-control addresses the process of accomplishing a task. Self-inspection is discussed in Chapter 16.

We now proceed with an examination of the steps in the control sequence.

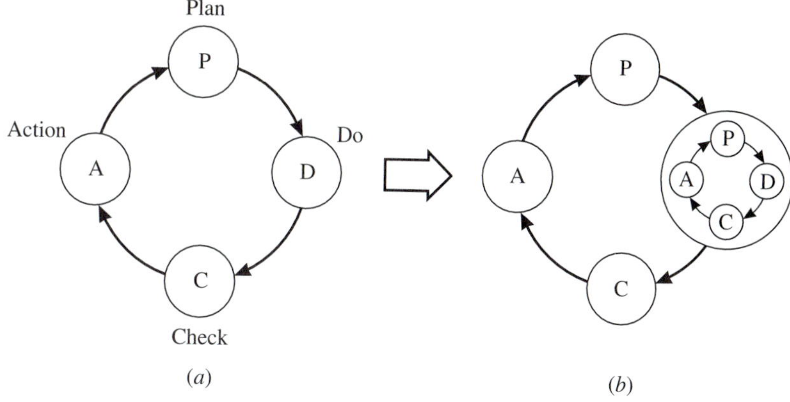

FIGURE 5.3
Deming's cycle. (*From JQH5, p. 41.4.*)

5.4 THE CONTROL SUBJECTS FOR QUALITY

Control subjects for quality are the critical parameters. At the technological level each division of a product—components, units, subsystems, and systems—has quality characteristics. Processing conditions (e.g., time to pay an insurance claim, oven temperature) and processing facilities also have quality characteristics. In addition, input materials and services have quality characteristics. Still more quality control subjects are imposed by external forces: clients, government regulations, and standardization bodies.

Beyond technological quality control subjects are managerial quality control subjects. These are mainly the performance goals for organization units and the associated managers. Managerial goals extend to nontechnological matters such as customer relations, financial trends (e.g., progress in reducing the cost of poor quality), employee relations, and community relations.

To identify and choose quality control subjects, several principles apply:

1. Quality control subjects should be aligned and linked with customer parameters. That is, the subjects should directly measure customer needs, satisfaction, and loyalty or measure product and process features that correlate with these customer parameters. Paramount are external customers, who affect sales income; equally important are internal customers, who affect internal costs such as the cost of poor quality. But let's face reality. Sometimes our control subjects are incomplete. For example, although advances have been made in measuring the quality of medical care, it is difficult to measure whether a physician detects a medical problem as early as possible.

   Table 5.2 shows examples of quality control subjects from different organizations. Later in this chapter we specify these categories further by defining the units of measure.

TABLE 5.2
Control subjects

Electronics manufacturer	A bank
Document quality	Operations—timeliness
Software quality	Retail banking—accuracy
Hardware quality	Commercial banking—loan payment posting
Process quality	Credit card and ATM cards—transactions
System quality	Financial and investments—transactions
	Human resources—personnel requisitions
	Information services—system downtime
	Administrative—work order status

2. Defining quality control subjects for work processes starts with defining work processes in terms of objectives, process steps, process customers, and customer needs.
3. Quality control subjects should recognize both components of the definition of quality, i.e., freedom from deficiencies and product features. The number of errors per thousand lines of computer code (KLOC) is important, but even perfect code does not mean that a customer will be satisfied with the software.
4. Potential quality control subjects can be identified by obtaining ideas from both customers and employees. Customers can be asked, "How do you evaluate the product or service that you receive from me?" A focus group of customers can provide valuable responses. Again, we are addressing both external and internal customers. All employees are sources of ideas, but employees who have direct contact with external customers can be a fertile source of imaginative ideas on quality control subjects.
5. Quality control subjects must be viewed by those who will be measured as valid, appropriate, and easy to understand when translated into numbers. These are nice notions, surely. But in the real world they can be pretty elusive.

Next we must establish the measurement process for these control subjects.

5.5 ESTABLISH MEASUREMENT

In order to quantify, we must create a system of measurement consisting of:

- *A unit of measure:* the unit used to report the value of a control subject, e.g., pounds, seconds, dollars
- *A sensor:* a method or instrument that can carry out the evaluation and state the findings in terms of the unit of measure.

Units of measure for product and process performance are usually expressed in technological terms; for example, fuel efficiency is measured in terms of distance traveled per volume of fuel; timeliness of service is expressed in minutes (hours, days, etc.) required to provide service.

Units of measure for product deficiencies usually take the form of a fraction:

$$\frac{\text{Number of occurrences}}{\text{Opportunity for occurrence}}$$

The numerator may be in such terms as defects per million, number of field failures, or cost of warranty charges. The denominator may be in such terms as number of units produced, dollar volume of sales, number of units in service, or length of time in service.

Units of measure for product features are more difficult to create. The number and variety of these features may be large. Sometimes inventing a new unit of measure is a fascinating technical challenge. In one example a manufacturer of a newly developed polystyrene product had to invent a unit of measure and a sensor to evaluate an important product feature. It was then possible to measure that feature both of the product and of competitors' products before releasing the product for manufacture. In another case the process of harvesting peas in the field required the development of a unit of measure for tenderness and the invention of a "tenderometer" gauge. A numerical scale was created, and measurements were taken in the field to determine when the peas were ready for harvesting.

Table 5.3 shows examples of units of measure for a manufacturing organization and for a service organization. It should be noted that for many service industries, the *time* taken to deliver a service to an external customer is the decisive control subject for measurement.

Often a number of important product features exist. To develop an overall unit of measure, we can identify the important product features and then define the relative importance of each feature. In subsequent measurement, each feature receives a score. The overall measure is calculated as the weighted average of the scores for all features. This approach is illustrated in Table 2.4. In using such an approach for periodic or continuous measurement, some cautions should be cited (Early, 1989). First, the relative importance of each feature is not precise and may change greatly over time. Second, improvement in certain features can result in an improved overall measure but can hide a deterioration in one feature that has great importance.

Measurement scales are part of a system of measurement. The most useful scale is the *ratio scale* in which we record the actual amounts of a parameter such as weight. An *interval scale* records ordered numbers but lacks an arithmetic origin such as zero—clock time is an example. An *ordinal scale* records information in ranked categories—an example is customer preference on the flavor of various soft drinks. Finally, the *nominal scale* classifies objects into categories without an ordering or origin point—an example is the count of population in each state. The type of measurement scale determines the statistical analysis that can be applied to the data. In this regard the ratio scale is the most powerful scale. For elaboration, see Emory and Cooper (1991).

The Sensor

The sensor is the means used to make the actual measurement. Most sensors are designed to provide information in terms of units of measure. For operational control subjects, the sensors are usually technological instruments or human beings employed

TABLE 5.3
Units of measure—examples

Electronics manufacturer	A bank
Document quality Defects per thousand formatted output pages	Operations Number of statements mailed late / Total number of statements processed
Software quality Defects corrected per thousand noncomment source statements	Retail banking Number of teller entry errors / Total number of teller entries
Hardware quality Field removal rate	Commercial banking Loan payments posted incorrectly / Total loan payments
Process quality Functional yields	Credit card and ATM cards Number of mispostings / Total number of transactions
System quality Total outages	Financial/investments Number of trading corrections / Number of trades made
	Human resources Requisitions not filled in 30 days / Total number of requisitions
	Information services Customer information system (CIS) downtime / Total CIS time
	Administrative Number of work orders not completed within 10 days / Number of work orders completed

as instruments (e.g., inspectors, auditors); for managerial subjects, the sensors are data systems. Choosing the sensor includes defining how the measurements will be made—how, when, and who will make the measurements—and the criteria for taking action. This information can be conveniently summarized in a control spreadsheet (see Figure 5.4).

There has been a continuing trend toward providing sensors with additional functions of the feedback loop: data recording, data analysis, comparison of performance with standards, and initiating corrective action.

Despite the large number of control subjects, relatively few human beings are needed to carry out the control process. Imagine a pyramid of control subjects: A few vital controls are carried out by supervisors and managers; another segment is carried out by the workforce; the remaining majority of control subjects is handled by nonhuman means (stable processes, automated processes, servomechanisms).

Clearly, sensors must be economical and easy to use. In addition, because sensors provide data that can lead to critical decisions on products and processes, sensors must be both accurate and precise. The meaning, measurement, and impact of accuracy and precision are discussed in Chapter 19.

Process control features / Control subject	Unit of measure	Type of sensor	Goal	Frequency of measurement	Sample size	Criteria for decision making	Responsibility for decision making	. . .
Wave solder conditions Solder temperature	Degree F (°F)	Thermo-couple	505 °F	Continuous	N/A	510 °F reduce heat 500 °F increase heat	Operator	. . .
Conveyor speed	Feet per minute (ft/min)	Timer	4.5 ft/min	1/hour	N/A	5 ft/min reduce speed 4 ft/min increase speed	Operator	. . .
Alloy purity	% total contaminates	Lab chemical analysis	1.5% max	1/month	15 grams	At 1.5%, drain bath, replace solder	Process engineer	. . .
	⋮	⋮	⋮	⋮	⋮	⋮	⋮	

FIGURE 5.4
Spreadsheet for who does what. (*Making Quality Happen, Juran Institute, Inc., senior executive workshop, p. F-8, Wilton, CT.*)

TABLE 5.4
Control subjects and goals

Control subject	Goals
Mean time between failures	Minimum of 5000 hours
Solder temperature of soldering process	500 degrees F
Overnight delivery	99.5% delivered prior to 10:30 A.M. next morning
Relative quality ranking	At least equal in quality to competitors A and B
Customer retention	95% of key customers from year to year

5.6 ESTABLISH STANDARDS OF PERFORMANCE

Each control subject must have a quality goal. Table 5.4 shows examples of control subjects and associated goals for a variety of control subjects ranging from those for products, processes, and departments to that of an entire organization.

This chapter concentrates on goals at operational levels; Section 7.5, "Development of Goals," discusses overall company quality goals.

To set operational goals, certain criteria must be met. The goals should be

- *Legitimate:* have official status.
- *Customer focused:* external and internal.
- *Measurable:* numbers.
- *Understandable:* clear to all.
- *In alignment:* integrated with higher levels.
- *Equitable:* fair for all individuals.

Goals for product features and process features are based on technological analysis. To encourage continuous improvement, goals should be based on high levels achieved by others (see Chapter 7 under "Benchmarking"). The deployment and alignment of company quality goals to operational goals is discussed in Chapter 7 under "Deployment of Goals."

5.7 MEASURE ACTUAL PERFORMANCE

In organizing for control, a useful technique is to establish a limited number of control stations for measurement. Each such control station is then given the responsibility for carrying out the steps of the feedback loop for a selected list of control subjects. A review of numerous control stations discloses that they are usually located at one of several principal junctures:

- At changes of jurisdiction, e.g., where products are moved between companies or between major departments.
- Before embarking on an irreversible path, e.g., setup approval before production.

- After creation of a critical quality.
- At dominant process variables, e.g., the vital few.
- At natural windows, for economical control.

The choice of control stations is aided by preparation of a flow diagram that shows the progression of events through which the product is produced.

It is essential to measure both the quality of the output going to the external customer ("final yield") and the quality at earlier points in the process, including the "first-time yield."

In Figure 5.5, 100 units of input enter a process. After operations A, B, and C, an inspection is conducted; 87 acceptable units continue on to operation D, 8 units are reprocessed at previous operations, and 5 units are discarded. The first-time yield is thus 87 percent. After operations D and E, a second inspection is conducted; 82 acceptable units (of the 87) are available for delivery, 2 units are reprocessed, and 3 units are discarded. Assuming that all reprocessed units are acceptable, the final yield is 92 (82 + 8 + 2), or 92 percent of the original input. Note how the measurement of yield at several places highlights several opportunities for improvement. This concept applies to both manufacturing and nonmanufacturing processes. Don't let different terminology (e.g., inspection versus checking) obscure the concept. For example, in a software development organization, the average number of software errors was about two errors per thousand lines of code, just before delivery to the customer. The average level of errors, however, when measured earlier in the development process, was 50 errors per thousand lines of code. Huge resources were needed to screen out these errors. Ironically, the head of the organization was unaware of this first-time yield until it was revealed by a consultant.

For each control station, it is necessary to define the work to be done: which control subjects are to be measured, goals and standards to be met, procedures, instruments to be used, the data to be recorded, and the decisions to be made including the criteria and responsibility for making each decision.

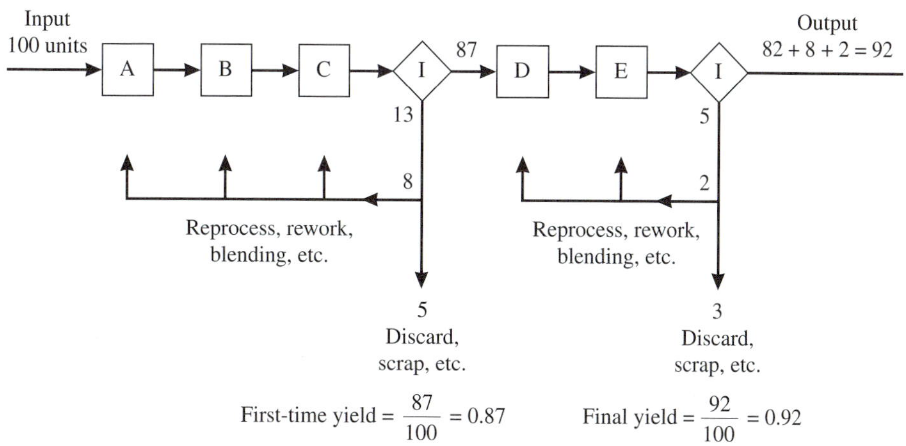

FIGURE 5.5
First-time yield and final yield (A, B, C, D, E = operations or tasks; I = inspections, checks, reviews).

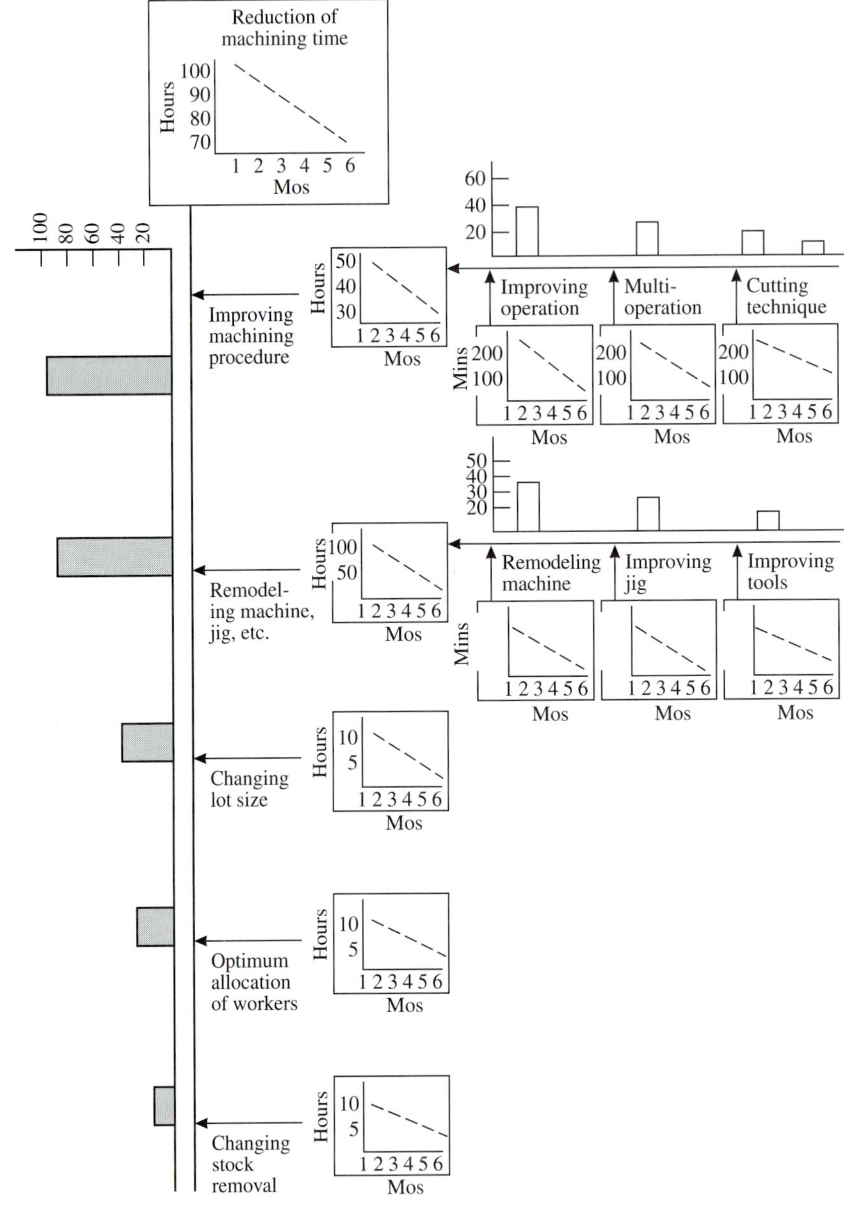

FIGURE 5.6
Example of a "flag diagram." (*Adapted from Kondo and Kano, 1999, p. 41.17.*)

See the example of a control spreadsheet in Section 5.5, "Establish Measurement." Keep in mind that control subjects include measurements both on product parameters and process variables. With all of this information, the feedback loop can function well.

The "flag diagram" (Figure 5.6) is an innovative illustration of how measurement can be combined with control subjects for tracking improvement. This diagram makes use of measurement data in combination with the Pareto concept and the cause-and-effect diagram (both discussed in Chapter 3).

The overall control subject (reduction of machining time) is divided into five major subjects, e.g., improving machining procedure. Each major subject is then further divided into secondary subjects, e.g., improving operation. Goals for each subject are shown as dotted lines on the charts and then performance is plotted on the same charts. The diagrams become a basis for review by the responsible manager and for action if there is a significant deviation from a goal.

5.8 COMPARE TO STANDARDS

This phase of the control process consists of comparing the measurement to the goal and deciding if any difference is significant enough to justify action. The criteria for taking action (or not taking action) should be numerically defined before measurements are taken, and training should be provided to ensure that the criteria are properly applied. Often the criteria can be simply stated: If a solder temperature exceeds 510 degrees F, decrease the heat; if the temperature is between 500 degrees and 510 degrees, then take no action on temperature. Other cases present a need to distinguish between real and apparent differences in measurements on a product or process. This task can be done by using the concept of statistical significance.

Statistical Significance

An observed difference between performance and a goal can be the result of (1) a real difference due to some cause or (2) an apparent difference arising from random variation. Further, differences between a measurement and a goal should not be viewed individually. Knowing the pattern of differences over time is essential to drawing correct conclusions. In Figure 5.7 the measurements at A and B and the trend at C represent real ("statistically significant") differences from the goal; the other measurements are due to random variation. Figure 5.7 is a statistical control chart—one of the elegant statistical tools used to evaluate statistical significance.

A control chart is a graphic comparison of process performance data to computed "control limits" drawn as limit lines on the chart. The process performance data usually consist of groups of measurements ("rational subgroups") selected in regular sequence of production.

A prime use of the control chart is to detect assignable causes of variation in the process. The term *assignable causes* has a special meaning, and understanding this meaning is a prerequisite to understanding the control chart concept (see Table 5.5).

Process variations are traceable to two kinds of causes: (1) random, i.e., due solely to chance, and (2) assignable, i.e., due to specific "special" causes. Ideally, only random (also called "common") causes should be present in a process. A process that is operating without assignable causes of variation is said to be "in a state of statistical control," which is usually abbreviated to "in control."

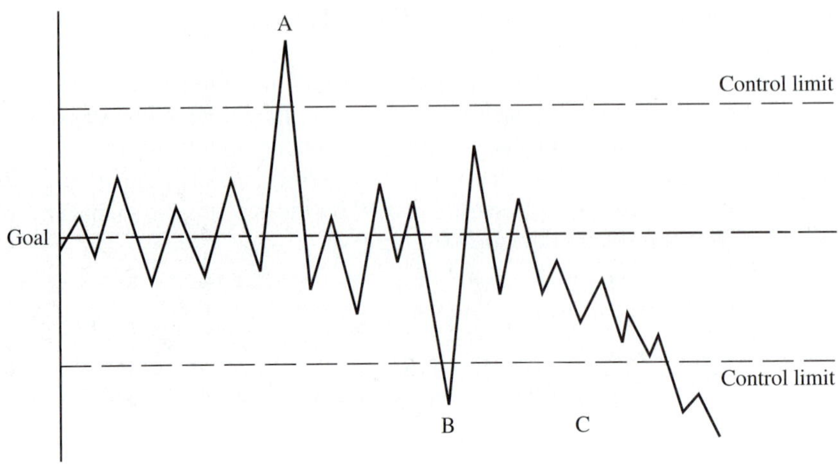

FIGURE 5.7
Control chart.

TABLE 5.5
Distinction between random and assignable causes of variation

Random (common) causes	Assignable (special) causes
Description	
Consists of many individual causes.	Consists of one or just a few individual causes.
Any one random cause results in a minute amount of variation (but many random causes act together to yield a substantial total).	Any one assignable cause can result in a large amount of variation.
Examples are human variation in setting control dials; slight vibration in machines; slight variation in raw material.	Examples are operator blunder, a faulty setup, or a batch of defective raw material.
Interpretation	
Random variation cannot economically be eliminated from a process.	Assignable variation can be detected; action to eliminate the causes is usually economically justified.
An observation within the control limits of random variation means the process should not be adjusted.	An observation beyond control limits means the process should be investigated and corrected.
With only random variation, the process is sufficiently stable to use sampling procedures to predict the quality of total production or make process optimization studies.	With assignable variation present, the process is not sufficiently stable to use sampling procedures for prediction.

The control chart distinguishes between random and assignable causes of variation through its choice of control limits. These are calculated from the laws of probability so that highly improbable random variations are presumed to be due not to random causes, but to assignable causes. When the actual variation exceeds the control limits, it is a signal that assignable causes entered the process and the process should be investigated. Variation within the control limits means that only random causes are present.

The important advantages of statistical control and the methodology of constructing and interpreting control charts are given in Chapter 18, "Statistical Process Control."

The control chart not only evaluates statistical significance but also provides an early warning of problems that could have major economic significance.

Random causes are usually chronic, associated with many minor variables, and thus difficult to diagnose and fix; assignable causes are typically sporadic and often have their origin in single variables, making diagnosis easier. A problem that exists when only random causes are present requires a basic analysis using quality improvement concepts (see Chapter 3, "Quality Improvement and Cost Reduction") or quality planning concepts (see Chapter 4, "Operational Quality Planning and Sales Income"). For example, a process may be in statistical control but does not have the inherent process capability (i.e., small variation) to meet a customer specification. A study is needed to improve the process capability. If a problem exists when assignable causes are present, then the quality control concepts in this section are appropriate. For example, a sudden increase in errors in processing insurance claims may be traced to one untrained person. We elaborate on these ideas in Chapter 18 under "Advantages of Decreasing Process Variability" and in Chapter 19 under "Conformance to Specification" and "Fitness for Use."

Economic Significance

Tools such as the statistical control chart serve several purposes—e.g., they document process performance and identify special situations such as assignable causes or trends. This type of tool provides an early warning of impending problems on the product. But identifying a statistically significant difference between a measurement and a goal does not always lead to corrective action. The presence of assignable causes does mean that the process is unstable, but sometimes assignable causes are so numerous that it is necessary to establish priorities for action based on economic significance and related parameters. When product problems are serious and/or frequent, then setting up a formal quality improvement project or taking other action is warranted.

5.9 TAKE ACTION ON THE DIFFERENCE

In the closing step of the feedback loop, action is taken to restore the process to a state of meeting the goal. Action may be needed for three types of conditions:

1. *Elimination of chronic sources of deficiency.* The feedback loop is not suitable for dealing with such chronic problems. Instead, the quality improvement process described in Chapter 3 or the operational quality planning process described in Chapter 4 should be employed.

2. *Elimination of sporadic sources of deficiency.* The feedback loop is well designed for this purpose. In these cases the cardinal issue is determining which changes caused the sporadic difference to arise. Discovery of those changes, plus action to restore control, can usually be carried out by local operating supervisors using troubleshooting procedures (see below).
3. *Continuous process regulation to minimize variation.* This situation requires linking each product characteristic to one or more process variables, providing a means for convenient adjustment of the setting for the process variables, and determining the relationship between the change in the setting of a process variable and the resulting effect on the product characteristic. These matters are discussed in Section 16.2 under "Correlation of Process Variables with Product Results" and in Chapter 18, "Statistical Process Control."

Section 16.10 provides guidance to operations on when to take action in the form of troubleshooting (quality control), quality improvement, or operational quality planning.

Troubleshooting

Troubleshooting (also called "firefighting") is the process of dealing with sporadic problems and restoring quality to the original level. For organizations that do not have a formal effort to reduce chronic and sporadic problems, operations managers often spend 30 percent of their time on troubleshooting; for the supervisors reporting to these managers, the time consumed frequently exceeds 60 percent. In a moment of jest, an executive once said: "Managers who are good at putting out fires can become heroes. I think some of our managers may be arsonists."

Troubleshooting is diagnostic and remedial action applied to sporadic problems and involves three steps:

1. *Identify the problem.* Identification means pinpointing the problem in terms of a single process indicator, the time of occurrence, and its effect. For example, the billing process at a hospital requires an average of 5.2 working days from patient discharge to mailing a final bill. For one week, the average time was 6.7 days, exceeding an upper control limit of 5.9 days on a control chart (Juran Institute, 1995).
2. *Diagnose the problem.* Diagnosis means investigating, developing, and testing theories on the cause of the problem. Analysis of the bills for the particular week revealed one specific set of bills that was delayed. The bills for that week were then classified by hospital department, payor, clerk preparing the bill, and discharging nursing unit. A Pareto diagram plotted percentage of all bills over seven days versus payer organization. This diagram showed that two-thirds of the delayed bills were for services to be paid for by a particular managed care plan. Further investigation revealed that the plan had just made significant revisions in procedures for submitting bills. These changes resulted in difficulty in the billing department and turned out to be the primary cause of the delayed bills for that week.
3. *Take remedial action.* Remediation requires taking steps to remove the cause identified in step 2. In this case immediate action was taken to modify the new procedures, change certain software, and identify a single point of contact with the insurance plan until the problem was resolved.

Example: billing department				
Process indicator: average days to bill				
Condition: weekly average exceeds 5.9 days				
Who	**What**	**Where**	**When**	**How**
Supervisor	If weekly bill volume is more than 800, add hours to part-time billing clerks	Weekly volume report	By 8:30 A.M. Monday	Inform both clerks and personnel
	Otherwise, convene troubleshooting team	Supervisor's office	By 11:00 A.M. Monday	By telephone
Troubleshooting team—supervisor, system-support technician, discharge planner	Troubleshoot the problem	—	Begin by 11:00 A.M. Monday	Standard methods

FIGURE 5.8
Contingency planning matrix.

Note that the diagnostic and remedial journeys for troubleshooting are similar to those for quality improvement (see Chapter 3, "Quality Improvement and Cost Reduction"). The approach with troubleshooting is usually less complex because the problem is localized to a specific sporadic time; in contrast, chronic problems are present for a sustained period of time.

Troubleshooting can be made more effective by anticipating problems and planning, in advance, for troubleshooting. A contingency planning matrix (see Figure 5.8) can be useful in this regard. Note how the planning tries to prevent problems and provide for action, when necessary, in a billing process. In manufacturing operations each product characteristic (quality control subject) must be linked to one or more process variables so that employees have a contingency plan to adjust the process when necessary. For elaboration, see Chapter 17 under "Review of Process Design."

5.10 PROCESS CONTROL SYSTEM THAT USES THE SIX-SIGMA CONCEPT

The GE Capital Mortgage Insurance Corporation provides us with an example of a process control system in the service industry. The company offers mortgage insurance to major lenders of mortgage funds for individual home buyers. Four key processes are involved: underwriting, billing, claims, and sales. The process control system employs both customer information and internal measurements. The elements of the process control system can be illustrated in the six-step feedback loop discussed in this chapter.

1. *Choose the control subjects.* Figure 5.9 calls these subjects "measurement categories"; e.g., a measurement category (control subject) for the underwriting process is turnaround time. A flow diagram documents the process and helps to identify process measures and outcome indications (process and product features).
2. *Establish measurement.* Nine units of measure (metrics) are employed, e.g., average turnaround time for underwriting.
3. *Establish standards of performance.* For each metric, customers provide input to establish a numerical specification, e.g., turnaround time for underwriting is four hours.
4. *Measure actual performance.* Data collection includes percentage of transactions meeting the specification, actual sigma, and a customer evaluation (on a scale of 5 to 1 with 5 meaning excellent). For turnaround time on underwriting, actual performance is 99.9 percent, which is at the 4.6 sigma level, with a customer evaluation rating of 5. At the 4.6 sigma level, the average opportunity for defects is about 970 defects per million opportunities; at the 6.0 sigma level, the average opportunity is only 3.4 defects (see Table 18.7).

Quality score card 1st quarter 1997

Measurement category	Customer specifications	Actual performed	Actual σ	Evaluation Excellent ⟷ Poor
Underwriting				
Turnaround time	4 hours	99.9%	4.6 σ	(5) 4 3 2 1
Accessibility	100%	99.5%	4.2 σ	5 (4) 3 2 1
Knowledgeable	Consistent application of guidelines	95.5%	3.2 σ	5 (4) 3 2 1
Billing				
Timeliness	3rd – 5th of month	99.9%	4.6 σ	(5) 4 3 2 1
Completeness	100%	98.9%	3.8 σ	5 (4) 3 2 1
Claims				
Timely payments	30 days	84%	2.5 σ	5 (4) 3 2 1
Work out cycle time	To guidelines 100%	95%	3.1 σ	5 (4) 3 2 1
Sales				
Meeting frequency	Monthly/ quarterly	100%	6+ σ	(5) 4 3 2 1
Knowledge	Answer questions when asked	86%	2.6 σ	5 (4) 3 2 1

FIGURE 5.9
Quality scorecard. (*Reprinted with permission by the ASQ.*)

5. *Compare to standards.* Control charts monitor the processes and provide the linkage between top-level measurements and lower-level process indicators. These charts along with the process flow diagram are displayed in the business area. A scorecard with data trends is presented to the customer (the major lenders).
6. *Take action on the difference.* Data are constantly reviewed to achieve process improvements aimed at a 6.0 sigma level by the year 2000. Periodic meetings with customers are held to review numerical performance and identify any changing customer needs.

For elaboration of this system, see Pautz (1998).

The elements of the feedback loop discussed in this chapter are universal. The concepts apply not only to both manufacturing and service industries but also to both executive and operational activities within all industries.

SUMMARY

- Control is the process we employ to meet standards.
- Control involves a universal sequence of steps: choosing the control subject, choosing a unit of measure, setting a goal, creating a sensor, measuring performance, interpreting the difference between actual performance and the goal, and taking action on the difference. Measurement is a quiet source of action.
- Self-control involves three elements: People must have knowledge of what they are supposed to do, knowledge of their performance, and means of regulating their performance.
- Troubleshooting is diagnostic and remedial action applied to sporadic troubles.

PROBLEMS

5.1. Select a specific task that you have regularly performed for an organization. Evaluate the degree to which this task meets the three criteria for self-control.

5.2. Interview someone who regularly performs a specific task for an organization. Explain the three criteria of self-control to that person and document the degree to which this task meets the criteria, as viewed by the person performing the task.

5.3. Place yourself in the role of a customer for any product—a good or a service. Identify at least four quality control subjects that are important to you as a customer and that the supplier should measure. For each quality control subject, propose a unit of measure.

5.4. Place yourself in the role of upper management for any organization producing goods or service. Identify at least four quality control subjects that are important to internal organizational performance and that the organization should measure. For each quality control subject, propose a unit of measure.

5.5. Select one process consisting of a series of tasks within an organization. Identify the location and the data that are collected on quality-related control subjects throughout the process.

5.6. Interview someone who regularly performs a manufacturing task that includes taking periodic measurements on a product or process characteristic and comparing the result to a specification. Determine how the person makes the following decisions:
(a) How large a deviation of a measurement from a specification is permitted before the person takes action to adjust the process?
(b) If a process adjustment is needed, what *amount* of adjustment is made?

5.7. You are designing process controls for the replacement of lost or stolen credit cards. Customers are concerned about responsibility for charges and quick replacement of cards. Identify two potential control subjects that will help manage the process while meeting customer needs. Then define a unit of measure and a sensor for each control subject.

5.8. A major hotel chain is the site for many business conferences. A key customer need is for comfortable meeting rooms that have adequate lighting, temperature control, and visual aid equipment. Identify two control subjects and a unit of measure and sensor for each subject.

REFERENCES

Early, J. F. (1989). "Strategies for Measuring Service Quality," *ASQC Quality Congress Transactions,* Milwaukee, pp. 2–9.

Emory, C. W. and D. R. Cooper (1991). *Business Research Methods,* Irwin, Homewood, IL.

Endres, A. C. (2000). *Implementing Juran's Roadmap for Quality Leadership,* John Wiley & Sons, New York.

Gitlow, H., A. Oppenheim, and R. Oppenheim (1995). *Quality Management—Tools and Methods for Improvement,* Irwin, Burr Ridge, IL.

Imai, M. (1986). *Kaizen,* McGraw-Hill, New York.

Juran Institute, Inc. (1995). *Work Team Excellence,* Wilton, CT.

Kondo, Y. and N. Kano (1999). "Quality in Japan," *JQH5,* Section 41.

Pautz, S. J. (1998). "Using Dashboards and Scorecards in a Service Industry," *ASQ Annual Quality Congress Proceedings,* Milwaukee, pp. 324–330.

Schonberger, R. J. (1999). "Economy of Control," *Quality Management Journal,* vol. 6, no. 1, pp. 10–18.

Zairi, M. (1994). *Measuring Performance for Business Results,* Chapman and Hall, London.

SUPPLEMENTARY READING

Measurement: Sink, D. S. (1991). "The Role of Measurement in Achieving World Class Quality and Productivity Management," *Industrial Engineering,* June, pp. 23–29.

Performance measurement: Zairi, M. (1994). *Measuring Performance for Business Results*, Chapman & Hall, London.

Quality control process: *JQH5,* Section 4.

Yield: Harry, M. and R. Schroeder (2000). *Six Sigma,* Doubleday, New York, pp. 83–91.

WEBSITE

Performance measures: www.zigonperf.com

6

PROCESS MANAGEMENT

6.1
FUNCTIONAL VERSUS PROCESS MANAGEMENT

In the future traditional organizational structure and concepts may see remarkable—indeed, extraordinary—change. Historically, most enterprises have organized around functional departments, e.g., design engineering, manufacturing (operations), marketing, service, and support (the columns in Figure 6.1). Management direction, objectives, and review come from the top and proceed downward through a vertical hierarchy. The functional departments ("silos") focus on achieving their functional (departmental) objectives. Thus a development function has an objective to develop a number of new products; a sales function has an objective in terms of sales quotas. The management of each function emphasizes priorities that will help meet the functional objectives. An examination of the overall objectives of the enterprise, however, reveals that success depends on certain critical results that are the outcome of cross-functional processes (the rows in Figure 6.1). The traditional emphasis on functional objectives can be a serious obstacle to achieving company business objectives that require cross-functional processes.

A *process* is a collection of activities that converts inputs into outputs or results. Thus a process may simply be several steps in a manufacturing or service area. But experience suggests that achieving business goals depends mostly on large, complex processes that go across functional departments. Examples of such cross-functional processes are billing, product development, and distribution. Other examples are materials procurement, hospital patient care, and insurance claims servicing.

We define a *primary process* as a collection of cross-functional activities that are essential for external customer satisfaction and achieving the mission of the organization. These activities integrate people, materials, energy, equipment, and information.

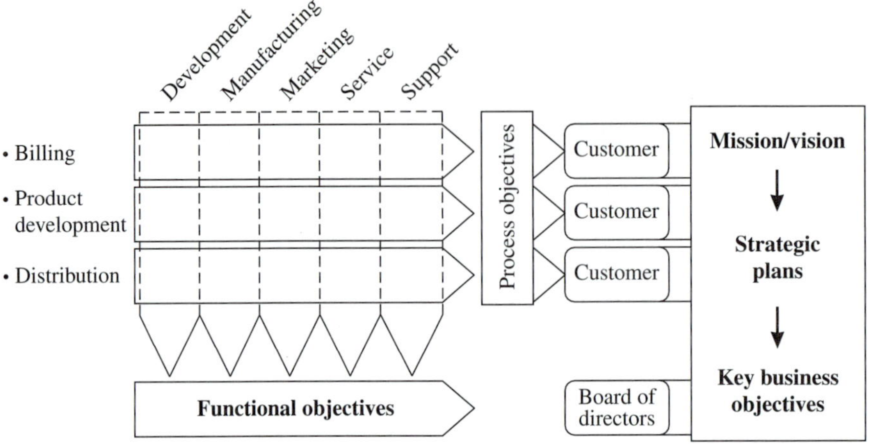

FIGURE 6.1
Work flow in a functional organization. *(From Juran Institute, Inc.)*

The products and services furnished to external customers are produced mostly by cross-functional primary processes. The managers of functional departments are responsible for functional pieces of the process, but no one is accountable for the whole process. Problems often arise because these managers focus on meeting functional objectives rather than process objectives. Problems frequently arise at the functional interfaces between departments (the "white space")—see Rummler and Brache (1995) in the Supplementary Reading.

The greatest opportunities for improvement exist at the cross-functional process level, which leads us to the concept of process management.

6.2 PROCESS MANAGEMENT

In this book process management is an approach for planning, controlling, and improving the primary processes in an organization by using permanent process teams. The distinguishing features of process management are

- Emphasis on customer needs rather than functional needs.
- Focus on a few key cross-functional processes.
- Process owners who are responsible for all aspects of the process.
- Permanent cross-functional process teams responsible for operating the process (permanent for the life of the process).
- Application at the process level of the trilogy of quality processes—quality planning, quality control, and quality improvement (see Chapters 3, 4, and 5).

An example of key processes within a research and development organization is described by Hildreth (1993). At Lederle–Praxis Biological, a division of American Cyanamid, six major processes were identified: prototype discovery research, candidate development research, good manufacturing practices (GMP) vaccine production, clinical research, product transfer into manufacturing, and research and development (R&D) regulatory. A process owner assigned to each process is responsible for improving customer satisfaction, increasing efficiency, decreasing average product R&D time, and improving the awareness and external recognition for scientific/technical excellence.

Process management really replaces the hierarchical vertical organization with a horizontal view of an organization. (See the final section of this chapter for a discussion of the impact on functional departments that still exist under process management.) Under process management, functions have extensive interactions with each other, which leads to a healthy understanding of the interdependencies among functions, i.e., a "systems viewpoint."

One of the early titles of this approach was business process quality management. Other names include process quality management and business process management. Part of the evolution includes an emphasis on both the effectiveness and efficiency aspects of performance (see below) rather than a narrower conformance to the requirements viewpoint of quality.

Process management is commanding much attention and increased usage (as contrasted to functional management), but full process management as described here is still a minority (perhaps 30 percent) form of management.

A road map for process management is shown in Figure 6.2. Process management starts when upper management selects key processes, appoints process owners and teams, and provides process mission statements and goals. We next consider, then, process selection.

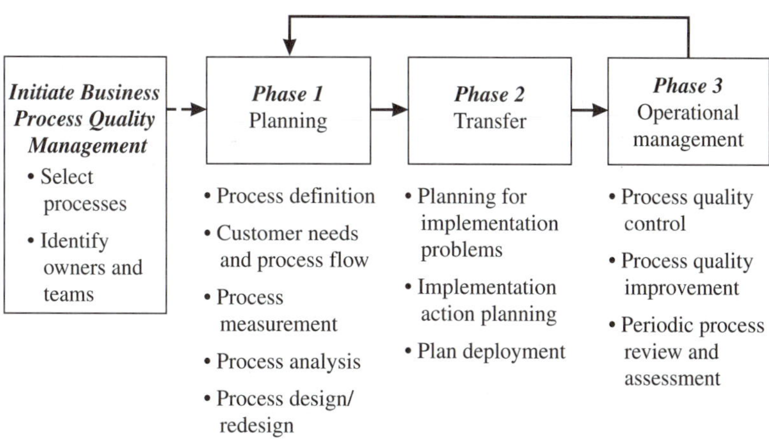

FIGURE 6.2
Road map for process management. (*From Juran Institute, Inc.*)

6.3 SELECTION OF PROCESSES

Organizations have many important cross-functional processes. From these, upper management should select a few primary processes for the process management approach. The processes selected should, of course, be aligned with the organization mission, strategic plans, and key business objectives (see Figure 6.1).

The selection of the processes is based on the critical success factors of the organization, i.e., the few events that must occur for the organization to be successful. Examples of critical success factors are reduction of product development times, increased perception of value, and higher yields. Candidate processes can then be ranked by assessing the importance of the process with regard to the critical factors and also the current process performance. This step is illustrated in Figure 6.3. The relevance of each factor is entered in the body of the matrix (e.g., 5 is high relevance of the process to the critical success factor, 3 is moderate, and 1 is low relevance). The *count* is simply the total of the relevance ratings. *Current performance* of a process is a rating of how the process is performing now (e.g., 1 is good performance, 3 is fair, and 5 is poor.) The count on relevance is then multiplied by the current performance rating to obtain a total score for the candidate process. Processes with the highest total would be likely selections for the formal process management approach. Hardaker and Ward (1987) provide a classic paper describing the use of critical success factors in process selection. *JQH5,* Section 6, describes other approaches for selecting processes for process management.

When a process is selected, the upper-management quality council should prepare a statement of the process mission and process goals. For the accounts payable process, the statement might be something like this:

Key business processes	Critical Success Factors (CSF)						Count	Current performance	**Total**
	Product quality 1	Supplier quality 2	Empowered, skilled employees 3	Customer satisfaction 4	Lowest cost of poor quality 5	Lowest delivered cost 6			
A									
B									
C									
D									
E									
. . .									

FIGURE 6.3
Critical success factors.

Process mission

The mission of the accounts payable process is to pay suppliers in a timely and accurate manner.

Process goals

The process owner and process management team will be responsible for increasing payment accuracy by 50 percent and for reducing the total process cost by 25 percent within two years.

This statement should be presented to the team for review and to provide direction.

6.4 ORGANIZE THE PROCESS TEAM

After selecting the processes, the quality council (see Section 8.3) appoints a process owner who is responsible for all aspects of process performance. Specifically:

- Be responsible for making the process effective, efficient, and adaptable (see below).
- Schedule, set agendas, and conduct process team meetings.
- Establish cooperative working relationships among all functions contributing to the process.
- Guide the process team in analyzing the current process and achieving improvement.
- Make assignments to team members.
- Resolve or escalate issues that may hinder improvement.
- Assure that team members receive training in process management.
- Manage the implementation of process changes.
- Schedule process reviews.
- Report progress of the process team to the quality council.

For critical cross-functional processes, the responsibility is indeed a heavy one because the owner does not have line responsibility and authority for all of the component activities of the process. But the owner is responsible to upper management for the overall performance of the process. In practice, the owner focuses on establishing working relationships through a process team, installing quality concepts, resolving or escalating cross-functional issues, and encouraging continuous progress. The typical owner is from a high level of management and is often either the manager with the most resources in the process or the one who is affected the most when problems occur. For example, the purchasing manager is a good choice for the purchasing process.

Some processes have an "executive owner" serving as a champion and a "working owner" responsible for day-to-day activities. This structure provides both the involvement and support of upper management and continuous management of the process details. The process owner (with these one or two tiers) is a permanent position.

The process team includes a manager or supervisor from each major function with work activities in the process. In contrast to the ad hoc teams described in Chapter 3 that solve specific quality improvement problems, the process team is permanent.

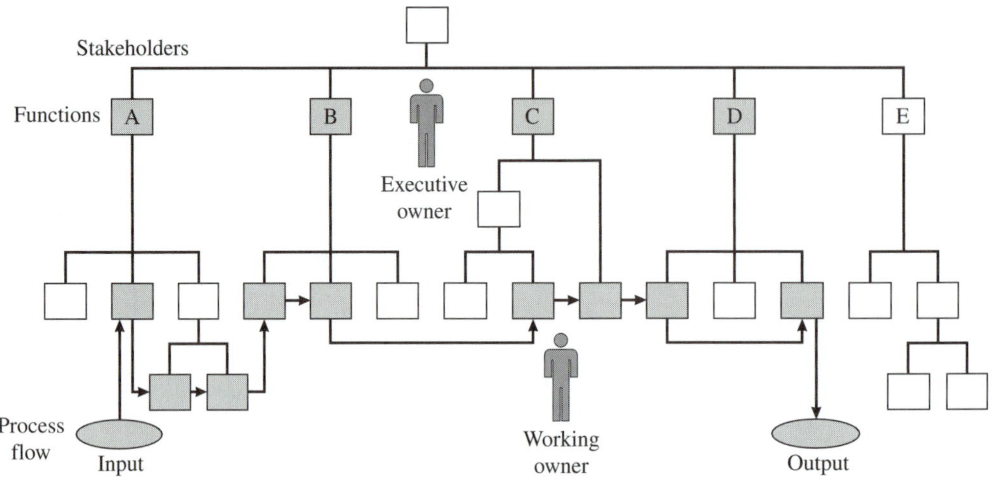

FIGURE 6.4
Organization infrastructure for process management in multifunctional organizations. (*From Juran Institute, Inc.*)

Typically the team has a maximum of eight members and a facilitator (see Chapter 8 under "Quality Project Team").

Figure 6.4 shows a multifunctional organization and one of its major processes. The shaded portions show the executive owner, the working owner, the process management team, and the functional managers who have work activities.

6.5 EXAMPLE OF PROCESS MANAGEMENT

An example of process management is provided by a process that involves preparing price quotations and delivery schedules for large orders that go beyond the quantities in standard price discount tables (Juran Institute, Inc., 1990). The process, called "contract management," required a review of the sales order, an analysis, and preparation of the final quotation. A process owner was selected, and a process team appointed. The previously used process, a manual one, is depicted in Figure 6.5. Note that the customer's request for a quotation was received at a branch office, traveled through various offices at different locations, and required 28 sequential approvals before the decision was given to the branch office, which notified the customer. Typically, customers waited 14 weeks for a quotation, and only 20 percent of these quotations resulted in a firm sales order.

Analysis of the process revealed that much of the delay time was due to nonconformance to *internal* customer requirements on the format and content of the proposal for the *external* customer. A redesigned electronic process eliminated many reasons for internal delays. Also, separate procedures were created to handle the "vital few"

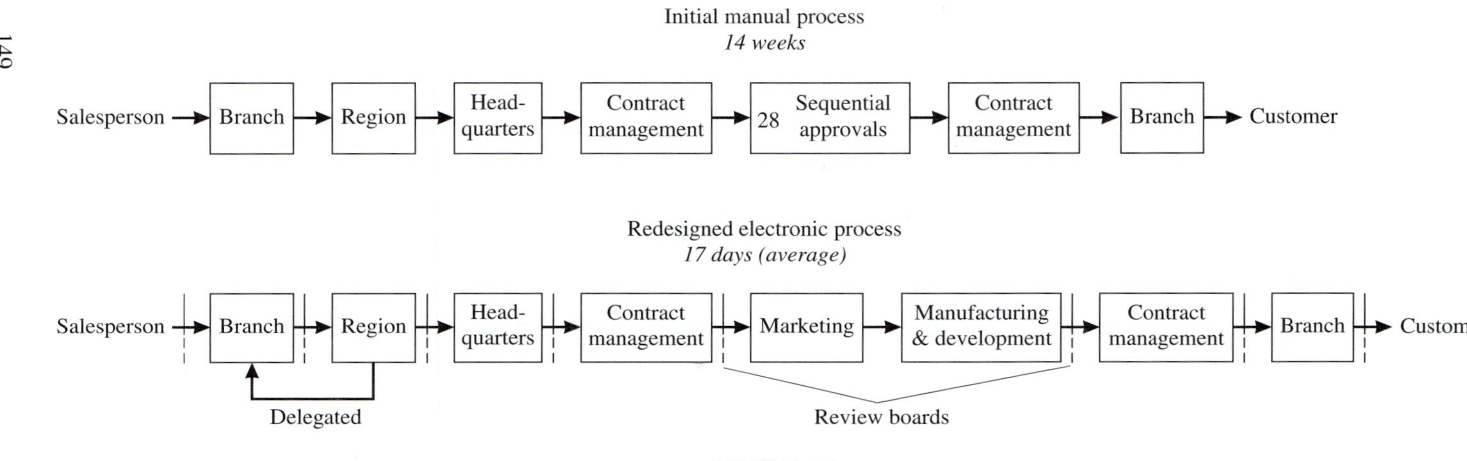

FIGURE 6.5
Contract management process. (*From Juran Institute, Inc., 1990.*)

and the "useful many" customer requests, with authority delegated to regional offices to make final decisions on the useful many. Finally, two review boards were set up to break the bottleneck of the 28 sequential reviews. The cycle time required to respond to the customer was reduced to an average of 17 days (later refinements achieved further savings). This shortened response time was instrumental in raising the yield of firm sales orders from 20 percent to a solid 60 percent.

For elaboration on this example, see *JQH5,* Section 6.

6.6 THE PLANNING PHASE OF PROCESS MANAGEMENT

The planning phase consists of five steps: (1) define the current process, (2) discover customer needs and flowchart the process, (3) establish process measurements, (4) analyze process data, and (5) design (or redesign) the process. Note the similarity of the planning phase with the six-sigma approach and the trilogy of quality planning, quality control, and quality improvement.

Define the Current Process

This process definition step establishes the process mission, goals, scope, and major subprocesses of the current (or "as is") process.

To start, the process team thoroughly reviews the process mission and goals statement provided by the quality council, along with other information such as strengths and weaknesses and performance history of the process. An example of a statement on mission and goals was given above for the accounts payable process.

The major subprocesses are described in a high-level or macro-flow diagram, e.g., Figure 6.5. Another example is provided in Figure 6.6 for the billings and collections process at the AAL insurance organization (Hooyman and Harshbarger, 1994). This step in process definition should "bound" the process in terms of where the process starts, which activities are included (and excluded), and where the process ends.

Discover Customer Needs and Flowchart the Process

In this step the team identifies the customers (i.e., all external and internal parties who are affected by the process), determines customer needs, and prioritizes those needs. The specifics of this process are described in Chapter 4, "Operational Quality Planning and Sales Income," and Chapter 12, "Understanding Customer Needs."

A more detailed flowchart (a "process map") is prepared showing the major activities and key customers and suppliers and their roles in the process. This flowchart creates an understanding between the process owner and team members of how the process works. The use of the flowchart to gain this understanding is important because initial discussions usually reveal disagreements on how the process really works. Creating the flowchart clarifies—often after much discussion—how the

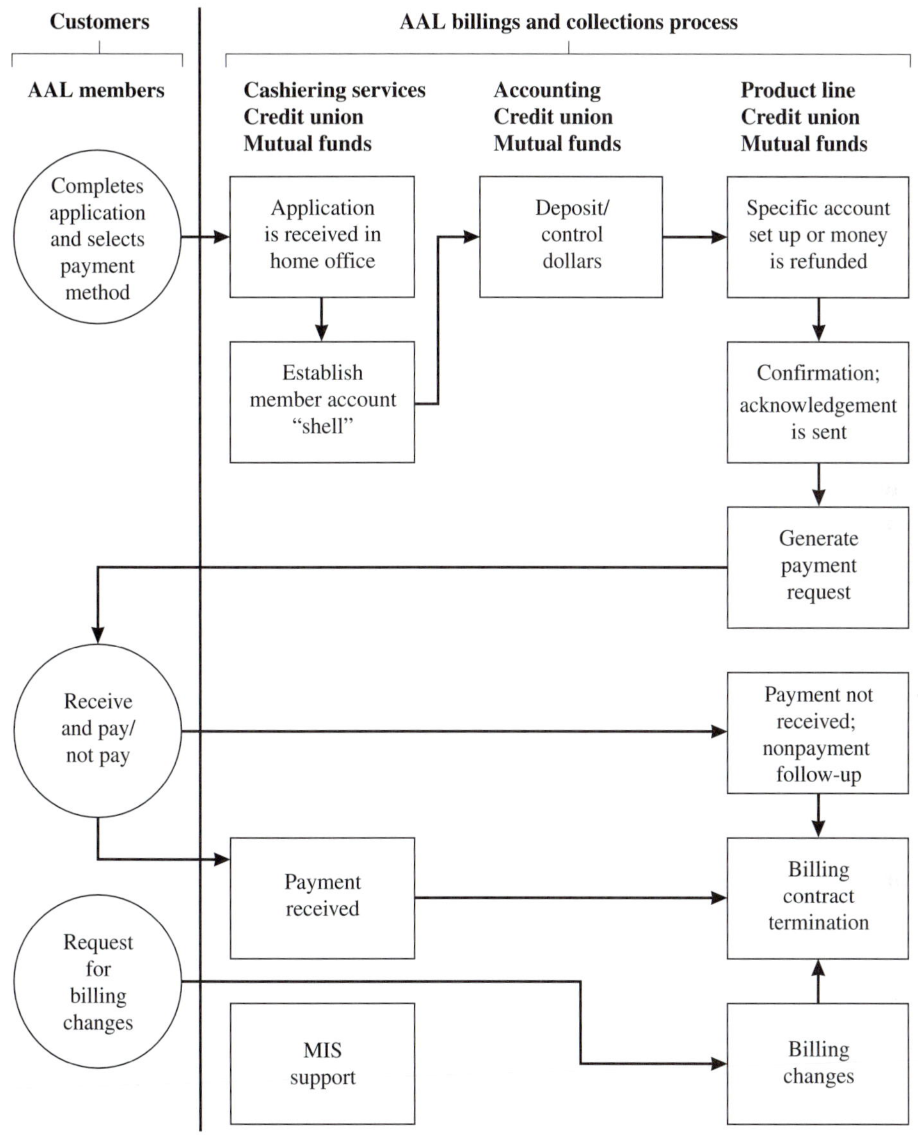

FIGURE 6.6
Macro-level process flow diagram for the billing and collections process.

process works. It is useful to schedule a work session of several hours in which the team discusses the process and prepares the flowchart. A facilitator describes how the work session will proceed, asks for inputs on the sequence of activities, and uses devices such as sticky notes to physically create the flowchart on a large board prepared for the purpose. The result is a starting point for analysis and improvement. An example of such a flowchart is given in Figure 6.7 (the measurement concept, M, is explained below).

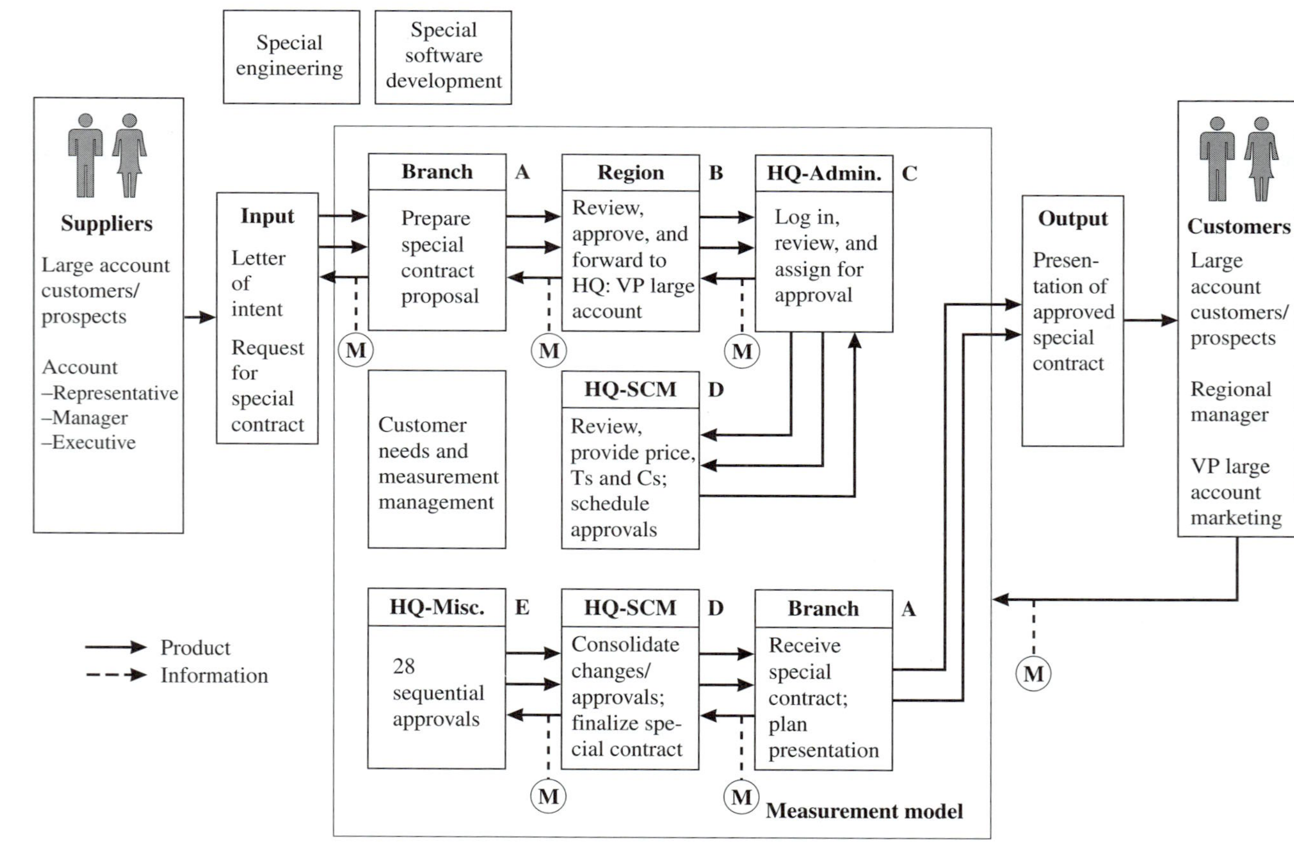

FIGURE 6.7
Flowchart of the special-contract management process, including process control measurement (M) points. (*From Juran Institute, Inc.*)

Van Aken and Hacker (1997) describe in seven steps how the work session and follow-up are conducted. After an initial flowchart is created by hand, software (such as Visio® or Optima!®) should be used to generate the flowchart.

Establish Process Measurements

Measurements from a process are initially needed to describe how well the process is doing and set the stage for process analysis and improvement. Later, measurements are employed to help control process performance and periodically determine process capability.

In deciding which measurements to collect from a process, the emphasis should be on the process mission statement, goals, and customer needs discussed above. In the framework shown in Figure 6.8, effectiveness focuses on meeting customer needs, efficiency addresses meeting the needs at least cost, and adaptability reflects the ability of the process to react positively when process or external conditions change. Although effectiveness and efficiency can be quantified easily, quantifying adaptability is more difficult. Snee (1993) discusses methods for making processes adaptable to changing conditions. The basic steps in quantifying parameters are discussed in Chapter 5, "Quality Control." Note, in Figure 6.7, how the measurement control points are indicated on the flowchart.

Process measurements should be linked to traditional business indicators. Some examples are shown in Figure 6.9.

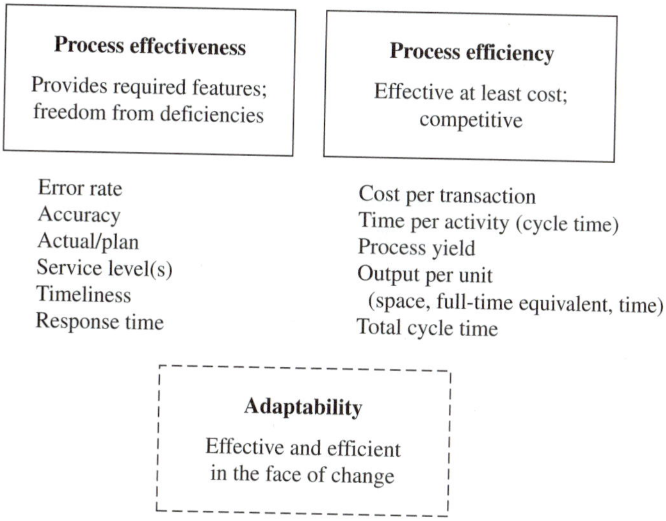

Process effectiveness

Provides required features; freedom from deficiencies

Error rate
Accuracy
Actual/plan
Service level(s)
Timeliness
Response time

Process efficiency

Effective at least cost; competitive

Cost per transaction
Time per activity (cycle time)
Process yield
Output per unit
(space, full-time equivalent, time)
Total cycle time

Adaptability

Effective and efficient in the face of change

FIGURE 6.8
Business process measurements.

Traditional business view		Process view	
Business objective	Business indicator	Key process	Process measure
Higher revenue	Percentage of sales quota achieved	Contract management	Contract close rate
		Product development	Development cycle time
	Percentage of revenue plan achieved	Account management	Backlog management and system assurance timeliness
	Value of orders canceled after shipment		Billing quality index
	Receivable days outstanding	Manufacturing	Manufacturing cycle time
Reduce costs	Inventory turns		

FIGURE 6.9
Linkages among business objectives, traditional business indicators, and process measures generated by the process-management approach—a few examples.
(*From Juran Institute, Inc.*)

Analyze Process Data

In this step we evaluate the process performance data, identify opportunities for improvement, and determine the causes of process problems. The specific approach and tools are described in Chapter 3, "Quality Improvement and Cost Reduction," Chapter 4, "Operational Quality Planning and Sales Income," and Chapter 5, "Quality Control," but with a special emphasis on the total process and subprocesses.

Performance data are evaluated for both process effectiveness and process efficiency (see above), and problems are identified using Pareto analysis, flow diagrams, and other means discussed in Chapter 3 under "Identify Potential Projects."

The high-level and detailed flow diagrams are key tools at this point. For example, at a telecommunications service company a team required four hours to construct a high-level flow diagram for a new service provided to corporate customers. The time consumed was a classic case of each member having a different view of how the process operated. The discussion and resulting high-level diagram exposed missing links, bottlenecks, unnecessary steps, and redundancies in the process. For the first time, all team members had a common understanding of the process.

The team then divided the high-level diagram into four segments and constructed detailed flow diagrams for each segment. A portion of one detailed diagram is shown in Figure 6.10. Examination of the four detached flow diagrams revealed 30 rework loops (and Pareto analysis showed that 6 of the 30 loops accounted for 82 percent of the total rework time). In Figure 6.10, two of the rework loops (26 and 27) are identified with a triangle symbol.

Flow diagrams often use four symbols for events: a diamond for a decision-making event, a small square for an activity, a group of consecutive squares for a

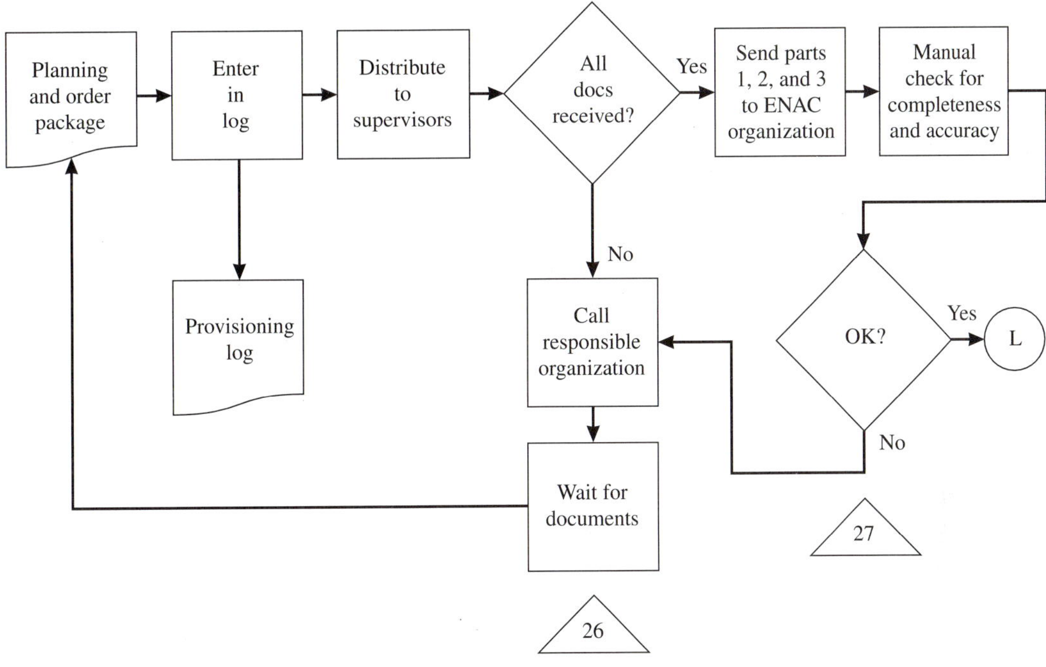

FIGURE 6.10
Flow diagram with rework loops (identified with triangles).
(*From Juran Institute, Inc., 1989.*)

rework loop, and a paper symbol for a document or database. Figure 6.11 provides a checklist of questions to ask for each of these symbols to help discover opportunities for improvement. The construction and analysis of flowcharts along with other simple and complex process analysis techniques was originated and refined by the industrial engineering profession. A useful reference is Salvendy (1992).

An example of a powerful (but more complex) technique is computer simulation of a process. Here a computer model is developed based on the logic sequence of process activities, along with data on the activities. The computer model then generates simulated results of process outputs. The time spent in developing the computer model is an investment to help reveal bottlenecks, underutilized process activities, and key causes of problems and to understand the process. Later the model can also help to evaluate potential solutions to problems. Batson and Williams (1998) describe seven cases, from both manufacturing and service industries, of using simulation in process improvement.

An excellent source of ideas to design or redesign a process is other organizations with similar processes. Some processes are common across many types of manufacturing and service industries (e.g., the hiring process), and the experience of other organizations can provide ideas that have been tested in practice. This approach is really an application of benchmarking (see Chapter 7 under "Competitive Benchmarking" and

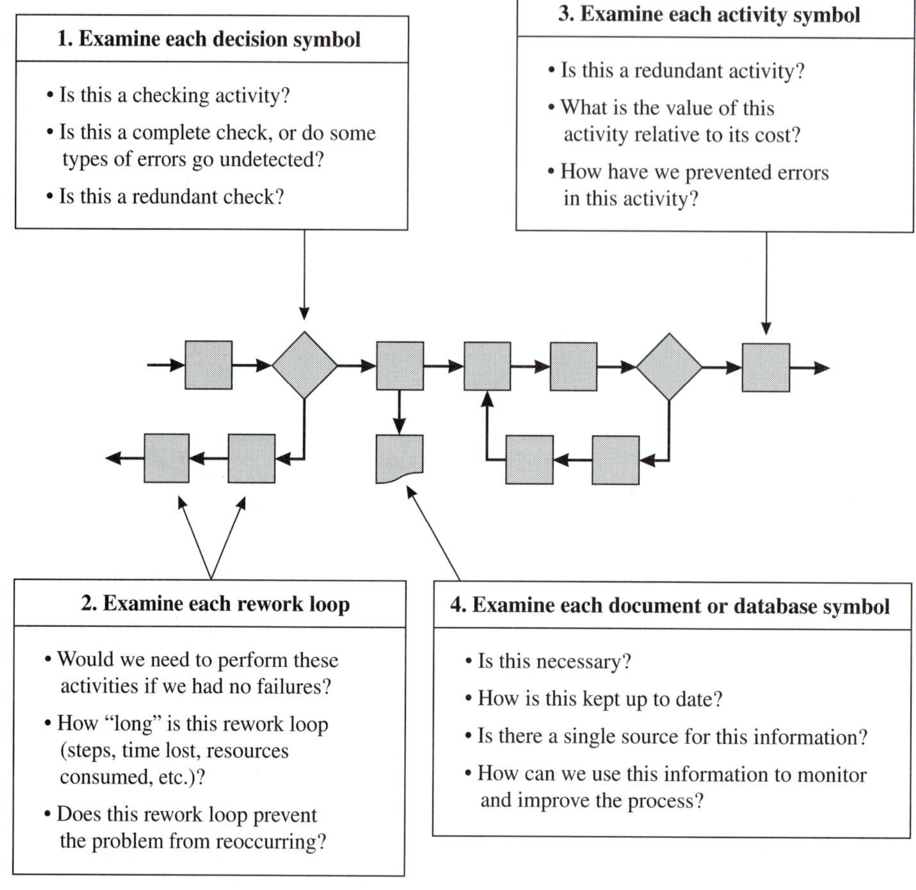

FIGURE 6.11
Analyzing a flow diagram. (*From Juran Institute, Inc., 1989.*)

also Camp, 1995). If it is possible to obtain a copy of process flow charts from another organization, so much the better.

Dow Corning now uses software to manage its process knowledge. This software creates interlinked maps of key processes, and the maps are stored on the company intranet. The "process repository" organizes information by process types and process parts. The Massachusetts Institute of Technology is licensing companies to use the process repository and software. A general process repository will be placed on the World Wide Web, giving managers access to a wealth of knowledge on process design (Carr, 1999).

At the end of the analyze phase, we have a clear understanding of the current process, have identified actual problems and causes, and have initial thoughts on types of improvement action necessary. This information should be reviewed by the executive process owner and other managers before proceeding to design/redesign.

Design (Redesign) the Process

This final step may involve radical change, incremental change, or both. The approach and tools of quality improvement, quality planning, and quality control are described in Chapters 3, 4, and 5, but these are applied with a special emphasis on the process and subprocesses. We start with the flowchart analysis of the current (as is) process and then redesign the process to create a flowchart for the revised (should) process. Design changes involve workflow, information and other technology, people, physical locations, and policies and regulations.

Radical redesign of a process is associated with the term *reengineering*. Hammer and Champy (1993) define *reengineering* as "the fundamental rethinking and radical redesign of business processes to achieve dramatic improvements in critical contemporary measures of performance, such as cost, quality, service, and speed." Clearly, the four key words are *fundamental, radical, dramatic,* and *processes.*

In reality, reengineering is a blend of the old and the new. Many concepts and techniques of industrial engineering, quality management, and other disciplines have long been an integral part of process improvement and redesign. Some key aspects of reengineering are

- Apply state-of-the-art information technology (see below).
- Analyze process activities for improvement; i.e., eliminate non-value-added work; simplify, combine, resequence activities; and minimize transfer of material and information especially across departments. This aspect of reengineering includes combining process steps so that an employee produces the final output rather than transferring the work for subsequent steps to other specialists.
- Benchmark against other organizations.
- Empower employees to make decisions to minimize time required for approvals.
- Remove causes of errors in processes to reduce rework and minimize checking and controls.
- Consider transfer of some activities upstream to suppliers or downstream to customers.
- Establish a single point of contact for customers so that one employee handles an entire service, rather than transfer the customer to other employees.
- Use creative thinking techniques including brainstorming.

This list presents highlights and is not all-inclusive.

A major contribution of reengineering is the emphasis on the application of the ever-increasing benefits of information technology on work. Hammer and Champy (1993) provide some illustrations:

- Information can appear simultaneously in as many places as it is needed.
- A generalist can do the work of an expert. For example, supported by integrated systems of information, a call center representative can handle all steps to service a customer.
- Businesses can simultaneously reap the benefits of centralization and decentralization. Decentralized divisions, at distant locations, can exchange information instantaneously with corporate headquarters and other divisions.

- Decision making is part of everyone's job. Database technology now provides many people with information that was previously available only to management.
- Field personnel can send and receive information wherever they are.
- The best contact with a potential buyer is effective contact (with or without human intervention). For example, banks, real estate brokers, and others can now provide customers with interactive video that provides a wealth of carefully prepared information.
- Things tell you where they are. For example, the locations of trucks and railroad cars can now easily be determined in real time.
- Plans get revised instantaneously. Instead of periodically revising plans and updating status, all work can now be done instantaneously and made available to all parties immediately.

The message is clear—information technology now has a major impact on how we organize and execute work.

In practice, design changes are both radical and incremental. Radical changes are dramatic, but incremental changes are collectively also essential.

Before the new design is placed into operation, conduct a design review and trial implementation. Typically, the process owner assembles a group of experts (from outside the process) to evaluate the design alternatives and the chosen design. Finally, the selected design should be tested under operating conditions using trial runs with regular operating personnel to predict both process effectiveness and process efficiency.

AT&T's similar approach to process management is summarized in Table 6.1. Note how many of the tools of quality improvement, quality planning, and quality control play an integral role in process management.

6.7 TRANSFERRING AND MANAGING THE NEW PROCESS

Transferring a new or revised process to operations involves carefully planning for the many aspects of the change. For a general discussion, see Chapter 4 under "Planning for Product Quality to Generate Sales Income." Managing the new process includes setting up appropriate process controls and planning for continuous improvement and periodic assessment. For elaboration, see Chapter 4, "Operational Quality Planning and Sales Income" and *JQH5,* pages 6.16–6.19.

Processes evolve into various phases of maturity. At General Electric five maturity classes are ad hoc, consistent, organized, systematic, and statistically stable (McNamara, 1997). McNamara reports that most important business processes are not statistically stable.

To summarize, process management includes some similarities to and some differences from the concepts discussed earlier in this book. The similarities include an emphasis-on-the-customer concept and the use of techniques from quality planning, quality control, and quality improvement. But important additions are provided by the process management concept:

TABLE 6.1
AT&T process quality management and improvement methodology

Steps	Activities	Tools
1. Establish process management responsibilities	Review owner selection criteria Identify owner and process members Establish/review responsibilities of owner and process members	Nominal group technique
2. Define process and identify customer requirements	Define process boundaries and major groups, outputs and customers, inputs and suppliers, and subprocesses and flows Conduct customer needs analysis Define customer requirements and communicate your own requirements to suppliers	Block diagram Survey Customer/supplier relations checklist Interview Benchmarking Affinity diagram Tree diagram
3. Define and establish measures	Decide on effective measures Review existing measures Install new measures and reporting system Establish customer satisfaction feedback system	Brainstorming Nominal group technique Survey Interview
4. Assess conformance to customer requirements	Collect and review data on process operations Identify and remove causes of abnormal variation Compare performance of stable process to requirements and determine chronic problem areas	Control chart Interview Survey Pareto diagram Cause-and-effect diagram Brainstorming Nominal group technique Trend chart
5. Investigate process to identify improvement opportunities	Gather data on process problems Identify potential process problem areas to pursue Document potential problem areas Gather data on subprocess problems Identify potential subprocess problems to pursue	Interview Flowcharting Brainstorming Pareto diagram Nominal group technique
6. Rank improvement opportunities and set objectives	Review improvement opportunities Establish priorities Negotiate objectives Decide on improvement projects	Pareto diagram Nominal group technique Trend chart
7. Improve process quality	Develop action plan Identify root causes Test and implement solution Follow through Perform periodic process review	Pareto diagram Nominal group technique Brainstorming Cause-and-effect diagram Cause-and-effect/force-field analysis Control chart Survey

Source: Shaw et al. (1988).

1. Emphasis is placed on the overall effectiveness, efficiency, and adaptability of a cross-functional *process*, rather than the output of individual functional departments.
2. The process is analyzed in an integrated manner not only for correcting defects, errors, and other problems but also for identifying and meeting customer needs.
3. Responsibility for the process is defined in terms of a process owner and a process team. The owner and the team are permanent. The process team may set up quality teams that operate temporarily to address specific problems within the total process.

Finally, the special features of process management require us to consider its impact on an organization.

6.8 IMPACT OF PROCESS MANAGEMENT ON AN ORGANIZATION

This chapter presents a concept of process management that identifies the key processes for organization success and selects process owners and process teams that are responsible for planning, controlling, and improving all aspects of processes. These teams are permanent, not ad hoc quality planning or quality improvement teams.

Although only a minority of organizations apply this full concept of process management, the application is growing. The changes are indeed profound: Work activities are focused in cross-functional process teams, not functional departments; the organization structure changes from hierarchical to flat; managers change from supervisors to coaches and leaders; employees change from people who perform narrow tasks and rigid procedures to empowered individuals who are trained in a broader work scope; and more.

The impact on the functional department organization can be dramatic. If an organization is run by processes, the processes will drive the organization, including the structure. When most work is done by permanent process teams, the role of functional departments becomes one of serving processes; i.e., a process is the internal customer of the functional departments. Functional departments will still be needed to do internal consulting, training, measurement, and research.

Perhaps process management will bury the silos, break down the walls between functional departments, and replace command and control management with empowered employees who management provides with capable processes. Perhaps.

SUMMARY

- A process is a collection of activities that convert inputs into outputs or results.
- A primary process is a collection of cross-functional activities that are essential for external customer satisfaction and achieving the mission of the organization.

- Process management is an approach for planning, controlling, and improving the primary processes in an organization by using permanent process teams.
- Process management is most effective when a process owner and a permanent process team manage and operate the process.

PROBLEMS

6.1. For an organization with which you are familiar, discuss the critical success factors, i.e., the few events that must occur for the organization to be successful. Then identify three cross-functional processes and rank each of them on the basis of (a) their importance with regard to the critical factors and (b) relative need for improvement.

6.2. For one of the processes in problem 6.1, prepare a process mission statement and a statement of process goals.

6.3. For one of your key business processes, propose a process owner and members of a process team.

6.4. For the process in problem 6.3, develop a flowchart of the current process.

6.5. For the process in problem 6.3, propose at least three process measurements that could be collected for process analysis and improvement.

6.6. For the process in problem 6.3, analyze the process and propose a redesign.

REFERENCES

Batson, R. G. and T. K. Williams (1998). "Process Simulation in Quality and BPR Teams," *Annual Quality Congress Proceedings,* ASQ, Milwaukee, WI, pp. 368–381.

Camp, R. C. (1995). *Business Process Benchmarking,* Quality Press, ASQ, Milwaukee, WI.

Carr, N. G. (1999). "A New Way to Manage Process Knowledge," *Harvard Business Review,* September–October, pp. 24–25.

Hammer, M. and J. Champy (1993). *Reengineering the Corporation,* HarperCollins, New York.

Hardaker, M. and B. K. Ward (1987). "Getting Things Done: How to Make a Team Work," *Harvard Business Review,* November–December, pp. 112–119.

Hildreth, S. W. (1993). "Rolling out BPQM in the Core R & D of Lederle–Praxis Biologicals, American Cyanamid," *Proceedings of R & D Quality Symposium,* Juran Institute, Inc., Wilton, CT, pp. 2A-1 to 2A-9.

Hooyman, J. A. and R. W. Harshbarger (1994). "Re-Engineering the Billing and Collections Process," *Impro 94,* Juran Institute, Inc., Wilton, CT, pp. 7A-3 to 7A-18.

Juran Institute, Inc. (1989). *Quality Improvement Tools—Flow Diagrams,* Wilton, CT.

McNamara, D. M. (1997). "Process Maturity Classes," *Annual Quality Congress Proceedings,* ASQ, Milwaukee, pp. 891–897.

Salvendy, G. (1992). *Handbook of Industrial Engineering,* 2nd ed., John Wiley & Sons, New York.

Shaw, G. T., E. Leger, and J. C. MacDorman (1988). "Process Quality at AT&T," *Impro Conference Proceedings*, Juran Institute, Inc., Wilton, CT, pp. 4D-5 to 4D-9.

Snee, R. D. (1993). "Creating Robust Work Processes," *Quality Progress,* February, pp. 37–41.

Van Aken, E. M. and S. K. Hacker (1997). "Enhancing Conventional Process Improvement with Systems Mapping," *Proceedings Annual Quality Congress,* ASQ, Milwaukee, pp. 84–94.

SUPPLEMENTARY READING

Process management: *JQH5,* Section 6.

AT&T (1988). *Process Quality Management and Improvement Guidelines*, AT&T Bell Laboratories, Indianapolis, IN.

Hammer, M. and S. Stanton (1999). "How Process Enterprises Really Work," *Harvard Business Review,* November–December, pp. 108–118.

Harrington, H. J. (1991). *Business Process Improvement*, McGraw-Hill, New York.

Melan, E. H. (1993). *Process Management*, McGraw-Hill, New York.

Pall, G. A. and L. A. Kelly (1999). *The Process Centered Enterprise*, St. Lucie Press, Boca Raton, FL.

Rummler, G. A. and A. P. Brache (1995). *Improving Performance,* 2nd ed., Jossey-Bass, San Francisco.

WEBSITES

Business process reengineering:
www.brint.com/BPR.htm
www.prosci.com

7

STRATEGIC QUALITY MANAGEMENT

7.1
ELEMENTS OF STRATEGIC QUALITY MANAGEMENT

Previous chapters present some specific concepts and techniques to achieve quality excellence. With that knowledge we are now ready to take a longer-range view of quality, i.e., strategies for quality. Strategies provide a game plan for the future.

Strategic quality management is the process of establishing long-range customer-focused goals and defining the approach to meeting those goals. Strategic quality management becomes an integral part of the overall strategic plan of an organization and is developed, implemented, and led by upper management. Strategic quality management (this chapter) is performed at the top organization level. Operational quality planning (Chapter 4) focuses at the product and process level with middle management.

We will examine the basic elements of strategic management and then address how the quality parameter can be integrated. The following elements provide a widely accepted framework:

- Define the mission and critical success factors.
- Study the internal and external environments and identify the strengths, weaknesses, opportunities, and threats for the organization.
- Define a long-term, ultimate goal (a "vision").
- Develop key strategies to achieve the vision.
- Develop strategic goals (long term and short term).
- Subdivide the goals and develop operational plans and projects ("deploy the goals") to achieve the goals.

- Provide executive leadership to implement the strategies.
- Review progress with measurements, assessments, and audits

Note that these elements cover both the development and the deployment of strategies.

Typically, strategy covers a five-year span in broad terms, with the first year in more detail, and with annual updating of the five-year strategy.

Thompson and Strickland (1998) is a helpful reference that elaborates on the general concept of strategic planning. *JQH5,* Section 13 on strategic deployment, discusses both development and deployment of quality strategies.

Specific approaches to strategic quality management are still evolving, but these are some necessary ingredients:

1. A focus on customer needs.
2. Continuous improvement for all processes in the organization (big Q).
3. Understanding the key customer, market, and operational conditions as input to setting strategic direction. This aspect links directly to four components of quality assessment described in Chapter 2. These components identify strengths, weaknesses, opportunities, and threats—in the language of strategic management, a "SWOT analysis." If a significant difference exists, then strategies, goals, and actions must be identified—call it a "gap analysis."
4. Leadership by upper management. Quality needs to be integrated into the eight elements of strategic management identified above. This ingredient of strategic quality management includes actions to set up organizational machinery to carry out improvement, empower the workforce to make improvements, train all levels to execute their quality responsibilities, establish measures and review progress against improvement goals, provide recognition for superior performance, and expand the reward system to reflect changes in job responsibilities.
5. Translation and deployment of strategies into annual business plans. This aspect requires a structured approach to align activities at three levels: organization level, key process level, work unit/individual level. Also, emphasis must be on actions by line departments instead of relying on a quality department.
6. Adequate resources "where people have the knowledge, skills, authority, and desire to decide, act, and take responsibility for the results of their actions and for the contribution to the success of the company" (the Eastman Chemical definition of empowerment). Also, certain resource capabilities serve as a source of competitive advantage over rival firms (e.g., product development, marketing). These "core competencies" require special attention and support.

The Japanese use a similar approach to strategic management, called "*hoshin* planning" (or "policy deployment"), which is built around the classic management cycle of plan, execute, and audit (or plan, do, check, act). Key aspects of *hoshin* planning include a focus on the planning process, company targets known by all employees, individual initiative, self-audit, and documentation and communication.

Next we present an overview of the strategic quality management process using four examples, and then we discuss the individual elements of the process.

7.2
INTEGRATING QUALITY INTO STRATEGIC MANAGEMENT

In a classic case General Electric conducted a full strategic planning study for one of its products. Product quality was one parameter addressed in the study (Utzig, 1980).

In this example General Electric was the minimum-unit-cost producer, but the company was losing market share. The strategic planning study posed several questions.

1. What are the financial goals for product X?

   The short-range financial goal was a return on investment of at least 25 percent; the 5-year goal on cumulative net income was $120 million.
2. What is our present quality goal with respect to the competition?

   The goal was to be equal to competitors A and B.
3. What are the key quality factors that influence the purchasing decision of potential customers?

   Marketing research determined that the factors for product X were reliability, efficient performance, durability, ease of inspection and maintenance, ease of wiring and installation, and product service.
4. How do we compare to the competition on each of the key factors?

   The surprising results are shown in Table 7.1. Note that GE was rated lowest on all six factors. Of course, the results were not believed—until the study was done three times. Can't you just hear the criticisms of the study?
5. Does anyone have a unique competitive advantage on quality?

   Table 7.1 shows that competitor A was best on all six factors.
6. What are the internal results on quality?

   Failure costs were low and complaints were low. Thus a traditional cost-of-poor-quality study would *not* have identified the seriousness of the problem.
7. What are alternative quality goals with respect to competition?

   Several alternatives were examined by studying the benefits and costs. It was decided that the present goal (to be equal in quality to competitors A and B) was to be retained.
8. For the chosen goal, what departmental goals are needed to achieve the changes of level in the key factors?

TABLE 7.1
Customer-based measurements

Product attributes	Mean importance rating	Competitor performance ratings		
		GE	A	B
Reliable operation	9.7	8.1	9.3	9.1
Efficient performance	9.5	8.3	9.4	9.0
Durability/life	9.3	8.4	9.5	8.9
Easy to inspect and maintain	8.7	8.1	9.0	8.6
Easy to wire and install	8.8	8.3	9.2	8.8
Product service	8.8	8.9	9.4	9.2

New departmental goals were needed in the design, manufacturing, service, and quality assurance departments.

9. What departmental plans must be developed?

   The goals were translated into specifics such as strengthening the magnetic structure to improve reliability, providing more uniform heat treatment, instituting a special training program for service technicians, and performing additional life testing on the product.

10. What resources are required?

   After the usual give-and-take of negotiation, appropriate resources were assigned.

Note that the driving force in this model (item 1) was a profit goal, not a quality goal. The competitive analysis, when viewed in light of the profit goal and regaining market share, was decisive in changing priorities and assigning the necessary resources. Business planning, involving items 8, 9, and 10, glued the quality strategy into place. This action led to a successful result.

Middle managers at Carolina Power and Light were the source of input in developing a strategy (Allen and Bailes, 1988). A series of meetings was held to identify critical issues that were obstacles to quality. More than 100 issues were identified. These were reduced to 22 issues, e.g., enhanced leadership, direction and control by senior management, and further meetings took place to develop goals to address the 22 issues. The result was 297 goals, e.g., sense of unity of purpose, visible commitment to corporate mission and beliefs. Finally, the goals were grouped into four "macrostrategies": communications, changing the culture and management style, recognition and reward, and education and training. For each of these strategies, brainstorming was employed to identify action items. Action items, providing a framework for each strategy, then led to the development of a five-year plan for quality.

Increasingly, organizations are finding that although upper-management leadership is essential, the development of strategy should involve both top-down and bottom-up viewpoints.

In another service industry example, Figure 7.1 shows the approach at Kelly Services, Inc. Note how the link between strategic planning and performance management begins with internal and external customer input.

A final example comes from Xerox. Figure 7.2 depicts the major steps in going from developing strategy at the corporate level and the linkage from corporate direction to annual objectives at the division level and then to the operations level.

In the next section we discuss how quality-related activities can be linked and integrated into the eight elements of strategic planning, resulting in a strategic quality management framework.

7.3 MISSION, ENVIRONMENT ANALYSIS, VISION

A *mission* is a statement of the organization's purpose and the scope of its operations, i.e., "the business we are in."

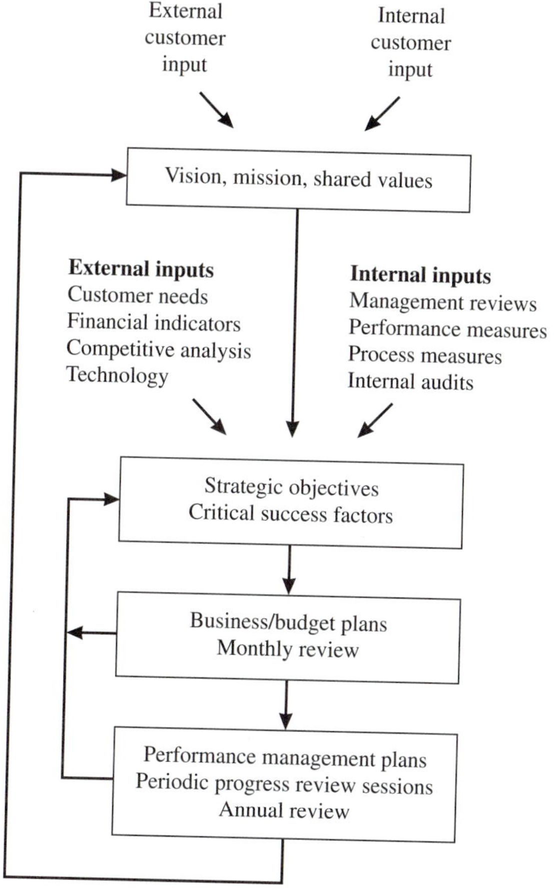

FIGURE 7.1
Kelly Services strategic planning process. (*From McCain, 1995.*)

Some examples:

Our mission is to provide any customer a means of moving people and things up, down, and sideways over short distances with higher reliability than any similar enterprise in the world. (Otis Elevator)

Our business is renting cars. Our mission is total customer satisfaction. (Avis Rent-a-Car)

We exist to create, make, and market useful products and services to satisfy the needs of our customers throughout the world. (Texas Instruments)

The mission of the American Red Cross is to improve the quality of human life; to enhance self-reliance and concern for others; and to help people avoid, prepare for, and cope with emergencies.

Note how all of these mission statements recognize quality.

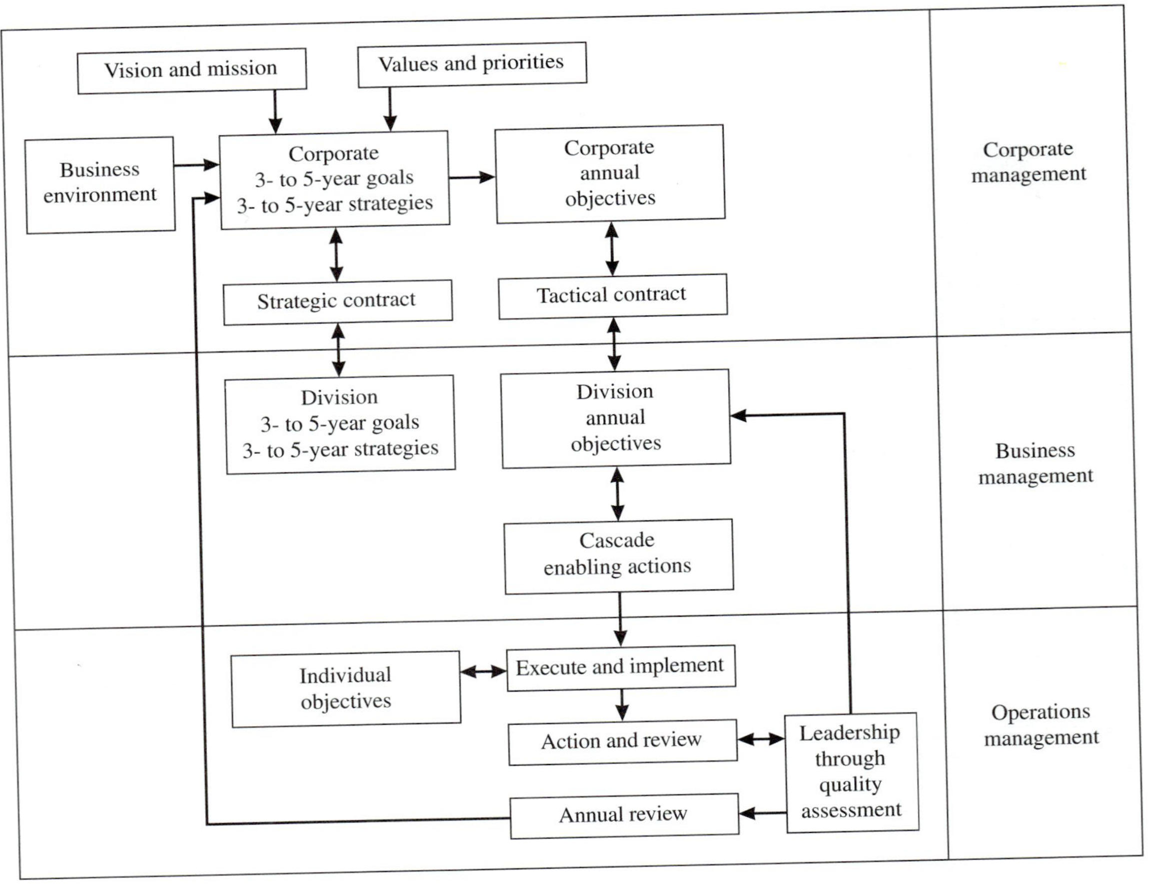

FIGURE 7.2
Xerox managing-for-results process. (*From Leo, 1994.*)

The mission statement sets forth what an organization is doing today, but we must then examine the external and internal environments and think strategically about the future. With respect to quality, the environment analysis should focus on four elements: cost of poor quality, market standing on quality, the current quality culture, and the current quality system. These elements, explained in Chapter 2, "Companywide Assessment of Quality," are essential because they identify the driving forces that must be addressed to define a vision and develop and deploy strategies.

A *vision statement* defines the desired future state of the organization. A *vision* can be viewed as the ultimate goal, one that may take five years or more to achieve. Some examples:

To be the leading supplier of PCs and PC servers in all customer segments. (Compaq Computers)

We will lead in delivering affordable, quality health care that exceeds the service and value of our customers' expectations. (Kaiser-Permanente)

To engineer, produce, and market the world's finest automobiles. (Cadillac Motor Car Division)

To be America's best quick service restaurant chain. We will provide each guest great tasting, healthful, reasonably priced fish, seafood, and chicken in a fast, friendly manner on every visit. (Long John Silver's)

Note how all these vision statements recognize quality.

In practice, mission statements focus on "what our business is now"; vision statements emphasize "what our business will be later." Sometimes one statement covers both the present and the future.

7.4 DEVELOPING STRATEGIES

A *strategy* is a guide on how to pursue the organization's mission and vision. Strategies set direction by identifying the key issues or activities that help develop specific goals and plans. Strategies must contribute significantly to the vision. In developing strategies and goals, we need the leadership of upper management and the participation of middle management. This scope provides a source of valuable ideas and also encourages ownership for implementation.

With respect to quality, there are usually a small number of strategies, say three to five. For example, the Cadillac Motor Car Division identified three strategies: a cultural change where teamwork and employee involvement are considered a competitive advantage, a focus on the customer with customer satisfaction in the master plan, and a more disciplined approach to planning that focuses all employees on quality goals.

Strategies can aim at both operational effectiveness and achieving a competitive advantage (Porter, 1996). *Operational effectiveness* means performing similar activities better than the competition does (e.g., conducting operations with less waste); *achieving a competitive advantage* means not performing the same activities as competitors

TABLE 7.2
Examples of quality strategies

Operational effectiveness	Competitive advantage
Reduce costs and cycle time through waste reduction using the six-sigma approach	Pursue customer loyalty and retention
Institute formal cross-functional process management	Formalize the product development process
Decentralize decision making through employees trained in problem solving	Pursue the Baldrige Award
Realign reward and recognition systems to focus on quality	Focus on quality/price ratio
Emphasize both incremental improvement and major breakthrough improvement	Build a learning organization
Pursue ISO certification	
Work jointly with suppliers to achieve quality excellence	

perform or performing similar activities in different ways (e.g., conducting a formal, intensified product development process to generate unique product features).

This distinction applies to developing quality strategies. Table 7.2 presents examples of some quality strategies for both operational effectiveness and competitive advantage.

The distinction reminds us that quality activities go far beyond reducing waste and the cost of poor quality. Strategies are the engine for quality success. These strategies must then be supported with goals, operational plans, and projects.

7.5
DEVELOPMENT OF GOALS; COMPETITIVE BENCHMARKING

A *goal* is a desired result to be achieved in a specified time. ("A goal is a dream with a deadline"—Anonymous.) Goals often have different names: The overall organization goal is sometimes called a "vision"; long-range company goals (say, five years) are called "strategic goals"; short-range goals (say, one year) are "tactical goals"; goals at various levels may be called "business goals," "objectives," or "targets." The terminology simply is not standardized.

DeFeo (1995) in *JQH5* identifies seven areas in which goals are minimally required:

- Product performance.
- Competitive performance.
- Quality improvement.
- Cost of poor quality.

- Performance of business processes.
- Customer satisfaction.
- Customer loyalty and retention.

Some examples of corporate quality goals prepared for a health products company for the coming year are

1. Reduce quality costs for the company by ______ percent.
2. Keep material loss for the company under ______.
3. Reduce the average leakage rate for product ______ to ______.
4. Employ ______ certified quality engineers.
5. Determine quality costs for at least one product.
6. Develop and implement a specific in-process quality data analysis technique for at least one product.
7. Define numerical reliability and maintainability objectives for at least one product.
8. Implement a procedure to assure that all corporate product specifications are reviewed by the plants before plant production planning starts.
9. Implement a procedure to assure that all specifications for suppliers are agreed to by the supplier before the purchasing contract is finalized.
10. Develop a quality procedures manual.
11. Require the president or senior vice president to make at least ______ visits to customers to review product quality.

Other examples of overall quality goals are

- Create the Taurus/Sable car model at a level of quality that is "best in class."
- Reduce by 50 precent the time required to resolve customer complaints.
- Increase the percentage of research results that become incorporated into products by ______ percent.

Note that these statements include a quantification in either terms of a product characteristic or a date (the end of the calendar year). These cover both product characteristics and tasks in the overall company quality program. Quality goals can also be created for individual departments.

Formulation of Quality Goals

Quality goals can be identified from several inputs. The most important source is the collection of four studies described in Chapter 2, "Companywide Assessment of Quality": cost of poor quality, market standing on quality, quality culture, and the quality system. These studies identify the strengths, weaknesses, opportunities, and threats. Other inputs to help formulate goals include

- Pareto analysis (see Chapter 3) of repetitive external alarm signals (field failures, complaints, returns, etc.).
- Pareto analysis of repetitive internal alarm signals (scrap, rework, sorting, 100 percent test, etc.).

- Proposals from key insiders—managers, supervisors, professionals, union shop stewards.
- Proposals from suggestion schemes.
- Field study of users' needs, costs.
- Data on performance of products versus competitors' (from users and from laboratory tests).
- Comments of key people outside the company (customers, vendors, journalists, critics).
- Findings and comments of government regulators, independent laboratories, reformers.

Analysis of these inputs requires, as in formulating policies, a mechanism that gives managers the opportunity to participate in setting goals without the burden of performing the detailed staff work. Quality engineers and other staff specialists are assigned the job of analyzing the available inputs and of creating any essential missing inputs. These analyses point to potential projects that are then proposed. The proposals are reviewed by managers at progressively higher organizational levels. At each level there is summary and consolidation until the corporate level is reached. The foregoing process is similar to that used in preparing the annual financial budget.

The process employed at a chemical plant of Union Carbide to set annual goals is shown in Figure 7.3. Note that this approach includes a vision, a strategy (long range), and tactical plans (short range).

Several alternative criteria may be used to define quality goals: historical performance, engineering analysis, competition (see "Competitive Benchmarking" below), or some absolute value (e.g., six sigma). Companies aspiring to excellence often set goals beyond those that are clearly attainable to encourage people to generate unusual approaches to achieving excellence. These are "stretch goals." Sometimes, the results can be remarkable, indeed extraordinary. All goals—but particularly stretch goals—require a strong follow-through of goal deployment and assignment of resources.

Finally, the nature of quality goals changes with the maturity of an organization's quality initiative, e.g., earlier goals for establishing infrastructure versus later goals that stem from quality assessments and benchmarking.

Competitive Benchmarking

Competitive benchmarking is "the continuous process of measuring products, services, and practices against the company's toughest competitors or those companies renowned as industry leaders" (Camp, 1994).

A benchmark is simply a reference point that is used as a standard of comparison for actual performance. Table 7.3 lists typical benchmarks.

Unfortunately, adherence to a specification may be woefully insufficient to generate sales. Also note that some of the other benchmarks listed in Table 7.3 go beyond competition. In this context *competition* means other organizations that compete to sell, e.g., office reproduction equipment. But a benchmark organization may be "best in our industry" or "best in any industry." For example, the Xerox Corporation used IBM and Kodak (direct competitors on some products) as benchmark organizations to

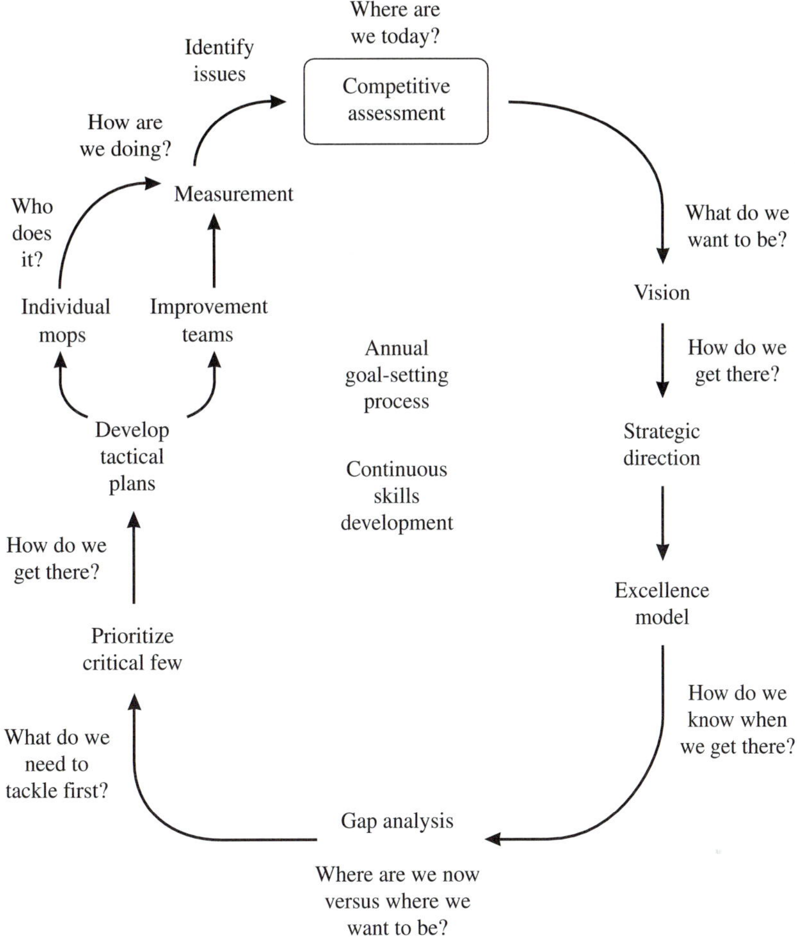

FIGURE 7.3
Annual goal-setting process. (*From Perry, 1989.*)

TABLE 7.3
Benchmarks

The specification
Customer desires
Competition
Best in our industry
Best in any industry

evaluate many Xerox operations. But for warehousing and distribution activities, Xerox chose as a benchmark the L.L. Bean Company, a catalog sales distributor of clothing and other consumer products. Benchmarks serve not only as a standard of comparison but also as a means of self-evaluation and subsequent improvement. The concept of searching for the best performer in *any* industry is a valuable contribution of the benchmarking approach.

The benchmarking process applies to subjects such as products, customer services, and internal processes. The steps in the process are

1. Identify the benchmark subjects.
2. Identify benchmark partners (organizations that will serve as benchmarks).
3. Determine the data collection method and collect the data.
4. Determine the competitive gap (comparing our company to the benchmark partners).
5. Project the future performance of the industry and our company.
6. Communicate the results.
7. Establish functional goals.
8. Develop action plans.
9. Implement plans and monitor results.
10. Recalibrate the benchmarks (repeat the benchmarking process, typically every three to five years depending on the subject).

Competitive benchmarking is an essential input to the formulation of goals. But benchmarking also provides sources of ideas for improvement along with the evidence that better practices not only exist but also must be instituted to be competitive. For a complete discussion on benchmarking, see *JQH5,* Section 12. Camp (1998) presents case studies of benchmarking in the manufacturing, service, nonprofit, government, and education sectors.

7.6 DEPLOYMENT OF GOALS

Broad goals do not lead directly to results; they must first be "deployed." *Deployment* means subdividing (aligning) the goals and allocating the goals to lower levels for the conversion into operational plans and projects (see Figure 7.4). Thus goals must be deployed from the organization level to the process level and to individual jobs.

For example, an organization had a strategy to be "known as the provider of the most responsive customer service." To support this strategy four goals were set (Juran Institute, 1992):

1. Reduce order lead times to two days by 1993.
2. Reduce by 50 percent the time interval to reply to customer complaints.
3. Decrease by 90 percent customer application development times.
4. Cut contract proposal times by 80 percent.

Note that each goal is specific, observable, and measurable. Specific projects were then set up to achieve these goals.

One hospital methodically defines the linkages among elements of its strategic planning process. The hospital has five long-term goals, e.g., improve customer success and satisfaction. Each goal is linked (in a matrix) to specific strategic objectives, key performance measures, annual targets, and 11 core processes, e.g., deliver ser-

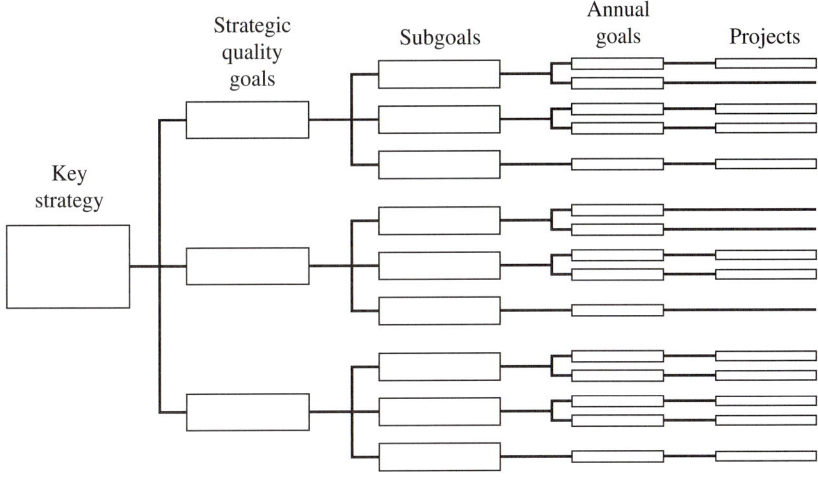

FIGURE 7.4
Deploying the vision. (*Juran Institute, Inc.*)

vices. The impact of a goal on a core process is further defined as high, medium, or low. Management "champions" then work with process owners to identify and pursue projects that will achieve the performance targets. This cascading process of deploying strategy and goals to lower levels is called the "golden thread."

Here is a key point: vision, strategies, and goals provide direction, but specific projects and other forms of action supply the methods to achieve results. Sometimes the number of projects can be dramatic, e.g., for General Electric's six-sigma goal, the number of projects grew from 200 in 1995 to about 47,000 in 1999 (Slater, 1999). For the Ford Taurus/Sable goal of becoming best in its class, more than 400 subgoals were defined, which led to more than 1500 project teams.

7.7 PROVIDE EXECUTIVE LEADERSHIP TO IMPLEMENT THE STRATEGIES

The most important factor in implementing quality strategies is the personal leadership of upper management. The organizational mechanism used is the quality council (see Chapter 8 under "Role of Upper Management").

Although upper-management leadership is essential, the development of strategy and plans should involve both top-down and bottom-up viewpoints. This process requires two-way communication between top management and lower levels—the Japanese call this concept "catch ball." The key elements are

- Communication of what top management proposes as the key focus areas for strategic planning.
- Nominations by managers at lower levels for additional areas for attention.

- Decisions on strategies, goals, and resources for deployment

At Fannie Mae, a financial services company, senior executives drafted the original vision, mission, and key strategies. Then directors and middle-level managers provided comments, which led to a draft of strategic goals. Next more than 600 managers and supervisors contributed ideas for deploying the goals to subgoals and projects. Senior management then created the final version of the key strategies and goals. Note the similarity to the approach employed at Carolina Power and Light (see Section 7.2).

To successfully implement the strategies requires upper management to provide an infrastructure for quality. For example:

- Build an organization with the necessary competencies and assign sufficient resources to the activities critical to the strategies.
- Emphasize continuous improvement.
- Install operating and information systems to support personnel, including quality policies and procedures.
- Create a positive quality culture, including rewards and recognition for achievement of superior performance.
- Exert personal leadership to drive implementation of the strategies.

This book explains these elements. One tool in providing direction is the concept of quality policies.

Quality Policies

A *policy* is a broad guide to action. It is a statement of principles or values. A policy differs from a procedure, which details how a given activity is to be accomplished. Thus a quality policy might state that quality costs will be measured. The corresponding procedure describes how the costs are to be measured.

Policies do not have to be vague. They can be specific enough to provide useful guidance. Here are two dramatic examples from different computer manufacturers:

> A new product must perform better than the product it replaces and better than the competition's, and this must be the case at the time of the first regular customer shipment.
>
> In selecting suppliers, decision makers are responsible for choosing the best source even if this means internal sources are not selected. (This policy rescinded a previous policy that placed a priority on buying from "sister divisions.")

The following corporate quality policies were prepared for discussion at a health products company.

1. At both the corporate and plant levels, the quality control department shall be independent of the production function.
2. The company shall place a new product on the market only if the overall quality is superior to the competition's.
3. All tasks necessary to achieve superior quality shall be taken, but each task shall be evaluated to assure that the investment has a tangible effect on quality.

4. Specific quality responsibilities for all company areas including top management shall be defined in writing.
5. Quality activities shall emphasize the prevention of quality problems rather than only detection and correction.
6. Quality and reliability shall be defined and measured in quantitative terms.
7. All quality parameters and tests shall reflect customer needs, usage conditions, and regulatory requirements.
8. Total company costs associated with achieving quality objectives shall be periodically measured.
9. Technical assistance shall be provided to suppliers to improve their quality control programs.
10. Each quality task responsibility defined for a functional department shall have a written procedure describing how to perform the task.
11. The company shall propose to regulating agencies or other organizations any additions or changes to industry practice that will ensure a minimum acceptable quality of products.
12. Each year, quality objectives shall be defined for corporate, division, and plant activities and shall include both product objectives and objectives on tasks in the company quality program.
13. All levels of management shall have a defined quality motivation program for the employees in their department.

These policies were prepared to provide guidelines for (1) planning the overall quality program and (2) defining the action to be taken when personnel request guidance.

Policies may also be needed within a functional department. Policies for use within a quality department might include the following statements:

1. The amount of inspection of incoming parts and materials shall be based on criticality and a quantitative analysis of supplier history.
2. The evaluation of new products for release to production shall include an analysis of data for compliance to performance requirements and shall also include an evaluation of overall fitness for use, including reliability, maintainability, and ease of user operation.
3. The evaluation of new products for compliance to performance requirements shall be made to defined numerical limits of performance.
4. Suppliers shall be supplied with a written statement of all quality requirements before a contract is signed.
5. Burn-in testing is not a cost-effective method of eliminating failures and shall be used only on the first units of a new product type (or a major modification of an existing product) to get rapid knowledge of problems.

Note that these examples of policies state (1) a principle to be followed or (2) *what* is to be done but not *how* it is to be done. The how is described in a procedure. Often it is best to have a policy instead of a procedure to provide flexibility for different situations.

A sensitive policy issue may arise as the result of improvement projects that reduce rework or reprocessing to correct errors. The people who have been doing the rework wonder: What will happen to me if this work is no longer necessary? This apprehension should be faced head on and a policy formulated. Several alternatives are possible:

- Guarantee that no employee will lose employment as a result of the quality effort. A few companies have issued such a policy statement.
- Rely on resignations and retirements as a source of new jobs for those whose jobs have been eliminated. Retrain affected workers to qualify them for the new jobs.
- Reassign affected employees to other areas. This approach can include creating positions for additional quality improvement work.
- Offer early retirement.
- If all else fails, offer termination assistance to help workers locate jobs in other companies.

Planning for this situation in advance can often lead to creative solutions; failure to plan in advance often leads to a minefield of problems, including serious disruptions of the lives of people. A compassionate approach is clearly a hallmark of true leadership.

7.8 REVIEW PROGRESS WITH MEASUREMENTS, ASSESSMENT, AND AUDITS; BALANCED SCORECARD

Once goals have been set and deployed into subgoals, business plans, and projects, then key measurements must be established. Various names apply, e.g., key results indicators, performance indicators. Readers are urged to review the basic elements of measurement—see Chapter 5 under "Measurement" and particularly Figure 5.2.

At the strategic level, measurements should be developed for each strategic goal defined in the strategic plan. The measurements usually include areas such as product performance, competitive performance, quality improvement, cost of poor quality, performance of business processes, customer satisfaction, and customer loyalty and retention (see above under Section 7.5, "Development of Goals"). Luther (1993) describes how Corning Inc. uses more than 200 key results indicators to integrate quality and business objectives.

Some organizations combine their measurements from financial, customer, internal processes, and learning and growth areas into a "balanced scorecard." Figure 7.5 illustrates the concept. The scorecard is used in conjunction with four management processes: translating the vision, communicating and linking strategy to departmental and individual objectives, integrating business and financial plans, and modifying strategies to reflect real-time learning.

For a discussion of broad assessments and audits, see Chapter 2, "Companywide Assessment of Quality." Examples of specific quality measures are presented in Chapters 13, 15, 16, 17, and 20 on functional activities. See also *JQH5,* Section 13, on strategic deployment.

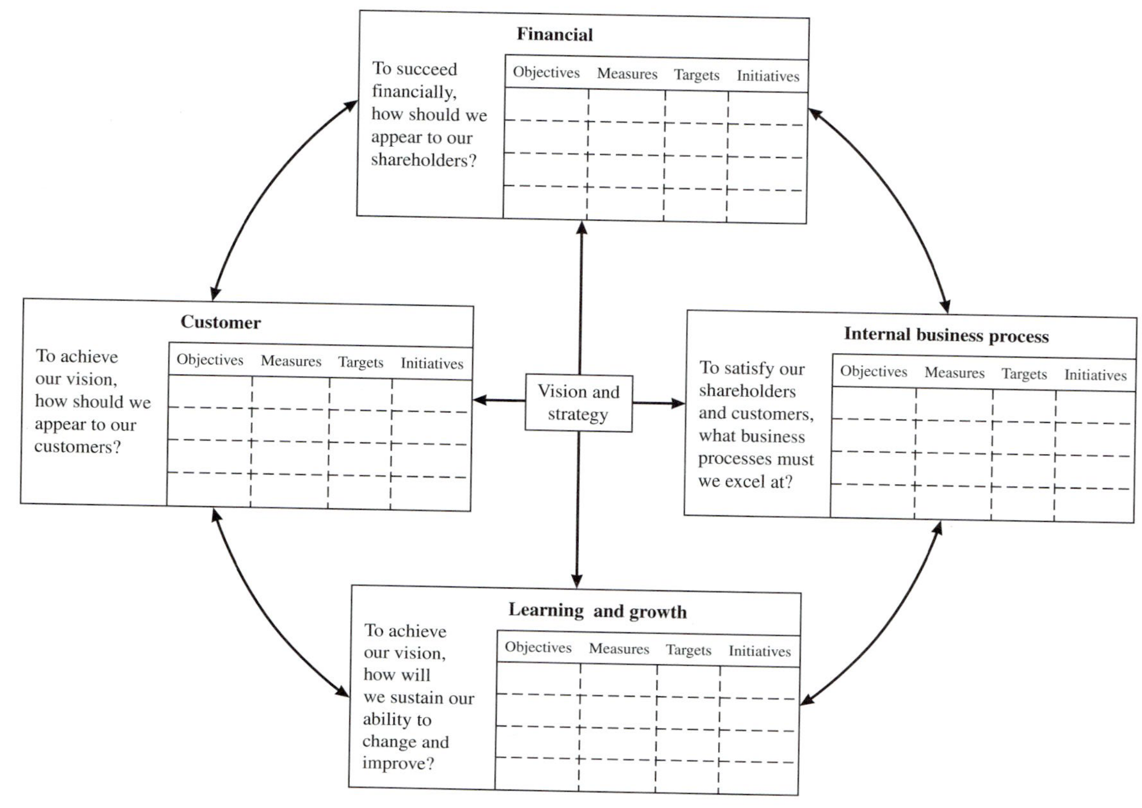

FIGURE 7.5
Translating vision and strategy: four perspectives. (*From Kaplan and Norton, 1996. Reprinted by permission of Harvard Business School Press.*)

TABLE 7.4
Learning disciplines and strategic management

Learning discipline	Relation to strategic management
Personal mastery—create an environment for learning	Encourage innovative thinking and strategies
Mental models—clarify situations to help shape actions	Develop strategies and strategic goals
Shared vision—build shared images of the future and how to get there	Participate in developing a vision and strategies
Team learning—develop intelligence and ability through teams	Use project and process teams
Systems thinking—understand the forces that shape complex activities	Study external and internal environments

7.9 THE LEARNING ORGANIZATION

"A learning organization is an organization skilled at creating, acquiring, and transferring knowledge, and at modifying its behavior to reflect new knowledge and insights" (Garvin, 1993).

The learning organization concept commands significant attention from organizations that are committed to continuous improvement. Examples are Xerox, GTE, Chaparral Steel, British Petroleum, General Electric, and AT&T.

The concept is based upon five "learning disciplines" that involve lifelong efforts of study and practice. These disciplines have linkages to strategic management (see Table 7.4).

The learning organization is an evolving concept. The basic reference is Senge (1990). Useful explanations including practical illustrations of the five disciplines are provided by Senge et al. (1994) and Garvin (1993).

7.10 IMPLEMENTING TOTAL QUALITY

Translating the elements of the management cycle for quality into reality can be addressed in five phases: decide, prepare, start, expand, and integrate.

Decide

In the *decide* phase, we face the question, Do we need a different approach to quality? Often, driving forces such as competitive pressures, customer dissatisfaction, and excessive costs of poor quality lead to a conclusion that the current quality system needs to be changed. This phase also examines some alternative approaches: statistical process control, six sigma, benchmarking, learning organization—the list goes on and on.

Prepare

In the *prepare* phase, training is given to upper managers and selected middle managers who then apply the training to develop initial goals, plans, and some assignments. The elements of this phase are shown in Figure 7.6.

Start

The *start* phase includes more training, pilot quality projects, and revision and expansion of various management systems to implement and sustain the new approach to quality (see Figures 7.7, 7.8, 7.9). Of cardinal importance are the pilot projects. Their success broadcasts a message to many in the organization who are watching with a wait-and-see attitude; any failures in pilot projects provide lessons for the next phase.

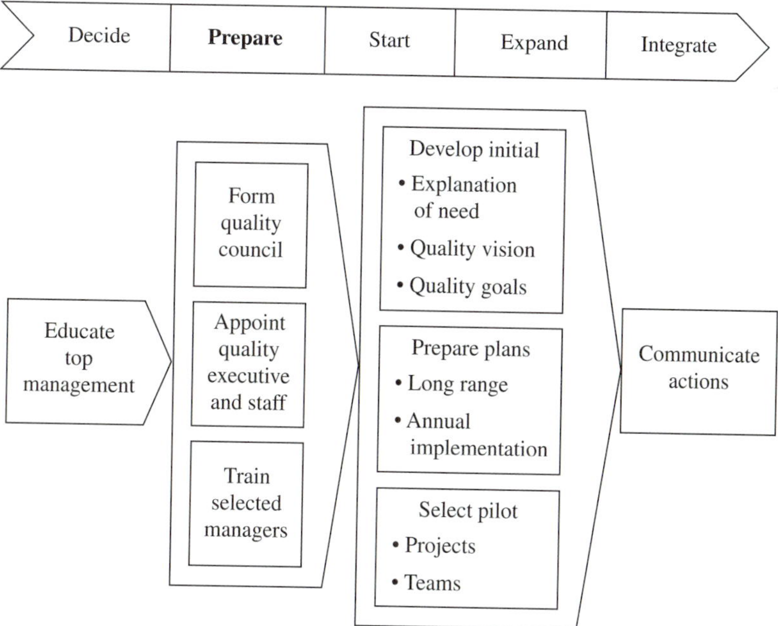

FIGURE 7.6
Prepare phase. (*From Juran Institute, Inc., 1991.*)

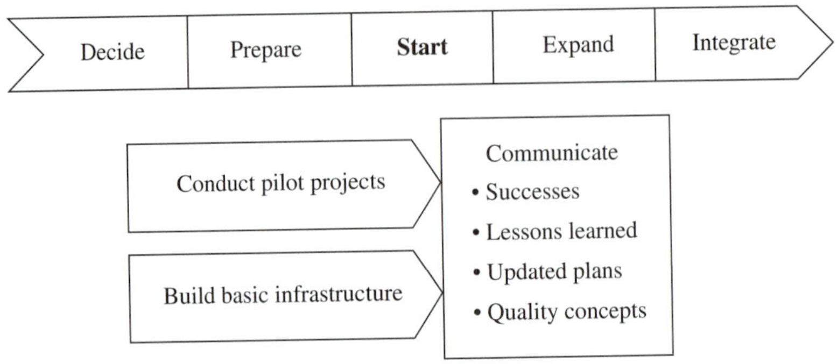

FIGURE 7.7
Start phase—overall. *(From Juran Institute, Inc., 1991.)*

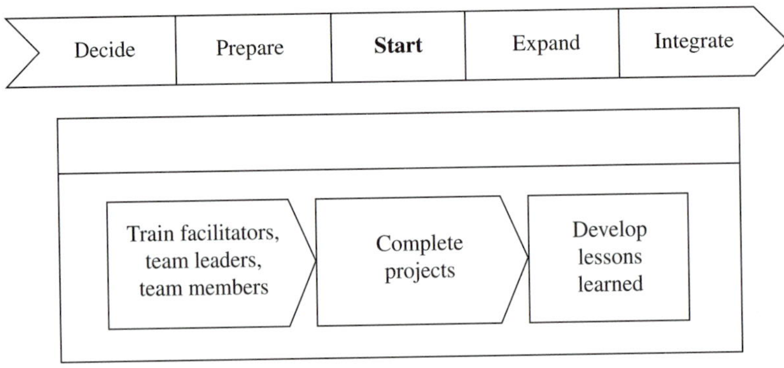

FIGURE 7.8
Start phase—pilot projects. (*From Juran Institute, Inc., 1991.*)

Expand

In the *expand* phase, the new approach is unfolded to other organizational units by setting up teams, measurement systems, individual quality initiatives, and additional training.

Integrate

Integrate is the ultimate phase, when quality becomes a way of life (Figure 7.10). Strategic goals are finalized and deployed to various levels; people are trained to participate on teams and to carry out individual quality roles; key business processes are identified and analyzed; assessments, reviews, and audits are in place. In summary, quality is no longer a "program"; it is now a part of business planning.

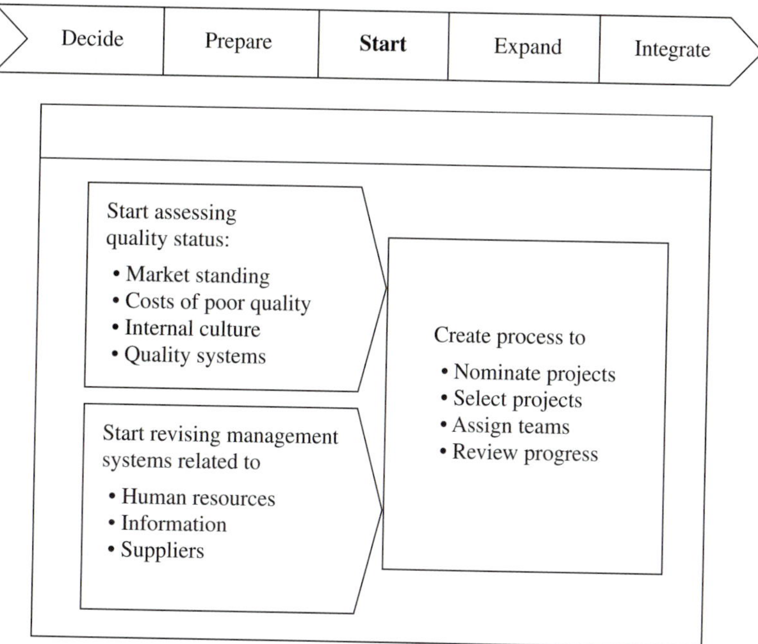

FIGURE 7.9
Developing the quality system infrastructure. (*From Juran Institute, Inc., 1991.*)

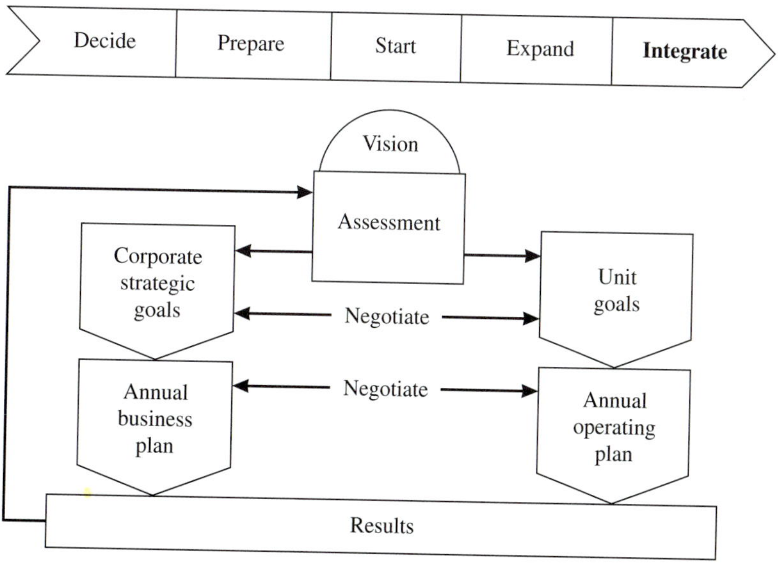

FIGURE 7.10
Integrate phase. (*From Juran Institute, Inc., 1991.*)

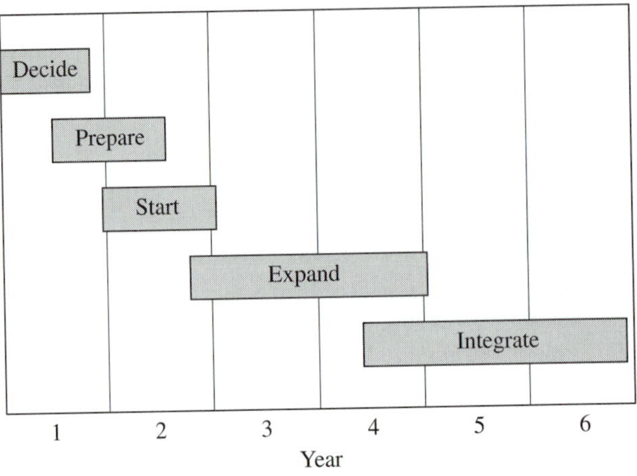

FIGURE 7.11
Timing of phases.

How long will all of this take? Tangible, measurable results will start to appear after about a year, but the heaven of quality truly being a way of life requires about six years. In most organizations the phases follow a fairly predictable calendar (Figure 7.11).

Endres (2000) provides a thorough description, with examples, of these five phases.

7.11 OBSTACLES TO ACHIEVING QUALITY GOALS

The reasons for failure are many, but seven stand out as important:

1. *Lack of leadership by upper management.* Without this leadership, quality success is likely to be doomed. Some in upper management may be "committed," but lack of *evidence* of this commitment has a damaging result on the rest of the organization. Chapter 8, "Organization for Quality," discusses the leadership role of upper management.
2. *Lack of an infrastructure for quality.* With other major activities, management has successfully delegated responsibility but only after evolving mechanisms that include clear goals, plans, organizational mechanisms for carrying out the plans, budgets, and provision for recognition and rewards. In contrast, these elements are usually vague or are missing with respect to quality.
3. *Failure to understand the skepticism about the "new quality program."* Many people have seen previous programs on quality quietly sink into oblivion. Unfortunately, the skepticism is not vocalized. The result is that management fails to present a convincing case of (a) "proof of the need" for the quality effort and (b) their determination to make the effort a success.
4. *An assumption by management that the exhortation approach will work.* This approach involves a strong presentation on proof of the need (i.e., convincing

everyone about the seriousness of the quality problem) coupled with relying principally on motivational techniques to inspire everyone to do better.

5. *Failure to "start small" and learn from pilot activities.* Sometimes, in a haste to achieve sizable results rapidly, the small pilot phase is omitted; instead, massive training takes place with the expectation that the troops can then simultaneously advance on all fronts. This approach doesn't work. A much better alternative uses a small number of pilot projects, with the scope of each project carefully defined so that completion is likely within six months. Perhaps the most common error in quality projects is failure to limit their scope to a digestible bite. People quickly grow tired of projects that seem to take forever.
6. *Reliance on specific techniques as the primary means of achieving quality goals.* Examples of such techniques are statistical process control, quality cost, quality circles, quality function deployment, etc. These techniques are valuable and often necessary, but they address only specific parts of the problem.
7. *Underestimating the time and resources required.* About 10 percent of the time of upper and middle management and professional specialists is required to achieve breakthroughs in quality. Typically, this time must be found without adding personnel. Thus priorities must be changed; i.e., other activities must be delayed or eliminated.

Organizations embarking on a significant new effort on quality are well advised to seek out the advice of other organizations and particularly to learn about their reasons for failure.

SUMMARY

- Strategic quality management is the process of establishing long-range customer-focused goals and defining the approach to meeting those goals.
- Upper management develops and implements strategic quality management.
- Approaches to strategic quality management have some common elements: a focus on customer needs, upper-management leadership, integrating quality strategies into business plans, and implementation by line departments.
- A policy is a broad guide to action.
- A goal is a statement of the desired result to be achieved within a specified time.
- Competitive benchmarking identifies key quality characteristics and uses other leading organizations as reference points to develop goals and strategies.
- We "deploy" quality goals by dividing them into specific deeds, allocating responsibility, and providing resources.
- Implementing a quality strategy involves five phases: decide, prepare, start, expand, and integrate.

PROBLEMS

7.1. Select one improvement program that you have observed in an organization. The program of improvement may focus on quality, safety, absenteeism, costs, or other matters. With respect to implementing the program, what were its strengths? What were the weaknesses in implementing? What recommendations would you make to implement an improvement program in the future?

7.2. Propose a list of five quality policies for a specific organization, e.g., your company, a supermarket.

7.3. Propose a list of five quality goals for a specific organization, e.g., your company, a supermarket.

7.4. Many organizations have a general statement called a "quality policy." This statement is often vague; e.g., the organization will supply the customer with "high quality." Obtain an example of such a statement and explain how it could be made more specific.

7.5. Refer to the list of 13 quality policies under "Quality Policies" in Section 7.7. Comment on the extent to which each policy applies to a specific organization.

7.6. A bank requires its branches to develop an annual quality plan that deploys corporate strategy into quality goals for five subject areas. One area is human resource utilization. Propose four goals to support this area.

7.7. Four months ago an organization decided that it needed a new approach to quality. One of the steps taken was the establishment of quality goals. As part of a "50 percent in 5" focus, each department now has a goal of reducing the cost of poor quality by 50 percent in 5 years. Comment on this approach.

7.8. For an organization with which you are familiar, define one strategic quality goal. For that goal, propose one specific quality project (improvement, planning, or control). For that project prepare a brief mission statement and a list of the functions that should be included on the project team.

REFERENCES

Allen, R. L. and C. V. Bailes (1988). "Managing the Startup of a Corporate Quality Improvement Effort—Translating Corporate Strategies into Field Operations," *Impro Conference Proceedings,* Juran Institute, Inc., Wilton, CT, pp. 6A-13 to 6A-18.

Camp, R. C. (1994). *Business Process Benchmarking: Finding and Implementing Best Practices,* Quality Press, ASQ, Milwaukee; Quality Resources, White Plains, NY.

Camp, R. C., ed. (1998). *Global Cases in Benchmarking: Best Practices from Organizations around the World,* ASQ Quality Press, Milwaukee.

Endres, A. (2000). *Implementing Juran's Roadmap for Quality Leadership,* John Wiley and Sons, New York.

Garvin, D. A. (1993). "Building a Learning Organization," *Harvard Business Review,* July–August, pp. 78–91.

Juran Institute, Inc. (1992). *Strategic Quality Planning,* Wilton, CT.

Kaplan, R. S. and D. P. Norton (1996). *The Balanced Scorecard,* Harvard Business School Press, Boston.

Leo, R. J. (1994). "A Corporate Business Excellence Process," *Impro Conference Proceedings,* Juran Institute, Inc., Wilton, CT, pp. 6A.1-1 to 6A.1-13.

Luther, D. B. (1993). "Advanced TQM: Measurements, Missteps, and Progress through Key Result Indicators at Corning," *National Productivity Review*, Winter, pp. 23–36.

McCain, C. (1995). "Successfully Solving the Quality Puzzle in a Service Company," *Impro Conference Proceedings,* Juran Institute, Inc., Wilton, CT, pp. 6A.1-1 to 6A.1-13.

Perry, A. C. (1989). "From Teams to Total Quality," *Impro Conference Proceedings,* Juran Institute, Inc., Wilton, CT, pp. 3A-7 to 3A-16.

Porter, M. E. (1996). "What Is Strategy?" *Harvard Business Review,* November–December, pp. 61–78.

Senge, P. M., A. Kleiner, C. Roberts, R. B. Ross, and B. J. Smith (1994). *The Fifth Discipline Fieldbook,* Currency Doubleday, New York.

Senge, P. M. (1990). *The Fifth Discipline,* Currency Doubleday, New York.

Slater, R. (1999). *Jack Welch and the GE Way,* McGraw-Hill, New York.

Thompson, A. A. Jr. and A. J. Strickland III (1998). *Strategic Management,* 10th ed., McGraw-Hill, New York.

Utzig, L. (1980). "Quality Reputation—A Precious Asset," *ASQ Technical Conference Transactions,* Milwaukee, pp. 145–154.

SUPPLEMENTARY READING

Benchmarking: AT&T (1992). *Benchmarking: Focus on World-Class Practices,* AT&T Benchmarking Team, Indianapolis, IN.

Burpo, T. R. (1998). "Benchmarking Start-Up: Pitfalls and Successes," *Annual Quality Congress Proceedings,* ASQ, Milwaukee, pp. 277–286.

Deployment: AT&T (1992). *Policy Deployment,* AT&T Bell Laboratories Quest Partnership, Indianapolis, IN.

Learning organization: Watson, G. H. (1998). "Bring Quality to the Masses: The Miracle of Loaves and Fishes," *Quality Progress,* June, pp. 29–32.

Davenport, T. and L. Prusak (1998). *Working Knowledge,* Harvard Business School Press, Boston.

Strategy: Berry, R. W. (1996). *An Investigation of the Relationship between World-Class Quality System Components and Performance,* Ph.D. dissertation, UMI Dissertation Services, Ann Arbor, MI.

WEBSITES

Benchmarking:

Benchmarking Exchange: www.benchnet.com

American Productivity and Quality Center: www.apqc.org

Learning organization: www.lguide.com

8

ORGANIZATION FOR QUALITY

8.1
EVOLUTION OF ORGANIZATION FOR QUALITY

Many organizations, particularly in manufacturing industries, have focused quality-related activities in a "quality department." Over the decades, the names and the scope of activities have changed—inspection, statistical quality control, quality control, reliability engineering, total quality control, quality assurance, total quality management (TQM).

During the 1980s and 1990s in the United States, five major trends emerged in organizing for quality:

1. Quality management tasks were assigned (or transferred) to functional line departments rather than to quality departments. For example, process capability studies were transferred from a quality department to a process engineering department.
2. The scope of quality management was broadened from operations only (little Q) to all activities (big Q) and from external customers to external and internal customers. Most organizations now train personnel in the functional departments in the tools of quality management and make employees responsible for implementing modern concepts of quality.
3. A major expansion occurred in the use of quality teams.
4. Authority to make decisions was delegated to lower levels.
5. Many companies are including key suppliers and customers in quality activities—improvement, planning, and control. The term employed is *partnering.* Sometimes partnering results in a "virtual corporation," i.e., a temporary network of functions from different companies that come together for a specific business purpose; the network then disbands when the purpose has been met. For elaboration, see Christie and Levary (1998).

These trends led to new approaches for coordinating and organizing quality activities. Becker et al. (1994) conducted research at 30 companies to learn the impact of TQM on the organization. The findings: TQM led to (1) flatter organizations, (2) an increase in liaison devices such as cross-functional teams, (3) a decrease in size, and an increase in output, (4) a shift to group reward systems, (5) participation in planning by lower-level employees, and (6) changing organizational boundaries.

8.2 COORDINATION OF QUALITY ACTIVITIES

The approach used to coordinate quality activities throughout an organization takes two major forms:

1. Coordination for *control* is achieved by the regular line and staff departments, primarily through employment of formal procedures and use of the feedback loop. Feedback loops take such forms as audit of execution versus plans, sampling to evaluate process and product quality, control charts, and reports on quality.
2. Coordination for *creating change* is achieved primarily through the use of quality project teams and other organizational forms for creating change.

Coordination for control is often the focus of a quality department; sometimes such a focus is so preoccupying that the quality department is unable to make major strides in coordination for change. As a result, some "parallel organizations" for creating change have evolved.

Parallel Organizations for Creating Change

All organizations are engaged in creating beneficial change as well as preventing adverse change ("control"). Much of the work of creating change consists of processing small, similar changes. An example is the continuing introduction of new products consisting of new colors, sizes, shapes, and so forth. Coordination for this level of change can often be handled by carefully prepared procedures.

Nonroutine, unusual programs of change usually require new organizational forms. These new forms are called "parallel organizations." *Parallel* means that these organizational forms exist in addition to and simultaneously with the regular "line" organization.

Examples of parallel organizations for achieving change in quality management are process teams, (see Chapter 6), quality councils, and various types of quality project teams (see below). Parallel organizations may be permanent or ad hoc (a process team versus a quality improvement team) and may be mandatory or voluntary (a quality council versus a workforce team). An early form of a parallel organization for product reliability is described by Gryna (1960).

A dramatically different type of parallel organization is the process management organization discussed in Chapter 6. Activities in any organization can be viewed as occurring at three levels: the organization level, the process level, and the individual job level. At the organization level, we are concerned with a broad view of the entire organization, a business unit, or a plant; at the process level, we address cross-functional work activities that produce a key output; at the job level, individuals perform specific activities that contribute to the output of processes. Rummler and Brache (1995) discuss these three levels in terms of goals, design, and management. The many activities at the three levels are interdependent and result in a limitless number of interactions to form a "system." Baker (1999) explains the importance of viewing an organization as a system of interdependent parts.

The full process management concept involves organizing by key processes; having permanent process owners and process teams; and making them responsible for managing all aspects of processes—planning, control, and improvement. The application of process management is still evolving, and only a few organizations have embraced the full concept.

Changing from a functional organization to a process organization is a major step that requires at least three years. Garvin (1995) presents the experiences of the CEOs of Xerox, USAA, SmithKline Beecham, and Pepsi in applying process management. Majchrozak and Wang (1996) describe the experiences of electronics manufacturers in changing from functional to process organizations. In both papers the message is clear: Management must be prepared to consistently emphasize a collaborative rather than a hierarchical culture.

Next we examine the roles of various constituencies in an overall quality effort. The key role is that of upper management.

8.3 ROLE OF UPPER MANAGEMENT

A company president said this: "My people have disappointed me. I clearly told them that quality is our first priority; I provided training; now two years later there is little evidence of improvement." He missed the point of active leadership. Of all the ingredients for successfully achieving quality superiority, one stands out: active leadership by upper management. Commitment to quality is assumed but is not enough.

Certain roles can be identified:

- Establish and serve on a quality council (see below).
- Establish quality strategies (see Chapter 7).
- Establish, align, and deploy quality goals (see Chapter 7).
- Provide the resources.
- Provide training in quality methodology (see Chapters 3, 4, and 5).
- Serve on upper-management quality improvement teams that address chronic problems of an upper-management nature (see Chapter 9).
- Review progress and stimulate improvement (see Chapter 9).
- Provide for reward and recognition (see Chapter 9).

In brief, upper management develops the strategies for quality and assures their implementation through personal leadership.

One effective example is the action taken by the head of a manufacturing division. He personally chairs an annual meeting at which improvement projects are proposed and discussed. By the end of the meeting, a list of approved projects for the coming year is finalized, and responsibility and resources assigned for each project.

Unfortunately, there is a price to be paid for this active leadership. The price is time. Upper managers will need to spend at least 10 percent of their time on quality activities—with other managers, with front-line employees, with suppliers, and with customers.

Providing Resources for Quality Activities

The modern approach to quality requires an investment of time and resources throughout the entire organization—for many people, the price is about 10 percent of their time. In the long run this investment yields time savings that then become available for quality activities or other activities; in the short run the investment of resources can be a problem.

Upper management has the key role in providing resources for quality activities. One alternative is to add resources, but in highly competitive times this approach may not be feasible. Often, time and resources can be found only by changing the priorities of both line and staff work units. Thus some work must be eliminated or delayed to make personnel available for the quality activities.

People who are assigned to quality teams should be aware of the amount of time that will be required while they are on a team. If time is a problem, they should be encouraged to propose changes in their other priorities (before the team activity starts). Resources for project teams will be made available only if the pilot teams demonstrate benefits by achieving tangible results—thus the importance of nurturing those pilot teams to yield results. As a track record builds for successful teams, the resource issue becomes less of a problem. One outcome can be for the annual budgeting process of an organization to include a list of proposed projects and necessary resources on the agenda for discussion.

In a dramatic action to provide resources for a comprehensive quality effort, upper management at a modem manufacturer assigned 3 percent of the total workforce (of 3000 people) to work full time on quality. The team included six upper-level managers who were responsible for designing and overseeing the effort. These managers returned to their regular jobs after one year.

Quality Council

A *quality council* (sometimes called a "leadership team") is a group of upper managers who develop the quality strategy and guide and support the implementation. Councils may be established at several levels—corporate, division, business unit. When multiple councils are established, they are usually linked together; i.e., members of high-level councils serve as chairpersons of lower-level councils. For any level, membership consists of upper managers—both line and staff. The chairperson is the manager having overall responsibility and authority for that level, e.g., the president for the corporate council and the division and site managers for their levels. One member of the council is the quality director whose role in the company is discussed below.

Each council should prepare a charter that includes responsibilities such as

- Formulate quality strategies and policies.
- Estimate major dimensions of the quality issue.
- Establish an infrastructure for selecting quality projects and assigning project team leaders and members.
- Provide resources including support for teams.
- Plan for training for all levels.
- Establish strategic measures of progress.
- Review progress and remove any obstacles to improvement.
- Provide for public recognition of teams.
- Revise the reward system to reflect progress in quality improvement.

At Kelly Services, Inc., (McCain, 1995), the quality council identified its key responsibilities as

- Setting and deploying vision, mission, shared values, quality policy, and quality goals.
- Reviewing progress against goals.
- Integrating quality goals into business plans and performance management plans.

Upper managers often ask: Isn't the quality council identical in membership to the regular top-management team? Usually, yes. If so, instead of having a separate council, why not add quality issues to the agenda of the periodic upper-management meetings? Eventually (when quality has become "a way of life") the two can be combined, *but not at the start.* The seriousness and complexity of the quality issues require a focus that is best achieved by meetings that address quality alone.

As a council works on its various activities, it often designates one or more full-time staff people to assist the council by preparing draft recommendations for council review. In another approach several council members serve on ad hoc task forces to investigate various issues for the council.

In addition, individual council members often serve as advocates ("champions") for key quality-related projects. In this role they continually monitor the project, use their executive position to remove obstacles for completion of the project, and recognize project teams for their efforts.

Providing evidence of upper-management leadership is clearly important in establishing a positive quality culture. For elaboration, see Section 9.6.

Finally, note that leadership is the first criterion of the Baldrige criteria. The leadership criterion covers organizational leadership and public responsibility and citizenship. Examination of the complete criteria reveals that leadership should be linked to other Baldrige criteria such as strategic planning, customer and market focus, information and analysis, and human resources.

Miller (1984) in the Supplementary Reading is a classic book on the new roles facing management in organizations.

8.4
ROLE OF THE QUALITY DIRECTOR

The quality director of the future is likely to have two primary roles—administering the quality department and assisting upper management with strategic quality management (Gryna, 1993).

The Quality Department of the Future

What will the future role of a quality department be? The major activities are indicated in Table 8.1. The table indicates some traditional activities of the quality department and some important departures from the current norm.

Note, for example, "transferring activities to line departments." Recent decades have shown that the best way by far to obtain *implementation* of quality methodologies is through line organizations rather than through a staff quality department. (Isn't it a shame that it took us so long to understand this point?) A few organizations have stressed this approach for several years, and some have been eminently successful in transferring many quality activities to the line organization.

To achieve success in such a transfer, the line departments must clearly and fully understand the activities for which they are held responsible. In addition, the line departments must be trained to execute these newly acquired responsibilities.

Examples of such transfer include shifting the sentinel (sorting) type inspection activity from a quality department to the workforce itself, transferring reliability engineering work from a quality department to the design engineering department, and transferring supplier quality activities from a quality department to the purchasing department.

A key role for the quality department is to help its internal customers achieve their quality objectives. In most organizations the key internal customer is the operations

TABLE 8.1
Functions of the quality department of the future

Companywide quality planning
Setting up quality measurement at all levels
Auditing outgoing quality
Auditing quality practices
Coordinating and assisting with quality projects
Participating in supplier partnerships
Training for quality
Consulting for quality
Developing new quality methodologies
Transferring activities to line departments

department (called "operations" in the service sector and "manufacturing" or "production" in the manufacturing sector). Clearly, quality and operations must be partners, not adversaries. As a quality director, the author each year had to obtain 50 percent of his departmental budget from the operations departments. This experience vividly causes a manager to become customer oriented. If a quality director truly wants to assist internal customers, a constructive step is to conduct internal "market research" on these customers. Bourquin (1995) describes how a unit of AT&T conducted a survey of internal customers (mainly manufacturing but also marketing, design, and other departments). The survey contained 10 questions reflecting both the relative importance of certain services and the degree of satisfaction with the services. The survey required little time or money but provided valuable quantitative feedback from the main internal customers.

The author believes that there will always be a need for a quality department to provide an independent evaluation of product quality and to furnish services to internal customers (see Table 8.1). Changing business conditions, however, make it essential that the role of the department be periodically reviewed and appropriate changes made. These changing business conditions include company mergers, changes in customer (external and internal) expectations, outsourcing, the information explosion, new communication technologies, the impact of global business and cultures, and the maturity of the quality effort within the organization. In addition, the emphasis on continuous improvement within organizations has resulted in internal competitors to a quality department for coordinating improvement activities. Such competitors include industrial engineering, human resources, financial audit, and information technology.

Assisting Upper Management with Strategic Quality Management (SQM)

A wonderful opportunity exists for a quality director to assist upper management to plan and execute the many activities of SQM. Some of these activities are shown in Table 8.2. (For elaboration, see Chapter 7.) Can a quality director achieve such an exalted role? It is useful to cite an analogy to the area of finance (see Table 8.3). Many organizations today have a chief financial officer (CFO). This officer is concerned with broad financial planning aspects, addressing questions such as, Where should our company be going with respect to finance? Other financial managers direct and manage detailed financial processes, such as accounts payable, accounts receivable, cash management, acquisitions, and budgeting. These roles are also vital in the organization but different from the broader role of the CFO. Finally, of course, line managers throughout the entire organization have specific activities that they must do to help meet the financial objectives of the company. Note that the quality function comprises both technical quality activities and strategic quality management activities. Provision must be made to staff a quality department with skills in these two areas. Larger organizations may benefit from having both a quality engineering function and a quality management function (just as accounting and financial planning are separate functions). These functions would then report to the quality director.

The quality director of the future could act as the right hand of upper management on quality in the same way that the chief financial officer acts as the right hand of upper management on finance. (Note that other essential quality activities such as

TABLE 8.2
Assisting upper management with strategic quality management

Assessing quality
Formulating goals and policies
Developing quality strategies to increase sales revenue and reduce internal costs
Delegating organizational responsibilities for quality
Carrying out reward and recognition
Reviewing progress
Determining personal roles for upper management
Acting as facilitator to the quality council
Integrating quality during strategic business planning cycle

TABLE 8.3
A contrast of roles

Finance (today)	Quality (the future)
Chief financial officer	Quality director
Other financial managers	Other quality managers
Line managers	Line managers

inspection, audit, and quality measurement must be directed and managed by other quality managers.)

Some of the scars of experience that have been accumulated within financial circles apply also to the quality function. Years ago, there was no such role as chief financial officer. The detailed financial processes were handled by one or several managers. As time went on, the need for a person with a broader financial viewpoint became apparent. In some companies the person who was then the "controller" was promoted to become the chief financial officer. Other companies felt that the controller did not have the breadth of vision and scope to assist upper management on broad financial planning, even though the controller was excellent in administering some of the detailed financial processes. Thus not every controller became a chief financial officer.

A similar issue arises in regard to the broad scope necessary for the quality director of the future. Is the present quality director prepared, or is that person willing to become prepared, for the broader business role with respect to quality? Quality directors who wish to assume this broader role in the future need to learn from the lessons of the financial controllers.

Table 8.4 shows some ingredients for success as a quality director of the type indicated here.

Clearly, this list goes far beyond the current scope of quality directors in many companies. In larger companies quality managers, including those who will manage some of the technical quality activities, will be needed at various levels. A quality director is likely to be involved in administering technical activities but will increasingly be called

TABLE 8.4
Ingredients for success as a quality director

Focus on customer orientation and customer advocacy
Ability to build collaborative relationships across functions
Strong oral and written communication skills to encourage sharing of information
Goal orientation
Ability to analyze complex issues and generate innovative solutions
Initiative, persistence, and self-confidence in gaining acceptance of new ideas
Ability to organize activities
Ability to provide for self-development of subordinates

Source: Adapted from Watson (1998). Used with permission of Business Systems Solutions, Inc.

upon to assist upper management as upper management—not the quality director—*leads* the company on quality. For elaboration on the future role of the quality department, see Gryna (1993).

Cinelli and Schein (1994) present the results of a survey on the profile of the U.S. senior quality executive. The survey examines the professional background, day-to-day activities, and responsibilities of the quality executive at 223 *Fortune* 500 firms. For the results of a survey of 38 U.K. companies on organizing for quality, see Groocock (1994).

8.5 ROLE OF MIDDLE MANAGEMENT

Middle managers, supervisors, professional specialists, and the workforce are the people who execute the quality strategy developed by upper management.

The roles of middle managers, supervisors, and specialists include

- Nominating quality problems for solutions.
- Serving as leaders of various types of quality teams.
- Serving as members of quality teams.
- Serving on task forces to assist the quality council in developing elements of the quality strategy.
- Leading the quality activities within their own area by demonstrating a personal commitment and encouraging their employees.
- Identifying customers and suppliers and meeting with them to discover and address their needs.

Increasingly, middle managers are asked to serve as team leaders as a continuing part of their job. For many of them, their role as a team leader requires special managerial skills. For a manager of a department directing the people in that department, a traditional hierarchical approach is common. The leader of a cross-functional quality improvement team faces several challenges; e.g., the leader usually has no hierarchical authority over anyone on the team because members come from various departments, serve part time, and have priorities back in their home department. The success of a team leader depends on technical competence, ability to get people to work together as a team, and a personal sense of responsibility for arriving at a solution to the assigned problem. Leading a team calls for quite an array of talents and willingness to assume responsibility. For some middle managers, the required change in managerial style is too much of a burden; for others, the role presents an opportunity. Dietch et al. (1989) identified and studied 15 characteristics of team leaders at the Southern California Edison Company. The research concluded that team leaders (as compared to team members) have a higher tolerance for handling setbacks, believe that they have great influence over what happens to them, exhibit a greater tolerance for ambiguity, are more flexible, and are more curious (see Section 9.6, "Provide Evidence of Management Leadership").

8.6
ROLE OF THE WORKFORCE

By workforce, we mean all employees except those in management and the professional specialists.

Recall that most quality problems are management or system controllable. Therefore management must (1) direct the steps necessary to identify and remove the causes of quality problems (see Chapter 3) and (2) provide the system that places workers in a state of self-control (see Section 5.3, "Self-Control"). Both inputs from and cooperation by the workforce are essential. The roles of the workforce include

- Nominating quality problems for solution.
- Serving as members of various types of quality teams.
- Identifying elements of their own jobs that do not meet the three criteria of self-control.
- Becoming knowledgeable as to the needs of their customers (internal and external).

At last we are starting to tap the potential of the workforce by using its experience, training, and knowledge. One plant manager says this: "No one knows a workplace and a radius of 20 feet around it better than the worker." Quality goals cannot be achieved unless we use the hands and the heads of the workforce. Period. Some of the workforce roles on teams are discussed below.

8.7
ROLE OF TEAMS—GENERAL

The "organization of the future" will be influenced by the interaction of two systems that are present in all organizations: the technical system (equipment, procedures, etc.) and the social system (people, roles, etc.); thus the name "sociotechnical systems" (STS).

Much of the research on sociotechnical systems has concentrated on designing new ways of organizing work, particularly at the workforce level. Team concepts play an important role in these new approaches. Some organizations now report that, within a given year, 40 percent of their people participate on a team; some organizations have a goal of 80 percent. A summary of the most common types of quality teams is given in Table 8.5. Quality project teams are discussed in Section 3.7, "Define Phase"; business management quality teams are described in Chapter 6; quality project teams, workforce teams, and self-managing teams are discussed below.

Aubrey and D. Gryna (1991) describe the experiences of more than 1000 quality teams during four years at 75 banking affiliates of Banc One. This effort yielded significant results: $18 million in cost savings and revenue enhancement; a 10 to 15 percent improvement in customer satisfaction rating; and a 5 to 10 percent reduction in costs, defects, and lost customers. On some teams, membership was assigned; on other teams, membership was voluntary. A summary of some of the results of an organizational nature is given in Table 8.6. The focus is to improve customer satisfaction, reduce cost and increase revenue, and improve communication between front-line employees and management.

Next we examine three types of teams: quality project teams, workforce teams, and self-managing teams.

TABLE 8.5
Summary of types of quality teams

	Quality project team	Workforce team	Business process quality team	Self-managing team
Purpose	Solve cross-functional quality problems	Solve problems within a department	Plan, control, and improve the quality of a key cross-functional process	Plan, execute, and control work to achieve a defined output
Membership	Combination of managers, professionals, and workforce from multiple departments	Primarily workforce from one department	Primarily managers and professionals from multiple departments	Primarily workforce from one work area
Basis and size of membership	Mandatory; 4–8 members	Voluntary; 6–12 members	Mandatory; 4–6 members	Mandatory; all members in the work area (6–18)
Continuity	Team disbands after project is completed	Team remains intact, project after project	Permanent	Permanent
Other names	Quality improvement team	Employee involvement group	Business process management team; process team	Self-supervising team; semiautonomous team

TABLE 8.6
Observations on organization of quality teams at a bank

Feature	Results of research
Team size	Average of 7 members, with a range of 2–11
Project selection	75% by management, 15% by quality council, 10% by individual team
Average savings related to project selection	Projects selected by management or the quality council achieved savings about twice as high as projects selected by the team
Project duration	Average of 3 months; 24 worker-hours per team member (excluding time spent outside team meetings)
Factors to maximize team success	Ideal team size of 4–5 employees; 75% officer/staff level, 25% nonexempt employees; members elected by management; project selected by management or quality council; project duration of 3–4 months with weekly 90-minute team meetings

8.8 QUALITY PROJECT TEAMS

A quality project team (often called a cross-functional team) usually consists of about six to eight persons who are drawn from multiple departments to address a selected quality problem. The job of the team is to bring the project to a successful conclusion as defined in the mission statement for the project (see Chapter 3 under "Problem and Mission Statement"). The project may be an improvement project (see Chapter 3), a planning project (see Chapter 4), or a control project (see Chapter 5).

The team meets periodically, and members serve part time in addition to performing their regular functional responsibilities. When the project is finished, the team disbands.

The project team consists of a sponsor (champion), a leader, a recorder, team members, and facilitators (other resources from disciplines such as accounting and information technology are invited to meetings when needed). This organization concept as applied in the six-sigma approach is shown in Figure 8.1. In 1996 the six-sigma approach at General Electric involved 3000 projects. Executive management provides the start by selecting projects, setting up project teams, providing resources, and encouraging the use of formal quality methodologies.

Project champion

The project champion (or sponsor) is someone from upper management who provides management, leadership, support, and mentoring to the team. The champion follows team progress and acts to remove obstacles to team progress.

Project team leader

The project team leader steers the team in its responsibility of carrying out the project. Successful leadership requires knowledge of the project area and skills in getting team members from several functional areas to work as a team. It is often useful for the team leader to come from the organizational unit most affected by the problem.

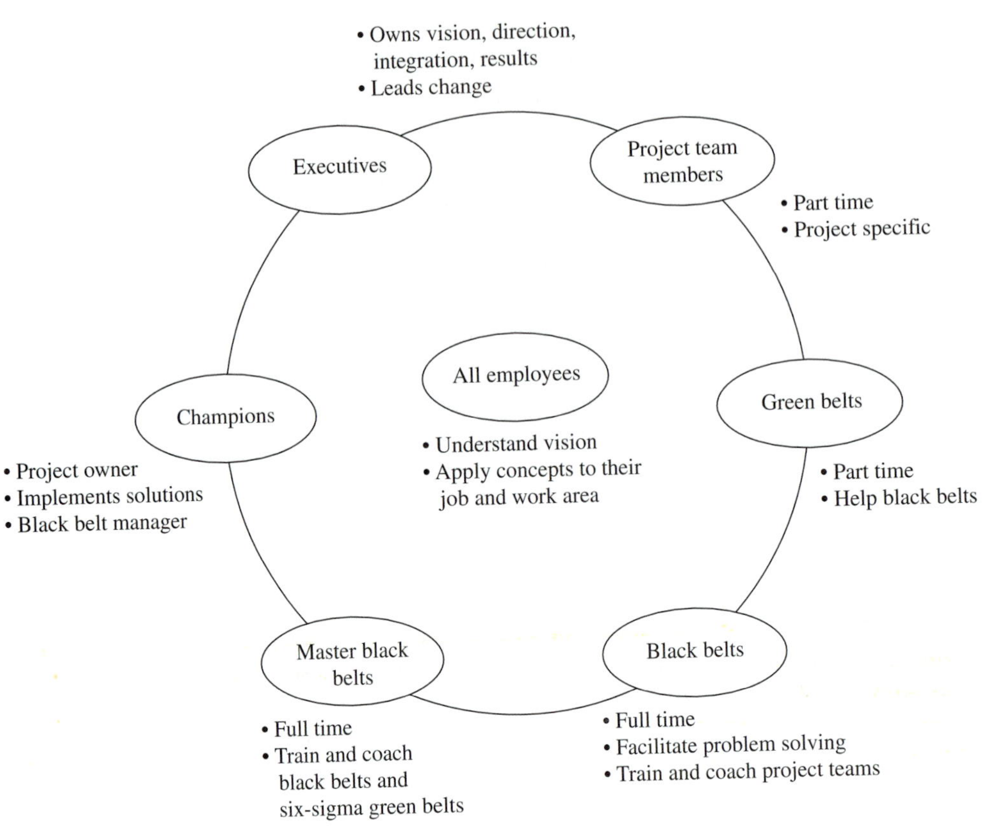

FIGURE 8.1
Six sigma requires *severe* coordination. (*©1996, 1997, 1998 by Six Sigma Academy, Inc. All rights reserved. Used with permission.*)

Project recorder

Each team requires a project recorder to handle documentation: agendas, minutes, reports, etc. The recorder should be a member of the project team.

Project team members

Team membership draws upon all of the skills and knowledge necessary for the project. For chronic problems the teams are usually cross functional and consist of middle management, professional, and workforce personnel. Surprisingly, some projects are relatively easy and can be handled with a minimum of skills and knowledge. (Such projects are often the result of a previous lack of a project approach.) Other projects are complex and require more depth in team membership, perhaps even including consulting specialists from within the company. Because team members come from various functional departments, there must be a clear understanding among team members and their functional managers on the priority and time required for team activity (meetings and other team assignments may take one day per week). When feasible, suppliers and customers should serve on the project team.

Facilitators

Supplementing the formal team membership is a "facilitator." Many companies have adopted the concept of using a facilitator to help project teams carry out their first project. Although not a member of the team, the facilitator can play an important role. The role of the facilitator consists of any or all of the following:

- Explaining the company's approach to quality improvement and how it differs from prior efforts at quality improvement.
- Providing assistance in team building.
- Assisting in the training of project teams.
- Assisting the project team leader to solve human relations problems among team members.
- Helping the team avoid a poor choice of project.
- Reporting progress on projects to management.
- Revitalizing a stalled project.
- Providing technical support in quality methodologies.
- Practicing the virtue of patience.

Organizations with numerous quality project teams often have several levels of facilitators. Thus in the six-sigma approach (Figure 8.1), three levels are employed: "master black belts," "black belts," and "green belts." A master black belt is usually a full-time facilitator and trains black belts and green belts; a black belt is full time and trains and assists the project team in problem solving; a green belt is part time and helps the black belts.

For elaboration on the roles on a quality project team, see *JQH5,* Section 5, under "Project Team."

An example of a special quality project team is the "blitz team." A *blitz team* is a project team that operates on an accelerated problem-solving schedule (several weeks for a solution rather than several months). The fast pace is accomplished by having the team meet frequently—several times a week, often for full days. A full-time facilitator may be assigned to the team. Additional help is provided for data collection and analysis. Most of the diagnostic work (data analysis, flowcharting) is done outside of the meetings. Skiba (1996) describes the application at the Mayo health care system.

Another special quality project team is the *virtual team.* A virtual team is a project team whose members reside in different geographical locations (even in different countries). The team employs one or more communication technologies—the telephone, overnight mail, fax, shared database, e-mail, the Internet, computer hookups, shared computer screens, and videoconferencing. For elaboration on the special issues involved and approaches for making virtual teams effective, see Henry and Hartzler (1998).

8.9 WORKFORCE TEAMS

One organizational mechanism for worker participation in quality are workforce teams, originally called quality circles. (Workers can also participate on other types of teams; see Table 8.5.) A workforce team is a group of people, usually from within one department, who volunteer to meet weekly (on company time) to address quality problems that occur within their department. Team members select the problems and are given training in problem-solving techniques. Note that this team is different from a quality project team where projects are cross functional and are selected by upper management based on priorities for overall company operations.

Where the introduction of workforce teams is carefully planned and where the company environment is supportive, they are highly successful. The benefits fall into two categories: measurable savings and improvement in the attitudes and behavior of people.

Workforce teams pursue two types of problems: those concerned with the personal well-being of workers and those concerned with the well-being of the company. Workers' problems are those of their environment, e.g., reducing the amount of junk food in vending machines, eliminating a draft in a work area, designing special work tables for worker convenience. These frustration-type problems are important to the workforce and therefore are frequently selected as early problems to address. Many of these problems can be solved in a short time with little or no investment. But management must be prepared to pay some dues by permitting workers to spend time on such frustration problems. Later, as teams start to address company problems, the focus is on products and processes. The benefits are improved quality for both internal and external customers and monetary savings. Savings on these types of projects typically range from $5000 to $25,000 per year with a benefit-to-cost ratio of at least four to one.

Perhaps the most important benefit of workforce teams is their effect on people's attitudes and behavior. The enthusiastic reactions of workers, sometimes streaked with emotion, are based on their personal involvement in solving problems. Beneficial effects fall into three categories (Gryna, 1981):

1. Effects on individuals' characteristics.

   Teams enable the individual to improve personal capabilities. Many members spoke of the benefits gained from group participation and learning specific problem-solving tools. One worker felt he had developed a better relationship with his wife as a result of his participation in a team because the team had improved his ability to interact with others.

   Teams increase the individual's self-respect. At Woodward Governor a worker spoke highly of teams because "the little guy can get in on things."

   Teams help workers change certain personality characteristics. Almost every organization reported at least one case of an extremely shy person who had become more outgoing through participation. Teams help workers develop the potential to become the supervisors of the future.

2. Effects on individuals' relations with others.

   Teams increase the respect of the supervisor for the workers. "As a result of circles, I find that I talk more with workers on the line."

   Teams increase workers' understanding of the difficulties faced by supervisors. As a result of problem selection, solving, and implementation, team members become aware for the first time of the supervisor's many burdens and demands on her or his time.

Teams increase management's respect for workers. "Some of the presentations by teams have been better than those of my staff people."

3. Effects on workers and their attitudes toward the company.

   Teams change some workers' negative attitudes. At one company a worker stated, "I always had a chip on my shoulder around here because I didn't think the company cared about the worker. As a result of some team projects, I've got a lot better attitude."

   Teams reduce conflict stemming from the working environment. The removal of these frustrations not only eliminates sources of conflicts but also encourages workers to think that they can deal with other frustrations as well.

   Teams help workers to understand why many problems cannot be solved quickly. For instance, certain process changes at the Henry, Illinois, facility of the BF Goodrich Chemical Group required approval of the technical function. Workers at the plant have learned that this approval process requires some time because of many other process changes being considered.

   Teams instill in the worker a better understanding of the importance of product quality. At Keystone Consolidated Industries, a team was started in the wire mill. At an early meeting the team was given the results of a study that showed the dollars per year that were being lost due to poor quality. At the next meeting a worker voluntarily showed some calculations he had made that indicated that the dollar value of the scrap he was personally producing was more than his salary.

With all the potential benefits of workforce teams, their success rate has been mixed. Baker (1988) provides some perceptive recommendations for management to support and sustain such teams.

1. Recognizing and rewarding (not necessarily monetarily) workers' efforts even if recommendations are not adopted. Giving workers increased discretion and self-control to act on their own recommendations is an excellent reward.
2. Offering monetary rewards through the suggestion program (which may have to be modified to accommodate joint submission by team members).
3. Providing sufficient training to expand worker skills to take on more complex projects.
4. Establishing a system for workforce teams to expand into cross-functional teams when it appears to be a logical step. Teams may become "fatigued" when they feel they have accomplished about all they can by themselves and see the need to work with their internal suppliers and customers.
5. Training middle managers in team tools and techniques so they can ask their subordinates the "right questions" and not be "outsiders." These tools are also useful for the managers' own processes.
6. Addressing middle-management resistance when diagnosed. Typically, management is concerned about a loss of authority and control.
7. Measuring effectiveness by focusing on the quality of the process—e.g., the training, group discussion process, interpersonal relationships, supervisory leadership

style—rather than outcomes such as reduction of scrap and costs. If the process is right, the outcomes will be also and that will reinforce employee involvement, as well as management commitment.

Workforce teams can improve product quality and work environment quality within a department. Such teams, however, cannot be the main driver for quality improvement within an organization because the problems with the greatest impact are cross functional and upper management must address them by setting up quality project teams.

Every organization must provide for the participation of the workforce in solving quality problems. One approach makes use of workforce teams. A more revolutionary approach involves self-managing teams.

8.10 SELF-MANAGING TEAMS

A *self-managing team* is a group of people who work together continuously and who plan, execute, and control their work to achieve a defined output. That definition starkly contrasts with the traditional Taylor system of work design. Basic to the Taylor system is the division of an overall task into narrow, specialized subtasks that a supervisor assigns to individuals. That supervisor then coordinates and controls the execution and handles the general supervision of the workers.

In contrast, here are two examples of self-managing work teams:

- In an electronics plant, an assembly team handles all aspects of a customer order: It receives the order, prepares the components, assembles and solders circuit boards, tests the boards, ships the boards, monitors inventory levels, and processes all paperwork.
- At an insurance company (the Aid Association for Lutherans), work was originally organized into three areas—life insurance, health insurance, and support services such as billing. Under the new organizational design, teams of 20 to 30 employees perform all 167 tasks that formerly were split among the three functional sections. Now field agents deal solely with one team. The result is a shorter processing time for cases.

Fishman (1999) describes self-managing teams at an aircraft engine facility—how the team decides who does what work; how to balance training, vacations, and overtime against work flow; how to make the process more efficient; how to handle team members who slack off. Hahn (1995) describes how team activities grow from basic steps (e.g., manage work flow) to intermediate (e.g., develop team members) to advanced (e.g., evaluate performance).

The concept of self-managing teams has been applied to many activities, e.g., manufacturing, accounting, customer service, purchasing, and sales.

The contrasting features of the traditional organization of the workforce and self-managing teams are shown in Table 8.7. The difference is revolutionary. Workers are empowered to make certain decisions previously reserved for a supervisor.

TABLE 8.7
Comparison of organizational forms

Feature	Traditional organization	Self-managing team
Scope of work	Each individual is responsible for a narrow scope	Team is responsible for a broad scope
Job categories for personnel	Many narrow categories	A few broad categories
Organizing, scheduling, and assigning work	Primarily by supervisor or staff	Primarily by team
Measuring and taking corrective action	Primarily by supervisor or staff	Primarily by team
Training provided	Training for task assigned to individual	Extensive training for multiple tasks plus interpersonal skill training
Opportunity for job rotation	Minimum	High because of extensive training
Reward system	Related to job, individual performance, and seniority	Related to team performance and scope of skills acquired by individual
Handling of personnel issues	Primarily by supervisory personnel or staff	Many issues handled by team
Sharing of business information	Limited to nonconfidential information	Open sharing of all information

The advantages of these teams include improvements in productivity, quality, customer satisfaction, and cost, as well as commitment of personnel. For example, a reduction of 30 percent in total production costs and turnover and absence rates of only 1/2 percent a year are now reported. Further, one layer of management can often be removed, thereby providing resources for quality improvement and quality planning projects.

Lawler (1986) provides a good perspective in terms of benefits and problems (see an adaptation in Table 8.8).

Clearly, the implementation of such a fascinating but radical approach will be like walking through a mine field. Some key steps involved are

1. Upper management commitment to undertake the approach and to accept some unknown risks.
2. In-depth orientation and participation of upper management, middle management, specialists, workforce, and union officers.
3. Analysis of production work flow to define logical segments for teams.
4. Definition of skills required, levels of skills, and requirements for certification.
5. Formation of teams and training for teams and individuals.
6. Development of production goals for teams and provision for continuous feedback of information to teams. Such feedback must have the content and timeliness needed to control the process.

TABLE 8.8
Self-managing teams

Benefits	Problems
Improved work methods	Salary costs go up
Helpful in recruiting	Training costs go up
Staffing flexibility	Personnel needed for training
Improved quality	Unmet expectations may occur
Output may improve	Resistance by middle management
Staff support reduced	Resistance by staff groups
Supervision reduced	Conflict between participants and nonparticipants
Improved decision making	Time lost in team meetings

Source: Adapted from Lawler (1986).

7. Changes to the compensation system to reflect the additional skills acquired by individuals.
8. Actions to develop trust between management and the workforce, e.g., the sharing of financial and other sensitive information on company performance.
9. An implementation plan spanning about three years and starting slowly, with a few pilot teams.

The most important—and most difficult—issue is how to guide middle management through the transition to this team concept. Some layers of management will no longer be needed because teams will plan and control their work and do much of their own supervision. As an example, one work activity, under traditional organization, involved 25 direct labor employees, 16 individual job descriptions, three supervisors, and a production manager. This changed to three working teams (and seven skill levels for individuals) that reported directly to the production manager, with no supervisors.

Thus some middle managers will have a new job. That new job may be working with a self-managing team but perhaps now as a member, facilitator, or technical consultant instead of the hierarchical supervisor. Such a role change affects power, knowledge, rewards, and status and thus will be threatening. An organization has a responsibility to explain the new roles for managers clearly and to provide the training, understanding, and patience to achieve success.

Self-managing work teams are not always successful. For some managers, supervisors, and workers, the demands of the concept are more than they are willing to accept. But self-managing teams can be highly effective if they fit the technology, they are implemented carefully, and the people in the organization are comfortable with the concept.

Although the terms are mostly used interchangeably, some organizations make a distinction between self-managing teams and self-directing teams. The self-managing team often includes a manager as a coach, and the self-directing team implies an autonomous team. Other titles include self-regulating teams, natural work groups, and high-performance teams.

For elaboration on self-managing teams (and other types of teams), see *JQH5,* Section 15. Piczak and Hauser (1996) provide a guide to implementation, including traditional issues such as management commitment and other matters such as sharing of company business data and obtaining union cooperation. Also the *Quality Management Journal,* 1997, vol. 4, no. 2, is a special issue on team research and includes a paper on self-managed teams in the sales function.

8.11 SELECTION, TRAINING, AND RETENTION OF PERSONNEL

The principles of selection, training, and retention of personnel are known but are not always practiced with sufficient intensity in many functions.

Selection of Personnel

Norrell, a human resource company, provides client companies with traditional temporary help, managed staffing, and outsourcing services. A survey of more than 1000 of their clients clarified the client definition of *quality* as "excellence of personnel." *Excellence* was defined in terms of eight criteria for clerical positions and nine criteria for technical-industrial positions. These criteria are used in the selection and training of personnel assigned to the client companies. In another example based on a survey of five service organizations and nine manufacturers, Jeffrey (1995) identified 15 competencies that these organizations and their customers viewed as important in customer service activities by front-line employees. Many of these competencies apply to both front-line personnel and back-office personnel.

A human resources firm that provides temporary personnel to clients performed research to select a series of questions for prospective employees. A battery of questions was developed and given to all employees. Employees who are rated superior by clients (based on marketing research studies) answer certain questions differently than employees who are not superior. (Other questions result in the same response from most employees.) Responses to these "differentiating questions" help to select new employees.

Personality is one important attribute for many (but not all) positions in the operations function. This attribute is increasingly important as organizing by teams becomes more prevalent. A chemical manufacturer even places job applicants in a team problem-solving situation as part of the selection process. Larry Silver of Raymond James Financial Services Inc. states the case well: "We need to recruit people who play well in the sandbox with others, i.e., don't throw sand in people's face and do work together to build castles."

One tool for evaluating personality types is the Myers-Briggs Type Indicator. This personality test describes 16 personality types that are based on four preference scales: extrovert or introvert, sensing or intuition, thinking or feeling, and judgment or perception. Thus one personality type is an extrovert, sensing, thinking, judgment person. Analyzing responses to test questions from prospective or current employees helps to determine the personality types of individuals. Organizations need many personality types, and the Myers-Briggs approach describes the contributions to the organization of each of the 16 types. By understanding the types and making job assignments accordingly, an organization can take advantage of all personality types to achieve high performance in the workplace. McDermott (1994) explains the 16 types and how the tool can help in recruiting new personnel and assigning current personnel. Anderson and Anderson (1997) describe a candidate assessment tool that uses 4 areas of abilities, 3 areas of motivation and interests, and 13 areas of personality.

Training

One essential ingredient of a broad quality program is extensive training. Table 8.9 identifies constituencies and subject matter.

Experience in training has identified the reasons that some training programs fail:

- *Failure to provide training at the time it will be used.* In too many cases training is given to employees who have little or no opportunity to use it until many months later (if ever). A much better approach schedules training for each group at the time it is needed—"just in time" training.
- *Lack of participation by line managers in designing training.* Without this participation, training is often technique oriented rather than problem and results oriented.
- *Reliance on the lecture method of training.* Particularly in the industrial world, training must be highly interactive; i.e., it must enable a trainee to *apply* the concepts during the training process.

TABLE 8.9
Who needs to be trained in what

Subject matter	Top management	Quality managers	Other middle managers	Specialists	Facilitators	Workforce
Quality awareness	X		X	X	X	X
Basic concepts	X	X	X	X	X	X
Strategic quality management	X	X				
Personal roles	X	X	X	X	X	X
Three quality processes	X	X	X	X	X	
Problem-solving methods		X	X	X	X	X
Basic statistics	X	X	X	X	X	X
Advanced statistics		X		X		
Quality in functional areas		X	X	X		
Motivation for quality	X	X	X		X	

- *Poor communication during training.* The technology of quality, particularly statistical methodology, can be mystifying to some people. Many benefits are possible if we emphasize simple language and graphical techniques.

Training programs are a failure if they do not result in a change in behavior. Applying these lessons can help to prevent such failures.

For elaboration on training, see *JQH5,* Section 16.

Retention

Investing increased resources in selection and training helps to retain skilled employees. Compensation, of course, is also an essential contributor to employee retention. Other factors, however, are also essential, including

1. Provide for career planning and development.
2. Design jobs for self-control (see Chapters 16 and 17).
3. Provide sufficient empowerment and other means for personnel to excel.
4. Remove the sources of job stress and burnout.
5. Provide continuous coaching for personnel.
6. Provide for participation in departmental planning.
7. Provide the opportunity to interact with customers (both external and internal).
8. Provide various forms of reward and recognition.

Performance appraisal has become a subject of intense debate (and confusion). The author believes that, when properly conducted, performance appraisal is a useful tool. For further discussion, see Section 9.7.

Retention of key employees depends on many issues. An information technology services firm conducted a survey of more than 200 employees to determine their reasons for leaving—with some surprising results. In rank order (based on frequency and intensity of their feeling), employees gave the following reasons for leaving: communicate with me more, train my boss, train me more, help me set goals and give me feedback, provide a higher salary, and provide better benefits. Many of these issues are addressed in Chapter 9, "Developing a Quality Culture."

Morgan and Smith (1996) is a useful reference for selection, training, and retention of personnel.

8.12 QUALITY PROJECT MANAGEMENT AND TRADITIONAL PROJECT MANAGEMENT

Earlier in this chapter we discussed the quality project team, i.e., a team formed to address a specific quality problem (in quality improvement, quality planning, or quality control). The main focus of such a team is quality in specific terms, e.g., a recurring defect in a current product. The term *project management* has a different meaning that can be traced to the 1950s. Project management typically means the planning and control of schedules, costs, and quality for a unique set of activities that will yield a defined outcome such as a complex product or service. Some early applications were to defense products and construction projects.

The Harvard Business School (1996) provides a project management model (Figure 8.2) in terms of (1) define and organize the project, (2) plan the project, and

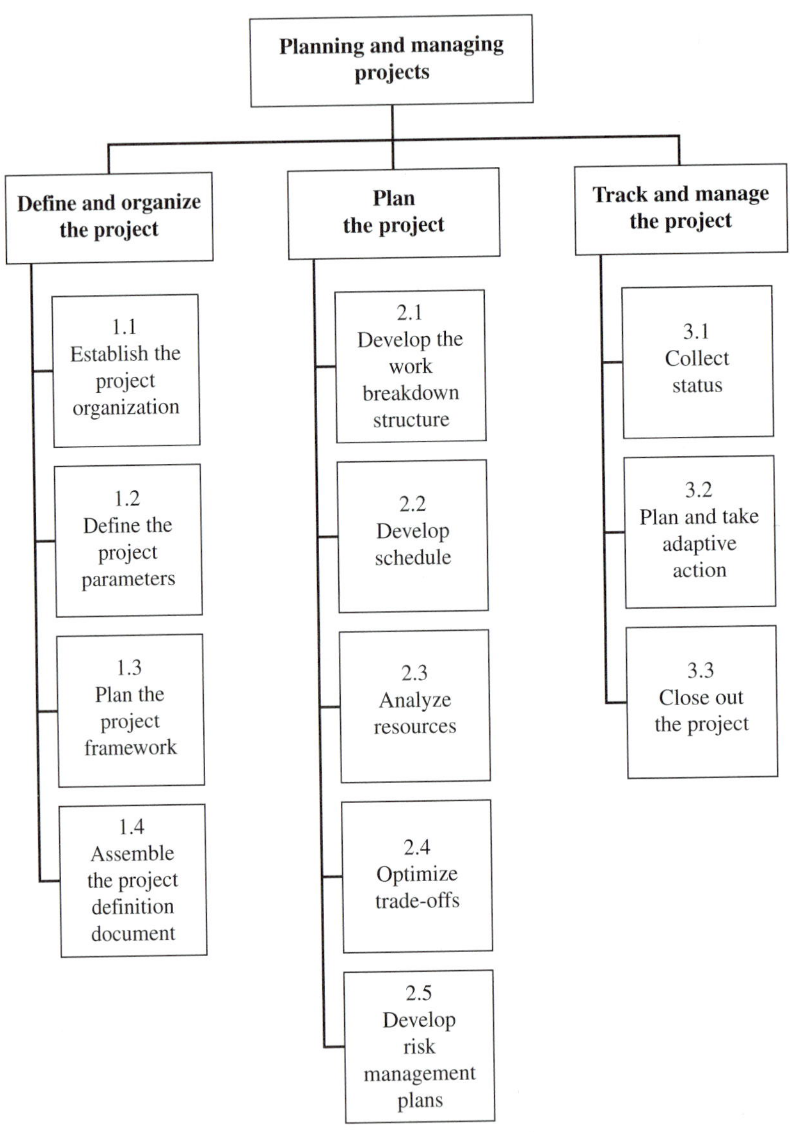

FIGURE 8.2
Project management process model. (*Reprinted by permission of Harvard Business School Publishing.*)

(3) track and manage the implementation of the project. Note how the model uses a framework of planning and control.

The concepts of traditional project management can be used to plan and control a complex quality activity. In an early application (Gryna, 1960), the author translated a vague concept of a "reliability program" into a project management document with 66 activities—each with a task statement, department responsible, other related departments, procedures, schedules, and controls to assure task completion (Figure 8.3).

We conclude this discussion of organizing for quality with a sober reminder: Quality superiority cannot be achieved without the active leadership of upper management. If such leadership is not present, then the first priority for the quality director must be to seek the key to obtain management leadership. The alternative is horrifying; i.e., wait until the quality problem gets so bad that management is forced to act.

SUMMARY

- Coordination of quality activities throughout an organization requires two efforts: coordination for control and coordination for creating change.
- Coordination for control is often the focus of a quality department; coordination for creating change often involves "parallel organizations" such as a quality council and quality project teams.
- New forms of organization aim to remove the barriers, or walls, between functional departments.
- To achieve quality excellence, upper management must lead the quality effort. The roles in this leadership can be identified.
- A quality council is a group of upper managers who develop quality strategy and guide and support its implementation.
- Middle management executes the quality strategy through a variety of roles.
- Inputs from the workforce are essential to identify the causes of quality problems and design work systems for self-control.
- Quality teams create change. Four important types of teams are quality project teams, workforce teams, process teams, and self-managing teams.
- The implementation of the quality strategy must occur through the line organization rather than through a staff quality department.
- The quality director of the future will have two roles: administering the quality department and assisting upper management with strategic quality management.

PROBLEMS

8.1. Conduct a research study and report on how any of the following institutions was organized for quality:
   (a) The Florentine Arte Della Lana (wool guild) of the 12th, 13th, and 14th centuries.
   (b) Venetian shipbuilding in the 14th century.

Task	Primary responsibility	Other functions concerned	Procedure approved		Controls	Schedule	
			Date	By		Start	Completion
Perform reliability predictions based on circuit and component analyses and use of reliability handbook. Compare to reliability budgets. If any design revisions are made, review reliability prediction and revise as required	Design engineer	Quality	Completed	Technical director	1. Reliability predictions to be included on page and line schedule 2. Analyses to be reviewed by group engineer 3. All analyses must be submitted to reliability engineer for approval		
Review all designs to determine component part and assembly limitations with respect to reliability requirements using economical processes and procedures	Electrical manufacturing	Quality		Manufacturing manager	1. All drawings must be reviewed by manufacturing representative prior to engineering release		
Establish failure reporting and problem resolution system from first development test through service use. 1. Establish and maintain failure reporting in test areas and service use 2. Establish and maintain failure reporting in selected vendors' plants	Quality	Engineering, manufacturing, procurement, logistics support		Quality manager	1. Assign specific reliability and assurance engineer and corrective action team to find problems and assure correction 2. Monitor reporting through assigned personnel in reporting areas		

FIGURE 8.3
Excerpts from Martin reliability management program for the ______ project.

(*c*) Construction of cathedrals in medieval Europe.
(*d*) The Gobelin tapestry industry in the 16th and 17th centuries.

8.2. Study the plan of organization for quality of any of the following institutions and report your conclusions:
(*a*) A hospital.
(*b*) A university.
(*c*) A chain supermarket.
(*d*) A chain of motels.
(*e*) A restaurant.
(*f*) A manufacturing company.

8.3. For any organization, make a list of the activities assigned to the quality department. Compare your list with Table 8.1 and explain why the two lists might be different.

8.4. For any organization, study the methods in use to attain
(*a*) Coordination for control.
(*b*) Coordination for change.
Report your findings.

8.5. Review the three elements of self-control in Chapter 5. Explain how each of these elements can be helpful in setting up a self-managing team.

8.6. An often-asked question is, Who is responsible for quality? As worded, the question is too vague. The question must be restated in terms of actions and decisions concerning quality. Choose one functional area of an organization. For this area, list in the rows of a table some key actions or decisions. Then set up a few columns showing organizational units that are candidates for responsibility. Fill in the table to show which unit should be responsible for each action or decision.

REFERENCES

Anderson, D. N. and T. Anderson (1997). "Finding the Right People With the Right Tools," *IIE Solutions Conference Proceedings,* Institute of Industrial Engineers, Norcross, GA.

Aubrey, C. A. II and D. S. Gryna (1991). "Revolution through Effective Improvement Projects," *ASQ Quality Congress Transactions,* Milwaukee, pp. 8–13.

Baker, E. M. (1988). "Managing Human Performance," in *Juran's Quality Control Handbook,* 4th ed., McGraw-Hill, New York, pp. 10.47, 10.48.

Baker, E. M. (1999). *Scoring a Whole in One,* Crisp Publications, Menlo Park, CA.

Becker, S., W. A. J. Golomski, and D. C. Lory (1994). "TQM and Organization of the Firm," *Quality Management Journal,* January, pp. 18–24.

Bourquin, C. R. (1995). "A Quality Department Surveys Its Customers (or, Shoes for the Cobbler's Children)," *Annual Quality Congress Proceedings,* ASQ, Milwaukee, pp. 970–976.

Christie, P. M. J. and R. R. Levary (1998). "Virtual Corporations: Recipe for Success," *Industrial Management,* July–August, pp. 7–11.

Cinelli, D. L. and L. Schein (1994). *A Profile of the U.S. Senior Quality Executive,* The Conference Board, New York.

Dietch, R., S. Tashian, and H. Green (1989). "Leadership Characteristics and Culture Change: An Exploratory Research Study," *Impro Conference Proceedings,* Juran Institute, Inc., Wilton, CT, pp. 3C-21 to 3C-29.

Fishman, C. (1999). "Engines of Democracy," *FAST COMPANY,* October, pp. 173–202.

Garvin, D. A. (1995). "Leveraging Processes for Strategic Advantage," *Harvard Business Review,* September–October, pp. 77–90.

Groocock, J. M. (1994). "Organizing for Quality—Including a Study of Corporate-Level Quality Management in Large U.K.-Owned Companies," *Quality Management Journal,* January, pp. 25–35.

Gryna, F. M. (1993). "The Role of the Quality Director—Revisited," *Impro Conference Proceedings,* Juran Institute, Inc., Wilton, CT, pp. 11.12–11.14.

Gryna, F. M. (1981). *Quality Circles,* Amacom, New York.

Gryna, F. M. (1960). "Total Quality Control through Reliability," *ASQ Quality Congress Transactions,* Milwaukee, pp. 746–754.

Hahn, J. R. (1995). "Self-Managing Teams at Aid Association for Lutherans," *Impro Conference Proceedings,* Juran Institute, Inc., Wilton, CT, pp. 7B.1-1 to 7B.1-24.

Harvard Business School (1996). *Project Management Manual,* Boston.

Henry, J. E. and M. Hartzler (1998). *Tools for Virtual Teams,* ASQ Quality Press, Milwaukee.

Jeffrey, J. R. (1995). "Preparing the Front Line," *Quality Progress,* February, pp. 79–82.

Lawler, E. E. III (1986). *Participative Strategies for Improving Organizational Performance,* Chapter 7, Jossey-Bass, San Francisco.

Majchrozak, A. and Q. Wang (1996). "Breaking the Functional Mindset in Process Organizations," *Harvard Business Review*, September–October, pp. 92–99.

McCain, C. (1995). "Successfully Solving the Quality Puzzle in a Service Company," *Impro Conference Proceedings,* Juran Institute, Inc., Wilton, CT, pp. 6A.1-1 to 6A.1-13.

McDermott, R. (1994). "The Human Dynamics of Total Quality," *Quality Congress Transactions*, ASQ, Milwaukee, pp. 225–233.

Morgan, R. B. and J. E. Smith (1996). *Staffing the New Workplace,* ASQ Quality Press, Milwaukee, and CCH Inc., Chicago.

Piczak, M. W. and R. Z. Hauser (1996). "Self-Directed Work Teams: A Guide to Implementation," *Quality Progress,* May, pp. 81–87.

Six Sigma Academy, Inc. (1998). *The Six Sigma Strategy for Breakthrough Performance,* Scottsdale, AZ.

Skiba, K. D. L. (1996). "Accelerated Continuous Improvement Methodology," *Impro Conference Proceedings,* Juran Institute, Inc., Wilton, CT, pp. 4E-l to 4E-7.

Watson, G. H. (1998). "The Emancipation of Quality: Building Bridges and Closing Gaps," *Quality Progress,* August, pp. 110–114.

SUPPLEMENTARY READING

Managements role: Miller, L. M. (1984). *American Spirit: Visions of a New Corporate Culture,* William Morrow, New York.

Teams: Katzenbach, J. R. and D. K. Smith (1993). *Wisdom of Teams,* Harvard Business School Press, Boston.

Project management: *JQH5,* Section 17.

Change management: Kanter, R. M. (1983). *The Change Masters,* Simon and Schuster, New York, Chapter 7.

Process management: *JQH5,* Section 6.

Rummler, G. A. and A. P. Brache (1995). *Improving Performance,* 2nd ed., Jossey-Bass, San Francisco.

Quality director: Shepherd, N. A. (1998). "Quality Manager 2000—Competitive Advantage: Mapping Change and the Role of the Quality Manager of the Future," *Annual Quality Congress Proceedings,* ASQ, Milwaukee, pp. 53–60.

9

DEVELOPING A QUALITY CULTURE

9.1
TECHNOLOGY AND CULTURE

To become superior in quality, we must pursue two courses of action:

1. *Develop technologies that meet customer needs.* Such technologies include those for developing new and improved products and those for quality processes and tools. We discuss these technologies throughout the book.
2. *Stimulate a "culture" throughout the organization that continually views quality as a primary goal.* Wouldn't it be great if we could identify techniques for creating a quality culture? Culture is not a technocratic issue. Nevertheless, certain approaches provide a path toward a quality culture. Such approaches are examined in this chapter.

Quality culture is the pattern of human habits, beliefs, values, and behavior concerning quality. Technology touches the head; culture touches the heart.

Recall that quality problems are mostly management controllable. Thus cultural issues apply to all levels—upper management, middle management, supervisors, technical specialists, business specialists, and the workforce. We first examine some classic theories of motivation.

9.2
THEORIES OF MOTIVATION

The professionals in this field are behavioral scientists. (Managers are merely experienced amateurs.) Studies by the behavioral scientists provide useful theories that help us to understand how human behavior responds to various stimuli.

TABLE 9.1
Hierarchy of human needs and forms of quality motivation

Maslow's list of human needs	Usual forms of quality motivation
Physiological needs: i.e., need for food, shelter, basic survival. In an industrial economy, this translates into minimum subsistence earnings.	Opportunity to increase earnings by receiving a bonus for good work.
Safety needs: i.e., once a subsistence level is achieved, the need to remain employed at that level.	Job security: e.g., quality makes sales; sales make jobs.
Belongingness and love needs: i.e., the need to belong to a group and be accepted.	Appeal to the employee as a member of the team—he or she must not let down the team.
Esteem needs: i.e., the need for self-respect and for the respect of others.	Appeal to pride of workmanship, to achieving a good score. Recognition through rewards, publicity, etc.
Self-actualization needs: i.e., the urge for creativity, for self-expression.	Opportunity to propose creative ideas, to participate in creative planning.

Hierarchy of Human Needs

Under the theory (Maslow, 1987), human needs fall into five fundamental categories in a predictable order of priorities. Table 9.1 shows this "hierarchy of human needs" together with the associated forms of motivation for quality.

Job Dissatisfaction and Satisfaction

Under this theory (Herzberg et al., 1959), job dissatisfaction and job satisfaction are not opposites. Job dissatisfaction is the result of specific dislikes—the pay is low, the working conditions are poor ("hygienic factors"). It is possible to eliminate these dislikes—raise the pay, change the working conditions. Companies that pay wages lower than the competition pays must face the harsh reality that they will not attract the best people and are probably doomed to quality mediocrity unless they have some miracle—but miracles don't come easy. Employee basic needs (e.g., wages) must be met before employee training needs (e.g., quality training) can be met successfully. For research on this and other employee issues, see Vass and Kincade (1999). But even when the revised hygienic conditions are accepted as normal, they do not motivate behavior.

In contrast, job satisfaction depends on what the worker *does.* Satisfaction comes from doing—motivation comes from factors such as job challenges, opportunities for creativity, identification with groups, and responsibility for planning. To illustrate: At the end of the day, an assembly line worker is happy to leave that monotonous job and go home to something more appealing. In the same company, a researcher may not leave precisely at closing time—the research project may be more fascinating than the outside hobby.

Theory X and Theory Y

Two theories bring us to a controversy about whether workers have lost their pride in their work. Is the change in the worker or in the work? These two alternatives have been given names—theory X and theory Y, respectively (McGregor, 1985).

Under theory X the modern worker has become lazy, uncooperative, etc. Managers must combat this decline in worker motivation through skillful use of incentives and penalties.

Under theory Y there has been no change in human nature. What has changed is the way in which work is organized. The solution is to create new job conditions that permit normal human drives to assert themselves.

Managers are not unanimous in adhering to one or the other of these theories. It is common to find, even within the same company, some managers who support theory X and others who support theory Y. This support is not merely philosophical—it is reflected in the operating conditions that prevail in different departments.

Both theory X and theory Y have their advocates. However, there seems to be no conclusive evidence that either can outperform the other in economic terms, i.e., productivity, cost, etc. But some evidence suggests that the theory Y approach makes for better human relations. Some studies have shown that certain workers regard repetitive, routine work as less stressful than the broader range of work with its demands for decision making and creativity.

9.3 CORPORATE CULTURE

Corporate culture consists of habits, beliefs, values, and behavior. Management needs to define and create the culture necessary for business success. In a classic book, Miller (1984) identifies eight "primary values" that promote employee loyalty, productivity, and innovation. His ideas reflect the thinking of many behavioral scientists and managers regarding the roles facing management. The eight values are

1. *Purpose.* Purpose is the vision stated in terms of product or service and benefit to the customer.
2. *Consensus.* Three decision-making styles—command, consultative, and consensus—should be matched to particular situations.
3. *Excellence.* Management creates an environment in which the pursuit of knowledge for improvement is pervasive.
4. *Unity.* The emphasis here is on employee participation and ownership of work.
5. *Performance.* Individual and team rewards are the focus along with performance measurements to tell individuals how they are doing.
6. *Empiricism.* Management by fact and the use of the scientific method form the basis of this value.
7. *Intimacy.* Intimacy relates to sharing ideas, feelings, and needs in an open and trusting manner without fear of punishment.
8. *Integrity.* The norm here is for managers to act as role models for ethical practices.

A discussion of culture quickly moves into an examination of the actions necessary for *change*. Miletich (1997) describes how Honeywell Space Systems identified seven "dimensions of organizational culture" and then developed a process for changing the culture. The seven dimensions of culture are risk orientation, relationships (of people), information, motivation, leadership, organizational structure, and organizational focus. For each of these dimensions, the executive team debated and documented a definition, determined the current state of the dimension, defined the desired state, developed an action plan to close the gap, and devised a method for measuring progress.

Mackin (1999) presents an organizational change curve (Figure 9.1) that shows how the process of change moves through time.

This curve is based on experiences at Merrill Lynch Credit Corp. When we make people aware of such a curve, they are personally able to move through the changes more successfully.

Wetlaufer (1999a) describes how the Ford Motor Company is driving change through a teaching initiative aimed at every Ford employee. Small-group discussions are held on strategy and competition, community service, and 360-degree feedback (see Section 9.7). The initiative also covers matters such as ideas for organizational success, personal values, how to motivate others, and thought processes for making difficult decisions.

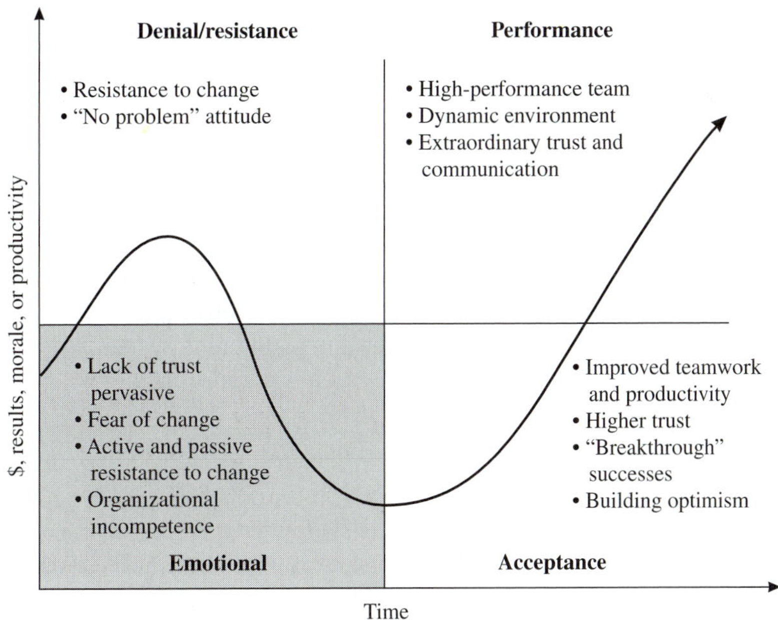

FIGURE 9.1
Organizational change curve.

In an unusual application, Bunn (1995) describes how a veterinary services laboratory measured organization culture and discovered a lack of collaboration, commitment, and creativity. He also reports on the results of actions taken.

An innovative approach to organization culture is called "appreciative inquiry." Here, discussions are held with employees to discover what practices in the organization are "most liberating to the human spirit." As these practices are identified, the focus then becomes planning and building the future to capitalize on these practices. For elaboration, see Hammond (1998) in the Supplementary Reading.

The subjects of corporate culture and the management of change are of great importance and have spawned much literature. Some recommended references are Kotter (1996), Baker (1999), and Covey (1989) in the Supplementary Reading.

9.4 QUALITY CULTURE

Quality culture is an integral part of corporate culture. Our discussion here concentrates on the aspects of corporate culture that relate to the quality of products and services.

Differences in quality culture can have extreme illustrations, both negative and positive. Two examples:

- *Negative quality culture ("hide the scrap" scenario).* The culture in a paint manufacturing plant put pressure on supervisors to avoid reporting any batch of paint that did not meet specifications. One supervisor resorted to hiding some paint; he directed his workers to dig a hole in the yard and bury the paint. In the service industries, reporting on the processing of transactions (e.g., insurance claims, mishandled baggage) is sometimes subject to data manipulation to yield good performance reports.
- *Positive quality culture ("climb the ladders to delight the customer" scenario).* The culture in a hotel resulted in taking an extraordinary step to please a customer—on short notice. About an hour before a seminar was to be presented, the seminar leader heard a steady, clear tinkling sound. The cause—small glass prisms in two chandeliers colliding due to air movement from the air conditioning system. He mentioned this distraction to the hotel management. Action was immediate—two work crews were assembled, ladders were set up, and every second prism was removed (about 100 in all).

Cameron and Sine (1999) discuss research on four different quality cultures and the quality tools associated with each culture. The four cultures are absence of quality emphasis, error detection, error prevention, and creative quality. Analysis of responses from managers in 68 organizations revealed that the more advanced levels of quality culture are, indeed, associated with higher levels of organization effectiveness.

An important starting point is to determine the current quality culture. This topic is discussed in Section 2.7. Watson and Gryna (1998) report on research in assessing quality culture in small manufacturing and service organizations. Knowing the quality culture enables us to implement a strategy in a way that encourages people to embrace the quality strategy and make it successful.

Culture can be changed. We need to provide goals and measurements, evidence of upper-management leadership, self-development and empowerment, participation, and recognition and rewards—critical success factors for achieving a positive quality culture. These paths are the means to drive changes in actions that lead to changes in attitude and finally to changes in quality culture. Above all, there must be a sense of urgency and initiative on quality throughout all levels of the organization. Instilling such a sense is easier said than done because "talk doesn't cook rice."

The paths for quality culture must be integrated with the methodologies and structure for quality (see Figure 9.2). The three elements of self-control are really a prerequisite for achieving a quality culture. Thus we must provide people with the knowledge of what they are supposed to do, provide feedback on how they are doing, and provide a means of regulating a capable process. Actions to "motivate" people will not be successful unless these basics of self-control are in place. Readers are urged to review Section 5.3, "Self-Control." Further elaboration is given in Chapters 16 and 17.

We proceed then to examine five key drivers for developing a quality culture.

FIGURE 9.2
Technology and culture.

9.5 PROVIDE QUALITY GOALS AND MEASUREMENTS AT ALL LEVELS

To ensure action on quality, a starting point is to provide quality goals and measurements at all levels (see Figure 9.3). In this process we develop goals and measures that are aligned with the mission, critical success factors, and quality strategy of the organization (see Section 7.6).

Clear quality goals for individuals are important stimuli for inspiring superiority in quality. Human beings commit themselves in two different ways: external and internal (Argyris, 1998). Under external commitment management defines the goals for employees and also the task required to achieve those goals. Under internal commitment management and employees jointly define goals, and the employees define the tasks to achieve the goals. Management must foster an environment of internal commitment.

Two important issues in disseminating information on quality are the language used and the content of the information.

We need to present information in different "languages" for different populations in an organization. The pyramid in Figure 9.4 depicts these populations and the corresponding languages. At the apex is upper management, usually the general manager and the top management team; at the base are first-line supervisors and the workforce; in between are middle managers and specialists.

These segments of the organization use different languages in everyday operations, and creating awareness of the need for quality must reflect this fact. Middle managers must not only understand their local dialects but also be fluent in the languages of the other levels (upper management and lower management and workforce). Thus middle managers must be "bilingual."

At the upper management level, creating an awareness of quality is best done in the language of money. Highlighting threats to sales income or opportunities for cost reduction are important. When quality can be related to either of these factors, an

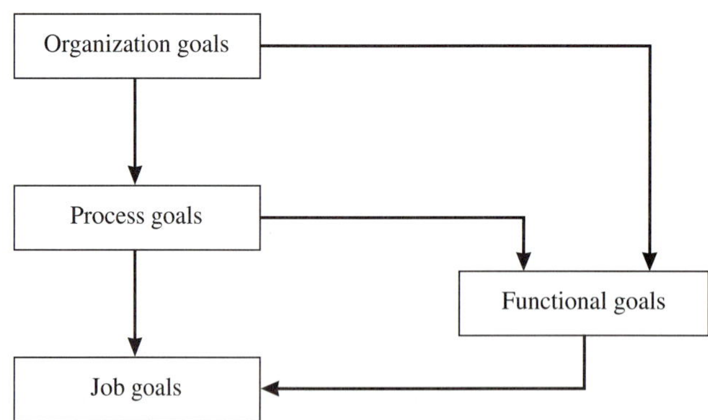

FIGURE 9.3
Goal alignment. (*Rummler and Brache, 1995. Reprinted by permission of Jossey-Bass, Inc., a subsidiary of John Wiley & Sons, Inc.*)

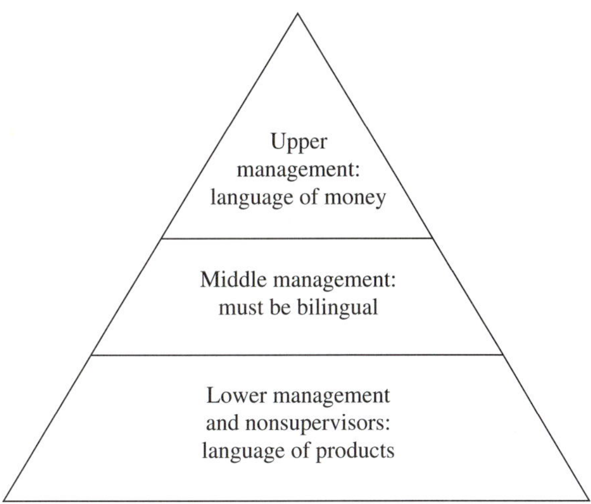

FIGURE 9.4
Common languages in the company.

essential step in inspiring upper management action has been taken. The reader is urged to review Chapter 2, "Companywide Assessment of Quality," which describes three studies. A study on marketplace standing will identify threats to sales income; a study on the cost of poor quality will highlight opportunities for cost reduction; a study on quality culture will help to identify some of the obstacles to inspiring action. Each of these assessments can be made for the total organization or for individual areas such as product development or operations. Note also that a presentation of the company's current status should be accompanied by an explanation of the *benefits* that can be expected from a new approach to quality. Demonstrations of these benefits are particularly useful (see Section 3.4 under "Use of a Bellwether Project"). Table 2.3 presents units of measure for both monetary and other languages to dramatize the importance of losses due to poor quality.

At middle management and lower levels, we sometimes can translate the impact of quality directly into the language of job security. When this step can be based on data, the result can be dramatic. A manufacturer of a consumer product was able to collect data on current customers who replaced the product made by the company. In one year about 430,000 customers replaced the product by purchasing a competing brand—and thus were lost customers. About one-third of these lost customers stated that the main reason for switching to another brand was "poor quality" (four times more important than price and other factors). The sales income lost due to poor quality was then calculated as $1.3 billion. This sales income would have provided about 2000 jobs—as many as were employed by one of the company's plants.

At Chase Manhattan Mortgage Corp., the link between strategy and performance goals and measures is formalized on an individual employee basis. As part of the implementation of strategic quality planning, each employee prepares a document listing information that includes primary job function, primary internal or external customer, and—for both department and the individual—productivity performance goals and measures and quality performance goals and measures. Later this information is used in the performance appraisal process, and team-based incentive compensation is aligned to achieve the goals.

Quality Measurements as a Continuous Focus

The message on quality must be sustained through continuous reinforcement. One form of reinforcement is quality measurement.

Throughout this book quality measurement is proposed for major functional activities, i.e., product development, purchasing, manufacturing, marketing and customer service, and administrative and support operations. These measurements become the "vital signs" that provide people with data not only to perform their tasks but also to maintain a continuing awareness of quality. The reader is urged to review the discussions on measurement and on self-control in Chapter 5, "Quality Control." In addition to continuous feedback to employees, some companies use a "chart room" to display key quality measurements as a dramatic way of showing the overall quality story. Lights on an operations floor provide a highly visible indicator of quality—green for good, yellow for marginal, and red for poor quality.

Units of measure must be carefully defined to inspire a positive priority for quality. An example that does just the opposite is a poorly defined measure of productivity. Measures of productivity are usually a ratio of product output to input resources. Some companies calculate productivity using *total* output (instead of output meeting specifications) divided by input resources. Although total output must be measured, a single productivity measure based only on total output sends a clear message that meeting goals and specifications is not important. Changing such deadly measures to count only good output provides continuing evidence that management sets a high priority on quality.

Reports and scoreboards of quality measurements can be highly effective, but caution must be exercised. Where the measurements show an unfavorable level of quality, the distinction between management-controllable and worker-controllable causes must be recognized. When the problem is mostly management controllable (the typical case), management must clearly be responsible for taking action. Otherwise, publication of the data implies that the low level of quality is the fault of the workers. Such an implication will be fiercely resented and will undermine a positive culture about quality (and even result in people hiding the defective output). For problems that are mostly worker controllable, the publishing or posting of the data must be accompanied by showing the workers exactly what steps they must personally take to improve their quality of output.

Steps that are "obvious" to management may not be obvious to employees.

Maintaining an awareness of quality can draw upon an array of ideas and techniques. These include quality newsletters, quality items on all meeting agendas, announcements on quality by key executives, conferences on quality, and "interest arousers" (e.g., letters from customers, quality puzzles in paycheck envelopes).

Sometimes a simple action stimulates an awareness of quality. At a private accounting firm, employees were shown a blank marketing research form used by a competing firm to obtain feedback on customer satisfaction. The form had specific questions about quality, and the employees were surprised that the competition was addressing quality in such a rigorous manner.

Human ingenuity provides an unending array of possibilities. But ideas for maintaining a focus on quality can *never* be a substitute for real action by management. Some management groups expect clever posters and other media to improve quality when the management-controllable causes of poor quality have not been corrected. If management has this expectation, the posters should be placed high off the ground; otherwise, vulgar comments may appear on them.

9.6
PROVIDE EVIDENCE OF MANAGEMENT LEADERSHIP

Management commitment is necessary but not sufficient. To inspire action within a company, the most important element is management leadership on quality—with the *evidence* to prove it.

The leadership role of upper management in strategic quality management is discussed in Section 8.3. This role includes establishing, aligning, and deploying quality goals and strategies and then serving on a quality council to lead the quality effort. In brief, upper management must initiate and support a vision of a total quality culture.

An example of the role of upper management is the Rank Xerox (RX), "quality policy deployment process" (Figure 9.5), described by Zairi (1994). Note how the process includes the use of the plan, do, check, act cycle (see Section 5.3) to plan, implement, and take necessary actions.

Upper-management quality-related activities will take about 10 percent of the managers' time—a heavy price to be paid by people who have many other demands on their time. (Establishing "proof of the need" is essential to convincing upper management to make the time investment.) When upper management spends time on these activities, it provides the evidence of leadership that inspires others to do their share.

Some upper management groups have chosen to be highly visible in the quality process by leading quality training. In such cases managers at a variety of levels personally conduct some of the managerial training for their subordinates. Sometimes, the concepts presented are emphasized by the manager/instructor in everyday practice.

A further form of evidence is upper-management quality improvement teams. Each team, consisting solely of upper-management members, addresses a problem that requires attention at its level. Examples include the effectiveness of the product development process, the quality of decision making in selecting new product managers, and the administrative aspects of high warranty costs.

The visibility of upper management taking such training and then conducting such projects sets an example for other levels to follow.

Occasionally, upper management has opportunities to take dramatic action to demonstrate its commitment to quality. Here are some examples:

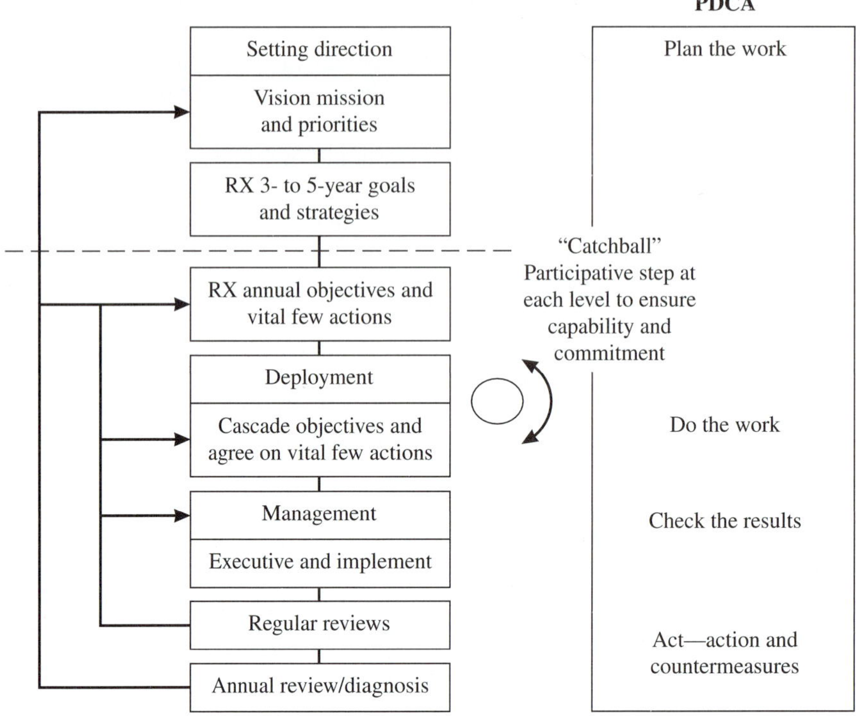

FIGURE 9.5
Quality policy deployment at Rank Xerox Ltd. (*Zairi, 1994. With kind permission from Kluwer Academic Publishers.*)

- A manufacturer of tires traditionally sold, at a discount, tires that had imperfections. (These imperfections had no impact on safety.) The sales income from these "seconds" made an important contribution to total sales income. A policy decision was made to discontinue all sales of tires with imperfections. The policy, one part of a broader company emphasis on quality, sent a strong message to all employees that quality was a top priority.
- A manufacturer of a small electronic product had always emphasized quality but found it necessary to rework about 8 percent of production. A potential customer objected to the rework concept, claiming that the rework might degrade the overall quality. (That customer was not satisfied that the product would be retested after rework.) The manufacturer announced that *no* rework would be permitted and that any nonconforming product would be discarded. In this case the strong message proclaimed by the policy helped to uncover many hidden causes of defects that had previously been tolerated and corrected by rework.
- One major utility requires department managers to submit and obtain approval each year of their departmental budget. This organization also requires each manager to

prepare an annual quality plan. Each *quality plan* must be approved before the annual budget is approved.

- A government agency had a reporting system on work output. Managers were evaluated on the quantity of output results versus goals. As the agency moved to a quality-oriented culture, the director took drastic action to demonstrate his feeling about the priority on quality. He discontinued the reports on output (even though he still had output goals from a higher level) and told his people that quality was the top priority. He explained that improvement in quality would contribute to meeting output goals by reducing the time spent on reprocessing activities due to poor quality.

Such dramatic actions, even though they are rare, are inspiring.

A particularly sensitive issue that has a major impact on quality culture is job security. The turmoil of downsizing and outsourcing understandably causes major apprehensions among employees. Employees correctly understand that improving quality to increase sales may create jobs, but they are concerned that reducing errors and other forms of waste may eliminate jobs. ("Am I working myself out of a job by participating in this quality improvement team?") How upper management handles this issue makes a deep impression on employees. Section 7.7 discusses this issue and presents alternatives for upper management.

Providing evidence of leadership may involve changes in the way management interacts with employees, i.e., the "style of leadership." A prerequisite to such change is understanding the present style. A division of the Rockwell Corporation decided that understanding the style was a key element in gaining employee support for quality. A survey was made to determine the management style of the president, his direct reports, and the people who reported to the direct reports (Warren, 1989). The management team exhibited six leadership styles: coercive, authoritative, affiliative, democratic, pace setting, and coaching. Figure 9.6 summarizes the results. The predominant style was pace setting ("The do-it-myself manager who performs many tasks personally, expects subordinates to follow his or her example and motivates by setting high standards and letting subordinates work on their own"). Although the pace-setting style had advantages, it was not deemed suitable for changing the organization or gaining employee support. (Pacesetters often take over a job themselves, have trouble delegating, are intolerant of mistakes, etc.) Further analysis revealed that a better management style is one where management prides a vision, sets clear standards and goals, shows individuals what is expected of them, lets employees do the job, and gives feedback along with rewards. A coaching process was developed to help managers change their style.

The matter of leadership for quality applies to all levels of management. Middle management and supervision must take the initiative to act as leaders on quality. In an innovative approach the CEO of SSM Health Care (a system that includes 21 hospitals) calls for everyone in her organization to be a leader for continuous improvement. Middle management, in particular, is encouraged and expected to create a work environment for improvement (Ryan, 1999). Some organizations tell personnel that they have two jobs—one is in their job description; their other job is improvement.

Kotter (1999) is an excellent reference in distinguishing between managers and leaders.

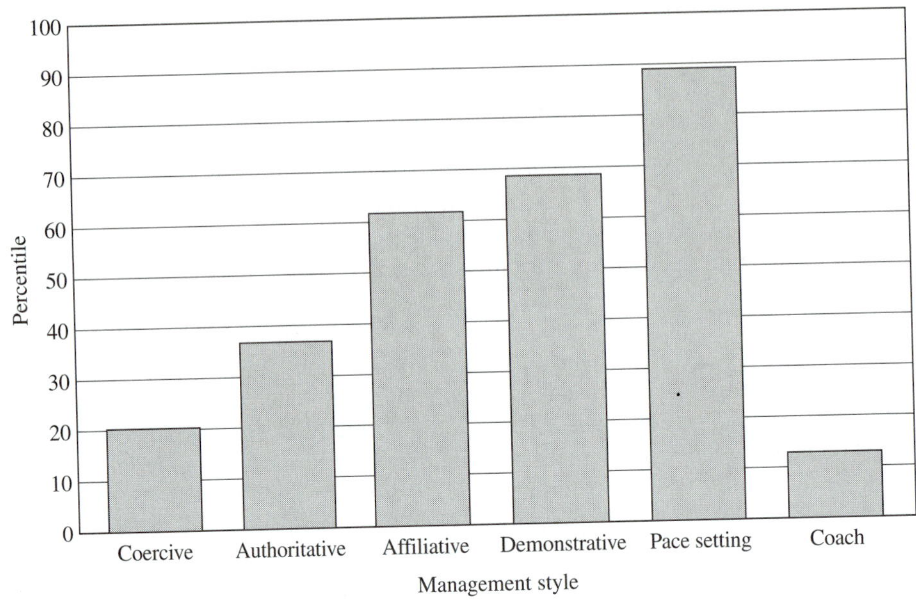

FIGURE 9.6
Composite of senior management. (*From Warren, 1989.*)

9.7 PROVIDE FOR SELF-DEVELOPMENT AND EMPOWERMENT

Inspiring people to take positive steps on quality is greatly influenced by the nature of the work performed by those people. We first provide some history and then address job content, empowerment, and performance appraisal.

The Taylor System

Frederick W. Taylor was a mechanical engineer who had worked as a machinist, foreman, and plant manager. He concluded from his experience that the supervisors and workers of his day (late 19th and early 20th centuries) lacked the education to make various essential decisions, e.g., which work methods to use, what constitutes a day's work.

Taylor's remedy was to separate planning from execution. He assigned engineers and specialists to do the planning, and he left to the supervisor and workers the job of executing the plans.

Taylor's system achieved spectacular increases in productivity. The resulting publicity stimulated further application of his ideas. The outcome was a widespread adoption of the concept of separating planning from execution. Through this adoption, the Taylor system (later called "scientific management") became widely used and deeply rooted in the United States and, to a lesser degree, among the industrialized countries of the West.

Originally the Taylor system, which is applied to both manufacturing and service industries, resulted in a rigid definition and specialization of work for individuals, particularly at the workforce level. Meanwhile, Taylor's major premise—lack of worker education—has been made obsolete by the remarkable rise in education at all levels including the workforce. As a result, organizations increasingly use the education, experience, and creativity of the workforce. Taylor's momentous contribution has been instrumental in achieving a surge of affluence during the 20th century in developing countries.

The Taylor system is based on the scientific analysis of work using knowledge and experimentation. The design of work can be further augmented with the concept of self-control.

Self-Control and Job Design

Recall the concept of self-control (see Section 5.3). People must have knowledge of what they are supposed to do, feedback on their performance, and the means of regulating their work if they are failing to meet the goals. Chapters 16 and 17 provide detailed checklists to evaluate adherence to these three elements of self-control. The lack of one or more of these three elements means that quality problems are management controllable. (Generally, at least 80 percent of quality problems are management controllable.) Placing workers in a state of self-control is a prerequisite to using behavioral approaches to motivate employees.

Some forms of job design place workers in a greater degree of self-control. These are examined below.

Job Characteristics

In a classic book, Hackman and Oldham (1980) describe five characteristics of jobs that provide more meaningful and satisfying ("enriched") jobs for workers. These characteristics and the actions needed to enrich jobs are shown in Table 9.2. Approaches to the redesign of jobs include several forms of job enlargement. In horizontal job enlargement, the scope of a job is increased by having workers perform a larger variety of tasks. The extreme of horizontal job enlargement is for each worker to produce a complete product unit. In vertical job enlargement, the job is enlarged by making workers responsible for tasks previously performed by others vertically higher in the organization (e.g., a supervisor).

Hackman and Oldham recommend caution on job redesign. They point out that most—but not all—people want a more demanding job. The opportunities are endless, but we must avoid placing employees in jobs to which they are not receptive or otherwise suited.

Self-Managing Teams

A special form of job enlargement is that applied to a group of workers, i.e., a self-managing team. Two elements are emphasized: (1) each worker is trained to have a

TABLE 9.2
Job enrichment characteristics and management actions

Characteristic	Definition	Action
Skill variety	Degree to which the job has a sufficient variety of activities to require a diversity of employee skills and talents	Combine sequential tasks to produce larger work modules (horizontal enlargement)
Task identity	Extent to which work requires doing a job from beginning to end and results in a completed visible unit of output	Arrange work into meaningful groups, e.g., by customer, by product
Task significance	Extent to which the job affects internal and external customers	Provide means of direct communication and personal contact with customer
Autonomy	Amount of employee self-control in planning and doing the work	Provide employee greater self-control for decision making (vertical enlargement)
Feedback	Degree to which direct knowledge of results is provided to employee	Create feedback systems to provide employees with information directly from the job

Source: Adapted from Hackman and Oldham (1980).

variety of skills, thereby permitting rotation of tasks, and (2) the team is given formal authority to execute certain job-planning and supervisory tasks (see Section 8.10, "Self-Managing Teams").

Empowerment

Empowerment is the process of delegating decision-making authority to lower levels within the organization. Particularly dramatic is empowerment of the workforce. But empowerment goes far beyond delegating authority and providing additional training. It means encouraging people to take the initiative and broaden their scope; it also means being supportive if mistakes are made.

As employees became more empowered in their work, the feeling of ownership and responsibility becomes more meaningful. Further, the act of empowering employees provides evidence of management's trust. Additional evidence is provided when management shares confidential business information with employees. For many organizations, such steps are clearly a change in the culture.

The concept of empowerment applies both to individuals and to groups of workers. Self-managed teams (see Section 8.10, "Self-Managing Teams") provide an illustration of empowerment for groups of workers.

With empowerment comes the need to redefine the basic roles of upper management, middle management, and the workforce. One model at a bank looks like this:

Upper management should act as shapers and coaches. As a shaper, it should create, communicate, and support the organization's mission. As a coach, it should help when asked but avoid entering into the day-to-day problems of middle management.

Middle management should not only run its area of responsibility but also work as a group to integrate all parts of the organization. In addition, it must support the workforce by eliminating obstacles to progress.

The *workforce* is the primary producer of the output for customers. Its closeness and knowledge about its work means that it should use its empowerment to determine how the work can best be done.

Note, under empowerment, how essential it is for management to provide employees with the information, feedback, and means of regulating their work; i.e., meeting the three elements of self-control is a prerequisite for empowerment and subsequent motivation. Self-control includes the training needed to make good decisions under empowerment. Empowerment requires employees to have the capability, the authority, and the desire to act.

Wetlaufer (1999b) describes a dramatic example of empowerment at a global electricity company. The characteristics include organization around teams to run operations and maintenance, elimination of functional departments, every person a generalist (a "mini-CEO"), and upper management acting as advisers.

But despite all the talk in many companies, "empowerment is still mostly an illusion" according to Argyris (1998). He believes that difficulties in achieving an empowerment environment are due to a failure by management to achieve an internal commitment from employees by jointly setting quality goals with employees rather than management unilaterally setting the goals (see Section 9.5).

Performance Appraisal

Performance appraisal is the process by which an organization periodically evaluates an employee's behavior and accomplishments. Performance appraisal has become a subject of intense debate (and confusion). The author believes that, when properly conducted, performance appraisal is a useful tool. Appraisal that coaches employees to a higher level of performance can be helpful; appraisal that is done only to rank employees for purposes of pay and advancement can be destructive. Graber, Breisch, and Breisch (1992) clarify some misunderstandings about the subject and provide constructive suggestions for beneficial appraisals. Eastman Chemical replaced its traditional performance appraisal system with an "employee development system" that focuses on employee development and coaching. Clearly, an appraisal process must deemphasize past performance and focus on assisting employees in their future job-related quality efforts.

An associated approach is the 360-degree appraisal process. In this approach several individuals who frequently interact with the employee participate in the appraisal process. Anyone who has significant contact or useful information about the employee's performance may be included. In the ideal 360-degree process, the individuals participate in both goal setting (at the beginning of a period) and performance appraisal (at the end of a period). Milliman and McFadden (1997) discuss the traditional view of human resource management versus the view of total quality management and then explain how the 360-degree system can contribute to more effective performance appraisals.

Section 5.7 covers the topic of continuous feedback on performance.

Selection and Training

Selection and training of personnel clearly have an important influence on people's development. Many of the principles are well known but are not always practiced with sufficient intensity. But this situation is changing in the United States. For example, some organizations are now making an annual investment in training of about 2 percent of sales income.

The Japanese have invested extensively in selection and training. Interviews and testing prior to employment assure compatibility of the candidate and the job. Rotational assignments then help to develop a broad base of technical skills, thus facilitating cooperation across departments. At the managerial level rotational assignments help to develop the individual's concern for the company as a whole.

For related matters on employee development, readers are urged to review Section 8.11, which discusses employee selection, training, and retention.

9.8
PROVIDE PARTICIPATION AS A MEANS OF INSPIRING ACTION

It is tempting to believe that, to inspire action on quality, we must start by changing the people's attitudes. A change in attitudes then should lead to a change in behavior. In reality, the opposite is true. If we first change people's behavior, then that will change their attitudes. Psychologists call this concept "cognitive dissonance."

An age-old principle that helps to change behavior is the concept of participation. By personally participating in quality activities, people acquire new knowledge, see the benefits of the quality disciplines, and obtain a sense of accomplishment by solving problems. This participation leads to lasting changes in behavior.

Various forms of participation are described throughout this book. Table 9.3 summarizes these forms for levels from upper management to first-line supervisors and the workforce.

Participation at *all* levels is decisive in inspiring action on quality. At the workforce level, however, participation can have impacts that border on the dramatic. In conducting research on workforce teams, some unforgettable events were observed (see Section 8.9, "Workforce Teams").

Participation should include the officers of labor unions. Competitive economic challenges faced by most organizations require management and unions to find ways to work together for their mutual benefit.

TABLE 9.3
Forms of participation

Form	Description	Level: Upper management	Middle management	Specialists	First-line supervisors/workers
Quality council	Serve on the council	X			
Quality improvement teams	Serve as a leader or member of a cross-functional improvement team	X	X	X	X
Quality workforce teams	Serve as a leader or member of a team within a department		X	X	X
Quality task forces	Serve on quality task forces appointed by the quality council	X	X	X	X
Process owner	Serve as owner of a business process	X	X		
Design review	Participate in design review meetings		X	X	
Process review	Participate in process review meetings		X	X	X
Provide planning	Identify obstacles to self-control		X	X	X
Set quality goals	Provide input or set goals	X	X	X	X
Plan own work	Handle all aspects of planning	X	X	X	X
Customer visits	Hold discussions on quality with customers	X	X	X	X
Supplier visits	Hold discussions on quality with suppliers	X	X	X	X
Meetings with management	Make presentations on quality activities		X	X	X
Visit other companies	Learn about quality activities	X	X	X	X
Job rotation	Work in the quality department or other departments	X	X	X	X
Conferences	Make presentations or chair sessions	X	X	X	X

For a discussion of how the quality profession and trade unions can benefit through a mutual interactive strategy, see Rubinstein and Ryan (1996).

9.9 PROVIDE RECOGNITION AND REWARDS

We define *recognition* as public acknowledgement of superior performance of specific activities. *Rewards* are benefits (such as salary increases, bonuses, and promotions) that are conferred for generally superior performance against goals.

Such expressions of esteem, which are discussed below, play an essential role in inspiring people on quality. An even more sustaining form is the positive feeling that people have internally when (1) their job has been designed to focus on self-development and (2) they are given opportunities to participate in planning and decision making. This environment tells employees that their skills, their judgment, and their integrity are trusted. Imagine their feeling when independent inspection is changed to self-inspection.

Recognition through public acknowledgment of superior activity can be provided at several levels—individual, team, and business unit. In planning for recognition, here are some questions to address:

- What type of activity will receive formal recognition, e.g., normal *participation* in an activity such as workforce teams, superior *effort,* or tangible *results?*
- Will recognition be given to individuals, to groups, or to both?
- Will selection of those to receive recognition be on a competitive or noncompetitive basis?
- What form will the recognition take, e.g., ceremonial, token award, or other?
- Who will decide on the form of recognition, e.g., a group of managers? Will others have input?
- Who will select the recipients, e.g., a management committee, peers of potential recipients, or someone else?
- How often will recognition be given? Many managers *overestimate* how often they provide recognition to employees.

As these questions apply to other activities (e.g., safety), experience is available in planning recognition.

Forms of Recognition

Forms of recognition range from a simple verbal message for a job well done (often overlooked in the rush of daily activities) to modest, or "token," awards. Token awards may be tangible (e.g., a savings bond, time off, a dinner) or intangible (sending a letter of praise, sending an employee to a seminar or conference, letting an employee be boss for a day).

Recognition must be genuine and must fit the local culture. Unfortunately, managers are sometimes naive about what best fits the culture. For example, each member of a quality team was given a shirt emblazoned with the name of the team. Some members, however, refused to accept the shirt—they viewed it as a "gimmick" that made a joke of their participation. In another case a banquet was held for members of teams. The plant manager made some brief remarks thanking the employees for their efforts. An enjoyable evening was had by all. Employees were quite appreciative, but some of them remarked (constructively) that it was the first time they had ever seen the plant manager: "Wouldn't it be nice if he occasionally toured the production floor and spoke with everyone?" In deciding on forms of recognition, managers should ask for suggestions from respected employees. Not only will their ideas fit the culture better, but the act of asking for suggestions shows recognition of their judgment.

Recognition can often be provided in an atmosphere of fun. Weinstein (1996) describes 52 ways (one for each week) to create an environment of fun and help create a positive corporate culture.

Sometimes programs of recognition are more useful than monetary rewards. One organization reviewed the results of its suggestion system. About 800 suggestions were received each year; about 25 percent were accepted and personnel received monetary rewards. But most people were dissatisfied—the winners said the decisions took too long and the monetary reward was too low; the others felt their suggestions should have been accepted. Under a new program monetary awards were eliminated and replaced by a simple "thanks." Also, decisions on suggestions must be made within a short time interval. Under the new program, 7700 ideas are received annually, and 60 percent are accepted.

Recognition systems at both the individual and team levels should reinforce both small wins and big victories.

Forms of Rewards

Rewards for quality-related activities are increasingly becoming part of the annual performance evaluation of upper managers, middle managers, specialists, and first-line supervisors. Forms of reward may include changes in base pay, merit increases, incentives, skill-based wages, a bonus, and stock plans.

The weakest area of quality motivation for managers is that of *improvement* of quality—for breakthrough to superior levels of performance. This weakness arises primarily because the problem of control—of meeting this year's goals—has a much higher priority.

Control sets its own priorities. When alarm bells ring, they demand corrective action then and there. The alarms must be heeded or the current goals will not be met. The manager wants to meet these current goals—managerial performance is judged mainly by measuring results against these goals. Firefighters get the best rewards—they are the heroes.

In contrast, improvement of quality is not needed to meet this year's goals—it is needed to attain leadership in some future year, or at least to remain competitive.

Hence improvement can be deferred, whereas control cannot. Moreover, improvement to new levels requires special organizational machinery such as is described in Chapter 3. Such special machinery is not needed to maintain current control.

One company has incorporated performance on improvement activities as a part of the annual appraisal of managers. A rating of less than adequate means that the manager will not receive a salary increase or a promotion for one year.

Texas Instruments measures the contribution to quality of every manager who has a profit-and-loss responsibility. Managers are evaluated annually using four measures: leading indicators, concurrent indicators, lagging indicators, and the cost of quality. The first three refer to quality measures before, during, and after creation of a product or service. For example, the field-complaint level is an example of a lagging indicator.

Upper management must change the reward system to inspire middle management to make breakthroughs to improved quality levels. A prerequisite, however, is for upper management to provide the infrastructure, resources, and training for such breakthroughs (see Chapter 3, "Quality Improvement and Cost Reduction").

A clear trend in management compensation is the increased use of variable compensation programs, i.e., monetary rewards above a base salary. These rewards include various incentive plans, gainsharing, pay for mastering additional skills, and profit sharing. Berman (1997) provides an exhaustive list and shows how the various rewards relate to team-based performance. Often these concepts can be applied to individuals or teams, but team-based programs are becoming increasingly popular.

In the service industry GTE California relates quality improvement to incentive compensation for both individuals and teams (Bowen, 1988). Team objectives that apply to key performance units or are companywide have both quality and cost objectives. The weights assigned to the different levels and to cost and quality are shown in Figure 9.7.

Kluge (1996) reports how Varian X-Ray Tube Products installed an incentive compensation plan on quality. Money is added or subtracted for every tube manufactured to a quality incentive pool—\$125 added for every good tube produced, \$500 subtracted for each tube scrapped, and \$600 subtracted for each dead-on-arrival tube that is returned from customers. At the end of each quarter, the money in the pool is divided equally among all people who influenced product quality. Previously, the culture was geared toward meeting production numbers at the end of each month. The incentive plan "radically changed the work ethic" and now employees focus on driving down quality costs. Based on an average salary, employees received a 13 percent bonus (and forfeited a 3 percent merit increase). All bonus money was funded from the reduced quality costs. Significant improvements were made in cost of scrap, dead-on-arrival tubes, yield, and market share.

Some organizations, large and small, use "gainsharing" to distribute savings from improvement activities. Typically, a formula defines the distribution to customers, employees, and the company. Gainsharing is a type of group incentive plan. Stratton (1998) describes how a gainsharing plan at a small manufacturing company reduced defects from 3.7 percent to 1.0 percent and generated an annual gainsharing of more than \$2500 per employee.

Survey after survey indicates that employees' primary concerns are lack of recognition and lack of involvement in decision making on the job. On the positive side, Lawler et al. (1995) show that people are motivated by success in making improve-

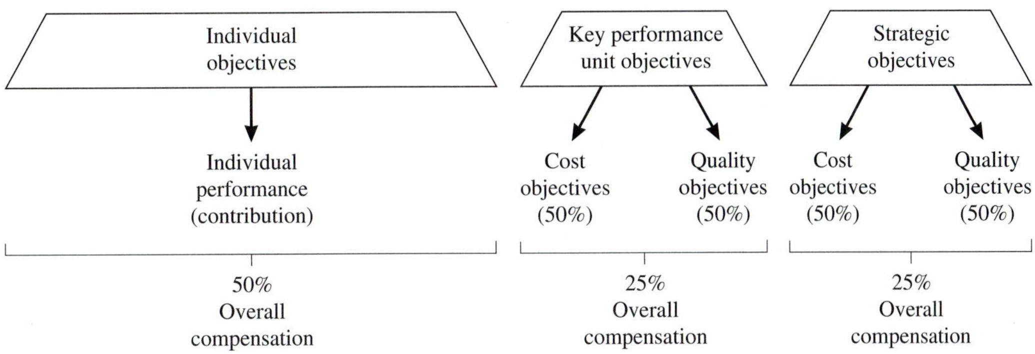

FIGURE 9.7
Incentive compensation. (*From Bowen, 1988.*)

ments and satisfying customers. These intrinsic motivators also include actions that make jobs more stimulating, build competence and pride, promote self-control, and result in a sense of achievement for employees. These matters provide the positive reinforcement of desired behavior that are the foundation of a strong quality culture. For elaboration, see *JQH5,* pages 15.26–15.28.

SUMMARY

- To become superior in quality, we need to (1) develop technologies to create products and processes that meet customer needs and (2) stimulate a culture that continually views quality as a primary goal.
- The culture for quality *can* be changed. We need to provide
  Goals and measurements.
  Evidence of upper-management leadership.
  Self-development and empowerment.
  Participation.
  Recognition and rewards.
- These elements must be integrated with the methodologies and structure for quality.
- To change culture requires years, not months; to change quality requires trust, not techniques.

PROBLEMS

9.1. Consider an individual sport such as golf or racquetball. Apply the three principles of self-control to such a sport and decide whether individuals are in a state of self-control. How does your conclusion relate to the design of jobs in an organization?

9.2. Prepare two lists to analyze the forces that have an impact on implementing quality improvement teams. One list should contain the restraining forces that are obstacles to implementing teams (e.g., lack of time); the other list should show the driving forces that can help (e.g., support by upper management). (This approach is part of a "force field analysis" or a "barriers and aids analysis.")

9.3. Look about your community to identify some of the ongoing drives or campaigns, such as for traffic safety, fund-raising, political election, or to keep your city clean. For any campaign, analyze and report on
(a) The methods used to secure attention.
(b) The methods used to secure interest and identification with the program.
(c) The methods used to secure action.

9.4. For any organization known to you, study the prevailing continuing program of motivation for quality. Report your findings on (1) the ingredients of the program and (2) the organization's effectiveness in carrying out the various aspects of the program.

9.5. Develop three quality indicators for the performance of a buyer in a purchasing department: a lagging indicator, a concurrent indicator, and a leading indicator.

9.6. Tektronix, Inc. (1985) describes "people involvement" as moving through nine types of management (a "continuum"). These are autocratic authoritarian management, directive command, selective information sharing, employee input, problem-solving groups, ad hoc task forces, participative decision making, work redesign and goal setting, and semiautonomous teams. For each type, describe in a few sentences the respective roles of managers and nonmanagers.

9.7. Speak with at least three people, preferably from different organizations. Identify three specific situations that contribute to a negative quality culture. Identify three situations that contribute to a positive quality culture.

REFERENCES

Argyris, C. (1998). "Empowerment: The Emperor's New Clothes," *Harvard Business Review,* May–June, pp. 98–105.

Berman, S. J. (1997). "Using the Balanced Scorecard and Variable Compensation as a Catalyst for Team Based Performance," *IIE Solutions Conference Proceedings*, Institute of Industrial Engineers, Norcross, GA.

Bowen, M. D. (1988). "Quality Improvement through Incentive Compensation," *Impro Conference Proceedings,* Juran Institute, Inc., Wilton, CT, pp. 3A-21 to 3A-24.

Bunn, T. O. (1995). "Changing the Culture of the NVSL," *Impro Conference Proceedings,* Juran Institute, Inc., Wilton, CT, pp. 2C.3-1 to 2C.3-10.

Cameron, K. and W. Sine (1999). "A Framework for Organizational Quality Culture," *Quality Management Journal,* vol. 6, no. 4, pp. 7–25.

Graber, J. M., R. E. Breisch, and W. E. Breisch (1992). "Performance Appraisals and Deming: A Misunderstanding," *Quality Progress,* June, pp. 59–62.

Hackman, J. R. and G. R. Oldham (1980). *Work Redesign.* Addison-Wesley, Reading, MA.

Herzberg, F., B. Mausman, and B. Synderman (1959). *The Motivation to Work,* 2nd ed., John Wiley and Sons, New York.

Kluge, R. H. (1996). "An Incentive Compensation Plan with an Eye on Quality," *Quality Progress,* December, pp. 65–68.

Kotter, J. P. (1999). *John P. Kotter on What Leaders Really Do,* Harvard Business School Press, Boston.

Lawler, E. E. III, S. A. Mohrmann, and G. E. Ledford Jr. (1995). *Creating High Performance Organizations: Practices and Results of Employee Involvement and Total Quality Management in Fortune 1000 Companies,* Jossey-Bass, San Francisco.

Mackin, T. (1999). "The Three Dimensions of Organizations of Quality," Florida Sterling Quality Conference, Orlando, FL.

Maslow, A. H. (1987). *Motivation and Personality,* 3rd ed., Harper & Row, New York.

McGregor, D. (1985). *The Human Side of Enterprise,* McGraw-Hill, New York.

Miletich, S. L. (1997). "Seven Cultural Dimensions of an Organization," Florida Sterling Quality Conference, Orlando, FL.

Miller, L. M. (1984). *American Spirit: Visions of a New Corporate Culture,* William Morrow, New York.

Milliman, J. F. and F. R. McFadden (1997). "Toward Changing Performance Appraisal to Address TQM Concerns: The 360-Degree Feedback Process," *Quality Management Journal,* vol. 4, no. 3, pp. 44–64.

Rubinstein, S. P. and J. Ryan (1996). "Survival for Quality and Unions," *Quality Progress,* July, pp. 50–53.

Rummler, G. A. and A. P. Brache (1995). *Improving Performance,* 2nd ed., Jossey-Bass, San Francisco.

Ryan, M. J. (1999). Keynote Address, Florida Sterling Conference, Orlando.

Strattton, B. (1998). "Texas Nameplate Company: All You Need Is Trust," *Quality Progress,* October, pp. 29–32.

Tektronix, Inc. (1985). "People Involvement: A Continuum," Tektronix, Inc., Beaverton, OR. All rights reserved. Reproduced with permission.

Vass, D. J. and D. H. Kincade (1999). "Relationship of TQM Implementation and Employee Opinion Survey: A Study of Three Manufacturers," *Quality Management Journal*, vol. 6, no. 1, pp. 60–73.

Warren, J. (1989). "We Have Found the Enemy. It Is Us," *Annual Quality Congress Transactions,* ASQ, Milwaukee, pp. 65–73.

Watson, M. A. and F. M. Gryna (1998). "Assessment of Quality Culture in Small Business: Exploratory Research," Report No. 905, College of Business, University of Tampa, Tampa, FL.

Weinstein, M. (1996). *Managing to Have Fun,* Simon and Schuster, New York.

Wetlaufer, S. (1999a). "Driving Change: An Interview with Ford Motor Company's Jacques Nasser," *Harvard Business Review,* March–April, pp. 76–88.

Wetlaufer, S. (1999b). "Organizing for Empowerment: An Interview with AES's Roger Sant and Dennis Bakke," *Harvard Business Review,* January–February, pp. 110–123.

Zairi, M. (1994). *Measuring Performance for Business Results,* Chapman and Hall, London.

SUPPLEMENTARY READING

Bringing about change: Kotter, J. P. (1996). *Leading Change,* Harvard Business School Press, Boston.

Covey, S. R. (1989). *The Habits of Highly Effective People,* Simon and Schuster, New York.

Corporate culture: Baker, E. M. (1999). *Scoring a Whole in One,* Crisp Publications, Menlo Park, CA.

Cameron, K. and W. Sine (1999). "A Framework for Organizational Quality Culture," *Quality Management Journal,* vol. 6, issue 4, pp. 7–25.

Collins, J. C. and J. I. Porras (1994). *Built to Last,* HarperBusiness, New York.

French, M., J. Pittman, R. Stacy, and J. Wetjen (1993). "Inova Quality Leadership—The Journey Continues," *Impro Conference Proceedings,* Juran Institute, Inc., Wilton, CT.

Hammond, S. A. (1998). *The Thin Book of Appreciative Inquiry,* The Thin Book Publishing Co., Plano, TX.

Managing human performance: *JQH5,* Section 15.

Herzberg, F. (1987). "One More Time—How Do You Motivate Employees?" *Harvard Business Review,* September–October, pp. 109–120.

10

BASIC CONCEPTS OF STATISTICS AND PROBABILITY

10.1 STATISTICAL TOOLS IN QUALITY

Statistics is the methodology used for the collection, organization, analysis, interpretation, and presentation of data. *Probability* is a measure that describes the chance that an event will occur. Experienced practitioners remind us that the data must be meaningful—not just easy to collect.

Since the beginning of the quality movement, practitioners have debated the relative importance of statistical methods versus managerial methods in achieving quality excellence. For an exchange of views, see Gunter (1998) and subsequent letters to the editor.

The author of this book views the knowledge necessary for quality this way: Knowledge of statistical methods is necessary but not sufficient; knowledge of managerial methods is necessary but not sufficient; knowledge of specific industry products and processes is necessary but not sufficient. Integration of these sources of knowledge is the key.

The body of knowledge of statistical methods is an essential tool of the modern approach to quality. Without statistics, drawing conclusions about data becomes lucky at best and disastrous in some cases.

10.2 THE CONCEPT OF STATISTICAL VARIATION

The concept of *variation* states that no two items will be perfectly identical. Variation is a fact of nature and a fact of industrial life. For example, even "identical" twins vary slightly in height and weight at birth.

The dimensions of a large-scale, integrated chip vary from chip to chip; cans of tomato soup vary slightly from can to can; the time required to assign a seat at an airline check-in counter varies from passenger to passenger. To disregard the existence of variation (or to rationalize falsely that it is small) can lead to incorrect decisions on major problems. Statistics helps to analyze data properly and draw conclusions, taking into account the existence of variation.

Statistical variation—variation due to random causes—is much greater than most people think. Often we decide what action to take based on the most recent data point, and we forget that the data point is part of a history of data. Malcolm Roberts once said: "Many managers run their systems by the last data point."

Data summarization can take several forms: tabular, graphical, and numerical. Sometimes one form will provide a useful, complete summarization. In other cases two or even three forms are needed for complete clarity.

10.3 TABULAR SUMMARIZATION OF DATA: FREQUENCY DISTRIBUTION

A *frequency distribution* is a tabulation of data arranged according to size. The raw data of the electrical resistance of 100 coils are given in Table 10.1. Table 10.2 shows the frequency distribution of these data with all measurements tabulated at their actual values. For example, there were 14 coils each of which had a resistance of 3.35 ohms (Ω); there were 5 coils each of which had a resistance of 3.30 Ω. The frequency distribution spotlights where most of the data are grouped (the data are centered about a resistance of 3.35) and how much variation there is in the data (resistance runs from 3.27 to 3.44 Ω). Table 10.2 shows the conventional frequency distribution and the cumulative frequency distribution in which the frequency values are accumulated to show the number of coils with resistances equal to or less than a specific value. The particular problem determines whether the conventional, cumulative, or both distributions are required.

TABLE 10.1
Resistance of 100 coils, Ω

3.37	3.34	3.38	3.32	3.33	3.28	3.34	3.31	3.33	3.34
3.29	3.36	3.30	3.31	3.33	3.34	3.34	3.36	3.39	3.34
3.35	3.36	3.30	3.32	3.33	3.35	3.35	3.34	3.32	3.38
3.32	3.37	3.34	3.38	3.36	3.37	3.36	3.31	3.33	3.30
3.35	3.33	3.38	3.37	3.44	3.32	3.36	3.32	3.29	3.35
3.38	3.39	3.34	3.32	3.30	3.39	3.36	3.40	3.32	3.33
3.29	3.41	3.27	3.36	3.41	3.37	3.36	3.37	3.33	3.36
3.31	3.33	3.35	3.34	3.35	3.34	3.31	3.36	3.37	3.35
3.40	3.35	3.37	3.35	3.32	3.36	3.38	3.35	3.31	3.34
3.35	3.36	3.39	3.31	3.31	3.30	3.35	3.33	3.35	3.31

TABLE 10.2
Tally of resistance values of 100 coils

Resistance, Ω	Tabulation	Frequency	Cumulative frequency
3.45			
3.44	𝍷	1	1
3.43			
3.42			
3.41	𝍷𝍷	2	3
3.40	𝍷𝍷	2	5
3.39	𝍷𝍷𝍷𝍷	4	9
3.38	𝍸 𝍷	6	15
3.37	𝍸 𝍷𝍷𝍷	8	23
3.36	𝍸 𝍸 𝍷𝍷𝍷	13	36
3.35	𝍸 𝍸 𝍷𝍷𝍷𝍷	14	50
3.34	𝍸 𝍸 𝍷𝍷	12	62
3.33	𝍸 𝍸	10	72
3.32	𝍸 𝍷𝍷𝍷𝍷	9	81
3.31	𝍸 𝍷𝍷𝍷𝍷	9	90
3.30	𝍸	5	95
3.29	𝍷𝍷𝍷	3	98
3.28	𝍷	1	99
3.27	𝍷	1	100
3.26			
Total		100	

TABLE 10.3
Frequency table of resistance values

Resistance, Ω	Frequency
3.415–3.445	1
3.385–3.415	8
3.355–3.385	27
3.325–3.355	36
3.295–3.325	23
3.265–3.295	5
Total	100

When there are a large number of highly variable data, the frequency distribution can become too large to serve as a summary of the original data. The data may be grouped into *cells* to provide a better summary. Table 10.3 shows the frequency distribution for these data grouped into six cells, each 0.03 Ω wide. Grouping the data into cells condenses the original data, and therefore some detail is lost.

The following is a common procedure for constructing a frequency distribution:

1. Decide on the number of cells. Table 10.4 provides a guide.
2. Calculate the approximate cell interval i. The cell interval equals the largest observation minus the smallest observation divided by the number of cells. Round this

TABLE 10.4
Number of cells in frequency distribution

Number of observations	Recommended number of cells
20–50	6
51–100	7
101–200	8
201–500	9
501–1000	10
Over 1000	11–20

result to some convenient number (preferably the nearest *uneven* number with the same number of significant digits as the actual data).

3. Construct the cells by listing cell boundaries.
   a. Each cell boundary should be to one more significant digit than the actual data and should end in a 5.
   b. The cell interval should be constant throughout the entire frequency distribution.
4. Tally each observation into the appropriate cell and then list the total frequency f for each cell.

This procedure should be adjusted when necessary to provide a clear summary of the data and to reveal the underling pattern of variation.

10.4 GRAPHICAL SUMMARIZATION OF DATA: THE HISTOGRAM

A *histogram* is a vertical bar chart of a frequency distribution. Figure 10.1 shows the histogram for the electrical resistance data. Note that as in the frequency distribution, the histogram highlights the center and amount of variation in the sample of data. The simplicity of construction and interpretation of the histogram makes it an effective tool in the elementary analysis of data.

Graphical methods are essential to effective data analysis and clear presentation of results. Many of these methods are used throughout this book. More are available. Experienced practitioners are fascinated by the graphical tools, and rightly so. The vividness of a picture when compared to the cold logic of numbers has practical benefits, e.g., identifying subtle relationships and presenting results in clear form. Experience dictates that the first step in data analysis is to plot the data.

An innovative variation of the histogram is the stem-and-leaf plot. Heyes (1985) presents data on wire break strength in grams (see Table 10.5) for supplier A. The corresponding stem-and-leaf plot is shown in Figure 10.2. Note that the stem is the first digit(s) of each value and the leaf is the remaining digits; e.g., for a value of 216, the stem is 2 and the leaf is 16. Note that this plot not only reveals the shape of the histogram but also makes it possible to regain the original values of the data.

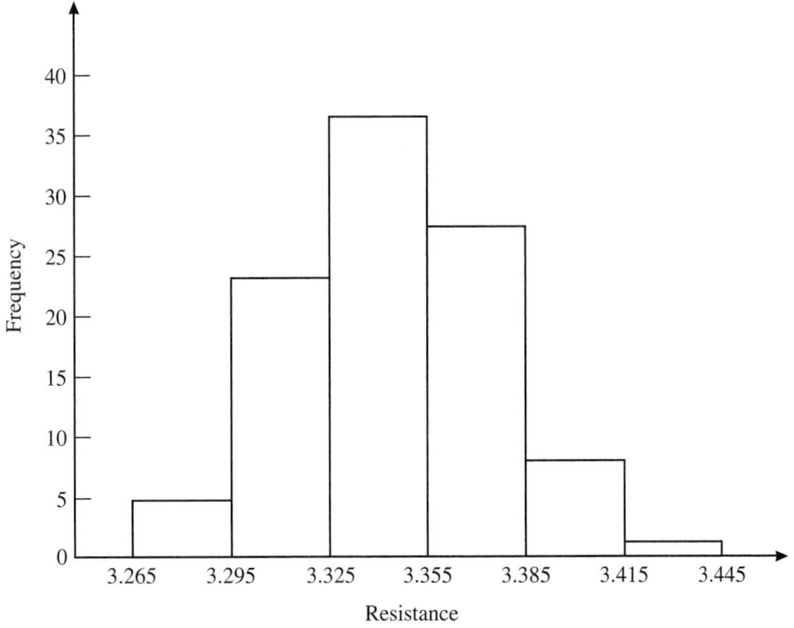

FIGURE 10.1
Histogram of resistance.

TABLE 10.5
Original data on wire break strength

1. 346	6. 402	11. 368
2. 338	7. 635	12. 376
3. 323	8. 281	13. 311
4. 438	9. 431	14. 379
5. 398	10. 390	15. 216

Stem	Leaf
2	16, 81
3	11, 23, 38, 46, 68, 76, 79, 90, 98
4	02, 31, 38
5	—
6	35

FIGURE 10.2
Stem-and-leaf plot. (*From Heyes, 1985.*)

10.5 BOX-AND-WHISKER PLOTS

A simple, clever, and effective way to summarize data is a box-and-whisker plot (usually called a "boxplot"). The *boxplot* is a graphical five-number summary of the data. In the basic (or "skeletal") boxplot, the five values are the median, maximum value, minimum value, first quartile, and third quartile. The quartiles are the values below which one-fourth and three-fourths of the observations lie.

Using the wire break strength data, the data are first arranged in rank order (see Table 10.6). The median is the middle value (the eighth rank, or 376). The extreme values are 216 and 635. The quartiles are 323 and 402 because those values divide the data into quarters. Figure 10.3 shows the resulting boxplot. The box, bounded by the two quartiles with the median inside the box, summarizes the middle part of the data. The lines extending out to the extreme values are the "whiskers." The longer whisker on the right suggests that the data include some values that are much larger than the other values. Also, the location of the median indicates that the values above the median are, as a group, closer to the median than the values below the median.

Innovative methods of graphical analysis and display of data are discussed in a now classic text by Tukey (1977). An excellent summary of graphical methods including the boxplot is presented in Wadsworth et al. (2001).

In another example a large service organization had a problem of poor reliability of copy machines. The copiers were repaired by two different firms (A and B). An improvement team conducted a study to see which contractor provided the faster and more consistent response to a repair call. The response time, in minutes, was recorded

TABLE 10.6
Ordered data on wire break strength

1. 216	6. 346	11. 398
2. 281	7. 368	12. 402
3. 311	8. 376	13. 431
4. 323	9. 379	14. 438
5. 338	10. 390	15. 635

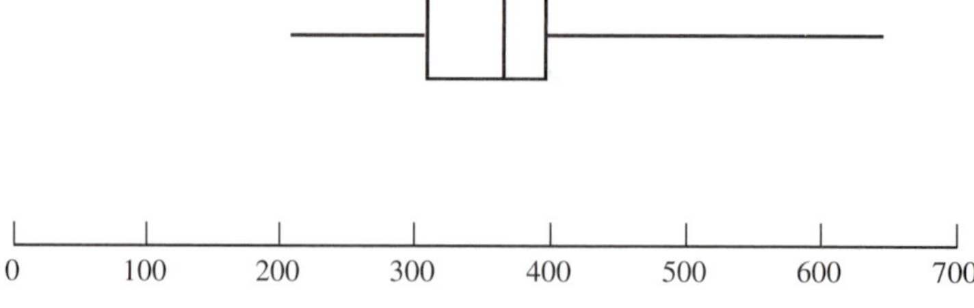

FIGURE 10.3
Boxplot. (*From Heyes, 1985.*)

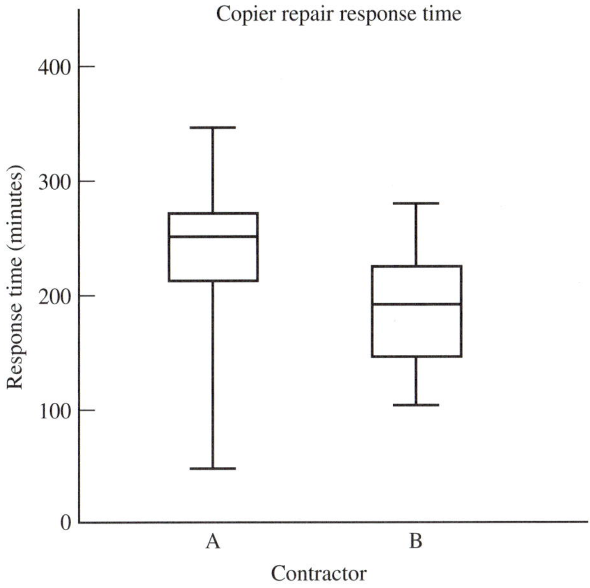

FIGURE 10.4
Basic boxplots of copier repair response time. (*Source: Juran Institute, Inc.*)

for a sample of 10 repair calls for each contractor. The resulting boxplots are shown in Figure 10.4. The team concluded that contractor B was superior because B was usually faster and more consistent than A.

10.6 GRAPHICAL SUMMARIZATION OF TIME-ORIENTED DATA: THE RUN CHART

The histogram is a simple and effective way to summarize data according to size. But it misses a dimension of data that is often important—time. The run chart is a plot of the data versus time. Such plots can reveal trends, cycles, and other changes over time. An example is presented in Figure 10.5.

More powerful plots of data over time are presented in Chapter 18, "Statistical Process Control."

Readers should be cautious in employing certain graphical methods of the popular process. Some of these (e.g., three-dimensional bar graphs) sacrifice clarity for glitz. An excellent summary of graphical methods in quality is presented in Wadsworth, et al. (2000). Harris (1996) provides a comprehensive reference on "information graphics." For unusually creative ideas on displaying all types of information, see Tufte (1997).

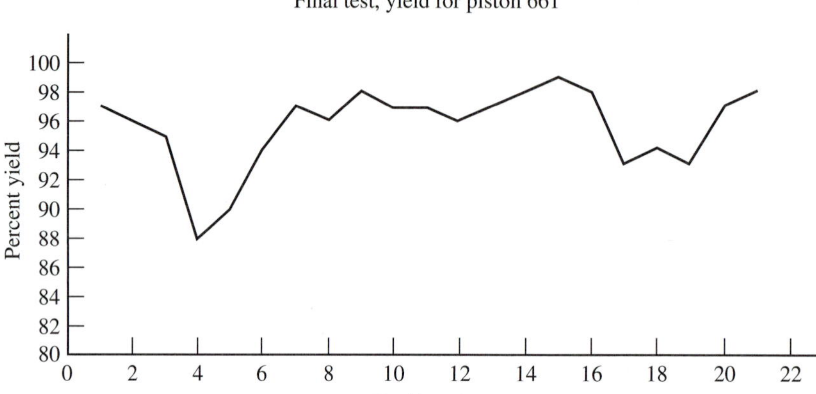

FIGURE 10.5
Run chart. (*Source: Juran Institute, Inc.*)

10.7 METHODS OF SUMMARIZING DATA: NUMERICAL

Data can also be summarized by computing (1) a measure of central tendency to indicate where most of the data are centered and (2) the measure of dispersion to indicate the amount of scatter in the data. Often these two measures provide an adequate summary.

The key measure of the central tendency is the *arithmetic mean,* or average. The definition of the *average* is

$$\bar{X} = \frac{\sum_{i=1}^{n} X_i}{n}$$

where $\bar{X}$ = sample mean
X_i = individual observations
n = number of observations
$\sum_{i=1}^{n}$ = summation of the X_i

Another measure of central tendency is the *median*—the middle value when the data are arranged according to size. The median is useful for reducing the effects of extreme values or for data that can be ranked but are not easily measurable, such as color or visual appearance.

Two measures of dispersion are commonly calculated. When the amount of data is small (10 or fewer observations), the range is useful. The *range* is the difference between the maximum value and the minimum value in the data. As the range is based on only two values, it is not as useful when the number of observations is large.

In general, the *standard deviation* is the most useful measure of dispersion. Like the mean, the definition of the standard deviation is a formula:

$$s = \sqrt{\frac{\sum_{i=1}^{n} (X - \bar{X})^2}{n - 1}}$$

where s is the sample standard deviation. The square of the standard deviation, s^2, is called the *variance*.

There is usually difficulty in understanding the "meaning" of the standard deviation. The only definition is a formula. There is no hidden meaning to the standard deviation, and it is best viewed as an index that shows the amount of variation in a set of data. Later applications of the standard deviation to making predictions will help clarify its meaning.

One useful technique is to calculate a relative measure of variation as the standard deviation divided by the mean (the coefficient of variation).

A problem that sometimes arises in the summarization of data is that one or more extreme values are far from the rest of the data. A simple (but not necessarily correct) solution is available: drop such values. The reasoning is that a measurement error or some other unknown factor makes the values "unrepresentative." Unfortunately, this approach may be rationalizing to eliminate an annoying problem of data analysis. The decision to keep or discard extreme values rests with the investigator. However, statistical tests are available to help make the decision (see *JQH5,* Section 44).

10.8
PROBABILITY DISTRIBUTIONS: GENERAL

A distinction is made between a sample and a population. A *sample* is a limited number of items taken from a larger source. A *population* is a large source of items from which the sample is taken. Measurements are made on the items. We take measurements from a sample and calculate a sample statistic, e.g., the mean. A *statistic* is a quantity computed from a sample to estimate a population parameter. It is usually assumed that the sample is a random one; i.e., each possible sample of n items has an equal chance of being selected (or the items are selected systematically from material that is itself random due to mixing during processing).

A *probability distribution function* is a mathematical formula that relates the values of the characteristic with their probability of occurrence in the *population.* The collection of these probabilities is called a probability distribution. The mean (μ) of a probability distribution is often called the expected value. Some distributions and their functions are summarized in Figure 10.6. Distributions are of two types:

1. *Continuous* (for "variables" data). When the characteristic being measured can take on any value (subject to the fineness of the measuring process), its probability distribution is called a "continuous probability distribution." For example, the probability

Distribution	Form	Probability function	Comments on application
Normal	μ	$y = \frac{1}{\sigma\sqrt{2\pi}} e^{-\frac{(X-\mu)^2}{2\sigma^2}}$ μ = Mean σ = Standard deviation	Applicable when there is a concentration of observations about the average and it is equally likely that observations will occur above and below the average. Variation in observations is usually the result of many small causes.
Exponential	μ	$y = \frac{1}{\mu} e^{-\frac{x}{\mu}}$	Applicable when it is likely that more observations will occur below the average than above.
Weibull	$\beta = 1/2$ $\alpha = 1$ $\beta = 1$ $\beta = 3$ X	$y = \alpha\beta(X-\gamma)^{\beta-1}e^{-\alpha(X-\gamma)^{\beta}}$ α = Scale parameter β = Shape parameter γ = Location parameter	Applicable in describing a wide variety of patterns in variation, including departures from the normal and exponential.
Poisson*	$p = 0.1$ $p = 0.3$ $p = 0.5$ r	$y = \frac{(np)^r e^{-np}}{r!}$ n = Number of trials r = Number of occurrences p = Probability of occurrence	Same as binomial but particularly applicable when there are many opportunities for occurrence of an event but a low probability (less than 0.10) on each trial.
Binomial*	$p = 0.1$ $p = 0.3$ $p = 0.5$ r	$y = \frac{n!}{r!(n-r)!} p^r q^{n-r}$ n = Number of trials r = Number of occurrences p = Probability of occurrence $q = 1 - p$	Applicable in defining the probability of r occurrences in n trials of an event that has a constant probability of occurrence on each independent trial.

FIGURE 10.6

Summary of common probability distributions. (Asterisks indicate that these are discrete distributions, but the curves are shown as continuous for ease of comparison with the continuous distributions. Strictly, the actual plots are discrete, not continuous.)

distribution for the resistance data of Table 10.1 is an example of a continuous probability distribution because the resistance could have any value, limited only by the fineness of the measuring instrument. Most continuous characteristics follow one of several common probability distributions, i.e., the normal distribution, the exponential distribution, and the Weibull distribution. These distributions find the probabilities associated with occurrences of the *actual values* of the characteristic. Other continuous distributions (e.g., *t, F,* and chi square) are important in data analysis but are not helpful in directly predicting the probability of occurrence of actual values.

2. *Discrete* (for "attributes" data). When the characteristic being measured can take on only certain specific values (e.g., integers 0, 1, 2, 3), its probability distribution is called a "discrete probability distribution." For example, the distribution for the number of defects r in a sample of five items is a discrete probability distribution because r can only be 0, 1, 2, 3, 4, or 5. The common discrete distributions are the Poisson and binomial (see Figure 10.6).

The following paragraphs explain how probability distributions can be used with a sample of observations to make predictions about the larger population. Such predictions assume that the data come from a process that is stable over time, which is not always the case. Plotting the data points in order of production provides a rough test for stability; plotting the data on a statistical control chart provides a rigorous test (see Chapter 18, "Statistical Process Control").

10.9 THE NORMAL PROBABILITY DISTRIBUTION

Many engineering characteristics can be approximated by the *normal distribution function:*

$$y = \frac{1}{\sigma\sqrt{2\pi}} e^{-(X-\mu)^2/2\sigma^2}$$

where $e = 2.718$
$\pi = 3.141$
μ = population mean
σ = population standard deviation

Problems are solved with a table, but note that the distribution requires only the average μ and standard deviation σ of the population.[1] The curve for the normal probability distribution is related to a frequency distribution and its histogram. As the sample becomes larger and larger and the width of each cell becomes smaller and smaller, the histogram approaches a smooth curve. If the entire population were measured and if it were normally distributed, the result would be as shown in Figure 10.6. Thus the *shape* of a histogram of sample data provides some indication of the probability distribution for the population. If the histogram resembles[2] the "bell" shape shown in Figure 10.6, this is a basis for assuming that the population follows a normal probability distribution.

[1]Unless otherwise indicated, Greek symbols will be used for population values and Roman symbols for sample values.

[2]The sample histogram need not look as if it came from a normal population. The assumption of normality is applied only to the population. Small deviations from normality are expected in random samples.

Using the Normal Probability Distribution to Make Predictions

Predictions require just two estimates and a table. The estimates are

$$\text{Estimate of } \mu \text{ is } \bar{X} \qquad \text{Estimate of } \sigma \text{ is } s$$

Sometimes $\hat{\sigma}$ is used to denote the estimate of σ. The calculations of the sample $\bar{X}$ and s are made by the methods previously discussed.

For example, from past experience a manufacturer concludes that the burnout time of a particular light bulb follows a normal distribution. A sample of 50 bulbs has been tested, and the average life is 60 days with a standard deviation of 20 days. How many bulbs in the entire population of light bulbs can be expected to be still working after 100 days of life?

The problem is to find the area under the curve beyond 100 days (see Figure 10.7). The area under a distribution curve between two stated limits represents the probability of occurrence. Therefore, the area beyond 100 days is the probability that a bulb will last more than 100 days. To find the area, calculate the difference Z between a particular value and the average of the curve in units of standard deviation:

$$Z = \frac{X - \mu}{\sigma}$$

In this problem $Z = (100 - 60) \div 20 = +2.0$. Table A in Appendix II shows a probability of 0.9773 for $Z = 2$. The statistical distribution tables provide probabilities that cover the span from $-\infty$ up to and including the value of X included in the formula (i.e., cumulative probabilities). Thus 0.9773 is the probability that a bulb will last 100 days or less. The normal curve is symmetrical about the average, and the total area is 1.000. The probability of a bulb lasting more than 100 days then is 1.0000 − 0.9773, or 0.0227, or 2.27 percent of the bulbs in the population will still be working after 100 days.

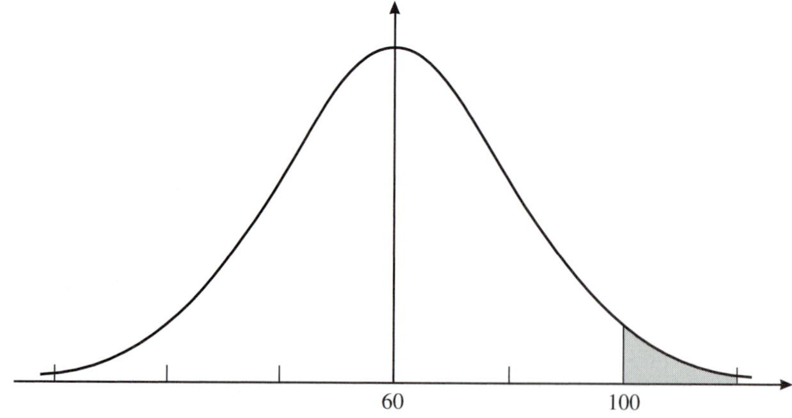

FIGURE 10.7
Distribution of light bulb life.

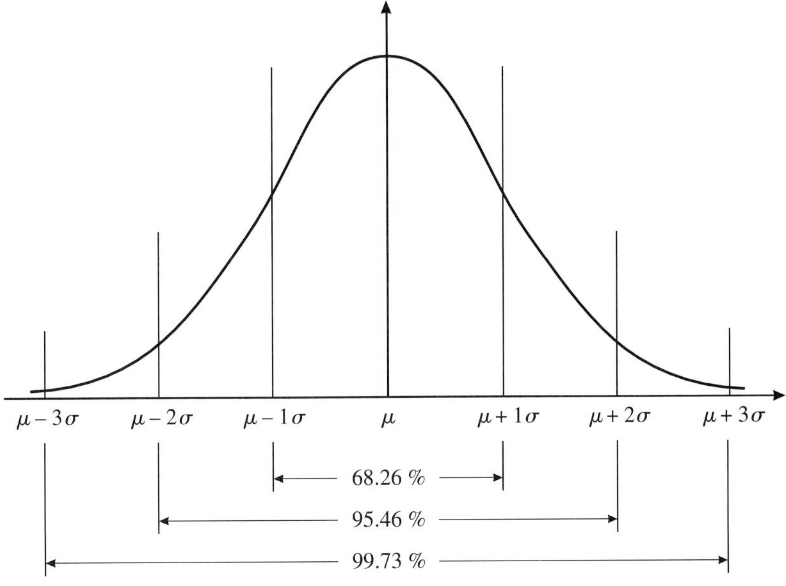

FIGURE 10.8
Areas of the normal curve (derived from Appendix II, Table A).

Similarly, if a characteristic is normally distributed and if estimates of the average and standard deviation of the population are obtained, this method can estimate the total percentage of production that will fall within engineering specification limits.

Figure 10.8 shows representative areas under the normal distribution curve. Thus 68.26 percent of the *population* will fall between the average of the population plus or minus 1 standard deviation of the population, 95.46 percent of the population will fall between the average and $\pm 2\sigma$, and finally, $\pm 3\sigma$ will include 99.73 percent of the population. The percentage of a *sample* within a set of limits can be quite different from the percentage within the same limits in the population.

10.10 THE NORMAL CURVE AND HISTOGRAM ANALYSIS

Because many processes produce results that reasonably follow a normal distribution, we can combine the histogram concept and the normal curve concept to yield a practical working tool known as *histogram analysis*.

A random sample is selected from the process, and measurements are made for the selected quality characteristics. A histogram is prepared and specification limits are added. Knowledge of the manufacturing process is then combined with insights provided by the histogram to draw conclusions about the ability of the process to meet the specifications.

Figure 10.9 shows 16 typical histograms. Readers are encouraged to interpret each of these pictures by asking two questions:

1. Does the process have the ability to meet the specification limits (process capability)?
2. What action on the process, if any, is appropriate?

These questions can be answered by analyzing the following characteristics:

1. *The centering of the histogram.* This defines the aim of the process.
2. *The width of the histogram.* This defines the variability about the aim.
3. *The shape of the histogram.* When a normal or bell-shaped curve is expected, then any significant deviation or other aberration is usually caused by a manufacturing (or other) condition that may be the root of the quality problem. For example, histograms with two or more peaks may reveal that several "populations" have been mixed together.

Note that for six-sigma quality, we need a process (1) centered between the limits, (2) with small variation so that each limit is six standard deviations from the mean, and (3) normally distributed. For elaboration, see Chapter 18.

An example of a histogram from the service industry is presented in Figure 10.10. Data were collected from four bank tellers on the time to conduct a transaction during the busy lunch-hour period. Note the similarity of the histogram to Figure 10.9*g*. The two peaks suggest a combination of two bell-shaped distributions, which in turn suggests two different work processes. Even when separate histograms by teller were plotted, the same double-peaked histograms resulted. Then the types of transactions were examined. Developing histograms by type of transaction revealed the reason for the double-peaked histograms. Simple transactions such as withdrawals and deposits resulted in short transaction times; complex transactions such as opening retirement accounts and obtaining certificates of deposit resulted in longer transaction times. Originally, management believed that two inexperienced tellers were the primary reason for delays in serving customers. Histogram analysis revealed the true causes as the extra steps required for complex transactions.

Histograms illustrate how variables data provide much more information than do attributes data. For example, Figures 10.9*b, d, g,* and *i* warn of potential trouble, even though all units in the sample are within the specification limits. With attributes measurement all the units would simply be classified as acceptable, and the inspection report would have stated "50 inspected, 0 defective"—therefore no problem. One customer had a dramatic experience based on a lot that yielded a sample histogram similar to Figure 10.9*i*. Although the sample indicated that the lot met quality requirements, the customer realized that the supplier must have made much scrap and screened it out before delivery. A rough calculation indicated that full production must have been about 25 percent defective. The histogram enabled the customer to deduce this *without ever having been inside the supplier's plant.* Note how the "product tells on the process." As the customer would eventually pay for this scrap (in the selling price), he wanted the situation corrected. The supplier was contacted, and advice was offered in a constructive manner.

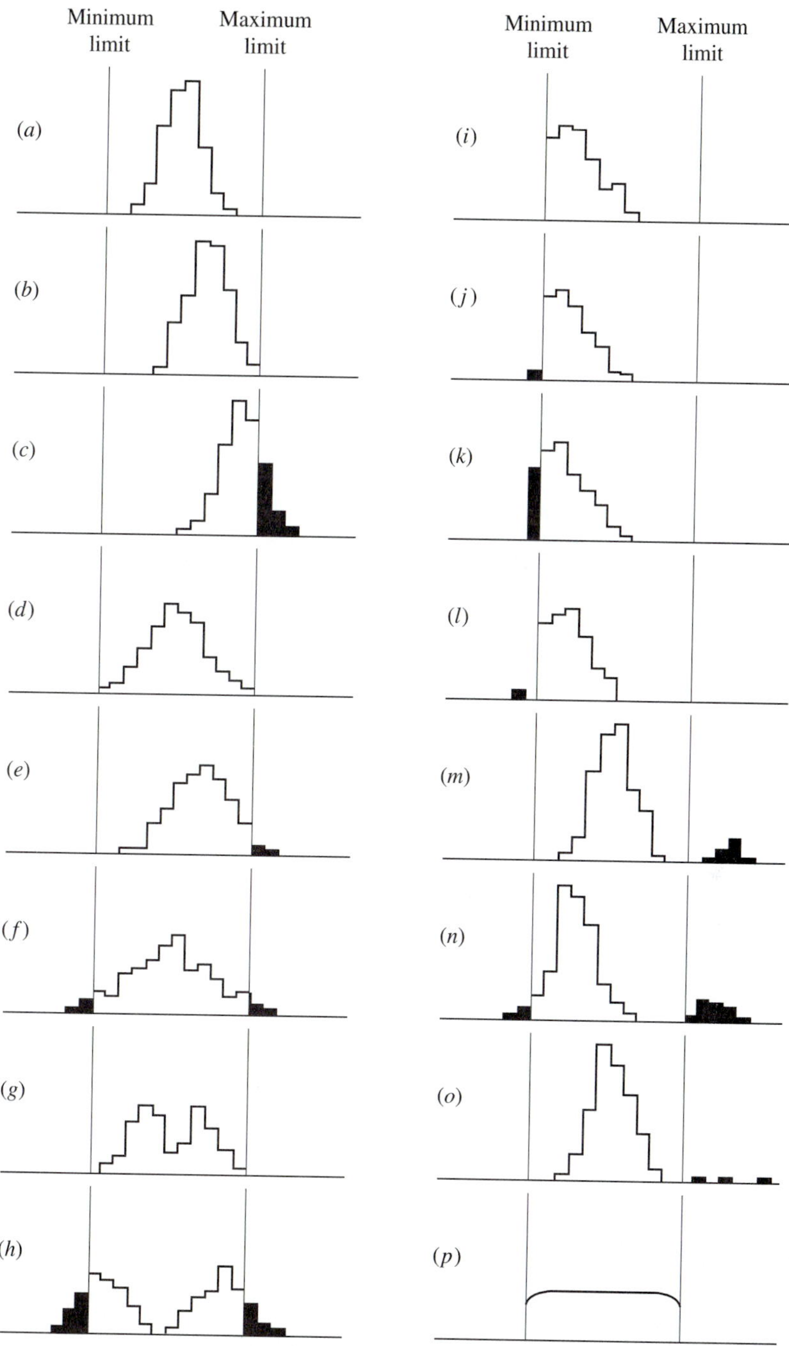

FIGURE 10.9
Distribution patterns related to tolerances. (*Adapted from Armstrong and Clarke, 1946.*)

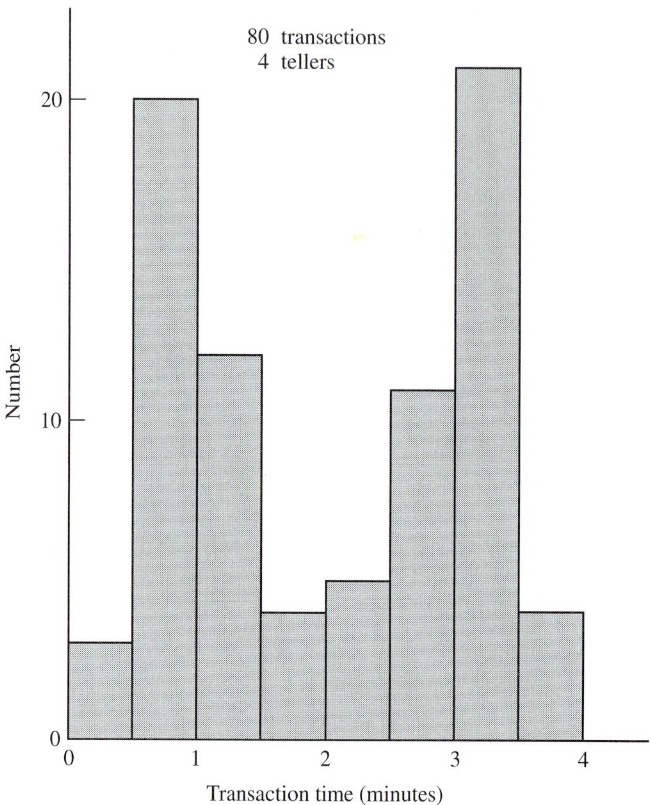

FIGURE 10.10
Histogram of bank teller transaction times. (*Juran Institute, Inc.*)

As a general rule, at least 50 measurements are needed for the histogram to reveal the basic pattern of variation. Histograms based on too few measurements can lead to incorrect conclusions because the shape of the histogram may be incomplete without the observer realizing it. Note that the discussion here is based on the assumption of normality.

Histograms have limitations. Because the samples are taken at random rather than in the order of manufacture, the time-to-time process trends during manufacture are not disclosed. Hence the seeming central tendency of a histogram may be illusory—the process may have drifted substantially. In like manner, the histogram does not disclose whether the supplier's process was operating at its best, i.e., whether it was in a state of statistical control (see Chapter 18).

In spite of these shortcomings, the histogram is an effective analytical tool. The key to its usefulness is its simplicity. It speaks a language that everyone understands—comparison of product measurements against specification limits. To draw useful conclusions from this comparison requires little experience in interpreting frequency distributions and no formal training in statistics. The experience soon expands to include applications in development, manufacturing, supplier relations, and field data.

10.11 THE EXPONENTIAL PROBABILITY DISTRIBUTION

The *exponential probability function is*

$$y = \frac{1}{\mu} e^{-X/\mu}$$

Figure 10.6 shows the shape of an exponential distribution curve. Note that the normal and exponential distributions have distinctly different shapes. An examination of the tables of areas shows that 50 percent of a normally distributed population occurs above the mean value and 50 percent below. In an exponential population, 36.8 percent is above the mean and 63.2 percent below the mean. This refutes the intuitive idea that the mean is always associated with a 50 percent probability. The exponential curve describes the loading pattern for some structural members because smaller loads are more numerous than are larger loads. The exponential curve is also useful in describing the distribution of failure times of complex equipments. Note the ski jump shape of the curve.

A fascinating property of the exponential distribution is that the standard deviation equals the mean.

Using the Exponential Probability Distribution to Make Predictions

Predictions based on an exponentially distributed population require only an estimate of the population mean. For example, the time between successive failures of a complex piece of repairable equipment is measured, and the resulting histogram is found to resemble the exponential probability curve. For the measurement made, the *mean time between failures* (commonly called MTBF) is 100 hours. What is the probability that the time between two successive failures of this equipment will be at least 20 hours?

The problem is one of finding the area under the curve beyond 20 hours (Figure 10.11). Table B in Appendix II gives the area under the curve beyond any particular value X that is substituted in the ratio X/μ. In this problem,

$$\frac{X}{\mu} = \frac{20}{100} = 0.20$$

From Table B the area under the curve beyond 20 hours is 0.8187. The probability that the time between two successive failures is greater than 20 hours is 0.8187; i.e., there is about an 82 percent chance that the equipment will operate without failure continuously

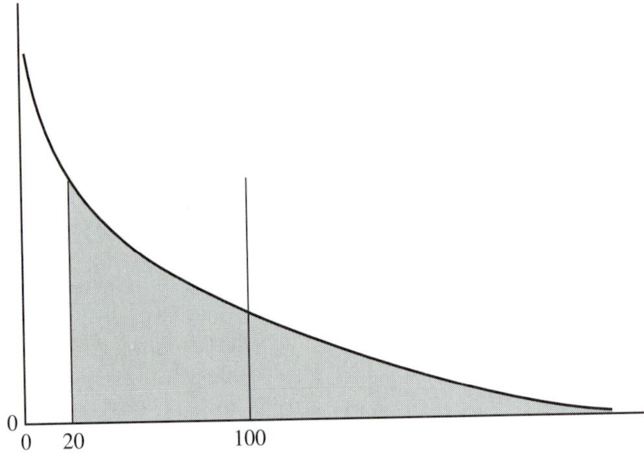

FIGURE 10.11
Distribution of time between failures.

for 20 or more hours. Similar calculations would give a probability of 0.9048 for 10 or more hours.

10.12 THE WEIBULL PROBABILITY DISTRIBUTION

The *Weibull distribution* is a family of distributions having the general function

$$y = \alpha\beta(X - \gamma)^{\beta-1}e^{-\alpha(X - \gamma)\beta}$$

where α = scale parameter
β = shape parameter
γ = location parameter

The curve of the function (Figure 10.6) varies greatly depending on the numerical values of the parameters. Most important is the shape parameter β, which reflects the pattern of the curve. Note that when β is 1.0, the Weibull function reduces to the exponential and that when β is about 3.5 (and $\alpha = 1$ and $\gamma = 0$), the Weibull closely approximates the normal distribution. In practice, β varies from about 1/3 to 5. The scale parameter α is related to the peakedness of the curve; i.e., as α changes, the curve becomes flatter or more peaked. The location parameter γ is the smallest possible value of X. This is often assumed to be 0, thereby simplifying the equation. It is often unnecessary to determine the values of these parameters because predictions are made directly from Weibull probability paper, but King (1981) gives procedures for graphically finding α, β, and γ. Table J in Appendix II provides a sample of Weibull paper. With this sample paper, β can be estimated by drawing a line parallel to the line of best fit and through the point circled on the vertical scale of 40.0. Then, the intersection with the arc gives an estimate of β. For an example, see Section 20.6, "Analyzing Field Data."

The Weibull covers many shapes of distributions. This feature makes the Weibull popular in practice because it reduces the problems of examining a set of data and deciding which of the common distributions (e.g., normal or exponential) fits best. Computer software such as Excel or MINITAB is useful in performing Weibull analyses.

Using the Weibull Probability Distribution to Make Predictions

An analytical approach for the Weibull distribution (even with tables) is cumbersome, and the predictions are usually made with Weibull probability paper. For example, seven heat-treated shafts were stress tested until each of them failed. The fatigue life (in terms of number of cycles to failure) was as follows:

11,251	40,122
17,786	46,638
26,432	52,374
28,811	

The problem is to predict the percentage of failure of the population for various values of fatigue life. The solution is to plot the data on Weibull paper, observe whether the points fall approximately in a straight line, and if so, read the probability predictions (percentage of failure) from the graph.

In a Weibull plot the original data are usually[3] plotted against *mean ranks.* (Thus the mean rank for the ith value in a sample of n ranked observations refers to the mean value of the percentage of the population that would be less than the ith value in repeated experiments of size n.) The mean ranks is calculated as $i/(n + 1)$. The mean ranks necessary for this example are based on a sample size of seven failures and are as shown in Table 10.7. The cycles to failure are now plotted on the Weibull graph paper against the corresponding values of the mean rank (see Figure 10.12). These

TABLE 10.7
Table of mean ranks

Failure number (i)	Mean rank
1	0.125
2	0.250
3	0.375
4	0.500
5	0.625
6	0.750
7	0.875

[3]There are other plotting positions (see e.g., *JQH5,* page 44.32).

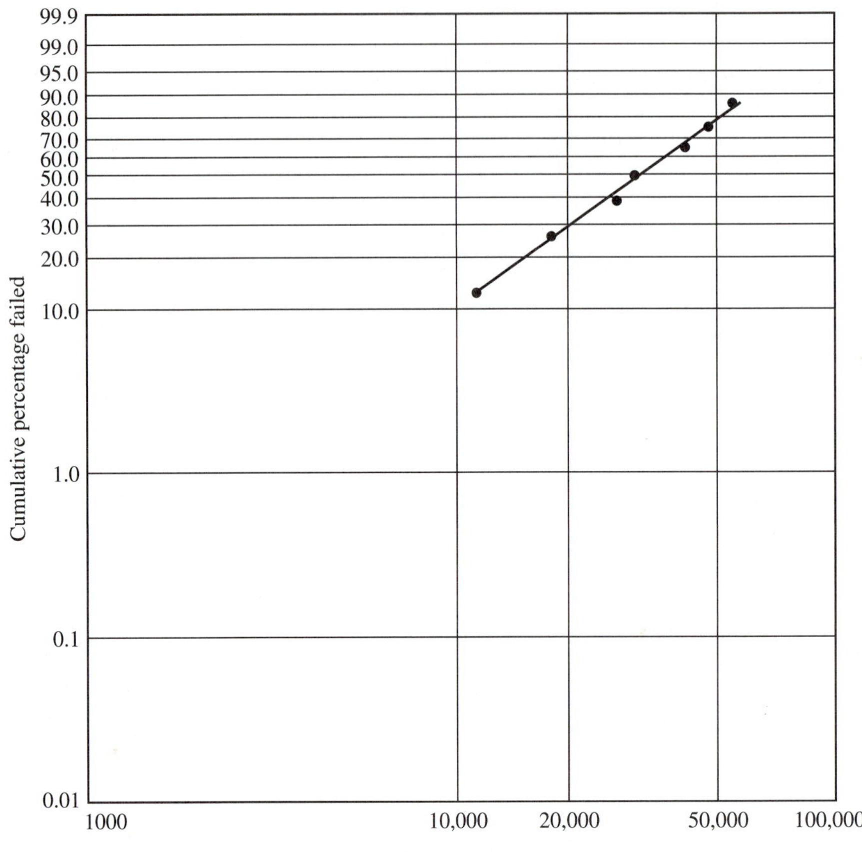

FIGURE 10.12
Distribution of fatigue life.

points fall approximately in a straight line, so it is assumed that the Weibull distribution applies. The vertical axis gives the cumulative percentage of failures in the population corresponding to the fatigue life shown on the horizontal axis. For example, about 50 percent of the population of shafts will fail in fewer than 32,000 cycles. About 80 percent of the population will fail in fewer than 52,000 cycles. By appropriate subtractions, predictions can be made of the percentage of failures between any two fatigue life limits.

It is tempting to extrapolate on probability paper, particularly to predict life. For example, suppose that the minimum fatigue life were specified as 8000 cycles, and the seven measurements displayed earlier were from tests conducted to evaluate the ability of the design to meet 8000 cycles. Because all seven tests exceeded 8000 cycles, the design seems adequate and should therefore be released for production. However, extrapolation on the Weibull paper predicts that about 5 percent of the *population* of shafts would fail in less than 8000 cycles. This information suggests a review of the design before release to production. Thus the small *sample* (all *within* specifications) gives a deceptive result, but the Weibull plot acts as an alarm signal by highlighting a potential problem.

Extrapolation can go in the other direction. Note that a probability plot of life-test data does *not* require that all tests be completed before the plotting starts. As each unit fails, the failure time can be plotted against the mean rank. If the early points appear to be following a straight line, it is tempting to draw in the line *before* all tests are finished. The line can then be extrapolated beyond the actual test data, and the life predictions can be made without accumulating a large amount of test time. The approach has been applied to predicting *early* in a warranty period the "vital few" components of a complex product that will be most troublesome. However, extrapolation has dangers. It requires the judicious melding of statistical theory, engineering experience, and judgment.

To make a valid Weibull plot, at least seven points are needed. Such small samples can have significant sampling variability, and it is wise to calculate confidence limits—see Wadsworth (1998) in the Supplementary Reading. Any fewer casts doubt on the ability of the plot to reveal the underlying pattern of variation. When the plot is prepared, it is hoped that the points will approximate a straight line. This configuration implies a single stable population, and the line can be used for predictions. However, non-straight-line plots are often extremely valuable (in the same way as nonnormal histograms) in suggesting that several populations have been mixed together.

Probability graph paper is available for the normal, exponential, Weibull, and other probability distributions. Although the mathematical functions and tables provide the same information, the graph paper reveals *relationships* between probabilities and values of X that are not readily apparent from the calculations. For example, the reduction in percentage defective in a population as a function of wider and wider specification limits can be easily portrayed by the graph paper.

Other continuous distributions include the continuous uniform (roughly speaking, all values having equal probabilities), log normal (logarithms of the original values are normally distributed), and the multinormal (e.g., a product with two measurement parameters, each normally distributed, is called a bivariate normal). For elaboration, see *JQH5,* Section 44.

The assumption of a particular distribution can be evaluated using a "goodness of fit" test. Such a test determines whether the deviation of a sample result from the theoretical population values is likely to be due to sampling variation (see *JQH5,* page 44.70).

10.13 THE POISSON PROBABILITY DISTRIBUTION

If the probability of occurrence p of an event is constant on each of n independent trials of the event, the probability of r occurrences in n trials is

$$\frac{(np)^r e^{-np}}{r!}$$

TABLE 10.8
Table of Poisson probabilities

r	Probability of r or fewer in sample
0	0.449
1	0.809
2	0.953
3	0.991
4	0.999
5	1.000

where n = number of trials
p = probability of occurrence
r = number of occurrences

Using the Poisson Distribution to Make Predictions

The Poisson is useful in calculating probabilities associated with sampling procedures. Table C in Appendix II directly gives cumulative Poisson probabilities, i.e., the probability of r or fewer occurrences in n trials of an event having probability p. For example, suppose that a lot of 300 units of product is submitted by a vendor whose past quality has been about 2 percent defective. A random sample of 40 units is selected from the lot. Table C in Appendix II provides the probability of r or fewer defectives in a sample of n units. Entering the table with a value of np equal to 40(0.02), or 0.8, for various values of r results in Table 10.8. Individual probabilities can be found by subtracting cumulative probabilities. Thus the probability of exactly two defectives is 0.953 − 0.809, or 0.144. Of course, the probabilities in Table 10.8 could also be found by substituting into the formula six times ($r = 0, 1, 2, 3, 4, 5$).

The Poisson is an approximation to more exact distributions and applies when the sample size is at least 16, the population size is at least 10 times the sample size, and the probability of occurrence p on each trial is less than 0.1. These conditions are often met.

The Poisson is not just an approximation. It can be used as an exact distribution when an event has many opportunities to occur but the probability of occurrence of any opportunity is extremely unlikely.

10.14
THE BINOMIAL PROBABILITY DISTRIBUTION

If the conditions of the Poisson distribution are not met, the binomial distribution may be applicable. If the probability of occurrence p of an event is constant on each of n independent trials of the event, then the probability of r occurrences in n trials is

$$\frac{n!}{r!(n-r)!}p^r q^{n-r}$$

where $q = 1 - p$.

TABLE 10.9
Table of binomial probabilities

r	P (Exactly r defectives in 6) $= [6!/r!\,(6-r)!](0.05)^r(0.95)^{6-r}$
0	0.7351
1	0.2321
2	0.0306
3	0.0021
4	0.0001
5	0.000
6	0.000

In practice, the assumption of a constant probability of occurrence is considered reasonable when the population size is at least 10 times the sample size.[4] (Note that the binomial has fewer conditions than the Poisson.)

Tables for the binomial are available (*JQH5,* Appendix II, pp. 14–15).

Using the Binomial Probability Distribution to Make Predictions

A lot of 100 units of product is submitted by a vendor whose past quality has been about 5 percent defective. A random sample of six units is selected from the lot. The probabilities of various sample results are given in Table 10.9.

In using the formula, note that $0! = 1$.

Other discrete distributions include the hypergeometric (used when the assumptions of the Poisson and binomial cannot be met), discrete uniform (all values have equal probabilities), and the multinomial (when two or more parameters are observed in a sample). See *JQH5,* Section 44.

10.15 BASIC THEOREMS OF PROBABILITY

Probability is expressed as a number that lies between 1.0 (certainty that an event will occur) and 0.0 (impossibility of occurrence).

A convenient definition of probability is one based on a frequency interpretation: If an event A can occur in s cases out of a total of n possible and equally probable cases, the probability that the event will occur is

$$P(A) = \frac{s}{n} = \frac{\text{Number of successful cases}}{\text{Total number of possible cases}}$$

[4]Under this condition, the change in probability from one trial to the next is negligible. If this condition is not met, the hypergeometric distribution should be used (*JQH5,* page 44.24, 44.25).

EXAMPLE 10.1 A lot consists of 100 parts. A single part is selected at random, and thus each of the 100 parts has an equal chance of being selected. Suppose that a lot contains a total of eight defectives. Then the probability of drawing a single part that is defective is 8/100, or 0.08.

The following theorems are useful in solving problems:

THEOREM 10.1 If $P(A)$ is the probability that an event A will occur, then the probability that A will not occur is $1 - P(A)$. This theorem is sometimes called the complementary rule of probability.

THEOREM 10.2 If A and B are two events, then the probability that either A or B will occur is

$$P(A \text{ or } B) = P(A) + P(B) - P(A \text{ and } B)$$

A special case of this theorem occurs when A and B cannot occur simultaneously (i.e., A and B are *mutually exclusive*). Then the probability that either A or B will occur is

$$P(A \text{ or } B) = P(A) + P(B)$$

EXAMPLE 10.2 The probability of r defectives in a sample of six units from a 5 percent defective lot was previously found by the binomial. The probability of zero defectives was 0.7351; the probability of one defective was 0.2321. The probability of zero or one defective is then $0.7351 + 0.2321$, or 0.9672.

THEOREM 10.3 If A or B are two events, then the probability that events A and B occur together is

$$P(A \text{ and } B) = P(A) \times P(B \mid A)$$

where $P(B \mid A)$ = probability that B will occur assuming that A has already occurred.

A special case of this theorem occurs when the two events are independent, i.e., when the occurrence of one event has no influence on the probability of the other event. If A and B are independent, then the probability of both A and B occurring is

$$P(A \text{ and } B) = P(A) \times P(B)$$

EXAMPLE 10.3 A complex system consists of two major subsystems that operate independently. The probability of successful performance of the first subsystem is 0.95; the corresponding probability for the second subsystem is 0.90. Both subsystems must operate successfully to achieve total system success. The probability of the successful operation of the total system is therefore $0.95 \times 0.90 = 0.855$.

The theorems above have been stated in terms of two events but can be expanded for any number of events.

10.16 ENUMERATIVE AND ANALYTIC STUDIES

Deming (1982) provides an important distinction between two types of statistical studies—enumerative and analytic. In an enumerative study, we measure a *sample* and then estimate the characteristics of the *population*. For example, we draw a random sample of 20 units from a lot of 100 units. We measure the 20 units and then make predictions about the population of 100 units. This approach implies a defined, existing population. Typically, in an enumerative study we do not document information on the order of production—a pity.

In an analytic study we measure periodic samples from a process that is continually manufacturing product, i.e., a changing population. Suppose we want to predict the results for the next batch of product from the process. Such a prediction requires two assumptions: (1) the process has stabilized to a set of conditions that will repeat in the future, and (2) any imbedded process changes with time such as trends or cyclical effects are known. To predict future process result, data on the order of production must be documented and used to analyze process stability and embedded time effects. Techniques such as statistical control charts (see Chapter 18, "Statistical Process Control") are helpful in such analysis. In practice, the prediction of future process results predominates and thus makes the analytic type of study of great importance.

10.17 STATISTICAL THINKING AT THREE LEVELS IN AN ORGANIZATION

Now that we understand the basic concepts of variation, it is useful to reflect on how this knowledge can be used throughout an organization.

Statistical thinking is not just applying formulas to data. Snee (1990) proposes that statistical thinking be applied at three levels: strategic, managerial, and operational. At the strategic level the focus is on concepts:

- Variation is present in all processes.
- All work is a series of interconnected processes.
- Reducing variation improves quality.

At the managerial level the focus is on systems:

- Statistical process control.
- Experimentation.
- Robustness of product and process design.

At the operational level the focus is on tools:

- Control charts.
- Formal design and analysis of experiments.
- Improvement, planning, and control tools.

At the upper-management level statistical thinking to reduce variation can take the form of helping all groups work toward a common goal, providing the workforce with the standards, feedback, and capable process that is needed for self-control, reducing the number of suppliers, providing sufficient training, and many other actions.

Statistical tools have been emphasized at the operation level with varying degrees of implementation. The major need is for upper management to understand statistical thinking at the strategic level and to integrate the concepts into systems at the managerial and operational levels. For elaboration, see Hoerl (1995).

10.18 COMPUTER SOFTWARE FOR STATISTICAL ANALYSIS

With the advent of statistical software packages, the practitioner can now use many statistical techniques that were not previously considered because of the difficulty in understanding the techniques or doing the calculations. (Historical note: The terms "software" and "bit" were first used by Dr. John Tukey, an eminent statistician.) Currently, most software packages provide a basic explanation of a technique, define the inputs required, and then present the results. But such accessibility has a danger. The practitioner must understand the assumptions behind the methods and what the final results do and do not mean. In the haste to obtain an answer and avoid tedious detail, there is a danger of applying a technique incorrectly or misunderstanding a result. Beware of these serious consequences.

Figure 10.13 illustrates the power of a software spreadsheet. Here the software supplied 1000 randomly generated normal numbers with a mean of 100 and a standard deviation of 10 and grouped these into cells of a frequency distribution. The theoretical normal distribution curve is also provided. Even with the large amount of data, the "fit" is not perfect, but the data clearly follow the normal distribution. Other examples of software output are presented in later chapters.

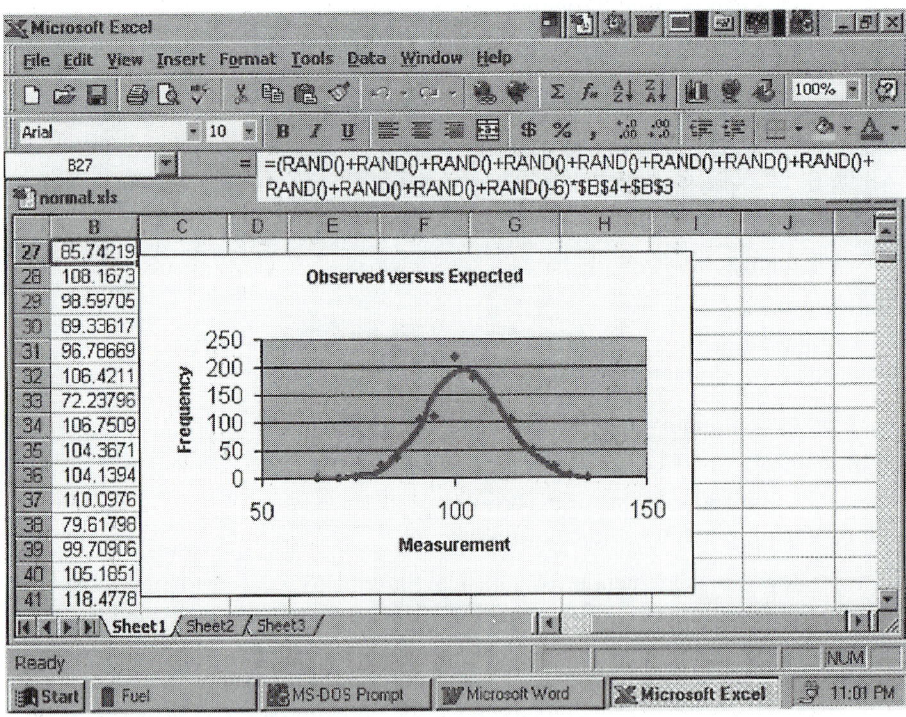

FIGURE 10.13
Graph of observed frequency counts versus expected frequencies for a normal distribution. *(Zimmerman and Icenogle, 1999.)*

Among the recommended statistical packages are MINITAB, SAS, BMDP, STATPRO, LABONE, SPSS, and SYSTAT. Some people prefer to use Microsoft Excel for spreadsheet analysis.

The statistical calculations presented in this book are available in statistical software packages. ASQ publishes an annual directory of software for quality assurance and quality control in *Quality Progress* magazine.

SUMMARY

- Statistical methods are essential in the modern approach to quality.
- Variation is a fact of nature and a fact of industrial life.
- In summarizing data, useful tabular and graphical tools include the frequency distribution, histogram, boxplot, and probability paper.
- In summarizing data, useful numerical indices include the average, median, range, and standard deviation.
- A sample is a limited number of items taken from a larger source called the population.
- A probability distribution function relates the values of a characteristic to their probability of occurrence in the population.
- The important continuous probability distributions are the normal, exponential, and Weibull; important discrete distributions are the Poisson and binomial.
- Three theorems of probability are basic in analyzing the probability of specific events.

PROBLEMS

Note: Many of the statistical problems in the book have intentionally been stated in business language. Thus the specific statistical technique required will often *not* be specified. With this approach the student should gain some experience in translating the business problem into a statistical formulation and then choosing the appropriate statistical technique.

10.1. The following data consist of 80 potency measurements of the drug streptomycin.

4.1	5.0	2.0	2.6	4.5	8.1	5.7	2.5
3.5	6.3	5.5	1.6	6.1	5.9	9.3	4.2
4.9	5.6	3.8	4.4	7.1	4.6	7.4	3.5
4.9	5.1	4.6	6.3	8.3	6.3	8.8	1.0
5.3	5.4	4.4	2.9	7.5	5.7	5.3	3.0
4.2	5.2	7.0	3.7	6.7	5.8	6.9	2.8
6.0	8.2	6.1	7.3	8.2	6.2	4.3	2.2
5.2	5.5	3.5	7.1	7.9	5.6	5.4	3.9
6.8	8.2	4.2	4.2	5.5	6.2	3.5	3.4
6.8	4.7	4.6	4.1	4.7	5.0	3.4	7.1

(*a*) Summarize the data in tabular form.
(*b*) Summarize the data in graphical form.

10.2. Compute a measure of central tendency and two measures of variation for the data given in Problem 10.1. Calculate the following three sets of limits: $\bar{X} \pm 1s$, $\bar{X} \pm 2s$,

$\bar{X} \pm 3s$. For each set, calculate the percentage of data values that fall within the limits. Compare these percentages to the theoretical percentages based on the normal distribution.

10.3. Examine the histograms in Figure 10.7. For each histogram, comment on *(a)* the ability of the process to meet specification limits and *(b)* what action, if any, on the process is appropriate.

10.4. Heyes (1985) presents the following data on the wire break strength for supplier B:

470	425	438	620	452
573	382	486	526	300
520	450	389	371	598

Prepare a boxplot for this data.

10.5. A company has a filling machine for low-pressure oxygen shells. Data collected over the past two months show an average weight after filling of 1.433 g with a standard deviation of 0.033 g. The specification for weight is 1.460 ± 0.085 g. Weight is normally distributed.
(a) What percentage will not meet the weight specification?
(b) Would you suggest a shift in the aim of the filling machine? Why or why not?

10.6. A company that makes fasteners has government specifications on a self-locking nut. The locking torque has both a maximum and a minimum specified. The offsetting machine used to make these nuts has been producing nuts with an average locking torque of 8.62 in-lb and a variance σ^2 of 4.49-squared in-lb. Torque is normally distributed.
(a) If the upper specification is 13.0 in-lb and the lower specification is 2.25 in-lb, what percent of these nuts will meet the specification limits?
(b) Another machine in the offset department can turn out the nuts with an average of 8.91 in-lb and a standard deviation of 2.33 in-lb. In a lot of 1000 nuts, how many would have too high a torque?
Answer: (a) 97.95 percent. *(b)* 40 nuts.

10.7. A power company defines service continuity as providing electric power within specified frequency and voltage limits to the customer's service entrance. Interruption of this service may be caused by equipment malfunctions or line outages due to planned maintenance or to unscheduled reasons. Records for the entire city indicate that there were 416 unscheduled interruptions in 1997 and 503 in 1996.
(a) Calculate the mean time between unscheduled interruptions, assuming that power is to be supplied continuously.
(b) What is the chance that power will be supplied to all users without interruption for at least 24 hours? For at least 48 hours? Assume an exponential distribution.

10.8. An analysis was made of repair time for an electrohydraulic servovalve used in fatigue test equipment. Discussions concluded that about 90 percent of all repairs could be made within six hours.
(a) Assuming an exponential distribution of repair time, calculate the average repair time.
(b) What is the probability that a repair would take between three and six hours?
Answer: (a) 2.6 hours. *(b)* 0.217.

10.9. Three designs of a certain shaft are to be compared. The information on the designs is summarized as follows:

	Design I	Design II	Design III
Material	Medium-carbon alloy steel	Medium-carbon unalloyed steel	Low-carbon special analysis steel
Process	Fully machined before heat treatment, then furnace-heated, oil-quenched, and tempered	Fully machined before heat treatment, then induction-scan-heated, water-quenched, and tempered	Fully machined before heat treatment, then furnace-heated, water-quenched, and tempered

	Design I	Design II	Design III
Equipment cost	Already available	$125,000	$500
Cost of finished shaft	$57	53	55

Fatigue tests were run on six shafts of each design with the following results (in units of thousands of cycles to failure):

I	II	III
180	210	900
240	360	1400
100	575	1500
50	330	340
220	130	850
110	575	600

(*a*) Rearrange the data in ascending order and make a Weibull plot for each design.
(*b*) For each design, estimate the number of cycles at which 10 percent of the population will fail. (This value is called the B_{10} life.) Do the same for 50 percent of the population.
(*c*) Calculate the average life of each design based on the test results. Then estimate the percentage of the population that will fail within this average life. Note that it is not 50 percent.
(*d*) Comment on replacing the current design I with II or III.

10.10. Life tests on a sample of five units were conducted to evaluate a component design before release to production. The units failed at the following times:

Unit number	Failure time, hours
1	1200
2	1900
3	2800
4	3500
5	4500

Suppose that the component was guaranteed to last 1000 hours. Any failures during this period must be replaced by the manufacturer at a cost of $200 for each component. Although the number of test data is small, management wants an estimate of the cost of replacements. If 4000 of these components are sold, provide a dollar estimate of the replacement cost.

10.11. The following data consist of 24 measurements (in seconds) on the time to answer a hot-line telephone inquiry at an AT&T call center (AT&T, 1990).

24	24	21	24	24	24
27	26	25	23	23	23
26	22	22	28	21	25
25	27	25	26	22	20

(a) Summarize the data in tabular form.
(b) Summarize the data in graphical form.

10.12. For the data of problem 10.11, perform the analyses stated in problem 10.2.

REFERENCES

Armstrong, G. R. and P. C. Clarke (1946). "Frequency Distribution vs Acceptance Table," *Industrial Quality Control,* vol. 3, no. 2, pp. 22–27.

AT&T (1990). *Analyzing Business Process Data: The Looking Glass,* AT&T Bell Laboratories, Indianapolis, IN.

Deming, W. E. (1982). *Out of the Crisis,* Massachusetts Institute of Technology, Cambridge, p. 132.

Gunter, B. (1998). "Farewell Fusillade," *Quality Progress,* April, pp. 111–119.

Harris, R. L. (1996). *Information Graphics,* Management Graphics, Atlanta, GA.

Heyes, G. B. (1985). "The Box Plot," *Quality Progress,* December, pp. 12–17.

Hoerl, R. W. (1995). "Enhancing the Bottom-Line Impact of Statistical Methods," *Quality Management Journal,* Summer, pp. 58–74.

Juran Institute, Inc. (1989). *Quality Improvement Tools—Box Plots,* Wilton, CT, p. 7.

King, J. R. (1981). *Probability Charts for Decision Making,* rev. ed., TEAM, Tamworth, NH.

Snee, R. D. (1990). "Statistical Thinking and Its Contribution to Total Quality," *The American Statistician,* May, pp. 116–121.

Tufte, E. R. (1997). *Visual Explanations,* Graphics Press, Cheshire, CT.

Tukey, J. W. (1977). *Exploratory Data Analysis,* Addison-Wesley, Reading, MA.

Wadsworth, H. M., K. S. Stephens, and A. B. Godfrey (2001). *Modern Methods for Quality Control and Improvement,* 2nd ed., John Wiley and Sons, New York.

Zimmerman, S. M. and M. L. Icenogle (1999). *Statistical Quality Control Using Excel,* Quality Press, ASQ, Milwaukee.

SUPPLEMENTARY READING

Basic concepts: *JQH5,* Section 44.

Basic, intermediate, and advanced concepts: Wadsworth, H. M. (1998). *Handbook of Statistical Methods for Engineers and Scientists,* 2nd ed., McGraw-Hill, New York.

International standards on statistical methods: Mundel, A. B. (1998). "Activities of ISOTC69: International Standards on Statistical Methods," *Proceedings of the Annual Quality Congress,* ASQ, Milwaukee, pp. 455–457.

WEBSITE

Statistics division of ASQ: www.asq.org/about/divisions/stats

11

STATISTICAL TOOLS FOR ANALYZING DATA

11.1 SCOPE OF DATA PLANNING AND ANALYSIS

Here are some types of problems that can benefit from statistical analysis:

1. Determining the usefulness of a limited number of test results in predicting the true value of a product characteristic.
2. Determining the number of tests required to provide adequate data for evaluation.
3. Comparing test data between two alternative designs.
4. Predicting the amount of product that will fall within specification limits.
5. Predicting system performance.
6. Controlling process quality by early detection of process changes.
7. Planning experiments to discover the factors influencing a characteristic of a product or process, i.e., exploratory experimentation.
8. Determining the quantitative relationship between two or more variables.

The central issue in all of these problems is the *prediction* of population parameters on the basis of sample results. Statistics ties in with the scientific method—for a perceptive discussion, see Box and Liu (1999) and Box (1999).

These statistical methods involve both simple and complex techniques. Fortunately, computer software has eliminated the drudgery of doing detailed calculations (and making errors in those calculations). Understanding the basic concepts is essential to correctly interpreting computer outputs, and this chapter focuses on both the concepts and the tools.

Providing training in statistical tools within an organization requires careful planning if actual implementation is to be achieved. Finn et al. (1995) identify common problems and provide recommendations for successful implementation.

11.2 STATISTICAL INFERENCE

Drawing conclusions from limited data is notoriously unreliable. The "gossip" of a small sample size can be dangerous. Examine the following problems concerned with the evaluation of test data. For each, give a yes or no answer based on your intuitive analysis of the problem. (Write your answers on a piece of paper *now* and then check for the correct answers at the end of this chapter.) Some of the problems are solved in the chapter.

Examples of Engineering Problems That Can Be Solved by Using the Concepts of Statistical Inference

1. A single-cavity molding process has been producing insulators with an average impact strength of 5.15 ft-lb [6.9834 Newton-meters (N-m)]. A group of 12 insulators from a new lot shows an average of 4.952 ft-lb (6.7149 N-m). Is this enough evidence to conclude that the new lot is lower in average strength?
2. Past data show the average hardness of brass parts to be 49.95. A new design claims to have higher hardness. A sample of 61 parts of the new design shows an average of 54.62. Does the new design actually have a different hardness?
3. Two types of spark plugs were tested for wear. A sample of 10 of design 1 showed an average wear of 0.0049 in (0.0124 cm). A sample of eight of design 2 showed an average wear of 0.0064 in (0.0163 cm). Are these enough data to conclude that design 1 is better than design 2?
4. Only 11.7 percent of the 60 new-alloy blades on a turbine rotor failed on test in a gas turbine where 20 percent have shown failures in a series of similar tests in the past. Are the new blades better than the old ones?
5. 1050 resistors supplied by one manufacturer were 3.71 percent defective. 1690 similar resistors from another manufacturer were 1.95 percent defective. Can one reasonably assert that the product of one plant is inferior to that of the other?

You probably had some incorrect answers. The statistical methods used to properly analyze these problems are called *statistical inference.* We start the chapter with the concept of sampling variation and sampling distributions.

11.3 SAMPLING VARIATION AND SAMPLING DISTRIBUTIONS

Suppose that a battery is to be evaluated to ensure that life requirements are met. A mean life of 30 hours is desired. Preliminary data indicate that life follows a normal distribution and that the standard deviation is equal to 10 hours. A sample of four batteries is selected at random from the process and tested. If the mean of the four is close to 30 hours, it is concluded that the battery meets the specification. Figure 11.1 plots the distribution of *individual* batteries from the population assuming that the true *mean* of the population is exactly 30 hours.

If a sample of four is life-tested, the following lifetimes might result: 34, 28, 38, and 24, giving a mean of 31.0. However, this random sample is selected from the many batteries made by the same process. Suppose that another sample of four is taken. The second sample of four is likely to be different from the first sample. Perhaps the results would be 40, 32, 18, and 29, giving a mean of 29.8. If the process of drawing many samples (with four in each sample) is repeated over and over, different results would be obtained in most samples. The fact that samples drawn from the *same* process can yield different sample results illustrates the concept of sampling variation.

Returning to the problem of evaluating the battery, a dilemma exists. In the actual evaluation, only one sample of four can be drawn (because of time and cost limitations). Yet the experiment of drawing many samples indicates that samples vary. The question is: How reliable is the sample of four that will be the basis of the decision? The final decision can be influenced by the luck of which sample is chosen. The key point is that the existence of sampling variation means that any one sample cannot be relied upon to always give an adequate decision. The statistical approach analyzes the results of the sample, *taking into account the possible sampling variation that could occur.* Formulas have been developed to define the expected amount of sampling variation. Knowing this, a valid decision can be reached by evaluating the one sample of data.

The problem, then, is to define how means of samples vary. If sampling is continued and for each sample of four the mean is calculated, these means could be compiled into a histogram. Figure 11.1 shows the resulting probability curve superim-

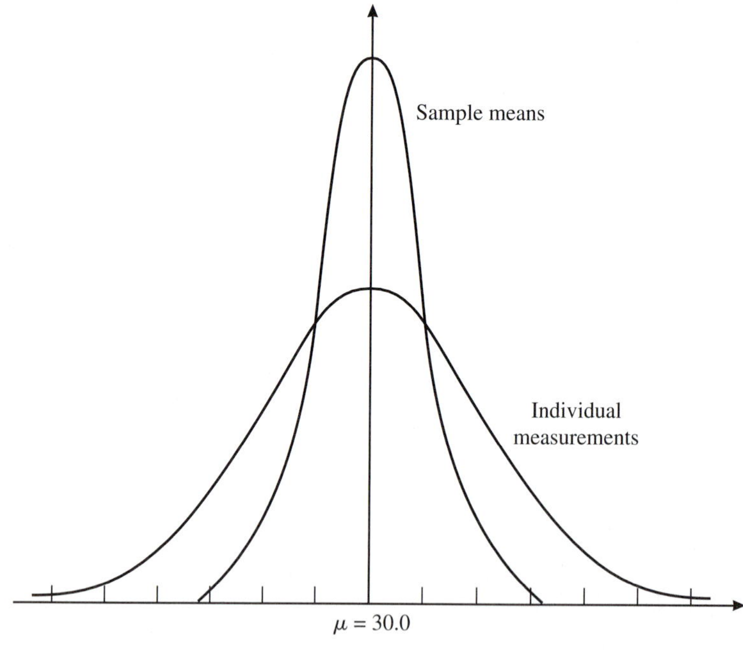

FIGURE 11.1
Distributions of individual measurements and sample means.

posed on the curve for the population. The narrow curve represents the distribution of life for the sample *means* (where each average includes four individual batteries). This is called the *sampling distribution of means.* The curve for means is narrower than the curve for individuals because in calculating means, extreme individual values are offset. The mathematical properties for the curve for averages have been studied and the following relationship developed:

$$\sigma_{\bar{x}} = \frac{\sigma}{\sqrt{n}}$$

where $\sigma_{\bar{x}}$ = standard deviation of means of samples (sometimes called the standard error of the mean)

σ = standard deviation of individual items

n = number of items in *each* sample

The mathematical justification for the relationship is the central limit theorem. This limit states that if $x_1, x_2, \ldots x_n$ are outcomes of a sample of n independent observations of a random variable X, then the mean of the samples of n will approximately follow a normal distribution with mean μ and standard deviation $\sigma_{\bar{x}} = \sigma/\sqrt{n}$.

The relationship is significant because if an estimate of the standard deviation of *individual* items can be obtained, then the standard deviation of sample means can be calculated from the foregoing relationship instead of running an experiment to generate sample averages. The problems of evaluating the battery can now be portrayed graphically (Figure 11.2).

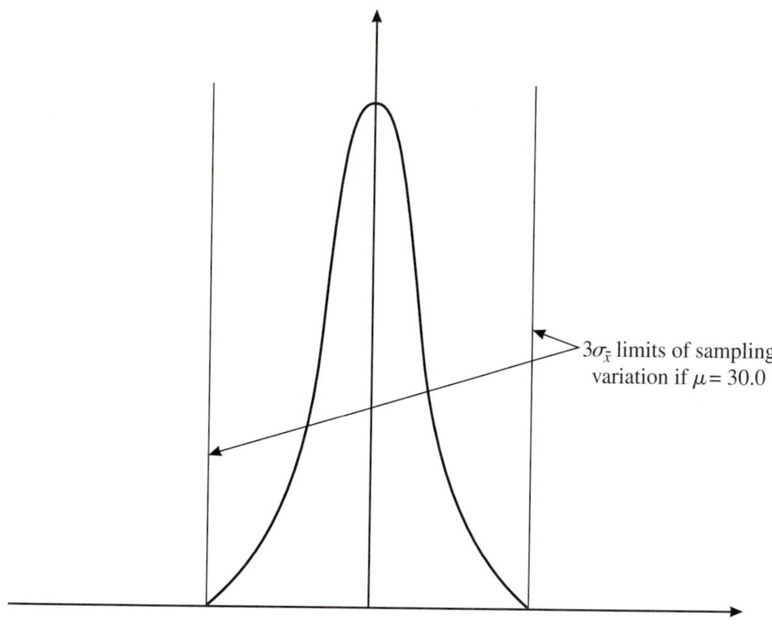

FIGURE 11.2
Distribution of sample means.

This concept of a sampling distribution is basic to the two major areas of statistical inference, i.e., estimation and tests of hypotheses, which are discussed next.

11.4
STATISTICAL ESTIMATION: CONFIDENCE LIMITS

Estimation is the process of analyzing a sample result in order to predict the corresponding value of the population parameter. For example, the sample of four batteries previously mentioned had a mean life of 31.0 hours. If this is a representative sample from the process, what estimate can be made of the true life of the entire population of batteries?

The estimation statement has two parts:

1. The *point estimate* is a single value used to estimate the population parameter. For example, 31.0 hours is the point estimate of the average life of the population. The sample mean has some important statistical properties: On the average, sample means will equal the population mean, and the variability of sample means will be less than other estimates of the population mean. More precisely, the point estimate for a population parameter should be unbiased; i.e., the expected value of the estimate should equal the population parameter. Also, the point estimate should have minimum variance compared to other estimators of the population parameter. Thus the sample mean is usually the best point estimate of the population mean.
2. The *confidence interval* is a range of values that include (with a preassigned probability called a *confidence level*) the true value of a population parameter. *Confidence limits* are the upper and lower boundaries of the confidence interval. A confidence level is the probability that an assertion about the value of a population parameter is correct.[1]

Confidence limits should not be confused with other limits, e.g., control limits, statistical tolerance limits (see Chapter 14, "Designing for Quality—Statistical Tools," for a distinction among several types of limits).

Table 11.1 summarizes confidence limit formulas for common parameters. The following examples illustrate some of these formulas.

EXAMPLE 11.1. Mean of a normal population. Twenty-five specimens of brass have a mean hardness of 54.62 and an estimated standard deviation of 5.34. Determine the 95 percent confidence limits on the mean.

Solution. Note that when the standard deviation is unknown and is estimated from the sample, the t distribution (Table D in Appendix II) must be used. The t value for 95 percent

[1]Confidence levels of 90, 95, or 99 percent are usually assumed in practice.

TABLE 11.1
Summary of confidence limit formulas

Parameters	Formulas
Mean of a normal population (standard deviation known)	$\bar{X} \pm Z_{\alpha/2} \frac{\sigma}{\sqrt{n}}$ where $\bar{X}$ = sample average Z = normal distribution coefficient σ = standard deviation of population n = sample size
Mean of a normal population (standard deviation unknown)	$\bar{X} \pm t_{\alpha/2} \frac{s}{\sqrt{n}}$ where t = distribution coefficient (with $n - 1$ degrees of freedom) s = estimated σ
Standard deviation of a normal population	Upper confidence limit $= s\sqrt{\frac{n-1}{\chi^2_{\alpha/2}}}$ Lower confidence limit $= s\sqrt{\frac{n-1}{\chi^2_{1-\alpha/2}}}$ where χ^2 = chi-square distribution coefficient with $n - 1$ degrees of freedom $1 - \alpha$ = confidence level
Population fraction defective	See Table F in Appendix II.
Difference between the means of two normal populations (standard deviations σ_1 and σ_2 known)	$(\bar{X}_1 - \bar{X}_2) \pm Z_{\alpha/2}\sqrt{\frac{\sigma_1^2}{n_1} + \frac{\sigma_2^2}{n_2}}$
Difference between the means of two normal populations ($\sigma_1 = \sigma_2$ but unknown)	$(\bar{X}_1 - \bar{X}_2) \pm t_{\alpha/2}\sqrt{\frac{1}{n_1} + \frac{1}{n_2}}$ $\times \sqrt{\frac{\Sigma(X - \bar{X}_1)^2 + \Sigma(X - \bar{X}_2)^2}{n_1 + n_2 - 2}}$
Mean time between failures based on an exponential population of time between failures	Upper confidence limit $= \frac{2rm}{\chi^2_{\alpha/2}}$ Lower confidence limit $= \frac{2rm}{\chi^2_{1-\alpha/2}}$ where r = number of occurrences in the sample (i.e., number of failures) m = sample mean time between failures DF $= 2r$

confidence is found by entering the table at 0.975 and 25 − 1, or 24, degrees of freedom[2] and reading a t value of 2.064.

$$\text{Confidence limits} = \bar{X} \pm t\frac{s}{\sqrt{n}},$$

$$= 54.62 \pm (2.064)\frac{5.34}{\sqrt{25}}$$

$$= 52.42 \text{ and } 56.82$$

There is 95 percent confidence that the true mean hardness of the brass is between 52.42 and 56.82.

EXAMPLE 11.2. Mean of an exponential population. A repairable radar system has been operated for 1200 hours, during which time eight failures occurred. What are the 90 percent confidence limits on the mean time between failures for the system?

Solution

$$\text{Sample MTBF} = \frac{1200}{8} = 150 \text{ h between failures}$$

$$\text{Upper confidence limit} = 2(1200)/7.962 = 301.4$$

$$\text{Lower confidence limit} = 2(1200)/26.296 = 91.3$$

The values 7.962 and 26.296 are obtained from the chi-square table (Table E in Appendix II). There is 90 percent confidence that the true mean time between failures is between 91.3 and 301.4 h.

Confusion has arisen on the application of the term *confidence level* to a reliability index such as MTBF. Using a different example, suppose that the numerical portion of a reliability requirement reads as follows:

"The MTBF shall be at least 100 hours at the 90 percent confidence level." This statement means that

1. The minimum MTBF must be 100 hours.
2. Actual tests shall be conducted on the product to demonstrate with 90 percent confidence that the 100-hour MTBF has been met.
3. The test data shall be analyzed by calculating the observed MTBF and the lower one-sided 90 percent confidence limit on MTBF. The true MTBF lies above this limit with 90 percent of confidence.
4. The lower one-sided confidence limit must be ≥100 hours.

The term *confidence level* from a statistical viewpoint has great implications on a test program. The observed MTBF must be *greater* than 100 if the lower confidence

[2] A mathematical derivation of degrees of freedom is beyond the scope of this book, but the underlying concept can be stated. *Degrees of freedom* (DF) is the parameter involved when, e.g., a sample standard deviation is used to estimate the true standard deviation of a universe. DF equals the number of measurements in the sample minus some number of constraints estimated from the data in order to compute the standard deviation. In this example it was necessary to estimate only one constant (the population mean) to compute the standard deviation. Therefore, DF = 25 − 1 = 24.

limit is to be ≥100. Confidence level means that sufficient tests must be conducted to demonstrate, with statistical validity, that a requirement has been met. Confidence level does *not* refer to the qualitative opinion about meeting a requirement. Also, confidence level does *not* lower a requirement; i.e., a 100-hour MTBF at a 90 percent confidence level does *not* mean that 0.90 × 100, or 90 hours, is acceptable. Such serious misunderstandings have occurred. When the term *confidence level* is used, a clear understanding should be verified and not assumed.

11.5 IMPORTANCE OF CONFIDENCE LIMITS IN PLANNING TEST PROGRAMS

Additional tests will increase the accuracy of estimates. Accuracy here refers to the agreement between an estimate and the true value of the population parameter. The increase in accuracy does not vary linearly with the number of tests—doubling the number of tests usually does *not* double the accuracy. Examine the graph (Figure 11.3) of the confidence interval for the mean against sample size (a standard deviation of 50.0 was assumed): When the sample size is small, an increase has a great effect on

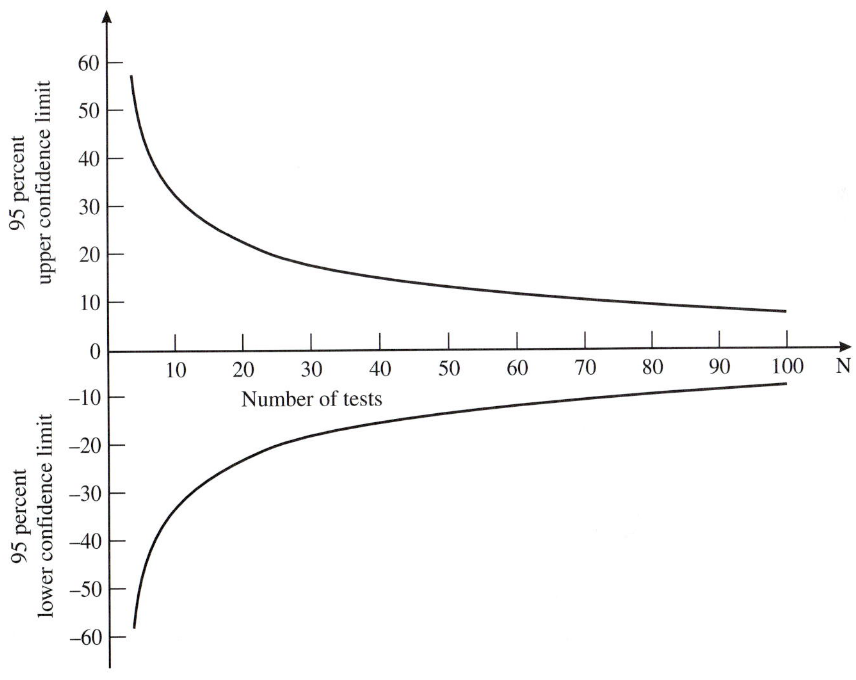

FIGURE 11.3
Width of confidence interval versus number of tests.

the width of the confidence interval; after about 30 units, an increase has a much smaller effect. The effect diminishes because of the square root in the formula for confidence limits. To double the accuracy requires a sample size four times larger. The inclusion of the cost parameter is vital here. The cost of additional tests must be evaluated against the value of the additional accuracy.

Further, if the sample is selected randomly and if the sample size is less than 10 percent of the population size, accuracy depends primarily on the absolute size of the sample rather than the sample size expressed as a percentage of the population size. Thus a sample size that is 1 percent of the population of 100,000 may be better than a 10 percent sample from a population of 1000.

11.6 DETERMINATION OF THE SAMPLE SIZE REQUIRED TO ACHIEVE A SPECIFIED ACCURACY IN AN ESTIMATE

Confidence limits can help to determine the size of a test program required to estimate a product characteristic within a specified accuracy. It is desired to estimate the true mean of the battery previously cited where $\sigma = 10$. The estimate must be within 2.0 hours of the true mean if the estimate is to be of any value. A 95 percent confidence level is desired on the confidence statement. The desired confidence interval is ± 2.0 hours, or

$$2.0 = \frac{(1.96)(10)}{\sqrt{n}} \qquad n = 96$$

A sample of 96 batteries will provide an average that is within 2.0 hours of the true mean (with 95 percent confidence). Notice the type of information required for estimating the mean of a normal population: (1) desired width of the confidence interval (the accuracy desired in the estimate), (2) desired confidence level, and (3) variability of the characteristic under investigation. The number of tests required cannot be determined until the engineer furnishes these items of information. Past information may also have a major role in designing a test program (see Section 11.11).

11.7 TEST OF HYPOTHESIS

Basic Concepts

A *hypothesis,* as used here, is an assertion made about a population. Usually, the assertion concerns the numerical value of some parameter of the population. For example, a hypothesis might state that the mean life of a population of batteries equals 30.0 hours, written as $H: \mu_0 = 30.0$. This assertion may or may not be correct. A *test of hypothesis* is a test of the validity of the assertion and is carried out by analysis of a sample of data.

Sample results must be carefully evaluated for two reasons. First, there are many other samples that, by chance alone, could be drawn from the population. Second, the numerical results in the sample actually selected can easily be compatible with several different hypotheses. These points are handled by recognizing the two types of sampling errors.

The Two Types of Sampling Errors.

In evaluating a hypothesis, two errors can be made:

1. *Reject* the hypothesis when it is *true*. This is called the *type I error* or the *level of significance*. The probability of the type I error is denoted by α.
2. *Accept* the hypothesis when it is *false*. This is called the *type II error* and the probability is denoted by β.

These errors are defined in terms of probability numbers and can be controlled to desired values. The results possible in testing a hypothesis are summarized in Table 11.2.

The type I error is shown graphically in Figure 11.4 for the hypothesis $H_0 : \mu_0 = 30.0$. The interval on the horizontal axis between the vertical lines represents the

TABLE 11.2
Type I (α) error and type II (β) error

	Suppose the H is:	
Suppose decision of analysis is:	**True**	**False**
Accept H	Correct decision $P = 1 - \alpha$	Wrong decision $P = \beta$
Reject H	Wrong decision $P = \alpha$	Correct decision $P = 1 - \beta$

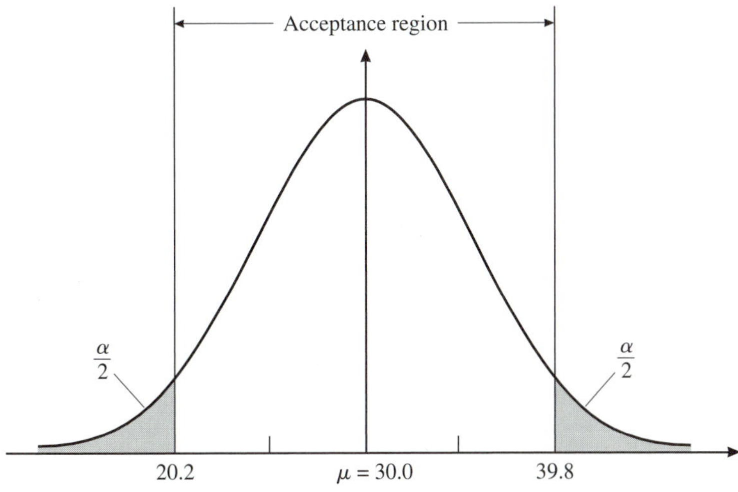

FIGURE 11.4
Acceptance region for $H_0 : \mu = 30.0$ and $H_1 : \mu \neq 30.0$.

acceptance region for the test of hypothesis. If the sample result (e.g., the mean) falls within the acceptance region, the hypothesis is accepted. Otherwise, it is rejected. The terms *accepted* and *rejected* require careful interpretation. The meanings are explained in Section 11.9 of this chapter. Notice that a small portion of the curve falls outside the acceptance region. This area (α) represents the probability of obtaining a sample result outside the acceptance region, even though the hypothesis is correct.

Suppose it has been decided that the type I error must not exceed 5 percent. This value is the probability of rejecting the hypothesis when, in truth, the true mean life is 30.0. The acceptance region can be obtained by locating values of mean life that have only a 5 percent chance of being exceeded when the true mean life is 30.0. Further, suppose a sample n of four measurements is taken and $\sigma = 10.0$.

Remember that the curve represents a population of sample means because the decision will be made on the basis of a sample mean. Sample means vary less than individual measurements according to the relationship $\sigma_{\bar{x}} = \sigma/\sqrt{n}$ (see Section 11.3, "Sampling Variation and Sampling Distributions").

Further, the distribution of sample means is approximately normal even if the distribution of the individual measurements (going into the means) is not normal. The approximation holds best for large values of n but is adequate for n as low as 4.

Table A in Appendix II shows that a 2.5 percent area in each tail is at a limit that is 1.96 standard deviations from 30.0. Then, under the hypothesis that $\mu_0 = 30.0$, 95 percent of sample means will fall within $\pm 1.96\sigma_{\bar{x}}$ of 30.0, or

$$\text{Upper limit} = 30.0 + 1.96\frac{10}{\sqrt{4}} = 39.8$$

$$\text{Lower limit} = 30.0 - 1.96\frac{10}{\sqrt{4}} = 20.2$$

The acceptance region is thereby defined as 20.2 to 39.8. If the mean of a random sample of four batteries is within this acceptance region, the hypothesis is accepted. If the mean falls outside the acceptance region, the hypothesis is rejected. This decision rule provides a type I error of 0.05.

The type II, or β, error, the probability of accepting a hypothesis when it is false, is shown in Figure 11.5 as the shaded area. Notice that it is possible to obtain a sample result within the acceptance region, even though the population has a true mean that is *not* equal to the mean stated in the hypothesis. The numerical value of β depends on the true value of the population mean (and also on n, σ, and α). The various probabilities are depicted by an *operating characteristic* (OC) *curve.*

The problem now is to construct an OC curve to define the magnitude of the type II (β) error. As β is the probability of *accepting* the original hypothesis ($\mu_0 = 30.0$) when it is *false,* the probability that a sample mean will fall between 20.2 and 39.8 must be found when the true mean of the population is something other than 30.0. This is done by finding the area under the normal curve for all possible values of the true mean of the population.

The results form the OC curve shown in Figure 11.6. The OC curve is a plot of the probability of accepting the original hypothesis as a function of the true value of the population parameter (and the given values of n, σ, and α). Note that for a mean

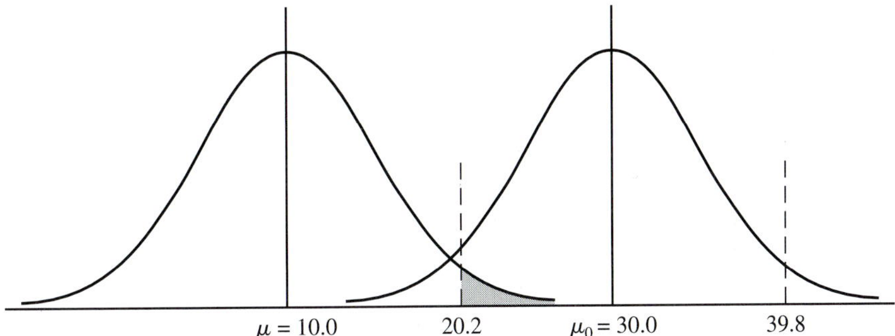

FIGURE 11.5
Type II, or β, error.

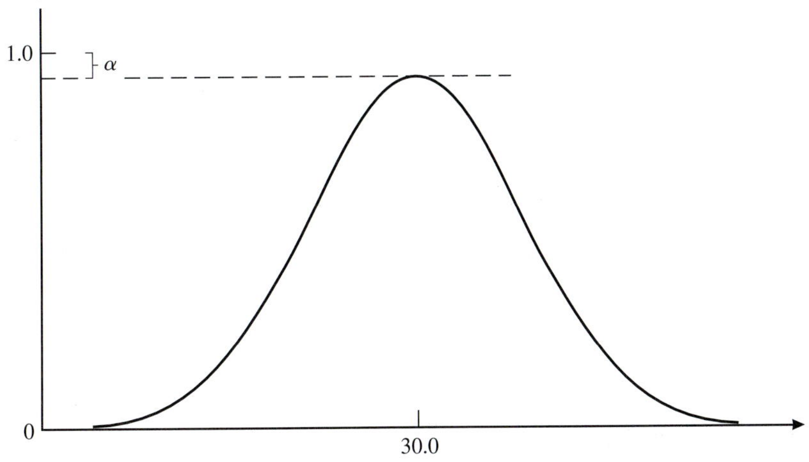

FIGURE 11.6
Operating characteristic curve.

equal to the hypothesis (30.0), the probability of acceptance is $1 - \alpha$. This curve should not be confused with that of a normal distribution of measurements. In some cases the shape is similar, but the meanings of an OC curve and a distribution curve are entirely different.

The Use of the Operating Characteristic Curve in Selecting an Acceptance Region

The acceptance region was determined by dividing the 5 percent allowable α error into two equal parts (see Figure 11.4). This process is called a *two-tail test.* The entire 5 percent error could also be placed at either the left or the right tail of the distribution curve. These are *one-tail tests.*

Operating characteristic curves for tests having these one-tail acceptance regions can be developed following the approach used for the two-tail region. Although the α error is the same, the β error depends on whether a one-tail or two-tail test is used.

In some problems knowledge is available to indicate that if the true mean of the population is *not* equal to the hypothesis value, then it is on one side of the hypothesis value. For example, a new material of supposedly higher mean strength will have a mean equal to or *greater than* that of the present material. Such information will help select a one-tail or two-tail test to make the β error as small as possible. The following guidelines are based on the analysis of OC curves:

Use a one-tail test with the entire α risk on the right tail if (1) it is suspected that (if μ_0 is not true) the true mean is $>\mu_0$ or (2) values of the population mean $<\mu_0$ are acceptable and we want to detect only a population mean $>\mu_0$.

Use a one-tail test with the entire α risk on the left tail if (1) it is suspected that (if μ_0 is not true) the true mean is $<\mu_0$ or (2) values of the population mean $>\mu_0$ are acceptable and we want to detect only a population mean $<\mu_0$.

Use a two-tail test if (1) there is no prior knowledge on the location of the true population mean or (2) we want to detect a true population mean less than or more than the μ_0 stated in the original hypothesis.[3]

The selection of a one- or two-tail test will be illustrated by some examples in Section 11.8. Every test of hypothesis has an OC curve. Duncan (1986) is a good source of OC curves. (Some references present "power" curves, but power is simply one minus the probability of acceptance, or $1 - \beta$.)

With this background, the discussion now proceeds to the steps for testing a hypothesis.

11.8 TESTING A HYPOTHESIS WHEN THE SAMPLE SIZE IS FIXED IN ADVANCE

Ideally, desired values for the type I and type II errors are defined in advance, and the required sample size determined (see Section 11.10). If the sample size is fixed in advance because of cost or time limitations, then usually the desired type I error is defined and this procedure is followed:

1. State the hypothesis.
2. Choose a value for the type I error. Common values are 0.01, 0.05, and 0.10.
3. Choose the test statistic for testing the hypothesis.
4. Determine the acceptance region for the test, i.e., the range of values of the test statistic that results in a decision to accept the hypothesis.

---

[3]With a two-tail test, the hypothesis is sometimes stated as the original hypothesis $H_0 : \mu_0 = 30.0$ against the alternative hypothesis $H_1 : \mu_0 \neq 30.0$. With a one-tail test, $H_0 : \mu_0 = 30.0$ is used against the alternative $H_1 : \mu_1 < 30.0$ if α is placed in the left tail, or $H_1 : \mu_1 > 30.0$ is used if α is placed in the right tail.

5. Obtain a sample of observations, compute the test statistic, and compare the value to the acceptance region to make a decision to accept or reject the hypothesis.
6. Draw an engineering conclusion.

Table 11.3 summarizes some common tests of hypotheses. The procedure is illustrated through the following examples. In these examples a type I error of 0.05 will be assumed.

TABLE 11.3
Summary of formulas on tests of hypotheses

Hypothesis	Test statistic and distribution
$H: \mu = \mu_0$ (the mean of a normal population is equal to a specified value μ_0; σ is known)	$Z = \dfrac{\bar{X} - \mu_0}{\sigma/\sqrt{n}}$ Normal distribution
$H: \mu = \mu_0$ (the mean of a normal population is equal to a specified value μ_0; σ is estimated by s)	$t = \dfrac{\bar{X} - \mu_0}{s/\sqrt{n}}$ t distribution with $n - 1$ degrees of freedom (DF)
$H: \mu_1 = \mu_2$ (the mean of population 1 is equal to the mean of population 2; assume that $\sigma_1 = \sigma_2$ and that both populations are normal)	$t = \dfrac{\bar{X}_1 - \bar{X}_2}{\sqrt{1/n_1 + 1/n_2}\sqrt{[(n_1 - 1)s_1^2 + (n_2 - 1)s_2^2]/(n_1 + n_2 - 2)}}$ t distribution with DF $= n_1 + n_2 - 2$
$H: \sigma = \sigma_0$ (the standard deviation of a normal population is equal to a specified value σ_0)	$\chi^2 = \dfrac{(n - 1)s^2}{\sigma_0^2}$ Chi-square distribution with DF $= n - 1$
$H: \sigma = \sigma_2$ (the standard deviation of population 1 is equal to the standard deviation of population 2; assume that both populations are normal)	$F = \dfrac{s_1^2}{s_2^2}$ F distribution with $DF_1 = n_1 - 1$ and $DF_2 = n_2 - 1$
$H: p = p_0$ (the fraction defective in a population is equal to a specified value p_0; assume that $np_0 \geq 5$)	$Z = \dfrac{p - p_0}{\sqrt{p_0(1 - p_0)/n}}$ Normal distribution
$H: p_1 = p_2$ (the fraction defective in population 1 is equal to the fraction defective in population 2; assume that n_1p_1 and n_2p_2 are each $\geq$5)	$Z = \dfrac{X_1/n_1 - X_2/n_2}{\sqrt{\hat{p}(1 - \hat{p})(1/n_1 + 1/n_2)}}$ $\quad \hat{p} = \dfrac{X_1 + X_2}{n_1 + n_2}$ Normal distribution

1. Test for a population mean, μ. (Standard deviation of the population is known.)

EXAMPLE 11.3. A single-cavity molding press has been producing insulators with a mean impact strength of 5.15 ft-lb (6.98 N-m) and with a standard deviation of 0.25 ft-lb (0.34 N-m). A new lot shows the following data from 12 specimens:

Specimen	Strength
1	5.02
2	4.87
3	4.95
4	4.88
5	5.01
6	4.93
7	4.91
8	5.09
9	4.96
10	4.89
11	5.06
12	4.85
	$\bar{X}$ = 4.95

Is the new lot from which the sample of 12 was taken different in mean impact strength from the past performance of the process?

Solution. $H_0 : \mu_0 = 5.15$ ft-lb (6.98 N-m). (The mean of the population from which the sample was taken is the same as the past process average.)

Test statistic

$$Z = \frac{\bar{X} - \mu_0}{\sigma/\sqrt{n}}$$

Acceptance region. Assuming no prior information and that a deviation on either side of the hypothesis average is important to detect, a two-tail test (Figure 11.7) is applicable. From Table A in Appendix II, the acceptance region is Z between -1.96 and $+1.96$.

Analysis of sample data

$$Z = \frac{4.95 - 5.15}{0.25/\sqrt{12}} = -2.75$$

Conclusion. Because Z is outside the acceptance region, the hypothesis is rejected. Therefore, sufficient evidence is present to conclude that the mean impact strength of the new process is significantly different from the mean of the past process. The answer to the first question in Section 11.2 is yes.

2. Test for two population means, μ_1 and μ_2, when the standard deviation is unknown but believed to be the same for the two populations. (This assumption can be evaluated by test 4.)

EXAMPLE 11.4. Two makes of spark plugs were operated in the alternate cylinders of an aircraft engine for 100 hours and the following data obtained:

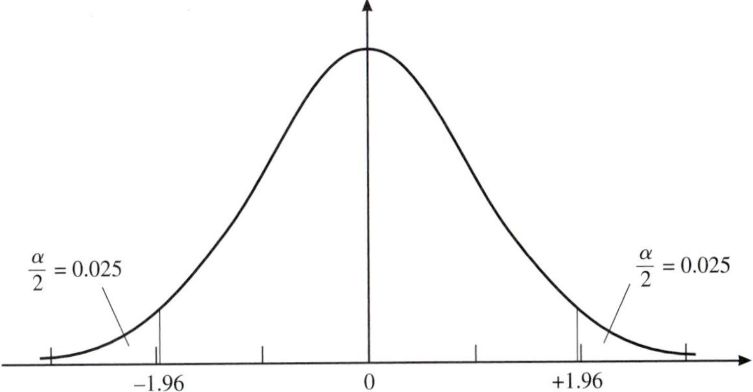

FIGURE 11.7
Distribution of Z (two-tail test).

	Make 1	Make 2
Number of spark plugs tested	10	8
Average wear per 100 h ($\bar{X}$), in	0.0049	0.0064
Variability (s), in	0.0005	0.0004

Can it be said that make 1 wears less than make 2?

Solution. $H : \mu_1 = \mu_2$.

Test statistic

$$t = \frac{\bar{X}_1 - \bar{X}_2}{\sqrt{1/n_1 + 1/n_2}\sqrt{[(n_1 - 1)s_1^2 + (n_2 - 1)s_2^2]/(n_1 + n_2 - 2)}}$$

with degrees of freedom $= n_1 + n_2 - 2$.

Acceptance region. We are concerned only with the possibility that make 1 wears less than make 2; therefore, use a one-tail test (Figure 11.8) with the entire α risk in the left tail. From Table D in Appendix II, the acceptance region is $t > -1.746$.

Analysis of sample data

$$t = \frac{0.0049 - 0.0064}{\sqrt{1/10 + 1/8}\sqrt{[(10 - 1)(0.0005)^2 + (8 - 1)(0.0004)^2]/(10 + 8 - 2)}} = -6.9$$

Conclusion. Because t is outside the acceptance region, the hypothesis is rejected. Therefore, sufficient evidence is present to conclude that make 1 wears less than make 2. The answer to the third question in Section 11.2 is yes.

A special case of the test for two population means is the paired-comparison test. Here the two samples of measurements come from a source that is thought to be homogeneous and thus greater precision can be obtained in comparing the two samples. The

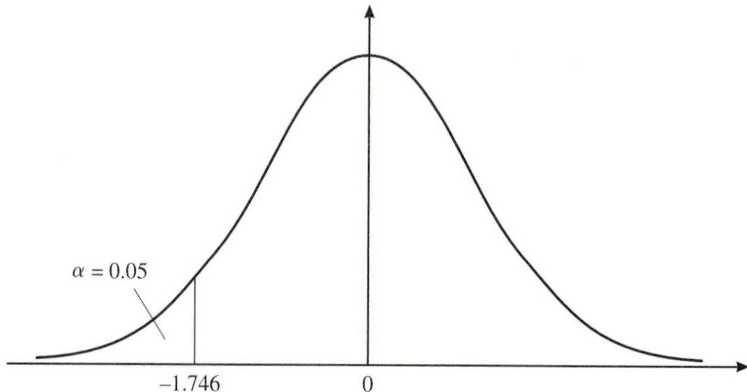

FIGURE 11.8
Distribution of t with α on left tail.

data are taken in pairs and the differences, *d,* within pairs, are analyzed, and a t test is run. For elaboration, see *JQH5,* page 44.65.

3. Test for a population standard deviation, σ.

EXAMPLE 11.5. For the insulator strengths tabulated in the first example, the sample standard deviation is 0.036 ft-lb (0.049 N-m). The previous variability, recorded over a period, has been established as a standard deviation of 0.25 ft-lb (0.34 N-m). Does the low value of 0.036 indicate that the new lot is significantly more uniform (i.e., standard deviation less than 0.25)?

Solution. $H : \sigma_0 = 0.25$ ft-lb (0.34 N-m).

Test statistic

$$\chi^2 = \frac{(n-1)s^2}{\sigma_0^2}$$

with degrees of freedom $= n - 1$.

Acceptance region. We believe that the standard deviation is smaller; therefore, we will use a one-tail test (Figure 11.9) with the entire risk on the left tail. From Table E in Appendix II the acceptance region is $\chi^2 \geq 4.57$.

Analysis of sample data

$$\chi^2 = \frac{(12-1)(0.078)^2}{(0.25)^2} = 1.08$$

Conclusion. Because χ^2 is outside the acceptance region, the hypothesis is rejected. Therefore, sufficient evidence is present to conclude that the new lot is more uniform than the old.

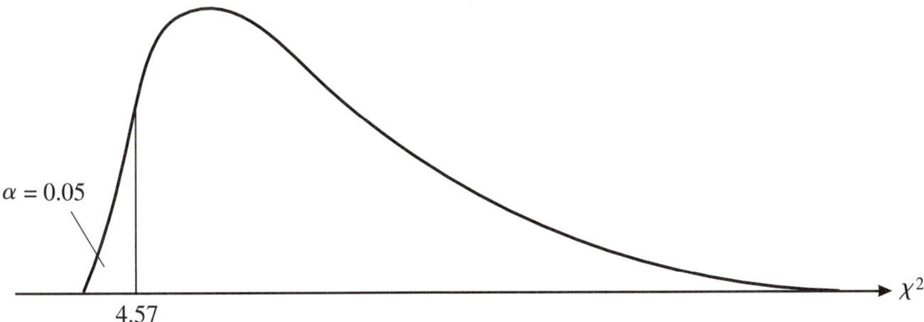

FIGURE 11.9
Distribution of χ^2 with α on left tail.

When the chi-square distribution is used to test a sample variance, the original data must come from a normally distributed population.

4. Test for the difference in variability (s_1 versus s_2) in two samples.

EXAMPLE 11.6. A materials laboratory was studying the effect of aging on a metal alloy. Researchers wanted to know whether the parts were more consistent in strength after aging than before. The following data were obtained:

	At start (1)	After 1 year (2)
Number of specimens (n)	9	7
Average strength ($\bar{X}$), psi	41,350	40,920
Variability (s), psi	934	659

Solution. $H : \sigma_1 = \sigma_2$.

Test statistic

$$F = \frac{s_1^2}{s_2^2} \quad \text{with} \quad DF_1 = n_1 - 1, \qquad DF_2 = n_2 - 1$$

Acceptance region. We are concerned with an improvement in variation; therefore, we will use a one-tail test (Figure 11.10) with the entire α risk in the right tail.

From Table G in Appendix II, the acceptance region is $F \leq 4.15$.

Analysis of sample data

$$F = \frac{(934)^2}{(659)^2} = 2.01$$

Conclusion. Because F is inside the acceptance region, the hypothesis is accepted. Therefore, there is not sufficient evidence to conclude that the parts were more consistent in strength after aging.

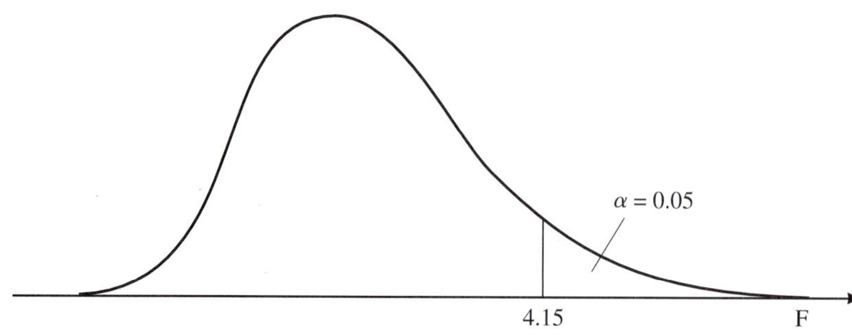

FIGURE 11.10
Distribution of F with α on right tail.

In this test and other tests that compare two samples, the samples must be independent to ensure valid conclusions.

11.9 DRAWING CONCLUSIONS FROM TESTS OF HYPOTHESES

The payoff for these tests of hypotheses comes from reaching useful conclusions. The meaning of *reject the hypothesis* or *accept the hypothesis* is shown in Table 11.4, together with some analogies to explain the subtleties of the meanings.

When a hypothesis is rejected, the practical conclusion is, The parameter value specified in the hypothesis is wrong. The conclusion is made with strong conviction—roughly speaking at a confidence level of $(1 - \alpha)$ percent. The key question then is: Just what is a good estimate of the value of the parameter for the population? Help can be provided on this question by calculating the confidence limits for the parameter (see Section 11.4).

When a hypothesis is accepted, the numerical value of the parameter stated in the hypothesis has not been proved, but it has not been disproved. It is not correct to say that the hypothesis has been proved as correct at the $(1 - \alpha)$ percent confidence level. Many other hypotheses could be accepted for the given sample of observations, and yet only one hypothesis can be true. Therefore, an acceptance does not mean a high probability of proof that a specific hypothesis is correct. (All other factors being equal, the smaller the sample size, the more likely it is that the hypothesis will be accepted. Less evidence certainly does not imply proof.)

With an acceptance of a hypothesis, a key question then is: What conclusion, if any, can be drawn about the parameter value in the hypothesis? Two approaches are suggested:

1. *Calculate confidence limits on the sample result.* These confidence limits define an interval within which the true population parameter lies. If this interval is small, an

TABLE 11.4
The meaning of a conclusion from tests of hypotheses

	If hypothesis is rejected	If hypothesis is accepted
Adequacy of evidence in the sample of observations	Sufficient to conclude that hypothesis is false	Not sufficient to conclude that hypothesis is false; hypothesis is a reasonable one but has not been proved to be true
Difference between sample result (e.g., $\bar{X}$) and hypothesis value (e.g., μ_0)	Unlikely that difference was due to chance (sampling) variation	Difference could easily have been due to chance (sampling) variation
Analogy of guilt or innocence in a court of law	Guilt has been established beyond a reasonable doubt	Have not established guilt beyond a reasonable doubt
Analogy of a batting average in baseball	If player got 300 base hits out of 1000 times at bat, this is sufficient to conclude that his overall batting average is about 0.300	If player got 3 hits in 10 times, this is not sufficient to conclude that his overall average is about 0.300

acceptance decision on the test of hypothesis means that the true population value is either equal to or close to the value stated in the hypothesis. Then it is reasonable to act as if the parameter value specified in the hypothesis is in fact correct. A relatively wide confidence interval is a stern warning that the true value of the population might be far different from that specified in the hypothesis. For example, the confidence limits of 21.2 and 40.8 on battery life in Section 11.4 would lead to an acceptance of the hypothesis of $\mu = 30.0$, but note that the confidence interval is relatively wide.

2. *Construct and review the operating characteristic curve for the test of hypothesis.* This curve defines the probability that other possible values of the population parameter could have been accepted by the test. Knowing these probabilities for values relatively close to the original hypothesis can help draw further conclusions about the acceptance of the original hypothesis. For example, Figure 11.6 shows the OC curve for a hypothesis that specified that the population mean is 30.0. Note that the probability of accepting the hypothesis when the population mean is 30.0 is 0.95 (or $1 - \alpha$). But also note that if μ really is 35.0, then the probability of accepting $\mu = 30.0$ is still high (about 0.83). If μ really is 42.0, the probability of accepting $\mu = 30.0$ is only about 0.33.

Care must always be taken in drawing business conclusions from the statistical conclusions, particularly when a hypothesis is accepted.

A test of hypothesis tests to see whether a statistically significant difference exists between the sample result and the value of the population parameter stated in the hypothesis. A decision to reject the hypothesis means that a statistically significant difference is present. However, the difference does not necessarily have practical significance. Large sample sizes, although not generally available, can detect small differences that may not have practical importance. Conversely, *accept hypothesis* means that a statistically significant difference was not found, but this outcome may have been due to a small sample size. A larger sample size could result in a "reject hypothesis" and thus detect a significant difference.

Instead of determining an acceptance region for a test of hypothesis and seeing whether the test statistic falls within the acceptance region, some people prefer to determine the significance probability (P). In this approach we calculate the test statistic and then look in the appropriate probability distribution table to determine the probability value (i.e., the smallest statistical significance level at which we can reject the hypothesis). Computer software often provides the P value as part of the data analysis. This P value is then compared to the significance level originally chosen: If P is equal to or less than the significance level, the hypothesis is rejected; if P is greater than the significance level, the hypothesis is accepted. Knowing the probability value does add information for decision making, but the same probability value could occur for other possible sample results (see *JQH5,* page 44.81 for elaboration). Thus, at least confidence limits on the sample result should be calculated (see above).

11.10
DETERMINING THE SAMPLE SIZE REQUIRED FOR TESTING A HYPOTHESIS

The preceding sections assume that the sample size is fixed by nonstatistical reasons and that only the type I error is predefined for the test. The ideal procedure is to predefine the desired type I and II errors and calculate the sample size required to cover both types of errors.

The sample size required will depend on (1) the sampling risks desired (α and β), (2) the size of the smallest true difference that is to be detected, and (3) the variation in characteristic being measured. The sample size can be determined by using the OC curve for the test (see *JQH5,* page 44.78).

The sample size can also be directly calculated:

$$n = \left[\frac{(Z_{\alpha/2} + Z_{\beta})\sigma}{\mu - \mu_0}\right]^2$$

Suppose that it was important to detect the fact that the mean life of the battery cited previously was 35.0 hours. Specifically, we want to be 80 percent sure of detecting this change ($\beta = 0.2$). Further, if the true mean was 30.0 hours (as stated in the hypothesis), we want to have only a 5 percent risk of rejecting the hypothesis ($\alpha = 0.05$). Then:

$$n = \left[\frac{(1.96 + 0.84)10}{35 - 30}\right]^2 = 31.4$$

The required sample size is 32.

11.11
THE DESIGN OF EXPERIMENTS

First, a bit of history. The body of knowledge called design of experiments (DOE) had early roots in the field of agriculture with the pioneering work of Sir Ronald A. Fisher. In the business world the application of DOE was focused in the manufacturing sector, particularly the chemical industry. Applications seemed to be inhibited because of the complexities of both the design and analysis and the practical difficulties of executing the experiments.

Several developments have sparked an increased attention to DOE. First, the six-sigma movement (see Chapter 3) integrates DOE as an important tool for preventing or solving problems and decreasing variation in processes. Second, several alternative adaptations of classical DOE have been developed that simplify the concepts and attempt to make applications more "practical." These alternatives include the Taguchi approach (see below). As with most attempts to simplify complex concepts, the adaptations have stimulated some controversies. Finally, the power of computer software in doing the statistical calculations eliminates another inhibiting burden of the past.

The future will bring exciting applications of DOE. Applications in the manufacturing sector will increase, and the service sector, with its strong emphasis on the human element in operations, will learn how to apply DOE. Some early applications in the service sector have included using DOE to reduce cycle time for mortgage application approval in financial services and improving quality and productivity in call center operations at a telecommunications organization.

Experiments can have numerous objectives, and the best strategy depends on the objective. In some experiments the objective is to find the most important variables affecting a quality characteristic. The plan for conducting such experiments is the *design of the experiment.* We will first cover an example that presents several alternative designs and defines the basic terminology and concepts.

Suppose that three detergents are to be compared for their ability in cleaning clothes in an automatic washing machine. The "whiteness" readings obtained by a special measuring procedure are the *dependent,* or *response, variable.* The independent variable under investigation (detergent) is a *factor,* and each variation of the factor is called a *level*; i.e., there are three levels. A treatment is a single level assigned to a single factor, detergent A. A treatment combination is the set of levels for all factors in a given experimental run. A factor may be qualitative (different detergents) or quantitative (water temperature). Finally, some experiments have a *fixed-effects model*; i.e., the levels investigated represent all levels of concern to the investigator (e.g., three brands of washing machines). Other experiments have a *random effects model*; i.e., the levels chosen are just a sample from a larger population (e.g., three operators of washing machines). A *mixed-effects model* has both fixed and random factors.

Figure 11.11 outlines six designs of experiments starting with the classical design in Figure 11.11*a*. Here all factors except detergent are held constant. Thus nine tests are run, i.e., three with each detergent with the washing time, make of machine, water temperature, and all other factors held constant. One drawback of this design is that the conclusions about detergent brands apply only to the specific conditions run in the experiment.

(*a*)

A	B	C
–	–	–
–	–	–
–	–	–

(*b*)

I	II	III
A	B	C
A	B	C
A	B	C

(*c*)

I	II	III
C	B	B
A	C	B
A	A	C

(*d*)

I	II	III
B	A	C
C	C	A
A	B	B

(*e*)

	I	II	III
1	C	A	B
2	B	C	A
3	A	B	C

(*f*)

	I	II	III
	ABC	ABC	ABC
1	--	--	--
2	--	--	--
3	--	--	--

FIGURE 11.11
Some experimental designs.

Figure 11.11*b* recognizes a second factor at three levels, i.e., washing machines brands I, II, and III. However, in this design it would not be known whether an observed difference was due to detergents or washing times.

In Figure 11.11*c*, the nine tests are assigned completely at random, thus the name *completely randomized design.* However, detergent A is not used with machine brand III, and detergent *B* is not used with machine brand I, thus complicating the conclusions.

Figure 11.11*d* shows the *randomized block design.* Here each block is a machine brand, and the detergents are run in random order within each block. This design guards against any possible bias due to the order in which the detergents are used and has advantages in the subsequent data analysis and conclusions. First, a test of hypothesis can be run to compare detergents and a separate test of hypothesis run to compare machines; all nine observations are used in both tests. Second, the conclusions concerning detergents apply for the three machines and vice versa, thus providing conclusions over a wider range of conditions.

Now suppose that another factor such as water temperature is also to be studied, using the *Latin square design* shown in Figure 11.11*e*. Note that this design requires that each detergent be used only once with each machine and only once with each temperature. Thus three factors can be evaluated (by three separate tests of hypothesis) with only nine observations. However, there is a danger. This design assumes no interaction among the factors. No interaction between detergent and machine means that the effect of changing from detergent A to B to C does not depend on which machine is used, and similarly for the other combinations of factors. The concept of interaction is shown in Figure 11.12.

Finally, the main factors and possible interactions could be investigated by the *factorial design* in Figure 11.11*f*. *Factorial* means that at least one test is run at every combination of the main factors, in this case 3 × 3 × 3 or 27 combinations. Separate tests of hypothesis can be run to evaluate the main factors and also the possible interactions. Again, all the observations contribute to each comparison. When there are many factors, a portion of the complete factorial (i.e., a "fractional factorial") is useful. (See Section 11.15.)

Several key tools used in this example are explained next.

Most problems can be handled with one of the standard experimental designs. Designs can be classified by the number of factors to be investigated, the structure of the experimental design, and the kind of information the experiment is intended to provide (Table 11.5).

For a description of both the design and analysis of various design structures, see *JQH5,* Section 47, "Design and Analysis of Experiments." A sequential approach to experimentation can often be helpful. For a series of four papers, see Carter (1996). Emanuel and Palanisamy (2000) discuss sequential experimentation at two levels and a maximum of seven factors.

Increasingly, experimental design is focusing on exploratory experimentation as a means of discovering new knowledge (Box, 1999). This emphasis may require the use of extensive designs such as full factorials, response surface methodology, and other designs. Yes, these are more complex than other designs, but the benefits are powerful. For example, a factorial design at two levels not only enables us to evaluate possible interactions of factors but also provides the data to develop a mathematical model of the response variable and the causal factors. Thus more complete knowledge of the most critical factors and the best levels of those factors becomes available.

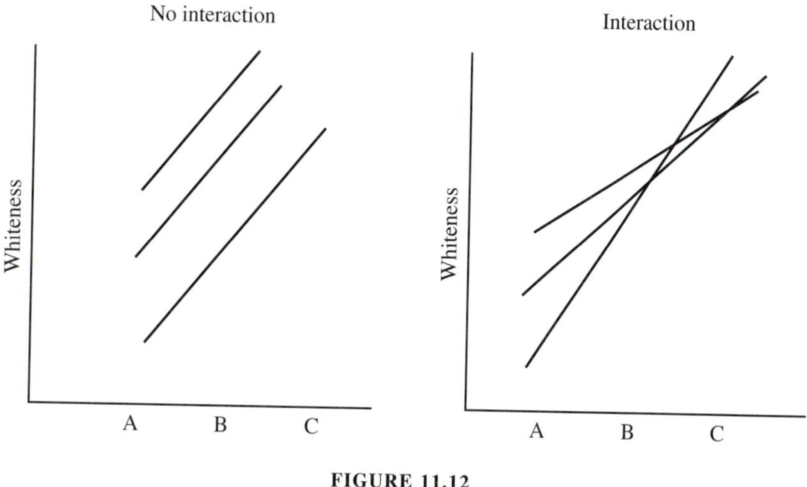

FIGURE 11.12
Interaction.

TABLE 11.5
Classification of designs

Design	Type of application
Completely randomized	Appropriate when only one experimental factor is being investigated.
Factorial	Appropriate when several factors are being investigated at two or more levels and interaction of factors may be significant.
Blocked factorial	Appropriate when number of runs required for factorial is too large to be carried out under homogeneous conditions.
Fractional factorial	Appropriate when many factors and levels exist and running all combinations is impractical.
Randomized block	Appropriate when one factor is being investigated and experimental material or environment can be divided into blocks or homogeneous groups.
Balanced incomplete block	Appropriate when all the treatments cannot be accommodated in a block.
Partially balanced incomplete block	Appropriate if a balanced incomplete block requires a larger number of blocks than is practical.
Latin square	Appropriate when one primary factor is under investigation and results may be affected by two other experimental variables or by two sources of nonhomogeneity. It is assumed that no interactions exist.
Youden square	Same as Latin square but number of rows, columns, and treatments need not be the same.
Nested	Appropriate when objective is to study relative variability instead of mean effect of sources of variation (e.g., variance of tests on the same sample and variance of different samples).
Response surface	Objective is to provide empirical maps (contour diagrams) illustrative of how factors under the experimenter's control influence the response.
Mixture designs	Same as factorial designs.

Source: Adapted from *JQH5,* Table 47.3.

11.12 SOME TOOLS FOR SOUND EXPERIMENTATION

Planned Grouping or Blocking

Beyond the factors selected for study, other uncontrollable variables may affect the outcome of the experiments. These variables are also called "noise," "experimental error," or "background variables." Where the experimenter is aware of these variables, it is often possible to plan the experiment so that

1. Possible effects due to uncontrollable variables do not affect information obtained about the factors of primary interest.
2. Some information about the effects of the uncontrollable variables can be obtained.

In designing experiments, wide use is made of the uniformity within blocks to minimize the effect of unwanted variables and to accentuate the effect of the variables under study. Designs that make use of this uniformity within blocks are called "block designs," and the process is called "planned grouping." A specific design ensures that different effects can be directly estimated without any entanglement ("confounding") of unwanted variables. In other words, no correlations exist among the factors under study. Designs that have this property are called "orthogonal" designs.

Randomization

The assignment of specimens to treatments in a purely chance manner is called "randomization" in the design of experiments. Such assignment increases the likelihood that the effect of uncontrolled variables will balance out. It also improves the validity of estimates of experimental error and makes possible the application of statistical tests of significance and the construction of confidence intervals.

Replication

Replication is the repetition of an observation or measurement. It is done to increase precision and to provide the means of measuring precision. (Some kinds of experiments have no outside source for measuring precision, so the measure must come from the experiment itself.) In addition, replication provides an opportunity for the effects of uncontrolled factors to balance out and thus aids randomization as a bias-decreasing tool. (In successive replications the randomization features must be independent.) Replication also helps to detect gross errors in the measurements.

In designing the experiment, some key questions that arise are

1. How large a difference in the conditions being compared is considered significant from an engineering point of view? (How large a difference do we want the experiment to detect?)
2. How much variation has been experienced in the quality characteristics under investigation?
3. What risk do we want to take that the experiment incorrectly concludes that a significant difference exists when the correct conclusion is that no significant difference exists? (This is the type I error.)
4. What risk do we want to take that the experiment fails to detect the difference that really does exist? (This is the type II error.)
5. Do we have any knowledge about possible interactions of the factors? Do we wish to test for these interactions?

11.13 CONTRAST BETWEEN THE CLASSICAL AND MODERN METHODS OF EXPERIMENTATION

The contrast between the classical method of experimentation (varying one factor at a time, holding everything else constant) and the modern approach is striking. Table 11.6 compares these two approaches for an experiment in which there are two factors

TABLE 11.6
Comparison of the classical and the modern methods of experimentation

Criteria	Classical	Modern
Basic procedure	Hold everything constant except the factor under investigation. Vary that factor and note the effect on the characteristic of concern. To investigate a second factor, conduct a separate experiment in the same manner.	Plan the experiment to evaluate both factors in one main experiment. Include, in the design, measurements to evaluate the effect of varying both factors simultaneously.
Experimental conditions	Care taken to have material, workers, and machine constant throughout the entire experiment.	Realizes difficulty of holding conditions reasonably constant throughout an entire experiment. Instead, experiment is divided into several groups or blocks of measurements. Within each block, conditions must be reasonably constant (except for deliberate variation to investigate a factor).
Experimental error	Recognized but not stated in quantitative terms.	Stated in quantitative terms.
Basis of evaluation	Effect due to a factor is evaluated with only a vague knowledge of the amount of experimental error.	Effect due to a factor is evaluated by comparing variation due to that factor with the quantitative measure of experimental error.
Possible bias due to sequence of measurements	Often assumed that sequence has no effect.	Guarded against by randomization.

(Continued)

(or variables) whose effects on a characteristic are being investigated. (The same conclusions hold for an experiment with more than two factors.)

This discussion has been restricted to the design or planning of the experiment. After the data are collected, the analysis phase beings. For simple experiments some of the basic tests of hypotheses and confidence limits (previously discussed) provide the tools of analysis. For more complex experiments we use additional tools such as the analytical analysis of variance (see below) and the graphical analysis of means (see Wadsworth, 1998).

For a perceptive discussion of the relation of design of experiments to the scientific method, see Box and Liu (1999).

11.14 ANALYSIS OF VARIANCE

Analysis of variance (ANOVA) is an important tool employed to analyze the results of an experiment. In this approach the total variation of all the measurements around the

TABLE 11.6 *(Continued)*

Criteria	Classical	Modern
Effect of varying both factors simultaneously ("interaction")	Not adequately planned into experiment. Frequently assumed that the effect of varying factor 1 (when factor 2 is held constant at some value) would be the same for any value of factor 2.	Experiment can be planned to include an investigation for interaction between factors.
Validity of results	Misleading and erroneous if interaction exists and is not realized.	Even if interaction exists, a valid evaluation of the main factors can be made.
Number of measurements	For a given amount of useful and valid information, more measurements needed than in the modern approach.	Fewer measurements needed for useful and valid information.
Definition of problem	Objective of experiment frequently not defined as necessary.	Designing the experiment requires defining the objective in detail (how large an effect do we want to determine, what numerical risks can be taken, etc.).
Application of conclusions	Sometimes disputed as applicable only to the controlled conditions under which the experiment was conducted.	Broad conditions can be planned into the experiment, thereby making conclusions applicable to a wider range of actual conditions.

overall mean is divided into sources of variation that are then analyzed for statistical significance. The following example explains this technique.

Four types of diets (1, 2, 3, 4) were fed to 24 animals. Blood coagulation time (the response variable) was recorded and is shown in Table 11.7.

The issue is whether diet has an effect on coagulation time. The null hypothesis is $\mu 1 = \mu 2 = \mu 3 = \mu 4$; i.e., the average coagulation time is the same for both diets.

The ANOVA analysis (using MINITAB®) is shown in Figure 11.13. The Total variation is divided into variation due to DietNum and Error (the residual when variation due to diet is subtracted from the total). The sum of squares (SS) for the total variation is the sum of the squared deviations of the 24 measurements around the overall average of the 24. The SS for diet is the sum of the squared deviations of the individual measurements for each diet around the averages for each diet. The SS for the error is found by subtracting the SS for diet from the SS for the total variation. The mean square (MS) is the sum of squares divided by the degrees of freedom (DF). The mean square value is essentially a variance (square of the standard deviation). The hypothesis is tested by comparing the MS for diet to the MS for error. This step is done by using the F test (Chapter 10) and calculating F as 76.00/5.60 or 13.57. If a significance level of 0.05 is used, the acceptance region on F is equal to or less than 3.10. The calculated F is much greater than 3.10, and thus the hypothesis is rejected; therefore, we conclude that diet does have an effect on coagulation time. Note that the MINITAB output shows

TABLE 11.7
Diet and coagulation time

Diet 1	Diet 2	Diet 3	Diet 4
62	63	68	56
60	67	66	62
63	71	71	60
59	64	67	61
	65	68	63
	66	68	64
			63
			59

Source: Juran Institute, Inc.

Source	DF	SS	MS	F	P
DietNum	3	228.00	76.00	13.57	0.000
Error	20	112.00	5.60		
Total	23	340.00			

```
                                     Individual 95% confidence intervals
                                     for mean based on pooled StDev
Level   N    Mean    StDev     ---+---------+---------+---------+---
1       4   61.000   1.826     (-------*-------)
2       6   66.000   2.828                         (------*------)
3       6   68.000   1.673                              (-----*-----)
4       8   61.000   2.619        (-----*-----)
                               ---+---------+---------+---------+---
Pooled StDev = 2.366             59.5      63.0      66.5      70.0
```

FIGURE 11.13
One-way analysis of variance and confidence intervals for CoagTime. (*From Juran Institute, Inc.*)

the significance probability as zero, meaning that it is highly unlikely that an F value of 13.57 could have occurred by chance if the means for the different diets were equal. This outcome is equivalent to rejecting the hypothesis. Also note that confidence intervals are provided for the diet means. Confidence intervals that do not overlap suggest that the diet means are equal, and we reject the hypothesis. Note how the confidence intervals provide useful information beyond the basic F test.

This example introduces only the basic concept of ANOVA and makes certain assumptions; e.g., the responses are independent and normally distributed and the population variances across all levels of the factor under study are equal. See *JQH5,* Section 47, for elaboration and for other examples of ANOVA for different experimental designs. Mazu (1996) describes the results of an experiment with seven variables that led to both a 40-pound decrease in variation and a 20 percent decrease in production costs. The ANOVA table quantified an amazing 32 sources of variation, including many interactions.

The analysis of variance determines which sources of variation (factors) contribute a significant amount of the total variation of the response variable, y. Knowing which factors are significant can lead to a mathematical model for y as a function of the dependent variables, x, found to be significant; i.e., $y = f(x_1, x_2, \ldots x_n)$ For elaboration, see DeVor et al. (1992), Section 18.

11.15 TAGUCHI APPROACH TO EXPERIMENTAL DESIGN

Professor Genichi Taguchi uses an approach to experimental design that has three purposes:

1. To design products and processes that perform consistently on target and are relatively insensitive ("robust") to factors that are difficult to control.
2. To design products that are relatively insensitive (robust) to component variation.
3. To minimize variation around a target value.

Thus the approach is meant to provide valuable information for product design and development. Taguchi views the engineering design process as having three phases: systems design, parameter design, and tolerance design. For a discussion of these phases and also the concept of robustness, see Section 13.4.

Taguchi proposes that the design of experiments can be particularly helpful in parameter design and tolerance design. He recommends that fractional factorial designs ("orthogonal arrays") are effective in obtaining useful information at minimum cost. Fractional factorial designs are appropriate when many factors and levels are present, but running all combinations in the experiment is impractical.

A fractional factorial design with seven factors (A, B, C, D, E, F, and G) each at two levels (1 and 2) is shown in Figure 11.14. This design requires only eight trials of the 128 combinations (the eight are indicated in black). The matrix shows the level (1 or 2) to be used for each of the seven factors. The column numbers (1 through 7) represent the factors (A through G). For example, trial number 1 runs each factor at level 1.

A fractional factorial requiring tests at only eight combinations of the seven variables clearly has time and cost advantages over a full factorial design. The information from such a small experiment is, however, substantially reduced from that of a full factorial experiment. Specifically, with the fractional design only certain interaction effects can be evaluated. Although interaction effects complicate an experiment, interaction of several variables is often the key to solving a problem. The selection of a specific orthogonal array design depends on the number of factors and levels, the interactions of interest, and the amount of confounding (mixing of variables) that can be permitted. For elaboration of the Taguchi approach and tables of orthogonal array designs, see Ross (1996).

For an extensive bibliography and a summary of some controversial aspects of the Taguchi approach, see *JQH5,* pages 47.58 and 47.59.

			A_1								A_2							
			B_1				B_2				B_1				B_2			
			C_1		C_2		C_1		C_2		C_1		C_2		C_1		C_2	
			D_1	D_2	D_1	D_2	D_1	D_2	D_1	D_2	D_1	D_2	D_1	D_2	D_1	D_2	D_1	D_2
E_1	F_1	G_1	■															
		G_2														■		
	F_2	G_1												■				
		G_2							■									
E_2	F_1	G_1								■								
		G_2											■					
	F_2	G_1													■			
		G_2		■														

Trial no.	Column no. 1	2	3	4	5	6	7
1	1	1	1	1	1	1	1
2	1	1	1	2	2	2	2
3	1	2	2	1	1	2	2
4	1	2	2	2	2	1	1
5	2	1	2	1	2	1	2
6	2	1	2	2	1	2	1
7	2	2	1	1	2	2	1
8	2	2	1	2	1	1	2

FIGURE 11.14
Fractional factorial design with matrix of levels. (*From Ross, 1996.*)

11.16 SHAININ AND THE RED X APPROACH TO EXPERIMENTAL DESIGN

Dorian Shainin has developed an approach called "statistical engineering," which is a combination of statistical and engineering techniques for solving engineering problems. The aim is to reduce variation by finding the largest cause of variation in a process. This cause—called the "Red X"— may be a single variable or an interaction of two or more variables.

The Red X approach uses techniques that are relatively simple and easily understood by operations personnel. Statistical engineering comprises 22 tools that combine selected DOEs, problem-solving tools, and statistical process control tools. The DOE-type tools range from multivari analysis to factorial designs. For a description of the tools, see Shainin (1993) and Bhote (1991).

11.17
PLANNING FOR EXPERIMENTAL DESIGN

Experienced statisticians have long known that formal design of experiments involves much more than selecting a design and making statistical calculations. A successful experiment requires careful planning of the activities before, during, and after the experiment. A classic paper by Bicking (1954) provides the definitive checklist of planning activities, reprinted in *JQH5,* pages 47.7 and 47.8. The checklist of 44 items covers the following items: obtain a clear problem statement, collect background information, design the experiment, carry out the experiment, analyze the data, interpret the results, and prepare the report. No experiment should be conducted without reviewing the checklist.

Bisgaard (1999) describes 11 points to be covered in a "preexperiment proposal" (reprinted by courtesy of Marcel Dekker, Inc.).

1. Statement of objective(s) of the experiment.
2. List of factors and their levels.
3. Description of how to measure the responses.
4. A table showing the design and possible confounding.
5. Layout of data collection sheets.
6. Description of the experimental procedure.
7. Schedule for the experiment.
8. An outline of how to analyze the data.
9. Budget for time, money, and other resources.
10. Anticipated problems and how to deal with them.
11. List of team members and responsibilities.

This list particularly applies to factorial experiments conducted in a sequential mode.

Barton (1997) describes some graphical methods useful in preexperiment planning. These methods show how graphical plots can be prepared based on preexperiment discussions with engineers who provide information on the product and process. These plots help to define the goals of the experiment; identify dependent, independent, and "nuisance" variables; and choose a mathematical relationship (a regression) between the independent and dependent variables. See also Coleman and Montgomery (1993) for further discussion of planning for experiments.

11.18
REGRESSION ANALYSIS

Quality problems sometimes require a study of the relationship between two or more variables, or *regression analysis.* The uses of regression analysis include forecasting and prediction, determining the important variables influencing some result, and locating optimum operating conditions.

The steps in a regression study are

1. Clearly define the objectives of the study. This step must include a definition of the dependent or response variable and the independent variables that are thought to be related to the dependent variable.
2. Collect data on the independent and dependent variables.
3. Prepare scatter diagrams (plots of one variable versus another).
4. Calculate the regression equation.
5. Study the equation to see how well it fits the data.
6. Provide measures of the precision of the equation.

The following example illustrates these steps.

Suppose that the life of a tool varies with the cutting speed of the tool and we want to predict life based on cutting speed. Thus life is the dependent variable (Y) and cutting speed is the independent variable (X). Data are collected at four cutting speeds (Table 11.8).

The plot of the data is a *scatter diagram* (Figure 11.15). This plot should always be prepared before making any further analysis. The graph alone may provide sufficient information on the relationship between the variables to draw conclusions on the

TABLE 11.8
Cutting speed (*X*, in feet per minute) versus tool life (*Y*, in minutes)

X	*Y*	*X*	*Y*	*X*	*Y*	*X*	*Y*
90	41	100	22	105	21	110	15
90	43	100	35	105	13	110	11
90	35	100	29	105	18	110	6
90	32	100	18	105	20	110	10

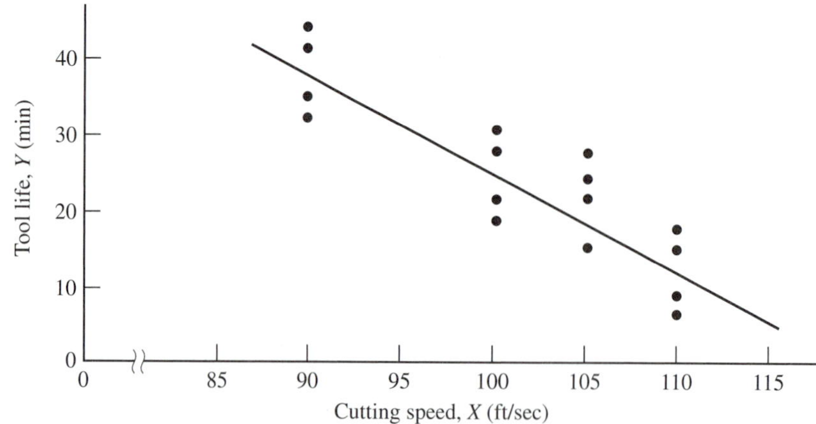

FIGURE 11.15
Tool life Y versus cutting speed X.

immediate problem, but the graph is also useful in suggesting possible forms of an estimating equation. Figure 11.15 suggests that life varies not only with cutting speed (i.e., life decreases with an increase in speed) but also in a linear manner (i.e., increases in speed result in a certain decrease in life that is the same over the range of the data). Note that the relationship is not perfect—the points scatter about the line.

Often it is valuable to determine a regression equation. For linear relationships, this can be done approximately by drawing a straight line by eye and then graphically estimating the Y intercept and slope. The linear regression model is

$$Y = \beta_0 + \beta_1 X + \varepsilon$$

where β_0 and β_1 are the unknown population intercept and slope and ε is a random-error term that may be due to measurement errors and/or the effects of other independent variables. This model is estimated from sample data by the form

$$\hat{Y} = b_0 + b_1 X$$

where $\hat{Y}$ is the predicted value of Y for a given value of X and b_0 and b_1 are the sample estimates of β_0 and β_1.

These estimates are usually found by the least-squares method, so named because it minimizes the sum of the squared deviations between the observed and predicted values of Y. The least-squares estimates are

$$b_1 = \frac{\Sigma(X_m - \bar{X})(Y_m - \bar{Y})}{\Sigma(X_m - \bar{X})^2} = \frac{\Sigma X_m Y_m - (\Sigma X_m \Sigma Y_m)/N}{\Sigma X_m^2 - (\Sigma X_m)^2/N}$$

$$b_0 = \bar{Y} - b_1 \bar{X}$$

The summations range from $m = 1$ to $m = N$, where N is the total number of sets of values of X and Y.

The detailed calculations are easily handled by a regression software program. For this data:

$$b_1 = \frac{-1191.25}{875} = -1.3614$$

$$b_0 = 23.06 - (-1.3614)(101.25) = 160.9018$$

and hence the prediction equation is

$$\hat{Y} = 106.90 - 1.3614X$$

After estimating the coefficients of the prediction equation, the equation should be plotted over the data to check for gross calculation errors. Roughly half the data points should be above the line and half below it. In addition, the equation should pass exactly through the point $\bar{X}, \bar{Y}$.

A number of criteria exist for judging the adequacy of the prediction equation. One common measure of the adequacy of the prediction equation is R^2, the proportion of variation explained by the prediction equation. R^2, or the *coefficient of determination,* is the ratio of the variation due to the regression, $\Sigma(\hat{Y}_m - \bar{Y})$, to the total variation, $\Sigma(Y_m - \bar{Y})^2$. $\hat{Y}_m$ is the predicted Y value for X_m. The calculation formula is

$$R^2 = \frac{b_1\Sigma(X_m - \bar{X})(Y_m - \bar{Y})}{\Sigma(Y_m - \bar{Y})^2}$$

$$= \frac{(-1.3614)(-1191.25)}{1958.94} = 0.828$$

Thus for this example the prediction equation explains 82.8 percent of the variation of the tool life. The coefficient of determination and *all* other measures of the precision of a regression relationship must be interpreted with great care. Regression analysis is not an area for the amateur.

An example of a MINITAB output for a regression analysis is shown in Figure 11.16. Note that the regression plot provides two sets of confidence limits. The inner set shows 95 percent confidence limits for the mean value (i.e., for predictions using the relationship represented by the line), and the outer set shows 95 percent prediction limits for individual values. Also note that the coefficient of determination ("R-sq") is 69.5 percent and the ANOVA table shows that the regression is significant with an F value of 43.29 (comparing the regression variation to the experimental error variation).

The above discussion covers simple *linear regression*—the prediction of a dependent variable, Y, from a single predictor variable, X. *Multiple regression* cases involve two or more predictor variables. For two predictor variables, the multiple linear regression model is

$$Y = \beta_0 + \beta_1 X_1 + \beta_2 X_2 + \varepsilon$$

For a discussion of how to estimate and examine the prediction equation, see *JQH5,* Section 44.

Another measure of the degree of association between two variables is the *simple linear correlation coefficient, r*. This coefficient varies between -1.0 and $+1.0$. Calculations for r are explained in *JQH5,* Section 44. Many software programs provide r as an output. Positive values indicate that as one variable increases, so does the other; negative values mean that as one variable decreases, so does the other. A coefficient of zero means that there is no association between the two variables; values of -1.0 or $+1.0$ indicate a perfect correlation.

Interpretation of the coefficient of correlation requires great care. First, r indicates the strength of a *linear* relationship. Thus, an r may be close to zero when there is actually a strong *nonlinear* relationship. Also, a high value of r does *not* necessarily imply that one variable *causes* the other. Variable X may contribute to variable Y but not be the sole cause, both X and Y may result from some common (but unknown) cause, or other reasons may explain the apparent correlation.

This brief treatment of regression is just an introduction to a complex subject. Further topics include confidence intervals and other measures of precision, multiple regression, and nonlinear regression. The literature provides more information (see *JQH5,* pages 44.89–44.108).

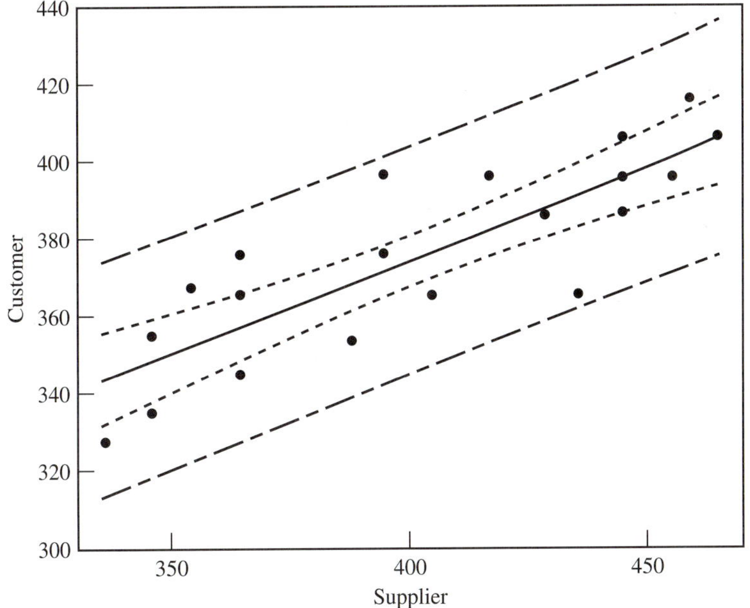

Regression

The regression equation is y = 183 + 0.476x

Predictor	Coef	StDev	T	P
Constant	182.81	29.36	6.23	0.000
x	0.47629	0.07239	6.58	0.000

S = 13.56   R-Sq = 69.5%   R-Sq(adj) = 67.9%

Source	DF	SS	MS	F	P
Regression	1	7955.9	7955.9	43.29	0.000
Error	19	3492.1	183.8		
Total	20	11448.0			

FIGURE 11.16

MINITAB output for regression analysis. (*From Juran Institute, Inc.*)

11.19 ADVANCED TOOLS OF DATA ANALYSIS

The growth of computer databases is resulting in large amounts of data that are easily accessible for analysis—*data mining* is the new term. This environment is leading to the use of various data analysis tools—some are new, some are old, some were developed by statisticians, some by other practitioners. The area of multivariate analysis includes tools such as principal component analysis, factor analysis, cluster analysis, discriminant analysis, multiple regression, neural networks, and time series analysis. Other tools include Bayesian analysis, fuzzy logic, image recognition, voice recognition, and artificial intelligence. An excellent source describing advanced tools is Wadsworth (1998).

Fortunately, computer software is available for many of these tools, and the time saved in making calculations can be invested in thoroughly understanding the meaning of the results.

Answers to questions at the beginning of the chapter:

1. Yes
2. Yes
3. Yes
4. No
5. Yes

SUMMARY

- Estimation is the process of analyzing a sample result in order to predict the corresponding value of the population parameter.
- The point estimate is a single value used to estimate the population parameter. The confidence interval is a range of values that include (with a preassigned probability called a confidence level) the true value of a population parameter.
- A hypothesis is an assertion made about a population. A test of hypothesis is a test of the validity of the assertion and is carried out by analysis of a sample of data.
- In evaluating a hypothesis, two types of errors can be made: type I errors (reject the hypothesis when it is true) and type II errors (accept the hypothesis when it is false).
- The statistical design of experiments provides plans for conducting experiments from which valid statistical analyses can be made.
- Randomization is the assignment of specimens to treatments in a purely chance manner.
- Replication is the repetition of an observation or measurement.
- Regression analysis is the study of the relationship between two or more variables.

PROBLEMS

Note: The specific questions have purposely been stated in nonstatistical language to provide the student with some practice in choosing techniques and making assumptions. When required, use a type I error of 0.05 and a confidence level of 95 percent. State any other assumptions needed.

11.1. In the casting industry the pouring temperature of metal is important. For an aluminum alloy, past experience shows a standard deviation of 15°. During a particular day, five temperature tests were made during the pouring time.

(*a*) If the average of these measurements was 1650°, make a statement about the average pouring temperature.

(*b*) If you had taken 25 measurements and obtained the same results, what effect would this have on your statement? Make such a revised statement.

11.2. At the casting firm above, a new aluminum alloy is being poured. During the first day of pouring, five pouring temperature tests were made, with these results:

1705° 1725° 1685° 1690° 1715°

Make a statement about the average pouring temperature of this metal.

11.3. A manufacturer pressure tests gaskets for leaks. The pressure at which this gasket leaked on nine trials was (in psi):

4000	3900	4500
4200	4400	4300
4800	4800	4300

Make a statement about the average "leak" pressure of this gasket.

11.4. The acceptable time for answering the phone at a call center is 15 seconds. In a sample of 500 calls, 427 were answered within 15 seconds. Make a statement concerning the true proportion that would be acceptable.

11.5. In a meat-packing firm, out of 600 pieces of beef, 420 were found to be Grade A. Make a statement about the true proportion of Grade A beef.

11.6. A specification requires that the average breaking strength of a certain material be at least 180 psi. Past data indicate the standard deviation of individual measurements to be 5 psi. How many tests are necessary to be 99 percent sure of detecting a lot that has an average strength of 170 psi?

11.7. Tests are to be run to estimate the average life of a product. Based on past data on similar products, it is assumed that the standard deviation of individual units is about 20 percent of the average life.

(*a*) How many units must be tested to be 90 percent sure that the sample estimate will be within 5 percent of the true average?

(*b*) Suppose funds are available to run only 25 tests. How sure would we be of obtaining an estimate within 5 percent?

Answer: (*a*) 44. (*b*) 78.8 percent.

11.8. A manufacturer of needles has a new method of controlling a diameter dimension. From many measurements of the present method, the average diameter is 0.076 cm with a standard deviation of 0.010 cm. A sample of 25 needles from the new process shows the average to be 0.071. If a smaller diameter is desirable, should the new method be adopted? (Assume that the standard deviation of the new method is the same as that for the present method.)

11.9. In the garment industry the breaking strength of cloth is important. A heavy cotton cloth must have at least an average breaking strength of 200 psi. From one particular lot of this cloth, these five measurements of breaking strength (in psi) were obtained:

206
194
203
196
192

Does this lot of cloth meet the requirement of an average breaking strength of 200 psi?

Answer: $t = -0.67$.

11.10. In a drug firm the variation in the weight of an antibiotic from batch to batch is important. With our present process, the standard deviation is 0.11 g. The research department has developed a new process that it believes will produce less variation. The following weight measurements (in grams) were obtained with the new process:

7.47
7.49
7.64
7.59
7.55

Does the new process have less variation than the old?

11.11. A paper manufacturer has a new method of coating paper. The less variation in the weight of this coating, the more uniform and better the product. The following 10 sample coatings were obtained by the new method:

Coating weights (in weight/unit area × 100)	
223	234
215	229
220	223
238	235
230	227

If the standard deviation in the past was 9.3, is this proposed method any better? Should the company switch to this method?

Answer: $\chi^2 = 5.43$.

11.12. A manufacturer of rubber products is trying to decide which "recipe" to use for a particular rubber compound. High tensile strength is desirable. Recipe 1 is cheaper to mix, but he is not sure if its strength is about the same as that of recipe 2. Five batches of rubber were made by each recipe and tested for tensile strength. These are the data collected (in psi):

Recipe 1	Recipe 2
3067	3200
2730	2777
2840	2623
2913	2044
2789	2834

Which recipe would you recommend be used?

11.13. Test runs with five models of an experimental engine showed that they operated, respectively, for 20, 18, 22, 17, and 18 min with 1 gal of a certain kind of fuel. A proposed specification states that the engine must operate for a mean of at least 22 min.
(*a*) What can we conclude about the ability of the engine to meet the specification?
(*b*) What is the probability that the sample mean could have come from a process whose true mean is equal to the specification mean?
(*c*) How low would the mean operating minutes (of the engine population) have to be to have a 50 percent chance of concluding that the engine does not meet the specification?

Answer: (*a*) $t = -3.4$. (*b*) Approx. 0.03. (*c*) 20.1.

11.14. A manufacturer claims that the average length in a large lot of parts is 2.680 in. A large amount of past data indicates the standard deviation of individual lengths to be 0.002 in. A sample of 25 parts shows an average of 2.678 in. The manufacturer says that the result is still consistent with his claim because only a small sample was taken.
(*a*) State a hypothesis to evaluate the claim.
(*b*) Evaluate his claim using the standard hypothesis testing approach.
(*c*) Evaluate his claim using the confidence limit approach.

11.15. An engineer wants to determine whether the type of test oven or temperature has a significant effect on the average life of a component. She proposes the following design of experiment:

	Oven 1	Oven 2	Oven 3
550°	1	0	1
575°	0	1	1
600°	1	1	0

The numbers in the body of the table represent the number of measurements to be made in the experiment. What are two reasons that interaction cannot be adequately evaluated in this design?

11.16. The molding department in a manufacturing plant has been making too many defectives. There are many opinions about the reasons. One opinion states that the molding time per record has a cause-and-effect relationship with the number of defectives produced per 100 units. Several trial lots of 100 units each were made with various mold times. The results were as follows:

Time, *s*	Number defective
2	16
4	13
5	8
7	8
10	4
11	6
13	5
17	3
17	5
20	3

Plot the data and graphically estimate the Y intercept and the slope.

Answer: The least-squares estimates are a Y intercept of 13.54 and a slope of -0.6076.

11.17. Assemble a team of three people to prepare a proposal for a designed experiment. Select a problem on a specific performance characteristic in a service industry. Using the 11 points in Section 11.17, develop an outline for the proposal.

REFERENCES

Barton, R. R. (1997). "Pre-Experiment Planning for Designed Experiments: Graphical Methods," *Journal of Quality Technology,* July, pp. 307–316.

Bhote, K. (1991). *World Class Quality,* AMACOM, America Management Association, New York.

Bisgaard, S. (1999). "Quality Quandries," *Quality Engineering,* vol. 11, no. 4, pp. 645–650.

Box, G. E. P. (1999). "Statistics as a Catalyst to Learning by Scientific Method, Part II—A Discussion," *Journal of Quality Technology,* vol. 31, no. 1, pp. 16–29.

Box, G. E. P. and P. Y. T. Liu (1999). "Statistics as a Catalyst to Learning by Scientific Method, Part I—An Example," *Journal of Quality Technology,* vol. 31, no. 1, pp. 1–15.

Carter, C. W. (1996). "Sequenced-Level Experimental Designs," *Quality Engineering,* vol. 8, no. 1 (pp. 181–188), no. 2 (pp. 361–366), no. 3 (pp. 499–504), no. 4 (pp. 695–698).

Coleman, D. E. and D. C. Montgomery (1993). "A Systematic Approach to Planning for a Designed Experiment" (with discussion), *Technometrics,* vol. 35, pp. 1–27.

Deming, W. E. (1982). *Out of the Crisis,* Massachusetts Institute of Technology, Cambridge, MA, p. 132.

DeVor, R. E., T.-H. Chang, and J. W. Sutherland (1992). *Statistical Quality Design and Control,* Prentice Hall, Upper Saddle River, NJ.

Emanuel, J. T. and M. Palanisamy (2000). "Sequential Experimentation Using Two-Level Fractional Factorials," *Quality Engineering,* vol. 12, no. 3, pp. 335–346.

Finn, L. M. (1995). "Getting Statistical Tools to the Masses: Common Problems in Training Design and Execution," *Annual Quality Congress Proceedings,* ASQ, Milwaukee, pp. 356–364.

Ross, P. J. (1996). *Taguchi Techniques for Quality Engineering,* McGraw-Hill, New York.
Shainin, R. D. (1993). "Strategies for Technical Problem Solving," *Quality Engineering,* vol. 5, no. 3, pp. 433–448.
Wadsworth, H. M. Jr. (1998). *Handbook of Statistical Methods for Engineers and Scientists,* 2nd ed., McGraw-Hill, New York.

SUPPLEMENTARY READING

Statistical methods: *JQH5,* Sections 44, 47.
Balestracci, D. (1998). "Data 'Sanity': Statistical Thinking Applied to Everyday Data," special publication of the ASQ statistics division.
ASQ "How To" booklets (series of 16).

WEBSITE

Statistics division of ASQ: www.asq.org/about/divisions/stats

12

UNDERSTANDING CUSTOMER NEEDS

12.1 QUALITY AND COMPETITIVE ADVANTAGE

The preceding chapters in this book present the key concepts to understand the dimensions of quality, determine current status, and start to achieve quality excellence. All departments in an organization (see Section 1.4, "The Quality Function") have a role in the quality effort. Starting with this chapter on understanding customer needs, the remainder of this book discusses the roles of these various departments.

In developing new products or services (or modifying existing ones), the operational quality planning process provides a useful framework. The following steps are involved: establish the project, identify the customers, discover customer needs, develop product, develop process, establish process controls, and transfer to operations. Chapter 4 describes this process. The current chapter provides elaboration on two of the early steps, i.e., identify the customers and discover customer needs. These two steps are part of the process of quality function deployment (see Section 4.10 and particularly Figure 4.5).

In the competitive world, all organizations aspire to have a unique competitive advantage. Such an advantage can be achieved by price, by ability to meet customer needs on short notice, and by quality. This chapter starts the journey to show how quality—both product features and freedom from deficiencies—can lead to a unique competitive advantage. By identifying customers, analyzing their needs, and understanding our quality status relative to competition, we can establish new product quality goals that will lead to a competitive advantage. We start with customers.

12.2 IDENTIFY THE CUSTOMERS

We define a *customer* as anyone who is affected by the product or process. Three categories of customers then emerge:

1. *External customers, both current and potential.* A multiplicity of these customers gives rise to a variety of influences, depending on whether the customer is economically powerful and on its technological sophistication. For service organizations the list of external customers may extend far in scope. For example, customers of the Internal Revenue Service include not only taxpayers but also the Treasury Department, Office of the President, Congress, accountants, lawyers, etc. Each customer has needs that must first be determined and then addressed in planning a product.
2. *Internal customers.* These customers include all functions affected by the product at both the managerial and workforce levels. Internal suppliers often view their internal customers as "captive" customers. Not so. Internal customers may have an alternative source; i.e., they may be able to purchase the product from an external supplier. For example, an engineering department procures research services from the company research department. The engineering department is an internal customer that may decide to use an external consultant to provide the services. An assembly plant purchases components from a sister plant within the company. That assembly department must be viewed as an internal customer that, in order to meet its own goals, could decide to go outside the company family to obtain the required quality on components.
3. *Suppliers as customers.* Suppliers should be viewed as extensions of internal customer departments such as manufacturing. Thus their needs must be understood and addressed during the planning for quality.

In identifying customers, some are obvious and some are not. An important tool for identifying all those who are affected (i.e., customers) is the flow diagram. Often no one individual or department is able to describe the total process; a cross-functional team can create the flow diagram. A flow diagram also helps to understand linkages that can reveal hidden customer needs. These linkages are information flows, work flows, and other interdependencies between operations.

For external customers a "cast of characters" is often the "customer." Thus in selling products and services to a hospital, a supplier must understand the needs of the hospital purchasing manager, a quality assurance manager, heads of hospital departments, physicians, nurses, and (of no little consequence) the patient. One manufacturer learns the needs of four levels of customers: those who approve the purchase, those who influence the decision, those who sign the purchase order, and those who are end users.

In practice, we must recognize that some customers are more important than others. It is typical that about 80 percent of the total sales volume comes from about 20 percent of the customers; these are the "vital few" customers who command priority. Within these key customer organizations are individual customers who also may have a hierarchy of importance; e.g., a surgeon at a hospital is a key customer for surgical needles.

Note that customers include both current and potential customers and that it is often useful to identify segments of customers (for an example of customer segmentation, see Section 4.9, "Spectrum of Customers"). Finally, a difficult but important decision involves which potential customers *not* to pursue. Restricting customer segments enables an organization to concentrate resources on the vital few customers and perform those activities that will lead to customer loyalty.

12.3 CUSTOMER BEHAVIOR

We will first define certain terms concerning customer behavior. The terms are *needs, expectations, satisfaction,* and *perception.*

Customer needs are the basic physiological and psychological requirements and desires for survival and well-being. A. H. Maslow is a primary source of information on both physiological and psychological needs. He identifies a hierarchy of such needs as physiological, safety, social, ego, and self-fulfillment (for elaboration, see Section 9.2).

Customer expectations are the anticipated characteristics and performance of the goods or service. Kano and Gitlow (1995) suggest that three levels of customer expectation are related to product attributes. The "expected" level of quality represents the minimum or "must be" attributes. We cannot drive up satisfaction with these attributes because they are taken for granted, but if performance of the basic attributes is poor, the strong dissatisfaction will result. At the "unitary" (or desired) level, better performance leads to greater satisfaction but (in a limited time period) usually in small increments. For the "attractive" (or surprising) level, better performance results in delighted customers because the attributes or the level of performance are a pleasant surprise to the customers. (For an application to hospitals, see Section 12.4.) Discovering and understanding customer needs and expectations is necessary to define specific product attributes for market research and product development.

Customer satisfaction is the degree to which the customer believes that the expectations are met or exceeded by the benefits received. Customer expectation has a strong influence on satisfaction. Thus a customer staying at a luxury hotel expects perfection, and even a minor inconvenience will result in low satisfaction. A budget hotel can have poor features, but if customers get a reasonable night's sleep, they will have high satisfaction because their expectation is low.

Customer perception is the impression made by the product. The perception occurs after a customer selects, organizes, and interprets information on the product. Customer perceptions are heavily based on previous experience. But other factors influence perception, and these factors can occur before the purchase, at the point of purchase, and after the purchase. For elaboration, see Onkvisit and Shaw (1994).

12.4 SCOPE OF HUMAN NEEDS AND EXPECTATIONS

In planning to collect information on customer needs, we must go beyond the search for obvious needs to the more subtle ones that present opportunities for innovative new-product designs.

First, let us focus on the distinction between stated needs and real needs. A consumer states a need for a "clothes dryer," but the real need is "to remove moisture"; a consumer wants a "lawn mower," but the real need is "to maintain height of lawn." In both cases, expressing the need in terms of a basic verb and noun can spawn new product ideas. One historical example is the replacement of hair nets by hair spray to satisfy the basic need of "secure hair." When a customer states: "I need an X," perhaps we should ask: "What would you use the X for?"

Some needs are disguised or even unknown to the customer at the time of purchase. Such needs often lead to the customer using the product in a manner different from that intended by the supplier—a telephone number intended for emergencies is used for routine questions, a hair dryer is used in winter weather to thaw a lock, a tractor is used in unusual soil conditions. Designers view such applications as misuse of the product but might better view them as new applications for their products.

Some of these applications are misuse, but such needs must be understood and, in some cases, alternative design concepts considered. Then there are other needs that go far beyond the utilitarian. Some needs may be perceptional (e.g., the now classic example of Stew Leonard's supermarket customers, who believe that only unwrapped fish could be fresh); some needs may be cultural (e.g., a process such as computer-aided design threatens to reduce the need for human expertise, and thus design engineers resist it).

The Hospital Corporation of America identifies several levels of customer expectations. At level 1, a customer *assumes* that a basic need will be met; at level 2, the customer will be *satisfied;* at level 3, the customer will be *delighted* with the service. For example, suppose a patient must receive 33 radiation treatments. Waiting time in the therapy area is one attribute of this outpatient service. At level 1, the patient *assumes* that the radiation equipment will be functioning each day for use; at level 2, the patient will be *satisfied* if the waiting time in the area is moderate, say, 15 minutes. At level 3, the patient will be *delighted* if the waiting time is short, say, one minute. To achieve a unique competitive advantage, we must focus on level 3.

Customer needs may be clear or disguised; they may be rational or less than rational. To create customers, we must discover and serve their needs.

Anderson and Narus (1998) view needs in terms of the worth, in monetary units, of the technical, economic, service, and social benefits a customer receives in exchange for the price paid.

12.5 SOURCES OF MARKET QUALITY INFORMATION

Market quality information includes quality alarm signals arising not only from a decline in sales but also from field failure reports, customer complaints, claims, lawsuits, etc.

Most alarm signals are poor measures of quality—rather, they are measures of expressed product dissatisfaction. A low level of alarm signals does not necessarily mean a high level of quality. Particularly for inexpensive products, complaint rates are a poor indicator of customer satisfaction. If customers are not satisfied, they simply switch brands—without submitting a complaint. Neither does the absence of product dissatisfaction mean that there is product satisfaction, since these two terms are not opposites. A product may be failure free and yet not be salable because a competitor's design is superior or has a lower price.

A second source of market quality information is the vast array of published data available relative to quality. Some of these data are internal to the company. Many are published external to the company. Such "field intelligence" includes databases on sales volume, price changes, success rate on bids, complaints, spare parts usage, salespersons' reports, ratings from customers and consumer journals, government reports, etc.

These two major sources of market quality information, while necessary, are not sufficient. A good deal of information is missing, and this data can be provided only through special marketing research studies.

12.6 MARKET RESEARCH IN QUALITY ("VOICE OF THE CUSTOMER")

The American Marketing Association defines *marketing research* (we use the term *market research*) as "the function which links the consumer, customer, and public to the marketer, through information—information used to identify and define marketing opportunities and problems; generate, refine, and evaluate marketing actions; monitor marketing performance; and improve understanding of marketing as a process."

Although market research includes the compilation and analysis of existing information (field failure reports, customer complaints, government reports), the exciting aspect concerns information that does not yet exist, i.e., the voice of the customer (VOC). VOC is a continuous process of collecting customer views on quality and can include customer needs, expectations, satisfaction, and perception. The emphasis is on in-depth observing, listening, and learning. This process addresses the three main purposes of market research for quality.

Purposes of Market Research in Quality

The broad purposes are mainly to

- Determine customer needs.
- Measure current customer satisfaction.
- Analyze customer retention and loyalty issues.

Determination of customer needs has several facets. These include both short-term needs to perform surgery on current products and longer-term needs to integrate quality into new-product development. We not only ask customers directly what their needs are but also methodically study how customers currently use the product; then we analyze their total system of use to identify hidden needs. This process provides specific ideas for modifying current products and helps to discover opportunities for new products. Needs must be periodically evaluated because today's new needs become a routine expectation tomorrow.

Measurement of current customer satisfaction also involves several elements. First, the term *quality* must be translated into specific attributes that customers say are important. These attributes should not be restricted to attributes on the core and auxiliary products provided to customers, but should also include quality attributes for the full life cycle from initial contact with a salesperson to customer service after the sale. We also need to learn the relative importance of these attributes to the customer and how our product compares to the competition (see Section 2.6 for an example). Finally, we must ask customers whether they would purchase from us again or recommend us to friends.

Using market research for customer retention and loyalty starts with the distinction between customer satisfaction and customer loyalty—two different concepts (see Section 4.4). Some levels of customer satisfaction do not achieve a high level of customer loyalty. The research must include finding the reasons for losing customers, i.e., customer defections. A framework for achieving high customer retention and loyalty is presented in Section 20.12.

Thus market research in quality looks for answers to some cardinal questions:

- What is the relative importance of various product qualities as seen by the user? The answers provided by market research are typically different from the prior beliefs of the manufacturer. Sometimes the difference is dramatic.
- For the more important qualities, how does our product compare with competitors' products, as seen by users?
- What is the effect of these competing qualities (including our own) on users' costs, well-being, and other aspects of fitness for use?
- What are users' problems about which they do not complain but which we might nevertheless be able to remedy?
- Do users have ideas that we might be able to utilize for their benefit?

In conducting market research, certain principles are basic:

1. Customer views on quality are most useful when they include views on competitors' products.
2. Market research should also capture the views of "noncustomers," i.e., prospective customers and lost customers.
3. In designing market research, operating managers within the organization should be asked what questions they need answers to in order to improve their operations.
4. The occasional survey is better than none, but periodic surveys provide data for analysis of trends and other matters.
5. The concept of sampling is important. Surveying a carefully designed sample of customers in depth is more useful than superficially surveying a large sample of customers.
6. The customer desires product performance during the entire life of a product, not just during the warranty period. Market research should go beyond the warranty period.

A useful reference for market research is Churchill (1991). Goodman et al. (1996) discuss eight common pitfalls that undermine the integrity and value of customer feedback.

In the following discussion the market research addresses customer needs for both components of quality—product features and freedom from deficiencies.

12.7
NEEDS RELATED TO PRODUCT FEATURES

To start, we need to identify the attributes that customers say are important in their purchasing decision. Table 12.1 shows the results of research in which customers (and potential customers) of floppy disks were asked to select important attributes. The relative importance was measured as the percentage of customers who selected each attribute. Note that the list of attributes focuses both on the disks and on specific and general aspects of service.

An important next step is to learn how our product compares with the competition's. This task can be accomplished by using a multiattribute study. Customers are asked to consider several product attributes and, for each attribute, to state the relative importance and a rating for our product and competing products. Section 2.6 presents three examples to illustrate customers' role in assessment. Additional examples of market research are given below.

In another approach for determining relative importance of service attributes, a bank asks customers to select the five most important services from a list of 12 (e.g., processing transactions without error, greeting you with a smile, knowledge of bank products and services).

Sometimes the list of attributes is lengthy, and the identification of the vital few can be extremely helpful (Mitsch, 1998). Thus at Unigroup (United Van Lines and Mayflower Transit), from a list of 71 attributes in moving household goods, customers selected 14 as most important (e.g., ability of salesperson, handling your goods with care—loading crew, keeping promises).

TABLE 12.1
Relative importance of attributes of floppy disks

Attribute	Relative importance, %
Quality of delivered products	98
Meeting delivery schedules	95
In-field reliability	95
Price	72
Provision of technical assistance	66
Handling large orders	64
Product sophistication	62
Meeting changing customer needs	51
Short delivery time	48
Capability of field service	45
Accessibility of suppliers' management	44
Specialization in floppy disk drive products	34

The relative importance of the product attributes can be determined by several methods. The simplest is to present a list to customers and ask them to select the most important. In another approach customers are asked to allocate 100 points over various attributes. In a more complicated method ("conjoint measurement"), customers are presented with combinations of product attributes and asked to indicate their preferences. The importance rating can then be calculated. Churchill (1991), Appendix 9B, describes this method. Peña (1999) presents a similar method for estimating relative importance of service quality attributes with applications to the measurement of quality at a university and to the Spanish railroad system.

The Shaving System Case

A leading manufacturer of razors and razor blades was confronted by a new competing model that featured an ingenious, improved blade-changing mechanism. When the competitor promoted the new model aggressively, the company marketing managers became apprehensive. They then succeeded in initiating a costly new-product development to bring out a model that could match the competition's. However, the top managers also authorized a field market research study to discover the reaction of consumers to the new development.

The planners of this classic research study identified the seven major quality characteristics of a shaving system. Next (through an intermediary market research company), the company provided each of several hundred consumers with all three principal shaving systems then on the market. These consumers were asked to use each shaving system for a month and to report:

- The ranking of the seven characteristics in their order of importance to the user.
- The ranking of the performance of each of the three systems for each of the seven quality characteristics.

Table 12.2 shows the plan of the market research study. The resulting data showed that, as seen by the users:

1. Ease of blade changing was the least important characteristic.

TABLE 12.2
Market research study—the shaving system case

	Users' rankings			
Quality characteristics	**Gillette**	**Gem**	**Schick**	**Importance**
1. Remove beard				
2. Safety				
3. Ease of cleaning				
4. Ease of blade changing				
5. . . .				
6. . . .				
7. . . .				

2. The competitor's new blade-changing mechanism had failed to create a user preference over competing forms of blade changing.

   These findings denied the beliefs of the company marketing managers and came as a welcome surprise. They enabled the company to terminate the costly new-product development then already in progress.
3. On one of the most critical qualities (product safety), the company's shaving system was inferior to both competing systems. This news came as an unwelcome surprise and stimulated the company to take steps to eliminate this weakness. Note that *this inferiority could not have been discovered from field complaints,* since no normal user would conduct such a comparative study on his or her own initiative. It was "an alarming situation for which our present alarm signals are silent."

Discovering Customer Needs and Marketing Opportunities

Market research in the field provides access to realities that cannot be discovered in the laboratory. The conditions of use can involve environments, loads, user training levels, misapplication, etc., all of which may be different from the conditions prevailing in the laboratory. The laboratory does provide a relatively prompt, inexpensive simulation that is most helpful for making many decisions. However, this simulation cannot fully disclose the needs of fitness for use under the actual conditions of use.

Field studies can provide access not only to the realities of the conditions of use but also to the users themselves. Through access to the user, it becomes possible to understand problems that are not a matter of poor quality but for which the manufacturer may nevertheless be able to provide a solution. For example, most users prefer to avoid disagreeable, time-consuming chores. The food-processing industry has successfully transferred many such chores from the household kitchen to the factory (e.g., soluble coffee, precooked foods) and incidentally greatly increased their sales.

The study of the user's operation can be aided by dissecting the process of use. This job is done by documenting all of the steps, analyzing them, and identifying opportunities for new-product development. Such information is valuable input to the product development function as that group embarks on a new design project.

EXAMPLE 12.1. For a certain health care product, a clamp controlled the amount of flow of fluid from an external source to a patient during a dialysis process. An analysis using a flow diagram showed that the proposed clamp was excellent in controlling the flow of fluid but was inconvenient in a later step that required the patient to wear the product under clothing. The original focus was on performance of the product, but the flow diagram pinpointed a later problem of inconvenience to the patient. See Section 13.4 for further discussion.

This approach gathers information by observation of customers rather than inquiry of customers. The approach is not new, but Leonard and Rayport (1997) suggest a useful framework of five steps, using a team to do the research:

1. *Observation.* Customers are observed carrying out normal routines.
2. *Capturing data.* Relatively few data are gathered through questions. Instead, photography and videography are often used.

3. *Reflection and analysis.* After gathering the data, the team reviews the visual and other data to identify customers' possible problems and needs.
4. *Brainstorming for solutions.* The observations are transformed into visual representations of possible solutions.
5. *Developing prototypes of possible solutions.* Prototypes clarify the concept of the new product or service and stimulate discussion with colleagues and customers.

Some researchers call this approach "stapling ourselves to the product."

Opportunities for improvement span the full range of customer use from initial receipt through operation. The primary stages and examples of ideas for improving fitness for use are given in Table 12.3. (Also see Table 4.10 for examples of hidden customer needs that presented opportunities for improvement.) These opportunities must then be translated into specific product quality goals, which, in turn, help to create a unique competitive advantage.

In another approach, Bovée and Thill (1992) suggest five ways of adding value through enhanced customer service with an application to financial services and retailing of compact discs (Table 12.4).

Sometimes information on customer needs is too broad to translate into specific product attributes and features. To obtain more precision, we can break down broadly stated needs into secondary and tertiary needs. This process provides a detailed quality function deployment matrix, i.e., a customer needs analysis matrix. An example is given in Table 12.5 for the scheduling of medical resources.

The methods to collect information on customer needs are many and varied. Some of the common methods include focus groups (see below), observations at customer sites (above), executive interactions with customers, special customer surveys, analysis of complaints, participation at trade shows, and comparing products with those of

TABLE 12.3
Opportunities for improving fitness for use of an industrial product

Stage	Opportunities
Receiving inspection	Provide data so incoming inspection can be eliminated
Material storage	Design product and packaging for ease of identification and handling
Processing	Do preprocessing of material (e.g., ready-mixed concrete); design product to maximize productivity when it is used in customer's manufacturing operation
Finished goods storage, warehouse and field	Design product and packaging for ease of identification and handling
Installation, alignment, and checkout	Use modular concepts and other means to facilitate setups by customer rather than manufacturer
Maintenance, preventive	Incorporate preventive maintenance in product (e.g., self-lubricated bearings)
Maintenance, corrective	Design product to permit self-diagnosis by user

TABLE 12.4
Adding value through customer service

Possible ways to add value (basic product)	Example from financial services (checking account)	Example from retailing (CDs)
Be flexible	Let customers design their own checks	Allow customers to return CDs
Tolerate customer errors	Cover overdrafts without charging a fee	Extend credit when customers forget to bring money
Give personal attention	Help customers with individual tax questions	Learn customers' musical tastes and suggest new CDs they might enjoy
Provide helpful information	Publish a brochure on financial planning	Distribute a newsletter that reviews new stereo equipment
Increase convenience	Install ATMs	Let customers order by phone

Source: Bovée and Thill (1992).

TABLE 12.5
Needs analysis spreadsheet

Primary needs	Secondary needs	Tertiary needs
Efficient use of clinical resources	No wasted time between patients	Procedure length adjustable to MD practice pattern
		No incompatible procedures scheduled
	Maximum number of appointments booked	Can easily find appointment records across facilities
		Accurate inventory of available resources—people, equipment, supplies
	⋮	⋮
Good clinical care	Right patient gets right procedure	Correct association of patient with MD orders
		Medical record information available
	Special patient needs known ahead and attended to	
	All locations have patient and MD information	

Source: Juran Institute, Inc.

competitors. A recent method is the monitoring of Internet messages to find out what customers are saying about a product (Finch, 1997).

JQH5, Section 3, provides a list of 13 methods for collecting information on customer needs. Plsek (1997) is an excellent reference on the application of creativity concepts to customer needs analysis.

Focus Groups

A focus group consists of about 8 to 14 current or potential customers who meet for about two hours to discuss a product. Here are some key features:

1. The discussion has a focus, hence the name.
2. The discussion can focus on current products, proposed products, or future products.
3. A moderator who is skilled in group dynamics guides the discussion.
4. The moderator has a clear goal on the information needed and a plan for guiding the discussion.
5. Often company personnel observe and listen in an adjacent room shielded by a one-way mirror.

Focus groups can cover many facets of a product or can discuss quality only. A discussion can be broad (e.g., obtaining views on customer needs) or can have a narrower scope (e.g., determining customer sensitivity to various degrees of surface imperfections on silverware). The groups might consist of customers from various market segments, or groups may represent noncustomer segments the firm wants to penetrate. Sessions can provide a depth of information on customer needs, expectations, perceptions, satisfaction, intentions, and reactions to new concepts or ideas. One useful approach is to ask customers to dream about the ideal product, e.g., the product that would overcome customer frustrations. The responses provide input for new-product development. Typically, focus group respondents are quite candid as they receive reinforcement by like-minded people in the group.

A key to a successful focus group is the qualification of the focus group moderator. Being a moderator is not a task for a well-intentioned amateur. The moderator must be a quick learner, a friendly leader, knowledgeable but not all-knowing, a good listener, a facilitator not a performer, flexible, empathic, a big-picture thinker, and a good writer; he or she must also have an excellent memory. For elaboration, see Greenbaum (1988).

Electronic meeting support (EMS) tools can help to record and summarize views during a focus group meeting. Typically, the participants meet in a room, with each participant sitting at a computer. Using the computers, the participants—simultaneously and anonymously—respond to the moderator's questions. Responses are collected and summarized by the computer network for all to see and discuss. For an application to the financial services sector, see Section 17.2.

12.8
NEEDS RELATED TO PRODUCT DEFICIENCIES

Understanding customer needs must also include the deficiencies side of the quality definition. Clearly, the emphasis during planning must be on *prevention* of deficiencies (defects, failures, errors, etc.).

With respect to prevention of deficiencies, the following chapters discuss the concepts and tools necessary in various functional areas. An important input to such prevention effort are the data on field complaints. Most organizations have systems of collecting and analyzing information on customer complaints. (For a discussion of these systems, see Section 20.10, "Processing and Resolution of Customer Complaints.") A Pareto analysis of field failures, complaints, product returns, etc., identifies the vital few quality problems to be addressed in both current and future product development. This type of analysis is in wide use. Table 12.6 shows a Pareto analysis of field complaints on fuel nozzles using several different measures. A conventional Pareto analysis would rank the various types of complaints according to the frequency of occurrence; but when those complaints are evaluated on an economic basis, the ranking changes. Note also that there are several possible economic indexes such as warranty costs and effect on user costs. The priorities for future product development can be quite different depending on the measure used in the Pareto analysis. Similarly, data on defects found internally must be analyzed on the proper basis to prioritize prevention efforts, many of which must be addressed during product development.

To prevent recurrence of specific complaints in future products requires a careful analysis of the words and phrases used to describe complaints.

EXAMPLE 12.2. An annual mail survey of automobile customers revealed a significant number of complaints on "fit of doors"; within the company, this term referred to the "margin" and "flushness." Margin is the space between the front door and fender or between the front and rear doors. Sometimes the margin was not uniform, e.g., a wider space at the top as compared to a narrower space at the bottom of the door. Flushness refers to the smoothness of fit of the door with the body of the car after the door is shut. Steps were taken (at a significant cost) during processing and assembly to correct the margin and flushness problems. But subsequent surveys again reported complaints on fit of doors.

TABLE 12.6
Pareto analysis of complaints on fuel nozzles

	Ranking based on various measures		
Failure mode	**Frequency**	**Warranty charges**	**Effect on user costs**
G	1	4	3
A	2	1	6
D	3	2	1
F	4	5	4
N	5	7	8
L	6	3	2
P	7	6	5
R	8	8	7

Fortunately, other market research input became available. The company held periodic focus group meetings at which customers answered a questionnaire and discussed selected issues in more detail. Fit of doors again was proclaimed to be a problem by the customers. On the spot, they were asked, "What do you mean by fit of doors?" Their answer was *not* margin or flushness. First, fit meant the amount of effort required to close the door. Customers complained that they had to "slam the door hard in order to get it to close completely the first time." Second, *fit* meant "sound." When they left the car and closed the door, they wanted to hear a solid, "businesslike" sound instead of "a metallic, loose sound" (that told them the door was not closed). The market research provided a correct understanding of the complaint, which subsequently required changes in the product design and the manufacturing process. Earlier, the company had acted on the complaint, but it had solved the wrong problem. It was a waste of resources—like spraying scrap with perfume.

When customer needs are determined, they must be related to internal business processes by defining metrics for each need. For an example, see Figure 17.3.

In analyzing customer needs related to both product features and to product deficiencies, an essential input is information on the current level of customer satisfaction.

12.9 MEASURING CUSTOMER SATISFACTION

In measuring customer satisfaction, terms like *customer satisfaction* and *quality* are too foggy. Instead, we must identify the attributes of the product that collectively define satisfaction. Examples are the accuracy of transactions at a bank or specific attributes of workmanship in a residential home. A sample of customers should be asked (in preliminary research) which attributes they think constitute high quality. Inputs from company managers, trade journals, and other sources should also be included. With these inputs the list of attributes is then finalized for research with the larger sample of customers. The list of attributes should span the entire cycle of customer contact from initial contact with a salesperson through use and servicing of the product and handling of complaints. Examples of customer satisfaction studies from both manufacturing and service industries are presented in Section 2.6.

It is possible to use a generic template of attributes. For example, in the area of customer service, the SERVQUAL model (Zeithaml et al., 1990) identifies five dimensions: tangibles, reliability, responsiveness, assurance, and empathy. Such broad terms need to be defined. Market research for the five areas is conducted by presenting customers with 22 statements that examine the difference between customer expectations and perceptions. These statements can be customized for each company.

Customer satisfaction measurement should also include asking customers about the relative importance of the various attributes (see above). When this information is combined with customer satisfaction data, the resulting quadrant diagram helps to direct future action on modifying a product to meet customer needs. For an example, see Figure 2.4.

Information about customer satisfaction relative to the competition can be sobering; this data may spur necessary action on a product. Data on competitor quality can be obtained from laboratory testing, direct questions to customers, mystery shoppers, focus groups, and benchmarking. When a noncustomer is purchasing from a competitor, it is useful to ask an open-ended question; e.g., what are the chief reasons that you buy from company X?

Another essential question to ask customers is, Would you purchase from us again, or would you recommend us to your friends? An example of this approach, along with numerical results at United Van Lines (UVL), is shown in Table 12.7.

Benchmarking sources report that world-class companies earn "definitely would recommend" scores 60 to 65 percent of the time. For quality superiority, organizations need to score in the top (not top two) response category.

Data on customer satisfaction can be collected by various methods, including market surveys, posttransaction interviews, focus groups, mystery shoppers, employee reports, surveys and interviews, call centers for complaints and other comments, customer advisory groups, and customer segment surveys. For further discussion, see *JQH5,* Section 18. Vavra (1997) and Churchill (1991) are also useful references.

The Web can also be used to collect customer satisfaction data through on-line customer surveys. Suppliers specializing in electronic surveys have assembled panels of ready respondents (see Websites at the end of the chapter). On-line research is faster than phone or mail surveys and can have higher response rates because it is more convenient and less intrusive than other methods. On the downside, on-line polling may not be representative because many consumers do not use the Internet regularly.

Customer satisfaction and needs data from various sources can be integrated into a database, and customer relationship management software can then be used to analyze and make the results available immediately to all functions in an organization. The potential for electronic commerce is exciting; e.g., data can be sliced by ZIP code to discover local needs.

Customer satisfaction data must be linked to customer loyalty analysis (see Section 20.11).

Collecting information on customers and their needs provides the input for the quality function deployment matrix on customers and customer needs (see Section 4.11). This data then helps us to translate customer needs into product features (discussed in Chapter 13).

Rhey and Gryna (2001) discuss the application of market research for quality to small business firms.

TABLE 12.7
Results of loyalty survey

	Current 4 months	
Willingness to recommend UVL	**Responses**	**Percentage**
Definitely would recommend	37	55.2
Would recommend	14	20.9
Might or might not recommend	9	13.4
Would not recommend	4	6.0
Definitely would not recommend	3	4.5

Source: United Van Lines, Inc. customer service survey.

12.10 MARKET RESEARCH FOR INTERNAL CUSTOMERS

Market research concepts apply to both internal and external customers. In the service sector a support department of AT&T Bell Laboratories conducted a survey of the needs of internal customers (Weimer, 1995). The result was a multilevel hierarchy of needs. Service quality was defined as one of the primary needs. Customers defined service quality with 10 secondary (tactical) needs such as "no red tape" and "access to service information." These needs were further defined in terms of 42 tertiary (operational) needs such as "deal with person providing service directly" and "know who supplies a service or owns a process."

For an application using EMS tools with a focus group of financial managers, see Section 17.2.

In another example, the central engineering staff of a large automotive manufacturer applied marketing research concepts to internal customers (Stevens, 1987). A survey questionnaire sent to internal customers first requested information about the frequency of contact with each of the seven staff areas. Respondents were then asked to state their degree of satisfaction with eight attributes of service: services meeting requirements, follow-up required, assistance available, competent people, cooperation, input requested, problem notification, and help in seeking contacts. An overall satisfaction rating was also requested. Finally, respondents were asked for details on any low ratings for the eight attributes and suggestions on how engineering could improve its service to help achieve customer quality goals.

Spending the time and effort to clearly understand customer needs provides the foundation for planning and executing the total work of the organization—product development, supplier relations, operations, marketing, field service, and support functions. That's the scenario for the rest of this book.

SUMMARY

- A customer is anyone who is affected by a product or process.
- Customers fall into three categories—external, internal, and suppliers.
- Customer needs may be clear or disguised; they may be rational or less than rational. But these needs must be discovered and served.
- A low level of customer dissatisfaction (e.g., complaints) does *not* necessarily mean that customers are satisfied.
- Market research for quality analyzes aspects of quality that are affected by forces in the marketplace.
- Customer satisfaction should be measured relative to the competition and should address both product features and freedom from deficiencies.
- Market research can help to discover opportunities for achieving a unique competitive advantage.

PROBLEMS

12.1. For a specific output of your department in an organization, sketch a simple flow diagram of the journey of the output throughout the organization. Then prepare a list of the internal and external customers.

12.2. Identify your own customers and suppliers for a representative 24-hour day.

12.3. For one of your important customers, define the steps you should take to discover the customer's primary needs. Emphasize the discovery of real needs rather than stated needs.

12.4. For any *consumer* product, prepare a plan of market research aimed at
(a) Identifying the principal product qualities.
(b) Ranking these qualities in their order of importance to users.
(c) Discovering the relative performance of competing companies with respect to the principal product qualities.

12.5. For any *industrial* product, prepare a plan of market research aimed at securing data similar to those set out in problem 12.4.

12.6. Walk through a supermarket and identify some products for which work has been transferred from the household kitchen to food-processing factories. In addition, identify some potential products for such transfer. Report your findings, with special emphasis on the quality problems involved.

12.7. You are participating in the preparation of a bid on a military optical system used for range finding. The military specification requires that the subsystems be interchangeable (so that in case of damage, the failed subsystem can be unplugged and replaced with a spare). You know from experience that this provision is wise. However, you are disturbed by the further requirement that the optical elements (lenses, prisms, etc.) within each subsystem should also be interchangeable. To your knowledge, this requirement is foolish because the military has never had the special repair and test facilities needed to reassemble optical elements in the field. To your knowledge, the military also lacks the means of testing for this interchangeability.

A procedure is available for proposing a change in such unrealistic specifications, but you once went through this procedure and found it to be shockingly long and difficult. In addition, your company has had poor success in securing military business because of being underbid in price by competitors. You have suspected for some time that one reason for this lack of success is that your company does its bidding on the basis of meeting all specification requirements (sensible or not), whereas the competitors may be doing their bidding on the basis of meeting only the requirements that make sense.

You are now faced with making a recommendation on how to structure the bid with respect to the requirement for the unneeded interchangeabilty. What do you propose? Why? If you propose to bid on the basis of not meeting the "foolish" requirement, what do you propose to do if the military later finds out what has happened?

REFERENCES

Anderson, J. C. and J. A. Narus (1998). "Business Marketing: Understand What Customers Value," *Harvard Business Review,* November–December, pp. 53–65.

Bovée, C. L. and J. V. Thill (1992). *Marketing,* McGraw-Hill, New York.

Churchill, G. A. Jr. (1991). *Marketing Research Methodological Foundations,* Dryden Press, Chicago.

Finch, B. J. (1997). "A New Way to Listen to the Customer," *Quality Progress,* May, pp. 73–76.

Goodman, J., D. DePalma, and S. Broetzmann (1996). "Maximizing the Value of Customer Feedback," *Quality Progress,* December, pp. 35–39.

Greenbaum, T. L. (1988). *The Practical Handbook and Guide to Focus Group Research,* Lexington Books, D.C. Heath and Company, Lexington, MA, pp. 50–54.

Kano, N. and H. Gitlow (1995). "The Kano Program," notes from seminar on May 4 and 5, University of Miami.

Leonard, D. and J. F. Rayport (1997). "Spark Innovation through Empathic Design," *Harvard Business Review,* November–December, pp. 102-113. Reprinted by permission of Harvard Business Review.

Mitsch, G. J. (1998). "Moving Experiences: How System-wide and Singular Customer Satisfaction Measurement Drives Quality at United and Mayflower," Customer Satisfaction and Quality Measurement Conference, American Marketing Association and American Society for Quality, Atlanta.

Onkvisit, S. and J. J. Shaw (1994). *Consumer Behavior,* Macmillan, New York.

Peña, D. (1999). "A Methodology for Building Service Quality Indices," *Annual Quality Congress Proceedings*, ASQ, Milwaukee, pp. 550–558.

Rhey, W. L. and F. M. Gryna (2001). *Market Research for Quality in Small Business, Quality Progress,* January.

Stevens, E. R. (1987). *Implementing an Internal-Customer Satisfaction Improvement Process,* Juran Report No. 8, pp. 140–145.

Vavra, T. G. (1997). *Improving Your Measurement of Customer Satisfaction,* Quality Press, ASQ, Milwaukee.

Weimer, C. K. (1995). "Translating R & D Needs into Support Measures," *Annual Quality Congress Proceedings*, ASQ, Milwaukee, pp. 113–119.

Zeithaml, V. A., A. Parasuraman, and L. L. Berry (1990). *Delivering Quality Service,* Free Press, New York.

SUPPLEMENTARY READING

Market research for quality: *JQH5,* Section 18.

Customer needs analysis: Plsek, P. E. (1997). *Creativity, Innovation, and Quality,* Quality Press, ASQ, Milwaukee.

Customer satisfaction and handling of complaints: American Productivity and Quality Center (1999). "White Paper on Complaint Management and Problem Resolution Consortium Benchmarking Study," Houston.

WEBSITES

Society of Consumer Affairs Professionals (SOCAP): www.socap.org

Guidestar Communications (on-line customer satisfaction surveys): www.guidestarco .com

13

DESIGNING FOR QUALITY

13.1 OPPORTUNITIES FOR IMPROVEMENT IN PRODUCT DESIGN

This step on the quality spiral translates the needs of the user into a set of product design requirements for manufacturing (or the operations function in a service organization). This process is called product development, research and development, engineering, or product design.

Many problems encountered by both external and internal customers can be traced to the design of the product.

EXAMPLE 13.1. In a classic study of 850 field failures of relatively simple electronic equipment, 43 percent of the failures were due to engineering design deficiencies.

EXAMPLE 13.2. In one chemical company a startling 50 percent of the product shipped was out of specification. Fortunately, the product was fit for use. A review concluded that many of the specifications were obsolete and had to be changed.

EXAMPLE 13.3. A study of health care products revealed that 34 percent of the product recalls were caused by faulty product or software design.

EXAMPLE 13.4. A manufacturer of terminals, modems, and other computer products analyzed the reasons for design changes. Local "wisdom" said that (1) about 10 percent of the changes were due to errors in the design and (2) the remaining changes were related to cost-reduction projects, requests from manufacturing, and changing customer requirements. But an in-depth analysis of design changes on four product lines reached a surprising conclusion: 78 percent of the changes were due to design errors.

For mechanical and electronic products of at least moderate complexity, errors during product development cause about 40 percent of fitness-for-use problems. When product development is responsible for both creating the formulation (design) of the product and developing the manufacturing process, as in chemicals, about 50 percent of the problems are due to development.

Product design people are highly educated and highly competent, particularly in quantitative skills. With respect to quality, a product design must provide the necessary product features to meet customer needs, and these features must be free of deficiencies. Poor designs are not the fault of individual designers—the cause is the *process* of design and development. We need, then, to identify some key issues concerning that process.

13.2 DESIGN AND DEVELOPMENT AS A PROCESS

A *process* is a collection of activities that produce an output or result. This concept is in contrast to viewing design and development as the job of a functional department, often called "Design and Development." When design and development is viewed as a process, several key issues emerge:

1. Design and development has "customers" (i.e., anyone affected by the design). Design and development departments frequently function as the brain trust for new product ideas with an emphasis on "come up with the new product and let other departments manufacture it and sell it." That thinking is becoming obsolete, but some residue remains and needs to be recognized. Clearly, design and development has customers—not only external customers who purchase the product but also internal customers who make, sell, and service the product. Thus a product design must use modern design concepts to satisfy the needs of customers—as defined by the customers.
2. For external customers the needs must be studied in detail and translated in a structured way into product features and design parameters—the goal being to knock a competitor off its pedestal. These needs are defined by customers, not by product development personnel. Competitive forces now dictate that customer needs must be analyzed in more detail than was done in the past (see Chapter 12, "Understanding Customer Needs"). An example of a new need is for products that are uniquely designed for one customer—in high or low volumes. Thus the emergence of "mass customization." This customization is often associated with faster order fulfillment. Feitzinger and Lee (1997) discuss the impact of mass customization on product design and process design.
3. For internal customers the development process must recognize that the design becomes a major determinant of costs in areas such as manufacturing, purchasing, and servicing. Thus the development function must share some of the responsibility for these costs—at least to the extent of including these other functions as part of the development process. A helpful concept in this regard is quality function deployment (see below).

4. Designers must recognize that product development is a cross-functional process (with subprocesses) involving the design department and other internal departments. A flow diagram of the development process (and key subprocesses) can be a useful starting point in identifying and understanding the roles of these departments. The leadership role of the design department in product development requires that designers become knowledgeable about the impact of the design on these other activities and that designers actively view these activities as part of the development process. It also means that the design function must gather and document information on the lessons learned during development for use in future designs.
5. Suppliers must become part of the development team. In this age of specialization, suppliers can provide a continuously updated expertise in designing certain subsystems of a system.
6. The product development process must address the issue of minimizing variation of product around the nominal values of design parameters. Traditionally, the design analysis sets the nominal (or desired average) value of a design parameter and then assigns (with less than a scientific approach) upper and lower specification limits around the nominal value. The implication is that all values within these limits are acceptable for performance. In practice, there are significant advantages to minimizing the variation around a nominal value (see Section 18.2). Design of experiments and other techniques can determine the optimum values of product and process parameters to minimize variation. Such approaches can help to achieve a six-sigma defect goal of 3.4 parts per million. See below under "Application of Design of Experiments (DOE) to Product and Process Design."
7. The product development process (for physical as well as service products) can benefit from the application of quality management concepts. Quality planning (Chapter 4) steps can provide a road map for developing a new or modified product; quality control (Chapter 5) can provide the measurements and feedback for the various steps in product development; quality improvement (Chapter 3) can provide the methodology for continuous improvement of the development process. See Section 13.14 for details.

This integrated view of the product development process is sometimes called "integrated product and process development" (IPPD). For elaboration on product development, see *JQH5,* Section 17 ("Project Management and Product Development") and Section 19 ("Quality in Research and Development"). Endres (1997) explains the use of the Juran approach in improving research and development. Sobek et al. (1998) explain how Toyota integrates product development using (1) "social processes" among engineers to foster in-depth technical knowledge and efficient communication and (2) design standards to speed development, allow flexibility, and build the company's base of knowledge. Brocato and Potocki (1998) explain a system model for quality management at the Johns Hopkins Applied Physics Laboratory. *JQH5,* pp. 19.3–19.5, discusses product development as a process in service industries.

We proceed, then, to examine the phases of product development and then specific techniques for design effectiveness.

TABLE 13.1
Forms of early warning of new-product problems

Phases of new-product progression	Forms of early warning of new-product troubles
Concept and feasibility study	Concept review
Prototype design	Design review, reliability and maintainability prediction, failure mode, effect and criticality analysis, safety analyses, value engineering
Prototype construction	Prototype test, environmental test, overstressing
Preproduction	Pilot production lots, evaluation of specifications
Early full-scale production	In-house testing (e.g., kitchen, road), consumer use panels, limited marketing area
Full-scale production, marketing, and use	Employees as test panels, special provisions for prompt feedback
All phases	Failure analysis, data collection and analysis

13.3 PHASES OF PRODUCT DEVELOPMENT AND THE EARLY WARNING CONCEPT

The process of developing modern products involves an evolution through distinct phases of development (see Table 13.1 for an example of phases).

The frequency and severity of problems caused by design have stimulated companies to develop more and better forms of early warning on impending troubles. These early warnings are available in various forms (Table 13.1). Special quality-oriented tools described in this chapter help to evaluate designs, and to improve the design process itself. Collectively, these early warnings and quality-oriented tools provide added assurance that the new designs will not create undue trouble as they progress around the spiral. Many forms of early warning are administered by specialists in reliability, maintainability, and other fields. The timing of their inputs is critical. Early timing can provide constructive help; late timing causes resistance to the warnings and often creates an atmosphere of blame. The cost of changes in a design can be huge; e.g., a design *change* during pilot production of a major electronics product can cost more than $1 million.

Next, we examine some techniques that help to ensure overall design effectiveness. These design assurance techniques address functional performance, reliability, maintainability, safety, manufacturability, and other attributes.

13.4 DESIGNING FOR BASIC FUNCTIONAL REQUIREMENTS; QUALITY FUNCTION DEPLOYMENT

Product development translates customer expectations for functional requirements into specific engineering and quality characteristics. For traditional products, this process is not complicated and can be achieved by experienced design engineers without using any special techniques. For modern products, it is useful to document and analyze the design logic. This means starting with the desired product attributes and then identifying the necessary characteristics for raw materials, parts, assemblies, and process steps. Such an approach goes under a variety of names, such as systems engineering, functional analysis systems technique, structured product/process analysis, and quality function deployment.

Quality Function Deployment

One technique for documenting overall design logic is quality function deployment (QFD). *QFD* is "a structured and disciplined process that provides a means to identify and carry the Voice of the Customer through each stage of product or service development and implementation. This process can be deployed horizontally through marketing, product planning, engineering, manufacturing, service and all other departments in an organization involved in product or service development" (ReVelle et al., 1998). QFD uses a series of interlocking matrixes that translates customer needs into product and process characteristics. (See Figure 13.1; it's identical to Figure 4.5 but is repeated here for convenience.)

Sometimes a matrix incorporates additional information or integrates information in an unusual and useful form. For example, Figure 13.2 is a matrix of customer needs ("customer requirements") and product features ("technical requirements") for paper being supplied to a commercial printer (Ernst and Young, 1990). Note the additional information on importance, weighting, correlations among requirements, units of target values (e.g., millimeters for width and thickness), and competitive evaluations. The "roof" of the diagram showing the correlations leads to a name for the diagram, i.e., the "house of quality."

Sanchez et al. (1993) provide a useful generic explanation of six steps to build a house of quality (Figure 13.3).

Much of the information for these matrixes is gathered by using the market research tools described in Chapter 12. Sometimes, even with simple products, the number of customer requirements and design requirements can become large, and then the number of relationships to investigate becomes unwieldy. For such cases, Shin and Kim (1997) propose the use of factor analysis. *Factor analysis* is a statistical technique that simplifies a large number of relationships among variables to a smaller number of parameters with a minimum loss of information.

An example of a QFD matrix for product development at a bank is shown in Table 13.2. This "product design matrix" translates customer needs into product features and product feature goals for service provided by an automated teller machine.

ReVelle et al. (1998) provide a thorough explanation of QFD (including the important use of supporting software) for both the manufacturing and service sectors. Comstock and Dooley (1998) dissect the application of QFD in two organizations (one commercial and one defense products) and identify the lessons learned for QFD effectiveness. Vonderembse et al. (1997) report on the results of a survey of 40 organizations using QFD. Overall, the organizations concluded that QFD helps achieve significantly

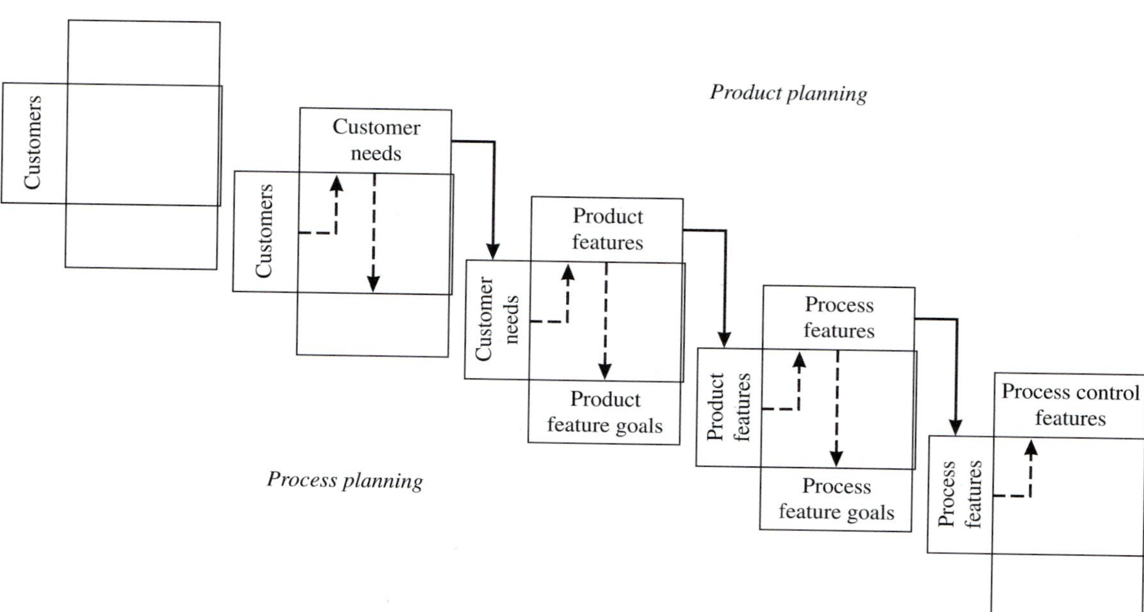

FIGURE 13.1
QFD spreadsheets in quality planning. (*Juran Institute, Inc. Copyright 1994. Used by permission.*)

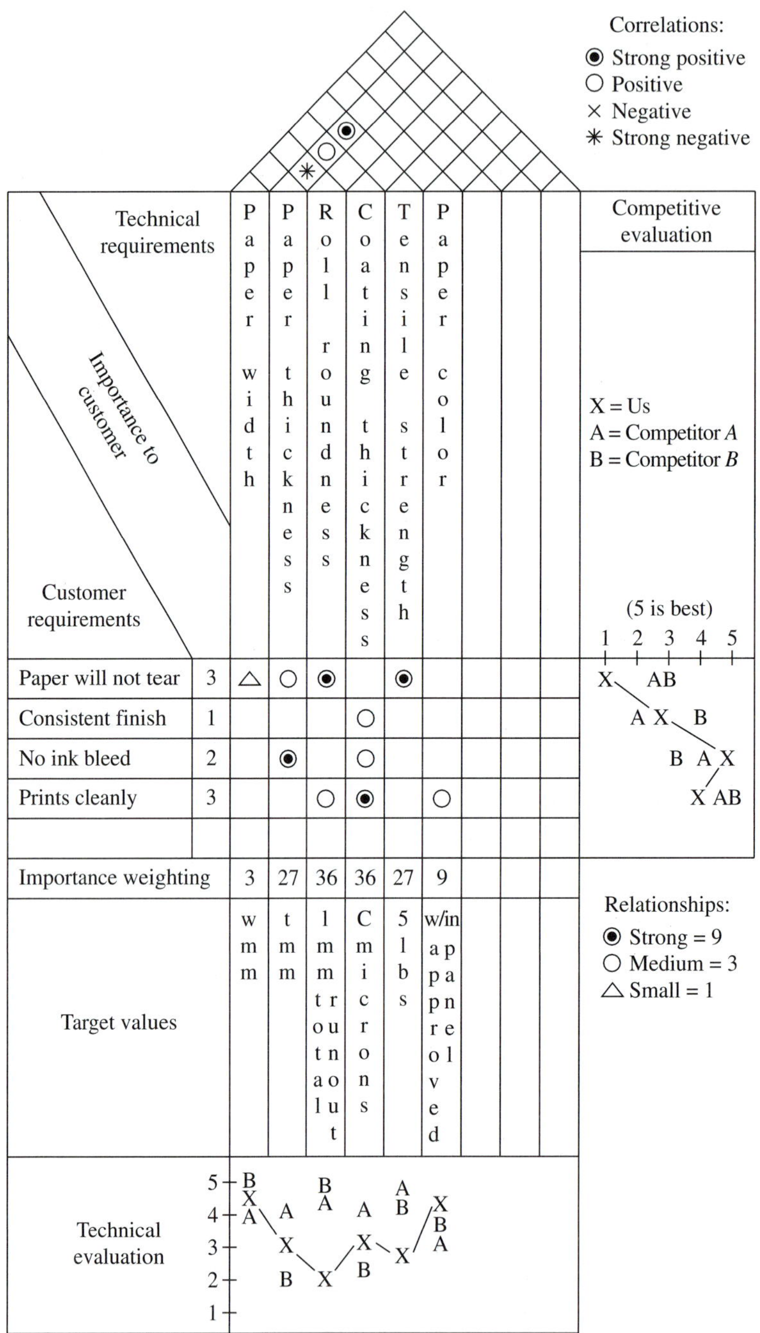

FIGURE 13.2
House of quality. (*From Ernst and Young, 1990.*)

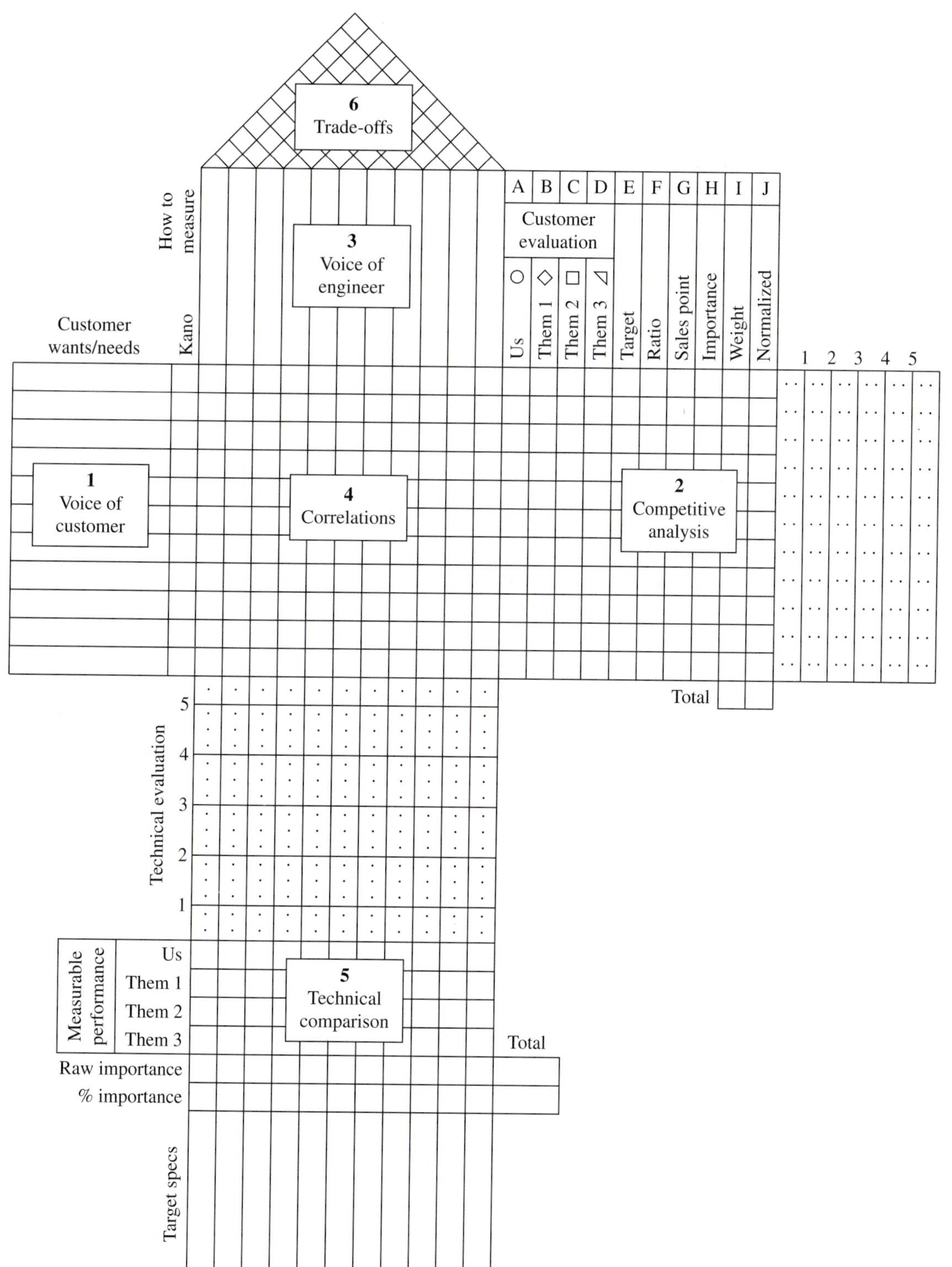

FIGURE 13.3

Six steps for building a house of quality. (*Reprinted by permission of John Wiley & Sons, Inc.*)

TABLE 13.2
Product design matrix for automated teller machine

Customers' needs	Product features	Product feature goals
Convenience	Hours of use	Available 24 hours a day, 99 percent of the time
Use by non-English speaking	Available in different languages	Customer can choose from four languages—English, Spanish, Japanese, and French
Easy access	Access to all accounts	Capable of accessing accounts and performing the following functions: get cash, make deposit, get information, transfer money between accounts
Speed/no lines	Multiple locations/machines available	Located at all 300 branches, with at least two machines available at all times
Sense of control	Customers can direct all transactions themselves	Directions can be followed by anyone with more than a sixth-grade education 90 percent of time
Privacy and safety	Location of machines; well lighted and secure	Machines securely located indoors with card access required

Source: Juran Institute, Inc.

better product designs than traditional practices and QFD creates an information-intensive atmosphere where communication increases and ideas are exchanged freely. Lesser benefits appear with respect to lowering product costs and reducing time to market.

Parameter Design and Robust Design

The most basic product feature is performance, i.e., the output—the color density of a television set, the turning radius of an automobile. To create such output, engineers use engineering principles to combine inputs of materials, parts, components, assemblies, liquids, etc. For each input the engineer identifies parameters and specifies numerical values to achieve the required output of the final product. For each parameter the specifications state a target (or nominal) value and a tolerance range around the target. The process is called "parameter and tolerance design."

In selecting these target values, it is useful to set values so that the performance of the product in the field is not affected by variability in manufacturing or field conditions. Then the design is said to be "robust." Robust designs provide optimum performance simultaneously with variation in manufacturing and field conditions.

Designers have always tried to create robust designs. But as products become more complex, with many factors affecting performance, it becomes difficult to know (1) what factors do affect performance and (2) what nominal values to set for each factor. Furthermore, some factors affect the mean value of an output parameter, whereas others affect the variation about the mean. One purpose of development testing is to investigate these matters. A powerful aid to planning development testing is the statistical design of experiments (see Section 11.11, "The Design of Experiments").

Application of Design of Experiments (DOE) to Product and Process Design

G. Taguchi has developed a method for determining the optimum target values of product and process parameters that will minimize variation while keeping the mean on target. This method is known as "parameter design;" setting limits around a target value is known as "tolerance design." The emphasis on setting target values on design parameters to minimize variation is an important contribution of this approach. In the Taguchi approach the design of experiments is employed to investigate control factors in the presence of noise factors. *Control factors* are factors that can be controlled in the design of a product or process. During design, we want to set their values for optimum product performance and minimum variation. *Noise factors* are factors that cannot be easily controlled (e.g., ambient temperature, variation and deterioration of materials), but we want to minimize them. Taguchi recommends using fractional factorial experiments (see Section 11.15) to identify the critical control factors and set their levels to minimize variation.

Phadke et al. (1983) describe an application of this approach to integrated circuits in the Bell System. The example concerns the dimensions of the contact window of a large-scale integrated chip. Windows that are not open or are too small result in a loss of contact to the devices, whereas excessively large windows lead to shorted device features. The steps in window forming and the factors critical to each step are shown in Table 13.3.

Levels of each of the nine critical factors were selected. Six of the factors have three levels each; three of the factors have only two levels each. A full factorial experiment to explore all possible factor-level combinations would require $3^6 \times 2^3$, or 5832, observations. (Twelve experiments were originally planned.) Instead, a fractional factorial design was chosen to investigate 18 combinations with a total of 34 measurements. Three dimensions of each window were measured. Transformed variables

TABLE 13.3
Fabrication steps and critical factors in window forming

Fabrication step	Critical factors
Applying photo resist	Photo-resist viscosity *(B)* and spin speed *(C)*
Baking	Baking temperature *(D)* and baking time *(E)*
Exposing	Mask dimension *(A)*, aperture *(F)*, and exposure time *(G)*
Developing	Developing time *(H)*
Plasma etching	Etching time *(I)*
Removing photo resist	Does not affect window size

instead of absolute values were used in the data analysis. The variables selected were the mean and the signal-to-noise ratio (s/n). The signal-to-noise ratio is defined as

$$s/n = \log_{10}\left(\frac{\text{mean}}{\text{standard deviation}}\right)$$

The problem was to determine the factor levels that yield a maximum s/n while keeping the mean on target. Two steps were needed:

1. Determining the factors that have a significant effect on the s/n. These are the factors that control process variability and are referred to as control factors. For each control factor, the level chosen is that with the highest s/n, thus maximizing the overall ratio.
2. Selecting the control factor that has the smallest effect on the s/n. This factor is called a "signal factor." The levels of the factors that are neither control factors nor signal factors are set at the nominal levels prior to the experiment. Finally, the level of the signal factor is set so that the mean response is on target.

The analysis of variance revealed the control factors to be factors *A, B, C, F, G,* and *H*. Based on an analysis of the data and engineering judgment, the signal factor was chosen to be the exposure time.

The levels selected for each factor are shown in Table 13.4. Using these levels, a sample of chips was manufactured. The resulting benefits were as follows:

	Old conditions	**Optimum conditions**
Standard deviation	0.29	0.14
Visual defects (window per chip)	0.12	0.04

After observing these improvements, the process engineers eliminated a number of in-process checks, thereby reducing the overall time spent by the wafers in photolithography by a factor of two.

Ross (1996) provides a complete discussion of the Taguchi approach. The Taguchi methodology has spawned some controversy. For a rigorous and practical account of

TABLE 13.4
Optimum factor levels

Label	Factor	Standard level	Optimum level
A	Mask dimension, μm	2.0	2.5
B	Viscosity	204	204
C	Spin speed, r/min	3000	4000
D	Bake temperature, °C	105	105
E	Bake time, min	30	30
F	Aperture	2	2
G	Exposure, PEP setting	Normal	Normal
H	Developing time, s	45	60
I	Plasma etch time, min	13.2	13.2

Source: Phadke et al. (1983, p. 1303).

the "triumphs and tragedies" of the methodology, see Pignatiello and Ramberg (1992). Montgomery (1997) and others agree that the concept of robust parameter design is good but are critical of the experimental designs and the data analysis methods (e.g., the signal-to-noise ratio) used in the Taguchi approach.

Design for Six Sigma (DFSS)

Section 3.6 describes the six-sigma approach in terms of five phases: define, measure, analyze, improve, and control (DMAIC). The DMAIC process is aimed mainly at reducing defects and errors in existing products, services, and processes. The importance of design has led to the adaptation of six sigma to design projects. Design for six sigma (DFSS) consists of five steps denoted as DMADV (Hahn et al., 2000):

1. *Define (D).* Identify the new (or modified) product to be designed and define a project team and project plan.
2. *Measure (M).* Plan and conduct research to understand customer needs and related requirements.
3. *Analyze (A).* Develop alternative design concepts, select a concept for high-level design, and predict the capability of the design to meet requirements.
4. *Design (D).* Develop the detailed design, evaluate its capability, and plan a pilot test.
5. *Verify (V).* Conduct the pilot test, analyze the results, and make design changes if needed.

Note the similarity to the quality planning road map (Section 4.11). In the six-sigma approach, a principal "critical to quality"(CTQ) characteristic is reliability—our next subject.

13.5 DESIGNING FOR TIME-ORIENTED PERFORMANCE (RELIABILITY)

Design engineers realize that a product should have a long service life with few failures. As products become more complex, failures increase with operating time. Traditional efforts of design, although necessary, are often not sufficient to achieve both the functional performance requirements and a low rate of failures with time. To prevent these failures, specialists have created a collection of tools called reliability engineering. Reliability is quality over time.

Reliability is the ability of a product to perform a required function under stated conditions for a stated period of time. (More simply, reliability is the chance that a product will work for the required time.) If this definition is dissected, four implications become apparent:

1. The quantification of reliability in terms of a probability.
2. A statement defining successful product performance.

3. A statement defining the environment in which the equipment must operate.
4. A statement of the required operating time between failures. (Otherwise, the probability is a meaningless number for time-oriented products.)

To achieve high reliability, it is necessary to define the specific tasks required. This task definition is called the *reliability program.* The early development of reliability programs emphasized the *design* phase of the product life cycle. However, it soon became apparent that the manufacturing and field-usage phases could not be handled separately. The result was reliability programs that spanned the full product life cycle, i.e., "cradle to grave."

A reliability program typically includes the following activities: setting overall reliability goals, apportionment of the reliability goals, stress analysis, identification of critical parts, failure mode and effect analysis, reliability prediction, design review, selection of suppliers (see Chapter 15), control of reliability during manufacturing (see Chapters 16 and 18), reliability testing, and failure reporting and corrective action system (see Chapter 20). Except as indicated, these activities are discussed in this chapter and Chapter 14.

Some elements of a reliability program are old (e.g., stress analysis, part selection). The significant new aspect is the *quantification* of reliability. The act of quantification makes reliability a design parameter just like weight and tensile strength. Thus reliability can be submitted to specification and verification. Quantification also helps to refine certain traditional design tasks such as stress analysis and part selection.

Before embarking on a discussion of reliability tasks, a remark on the application of the tasks is in order.

- These tasks are clearly not warranted for simple products. However, many products that were originally simple are now becoming more complex. In this case the various reliability tasks should be examined to see which, if any, might be justified.
- Reliability techniques were originally developed for electronic and space products, but their adaptation to mechanical products has now achieved success.
- These techniques apply not only to products a company markets but also to capital equipment a company purchases, e.g., numerically controlled machines. A less obvious application is to chemical-processing equipment.

For an overview of key reliability activities, see Figure 13.4 as well as Section 8.12, including Figure 8.3.

Setting Overall Reliability Goals

The original development of reliability quantification consisted of a probability and a mission time along with a definition of performance and usage conditions. This definition proved confusing to many people, so the index was abbreviated (using a mathematical relationship) to mean time between failures (MTBF). Many people believe

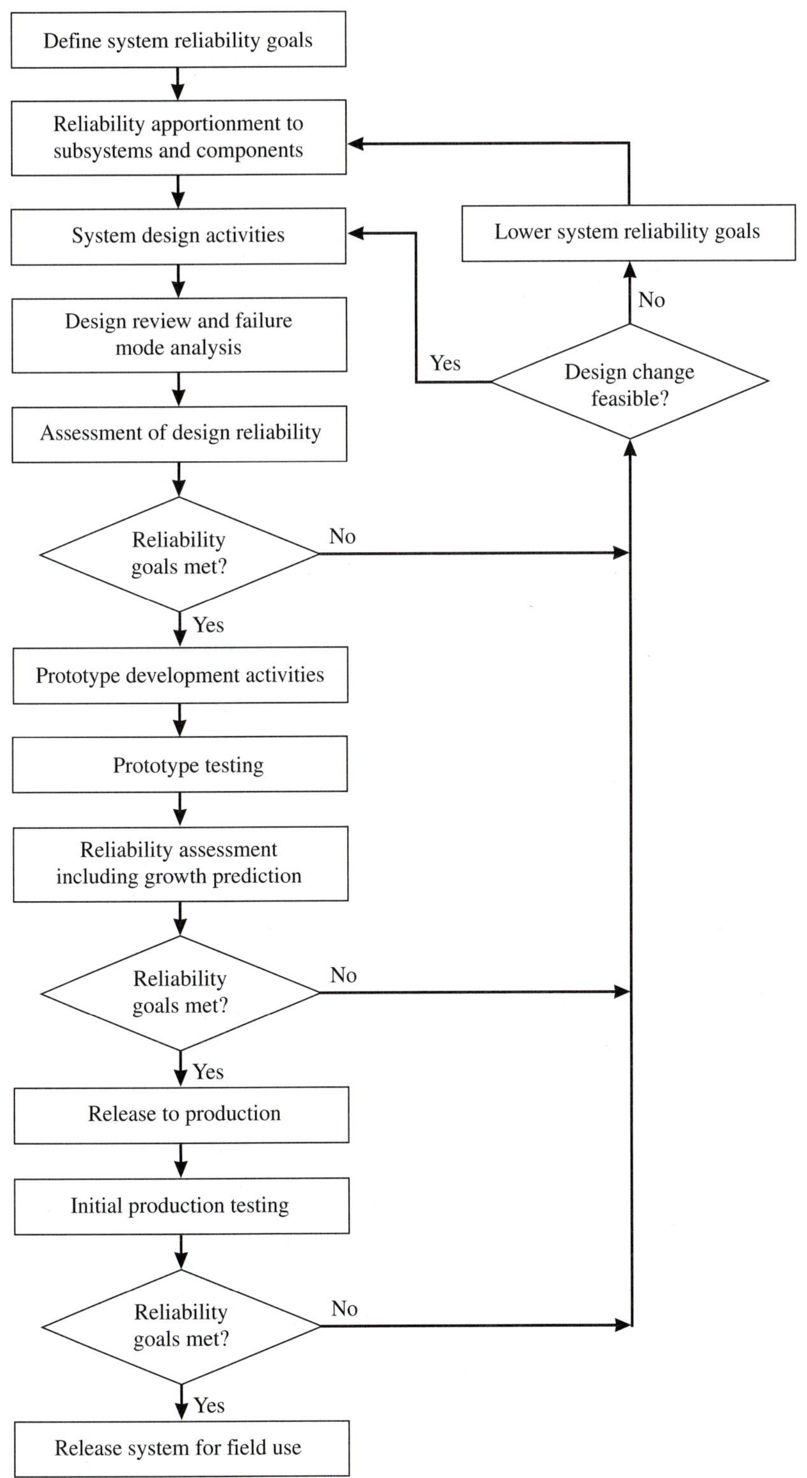

FIGURE 13.4
Some key reliability activities. (*From Ireson et al., 1996, p. 24.18.*)

TABLE 13.5
Reliability figures of merit

Figure of merit	Meaning
Mean time between failures (MTBF)	Mean time between successive failures of a repairable product
Failure rate	Number of failures per unit time
Mean time to failure (MTTF)	Mean time to failure of a nonrepairable product or mean time to first failure of a repairable product
Mean life	Mean value of life ("life" may be related to major overhead, wear-out time, etc.)
Mean time to first failure (MTFF)	Mean time to first failure of a repairable product
Mean time between maintenance (MTBM)	Mean time between a specified type of maintenance action
Longevity	Wear-out time for a product
Availability	Operating time expressed as a percentage of operating and repair time
System effectiveness	Extent to which a product achieves the requirements of the user
Probability of success	Same as reliability (but often used for "one shot" or non-time-oriented products)
b_{10} life	Life during which 10% of the population would have failed
b_{50} life	Median life, or life during which 50% of the population would have failed
Repairs/100	Number of repairs per 100 operating hours

MTBF to be the only reliability index, but no single index applies to most products. A summary of common indexes (often called *figures of merit*) is presented in Table 13.5.

As experience is gained in quantifying reliability, many companies are learning that it is best to create an index that uniquely meets the needs of those who will use the index. Users of the index include not only internal technical personnel but also marketing personnel and users of the product. Two examples of reliability indexes and goals follow:

- *For a telephone system.* The downtime of each switching center should be a maximum of 24 hours per 40 years.
- *For an engine manufacturer.* Seventy percent of the engines produced should pass through the warranty period without generating a claim. The number of failures per failed engine should not exceed one.

Note that both examples quantify reliability.

Setting overall reliability goals requires a meeting of the minds on (1) reliability as a number, (2) the environmental conditions to which the numbers apply, and (3) a definition of successful product performance. This task is not easily accomplished. However, the act of requiring designers to define both environmental conditions and successful product performance with precision forces the designers to understand the design in greater depth.

Reliability Apportionment, Prediction, and Analysis

The process of reliability quantification involves three phases:

1. *Apportionment (or budgeting).* The process of allocating (aligning) reliability objectives among various elements that collectively make up a higher-level product.
2. *Prediction.* The use of prior performance data plus probability theory to calculate the expected failure rates for various circuits, configurations, etc.
3. *Analysis.* The identification of the strong and weak portions of the design to serve as a basis for improvements, trade-offs, and similar actions.

These phases are illustrated in Table 13.6, which was one of the first published examples.

In the top section of the table, an overall reliability requirement of 95 percent for 1.45 hours is apportioned to the six subsystems of a missile. The second section of the table apportions the budget of the explosive subsystem to the three units within the subsystem. The allocation for the fusing circuitry is 0.998 or, in terms of MTBF, 725 hours. In the final section of the table, the proposed design for the circuitry is analyzed and a reliability prediction made, using the method of adding failure rates. As the prediction indicates an MTBF of 1398 hours as compared to a budget of 725 hours, the proposed design is acceptable. The prediction technique not only provides a quantitative evaluation of a design or a design change but also can identify design areas having the largest potential for reliability improvement. Thus the vital few will be obvious by noting the components with the highest failure rates. In this example the transistors, diodes, and tantalum capacitors account for about 70 percent of all the unreliability.

The approach of adding failure rates to predict system reliability is analogous to the control of weight in aircraft structures, where a running record is kept of weight as various parts are added to the design.

Reliability prediction is a continuous process starting with paper *predictions* based on a design analysis, plus historical failure-rate information (see Section 14.5). The evaluation ends with reliability *measurement* based on data from customer use of the product. Table 13.7 lists some characteristics of the various phases.

Although the visible result of the prediction procedure is to quantify the reliability numbers, the *process* of prediction is usually as important as the resulting numbers. This is so because the prediction cannot be made without obtaining rather detailed information on product missions, environments, critical component histories, etc. Acquiring this information often gives the designer knowledge previously unavailable.

TABLE 13.6
Establishment of reliability objectives*

System breakdown

Subsystem	**Type of operation**	**Reliability**	**Unreliability**	**Failure rate per hour**	**Reliability objective***
Air frame	Continuous	0.997	0.003	0.0021	483
Rocket motor	One shot	0.995	0.005		1/200 operations
Transmitter	Continuous	0.982	0.018	0.0126	80.5 h
Receiver	Continuous	0.988	0.012	0.0084	121 h
Control system	Continuous	0.993	0.007	0.0049	207 h
Explosive system	One shot	0.995	0.005		1/200 operations
System		0.95	0.05		

Explosive subsystem breakdown

Unit	**Operating mode**	**Reliability**	**Unreliability**	**Reliability objective**
Fusing circuitry	Continuous	0.998	0.002	725 h
Safety and arming mechanism	One shot	0.999	0.001	1/1000 operations
Warhead	One shot	0.998	0.002	2/1000 operations
Explosive subsystem		0.995	0.005	

Unit breakdown

Fusing circuitry component part classification	**Number used, n**	**Failure rate per part, λ (%/1000 h)**	**Total part failure rate, $n\lambda$ (%/1000 h)**
Transistors	93	0.30	27.90
Diodes	87	0.15	13.05
Film resistors	112	0.04	4.48
Wirewound resistors	29	0.20	5.80
Paper capacitors	63	0.04	2.52
Tantalum capacitors	17	0.05	8.50
Transformers	13	0.20	2.60
Inductors	11	0.14	1.54
Solder joints and wires	512	0.01	5.12
			71.51

$$\text{MTBF} = \frac{1}{\text{failure rate}} = \frac{1}{\Sigma n\lambda} = \frac{1}{0.0007151} = 1398 \text{ h}$$

Source: Adapted by F. M. Gryna, Jr., from Beaton (1959, p. 65).
*For a mission time of 1.45 hours.

Even if the designer is unable to secure the needed information, this inability nevertheless identifies the areas of ignorance in which the designer is forced to work.

The use of computer software improves the effectiveness of reliability predictions and reliability analyses. Such software not only handles the detailed calculations but also makes it feasible to identify and assess many alternatives before finalizing a design. The "Directory of Software" published periodically in *Quality Progress* includes a description of software for reliability.

TABLE 13.7
Stages of reliability prediction and measurement

	1. Start of design	2. During detailed design	3. At final design	4. From system tests	5. From customer usage
Basis	Prediction based on approximate part counts and part failure rates from previous product usage; little knowledge of stress levels, redundancy, etc.	Prediction based on quantities and types of parts, redundancies, stress levels, etc.	Prediction based on types and quantities of part failure rates for expected stress levels, redundancies, external environments, special maintenance practices, special effects of system complexity, cycling effects, etc.	Measurement based on the results of tests of the complete systems: appropriate reliability indexes are calculated from the number of failures and operating time	Same as step 4 except calculations are based on customer usage data
Primary uses	Evaluate feasibility of meeting a proposed numerical requirement Help in establishing a reliability goal for design	Evaluate overall reliability Define problem areas	Evaluate overall reliability Define problem areas	Evaluate overall reliability Define problem areas	Measure achieved reliability Define problem areas Obtain data for future designs

Note: System tests in steps 4 and/or 5 may reveal problems that result in a revision of the "final" design. Such changes can be evaluated by repeating steps 3, 4, and 5.

Parts Selection and Control

Table 13.6 shows how system reliability rests on a base of reliability of the component parts.

The vital role played by parts reliability has resulted in programs for thorough selection, evaluation, and control of parts. These programs include parts application studies, approved parts lists, identification of critical components, and use of derating.

Critical components list

A component part is considered "critical" if any of the following conditions apply:

- It has a high population in the equipment.
- It has a single source of supply.
- It must function to special, tight limits.
- It has not been proved to the reliability standard; i.e., no test data are available, or usage data are insufficient.

The critical components list should be prepared early in the design effort. It is common practice to formalize these lists, showing, for each critical component, the nature of the critical features, the plan for quantifying reliability, the plan for improving reliability, etc. The list becomes the basic planning document for (1) test programs for qualifying parts, (2) design guidance in application studies and techniques, and (3) design guidance for application of redundant parts, circuits, or subsystems.

Derating practice

Derating is the assignment of a product to operate at stress levels below its normal rating, e.g., a capacity rated at 300 V is used in a 200-V application. For many components, data are available showing failure rate as a function of stress levels. The conservative designer will use such data to achieve reliability by using the parts at low power ratios and low ambient temperatures.

Some companies have established internal policies with respect to derating. Derating is also a form of quantifying the factor of safety and hence lends itself to setting guidelines as to the design margins to be used. Derating may be considered a method of determining more scientifically the factor of safety that engineers have long provided on an empirical basis. For example, if the calculated load of a structure is 20 tons, engineers might design the structure to withstand 100 tons as a protection against unanticipated loads, misuse, hidden flaws, deterioration, etc.

Failure Mode, Effect, and Criticality Analysis

Two techniques provide a methodical way of examining a design for possible ways in which failures can occur. In the *failure mode, effect, and criticality analysis* (FMECA), a product is examined at the system and/or lower levels for all the ways in which a failure may occur. For each potential failure, an estimate is made of its effect on the total system and of its seriousness. In addition, a review is made of the action being taken (or planned) to minimize the probability of failure or to minimize the

1 = Very low (<1 in 1000)	T = Type of failure	Product HRC-1
2 = Low (3 in 1000)	P = Probability of occurrence	Date Jan. 14, 2000
3 = Medium (5 in 1000)	S = Seriousness of failure to system	By S.M.
4 = High (7 in 1000)	H = Hydraulic failure	
5 = Very high (>9 in 1000)	M = Mechanical failure	
	W = Wear failure	
	C = Customer abuse	

Component part number	Possible failure	Cause of failure	T	P	S	Effect of failure on product	Alternatives
Worm bearing 4224	Bearing worn	Not aligned with bottom housing	M	1	4	Spray head wobble or slowing down	Improve inspection
Zytel 101		Excessive spray head wobble	M	1	3	Ditto	Improve worm bearing
Bearing stem 4225	Excessive wear	Poor bearing/ material combination	M	5	4	Spray head wobbles and loses power	Change stem material
Brass		Dirty water in bearing area	M	5	4	Ditto	Improve worm seal area
		Excessive spray head wobble	M	2	3	Ditto	Improve operating instructions
Thrust washer 4226	Excessive wear	High water pressure	M	2	5	Spray head will stall out	Inform customer in instructions
Fulton 404		Dirty water in washers	M	5	5	Ditto	Improve worm seal design
Worm 4527	Excessive wear in bearing area	Poor bearing/ material combination	M	5	4	Spray head wobbles and loses power	Change bearing stem material
Brass		Dirty water in bearing area	M	5	4	Ditto	Improve worm seal design
		Excessive spray head wobble	M	2	3	Ditto	Improve operating instructions

FIGURE 13.5
Failure mode, effect, and criticality analysis.

effect of failure. Figure 13.5 shows a portion of an FMECA for a traveling lawn sprinkler. Each hardware item is listed on a separate line. Note that the failure "mode" is the symptom of the failure, as distinct from the cause of failure, which consists of the proved reasons for the existence of the symptoms. The analysis can be elaborated to include such matters as

- *Safety.* Injury is the most serious of all failure effects. In consequence, safety is handled through special programs.
- *Effect on downtime.* Must the system stop until repairs are made, or can repairs be made during an off-duty time?
- *Access.* What hardware items must be removed to get at the failed component?
- *Repair planning.* What is the anticipated repair time? What special repair tools are needed?
- *Recommendations.* What changes in designs or specifications should be made? What tests should be added? What instructions should be included in manuals of inspection, operation, or maintenance?

In Figure 13.5, a ranking procedure has been applied to assign priorities to the failure modes for further study. The ranking is twofold: (1) the probability of occurrence of the failure mode and (2) the severity of the effect. For each of these, a scale of 1 to 5 is used. If desired, a risk priority number can be calculated as the product of the ratings. Priority is then assigned to investigating failure modes with high risk priority numbers.

In this example the analysis revealed that about 30 percent of the expected failures were in the worm-and-bearing-stem area and that a redesign could easily be justified.

For most products, conducting the analysis of failure mode and failure effect for each component is not economical. Instead, engineering judgment is used to single out items that are critical to the operation of the product. As the FMECA proceeds for these selected items, the designer will discover that ready answers for some of the failure modes are lacking and that further analysis is necessary.

Generally, FMECA on one item is helpful to designers of other items in the system. In addition, the analyses are useful in planning for inspection, assembly, maintainability, and safety.

Although the FMECA was originally developed for analyzing design of physical products, the concept also applies to the service sector. Figure 13.6 shows an application at a bank in which the "product" is a new checking account and the "component" is printing new checks.

Fault tree analysis (FTA) is another method of studying potential failures in a product. Whereas FMECA usually studies all potential modes of failure, FTA studies are usually applied only to failures that are considered serious enough to warrant detailed analysis. For an example, see below under "Fault Tree Analysis."

The FMECA should not be developed by an individual designer, but by a team of designers, operations, quality, and other personnel who examine the proposed design from the several perspectives necessary for external and internal customer satisfaction. Perkins (1996) discusses this team approach in a resource-limited design community. Hatty and Owens (1994) explain how training and a multidisciplined team are important prerequisites for an effective analysis. The FMECA can also be applied to analyze proposed processes (rather than products). See Section 16.5.

Hoyland and Rausand (1994) list several software programs designed to facilitate FMECA. Ireson et al. (1996), Chapter 6, provides excellent detail on FMECA and associated techniques.

Product: New checking account

Component: Printing new checks

1	2	3	4	5	6	7	8	9
Mode of failure	Cause of failure	Effect of failure	Frequency of occurrence (1–10)	Degree of severity (1–10)	Chance of detection (1–10)	Risk priority (1–1000) [4] × [5] × [6]	Design action	Design validation
Checks being printed incorrectly	Incorrect information on application form	Checks have to be reissued	4	6	8	192	Clerk reviews information with customer	
	Data entry error	Ditto	8	6	5	240	Review step in software	
	Information entered in wrong field on application form	Ditto	5	6	2	60	—	
Incorrect account number								

FIGURE 13.6
Failure mode and effect analysis: checking account. (*From Juran Institute, Inc.*)

Evaluating Designs by Testing

Although reliability prediction, design review, FMECA, and other techniques are valuable as early-warning devices, they cannot be a substitute for the ultimate proof, i.e., use of the product by the customer. However, field experience comes too late and must be preceded by a substitute—various forms of testing the product to simulate field use.

Long before reliability technology was developed, several types of tests (performance, environmental, stress, life) were made to evaluate a design. The advent of reliability, maintainability, and other parameters resulted in additional types of tests.

A summary of types of tests for evaluating a design is given in Table 13.8. Test programs often use one type of test for more than one purpose, e.g., evaluating both performance and environmental capabilities.

All tests provide some degree of design assurance. They also involve a risk of leading one astray. The principal sources of risk follow.

Intended use versus actual use. The designer typically aims to attain fitness for the intended use. However, actual use can differ from the designer's concept because of variations in environment and other conditions of use. In addition, some users will misapply or misuse the product.

Model construction versus subsequent production. Models are usually built by skilled specialists under the supervision of designers. Subsequent production is carried out by less skilled factory workers under supervisors who must meet standards of productivity as well as quality. In addition, factory processes seldom possess the adaptive flexibility available in the model shop.

Variability due to small numbers. The number of models built is usually small. (Often there is only one.) Yet tests on these small numbers are used to judge the adequacy of the design for making many production units, sometimes running into the thousands and even millions.

TABLE 13.8
Summary of tests used to evaluate a design

Type of test	Purpose
Performance	Determine ability of product to meet basic performance requirements
Environmental	Evaluate ability of product to withstand defined environmental levels; determine interval environments generated by product operation; verify environmental levels specified
Stress	Determine levels of stress that a product can withstand in order to determine the safety margin inherent in the design; determine modes of failure that are not associated with time
Reliability	Determine product reliability and compare to requirements; monitor for trends
Maintainability	Determine time required to make repairs and compare to requirements
Life	Determine wear-out time for a product and failure modes associated with time
Pilot run	Determine whether fabrication and assembly processes are capable of meeting design requirements; determine whether reliability will be degraded
Customer ("beta")	Determine whether product functions properly under customer usage conditions

Evaluation of test results. Pressures to release a design for production can result in test plans and evaluations that do not objectively evaluate conformance to performance requirements, let alone complete fitness for use. One organization studied the process of qualifying a new design by having an independent review team analyze a sample of "qualification test" results on designs that had been approved for release. Two conclusions were that

- Pressures to release designs caused design approvals that later resulted in field problems. The situation was traced to (1) inadequate initial testing or (2) lack of verification of design changes made to correct failures that surfaced on the initial test.
- More than 50 percent of the approved test results were rejected because the test procedure was unable to evaluate the requirements set by the product development team.

As a homework problem, readers are asked to recommend action to avoid these risks.

Methods for Improving Reliability during Design

The general approach to quality improvement (see Chapter 3) is widely applicable to reliability improvement as far as economic analysis and managerial tools are concerned. The differences are in the technological tools used for diagnosis and remedy. Projects can be identified through reliability prediction, design review, FMECA, or other reliability evaluation techniques.

Action to improve reliability during the design phase is best taken by the designer. The designer understands best the engineering principles involved in the design. The reliability engineer can help by defining areas needing improvement and by assisting in the development of alternatives. The following actions indicate some approaches to improving a design:

1. *Review the users' needs* to see whether the *function* of the unreliable parts is really necessary to the user. If not, eliminate those parts from the design. Alternatively, determine whether the reliability index (figure of merit) correctly reflects the real needs of the user. For example, availability (see below) is sometimes more meaningful than reliability. If so, a good maintenance program might improve availability and hence ease the reliability problem.
2. *Consider trade-offs* of reliability for other parameters, e.g., functional performance, weight. Here again, you may find that customers' real needs may be better served by such a trade-off.
3. *Use redundancy* to provide more than one means of accomplishing a given task in such a way that all the means must fail before the system fails. This technique is discussed in Section 14.4, "The Relationship between Part and System Reliability."
4. *Review the selection of any parts* that are relatively new and unproven. Use standard parts whose reliability has been proven by actual field usage. (However, be sure that the conditions of previous use are applicable to the new product.)
5. *Use derating* to ensure that the stresses applied to the parts are lower than the stresses the parts can normally withstand.

6. *Use "robust" design methods* that enable a product to handle unexpected environments.
7. *Control the operating environment* to provide conditions that yield lower failure rates. Common examples are potting electronic components to protect them against climate and shock and the use of cooling systems to keep down ambient temperatures.
8. *Specify replacement schedules* to remove and replace low-reliability parts before they reach the wear-out stage. In many cases the replacement is made contingent on the results of checkouts or tests that determine whether degradation has reached a prescribed limit.
9. *Prescribe screening tests* to detect "infant mortality" failures and to eliminate substandard components. The tests take various forms, e.g., bench tests, "burn-in," accelerated life tests.
10. *Conduct research and development* to improve the basic reliability of components that contribute most of the unreliability. Although such improvements avoid the need for subsequent trade-offs, they may require advancing the state of the art and hence making an investment of an unpredictable size.

Although none of the foregoing actions provides a perfect solution, the range of choice is broad. In some instances the designer can arrive at a solution independently. More usually, collaboration with other company specialists will be necessary. In still other cases, the customer and/or company management will have to be adaptable because of the broader considerations involved.

13.6 AVAILABILITY

One major parameter of fitness for use is availability. *Availability* is the ability of a product, when used under given conditions, to perform satisfactorily when called upon. The total time in the operative state (also called uptime) is the sum of the time spent in active use and in the standby state. The total time in the nonoperative state (also called downtime) is the sum of the time spent under active repair and waiting for spare parts, paperwork, etc. The quantification of both availability and unavailability dramatizes the extent of the problems and areas for potential improvements. Formulas for quantifying availability are presented in Section 14.9, "Availability."

The proportion of time that a product is available for use depends on (1) freedom from failures, i.e., reliability, and (2) the ease with which service can be restored after a failure. The latter factor brings us to the subject of maintainability.

Designing for Maintainability

The tools for ensuring maintainability follow the same basic pattern as those for ensuring reliability; i.e., there are tools for specifying, predicting, analyzing, and measuring maintainability. The tools apply to both preventive maintenance (to reduce the number of failures) and corrective maintenance (to restore a product to operable condition). Some of the basic tools are discussed below.

Maintainability is often specified in quantitative form, such as the mean time to repair (MTTR). MTTR is the mean time needed to perform repair work assuming that a spare part and technician are available, e.g., a test set may have a specified MTTR of 2.5. As with reliability, no one maintainability index applies to most products. Other examples of indexes are percentage of downtime due to hardware failures, percentage of downtime due to software errors, and mean time between preventive maintenance actions.

Following the approach used in reliability, a maintainability goal for a product can be apportioned to the various components of the product. Also, maintainability can be predicted based on an analysis of the design. For elaboration, see Ireson et al. (1996), Chapter 15.

Approaches to improving maintainability of a design are both general and specific. General approaches include

- *Reliability versus maintainability.* For example, given an availability requirement, should the response be an improvement in reliability or in maintainability?
- *Modular versus nonmodular construction.* Modular design requires added design effort but reduces the time required for diagnosis and remedy in the field. The fault need only be localized to the module level, after which the defective module is simply unplugged and replaced. This concept also applies to consumer products such as television sets.
- *Repairs versus throwaway.* For some products or modules, the cost of field repair exceeds the cost of making new units in the factory. In such cases design for throwaway is an economic improvement in maintainability.
- *Built-in versus external test equipment.* Built-in test features reduce diagnostic time but usually at an added investment.
- *Person versus machine.* For example, should the operation/maintenance function be highly engineered with special instrumentation and repair facilities, or should it be left to skilled technicians with general-use equipment?

Specific approaches are based on detailed checklists that are used as a guide to good design for maintainability.

Maintainability can also be demonstrated by testing. The demonstration consists of measuring the time needed to locate and repair malfunctions or to perform selected maintenance tasks.

13.7 DESIGNING FOR SAFETY

Safety analysis tools include hazard quantification, designation of safety-oriented characteristics and components, fault tree analysis, fail-safe concepts, in-house and field testing, and publication of product ratings.

Quantification of safety

Generally, quantification of safety has been time related. Industrial injury rates are quantified on the basis of lost-time accidents per million labor-hours of exposure. (Note that this value expresses the frequency of occurrence, but does not indicate the severity of the accidents.) Motor vehicle injury rates are on the basis of injuries per 100 million miles. School injury rates are on the basis of injuries per 100,000 student days.

Product designers have tended to quantify safety in two ways:

1. *Hazard frequency.* A hazard is any combination of parts, components, conditions, or changing set of circumstances that present an injury potential. Hazard frequency takes the form of frequency of occurrence of an unsafe event and/or injuries per unit of time, e.g., per million hours of exposure. MIL-STD-882D has established categories of probability levels for hazards ranging from frequent to improbable. Such probabilities are sometimes referred to as "risk."
2. *Hazard severity.* MIL-STD-882D recognizes four levels of severity:
   - *Category I—catastrophic:* may cause death or system loss.
   - *Category II—critical:* may cause severe injury, severe occupational illness, or major system damage.
   - *Category III—marginal:* may cause minor injury, minor occupational illness, or minor system damage.
   - *Category IV—negligible:* will not result in injury, occupational illness, or system change.

MIL-STD-882C identifies 22 tasks to eliminate hazards.

Hazard analysis

Hazard analysis is similar to FMECA—but the failure event is one that causes an injury. Three forms of hazard analysis can be prepared: design concept, operating procedures, and hardware failures.

Fault tree analysis

This top-down approach starts by supposing that an accident takes place. It then considers the possible direct causes which could lead to this accident. Next, it looks for the origins of these causes. Finally, it looks for ways to avoid these origins and causes. The branching out of origins and causes is what gives the technique the name of "fault tree" analysis. The approach is the reverse of FMECA, which starts with origins and causes and looks for any resulting bad effects.

Hammer (1980) presents a fault tree analysis for an interlock safety circuit (Figure 13.7).

Based on field experience with specific products, detailed checklists are often developed to provide the designer with information on potential hazards, the injuries that can result, and specific types of design actions that can be taken to minimize the risk.

As products have become more complex, the interaction of products with the human beings operating the product has assumed increasing importance. The evaluation of the design of a product to ensure compatibility with the capabilities of human beings is referred to as "ergonomics" or "human-centered design." For elaboration, see Ireson et al. (1996), Chapter 9. For a discussion of the legal aspects of safety, see Section 20.7.

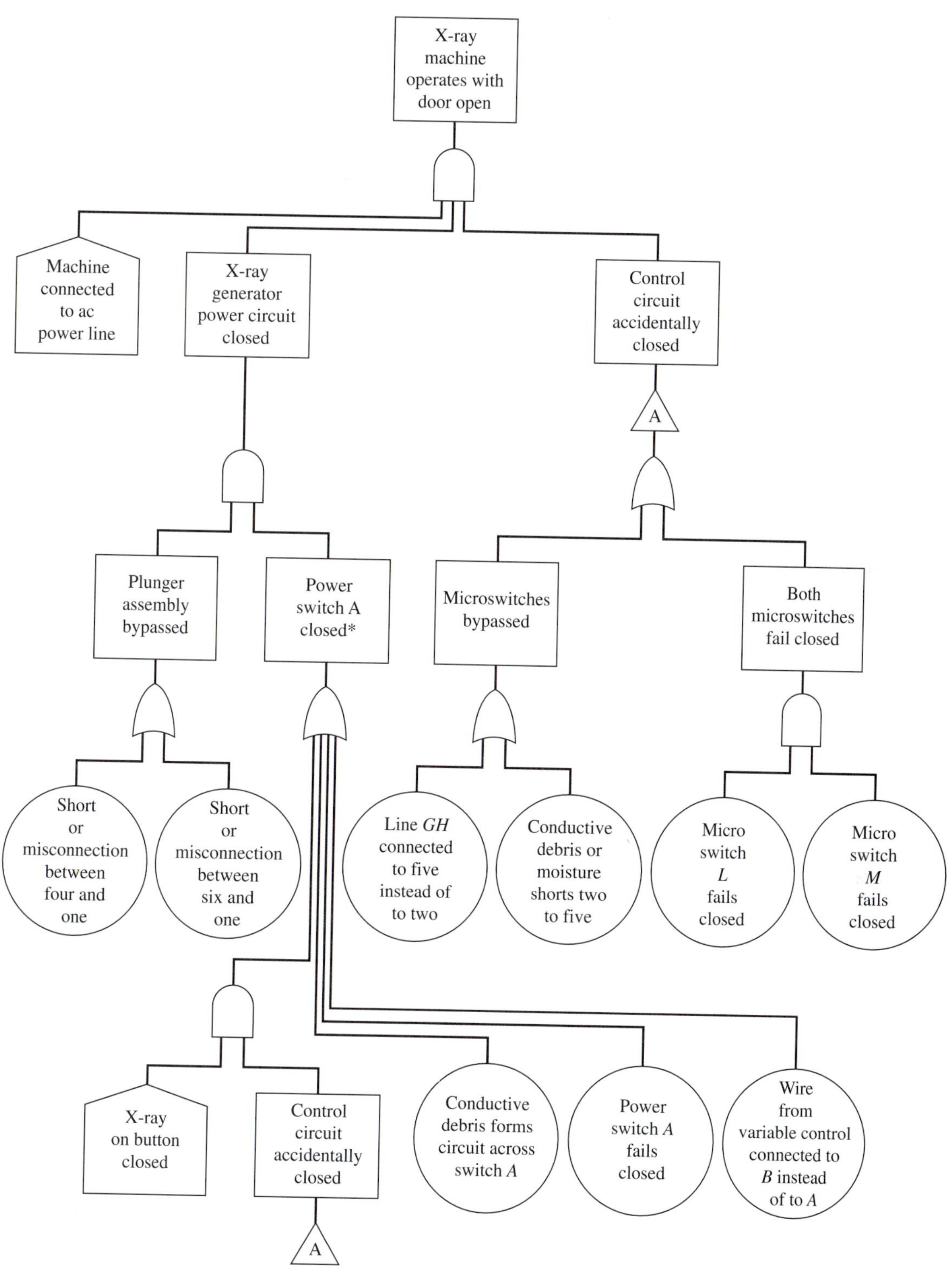

FIGURE 13.7
Fault tree analysis of an interlock safety circuit. (*From Hammer, 1980.*)

13.8 DESIGNING FOR MANUFACTURABILITY

Decisions made during design are the dominant influence on product costs, ability to meet specifications, and the time required to bring a new product to the marketplace. Furthermore, once these decisions are made (in hardware or software), the cost of changes in the design can be huge; e.g., a design *change* during pilot production of a major electronics product can cost more than $1 million.

An important set of decisions is the selection of specifications for product characteristics to be controlled during production. *Specification limits* specify the allowable limits of variability above and below the nominal value set by the designer.

The selection of limits has a dual effect on the economics of quality. The limits affect:

- Fitness for use and hence the salability of the product.
- Costs of manufacture (facilities, tooling, productivity) and quality (equipment, inspection, scrap, rework, material review, etc.).

Chapter 14 explains some quantitative techniques for setting specification limits.

A technique called design for manufacturability focuses on simplifying a design to make it more producible. The emphasis is on reducing the total number of parts, the number of different parts, and the total number of manufacturing operations. This type of analysis is not new—"value engineering" tools have been useful in achieving design simplification (see Section 13.9). What is new, however, is the computer software available for analyzing a design and identifying opportunities for simplifying assembly products. Such software dissects the assembly step by step; poses questions concerning parts and subassemblies; and provides a summary of the number of parts, the assembly time, and the theoretical minimum number of parts or subassemblies. Use of such software enables designers to learn the principles for ease of manufacturing analogous to reliability, maintainability, and safety analyses. In one example the proposed design of a new electronic cash register was analyzed with design for manufacturability (DFM) software. As a result, the number of parts was reduced by 65 percent. A person using no screws or bolts can assemble the register in less than two minutes—blindfolded. This simplified terminal was put onto the marketplace in 24 months—a record. Such design simplification reduces assembly errors and other sources of quality problems during manufacture.

Planning for inspection, packaging, transportation, and storage is discussed in Chapters 19 and 20. Some of this planning also has an impact on the product design.

In organizations using the six-sigma approach, the extremely low defect levels desired (3.4 defects per million) has an impact on product development. Product designers must understand and study manufacturing process capability before the design is released. Kodak designers use an approach that starts with the identification of critical parts and critical characteristics. The measurement process is then defined, including establishing measurement capability. The manufacturing process is then studied first for stability and then for process capability. For elaboration, see Turmel and Gartz (1997). Process stability and process capability are covered in this book in Chapter 18, "Statistical Process Control."

13.9
COST AND PRODUCT PERFORMANCE

Designing for reliability, maintainability, safety, and other parameters must be done with a simultaneous objective of minimizing cost. Formal techniques of achieving an optimum balance between performance and cost include both quantitative and qualitative approaches.

The quantitative approach makes use of a ratio relating performance and cost. Such a ratio tells "what we get for each dollar we spend." The ratio is particularly useful in comparing alternative design approaches to accomplishing a desired function.

A cost-effectiveness comparison of four alternative designs is shown in Table 13.9. Note that design 3 is the optimum design, even though design 4 has higher availability.

Another approach to comparing several different designs over a number of attributes is shown in Table 13.10. A design for a food-waste disposer is compared to the

TABLE 13.9
Cost-effectiveness comparison of alternative designs

	Design			
	1	**2**	**3**	**4**
Mean time between failures (MTBF)	100	200	500	500
Mean downtime (MDT)	18	18	15	6
Availability*	0.847	0.917	0.971	0.988
Life cycle cost (†)	51,000	49,000	50,000	52,000
Number of effective hours †	8,470	9,170	9,710	9,880
Cost/effective hour ($)	6.02	5.34	5.15	5.26

*Availability $= \dfrac{\text{MTBF}}{\text{MTBF} + \text{MDT}}$.

†Number of effective hours = 10,000 h of life × availability.

TABLE 13.10
Value comparison of food-waste disposers

		Competitor designs	
	Company design	***B***	***G***
Grinding time	9/1.73 = 5.2	9/0.87 = 10.3	6/2.14 = 2.8
Fineness of grind	4/9.18 = 0.4	4/7.82 = 0.5	4/11.88 = 0.3
Frequency of jamming	9/2.25 = 4.0	9/1.98 = 4.6	9/2.46 = 3.7
Noise	4/0.40 = 10.0	4/0.45 = 8.9	4/0.52 = 7.7
Self-cleaning ability	4/0.62 = 6.5	2/0.49 = 4.1	4/0.58 = 6.9
Electrical safety	16/0.58 = 27.6	16/0.52 = 30.8	16/0.43 = 37.2
Particle protection	6/0.29 = 20.7	6/0.30 = 20.0	2/0.37 = 5.4
Ease of servicing	6/0.70 = 8.6	4/0.52 = 7.7	6/0.98 = 6.1
Cutter life	9/0.96 = 9.4	9/0.83 = 10.8	9/1.32 = 6.8
Ease of installation	9/0.54 = 16.7	9/0.33 = 27.3	9/0.70 = 11.8
Total	76/17.25 = 4.4	72/14.11 = 5.1	69/21.44 = 3.2

Note: Value $= \dfrac{\text{Design score}}{\text{Cost of achieving}}$.

designs of two competitor models for 10 attributes describing fitness for use. For each combination of attribute and design, an effectiveness cost ratio is calculated. For example, for design *G* and the grinding-time characteristic, the ratio is 6/2.14, or 2.8. The value of 6 is the product of a weighting factor of 3 for grinding time and a score of 2 for design *G* on grinding time. The value of $2.14 is the estimated cost of achieving the grinding time of design *G* using the design *G* concept in production. The total of the ratios for each company provides a cost-effectiveness type of index.

Several approaches to achieving a balance between performance and cost have been developed. Value engineering is a technique for evaluating the design of a product to assure that the essential functions are provided at minimal overall cost to the manufacturer or user. A useful reference is Park (1999). A complementary technique is the "design to cost" approach. It starts with a definition of (1) a cost target for the product and (2) the function desired. Alternative design concepts are then developed and evaluated.

During the development cycle, the product undergoes several reviews. One form is a business review in which the current results of the development effort are summarized and a decision is made on whether to proceed. Another type of review is technical in nature and is usually called "design review."

13.10 DESIGN REVIEW

Design review is a formal, documented, comprehensive, and systematic examination of a design to evaluate the design requirements and the capability of the design to meet these requirements and to identify problems and propose solutions.

Design review is not new. However, in the past the term has referred to an informal evaluation of the design. Modern products often require a more formal program. A formal design review recognizes that many individual designers do not have specialized knowledge in reliability, maintainability, safety, producibility, and the other parameters that are important in achieving an optimum design. The design review aims to provide such knowledge.

For modern products, design reviews are based on the following concepts:

1. Design reviews become mandatory because of either customer demand or upper-management policy declaration.
2. Design reviews are conducted by a team consisting mainly of specialists who are not directly associated with the development of the design. These specialists must be highly experienced and must bring with them the reputation of being objective. The combination of competence, experience, and objectivity is present in some people, but these people are in great demand. The success of design reviews largely depends on the degree to which management supports the program by insisting that the best specialists be made available for design review work. The program deteriorates into superficial activity if (1) inexperienced people are assigned to design reviews or (2) those on the design review team are not given sufficient time to study product information prior to design review meetings.

3. Design reviews are formal. They are planned and scheduled like any other legitimized activity. The meetings are built around prepared agendas and documentation sent out in advance. Minutes of meetings are prepared and circulated. Follow-ups for action are likewise formalized.
4. Design reviews cover all quality-related parameters and others as well. The parameters can include reliability, maintainability, safety, producibility, weight, packaging, appearance, cost, etc.
5. As much as possible, design reviews are made based on defined criteria. Such criteria may include customer requirements, internal goals, and experience with previous products.
6. Design reviews are conducted at several phases of the progression of the design, such as design concept, prototype design and test, and final design. Reviews are made at several levels of the product hierarchy, such as system and subsystem.
7. The ultimate decision on inputs from the design review rests with the designer. The designer must listen to the inputs, but on matters of structural integrity and other creative aspects of the design, the designer retains the monopoly on decisions. The control and publication of the specification remain with the designer.

A universal obstacle to design review is the resistance of the design department. The common practice has been for this department to hold a virtual monopoly on design decisions, i.e., these decisions have historically been immune from challenge unless actual product trouble is encountered. With such a background, it is not surprising that designers have resisted the use of design reviews challenging their designs. Designers have contended that such challenges are based purely on grounds of theory and analysis (at which they regard themselves as the top experts) rather than on the traditional grounds of "failed hardware." This resistance is further aggravated in companies that permit reliability engineers to propose competing designs. Designers have resisted the idea of having competitors even more than they have resisted the idea of design review. An exciting concept that can overcome some of these cultural obstacles is concurrent engineering.

13.11 CONCURRENT ENGINEERING

Concurrent engineering, also called simultaneous engineering, is the process of designing a product using all inputs and evaluations *simultaneously and early* during design to ensure that internal and external customers' needs are met. The aim is to reduce the time from product concept to market, prevent quality and reliability problems, and reduce costs. An example is the approach used on the Taurus automobile (see Section 4.11).

Traditionally, activities during product development are handled sequentially, not concurrently. Thus, a marketing or research department identifies a product idea; next design engineering creates a design and builds a few prototype units; the purchasing department then calls for bids from suppliers; after that the manufacturing department produces units, etc. At each step the output of one department is "thrown over the

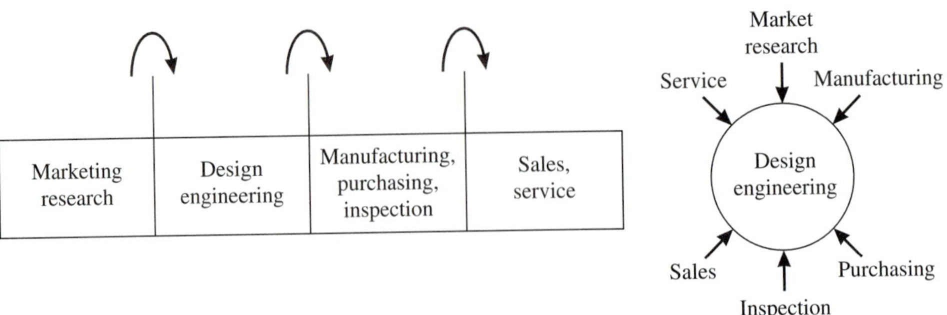

FIGURE 13.8
Walls of functional departments versus teamwork.

TABLE 13.11
Design engineering—contrast of traditional and concurrent

	Traditional	Concurrent
Organization	Engineering is separate from manufacturing and other functions; emphasis is on functional objectives	Multifunctional team with emphasis on team objective
Timing of inputs from other functions and suppliers	Mostly after design has been finalized by engineering	Simultaneous with creation of performance characteristics of design
Frequency and timing of design changes	Large number of changes, many of which occur after testing or during production	Smaller number of changes, most of which occur early, before the design is finalized
Information systems	Transfer of knowledge among functions and design changes are subject to delays of paperwork systems	Computer systems interface so that all functions have immediate access to design changes and other information
Physical location of functions	Usually in separate locations	Often located in one area

wall" to the next department; i.e., there is little input during design from functions that are impacted by the design (Figure 13.8).

A further contrast of traditional and concurrent engineering is shown in Table 13.11. Concurrent engineering is not a set of techniques; it is a concept that enables all who are impacted by a design to (1) have *early access* to design information and (2) have the ability to influence the final design to identify and prevent future problems. All design parameters discussed in this chapter—basic functional requirements, reliability, maintainability, safety, human factors, manufacturability, inspection, packaging, transportation, and storage—can be addressed during concurrent engineering.

Dramatic benefits are reported from concurrent engineering, e.g., 75 percent fewer engineering changes, 55 percent less time from product concept to market. Fleischer and Liker (1997) provide case studies at seven manufacturing organizations.

13.12 SOFTWARE DEVELOPMENT

The software development life cycle can be divided into six phases:

1. Requirements analysis.
2. Preliminary design.
3. Detailed design.
4. Coding (programming).
5. Testing and installation.
6. Maintenance.

For elaboration of these steps, see *JQH5,* Section 20.

Experience with software development provides some surprising statistics (Table 13.12). Note that 67 percent of the total effort is spent on *maintenance* that involves modifications to the software due to (1) changes in the initial requirements and (2) errors not previously detected. Thus we have an opportunity for improvement.

Also note that 46 percent of the basic development effort goes for testing and installation. Sometimes the extent of testing (and review) during software development is great.

EXAMPLE 13.5. In one software development organization, the number of errors per thousand lines of code (KLOC) measured at the *final* stage was 2/KLOC. At the *first* stage of review/inspection during development, the number was 50/KLOC. The intermediate stages included various forms of inspection, review, and testing to weed out errors. There was an opportunity for improvement.

According to the Pentagon and the Software Engineering Institute at Carnegie Mellon University, every 1000 lines of code have 5 to 15 bugs, costing up to $30,000 to correct (*Business Week,* December 6, 1999, p. 108).

TABLE 13.12
Typical resource requirements for phases of a software development life cycle

	Total effort, %	Development effort, %
Requirements definition	3	9
Preliminary design	3	9
Detailed design	5	15
Coding	7	21
Testing and installation	15	46
Maintenance	67	

Source: JQH4, p. 14.7.

For a description of diagnosis involving software errors, see Levenson and Turner (1993) in the Supplementary Reading. In testing software for use with medical equipment that delivers radiation to cancer patients, certain lines of code (used for the test only) were inadvertently left in the software. This error resulted in excessive radiation delivery and led to 13 deaths.

But errors in software are only one part of the quality dimension. As with all products, quality must address both product features and freedom from deficiencies (errors). Prahalad and Krishnan (1999) explain the need to consider these several aspects of quality in information technology.

One framework for developing product features to satisfy customer needs in software is the quality planning road map presented in Section 4.11. Chapter 22, "Quality Information Systems," presents the results of research to identify preferred practices in software development as applied to developing an information system for quality. Those preferred practices directly relate to the steps in the quality planning road map.

13.13
QUALITY MEASUREMENT IN DESIGN

The management of quality-related activities in design—as in all functional activities—must include provision for measurement. A popular rallying cry is, "What gets measured, gets done." Readers are urged to review the 10 basic principles of quality measurement in Section 5.2.

Table 13.13 shows units of measure for various subject areas of product design.

Endres (1997), Chapters 3 and 4, provides extensive discussion of measuring quality in design, research, and development.

13.14
IMPROVING THE EFFECTIVENESS OF PRODUCT DEVELOPMENT

Chapter 2, "Companywide Assessment of Quality," explains four studies that provide a thorough picture of the current status of quality in an organization. These studies are cost of poor quality, standing in the marketplace, quality culture, and assessment of current quality activities (quality system). Such studies can also be applied to assess the current status of the product development process with respect to quality. Endres (1997), Chapters 3 and 4, presents examples of such studies in product development by companies including Alcoa, Allied Signal, AT&T, Corning, Duracell, Eastman Chemical, IBM, and Olin. These assessments help define the actions needed to increase the effectiveness—pack more punch—of product development.

The trilogy of quality processes provides a useful model for product development. Thus the quality planning process (Chapter 4) provides the steps for effectively developing new products. The steps are

- Establish the project.
- Identify customers.

TABLE 13.13
Examples of quality measurement in design

Subject	Unit of measure
Overall design process	Cost of poor quality Number of months from first pilot unit to steady-state products
Design changes	Number of design changes (1) at design review, (2) at development testing, (3) after design release for steady-state production Number of design changes (1) to meet requirements, (2) to improve performance, (3) to facilitate manufacture Number of design changes requested by customer Number of waivers to specifications Number of drawing errors found by checkers on first check
Reliability, maintainability	Ratio of predicted reliability to actual reliability Ratio of actual reliability to reliability requirement Maintainability index compared to prior design
Software	Number of software errors per KLOC—first internal review Number of software errors per KLOC—final internal review Number of software errors per KLOC—discovered by customer Average score given by customers on overall quality of software
Ease of manufacture	Ratio of number of parts to theoretical minimum number Total assembly time Total number of operations

- Discover customer needs.
- Develop the product.
- Develop the process.
- Establish process controls; transfer to operations.

The quality control process (Chapter 5) measures performance of the product development function. The steps are

- Choose control subjects.
- Establish measurement.
- Set standards.
- Measure performance.
- Compare to standards.
- Take action on the difference.

The quality improvement process (Chapter 3) identifies and solves chronic problems within product development. The steps are

- Prove the need.
- Identify projects.
- Organize project teams.
- Verify the project need and mission.

- Diagnose causes.
- Provide a remedy.
- Deal with resistance to change.
- Institute controls to hold the gains.

Applying the quality planning, quality control, and quality improvement processes to the product development process is a major undertaking that is likely to occur only if one or more of the four components of assessment clearly establish the need for action (see Section 13.1). Using the trilogy of quality processes, Bailey et al. (1999) examine the research processes in the semiconductor industry.

A historical review of past product design changes can be a useful starting point for improvement. A study of 24 design changes revealed the following:

- Eleven of the changes were made to correct performance, reliability, or safety weaknesses; eight changes were made to correct administrative or paperwork errors; five changes were necessary to facilitate the manufacture of the product.
- Of the problems associated with these design changes, 23 were first found during production and one was found during field testing.
- In all 24 cases notification of the change was given only to the original designer.

Two conclusions emerged from the study: The process of product development was finding problems too late, and feedback on problems was not shared with all designers.

The concept of self-control provides a framework for analyzing the work of designers. Universal in application, the concept holds that three criteria (see Section 5.3, "Self-Control") must be met before a person can be held responsible for controlling the quality of his or her activities. The designer is said to be in a state of self-control only if all three criteria are fully met. A weakness in any of the three criteria requires the *process of product development* to be analyzed and corrected, instead of looking to the individual designer for improvement. Problem 13.14 asks readers to identify questions for each of the criteria for self-control.

SUMMARY

- For complex products, errors during product development cause about 50 percent of fitness-for-use problems.
- Product development is a process having distinct phases that can incorporate forms of early warning of new-product troubles.
- Quality function deployment is a technique consisting of interlocking matrixes that translate customer needs into product and process characteristics.
- Robust designs provide optimum performance regardless of variation in manufacturing and field conditions.
- Taguchi methods determine optimum values of product and process parameters that minimize variation while keeping the mean on target.
- Design for six sigma has five steps: define, measure, analyze, design, and verify.

- Reliability is the ability of an item to perform a required function under stated conditions for a stated period of time.
- Reliability quantification involves three phases: apportionment, prediction, and analysis. Maintainability quantification follows a familiar approach.
- Failure mode, effect, and criticality analysis and fault tree analysis are helpful qualitative tools for design assurance.
- The design-for-manufacturability technique aims to simplify a product design to facilitate manufacture.
- Value engineering and design-to-cost techniques analyze designs to achieve an optimum balance between performance and cost.
- Design review is a systematic examination of design requirements and the capability of the design to meet the requirements.
- Concurrent engineering is the process of designing a product using all inputs and evaluations simultaneously and early during design to assure that both internal and external customers' needs are met.
- The process of product development can be examined by using the elements of quality planning, control, and improvement.
- In software development, 67 percent of the effort involves making changes to the initial software—thus there is an opportunity for improvement.
- The management of quality in design must include provision for measurement.

PROBLEMS

13.1. Make a failure mode, effect, and criticality analysis for one of the following products:
(*a*) A product acceptable to the instructor.
(*b*) A flashlight.
(*c*) A toaster.
(*d*) A vacuum cleaner.

13.2. Make a fault tree analysis for one of the products mentioned in problem 13.1.

13.3. Visit a local plant and determine whether any formal or informal numerical reliability and maintainability goals are issued to the design function for its guidance in designing new products.

13.4. Obtain a schematic diagram on a product for which you can also obtain a list of the components that fail most frequently. Show the diagram to a group of engineering students most closely associated with the product (e.g., a mechanical product would be shown to mechanical engineering students). Have the students *independently* write their opinion of the most likely components to fail by ranking the top three.
(*a*) Summarize the results and comment on the agreement, or lack thereof, among the students.
(*b*) Comment on the students' opinions versus actual product history.

13.5. A reliability prediction can be made by the designer of a product or by an engineer in a staff reliability department that might be a part of the design function. One advantage

of the designer making the prediction is that his or her knowledge of the design will likely make possible a faster and more thorough job.
(*a*) What is one other advantage in having the designer make the prediction?
(*b*) Are there any disadvantages to having the designer make the prediction?

13.6. Prepare a formal presentation to gain adoption of one of the following:
(*a*) Quantification of reliability goals, apportionment, and prediction.
(*b*) Formal design reviews.
(*c*) Failure mode, effect, and criticality analysis.
(*d*) Critical components program.

You will make the presentation to one or more people who will be invited into the classroom by the instructor. These people may be from industry or may be other students or faculty. (The instructor will announce time and other limitations on your presentation.)

13.7. Outline a reliability test for one of the following products:
(*a*) A product acceptable to the instructor.
(*b*) A household clothes dryer.
(*c*) A motor for a windshield wiper.
(*d*) An electric food mixer.
(*e*) An automobile spark plug.

The testing must cover performance, environmental, and time aspects.

13.8. Speak with some practicing design engineers and learn the extent of feedback of field information to them on their own design work.

13.9. You are the design engineering manager for a refrigerator. Most of your day is spent on administrative work. You do not have time to get into details on new or modified designs. However, you must approve (by sign-off) all new designs or changes. In reality, your sign-off consists of a brief review of the design, but you rely basically on the competence of your individual designers. You do not want to institute a formal reliability program for designers or set up a reliability group at this time. What action could you take to give yourself some assurance that a design presented to you has been adequately examined by the designer with respect to reliability? It is not possible to increase the testing, and any action you take must involve a minimum of additional costs.

13.10. You work for a public utility and your department employs outside contractors who design and build various equipment and building installations. You have just finished a value engineering seminar at a university, and you wonder whether your utility should consider establishing such a function to evaluate contractor designs. Someone in your management group has heard that value engineering can reduce costs but "by lowering the performance or reliability of a design." Comment.

13.11. The chapter discusses a number of concepts (e.g., reliability prediction, design review). Select several concepts and outline a potential application of the concept to an actual problem on a specific product. The outline should include the following items:
(*a*) The name of the concept.
(*b*) A brief statement of the problem.
(*c*) Application of the concept to the problem.

(*d*) Potential advantages.
(*e*) Obstacles to actual implementation.
(*f*) An approach to use to overcome each obstacle.

The outline for each concept should be about one page.

13.12. The section "Evaluating Designs by Testing" stated four risks. For each risk, recommend one or more preventive actions.

13.13. In designing the SX-70 camera system, the Polaroid Corporation first determined the top customer complaints on previous models. For example, customers had trouble getting the correct focus; customers forgot to change the batteries; customers did not like changing lenses; customers were not sure when to use the flash. Propose a design feature to prevent each complaint.

13.14. Review the concept of self-control in Chapter 5. Apply the concept to design engineering by creating three questions for each of the three criteria of self-control.

13.15. For an organization that has a system of documenting design changes, make a historical review of the documentation that initiated about 25 design changes. Draw some initial conclusions about the *process* of product development.

REFERENCES

Bailey, D. E., F. S. Settles, and D. Sanrow (1999). "Applying Continuous Quality Improvement Techniques to a Research Environment," *Quality Management Journal,* vol. 6, no. 2, pp. 62–77.

Beaton, G. N. (1959). "Putting the R&D Reliability Dollar to Work," *Proceedings of the Fifth National Symposium on Reliability and Quality Control,* Institute of Electrical and Electronics Engineers, New York, p. 65.

Breyfogle, F. W. III (1992). *Statistical Methods for Testing, Development, and Manufacturing,* John Wiley & Sons, New York.

Brocato, R. C. and K. A. Potocki (1998). "A System Model for Quality Management Principles: Case Study Application to Technical Staff at an R&D Organization," *Annual Quality Congress Proceedings,* ASQ, Milwaukee, pp. 190–198.

Comstock, T. N. and K. Dooley (1998). "A Tale of Two QFDs," *Quality Management Journal,* vol. 5, no. 4, pp. 32–45.

Endres, A. (1997). *Improving R&D Performance the Juran Way,* John Wiley & Sons, New York.

Ernst and Young Quality Improvement Consulting Group (1990). *Total Quality,* Dow Jones-Irwin, Homewood, IL, p. 121

Feitzinger, E. and H. L. Lee (1997). "Mass Customization at Hewlett-Packard: The Power of Postponement," *Harvard Business Review,* January–February, pp. 116–121.

Fleischer, M. and J. K. Liker (1997). *Concurrent Engineering Effectiveness,* Hanser Publications, Cincinnati, OH.

Hahn, G. J., N. Doganaksoy, and R. Hoerl (2000). "The Evolution of Six Sigma," *Quality Engineering,* vol. 12, no. 3, pp. 317–326.

Hammer, W. (1980). *Product Safety Management and Engineering,* Prentice Hall, Englewood Cliffs, NJ.

Hatty, M. and N. Owens (1994). "Potential Failure Modes and Effects Analysis: A Business Perspective," *Quality Engineering,* vol. 7, no. 1, pp. 169–186.

Hoyland, A. and M. Rausand (1994). *System Reliability Theory: Models and Statistical Methods,* John Wiley & Sons, New York.

Ireson, W. G., C. F. Coombs Jr., and R. Y. Moss (1996). *Handbook of Reliability Engineering and Management,* 2nd ed., McGraw-Hill, New York.

Montgomery, D. C. (1997). *Introduction to Statistical Quality Control,* 3rd ed., John Wiley & Sons, New York.

O'Connor, P. D. T. (1995). *Practical Reliability Engineering,* 3rd ed. rev., John Wiley & Sons, New York.

Park, R. (1999). *Value Engineering,* St. Lucie Press, Boca Raton, FL.

Perkins, D. (1996). "FMEA for a Real (Resource Limited) Design Community," *Annual Quality Congress Proceedings,* ASQ, Milwaukee, pp. 435–442.

Phadke, M. S., R. R. Kackar, D. V. Speeny, and M. J. Grieco (1983). "Off-Line Quality Control in Integrated Circuit Fabrication Using Experimental Design," *The Bell System Technical Journal,* vol. 62, no. 5, pp. 1273–1309.

Pignatiello, J. J. and J. S. Ramberg (1992). "Top Ten Triumphs and Tragedies of Genichi Taguchi," *Quality Engineering,* vol. 4, no. 2, pp. 211–225.

Prahalad, C. K. and M. S. Krishnan (1999). "The New Meaning of Quality in the Information Age," *Harvard Business Review,* September–October, pp. 109–118.

ReVelle, J. B., J. W. Moran, and C. A. Cox (1998). *The QFD Handbook,* John Wiley & Sons, New York.

Ross, P. J. (1996). *Taguchi Techniques for Quality Engineering,* McGraw-Hill, New York.

Sanchez, S. M., J. S. Ramberg, J. Fiero, and J. J. Pignatiello, Jr. (1993). "Quality by Design," Chapter 10 in *Concurrent Engineering: Automation, Tools, and Techniques,* edited by A. Kusiak, John Wiley & Sons, New York.

Shin, J-S. and K-J. Kim (1997). "Restructuring a House of Quality Using Factor Analysis," *Quality Engineering,* vol. 9, no. 4, pp. 739–746.

Sobek, D. K. II, J. K. Liker, and A. C. Ward (1998). "Another Look at How Toyota Integrates Product Development," *Harvard Business Review,* July–August, pp. 36–49.

Turmel, J. and L. Gartz (1997). "Designing in Quality Improvement: A Systematic Approach to Designing for Six Sigma," *Annual Quality Congress Proceedings,* ASQ, Milwaukee, pp. 391–398.

Vonderembse, M., T. Van Fossen, and T. S. Raghunathan (1997). "Is Quality Function Deployment Good for Product Development? Forty Companies Say Yes," *Quality Management Journal,* vol. 4, no. 3, pp. 65–79.

SUPPLEMENTARY READING

Product development: *JQH5,* Sections 17 and 19.

Bieda, J. (1997). *Practical Product Assurance Management,* ASQ Quality Press, Milwaukee.

Software development: *JQH5,* Section 20.

Levenson, N. G. and C. S. Turner (1993). "An Investigation of the Therac-25 Accidents," *Computer,* July, pp. 18–41.

History of design problems: Mundel, A. B. (1991). *Ethics in Quality,* ASQ Quality Press, Milwaukee, Chapter 6.

WEBSITES

ASQ design and construction division: www.envet.org/asq/DCD
ASQ reliability division: www.asq-rd.org
QFD Institute: www.nauticom.net/www.qfdi

14

DESIGNING FOR QUALITY—STATISTICAL TOOLS

14.1 THE STATISTICAL TOOLKIT FOR DESIGN

Statistical tools for quality in the design and development process include techniques such as graphical means of summarizing data, probability distributions, theorems of probability, confidence limits, tests of hypotheses, design of experiments, and regression analysis. These are explained in Chapter 10, "Basic Concepts of Statistics and Probability," and Chapter 11, "Statistical Tools for Analyzing Data."

To supplement these techniques, this chapter explains some statistical tools for reliability and availability and tools for setting specification limits on product characteristics.

We proceed to examine some statistical tools that are useful in design.

14.2 FAILURE PATTERNS FOR COMPLEX PRODUCTS

Methodology for quantifying reliability was first developed for complex products. Suppose that a piece of equipment is placed on test, is run until it fails, and the failure time is recorded. The equipment is repaired and again placed on test, and the time of the next failure is recorded. The procedure is repeated to accumulate the data shown in Table 14.1. The failure rate is calculated, for equal time intervals, as the number of failures per unit of time. When the failure rate is plotted against time, the result (Figure 14.1) often follows a familiar pattern of failure known as the *bathtub curve*. Three periods are apparent. These periods differ in the frequency of failure and in the failure causation pattern:

TABLE 14.1
Failure history for a unit of electronic ground support equipment

Time of failure, infant mortality period		Time of failure, constant-failure-rate period		Time of failure, wear-out period	
1.0	7.2	28.1	60.2	100.8	125.8
1.2	7.9	28.2	63.7	102.6	126.6
1.3	8.3	29.0	64.6	103.2	127.7
2.0	8.7	29.9	65.3	104.0	128.4
2.4	9.2	30.6	66.2	104.3	129.2
2.9	9.8	32.4	70.1	105.0	129.5
3.0	10.2	33.0	71.0	105.8	129.9
3.1	10.4	35.3	75.1	106.5	
3.3	11.9	36.1	75.6	110.7	
3.5	13.8	40.1	78.4	112.6	
3.8	14.4	42.8	79.2	113.5	
4.3	15.6	43.7	84.1	114.8	
4.6	16.2	44.5	86.0	115.1	
4.7	17.0	50.4	87.9	117.4	
4.8	17.5	51.2	88.4	118.3	
5.2	19.2	52.0	89.9	119.7	
5.4		53.3	90.8	120.6	
5.9		54.2	91.1	121.0	
6.4		55.6	91.5	122.9	
6.8		56.4	92.1	123.3	
6.9		58.3	97.9	124.5	

1. *The infant mortality period.* This period is characterized by high failure rates that show up early in use (see the lower half of Figure 14.1). Commonly, these failures are the result of blunders in design or manufacture, misuse, or misapplication. Usually, once corrected, these failures do not occur again, e.g., an oil hole that is not drilled. Sometimes it is possible to "debug" the product by a simulated use test or by overstressing (in electronics this is known as burn-in). The weak units still fail, but the failure takes place in the test rig rather than in service. O'Connor (1995), Chapter 9, explains the use of burn-in tests and environmental screening tests.
2. *The constant-failure-rate period.* Here the failures result from the limitations inherent in the design, changes in the environment, and accidents caused by use or maintenance. The accidents can be held down by good control on operating and maintenance procedures. However, a reduction in the failure rate requires a basic redesign.
3. *The wear-out period.* These failures are due to old age; e.g., the metal becomes embrittled or the insulation dries out. A reduction in failure rates requires preventive replacement of these dying components before they result in catastrophic failure.

The top portion of Figure 14.1 shows the corresponding Weibull plot when $\alpha = 2.6$ was applied to the original data (see Section 10.12, "The Weibull Probability Distribution"). The values of the shape parameter, β, were approximately 0.5, 1.0, and 6.0, respectively. A shape parameter of less than 1.0 indicates a decreasing failure rate,

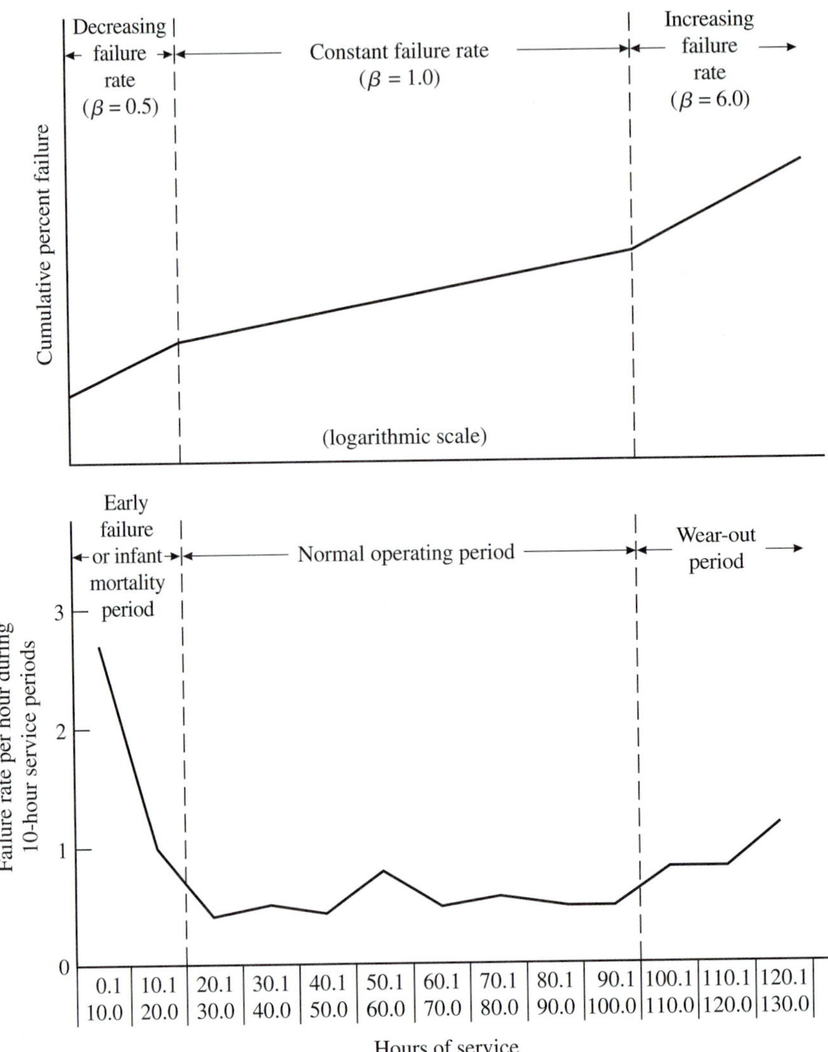

FIGURE 14.1
Failure rate versus time.

a value of 1.0 a constant failure rate, and a value greater than 1.0 an increasing failure rate (see Figure 14.1).

The Distribution of Time between Failures

Together with low failure rates during the infant mortality period, users are concerned with the length of time that a product will run without failure. Thus, for repairable products the *time between failures* (TBF) is a critical characteristic. The variation in time between failures can be studied statistically. The corresponding characteristic for nonrepairable products is usually called the *time to failure.*

When the failure rate is constant, the distribution of time between failures is distributed exponentially. Consider the 42 failure times in the constant-failure-rate portion of Table 14.1. The time between failures for successive failures can be tallied, and the 41 resulting TBFs can be formed into the frequency distribution shown in Figure 14.2*a*. The distribution is roughly exponential in shape, indicating that when the failure rate is constant, the distribution of time between failures (not *mean time* between failures) is exponential. This distribution is the basis of the *exponential formula for reliability.*

14.3
THE EXPONENTIAL FORMULA FOR RELIABILITY

The distribution of TBF indicates the chance of failure-free operation for the specified time period. The chance of obtaining failure-free operation for a specified time period *or longer* can be shown by changing the TBF distribution to a distribution showing the number of intervals equal to or greater than a specified time length (Figure 14.2*b*). If the frequencies are expressed as relative frequencies, they become estimates of the probability of survival. When the failure rate is constant, the probability of survival (or reliability) is

$$P_s = R = e^{-t/\mu} = e^{-t\lambda}$$

where $P_s = R$ = probability of failure-free operation for a time period equal to or greater than t
e = 2.718
t = specified period of failure-free operation
μ = mean time between failures (the mean of TBF distribution)
λ = failure rate (the reciprocal of μ)

Note that this formula is simply the exponential probability distribution rewritten in terms of reliability.

EXAMPLE 14.1. A washing machine requires 30 min to clean a load of clothes. The mean time between failures of the machine is 100 h. Assuming a constant failure rate, what is the chance of the machine completing a cycle without failure?

$$R = e^{-t/\mu} = e^{-0.5/100} = 0.995$$

There is a 99.5 percent chance of completing a washing cycle.

How about the assumption of a constant failure rate? In practice, sufficient data are usually not available to evaluate the assumption. However, experience suggests that this assumption is often true, particularly when (1) infant mortality types of failures have been eliminated before delivery of the product to the user and (2) the user replaces the product or specific components before the wear-out phase begins.

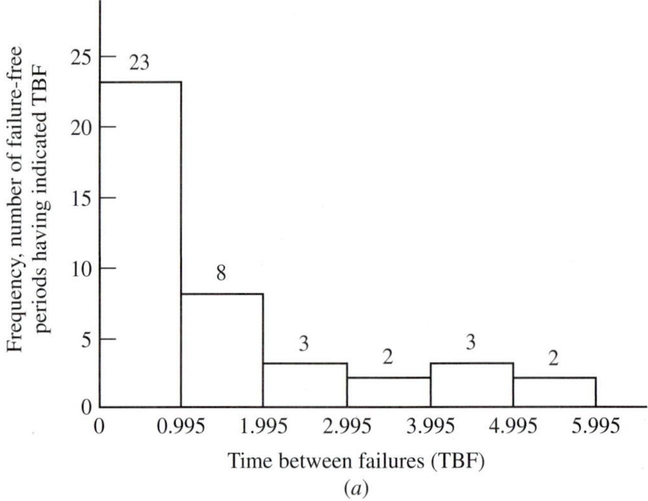

(*a*)

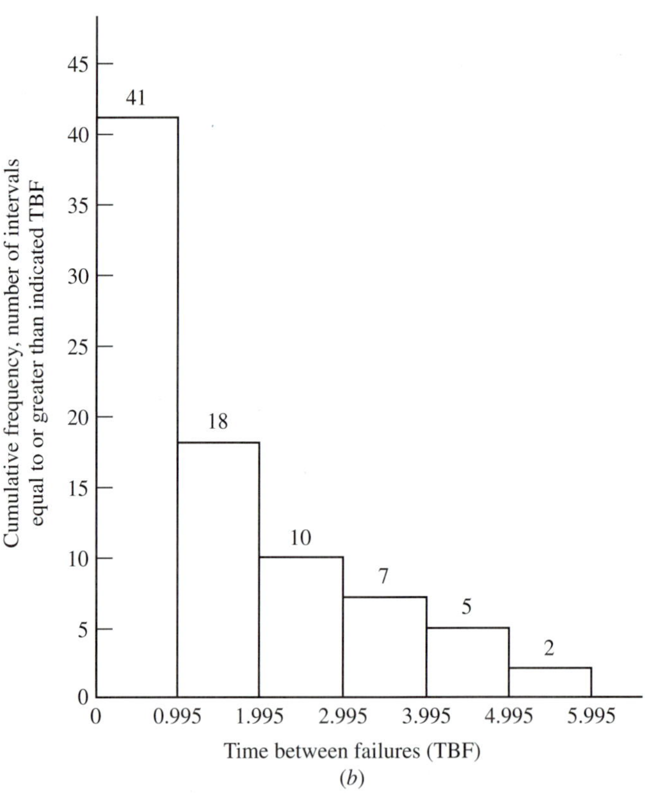

(*b*)

FIGURE 14.2
(*a*) Histogram of TBF. (*b*) Cumulative histogram of TBF.

The Meaning of *Mean Time between Failures*

Confusion surrounds the meaning of *mean time between failures* (MTBF). Further explanation is warranted:

1. The *MTBF* is the mean (or average) time between successive failures of a product. This definition assumes that the product in question can be repaired and placed back into operation after each failure. For nonrepairable products, the term *mean time to failure* (MTTF) is used.
2. If the failure rate is constant, the probability that a product will operate without failure for a time equal to or greater than its MTBF is only 37 percent. This outcome is based on the exponential distribution. (R is equal to 0.37 when t is equal to the MTBF.) This result is contrary to the intuitive feeling that there is a 50-50 chance of exceeding an MTBF.
3. MTBF is not the same as "operating life," "service life," or other indexes, which generally connote overhaul or replacement time.
4. An increase in an MTBF does not result in a proportional increase in reliability (the probability of survival). If $t = 1$ hour, the following table shows the MTBF required to obtain various reliabilities:

MTBF	R
5	0.82
10	0.90
20	0.95
100	0.99

A fivefold increase in MTBF from 20 to 100 hours is necessary to increase the reliability by 4 percentage points as compared with a doubling of the MTBF from 5 to 10 hours to get 8 percentage points' increase in reliability.

MTBF is a useful measure of reliability, but it is *not* correct for all applications. Section 13.5, "Designing for Time-Oriented Performance (Reliability)," includes a list of other reliability indexes.

14.4
THE RELATIONSHIP BETWEEN PART AND SYSTEM RELIABILITY

It is often assumed that system reliability (i.e., the probability of survival P_s) is the product of the individual reliabilities of the n parts within the system:

$$P_s = P_1 P_2 \ldots P_n$$

For example, if a communications system has four subsystems with reliabilities of 0.970, 0.989, 0.995, and 0.996, the system reliability is 0.951. The formula assumes that (1) the failure of any part causes failure of the system and (2) the reliabilities of the parts are independent of one another, i.e., the reliability of one part does not depend on the functioning of another part.

These assumptions are *not* always true, but in practice the formula serves two purposes. First, it shows the effect of increased complexity of equipment on overall reliability. As the number of parts in a system increases, the system reliability decreases dramatically (see Figure 14.3). Second, the formula is often a convenient approximation that can be refined as information on the interrelationships of the parts becomes available.

Pyzdek (1994) discusses the design of an ambulance service to handle emergency calls. The reliability block diagram for the rescue system is shown in Figure 14.4. For this example the reliability of the system is (0.998)(0.999) . . . (0.994) = 0.9195; i.e., the system would succeed in its mission about 92 percent of the time.

When it can be assumed that (1) the failure of any part causes system failure, (2) the parts are independent, and (3) each part follows an exponential distribution, then

$$P_s = e^{-t_1\lambda_1}e^{\lambda_1}e^{-t_2\lambda_2} \ldots e^{-t_n\lambda_n}$$

Further, if t is the same for each part,

$$P_s = e^{-t\Sigma\lambda}$$

Thus when the failure rate is constant (and therefore the exponential distribution can be applied), a reliability prediction of a system can be made based on the addition of the part failure rates (see Section 14.5).

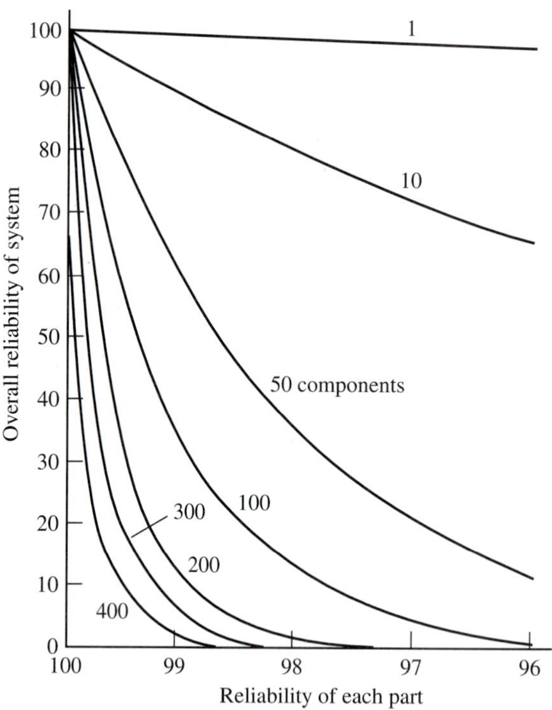

FIGURE 14.3
Relationship of part and system reliability.

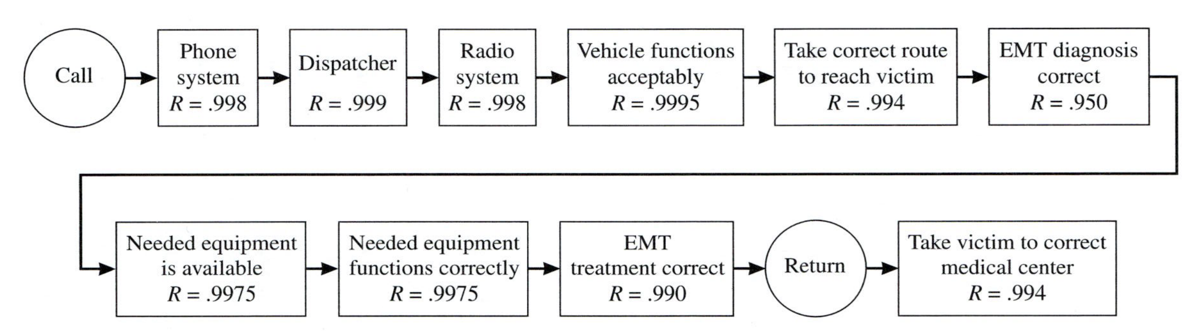

FIGURE 14.4
Reliability block diagram of emergency rescue system. (*Reprinted with permission by the ASQ.*)

Sometimes designs are planned with redundancy so that the failure of one part will not cause system failure. Redundancy is an old design technique invented long before the advent of reliability prediction techniques. However, the designer can now predict in *quantitative* terms the effect of redundancy on system reliability.

Redundancy is the existence of more than one element for accomplishing a given task, where all elements must fail before there is an overall failure to the system. In *parallel redundancy* (one of several types of redundancy), two or more elements operate at the same time to accomplish the task, and any single element is capable of handling the job itself in case of failure of the other elements. When parallel redundancy is used, the overall reliability is calculated as follows:

$$P_s = 1 - (1 - P_1)^n$$

where P_s = reliability of the system
P_1 = reliability of the individual elements in the redundancy
n = number of identical redundant elements

EXAMPLE 14.2. Suppose that a unit has a reliability of 99.0 percent for a specified mission time. If two identical units are used in parallel redundancy, what overall reliability will be obtained?

$$R = 1 - (1 - 0.99)(1 - 0.99) = 0.9999, \text{ or } 99.99 \text{ percent}$$

14.5 PREDICTING RELIABILITY DURING DESIGN

In Section 13.5 we introduced the reliability prediction method. Reliability prediction methods are still evolving. Several methods will be discussed in this chapter. Ireson et al. (1996) provide an extensive discussion of reliability prediction.

The following steps make up a reliability prediction method:

1. *Define the product and its functional operation.* The system, subsystems, and units must be precisely defined in terms of their functional configurations and boundaries. This precise definition is aided by preparation of a functional block diagram that shows the subsystems and lower-level products, their interrelationships, and the interfaces with other systems.

   Given a functional block diagram and a well-defined statement of the functional requirements of the product, the conditions that constitute failure or unsatisfactory performance can be defined.
2. *Prepare a reliability block diagram.* For systems in which there are redundancies or other special interrelationships among parts, a reliability block diagram is useful. This diagram is similar to a functional block diagram, but the reliability block diagram shows exactly what must function for successful operation of the system. The diagram shows redundancies and alternative modes of operation. The reliability block diagram is the foundation for developing the probability model for reliability. O'Connor (1995) provides further discussion.
3. *Develop the probability model for predicting reliability.* A simple model may only add failure rates; or a complex model can account for redundancies and other conditions.

4. *Collect information relevant to parts reliability.* This data includes information such as parts function, parts ratings, stresses, internal and external environments, and operating time. Many sources of failure-rate information state failure rates as a function of operating parameters. For example, failure rates for fixed ceramic capacitors are stated as a function of (1) expected operating temperature and (2) the ratio of the operating voltage to the rated voltage. Such data show the effect of derating (assigning a part to operate below its rated voltage) on reducing the failure rate.
5. *Select parts reliability data.* The required parts data consist of information on catastrophic failures and on tolerance variations with respect to time under known operating and environmental conditions. Acquiring these data is a major problem for the designer, since there is no single reliability data bank comparable to handbooks such as those that are available for physical properties of materials. Instead, the designer must build a data bank by securing reliability data from a variety of sources:
   - Field performance studies conducted under controlled conditions.
   - Specified life tests.
   - Data from parts manufacturers or industry associations.
   - Customers' parts qualification and inspection tests.
   - Government agency data banks such as the Government Industry Data Exchange Program (GIDEP) and by the Reliability Analysis Center (RAC).
6. *Combine all of the above to obtain the numerical reliability prediction.* An example of a relatively simple reliability prediction method is shown in Section 13.5. Other prediction methods are based on various statistical distributions, as explained in the following sections.

   A technique for predicting future performance from tests and field data is the concept of reliability growth. The concept assumes that a product is undergoing continuous improvements in design and refinements in operating and maintenance procedures. Thus the product performance will improve ("grow") with time. See below.

Ireson et al. (1996) and O'Connor (1995) are excellent references for reliability prediction. Included are the basic methods of prediction, repairable versus nonrepairable systems, electronic and mechanical reliability, reliability testing, and software reliability. *JQH5,* Section 48, provides extensive discussion of reliability data analysis, including topics such as censored life data (not all test units have failed during the test) and accelerated life test data analysis. Dodson (1999) explains how the use of computer spreadsheets can simplify reliability modeling using various statistical distributions.

Reliability prediction techniques based on component failure data to estimate system failure rates have generated controversy. Jones and Hayes (1999) present a comparison of predicted and observed performance for five prediction techniques using parts count analyses. The predictions differed greatly from observed field behavior and from each other. ANSI/IEC/ASQ D60300-3-1 (1997) compares five analysis techniques—FMEA/FMECA, fault tree analysis, reliability block diagram, Markov analysis, and parts count reliability prediction.

Reliability of a system evolves during design, development, testing, production, and field use. The concept of *reliability growth* assumes that the causes of product failures are discovered and action taken to remove the causes, thus resulting in improved

TABLE 14.2
Example of mechanical parts and subsystem failure rates

Part description	Quantity	Generic failure rate per million hours	Total failure rate per million hours
Heavy-duty ball bearing	6	14.4	86.4
Brake assembly	4	16.8	67.2
Cam	2	0.016	0.032
Pneumatic hose	1	29.28	29.28
Fixed displacement pump	1	1.464	1.464
Manifold	1	8.80	8.80
Guide pin	5	13.0	65.0
Control valve	1	15.20	15.20
Total assembly failure rate			273.376

MTBF = 1/0.000273376 = 3657.9 h
Source: Adapted from Ireson et al., p. 19.9.

reliability of future units ("test, analyze, and fix"). Reliability growth models provide predictions of reliability due to such improvements. For elaboration, see O'Connor (1995). Also, ANSI/IEC/ASQ D 1164-1997 describes several methods of estimating reliability growth.

14.6 PREDICTING RELIABILITY BASED ON THE EXPONENTIAL DISTRIBUTION

When the failure rate is constant and when study of a functional block diagram reveals that all parts must function for system success, then reliability is predicted to be the simple total of failure rates. An example of a subsystem prediction is shown in Table 14.2. The prediction for the subsystem is made by adding the failure rates of the parts; the MTBF is then calculated as the reciprocal of the failure rate.

For further discussion of reliability prediction, including an example for an electronic system, see Section 13.5.

14.7 PREDICTING RELIABILITY BASED ON THE WEIBULL DISTRIBUTION

Prediction of overall reliability based on the simple addition of component failure rates is valid only if the failure rate is constant. When this assumption cannot be made, an alternative approach based on the Weibull distribution can be used.

1. Graphically, use the Weibull distribution to predict the reliability R for the time period specified. $R = 100 - \%$ failure. Do this for each component. (See Section 10.12 under "Using the Weibull Probability Distribution to Make Predictions.")

2. Combine the component reliabilities using the product rule and/or redundancy formulas to obtain the prediction of system reliability.

Predictions of reliability using the exponential distribution or the Weibull distribution are based on reliability as a function of time. We next consider reliability as a function of stress and strength.

14.8 RELIABILITY AS A FUNCTION OF APPLIED STRESS AND STRENGTH

Failures are not always a function of time. In some cases a part will function indefinitely if its strength is greater than the stress applied to it. The terms *strength* and *stress* here are used in the broad sense of inherent capability and operating conditions applied to a part, respectively.

For example, operating temperature is a critical parameter, and the maximum expected temperature is 145°F (63°C). Further, capability is indicated by a strength distribution having a mean of 172°F (78°C) and a standard deviation of 13°F (7°C) (see Figure 14.5). With knowledge of only the maximum temperatures, the safety margin is

$$\frac{172 - 145}{13} = 2.08$$

The safety margin says that the average strength is 2.08 standard deviations above the maximum expected temperature of 145°F (63°C). Table A in Appendix II can be used to calculate a reliability of 0.981 (the area beyond 145°F [63°C]).

This calculation illustrates the importance of *variation* in addition to the *average* value during design. Designers have always recognized the existence of variation by using a *safety factor* in design. However, the safety factor is often defined as the ratio of average strength to the worst stress expected.

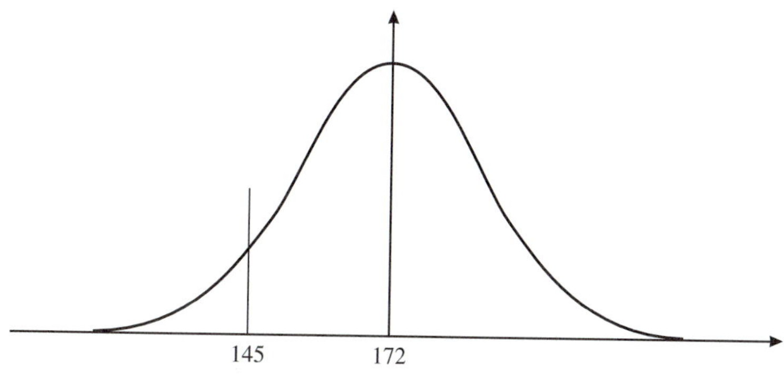

FIGURE 14.5
Distribution of strength.

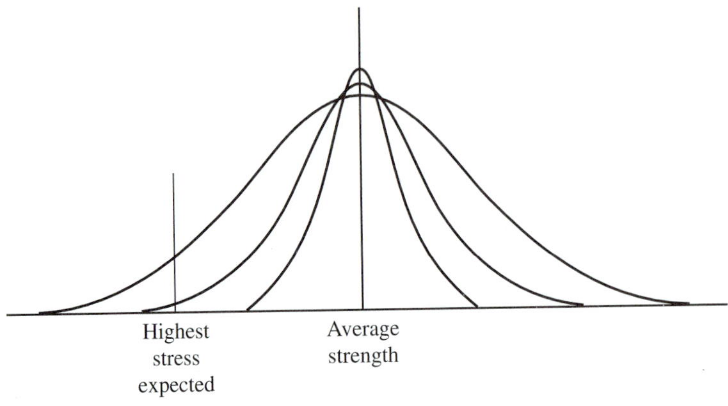

FIGURE 14.6
Variation and safety factor.

Note that in Figure 14.6, all the designs have the same safety factor. Also note that the reliability (probability of a part having a strength greater than the stress) varies considerably. Thus the uncertainty often associated with this definition of safety factor is in part due to its failure to reflect the *variation* in both strength and stress. Such variation is partially reflected in a safety margin, defined as

$$\frac{\text{Average strength} - \text{Worst stress}}{\text{Standard deviation of strength}}$$

This recognizes the variation in strength but is conservative because it does not recognize a variation in stress.

The recent emphasis of the six-sigma approach relates to a fascinating historical event. Almost half a century ago, Lusser (1958) proposed the use of safety margins as a way to design-in high reliability for critical products such as guided missiles. He suggested safety margins for both strength and stress. Specifically, the reliability boundary (maximum stress) is defined as six standard deviations (of stress) above the average stress. He also proposed that the average strength should be at least five standard deviations (of strength) above the reliability boundary. High reliability is provided because it is unlikely that a high stress (with extremely low probability of occurrence) would combine with a low strength (also with low probability of occurrence). Perhaps Lusser was an early pioneer of the six-sigma approach.

14.9 AVAILABILITY

Availability has been defined as the probability that a product, when used under given conditions, will perform satisfactorily when called upon. Availability considers the operating time of the product and the time required for repairs. Idle time, during which the product is not needed, is excluded.

Availability is calculated as the ratio of operating time to operating time plus downtime. However, downtime can be viewed in two ways:

1. *Total downtime.* This period includes the active repair (diagnosis and repair time), preventive maintenance time, and logistics time (time spent waiting for personnel, spare parts, etc.). When total downtime is used, the resulting ratio is called *operational availability* (A_0).
2. *Active repair time.* The resulting ratio is called *intrinsic availability* (A_i). Under certain conditions, availability can be calculated as

$$A_0 = \frac{\text{MTBF}}{\text{MTBF} + \text{MDT}} \quad \text{and} \quad A_i = \frac{\text{MTBF}}{\text{MTBF} + \text{MTTR}}$$

where MTBF = mean time between failures
MDT = mean downtime
MTTR = mean time to repair

This is known as the steady state formula for availability.

Garrick and Mulvihill (1974) present data on certain subsystems of a mechanized bulk mail system (see Table 14.3). If estimates of reliability and maintainability can be made during the design process, availability can be evaluated before the design is released for production.

The steady state formula for availability has the virtue of simplicity. However, the formula has several assumptions that are not always met in the real world. The assumptions are

- The product is operating in the constant-failure-rate period of the overall life. Thus the failure-time distribution is exponential.
- The downtime or repair-time distribution is exponential.
- Attempts to locate system failures do not change the overall system failure rate.
- No reliability growth occurs. (Such growth might be due to design improvements or through debugging of bad parts.)
- Preventive maintenance is scheduled outside the time frame included in the availability calculation.

More precise formulas for calculating availability depend on operational conditions and statistical assumptions. These formulas are discussed by Ireson et al. (1996).

TABLE 14.3
Availability data for mail system equipment

Equipment	MTBF, h	MTTR, h	Availability (%)
Sack sorter	90	1.620	98.2
Parcel sorter	160	0.8867	99.4
Conveyor, induction	17,900	1.920	100.0
Deflector, traveling	3,516	3.070	99.9

14.10 SETTING SPECIFICATION LIMITS

A major step in the development of physical products is the conversion of product *features* into dimensional, chemical, electrical, and other *characteristics* of the product. Thus a heating system for an automobile will have many characteristics on the heater, air ducts, blower assembly, engine coolant, etc.

For each characteristic the designer must specify (1) the desired average (or "nominal value") and (2) the specification limits (or "tolerance limits") above and below the nominal value that individual units of product must meet. These two elements relate to parameter design and tolerance design discussed in Section 13.4, "Designing for Basic Functional Requirements."

The specification limits should reflect the functional needs of the product, manufacturing variability, and economic consequences. These three aspects are addressed in the next three sections.

Anand (1996) describes several cases on the role of statistics in determining specification limits.

14.11 SPECIFICATION LIMITS AND FUNCTIONAL NEEDS

Sometimes data can be developed to relate product performance to measurements on a critical component. For example, a thermostat may be required to turn on and shut off a power source at specified low and high temperature values, respectively. A number of thermostat elements are built and tested. The prime recorded data are (1) turn-on temperature, (2) shutoff temperature, and (3) physical characteristics of the thermostat elements. We can then prepare scatter diagrams (Figure 14.7) and regression equations to help establish critical component tolerances on a scientific basis within the confidence limits for the numbers involved. Ideally, the sample size is sufficient, and the data come from a statistically controlled process—two conditions that are both rarely achieved.

O'Connor (1995), Chapter 7, explains how this approach can be related to the Taguchi approach to develop a more robust design.

14.12 SPECIFICATION LIMITS AND MANUFACTURING VARIABILITY

Generally, designers will not be provided with information on process capability. Their problem will be to obtain a sample of data from the process, calculate the limits that the process can meet, and compare these to the limits they were going to specify. (If they do not have any limits in mind, the capability limits calculated from process data provide them with a set of limits that are realistic from the viewpoint of

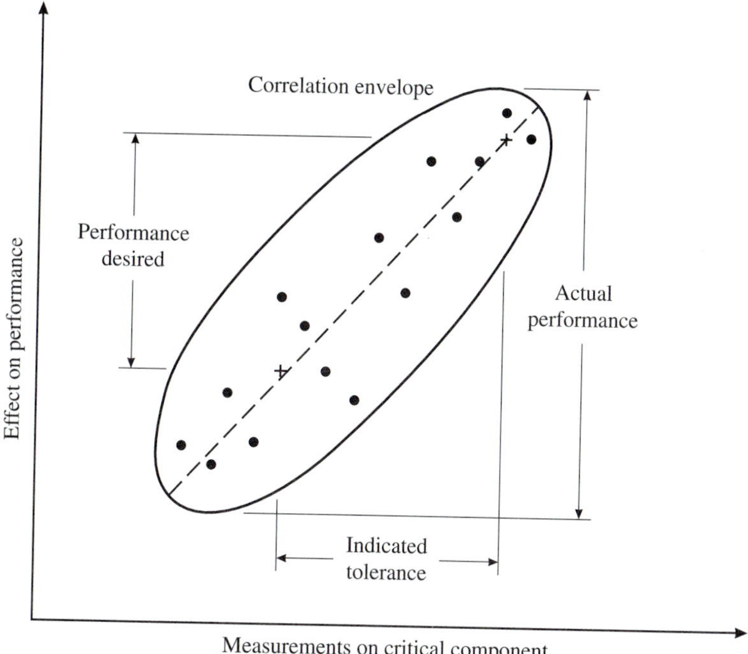

FIGURE 14.7
Approach to functional tolerancing.

producibility. These limits must then be evaluated against the functional needs of the product.)

Statistically, the problem is to predict the limits of variation of individual items in the total population based on a sample of data. For example, suppose that a product characteristic is normally distributed with a population average of 5.000 in (12.7 cm) and a population standard deviation of 0.001 in (0.00254 cm). Limits can then be calculated to include any given percentage of the population. Figure 14.8 shows the location of the 99 percent limits. Table A in Appendix II indicates that ±2.575 standard deviations will include 99 percent of the population. Thus, in this example, a realistic set of tolerance limits would be

$$5.000 \pm 2.575(0.001) = \begin{matrix} 5.003 \\ 4.997 \end{matrix}$$

Ninety-nine percent of the individual pieces in the population will have values between 4.997 and 5.003.

In practice, the average and standard deviation of the population are not known, but must be estimated from a sample of product from the process. As a first approximation, tolerance limits are sometimes set at

$$\bar{X} \pm 3s$$

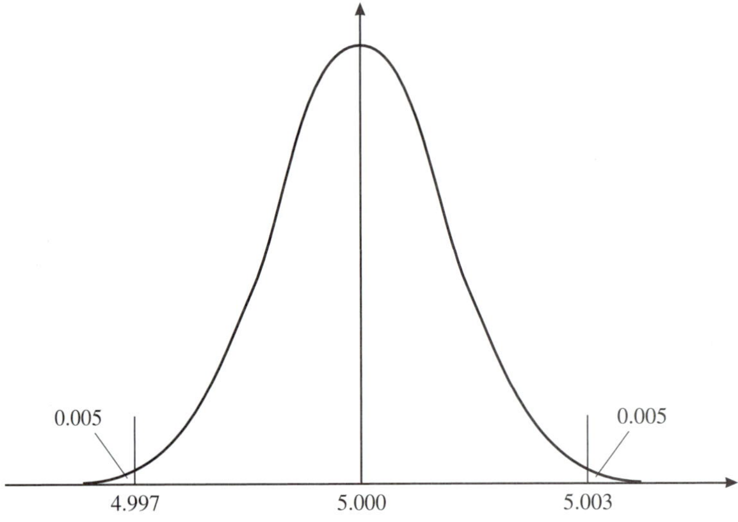

FIGURE 14.8
Distribution with 99 percent limits.

Here the average $\bar{X}$ and standard deviation s of the sample are used directly as estimates of the population values. If the true average and standard deviation of the population happen to be equal to those of the sample and if the characteristic is normally distributed, then 99.73 percent of the pieces in the population will fall within the limits calculated above. These limits are frequently called *natural tolerance limits* (limits that recognize the actual variation of the process and therefore are realistic). This approximation ignores the possible error in both the average and standard deviation as estimated from the sample.

Methodology has been developed for setting tolerance limits in a more precise manner. For example, formulas and tables are available for determining tolerance limits based on a normally distributed population. Table H in Appendix II provides factors for calculating tolerance limits that recognize the uncertainty in the sample mean and sample standard deviation. The tolerance limits are determined as

$$\bar{X} \pm Ks$$

The factor K is a function of the confidence level desired, the percentage of the population to be included within the tolerance limits, and the number of data values in the sample.

For example, suppose that a sample of 10 resistors from a process yielded an average and standard deviation of 5.04Ω and 0.016Ω, respectively. The tolerance limits are to include 99 percent of the population, and the tolerance statement is to have a confidence level of 95 percent. Referring to Table H in Appendix II, the value of K is 4.433, and tolerance limits are then calculated as

$$5.04 \pm 4.433(0.016) = \begin{matrix} 5.11 \\ 4.97 \end{matrix}$$

We are 95 percent confident that at least 99 percent of the resistors in the population will have a resistance between 4.97Ω and 5.11Ω. Tolerance limits calculated in this manner are often called statistical tolerance limits. This approach is more rigorous than the $\pm 3s$ natural tolerance limits, but the two percentages in the statement are a mystery to those without a statistical background.

For products in some industries (e.g., electronics), the number of units outside of specification limits is stated in terms of parts per million (PPM). Thus if limits are set at ± 3 standard deviations, 2700 PPM (100% − 99.73%) will fall outside the limits. For many applications (e.g., a personal computer with many logic gates), such a level is totally unacceptable. Table 14.4 shows the PPM for several standard deviations. These levels of PPM assume that the process average is constant at the nominal specification. A deviation from the nominal value will result in a higher PPM value. To allow for modest shifts in the process average, some manufacturers follow a guideline for setting specification limits at $\pm 6\sigma$. See Section 18.15 under "Six-Sigma Concept of Process Capability."

Turmel and Gartz (1997) describe how Kodak integrates six-sigma concepts during product development. As part of this process, product development engineers must focus on manufacturing process capability measures (see Section 18.10) to optimize design features and prevent defects. In addition, designers must study process data (from internal Kodak processes and supplier processes) to assure that the manufacturing processes have both the capability and the stability to achieve six-sigma defect levels. This analysis involves both critical product characteristics and critical process variables.

Designers must often set tolerance limits with only a few measurements from the process (or more likely from the development tests conducted under laboratory conditions). In developing a paint formulation, the following values of gloss were obtained: 76.5, 75.2, 77.5, 78.9, 76.1, 78.3, and 77.7. A group of chemists was asked where they would set a minimum specification limit. Their answer was 75.0—a reasonable answer for those without statistical knowledge. Figure 14.9 shows a plot of the data on normal probability paper. If the line is extrapolated to 75.0, the plot predicts that about 11 percent of the population will fall below 75.0—even though all of the sample data exceed 75.0 Of course, a larger sample size is preferred and further statistical analyses could be made, but the plot provides a simple tool for evaluating a small sample of data.

All methods of setting tolerance limits based on process data assume that the sample of data represents a process that is sufficiently stable to be predictable. In practice, the assumption is often accepted without any formal evaluation. If sufficient data are available, the assumption should be checked with a control chart.

TABLE 14.4
Standard deviations and PPM (centered process)

Number of standard deviations	Parts per million (PPM)
$\pm 3\sigma$	2700
$\pm 4\sigma$	63
$\pm 5\sigma$	0.57
$\pm 6\sigma$	0.002

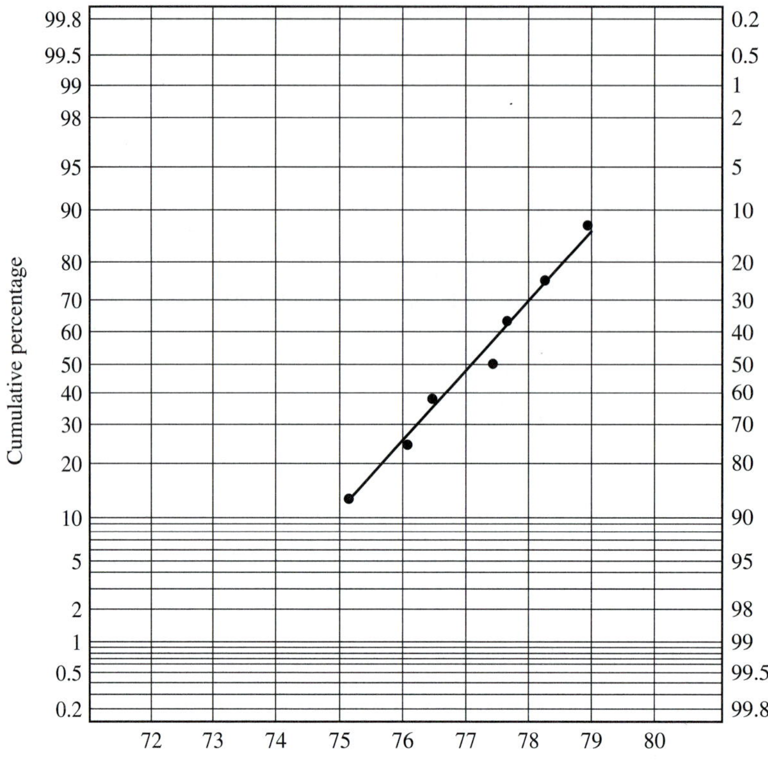

FIGURE 14.9
Probability plot of development data.

Statistical tolerance limits are sometimes confused with other limits used in engineering and statistics. Table 14.5 summarizes the distinctions among five types of limits (see also *JQH5,* pages 44.47–44.58).

14.13 SPECIFICATION LIMITS AND ECONOMIC CONSEQUENCES

In setting traditional specification limits around a nominal value, we assume that for product falling within specification limits there is no monetary loss. For product falling outside the specification limits, the loss is the cost of replacing the product.

Another viewpoint holds that *any* deviation from the nominal value causes a loss. Thus there is an ideal (nominal) value that customers desire, and any deviation from this ideal results in customer dissatisfaction. This loss can be described by a loss function (Figure 14.10).

TABLE 14.5
Distinctions among limits

Name of limit	Meaning
Tolerance	Set by the engineering design function to define the minimum and maximum values allowable for the product to work properly
Statistical tolerance	Calculated from process data to define the amount of variation that the process exhibits; these limits will contain a specified proportion of the total population
Prediction	Calculated from process data to define the limits which will contain all of k future observations
Confidence	Calculated from data to define an interval within which a population parameter lies
Control	Calculated from process data to define the limits of chance (random) variation around some central value

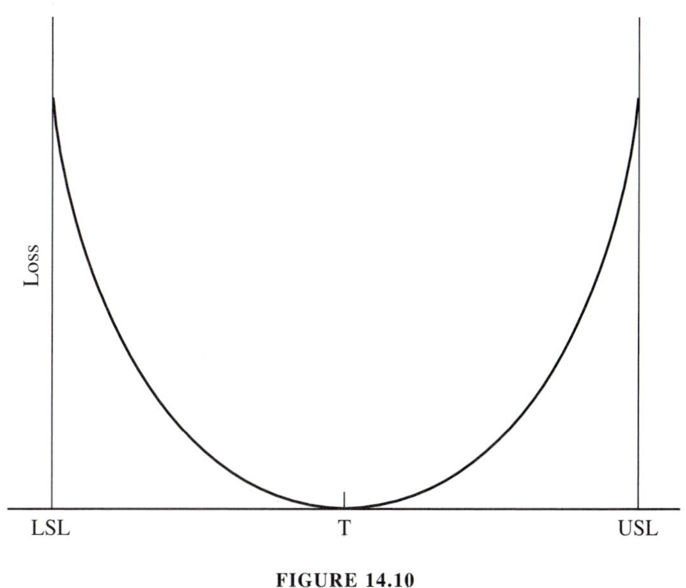

FIGURE 14.10
Loss function.

Many formulas can predict loss as a function of deviation from the target. Taguchi proposes the use of a simple quadratic loss function:

$$L = k(X - T)^2$$

where L = loss in monetary terms
k = cost coefficient
X = value of quality characteristic
T = target value

Ross (1996) provides an example to illustrate how the loss function can help to determine specification limits. In automatic transmissions for trucks, shift points are designed to occur at a certain speed and throttle position. Suppose it costs the producer \$100 to adjust a valve body under warranty when a customer complains of the shift point. Research indicates that the average customer would request an adjustment if the shift point is off from the nominal by 40 rpm transmission output speed on the first-to-second gear shift. The loss function is then:

$$\text{Loss} = k(X - T)^2$$

$$100 = k(40)^2$$

$$k = \$0.0625$$

This adjustment can be made at the factory at a lower cost, about \$10. The loss function is now used to calculate the specification limits:

$$\$10 = 0.0625(X - T)^2$$

$$(X - T) = \pm 12.65 \text{ or } \pm 13 \text{ rpm}$$

The specification limits should be set at ±13 rpm around the desired nominal value. If the transmission shift point is further than 13 rpm from the nominal, making the adjustment at the factory is less expensive than waiting for a customer complaint and making the adjustment under warranty in the field. Ross discusses how the loss function can be applied to set one-sided specification limits, e.g., a minimum value or a maximum value.

14.14 SPECIFICATION LIMITS FOR INTERACTING DIMENSIONS

Interacting dimensions mate or merge with other dimensions to create a final result. Consider the simple mechanical assembly shown in Figure 14.11. The lengths of components A, B, and C are interacting dimensions because they determine the overall assembly length.

Suppose the components were manufactured to the specifications indicated in Figure 14.11. A logical specification on the assembly length would be 3.500 ± 0.0035, giving limits of 3.5035 and 3.4965. This logic may be verified from the two extreme assemblies:

Maximum	Minimum
1.001	0.999
0.5005	0.4995
2.002	1.998
3.5035	3.4965

The approach of adding component tolerances is mathematically correct but is often too conservative. Suppose that about 1 percent of the pieces of component A are

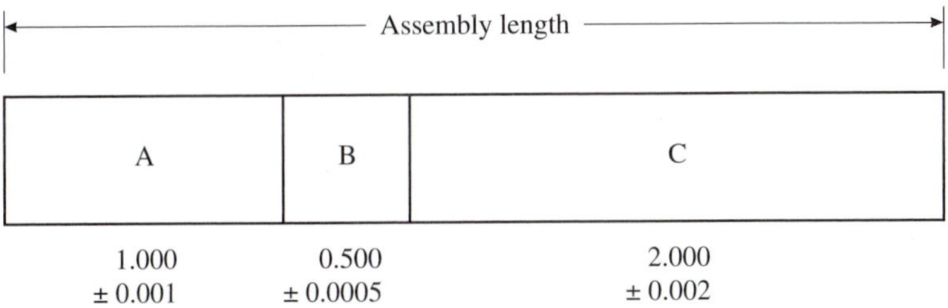

FIGURE 14.11
Mechanical assembly.

expected to be below the lower tolerance limit for component A and suppose the same for components B and C. If a component A is selected at random, there is, on average, 1 chance in 100 that it will be on the low side, and similarly for components B and C. The key point is this: If assemblies are made at random and if the components are manufactured independently, then the chance that an assembly will have all *three* components simultaneously below the lower tolerance limit is

$$\frac{1}{100} \times \frac{1}{100} \times \frac{1}{100} = \frac{1}{1{,}000{,}000}$$

There is only about one chance in a million that all three components will be too small, resulting in a small assembly. Thus setting component and assembly tolerances based on the simple addition formula is conservative in that it fails to recognize the extremely low probability of an assembly containing all low (or all high) components.

The statistical approach is based on the relationship between the variances of a number of independent causes and the variance of the dependent or overall result. This may be written as

$$\sigma_{\text{result}} = \sqrt{\sigma^2_{\text{cause } A} + \sigma^2_{\text{cause } B} + \sigma^2_{\text{cause } C} + \cdots}$$

In terms of the assembly example, the formula is

$$\sigma_{\text{assembly}} = \sqrt{\sigma_A^2 + \sigma_B^2 + \sigma_C^2}$$

Now suppose that, for each component, the tolerance range is equal to ± 3 standard deviations (or any constant multiple of the standard deviation). As σ is equal to T divided by 3, the variance relationship may be rewritten as

$$\frac{T}{3} = \sqrt{\left(\frac{T_A}{3}\right)^2 + \left(\frac{T_B}{3}\right)^2 + \left(\frac{T_C}{3}\right)^2}$$

or

$$T_{\text{assembly}} = \sqrt{T_A^2 + T_B^2 + T_C^2}$$

Thus the squares of tolerances are added to determine the square of the tolerance for the overall result. This formula compares to the simple addition of tolerances commonly used.

The effect of the statistical approach is dramatic. Listed below are two possible sets of component tolerances that when used with the formula above will yield an assembly tolerance equal to ±0.0035.

Component	Alternative 1	Alternative 2
A	±0.002	±0.001
B	±0.002	±0.001
C	±0.002	±0.003

With alternative 1 the tolerance for component A has been doubled, the tolerance for component B has been quadrupled, and the tolerance for component C has been kept the same as the original component tolerance based on the simple addition approach. If alternative 2 is chosen, similar significant increases in the component tolerances may be achieved. This formula, then, may result in a larger component tolerance with *no* change in the manufacturing processes and *no* change in the assembly tolerance.

The risk of this approach is that an assembly may fall outside the assembly tolerance. However, this probability can be calculated by expressing the component tolerances as standard deviations, calculating the standard deviation of the result, and finding the area under the normal curve outside the assembly tolerance limits. For example, if each component tolerance is equal to 3σ, then 99.73 percent of the assemblies will be within the assembly tolerance, i.e., 0.27 percent, or about 3 assemblies in 1000 taken at random, would fail to meet the assembly tolerance. The risk can be eliminated by changing components for the few assemblies that do not meet the assembly tolerance.

The tolerance formula is not restricted to outside dimensions of assemblies. Generalizing, the left side of the equation contains the dependent variable or *physical result,* while the right side of the equation contains the independent variables of *physical causes.* If the result is placed on the left and the causes on the right, the formula always has *plus* signs under the square root—even if the result is an internal dimension (such as the clearance between a shaft and hole). The causes of variation are *additive* wherever the physical result happens to fall.

The formula has been applied to a variety of mechanical and electronic products. The concept may be applied to several interacting variables in an engineering relationship. The nature of the relationship need not be additive (assembly example) or subtractive (shaft-and-hole example). The tolerance formula can be adapted to predict the variation of results that are the product and/or the division of several variables.

Assumptions of the Formula

The formula is based on several assumptions:

- The component dimensions are independent and the components are assembled randomly. This assumption is usually met in practice.

- Each component dimension should be normally distributed. Some departure from this assumption is permissible.
- The actual average for each component is equal to the nominal value stated in the specification. For the original assembly example, the actual averages for components A, B, and C must be 1.000, 0.500, and 2.000, respectively. Otherwise, the nominal value of 3.500 will not be achieved for the assembly and tolerance limits set about 3.500 will not be realistic. Thus it is important to control the average value for interacting dimensions. Consequently, process control techniques are needed using variables measurement.

Use caution if any assumption is violated. Reasonable departures from the assumptions may still permit the concept of the formula to be applied. Notice that in the example the formula resulted in the doubling of certain tolerances. This much of an increase may not even be necessary from the viewpoint of process capability.

Bender (1975) has studied these assumptions for some complex assembly cases and concluded, based on a "combination of probability and experience," that a factor of 1.5 should be included to account for the assumptions, i.e.,

$$T_{\text{result}} = 1.5\sqrt{T_A^2 + T_B^2 + T_C^2 + \cdots}$$

Graves (1997) suggests that different factors be developed for initial versus mature production, high versus low volume production, and mature versus developing technology and measurement processes.

Finally, variation simulation analysis is a technique that uses computer simulation to analyze tolerances. This technique can handle product characteristics having either normal or nonnormal distributions.

Dodson (1999) describes the use of simulation in the tolerance design of circuits; Gomer (1999) demonstrates simulation to analyze tolerances in engine design.

SUMMARY

- Complex products typically experience three periods of product life: infant mortality, constant failure rate, and wear out.
- When the failure rate is constant, the distribution of time between failures is exponential. This distribution is the basis of the exponential formula for reliability.
- If failure of any part causes failure of the system and if the reliabilities of the parts are independent, then system reliability is the product of the parts reliabilities.
- Redundancy is the existence of more than one element for accomplishing a given task.
- For time-oriented products, reliability can be predicted during design by using the exponential or Weibull distribution.
- For non-time-oriented products, reliability can be predicted as a function of stress and strength.
- Availability is the probability that a product, when used under given conditions, will perform satisfactorily when called upon.

- Specification limits must be based on the functional needs of the product and the variability of the manufacturing process. Statistical analysis quantifies process variability for the designer.
- Specification limits for interacting dimensions should recognize the probabilistic features of forming the final result.

PROBLEMS

14.1. A radar set has a MTBF of 240 h based on the exponential distribution. Suppose that a certain mission requires failure-free operation of the set for 24 h. What is the chance that the set will complete a mission without failure?

Answer: 0.91

14.2. A piece of ground support equipment for a missile has a specified MTBF of 100 h. What is the reliability for a mission time of 1 h? 10 h? 50 h? 100 h? 200 h? 300 h? Graph these answers by plotting mission time versus reliability. Assume an exponential distribution.

14.3. The average life of subassembly A is 2000 h. Data indicate that this life characteristic is exponentially distributed.

(a) What percentage of the subassemblies in the population will last at least 200 h?

(b) The average life of subassembly B is 1000 h, and the life is exponentially distributed. What percentage of the subassemblies in the population will last at least 200 hours?

(c) These subassemblies are independently manufactured and then connected in series to form the total assembly. What percentage of assemblies in the population will last at least 200 hours?

Answer: (a) 90.5 percent. *(b)* 81.9 percent. *(c)* 74.1 percent.

14.4. It is expected that the average time to repair a failure on a certain product is 4 h. Assume that repair time is exponentially distributed. What is the chance that the time for a repair will be between 3 h and 5 h?

14.5. The following table summarizes basic failure-rate data on components in an electronic subsystem:

Component	Quantity	Failure rate per hour
Silicon transistor	40	74.0×10^{-6}
Film resistor	100	3.0×10^{-6}
Paper capacitor	50	10.0×10^{-6}

Estimate the MTBF. (Assume an exponential distribution. All components are critical for subsystem success.)

Answer: 267 h.

14.6. A system consists of subsystems A, B, and C. The system is primarily used on a certain mission that lasts 8 h. The following information has been collected:

Subsystem	Required operating time during mission, h	Type of failure distribution	Reliability information
A	8	Exponential	50% of subsystems will last at least 14 h
B	3	Normal	Average life is 6 h with a standard deviation of 1.5 h
C	4	Weibull with $\beta = 1.0$	Average life is 40 h

Assuming independence of the subsystems, calculate the reliability for a mission.

14.7. A hydraulic subsystem consists of two subsystems in parallel, each having the following components and characteristics:

Components	Failures/10^6 h	Number of components
Pump	23.4	1
Quick disconnect	2.4	3
Check valve	6.1	2
Shutoff valve	7.9	1
Lines and fittings	3.13	7

The components within each subsystem are all necessary for subsystem success. The two parallel subsystems operate simultaneously, and either can perform the mission. What is the mission reliability if the mission time is 300 h? (Assume an exponential distribution.)

14.8. The following estimates, based on field experience, are available on three subsystems:

Subsystem	Percent failed at 1000 mi	Weibull β value
A	0.1	2.0
B	0.2	1.8
C	0.5	1.0

If these estimates are assumed to be applicable for similar subsystems that will be used in a new system, predict the reliability (in terms of percentage successful) at the end of 3000 mi, 5000 mi, 8000 mi, and 10,000 mi.

14.9. It is desired that a power plant be in operating condition 95 percent of the time. The average time required for repairing a failure is about 24 h. What must the MTBF be for the power plant to meet the 95 percent objective?

Answer: 456 h.

14.10. A manufacturing process runs continuously 24 h per day and 7 days a week (except for planned shutdowns). Past data indicate a 50 percent probability that the time between successive failures is 100 h or more. The average repair time for failures is 6 h. Failure times and repair times are both exponentially distributed. Calculate the availability of the process.

14.11. The following table summarizes data on components in a hydraulic system:

Component	Quality	Failure rate per hour
Relief valve	1	200×10^{-6}
Check valve	1	150×10^{-6}
Filter	1	100×10^{-6}
Cylinder	1	50×10^{-6}

Assume that all components must operate for system success and that the system is used continuously throughout the 8760 h in a year with no shutdowns except for failures. Repair time varies with the type of failure, but 50 percent of the repairs require 3 h or more. The following average cost estimates apply:

$$\text{Material cost/failure} = \$400.00$$

$$\text{Repair labor cost/hour} = \$20.00$$

Assume that failure time and repair time are exponentially distributed. Calculate the average cost of repairing failures per year.

Answer: \$2029.

14.12. Measurements were made on the bore dimension of an impeller. A sample of 20 from a pilot run production showed a mean value of 25.038 cm and a standard deviation of 0.000381 cm. All the units functioned properly, so it was decided to use the data to set specification limits for regular production.

(a) Suppose it was assumed that the sample estimates were exactly equal to the population mean and standard deviation. What specification limits should be set to include 99 percent of production?

(b) There is uncertainty that the sample and population values are equal. Based on the sample of 20, what limits should be set to be 95 percent sure of including 99 percent of production?

(c) Explain the meaning of the difference in *(a)* and *(b)*.

(d) What assumptions were necessary to determine both sets of limits?

14.13. A circuit contains three resistors in series. Past data show these data on resistance:

Resistor	Mean, Ω	Standard deviation, Ω
1	125	3
2	200	4
3	600	12

(a) What percentage of circuits would meet the specification on total resistance of 930 ± 30Ω?

(b) Ask a local distributor whether it is reasonable to assume that the resistance of a resistor is normally distributed.

14.14. A manufacturer of rotary lawn mowers received numerous complaints concerning the effort required to push its product. Studies soon found that the small clearance between the wheel bushing and shaft was the cause. Designers chose to make the clearance large enough for easy rotation of the wheel (or for a heavy grease coating) but still "tight enough" to prevent wobbling. Because of a large inventory of wheels and shafts, a decision was made to ream the bushings to a larger inside diameter (ID) and retain the shafts. The following specifications were proposed:

$$\text{Shaft diameter} = 0.800 \pm 0.002 \text{ in}$$

$$\text{New clearance} = 0.800 \pm 0.003 \text{ in}$$

$$\text{Bushing ID} = 0.800 \pm 0.001 \text{ in}$$

Production people claimed that they could not economically hold the tolerance on the bushing ID. What comment would you make about this claim?

Answer: ID tolerance of ± 0.0022 could be allowed.

14.15. An assembly consists of two parts (*A* and *B*) that mate together "end to end" to form an overall length, *C*. It is desired that the overall length, *C*, meet a specification of 3.000 $\pm$ 0.005 cm. The nominal specification on *A* is 2.000, and on *B* it is 1.000. The manufacturing process for *B* has much more variability than the process for *A*. Specifically, the tolerance for part *B* should be twice as large as for part *A*. Assemblies are to be made at random, and parts *A* and *B* are independently manufactured. Assuming that we want only a small risk of not meeting the specification on *C*, what tolerances should be set on *A* and *B*?

14.16. A canning factory decided to set tolerance limits for filling cans of a new product by sampling a pilot run of 30 cans. The results of that run yielded an average of 446 g with a standard deviation of 1.25 g. What tolerance limits would be 95 percent certain of including 99 percent of production? The label on the can states that it contains 453 g of the product. How many grams of product should the average can contain if the company is to be 95 percent certain that 99 percent of production contains at least 453 g?

REFERENCES

Anand, K. N. (1996). "The Role of Statistics in Determining Product and Part Specifications: A Few Indian Experiences," *Quality Engineering,* vol. 9, no. 2, pp. 187–193.

ANSI/IEC/ASQ D60300-3-1-1997 (1997). *Dependability management Part 3: application guide—Section 1: analysis techniques for dependability: Guide on methodology,* ASQ, Milwaukee.

Bender, A. (1975). "Statistical Tolerancing as It Relates to Quality Control and the Designer," *Automotive Division Newsletter of ASQC,* April, p. 12.

Dodson, B. (1999). "Reliability Modeling with Spreadsheets," *Proceedings of the Annual Quality Congress,* ASQ, Milwaukee, pp. 575–585.

Endres, A. (1997). *Improving R&D Performance the Juran Way,* John Wiley & Sons, New York.

Garrick, J. B. and R. J. Mulvihill (1974). "Reliability and Maintainability of Mechanized Bulk Mail Systems," *Proceedings of Annual Reliability and Maintainability Symposium,* Institute of Electrical and Electronics Engineers, New York.

Gomer, P. (1998). "Design for Tolerance of Dynamic Mechanical Assemblies," *Annual Quality Congress Proceedings,* ASQ, Milwaukee, pp. 490–500.

Graves, S. B. (1997). "How to Reduce Costs Using a Tolerance Analysis Formula Tailored to Your Organization," *Report No. 157,* Center for Quality and Productivity Improvement, University of Wisconsin, Madison.

Ireson, W. G., C. F. Coombs Jr., and R. Y. Moss (1996). *Handbook of Reliability Engineering and Management,* 2nd ed., McGraw-Hill, New York.

Jones, J. and J. Hayes (1999). "A Comparison of Electronic Reliability Prediction Models," *IEEE Transactions on Reliability,* vol. 48, no. 2, pp. 127–134.

Lusser, R. (1958). *Reliability through Safety Margins,* United States Army Ordnance Missile Command, Redstone Arsenal, AL.

O'Connor, P. D. T. (1995). *Practical Reliability Engineering,* 3rd ed. rev., John Wiley & Sons, New York.

Pyzdek, T. (1994). "Service Systems Reliability Engineering," *Proceedings of the Annual Quality Congress,* ASQ, Milwaukee, pp. 174–187.

Ross, P. J. (1996). *Taguchi Techniques for Quality Engineering,* McGraw-Hill, New York.

Turmel, J. and L. Gartz (1997). "Designing in Quality Improvement: A Systematic Approach to Designing for Six Sigma," *Proceedings of the Annual Quality Congress,* ASQ, Milwaukee, pp. 391–398.

SUPPLEMENTARY READING

Reliability, maintainability, and availability quantification: *JQH5,* Sections 19 and 48.
McLinn, J. A. (1997). *Weibull Analysis Primer,* 3rd ed., Williams Enterprises, Cool, CA.
Raheja, D. G. (1991). *Assurance Technologies,* McGraw-Hill, New York.
Statistical tolerance limits: *JQH5,* pp. 44.47–44.54.

WEBSITES

ASQ reliability division: www.asq-rd.org
ASQ statistics division: www.asq.org/about/divisions/stats
Government Industry Data Exchange Program (GIDEP): www.gidep.corona.navy.mil/data_inf/opscntr.htm
Reliability Analysis Center (RAC): www.rac.iitri.org

15

SUPPLY CHAIN MANAGEMENT

15.1
SUPPLIER RELATIONS—A REVOLUTION

This step on the spiral of quality concerns the purchase of goods or services from suppliers, or vendors.

For many companies, purchases account for 60 percent of the sales dollar and are the source of half of the quality problems. Poor quality of supplier items results in extra costs for the purchaser; e.g., for one appliance manufacturer, 75 percent of all warranty claims were traced to purchased components for the appliances.

Current emphasis on inventory reduction provides a further focus on quality. Under the just-in-time inventory concept, goods are received from suppliers only in the quantity and at the time that they are needed for production. The buyer stocks no inventories. If a portion of the purchased product is defective, production at the buyer's plant is disrupted because of the lack of a backup inventory. With conventional purchasing, supplier quality problems can be hidden by excess inventory; with the just-in-time concept, purchased product must meet quality requirements.

The interdependence of buyers and suppliers has increased dramatically. Sometimes the interdependence takes the form of integrated facilities, e.g., a can manufacturer locates next door to a brewery; sometimes technological skills are involved, e.g., an automobile manufacturer asks a supplier to propose a design for a purchased item. The supplier becomes an extension of the buyer's organization—a virtual department.

These circumstances have lead to a revolution in the relationship between buyers and suppliers. In the past the parties were often adversaries; some purchasers viewed suppliers as potential criminals who might try to sneak some defective product past the purchaser's incoming inspection. Today the key phrase is *partnership alliance,* i.e., working closely together for the mutual benefit of both parties.

This new viewpoint of supplier relations requires changing the purchasing process from a traditional view to a strategic view. An overview of some changes is shown in Table 15.1. For elaboration, see *JQH5,* page 21.6.

Part of the revolution in supplier relations is the expansion of the traditional supplier concept to the broader supply chain concept (see Figure 15.1). Donovan and Maresca (in *JQH5*) define the supply chain as the tasks, activities, events, processes, and interactions undertaken by *all* suppliers and *all* end users in the development, procurement, production, delivery, and consumption of a specific good or service. Note that this definition includes end users, prime suppliers or distributors, and multiple tiers of suppliers to the prime manufacturing or service organizations. Supply chain management is reserved for items that are of strategic importance to an organization.

The purchasing function has the primary role of managing the supply chain to achieve high quality and value throughout the supply chain. Admittedly, this ideal is lofty, but it highlights a new focus—*from managing purchasing transactions and troubleshooting to managing processes and supplier relationships.* Under supply chain management, mechanisms must be put in place to assure adequate linkages among parties in the supply chain. Such mechanisms include clear contractual requirements and continuous feedback and communication. For further discussion of the supply chain concept see *JQH5,* pages 21.4–21.9. Of course, managing supply chains is difficult. Fisher (1997) describes some of these difficulties and suggests a framework for supply chains based on the nature of the product demand and whether the products are primarily functional or primarily innovative. Supply chains also apply to the service sector. For an example involving physicians in a hospital and health care system, see Zimmerli (1996). This example also illustrates the use of internal customer surveys for the major processes of distribution, purchasing, and sterile processing.

TABLE 15.1
Traditional versus strategic view of the purchasing process

Aspect in the purchasing process	Traditional view	Strategic view
Supplier relationship	Adversarial, competitive, distrusting	Cooperative, partnership, based on trust
Length of relationship	Short term	Long term; indefinite
Quality assurance	Inspection upon receipt	No incoming inspection necessary
Supplier base	Many suppliers, managed in aggregate	Few suppliers, carefully selected and managed
Purchasing business plans	Independent of end-user organization business plans	Integrated with end-user organization business plans
Focus of purchasing decisions	Price	Total cost of ownership

Source: Adapted from *JQH5,* p. 21.6.

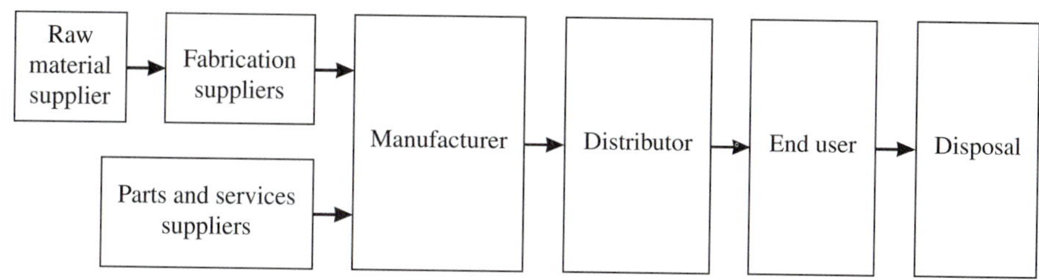

FIGURE 15.1
Elements of a supply chain. (*Source: JQH5, p. 21.4.*)

Much has been written about the importance of developing trust to replace the adversarial relationships of the past. Significant progress has been made, but in practice, suppliers (both large and small) still report pockets of arrogance exhibited by some purchasers.

15.2 SCOPE OF ACTIVITIES FOR SUPPLIER QUALITY

A purchasing system includes three key activities: specification of requirements, selection of a supplier, and supply chain management. The overall quality objective is to meet the needs of the purchaser (and the ultimate user) with a minimum of incoming inspection or later corrective action; this objective in turn leads to minimizing overall cost.

To achieve this quality objective, certain primary activities must be identified and responsibilities assigned. Table 15.2 shows a typical list of responsibilities as assigned in one company. These activities are discussed in this chapter. Further elaboration is provided in *JQH5,* Section 21.

The responsibility matrix in Table 15.2 shows the quality department as having the principal responsibility for many supplier quality activities. Under an alternative policy, the purchasing department has the principal responsibility for quality while others (e.g., product development and quality) have collateral responsibility. Such a shift in responsibility places a stronger focus on quality in setting priorities on delivery schedules, price, and quality. To meet this responsibility, most purchasing departments would need to supplement their technical capabilities. Some organizations have met that need by transferring technical specialists into the purchasing department.

To reflect a broad view of suppliers and the supply chain concept, some organizations are shifting from a function-based organization for purchasing transactions to a process-based organization for managing the supply chain (see *JQH5,* page 21.10). The process-based organization (see Chapter 6, "Process Management") uses a cross-functional team and process owner to focus on the total cost of ownership (rather than initial price of a purchased item), identify opportunities for increased value, and achieve a competitive advantage.

TABLE 15.2
Responsibility matrix—supplier relations

	Participating departments		
Activity	**Product development**	**Purchasing**	**Quality**
Defining product and program quality requirements	××		×
Evaluating alternative suppliers	×	×	××
Selecting suppliers		××	
Conducting joint quality planning	×		××
Cooperating with the supplier during the execution of the contract	×	×	××
Obtaining proof of conformance to requirements	×		××
Certifying qualified suppliers	×	×	××
Conducting quality improvement programs as required	×	×	××
Creating and utilizing supplier quality ratings		××	×

Note: ××, principal responsibility; ×, collateral responsibility.

Next we examine how quality relates to three key activities of supply management: specification of requirements, selection of suppliers, and management of the supply chain.

15.3 SPECIFICATION OF QUALITY REQUIREMENTS FOR SUPPLIERS

Goals and requirements for suppliers must be aligned with those for each link in the supply chain, particularly the end user and the purchasing organization. These goals and requirements include quality parameters and general business issues.

For modern products quality planning starts before a contract is signed. Such planning must recognize two issues:

1. The buyer must transmit to the supplier a full understanding of the use to be made of the product. Communicating usage requirements can be difficult even for a simple product.
2. The buyer must obtain information to be sure that the supplier can provide a product that meets all fitness-for-use requirements.

The complexity of many modern products makes it difficult to communicate usage needs to a supplier in a specification. Not only are the field usage conditions of a complex product sometimes poorly known, but the internal environments surrounding a particular component may not be known until the complete product is designed and tested. For example, specifying accurate temperature and vibration requirements to a supplier of an electric component may not be feasible until the complete system is developed. Such cases require, at the least, continuous cooperation between supplier and buyer. In special cases it may be necessary to award separate development and production contracts to discover how to finalize requirements.

Circumstances may require two kinds of specifications:

1. Specifications defining the product requirements.
2. Specifications defining the quality-related activities expected of the supplier, i.e., the supplier's quality system.

Definition of Numerical Quality and Reliability Requirements for Lots

Beyond the quality and reliability requirements imposed on individual units or products, there is usually a need for added numerical criteria to judge conformance of *lots* of products.

These criteria are typically needed in acceptance sampling procedures (see Chapter 19), which makes it possible to accept or reject an entire lot of product based on the inspection and test result of a random sample from the lot. The application of sampling procedures is facilitated if lot quality requirements are defined in numerical terms. Examples of numerical indexes are shown in Table 15.3.

The selection of numerical values for these criteria depends on several factors and also on probability considerations. These matters are discussed in Chapter 19. These

TABLE 15.3
Forms of numerical sampling criteria

Quality index	Meaning	Typical values, %	Common misinterpretation
Parts per million (PPM)	Number of defects per million items	5–1000	—
Acceptable quality level (AQL)*	Percentage defective that has a high probability (say ≥0.90) of being accepted by the sampling plan	0.01–10.0	All accepted lots are at least as good as the AQL; all rejected lots are worse than the AQL
Lot tolerance percentage defective (LTPD)	Percentage defective that has a low probability (say ≤0.10) of being accepted by the sampling plan	0.5–10.0	All lots better than the LTPD will be accepted; all lots worse than the LTPD will be rejected
Average outgoing quality limit (AOQL)	Worse average percentage defective over many lots after sampling inspection has been performed and rejected lots 100% inspected	0.1–10.0	All accepted lots are at least as good as the AOQL; all rejected lots are worse than the AOQL

*Some sampling tables and other sources define AQL as the maximum percentage defective considered satisfactory as a process average.

criteria are also a means of indexing sampling plans developed from statistical concepts. Unfortunately, many suppliers do not understand the statistical concepts and make incorrect interpretations of the quality-level requirement and also the results of sampling inspection (see Table 15.3). Also, these criteria can be a source of confusion in product liability discussions. Suppliers must understand that *all* product submitted is expected to meet specifications.

For complex and/or time-oriented products, numerical reliability requirements can be defined in supplier purchasing documents. Sometimes such requirements are stated in terms of mean time between failures. Numerical reliability requirements can help to clarify what a customer means by "high reliability."

EXAMPLE 15.1. A capacitor manufacturer requested bids on a unit of manufacturing equipment that was to perform several manufacturing operations. Reliability of the equipment was important to maintaining production schedules, so a numerical requirement on "mean time between jams" (MTBJ) was specified to prospective bidders. (Previously, reliability had not been treated quantitatively. Equipment manufacturers had always promised high reliability, but results had been disappointing.) After several rounds of discussion with bidders the manufacturer concluded that the desired level of reliability was unrealistic if the machine were to perform several operations. The capacitor manufacturer finally decided to revise the requirement for several operations and thereby reduce the complexity of the equipment. The effort to specify a numerical requirement in the procurement document forced a clear understanding of reliability. Suppliers can also be required to demonstrate, by test, specified levels of reliability.

Definition of the Supplier Quality System

The second type of specification is a departure from the traditional practice of not telling a supplier how to run his or her plant. Defining required activities within a supplier's plant is sometimes necessary to ensure that a supplier has the expertise to conduct the full program needed to result in a satisfactory product. For some products, government regulations require that a buyer impose certain processing requirements (e.g., sanitary conditions for manufacturing pharmaceutical products) on suppliers. For other products, such as a complex mechanical or electronic subsystem, the overall system requirements may result in a need for a supplier to meet a numerical reliability or maintainability requirement and to conduct certain activities to ensure that such requirements are met (see Chapter 13). For still other products, suppliers are required to use statistical process control techniques on selected product characteristics or process parameters. Documents such as the ISO 9000 series and QS 9000, which define the elements of quality programs, can be cited as requirements in a contract with a supplier.

EXAMPLE 15.2. Several suppliers were asked to submit bids on a battery needed in a space program. They were given a numerical reliability goal and asked to include in their bid proposal a description of the reliability activities that would be conducted to help meet the goal. Most of the prospective suppliers included a reliability program consisting of appropriate reliability activities for a battery. However, one supplier apparently had no expertise in formal reliability methodology and submitted a surprising write-up. That supplier made a word-for-word copy of a reliability program write-up previously published for a missile system (the word *battery* was substituted for *missile*). This led to a suspicion, later confirmed, that the supplier knew little about reliability programs.

For complex products for which a supplier is asked to design and manufacture a product, the supplier can be required to include in the proposal a preliminary reliability prediction; a failure mode, effect, and criticality analysis; a reliability test plan; or other reliability analyses (see Chapter 13). The supplier's response not only provides some assurance on the design concept but also shows that the supplier has the reliability expertise to conduct the program and has included the funds and schedule time in the proposal.

We proceed next to the selection of suppliers.

15.4 SUPPLIER SELECTION; OUTSOURCING

Should we make or buy? This decision requires an analysis of factors such as the skills and facilities needed, available internal capacity, ability to meet delivery schedules, expected costs of making or buying, and other matters. This question brings us to the issue of outsourcing.

Outsourcing

Outsourcing is the process of subcontracting to a supplier external to the organization an activity that is currently conducted in house. Outsourcing is undertaken to reduce costs (the primary impetus), reduce cycle time, or improve quality. Estimates suggest that at least 85 percent of major corporations now outsource at least some activities. A trade association, the Outsourcing Institute, now exists.

But what activities should be outsourced? One principle holds that outsourcing should be confined to activities that are required but do not provide a competitive advantage, e.g., security, facility maintenance, administration of health benefits. Activities that are strategic and involve core competencies should not be outsourced. In practice, some organizations outsource significant (core) functional activities such as customer service, marketing, product design, and information technology.

Many important business issues enter into decisions about outsourcing. Bettis et al. (1992) offer important cautions about outsourcing core activities such as design and manufacturing. The more that outsourcing results in a supplier obtaining technical knowledge and market knowledge, the higher the risks to the company doing the outsourcing.

Outsourcing reduces internal costs by reducing personnel because the outsourcer (supplier) companies have the technology and knowledge to perform certain tasks more efficiently than some companies can do internally. But there can be a serious impact on product quality if the supplier does not assign a high priority to quality. Outsourcing can also undermine employee morale and loyalty by creating fear that other activities will also be outsourced, resulting in a further loss of jobs. Some forward-looking companies like Eastman Chemical follow a policy that if an activity is outsourced no one will lose a job—people are retrained to take the positions of those who are normally retiring. Outsourcing also assumes that a capable supplier can be found and that adequate monitoring of the contract will assure high quality. Sometimes these issues are glossed over in the zeal to reduce costs, and the result can be significant quality problems on the purchased items. Peterson (1998) analyzes some potential shortcomings when dealing with contract manufacturers. Bossert (1994) provides a checklist of 11 elements (e.g., inspection instructions, sufficient manufacturing controls) to compare contract manufacturing services.

Once these skills are lost through outsourcing, it is difficult to reverse the process if later events require that the activity be brought back into an organization. This situation could be devastating for activities such as product design and selected operations activities.

Clearly, outsourcing can be a sensible, viable business decision—after all, the subcontracting of selected manufacturing activities has been a part of manufacturing history. But the desire for cost reduction may be taking outsourcing too far. Perhaps we should first study the entire activity to be outsourced as a quality improvement project using the road map provided in Chapter 3, "Quality Improvement and Cost Reduction." Thus, suppose customer service is a candidate for outsourcing. The *process* of customer service would be studied for both effectiveness and efficiency, and the necessary internal changes would be made. The result (quality and costs) could then be compared to those of outside suppliers. Some organizations even set up the activity under question as a separate profit center (to compete against outside suppliers) as a means to spur internal improvement—the threat of job loss is a powerful spur.

To sum up on a sensitive issue: Outsourcing can provide superior quality and lower costs for an activity that a company cannot easily develop and maintain on its own, e.g., information technology. Outsourcing can also enable a company to focus resources on the core competencies that are important for competitive advantage, e.g., product design, operations, marketing. But these core activities vary by organization. The author believes that the core competencies must be carefully identified within each organization, and once identified they should be performed internally and not be outsourced.

Multiple Suppliers versus Single Source

Multiple sources of supply have advantages: Competition can result in better quality, lower costs, better service, and minimum disruption of supply due to strikes or other catastrophes.

A single source of supply also has advantages: The size of the contract given to a single source will be larger than with multiple sources, and the supplier will attach more significance to the contract. With a single source, communications are simplified and more time is available for working closely with the supplier. The most dramatic examples of single sources are multidivisional companies in which some divisions are suppliers to others.

A clear trend has emerged: Organizations are significantly reducing the number of multiple suppliers. Since about 1980 reductions of 50 to 70 percent in the supplier base have become common. This trend does *not* necessarily mean that businesses are going to single source for all purchases; it *does* mean a single source for some purchases and fewer multiple suppliers for other purchases. Working with a smaller number of suppliers helps to achieve useful partnerships by providing the time and skills necessary to facilitate in-depth cooperation. The forms of cooperation are discussed later in this chapter.

Whether a single source or multiple suppliers, selection must be based on the reputation of the supplier, qualification tests of the supplier's design, survey of the supplier's manufacturing facility, and information from data banks and other sources on supplier quality.

15.5 ASSESSMENT OF SUPPLIER CAPABILITY

Evaluating supplier quality capability involves one or both of the following actions:

1. Qualifying the supplier's design through the evaluation of product samples.
2. Qualifying the supplier's capability to meet quality requirements on production lots, i.e., the supplier's quality system.

Qualifying the Supplier's Design

In some cases the supplier is asked to create a new design to meet the functions desired by the purchaser. In these cases the supplier makes samples based on the proposed design. (Such samples are often made in an engineering model shop because a manufacturing process for the new design has not yet been created.) The samples are tested (the "qualification test") either by the purchaser or by the supplier, who then submits the results to the purchaser. Qualification test results are often rejected. Two reasons are common: (1) The test results show that the design does not provide the product functions desired, or (2) the test procedure is not adequate to evaluate the performance of the product. Such rejections (and ensuing delays in shipments) can be prevented by starting with a rigorous definition of product requirements and by requiring an approval of the test procedure before the tests commence.

Qualification test results do show whether the supplier has created a design that meets the performance requirements; such test results do *not* show whether the supplier is capable of manufacturing the item under production conditions.

A supplier may be required to submit a failure mode, effects, and criticality analysis as evidence of analyses to prevent product or process failures. Increasingly, this requirement is part of the six-sigma approach to quality during design.

Qualifying the Supplier Manufacturing Process

Evaluation of the supplier's manufacturing capability can be done by reviewing past data on similar products, performing process capability analysis, or evaluating the supplier's quality system through a quality survey.

Data showing the supplier's past performance on the same or similar products may be available within the local buyer's organization, other divisions of the same corporation, government data banks, or industry data banks.

With the process capability analysis approach, data on key product characteristics are collected from the process and evaluated by using statistical indexes for process capability (see Chapter 18). All evaluation occurs before the supplier is authorized to proceed with full production. Typically, process capability analysis of a supplier process is reserved for significant product characteristics, safety-related items, or products requiring compliance with government regulations.

The third approach, a quality survey, is explained below.

When all three approaches can be used, the information collected can provide a sound prediction of supplier capability.

Supplier Quality Survey (Supplier Quality Evaluation)

A supplier quality survey is an evaluation of the ability of a supplier's quality system to meet quality requirements on production lots, i.e., to prevent, identify, and remove any product that does not meet requirements. The results of the survey are used in the supplier selection process, or, if the supplier has already been chosen, the survey alerts the purchaser to areas where the supplier may need help in meeting requirements. The survey can vary from a simple questionnaire mailed to the supplier to a visit to the supplier's facility.

The questionnaire poses explicit questions such as these submitted to suppliers of a manufacturer of medical devices:

- Has your company received the quality requirements on the product and agreed that they can be fully met?
- Are your final inspection results documented?
- Do you agree to provide the purchaser with advance notice of any changes in your product design?
- What protective garments do your employees wear to reduce product contamination?
- Describe the air-filtration system in your manufacturing areas.

The more formal quality survey consists of a visit to the supplier's facility by a team of observers from departments such as quality, engineering, manufacturing, and purchasing. Such a visit may be part of a broader survey of the supplier covering financial, managerial, and technological competence. Depending on the product involved, the activities included in the quality portion of the survey can be chosen from the following list:

- *Management:* philosophy, quality policies, organization structure, indoctrination, commitment to quality.
- *Design:* organization, systems in use, caliber of specifications, orientation to modern techniques, attention to reliability, engineering change control, development laboratories.
- *Manufacture:* physical facilities, maintenance, special processes, process capability, production capacity, caliber of planning, lot identification and traceability.
- *Purchasing:* specifications, supplier relations, procedures.
- *Quality:* organization structure, availability of quality and reliability engineers, quality planning (materials, in-process, finished goods, packing, storage, shipping, usage, field service), audit of adherence to plan.
- *Inspection and test:* laboratories, special tests, instruments, measurement control.
- *Quality coordination:* organization for coordination, order analysis, control over subcontractors, quality cost analysis, corrective action loop, disposition of nonconforming product.
- *Data systems:* facilities, procedures, effective use reports.
- *Personnel:* indoctrination, training motivation.
- *Quality results:* performance attained, self-use of product, prestigious customers, prestigious subcontractors.

Following the survey, the team reports its findings. These consist of (1) some objective findings as to the supplier's facilities (or lack of facilities), (2) subjective judgments on the effectiveness of the supplier's operations, (3) a further judgment on the extent of assistance needed by the supplier, (4) a highly subjective prediction as to whether the supplier will deliver a good product if awarded a contract.

The quality survey is a technique for evaluating the supplier's ability to meet quality requirements on production lots. The evaluation of various quality activities can be quantified by a scoring system.

A scoring system that includes importance weights for activities is illustrated in Table 15.4. This system is used by a manufacturer of electronic assemblies. In this case the importance weights (W) vary from 1 to 4 and must total to 25 for each of the three areas surveyed. The weights show the relative importance of the various activities in the overall index. The actual ratings (R) of the activities observed are assigned as follows:

10: The specific activity is satisfactory in every respect (or does not apply).
8: The activity meets minimum requirements but improvements could be made.
0: The activity is unsatisfactory.

Supplier quality surveys have both merits and limitations. On the positive side, such surveys can identify important weaknesses such as a lack of special test equipment or an absence of essential training programs. Further, the survey opens up lines of communication and can stimulate action on quality by the supplier's upper management. On the negative side, surveys that emphasize the supplier's organization, procedures, and documentation have had only limited success in predicting future performance of the product.

TABLE 15.4
Scoring of a supplier quality survey

	Receiving inspection			Manufacturing			Final inspection		
Activity	*R*	*W*	*R* × *W*	*R*	*W*	*R* × *W*	*R*	*W*	*R* × *W*
Quality management	8	3	24	8	3	24	8	3	24
Quality planning	8	4	32	8	4	32	10	4	40
Inspection equipment	10	3	30	10	3	30	10	3	30
Calibration	0	3	0	10	3	30	0	3	0
Drawing control	0	3	0	10	2	20	10	2	20
Corrective action	10	3	30	8	3	24	8	3	24
Handling rejects	10	2	20	8	2	16	10	3	30
Storage and shipping	10	1	10	10	1	10	10	1	10
Environment	8	1	8	8	1	8	8	1	8
Personnel experience	10	2	20	10	3	30	10	2	20
Area total			174			224			206

Note: R, rating; *W*, weight.
Interpretation of area totals:
Fully approved: Each of the three area totals is 250.
Approved: None of the three area totals is less than 200.
Conditionally approved: No single total is less than 180.
Unapproved: One or more of the area totals is less than 180.

Suppliers in some industries have been burdened with quality surveys from many purchasers. These repeat surveys (called "multiple assessment") are time-consuming for suppliers. In another approach, a standard specification of the elements of a quality system (e.g., the ISO 9000 series) is created and assessors are trained to use the specification to evaluate supplier capability. A list of suppliers that have passed the assessment is published, and other purchasers are encouraged to use these results instead of making their own assessment of a supplier. The assessors are independent of the supplier or purchasing organization—thus the term *third-party assessment.* In some countries a national standards organization acts in this role.

Bossert (1998) describes how supplier evaluation can start with an ISO 9000 assessment and then be supplemented by a quality survey including a supplier visit. The supplier visit covers contract and specification review, process audit, process risk analysis (using a failure mode and effects analysis), and statistical techniques (including the measurement process).

On to the third phase of supplier relations—management of the supply chain through quality planning, quality control, and quality improvement.

15.6 SUPPLY CHAIN QUALITY PLANNING

Donovan and Maresca (in *JQH5*) suggest the following steps for a purchasing process that involves the purchasing organization, suppliers, and end users. This approach is sometimes called a "sourcing process."

1. Document the organization's historic, current, and future procurement activity.
2. Identify a commodity from the procurement activity that represents both high expenditure and high criticality to the business.
3. For this commodity, assemble a cross-functional team.
4. Determine the sourcing needs of the customer through data collection, survey, and other activities.
5. Analyze the supply industry's structure, capabilities, and trends.
6. Analyze the cost components of the commodity's total cost of ownership.
7. Translate the customer needs into a sourcing process that will satisfy the customer and provide the opportunity to manage and optimize the total cost of ownership.
8. Obtain management endorsement to transfer the sourcing strategy into operation. Implement it. For elaboration, see *JQH5,* pages 21.18–21.20.

In doing the detailed quality planning with suppliers, three approaches emerge:

- *Inspection.* The focus is on various forms of product inspection.
- *Prevention.* The premise is that quality must be built in by the supplier, with the purchaser's help. But there is still an arm's-length relationship between purchaser and supplier.
- *Partnership.* Suppliers are offered the financial security of a long-term relationship in exchange for a supplier's commitment to quality that includes a strong teamwork relationship with the buyer.

Partnership—involving not just quality but also other business issues—is clearly the wave of the future. Teamwork actions vary greatly, e.g., training a supplier's staff in quality techniques, including suppliers in a design review meeting to gain ideas on how supplier parts can best be used, sharing confidential sales projections with suppliers to assist in supplier production scheduling. Such partnerships often lead to formation of supplier quality councils, which help provide new approaches for the benefits of both the buyer and suppliers. Various opportunities for teamwork are discussed below. But such teamwork depends on truly open communication between buyers and suppliers.

Such cooperation can best be achieved by setting up multiple channels of communication: Designers must communicate directly with designers, quality specialists with quality specialists, etc. These multiple channels are a drastic departure from the single channel, which is the method in common use for purchase of traditional products. In the single-channel approach, a specialist in the buyer's organization must work through the purchasing agent, who in turn speaks with the salesperson in the supplier's organization, to obtain information. Of course, the concept of multiple channels seems sensible, but wouldn't it be useful to determine whether multiple channels yield better results on quality? Carter and Miller (1989) did just that.

EXAMPLE 15.3. In an innovative research study, they compared quality levels for two communication structures: serial (single channel) and parallel (multiple channel). At a manufacturer of mechanical seals, one section of a plant followed the serial communication concept while a second area used parallel communication. Over a 19-month period, the section using parallel communication improved the average percentage of items rejected from 30.3 percent to 15.0 percent, a statistically significant difference; the section with serial communication had no such improvement—in fact, its rejection percentage increased slightly.

We next address how partnership can be achieved through joint economic planning, joint technological planning, and cooperation during contract execution.

Joint Economic Planning

The economic aspects of joint quality planning concentrate on two major approaches:

- *Value rather than conformance to specification.* The technique used is to analyze the value of what is being bought and to try to effect an improvement. The organized approach is known as value engineering (see Section 13.9, "Cost and Product Performance"). Applied to supplier quality relations, value engineering looks for excessive costs due to (1) overspecification for the use to which the product will be put, e.g., a special product ordered when a standard product would do, (2) emphasis on original price rather than on cost of use over the life of the product, and (3) emphasis on conformance to specification, not fitness for use. Suppliers are encouraged to make recommendations on design or other requirements that will improve or maintain quality at a lower cost.
- *Total cost of ownership.* To the purchase price, the buyer must add a whole array of quality-related costs: incoming inspection, materials review, production delays, downtime, extra inventories, etc. However, the supplier also has a set of costs it is trying to optimize. The buyer should put together the data needed to understand the life cycle costs or the cost of use and then press for a result that will optimize these.

EXAMPLE 15.4. A heavy-equipment manufacturer bought 11,000 castings per year from several suppliers. It was decided to calculate the total cost of the purchased casting as the original purchase price plus incoming inspection costs plus the costs of rejections detected later in assembly. The unit purchase price on a contract given to the lowest bidder was $19. The inspection and rejection costs amounted to an additional $2.11. The variation among bid prices was $2. Thus the lowest bid does not always result in the lowest total cost.

Joint Technological Planning

The standard elements of such planning include

1. Agreement on the meaning of performance requirements in the specifications.
2. Quantification of quality, reliability, and maintainability requirements.

EXAMPLE 15.5. A supplier was given a contract to provide an air-conditioning system with a mean time between failures of at least 2000 hours. As part of joint planning, the supplier was required to submit a detailed reliability program early in the design phase. The program write-up was submitted and included a provision to impose the same 2000-hour requirement on each supplier of parts for the system. This revealed a complete lack of understanding by the supplier of the multiplication rule (see Section 14.4, "The Relationship between Part and System Reliability").

3. Definition of reliability and maintainability tasks to be conducted by the supplier.
4. Preparation of a process control plan for the manufacturing process. The supplier can be asked to submit a plan summarizing the specific activities which will be conducted during the manufacture of the product. Typically, the plan must include statistical process control techniques to prevent defects by detecting problems early.
5. Definition of special tasks required of the supplier. These may include activities to ensure that good manufacturing practices are met, special analyses are prepared for critical items, etc.
6. Seriousness classification of defects to help the supplier understand where to concentrate efforts.
7. Establishment of sensory standards for qualities that require use of the human being as an instrument.

EXAMPLE 15.6. The federal government was faced with the problem of defining the limits of color on a military uniform. It was finally decided to prepare physical samples of the lightest and darkest acceptable colors. Such standards were then sent out with the provision that the standards would be replaced periodically because of color fading.

8. Standardization of test methods and test conditions between supplier and buyer to ensure their compatibility.

EXAMPLE 15.7. A carpet manufacturer repeatedly complained to a yarn supplier about yarn weight. The supplier visited the customer to verify the test methods. Their test methods were alike. Next an impartial testing lab verified the tests at the carpet plant. Finally, the mystery was solved. The supplier was spinning (and measuring) the yarn at bone-dry conditions, but the carpet manufacturer measured at standard conditions. During this period, $62,000 more was spent for yarn than if it had been purchased at standard weight.

9. Establishment of sampling plans and other criteria relative to inspection and test activity. From the supplier's viewpoint, the plan should accept lots having the usual process average. For the buyer the critical factor is the amount of damage caused by one defect getting through the sampling screen. Balancing the cost of sorting versus sampling can be a useful input in designing a sampling plan (see Chapter 19). In addition to sampling criteria, error of measurement can also be a problem (see Section 19.8, "Errors of Measurement").
10. Establishment of quality levels. In the past suppliers were often given "acceptable quality levels" (AQL). The AQL value was just one point on the "operating characteristic" curve that described the risks associated with sampling plans. A typical AQL value might be 2.0 percent. Many suppliers interpreted this to mean that product which included 2 percent defective was acceptable. It is best to make clear to the supplier through the contract that *all* product submitted is expected to meet specifications and that any nonconforming product may be returned for replacement. In many industries the unit of measurement is defects per million (DPM).
11. Establishment of a system of lot identification and traceability. This concept has always been present in some degree, e.g., heat numbers of steel, lot numbers of pharmaceutical products. More recently, with intensified attention to product reliability, this procedure is more acutely needed to simplify the localization of trouble, to reduce the volume of product recall, and to fix responsibility. These traceability

systems, while demanding some extra effort to preserve the order of manufacture and identify the product, make greater precision in sampling possible.

12. Establishment of a system of timely response to alarm signals resulting from defects. Under many contracts, the buyer and supplier are yoked to a common timetable for completion of the final product. Usually, a separate department (e.g., materials management) presides over major aspects of scheduling. However, upper management properly looks to the people associated with the quality function to set up alarm signals to detect quality failures and to act positively on these signals to avoid deterioration, whether in quality, cost, or delivery.

Such depth of joint technological planning bears no resemblance to the old approach of sending a supplier a blueprint with a fixed design and a schedule.

15.7 SUPPLY CHAIN QUALITY CONTROL

Donovan and Maresca (*JQH5,* Section 21) suggest these steps for successful supplier control:

1. Create a cross-functional team.
2. Determine critical performance metrics.
3. Determine minimum standards of performance.
4. Reduce the supplier base to those able to meet minimum performance requirements.
5. Assess supplier performance:
   a. Supplier quality systems assessment.
   b. Supplier business management.
   c. Supplier product fitness for use.

For elaboration, see *JQH5,* pages 21.20–21.23.

The detailed quality control activities focus on cooperation during contract execution, supplier certification, supplier rating, and quality measurement for supplier relations. These activities emphasize continuous feedback to suppliers.

Cooperation during Contract Execution

This cooperation usually concentrates on the following activities.

Evaluation of initial samples of product

Under many circumstances the supplier must submit test results of a small initial sample produced from production tooling and a sample from the first production shipment before the full shipment is made. The latter evaluation can be accomplished by having a buyer's representative visit the supplier's plant and observe the inspection of a random sample selected from the first production lot. A review can also be made of process capability of process control type data from that lot.

Design information and changes

Design changes may take place at the initiative of either the buyer or the supplier. Either way, the supplier should be treated like an in-house department when developing procedures for processing design changes. This need is especially acute for modern products, for which design changes can affect products, processes, tools, instruments, stored materials, procedures, etc. Some of these effects are obvious, but others are subtle, requiring a complete analysis to identify the effects. Failure to provide adequate design change information to suppliers has been a distinct obstacle to good supplier relations.

Surveillance of supplier quality

Quality surveillance is the continuing monitoring and verification of the status of procedures, methods, conditions, processes, products, services, and analysis of records in relation to stated references to ensure that specified requirements for quality are being met (ISO 8402). Surveillance by the buyer can take several forms: inspecting the product, meeting with suppliers to review quality status, auditing elements of the supplier quality program, monitoring of the manufacturing practices of the supplier, reviewing statistical process control data, and witnessing specific operations or tests. Major or critical contracts require on-site presence or repeat visits.

Evaluating delivered product

Evaluation of supplier product can be achieved by using one of the methods listed in Table 15.5.

TABLE 15.5
Methods of evaluating supplier product

Method	Approach	Application
100 percent inspection	Every item in a lot is evaluated for all or some of the characteristics in the specification	Critical items where the cost of inspection is justified by the cost of risk of defectives; also used to establish quality level of new suppliers
Sampling inspection	A sample of each lot is evaluated by a predefined sampling plan and a decision is made to accept or reject lot	Important items where the supplier has established an adequate quality record by the prior history of lots submitted
Identifying inspection	The product is examined to ensure that the supplier sent the correct product; no inspection of characteristics is made	Items of less importance where the reliability of the supplier laboratory has been established in addition to the quality level of the product
No inspection	The lot is sent directly to a storeroom or processing department	For purchase of standard materials or goods not used in the product, e.g., office supplies
Using supplier data (supplier certification)	Data of the supplier inspection is used in place of incoming inspection	Items for which a supplier has established a strong quality record

In previous decades incoming inspection often consumed a large amount of time and effort. With the advent of modern complex products, many companies have found that they do not have the necessary inspection skills or equipment. This situation has forced them to rely more on the supplier's quality system or inspection and test data, as discussed later in this chapter.

The choice of evaluation method depends on a variety of factors:

- Prior quality history on the part and supplier.
- Criticality of the part on overall system performance.
- Criticality on later manufacturing operations.
- Warranty or use history.
- Supplier process capability information.
- The nature of the manufacturing process. For example, a press operation depends primarily on the adequacy of setup. Information on the first few pieces and last few pieces in a production run is usually sufficient to draw conclusions about the entire run.
- Product homogeneity. For example, fluid products are homogeneous, and the need for large sample sizes is thus less.
- Availability of required inspection skills and equipment.

A useful tool for learning about a supplier's process and comparing several suppliers' manufacturing product to the same specification is the histogram (see Section 10.10, "The Normal Curve and Histogram Analysis"). A random sample is selected from a lot, and measurements are made on the selected quality characteristics. The data are charted as frequency histograms. The analysis consists of comparing the histograms to the specification limits.

An application of histograms to evaluating the hardenability of a particular grade of steel from three suppliers is shown in Figure 15.2. The specification was a maximum Rockwell C reading of 43 measured at Jominy position J8. Histograms were also prepared for carbon, manganese, nickel, and chromium content. Analysis revealed:

- Supplier 46 had a process without any strong central tendency. The histogram on nickel for this supplier was also rectangular in shape, indicating a lack of control of the nickel content and resulting in several heats of steel with excessively high Rockwell values.
- Supplier 27 had several heats above the maximum, although the process had a central value of about 28. The histograms for manganese, nickel, and chromium showed several values above and apart from the main histogram.
- Supplier 74 showed much less variability than the others. Analysis of other histograms for this supplier suggested that about half of the original heats of steel had been screened out and used for other applications.

Note how these analyses can be made without visiting the supplier plants, i.e., "the product tells on the process." Histograms have limitations (see Chapter 10), but they are an effective tool for incoming inspection.

Action on nonconforming product

During the performance of the contract, there will arise instances of nonconformance. These may be on the product itself or on process requirements or procedural requirements. Priority effort should go to cases where a product is unfit for use.

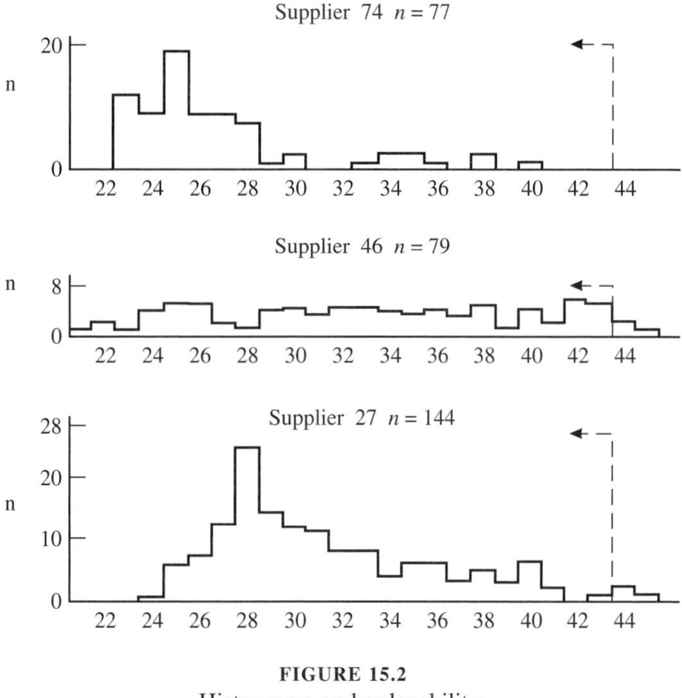

FIGURE 15.2
Histograms on hardenability.

Communications to the supplier on nonconformance must include a precise description of the symptoms of the defects. The best description is in the form of samples, but if this is not possible, the supplier should have the opportunity to visit the site of the trouble. There are numerous related questions: What disposition is to be made of the defective items? Who will sort or repair them? Who will pay the costs? What were the causes? What steps are needed to avoid a recurrence? These questions are outside the scope of pure defect detection; they require discussion among departments within each company and further discussions between buyer and supplier.

Supplier Certification

A "certified" supplier is one whose quality data record establishes that it is not necessary to perform routine inspection and test on each lot or batch received. A "preferred" supplier produces better quality than the minimum. An "approved" supplier meets minimum requirements. Some organizations use different terms and even different rankings but usually certified suppliers are the ideal. Unfortunately, they are in the minority. Spooner and Collins (1995) describe how criteria were developed at Walker Manufacturing to define these categories; e.g., a certified supplier has performed "at an overall 90 percent compliance level and met the individual rating component requirements for four consecutive quarters within two years."

ASQ recommends eight criteria for certification. These are summarized in Table 15.6.

TABLE 15.6
Criteria for supplier certification

Criteria	Examples
No product-related lot rejections for at least one year	An alternative is volume related, e.g., no rejects in 20 consecutive lots
No non-product-related rejections for at least six months	The marking on a container or the timeliness of an analysis document
No production-related negative incidents for at least six months	Ease with which the supplier's product can be used in the buyer's process or product
Passed a recent on-site quality system evaluation	A supplier survey on defined criteria
Has a totally agreed on specification	No ambiguous phrases like "characteristic odor" or "clear of contamination"
Fully documented process and quality system	The system must include plans for continuous improvement
Timely copies of inspection and test data	Real-time availability of data
Process is stable and in control	Statistical control and process capability studies

Source: Adapted from Maass et al. (1990).

Supplier certification provides a model for the low DPM levels necessary for just-in-time manufacture, drastically reduces buyer inspection costs, and identifies suppliers for partnerships. Certified suppliers receive preference in competitive bidding and achieve industry recognition by their certified status.

Schneider et al. (1995) explain how process capability indexes are used as part of the certification process at Dow Chemical. The concept of supplier certification equally applies to the service sector. Brown (1998) describes the approach used by a telecommunications company for suppliers of building leasing, maintenance, security, and food service.

Supplier Quality Rating

Supplier quality rating provides a quantitative summary of supplier quality over a period of time. This type of rating is useful in deciding how to allocate purchases among suppliers. Rating furnishes both buyer and supplier with common factual information that becomes a key input for identification and tracking of improvement efforts and for allocating future purchases among suppliers.

To create a single numerical quality score is difficult because there are several units of measure, such as

- The quality of multiple lots expressed as lots rejected versus lots inspected.
- The quality of multiple parts expressed as percentage nonconforming.
- The quality of specific characteristics expressed in numerous natural units, e.g., ohmic resistance, percentage of active ingredient, mean time between failures.
- The economic consequences of bad quality, expressed in dollars.

Because these units of measure vary in importance among different companies, published rating schemes differ markedly in emphasis.

Measures in use

Supplier quality rating plans are based on one or more of the following measures:

Product percentage nonconforming. This measure is a ratio of the amount of defective items received to the total number of items received. On a lot-by-lot basis, the formula is number of lots rejected divided by number of lots received; on an individual piece basis, the formula is number of individual pieces rejected divided by the number of individual pieces received.

Overall product quality. This plan summarizes supplier performance at incoming inspection and later phases of product application. Points are assigned for each phase with the maximum number of points given when no problems are encountered. Table 15.7 shows an example from AT&T. Note that the phases are incoming inspection, production failures, vendor response to problems, and AT&T customer complaints. Each rating element has further detailed criteria that are used to assign points for the element; e.g., if 3 percent of lots are rejected for "visual/mechanical" reasons in a rating period, then one point is deducted from the maximum of five for that element. Note that the overall rating evaluates the supplier response to problems, whereas detailed criteria include both timeliness and adequacy of the response.

Economic analysis. This type of plan compares suppliers on the total dollar cost for specific purchases. The total dollar cost includes the quoted price plus quality costs associated with defect prevention, detection, and correction.

Composite plan. Supplier performance is not limited to quality. It includes delivery against schedule, price, and other performance categories. These multiple needs suggest that supplier rating should include overall supplier performance

TABLE 15.7
AT&T quality performance rating

Rating element	Maximum points
Incoming inspection	
Visual mechanical PPM	10
Visual/mechanical—percentage of lot rejections	5
Testing PPM	10
Testing—lot rejections	5
Ship-to-stock credit	—
Production failures	
Shop complaints	20
Quality appraisal	10
Vendor response	
Response to problems	10
Failure analysis response	20
AT&T customer complaints	10
Total	100

Source: Nocera et al. (1989).

TABLE 15.8
Supplier rating report

Overall combined rating	92.46
Total quality rating	99.05
Total delivery rating	95.58
Total cost rating	79.22
Total response rating	83.30
Total lots received	18
Total parts received	398,351
Total parts rejected	3,804

Source: Wind (1991).

rather than just supplied quality performance. The purchasing department is a strong advocate of this principle and has valid grounds for this advocacy. Table 15.8 illustrates this approach with an example from the Tecumseh Products Company. The overall rating of 92.46 is calculated by combining the four ratings using a weight of 40 percent for quality, 30 percent for delivery, 20 percent for cost, and 10 percent for responsiveness to problems. Walker Manufacturing uses categories and weights of 35 percent for quality, 35 percent for delivery, 20 percent for price, and 10 percent for supplier support (Spooner and Collins, 1995).

Some organizations use periodic supplier rating to determine the share of future purchases given to each supplier. The rating system and its effect on market share are fully explained to all suppliers. The approach has been used successfully by both automotive and appliance manufacturers to highlight the importance of quality to their suppliers.

Quality Measurement in Supplier Relations

The management of quality-related activities in supplier relations must include provision for measurement. Readers are urged to review the 10 basic principles of quality measurement in Section 5.2, "Measurement."

Table 15.9 shows units of measure for various subject areas of supplier relations. Klenz (2000) discusses the use of a data warehouse for supplier quality analysis.

15.8 SUPPLY CHAIN QUALITY IMPROVEMENT

Donovan and Maresca (*JQH5,* Section 21) propose a sequence of five tiers of progression for improvement:

1. Create a joint team of the end user and supplier to align goals, analyze the supply chain business process, and work on chronic problems.
2. Focus on cost reduction including the cost of poor quality.

TABLE 15.9
Examples of quality measurement in supplier relations

Subject	Units of measure
Quality of submitted lots	Percentage of lots rejected Cost of poor quality Percentage of lots accepted on waiver Number of rejected lots classified "use as is"
Supplier relations program	Percentage of suppliers certified Percentage of suppliers classified acceptable as a result of a supplier survey Percentage of qualification test *procedures* approved on first submission Percentage of qualification test *results* approved on first submission Percentage of initial product samples approved on first submission Percentage of first production shipments approved on first submission Percentage of suppliers submitting data Average time to resolve problems
Business relationships	Average number of multiple suppliers per item Percentages of purchases as single source Percentage of purchases to lowest bidder Average time to secure bids Average time to secure answers to technical inquiries
Adequacy of inventory	Percentage of stockouts
Service to suppliers	Average number of days to pay supplier invoice Number of accounts payable beyond *X* days

3. Evaluate the value added by each link in the supply chain.
4. Exchange information and ideas on a routine basis throughout the chain.
5. Have the supply chain work as a single process with all parties routinely collaborating on improvement opportunities to generate value for customers as well as suppliers.

For elaboration, see *JQH5,* pp. 21.23–21.25.

The general approach to handling chronic supplier problems follows the step-by-step approach to improvement explained in Chapter 3, "Quality Improvement and Cost Reduction." This process includes the early steps of establishing the proof of the need for the supplier to take action and the application of Pareto analysis to identify the vital few problems. The following section on Pareto analysis of suppliers explains the form of such analyses of suppliers' problems.

Cooperation often requires that technical assistance be provided to suppliers. Miller and Kegaris (1986) describe how businesses may need to share proprietary information on a "need to know" basis. This often represents a major breakthrough in communications.

Sometimes upper management must provide the leadership in obtaining action from suppliers. Amazing results can be achieved when the initial step in an improvement program is a meeting of both the buyer's and supplier's upper-management teams, who plan the action steps for improvement together. Such discussions have much more impact than a meeting of the two quality managers does.

EXAMPLE 15.8. For an appliance manufacturer, 75 percent of the warranty costs were due to suppliers' items. The president and his staff met individually with the counterpart team from each of 10 key suppliers. Warranty data were presented to establish the "proof of the need." A goal was set for a 50 percent reduction in warranty costs over a five-year period. Each supplier was asked to develop a quality improvement program. The purchaser provided an eight-hour training session for the president and staff members of the key suppliers. Follow-up meetings were held. A system of supplier recognition awards was set up and purchasing practices were changed to transfer business to the best suppliers. The result: a decline in service calls from 41 to 13 calls per 100 products and a saving of $16 per unit in warranty costs.

Pareto Analysis of Suppliers

Supplier improvement programs can fail because the vital few problems are not identified and attacked. Instead, the programs consist of broad attempts to tighten up all procedures. The Pareto analysis (see Section 3.7 under "The Pareto Principle") can be used to identify the problem in a number of forms:

1. *Analysis of losses (defects, lot rejections, etc.) by material number* or *part number.* Such analysis serves a useful purpose as applied to catalog numbers involving substantial or frequent purchases.
2. *Analysis of losses by product family.* This process identifies the vital few product families present in small but numerous purchases of common product families, e.g., fasteners, paints.
3. *Analysis of losses by process,* i.e., classification of the defects or lot rejections in terms of the processes to which they relate, e.g., plating, swaging, coil winding.
4. *Analysis by supplier across the entire spectrum of purchases.* This process can help to identify weaknesses in the supplier's managerial approach as contrasted to the technological, which is more usually correlated with products and processes. One company had 222 suppliers on the active list. Of these, 38 (or 17 percent) accounted for 53 percent of the lot rejections and 45 percent of the bad parts.
5. *Analysis by total cost of the parts.* In one company 37 percent of the part numbers purchased accounted for only 5 percent of the total dollar volume of purchases but for a much higher percentage of the total incoming inspection cost. The conclusion was that these "useful many" parts should be purchased from the best suppliers, even at top prices. The alternative of relying on incoming inspection would be even more costly.
6. *Analysis by failure mode.* This technique is used to discover major defects in the management system. For example, suppose that studies disclose multiple instances of working to the wrong issue of the specification. In such cases the system used for specification revision should be reexamined. If value analysis discovers multiple instances of overspecification, the design procedures for choosing components

should be reexamined. These analyses by failure mode can reveal how the buyer is contributing to his or her own problems.

The cross-functional team approach for quality improvement described in Chapter 3 also applies to supplier quality. This means that there should be joint customer supplier teams and also that suppliers must be encouraged to set up an infrastructure (quality council, formation of teams, identification of projects, execution of projects) internally to address quality. Chen and Batson (1996) describe how Johnson & Johnson Consumer Products use 17 steps in this approach to supplier quality improvement. *JQH5,* Section 29, presents quality improvement in the automotive industry. In the service sector, Sun Health Alliance employs an innovative approach to stimulate improvement (Nussman, 1993). Sun provides grant funding to partner hospitals and corporate partners to support quality improvement demonstration projects. The projects include patient care topics (e.g., establishing clinical pathways for specific diagnoses); nonclinical topics (e.g., reducing turnaround time for lab results); employee-specific topics (e.g., reducing turnover or employee "needle sticks").

Handfield et al. (2000) discuss the results of research to identify "pitfalls" in supplier development. The research involved 84 companies in the fields of telecommunications, automobiles, electronics, computers, services, chemicals, consumer nondurable goods, and aerospace. The pitfalls were mainly concerned with identifying key projects, defining the details of the agreement between the buyer and supplier organizations, and monitoring the status and modifying strategies when necessary.

Some of the pitfalls were supplier specific; some were buyer specific; some were specific to the supplier-buyer relationship.

The supplier-specific pitfalls stemmed chiefly from the suppliers' lack of commitment and lack of technical or human resources. To avoid these pitfalls, companies took these actions:

1. Show suppliers where they stand.
2. Tie business relationships to performance improvement.
3. Illustrate supplier benefits clearly.
4. Ensure follow-up through a supplier champion (a supplier employee).
5. Keep initial improvements simple.
6. Draw on buyer's resources.
7. Offer personnel support
8. Build training centers.

The buyer-specific pitfalls occur when buyers see no obvious potential benefits from working on supplier development. To avoid this situation, companies found these tactics helpful:

1. Consolidate to fewer suppliers.
2. Keep a long-term focus.
3. Determine the total cost of ownership.
4. Set small goals.
5. Make executive commitment in the buyer organization a priority.

The supplier-buyer relationship pitfalls involved lack of trust between the organizations, poor alignment of cultures, and insufficient inducements to suppliers. Constructive solutions included

1. Delegate an ombudsman from the buyer organization.
2. Make provisions for handling confidential information.
3. Spell out clearly a cooperative purchasing relationship with well-defined objectives beyond purchase price.
4. Minimize legal involvement.
5. Adapt to local cultures.
6. Create a road map that defines responsibilities and expectations for both organizations.
7. Offer financial incentives.
8. Show suppliers how they can become "designed in" to buyer products and thus have greater potential for future business.
9. Offer repeat business as an incentive.

Supplier quality improvement needs upper management at all links in the supply chain to provide a structured approach (see Chapter 3) to improvement. Cheerleading and flag waving will not work.

SUMMARY

- A revolution in the relationship between buyers and suppliers has emerged in the form of supplier partnerships and the supply chain.
- Quality specifications often define requirements for both the product and the quality system.
- Organizations are significantly reducing the number of multiple suppliers.
- Outsourcing has both benefits and risks.
- Core competencies must be identified and performed internally.
- Evaluating supplier quality capability involves qualifying the supplier's design and the manufacturing process.
- Supplier partnerships require joint economic planning, joint technological planning, and cooperation during contract execution.
- A certified supplier is one that, after extensive investigation, is found to supply material of such quality that it is not necessary to perform routine testing on each lot received.
- Measurements for supplier relations should be based on input from customers; provide for both evaluation and feedback; and include early, concurrent, and lagging indicators of performance.
- Quality and reliability requirements should be stated in quantitative terms.
- Suppliers must understand that *all* product submitted is expected to meet specifications.
- The results of supplier surveys can be stated in quantitative terms.

- Histogram analyses of supplier data can reveal much information about the supplier's process.
- Pareto analysis of supplier data helps to establish priorities for improvement efforts.
- Supplier quality rating provides a quantitative summary of supplier quality over a period of time.
- Supplier quality improvement needs upper management involvement at all links in the supply chain.

PROBLEMS

15.1. Visit the purchasing agent of some local institution to learn the overall approach to supplier selection and the role of supplier quality performance in this selection process. Report your findings.

15.2. Visit a sampling of local suppliers (printer, merchant, repair shop, etc.) to learn the role of quality performance in their relationship with their clients. Report your findings.

15.3. A government agency contracted with a company to design and build a satellite system. Months after the contract was signed, the company discovered that the design would not be immune to certain types of radar interference. The agency claimed that it had described the performance desired for the satellites. The company disagreed (with respect to the radar interference). If this need had been realized at the start of the project, creating an appropriate design would have been relatively simple. The satellites are in an advanced stage of design and construction, and the necessary changes would cost \$100 million. There was further confusion. The company had chosen a supplier to manufacture the satellites. This supplier had previous experience on such products, and some people claimed that the supplier should have been aware of the radar interference problem. Comment on the actions that should be taken by three such organizations to prevent such a situation on a future project (*Business Week,* 1978).

15.4. During World War II, many manufacturers made products that were totally new to them. For example, the Ford Motor Company was asked to produce fuselage sections for B-24 aircraft. To do this, Ford had to work closely with the Consolidated Company, which was responsible for the manufacture of the entire aircraft. Thus Ford was a supplier for Consolidated. There was much friction between the companies. Lindbergh (1970, pages 644–676) describes the background of this classic case:

> In short, if the Consolidated men were carrying a chip on one shoulder, the Ford men arrived with a chip on each shoulder. Instead of taking the attitude that they had come to San Diego to learn how to build Consolidated bombers from the company that had developed those bombers, they took the attitude that they were there only as a preliminary to showing Consolidated how to build Consolidated bombers better and on mass production. The inevitable result was a deep-rooted antagonism which still exists.

The first article delivered by Ford was "not only as bad but considerably worse than the aviation people said it would be—rivets missing . . . badly formed skin . . . cracks *already* started . . . etc." However, this article had been passed both by Ford inspection and by the Army inspector stationed at Ford. Lindbergh concluded:

> What has happened is clear enough: under pressure, and encouraged by the desire to get production under way at Willow Run, and more than a little due to lack of experience, both Army and Ford inspection passed material that should have been rejected (and which was rejected by the more experienced and impartial inspectors at Tulsa).

Describe the *specific* actions that you would recommend to correct the immediate problem and prevent a recurrence in the future.

15.5. Apply the composite plan of supplier rating (Section 15.7) to compare three suppliers for one of the following: *(a)* any product or service acceptable to the instructor; *(b)* an automatic washing machine; *(c)* a new automobile; *(d)* a lawn mower.

15.6. Can you think of a situation other than 100 percent screening inspection that would result in the histogram shown in Figure 15.3?

15.7. Can you describe what caused the unusual histogram plots in Figure 15.4?

15.8. You have been asked to propose a specific quality rating procedure for use in one of the following types of organizations: *(a)* a company acceptable to the instructor; *(b)* a large municipal government; *(c)* a manufacturer of plastic toys; *(d)* a bank; *(e)* a manufacturer of whisky. Research the literature for specific procedures and select (or create) a procedure for the organization.

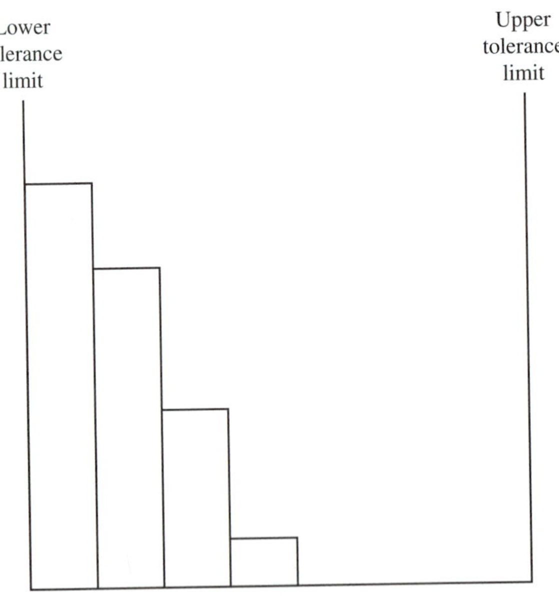

FIGURE 15.3
Sample histogram.

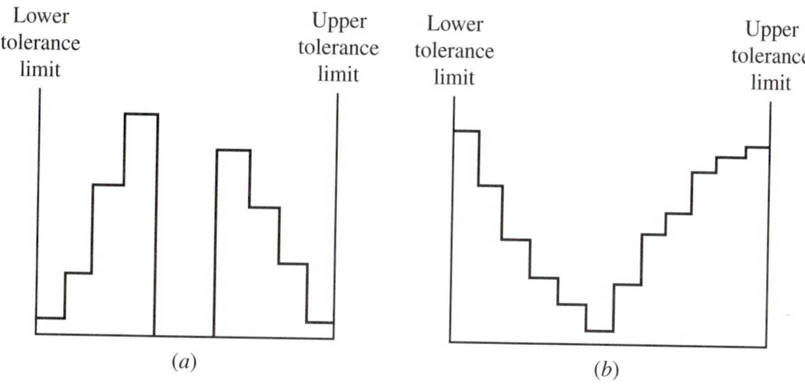

FIGURE 15.4
Sample histograms.

15.9. Visit a local organization and learn how it determines the quality of purchased items. Define the specific procedures used and what use is made of the information compiled.

15.10. Outline a potential application of several concepts in this chapter. Follow the instructions given in problem 13.11.

15.11. In an organization with which you are familiar, identify one activity as a candidate for outsourcing. Analyze the potential benefits and disadvantages of outsourcing that activity. Also, discuss how you might address the disadvantages.

REFERENCES

Bettis, R. A., S. P. Bradley, and G. Hamel (1992). "Outsourcing and Industrial Decline," *Academy of Management Executive,* vol. 6, no. 1, pp. 7–22.

Bossert, J. L. (1998). "Considerations for Global Supplier Quality," *Quality Progress,* January, pp. 29–32.

Bossert, J. L., ed. (1994). *Supplier Management Handbook,* ASQ Quality Press, p. 212.

Brown, J. O. (1998). "A Practical Approach to Service-Supplier Certification," *Quality Progress,* January, pp. 35–39.

Business Week (1978). "A $100 Million Satellite Error," August 7, p. 52.

Carter, J. R. and J. G. Miller (1989). "The Impact of Alternative Vendor/Buyer Communication Structures on the Quality of Purchased Materials," *Decision Sciences,* Fall, pp. 759–776.

Chen, B. A. and R. G. Batson (1996). "A Team Based Supplier Quality Improvement Process," *Annual Quality Congress Proceedings,* ASQ, Milwaukee, pp. 537–544.

Fisher, M. L. (1997). "What Is the Right Supply Chain for Your Product?" *Harvard Business Review,* March–April, pp. 105–116.

Handfield, R. B., D. R. Krause, T. V. Scannell, and R. M. Monczka (2000). "Avoid the Pitfalls in Supplier Development," *Sloan Management Review*, vol. 41, no. 2, pp. 37–49. By permission of the publisher. All rights reserved.

Klenz, B. W. (2000). "Leveraging the Data Warehouse for Supplier Quality Analysis," *Annual Quality Congress Proceedings,* ASQ, Milwaukee, pp. 519–528.

Lindbergh, C. A. (1970). *The Wartime Journals of Charles A. Lindbergh,* Harcourt Brace Jovanovich, New York.

Maass, R. A., J. O. Brown, and J. L. Bossert (1990). *Supplier Certification—A Continuous Improvement Strategy,* ASQ Quality Press, Milwaukee.

Miller, G. D. and R. J. Kegaris (1986). "An Alcoa-Kodak Joint Team," *Juran Report Number Six,* Juran Institute, Inc., Wilton, CT, pp. 29–34.

Nocera, C. D., M. K. Foliano, and R. E. Blalock (1989). "Vendor Rating and Certification," *Impro Conference Proceedings,* Juran Institute, Inc., Wilton, CT, pp. 9A-29 to 9A-38.

Nussman, H. B. (1993). "The Sun Health Alliance for Quality—A Unique Customer-Supplier Partnership," *Impro Conference Proceedings,* Juran Institute, Inc., Wilton, CT, pp. 3A.1-1 to 3A.1-4.

Peterson, Y. S. (1998). "Outsourcing: Opportunity or Burden," *Quality Progress,* June, pp. 63–64.

Schneider, H., J. Pruett, and C. Lagrange (1995). "Uses of Process Capability Indices in the Supplier Certification Process," *Quality Engineering,* vol. 8, no. 1, pp. 225–235.

Spooner, G. R. and D. W. Collins (1995). "A Cross Functional Approach to Supplier Evaluation," *Proceedings of the Annual Quality Congress,* ASQ, Milwaukee, pp. 825–832.

Wind, J. F. (1991). "Revolutionize Supplier Rating by Computerization," *Quality Congress Transactions,* ASQ, Milwaukee, pp. 556–564.

Zimmerli, B. (1996). "Re-engineering the Supply Chain," *Impro Conference Proceedings,* Juran Institute, Inc., Wilton, CT, pp. 4F-1 to 4F-18.

SUPPLEMENTARY READING

Purchasing and quality: *JQH5,* Sections 21 and 29.

Johnson, R. H. and R. T. Weber (1985). *Buying Quality,* Watts Publications, New York.

Pyzdek, T. and R. W. Berger (1992). *Quality Engineering Handbook,* Marcel Dekker, New York; ASQ Quality Press, Milwaukee, Chapter 7.

WEBSITE

ASQ customer supplier division: www.asqcd.org

16

OPERATIONS—MANUFACTURING SECTOR

16.1 QUALITY IN MANUFACTURING IN THE 21ST CENTURY

Operations is the nerve center of an organization—where the action is. This chapter covers operations in the manufacturing sector; Chapter 17 covers operations in the service sector.

In manufacturing industries, operations are activities, typically carried out in a factory, that transform material into the final product. Before we consider the planning, control, and improvement of manufacturing activities, we must recognize four important issues that will transform traditional manufacturing of the 20th century to a different manufacturing in the 21st century. These issues are discussed below.

Customer Demands for Higher Quality, Reduced Inventories, and Faster Response Time

As products and processes become more complex, new "world class" quality levels are now common. For many products, quality levels of 1 to 3 percent defective are being replaced by 1 to 10 defects per million parts (or 3.4 defects per million as in the six-sigma approach). Customers demand reduced inventory levels based on the "just in time" (JIT) production system. Under JIT, the concept of large lot sizes is challenged by reducing setup time, redesigning processes, and standardizing jobs. The results are smaller lot sizes and lower inventory. But JIT works only if product quality is high because little or no inventory exists to replace defective product. Finally, customers want faster response time from suppliers—to both develop and manufacture new products. That faster response time puts pressure on the product development process and can result in inadequate review of new designs for product performance and for manufacturability. Collectively, these three parameters (quality, inventories, response time) place a heavy burden on operations.

Lean Manufacturing

Lean manufacturing is the process of designing manufacturing systems to reduce costs by eliminating product and process waste. The emphasis is on eliminating non-value-added activities such as producing defective product, excess inventory charges due to work-in-process and finished goods inventory, excess internal and external transportation of product, excessive inspection, and idle time of equipment or workers due to poor balance of work steps in a sequential process. The goal of lean manufacturing has long been one of the goals of industrial engineering. Shuker (2000) provides a useful introduction to lean manufacturing based on the Toyota production system.

Agile Competition

An agile organization is able to respond to constantly changing customer opportunities. This characteristic means changing over from one product to another quickly, manufacturing goods to customer order in small lot sizes, customizing goods for individual customers, and using the expertise of people and facilities not only within the company but also among groups of cooperating companies (partners). Goldman, Nagel, and Preiss (1995) describe the concept and include examples.

This concept includes the "virtual" organization—a group of companies linked by an electronic network to enable the partners to satisfy a common customer objective. The virtual organization may be partially created by transferring complete functions to a supplier—outsourcing. See Section 15.4 for a discussion of the pros and cons of outsourcing.

Impact of Technology

Technology (including computer information systems) is clearly improving quality by providing a wider variety of outputs and also more consistent output. The infusion of technology makes some jobs more complex, thereby requiring extensive job skills and quality planning; technology also makes other jobs less complex but may contribute to job monotony.

These four issues suggest that quality during operations can no longer focus on inspection and checking but must respond to ever-increasing customer demands and changing competitive conditions. Skrabec (1997) in the Supplementary Reading describes the major change in manufacturing as a "paradigm shift."

We proceed now to examine specific methods for planning, controlling, and improving quality during manufacturing operations.

16.2 INITIAL PLANNING FOR QUALITY

Planning starts with a review of product designs (see Section 13.10). Then we review the process designs to identify key product and process characteristics, determine the importance of product characteristics, analyze the process flow diagram, error-proof the process, plan for a neat and clean workplace, validate the measurement process, and plan for operator self-control. These elements are discussed below. In addition, the all important topic of process capability is covered in Section 18.10.

Review of Product Designs

There is a clear advantage to having a new product design reviewed by operations personnel before the design is finalized for the marketplace. In practice, the extent of such review varies greatly—from essentially nothing ("tossing it over the wall" to the operations people) to a structured review using formal criteria and follow-up on open issues. Although a product design review often occurs during the design and development process (see Section 13.10), the emphasis is on the adequacy of field performance.

Review of product designs prior to release to operations must include an evaluation of producibility. This evaluation includes the following issues:

1. Identification of key product and process characteristics (see below).
2. Relative importance of various product characteristics (see below).
3. Design for manufacturability (see Section 13.8).
4. Process robustness. A process is robust if it is flexible, easy to operate, error proof, and its performance will tolerate uncontrollable variations in factors internal and external to the process. Such an ideal is approached by careful planning of all process elements, e.g., cross-functional training of personnel to cover vacations. For a discussion of robustness, see Snee (1993).
5. Availability of capable manufacturing processes to meet product requirements, i.e., processes that not only meet specifications but do so with minimum variation. The important subject of process capability is covered in Chapter 17 under "Process Capability."
6. Availability of capable measurement processes. This is discussed below and in Section 19.8.
7. Identification of special needs for the product, e.g., handling, transportation, and storage during manufacture.
8. Material control, e.g., identification, traceability, segregation, contamination control.
9. Special skills required of operations personnel.

This review of the product design must be supplemented by a review of the process design. The process review (discussed below) includes producibility issues initially raised in the product design review.

Identification of Key Product and Process Characteristics

Key *product* characteristics are the features that a product contains to meet customer needs. Key *process* characteristics are those that will create the key product characteristics. Product and process characteristics can be identified by using inputs from market research, quality function deployment, design review, and failure mode and effect analysis. Somerton and Mlinar (1996) describe the use of these and other tools to identify key characteristics.

Relative Importance of Product Characteristics

Planners are better able to allocate resources when they know the relative importance of the many product characteristics.

One technique for establishing the relative importance is the identification of critical items (see Section 13.5). Critical items are the product characteristics that require a high level of attention to ensure that all requirements are met. One company identifies "quality sensitive parts" by using criteria such as part complexity and high-failure-rate parts. For such parts special planning includes supplier involvement before and during the contract, process capability studies, reliability verification, and other activities.

Another technique is the classification of characteristics. Under this system the relative importance of characteristics is determined and indicated on drawings and other documents. The classification can be simply "functional" (or "critical to quality") or "nonfunctional." Another system uses several degrees of importance such as critical, major, minor, and incidental. The classification uses criteria that reflect safety, operating failure, performance, service, and manufacture. For elaboration, see Section 19.5 and *JQH5,* pages 22.6–22.7.

Analysis of the Process Flow Diagram

A process design can be reviewed by laying out the overall process in a flow diagram. Several types are useful. One type shows the paths followed by materials through their progression into a finished product. An example for a coating process at the James River Graphics Company is shown in Figure 16.1. Planners use such a diagram to divide the flow into logical sections called workstations. For each workstation they prepare a formal document listing such items as operations to be performed, sequence of operations, facilities and instruments to be employed, and process conditions to be maintained. This formal document becomes the plan to be carried out by the production supervisors and workforce. The document serves as the basis for control activities by the inspectors. It also becomes the standard against which the process audits are conducted.

Correlation of process variables with product results

A critical aspect of planning during manufacture is to discover, by data collection and analysis, the relationships between process features or variables and product features

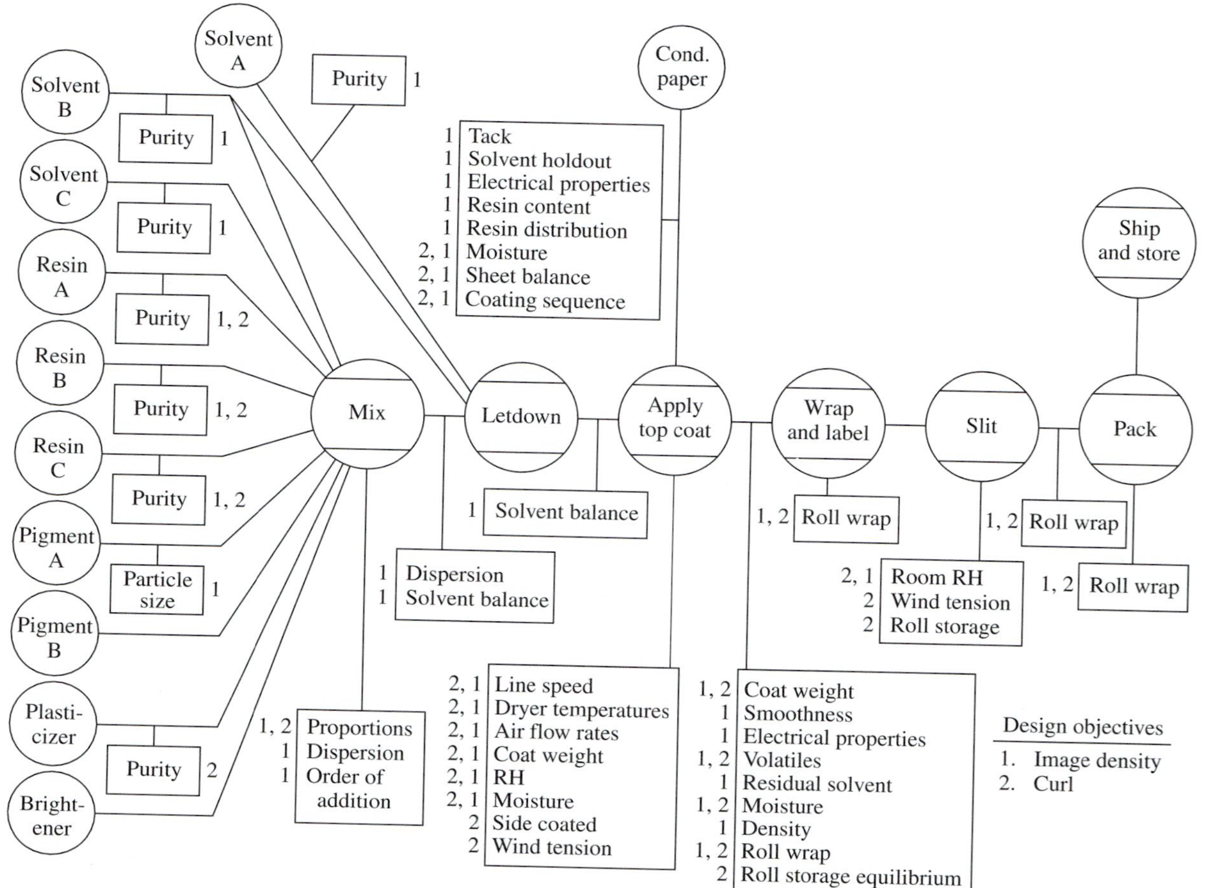

FIGURE 16.1
Product and process analysis chart. (*From Siff, 1984.*)

or results. Such knowledge enables a planner to create process control features, including limits and regulating mechanisms on the variables, to keep the process in a steady state and achieve the specified product results. In Figure 16.1 each process variable is shown in a rectangle attached to the circle representing an operation; product results are listed in rectangles between operations, at the point where conformance can be verified. Some characteristics (e.g., coat weight) are both process variables and product results.

For each control station in a process, designers identify the numerous control subjects over which control is to be exercised. Each control subject requires a feedback loop made up of multiple process control features. A process control spreadsheet helps to summarize the detail—as an example, Figure 5.4 shows, for each control subject, the unit of measure, type of sensor, goal, frequency of measurement, sample size, and the criteria and responsibility for decision making. For elaboration, see Juran (1992, page 286).

Determining the optimal settings and tolerances for process variables sometimes requires much data collection and analysis. Eibl et al. (1992) discuss such planning and analysis for a paint-coating process for which little information was available about the relationship between process variables and product results.

Many companies have not studied the relationship between process variables and product results. The consequences of this lack of knowledge can be severe. In the electronic component manufacturing industry, some yields are shockingly low and will likely remain that way until the process variables are studied in depth. In all industries the imposition of new quality demands such as six sigma requires a much deeper understanding of product results and process variables than in the past.

To understand fully the relationship between process variables and product results, we often need to apply the concept of statistical design of experiments (see Section 11.11). See also the discussion of the Taguchi approach (Section 13.4). Under the six-sigma approach, factorial experiments are becoming necessary to understand interaction between several variables and product results. But upper management must supply the missing elements, i.e., the resources for full-time personnel to design and analyze the experiments and the training of process engineers to integrate these concepts in process planning.

Error-Proofing the Process

An important element of prevention is the concept of designing the process to be error free through "error proofing" (the Japanese call it *poka-yoke*).

A widely used form of error-proofing is the design (or redesign of the machines and tools (the "hardware") so as to make human error improbable or even impossible. For example, components and tools may be designed with lugs and notches to achieve a lock-and-key effect, which makes it impossible to misassemble them. Tools may be designed to sense the presence and correctness of prior operations automatically or to stop the process on sensing depletion of the material supply. For example, in the textile industry a break in a thread releases a spring-loaded device that stops the machine. Protective systems, e.g., fire detection, can be designed to be "fail-safe" and to sound alarms as well as all-clear signals.

TABLE 16.1
Summary of error-proofing principles

Principle	Objective	Example
Elimination	Eliminating the possibility of error	Redesigning the process or product so that the task is no longer necessary
Replacement	Substituting a more reliable process for the worker	Using robotics (e.g., in welding or painting)
Facilitation	Making the work easier to perform	Color-coding parts
Detection	Detecting the error before further processing	Developing computer software that notifies the worker when a wrong type of keyboard entry is made (e.g., alpha versus numeric)
Mitigation	Minimizing the effect of the error	Utilizing fuses for overloaded circuits

In a classic study, Nakojo and Kume (1985) discuss five fundamental principles of error-proofing developed from an analysis of about 1000 examples collected mainly from assembly lines. These principles are elimination, replacement, facilitation, detection, and mitigation (see Table 16.1).

See *JQH5,* pages 22.24–22.26, for further examples of error-proofing.

An adjunct to error-proofing is the use of *poka-yoke* systems. These are control systems built into the process that stop the equipment when some type of irregularity occurs and signal the operator to address the problem (see Section 16.7).

Plan for Neat and Clean Workplaces

How obvious, but the reality is that many workplaces are dirty and are disorganized. The benefits of a good workplace include the prevention of defects; prevention of accidents; and the elimination of time wasted searching for tools, documentation, and other ingredients of manufacture. A simple body of knowledge now provides us with a framework to create the neat and clean workplace. The approach is called 5S for sort, set in order, shine, standardize, and sustain (see Figure 16.2). The steps are as follows:

1. *Sort.* Remove all items from the workplace that are not needed for current operations.
2. *Set in order.* Arrange workplace items so that they are easy to find, to use, and to put away.
3. *Shine.* Sweep, wipe, and keep the workplace clean.
4. *Standardize.* Make "shine" become a habit.
5. *Sustain.* Create the conditions (e.g., time, resources, rewards) to maintain a commitment to the 5S approach.

Decades ago industries producing critical items (health care, aerospace) learned that clean and neat workplaces are essential to achieve extremely low levels of defects. The quality levels demanded by the six-sigma approach now provide the same impetus.

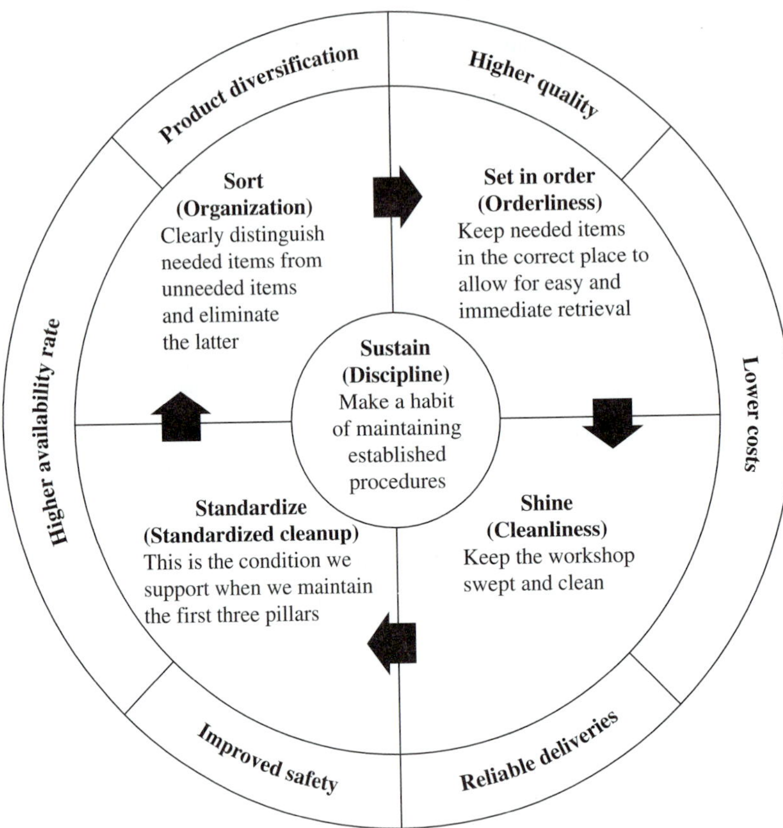

FIGURE 16.2

The 5S concept. (*Reprinted by permission. Productivity Press, a division of Productivity, Inc. PO Box 13390, Portland, OR 97213. 800-394-6868. www.productivityinc.com*)

Perhaps the significance of the 5S approach is the simplicity. The benefits are obvious; the tools are the simplest work-simplification tools; the tools are easy to understand and apply. Simple tools sometimes get dramatic results, and that's what has happened with 5S. For elaboration of the five steps, see The Productivity Process Development Team (1996).

Validate the Measurement System

Particularly with the low defect levels demanded under the six-sigma approach, it is important to understand not only the capability of the manufacturing process but also the capability of the measurement process. Thus planning and control of the measurement process become part of the six-sigma approach. Previous studies assumed that variation of the measurement process was small compared to variation caused by the

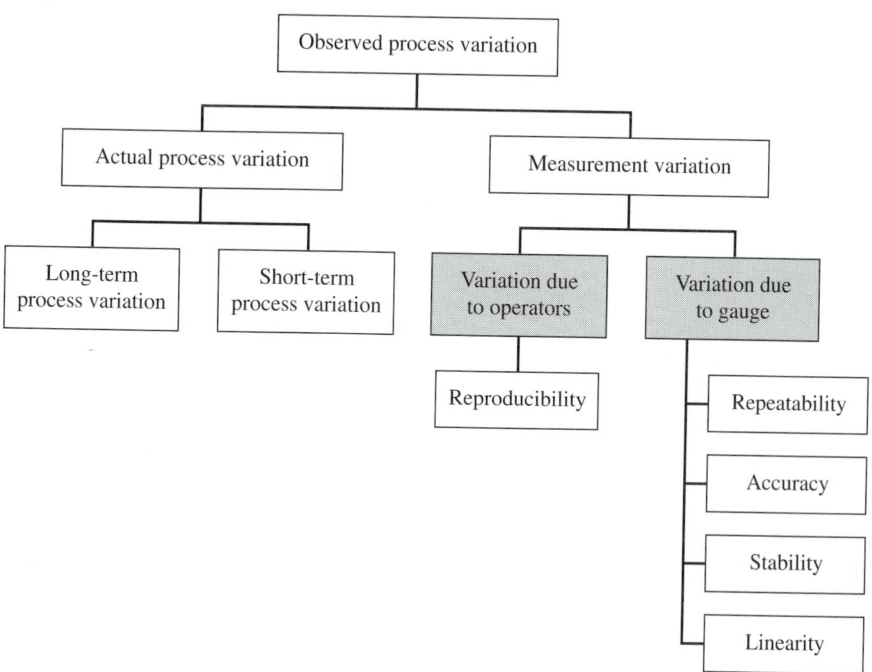

FIGURE 16.3
Possible sources of variation. (*Source: Juran Institute, Inc.*)

manufacturing process and thus could be ignored (in practice, the assumption was rarely tested in most industries). When variation due to the measurement process alone is even moderately large, the result will be mistakes in determining whether product meets specifications—some "good" product will be incorrectly classified as defective, and some "bad" product will be incorrectly classified as good. To quantify the measurement process, see Section 19.8. Thus the time has come to evaluate measurement capability and to determine whether measuring equipment is accurately measuring process output.

Figure 16.3 provides perspective on the measurement issue. Note that the observed process variation (i.e., the variation of recorded measurements) is due to two sources: variation of the process manufacturing the product and variation of the measurement process. Further, these two sources contain various components. The components shown in Figure 16.3 can be quantified and analyzed to determine measurement capability. The variation due to the process manufacturing the product can also be quantified—see Chapter 17 under "Process Capability."

Even in consumer product industries such as the manufacture of razor blades, tolerances can be on the order of the wavelength of visible light. Such tolerances are a long way from the days of tolerances in thousands or 10 thousands and using measuring instruments like micrometers and even supermicrometers. The measurement process must be capable of handling these conditions.

16.3
CONCEPT OF CONTROLLABILITY: SELF-CONTROL

The concept of self-control was introduced in Section 5.3. An ideal objective for manufacturing planning is to place human beings in a state of self-control, i.e., to provide them with all they need to meet quality objectives. To do so, we must provide people with the following:

1. Knowledge of what they are supposed to do.
   - Clear and complete work procedures.
   - Clear and complete performance standards.
   - Adequate selection and training of personnel.
2. Knowledge of what they are actually doing (performance).
   - Adequate review of work.
   - Feedback of review results.
3. Ability and desire to regulate the process for minimum variation.
   - A process and job design capable of meeting quality objectives.
   - Process adjustments that will minimize variation.
   - Adequate worker training in adjusting the process.
   - Process maintenance to maintain the inherent process capability.
   - A strong quality culture and environment.

As we will see, most organizations do not adhere to the three elements and ten subelements of self-control.

The concept of self-control has objectives that are similar to those of the Toyota production system. For a perceptive dissection of the Toyota system into four basic rules, see Spear and Bowen (1999).

The three basic criteria for self-control make possible a separation of defects into categories of "controllability," of which the most important are

1. *Worker controllable.* A defect or nonconformity is worker controllable if all three criteria for self-control have been met.
2. *Management controllable.* A defect or nonconformity is management controllable if one or more of the criteria for self-control have not been met.

Only management can provide the means of meeting the criteria for self-control. Hence, any failure to meet these criteria is a failure of management, and the resulting defects are therefore beyond the control of the workers. This theory is not 100 percent sound. Workers commonly have a duty to call management's attention to deficiencies in the system of control, and sometimes they do not do so. (Sometimes they do, and it is management that fails to act.) However, the theory is much more right than wrong.

Whether the defects or nonconformities in a plant are mainly management controllable or worker controllable is of the highest order of importance. To reduce the former requires a program in which the main contributions must come from managers, supervisors, and technical specialists. To reduce the latter requires a different kind of program in which much of the contribution comes from workers. The great difference between these two kinds of programs suggests that managers should quantify their knowledge of the state of controllability before embarking on major programs.

TABLE 16.2
Controllability study in a machine shop

Category	Percent
Management controllable	
Inadequate training	15
Machine inadequate	8
Machine maintenance inadequate	8
Other process problems	8
Material handling inadequate	7
Tool, gage, fixture (TGF) maintenance inadequate	6
TGF inadequate	5
Wrong material	3
Operation run out of sequence	3
Miscellaneous	5
Total	68
Worker controllable	
Failure to check work	11
Improper operation of machine	11
Other (e.g., piece mislocated)	10
Total	32

An example of a controllability study is given in Table 16.2. A diagnostic team was set up to study scrap and rework reports in six machine shop departments for 17 working days. The defect cause was entered on each report by a quality engineer who was assigned to collect the data. When the cause was not apparent, the team reviewed the defect and, when necessary, contacted other specialists (who had been alerted by management about the priority of the project) to identify the cause. The purpose of the study was to resolve a lack of agreement on the causes of chronically high scrap and rework. It did the job. The study was decisive in obtaining agreement on the focus of the improvement program. In less than one year, more than $2 million was saved, and important strides were made in reducing production backlogs.

Controllability can also be evaluated by posing specific questions for each of the three criteria of self-control. (Typical questions that can be posed are presented below.) Although this approach does not yield a quantitative evaluation of management and worker-controllable defects, it does show whether the defects are primarily management or worker controllable.

In the experience of the author, defects are about 80 percent management controllable. This figure does not vary much from industry to industry, but it does vary greatly among processes. Other investigators in Japan, Sweden, the Netherlands, and Czechoslovakia have reached similar conclusions.

While the available quantitative studies make clear that defects are mainly management controllable, many industrial managers do not know this or are unable to accept the data. Their long-standing beliefs are that most defects are the result of worker carelessness, indifference, and even sabotage. Such managers are easily persuaded to embark on worker-motivation schemes that, under the usual state of affairs, aim at a small minority of the problems and hence are doomed to achieve minor result at best. The issue is not whether quality problems *in an industry* are management controllable. The need is to determine the answer *in a given plant.* This answer requires solid facts, preferably through a controllability study of actual defects, as shown in Table 16.2.

We now discuss the three main criteria for self-control.

Criterion 1: Knowledge of "Supposed to Do"

This knowledge commonly consists of the following:

1. The product standard, which may be a written specification, a product sample, or other definition of the end result to be attained.
2. The process standard, which may be a written process specification, written process instructions, an oral instruction, or other definition of "means to an end."
3. A definition of responsibility, i.e., what decisions to make and what actions to take.

In developing product specifications, some essential precautions must be observed.

Unequivocal information must be provided

Specifications should be quantitative. If such specifications are not available, physical or photographic standards should be provided. But beyond the need for *clear* product specifications, there is also a need for *consistent* and *credible* specifications. In some organizations, production supervisors have a secret "black book" that contains the "real" specification limits used by inspectors for accepting product. A further problem is communicating changes to specifications, especially when there is a constant parade of changes.

Information on seriousness must be provided

All specifications contain multiple characteristics, and these are not equally important. Production personnel must be guided and trained to meet all specification limits. But they must also be given information on the relative importance of each characteristic in order to focus on priorities. Section 19.5 explains methods of defining relative seriousness.

Reasons must be explained

Explanation of the purposes served by both the product and the specification helps workers to understand why both the nominal specification value and the limits must be met.

Process specifications must be provided

Work methods and process conditions (e.g., temperature, pressure, time cycles) must be made unequivocally clear.

LTV, a steel manufacturer, uses a highly structured system of identifying key process variables, defining process control standards, communicating the information to the workforce, monitoring performance, and accomplishing diagnosis when problems arise. The process specification is a collection of process control standard procedures (Figure 16.4). A procedure is developed for controlling the key process variables

LTV Steel

INTEGRATED PROCESS CONTROL
Standard Procedures
Process Control

Plant Indiana Harbor

Dept No. 3 Sheet Mill

File No. 716-2.2.2

Date Orig Issue

Revision No. 1

Date Revised

Control Area	Control Point	Control Element	No.
Tandem Mill	Rolling	Rolling Solutions	2.2.2

Control Task
To maintain rolling solution characteristics at the proper levels.

Responsible for Control
Solution Attendant

Process Standard
- Oil concentration must be 2.5% to 3.5%
- Solution temperature must be 110°F - 120°F
- SAP value must be above 120
- Iron fines must be less than 600 ppm

Reason for Control
- To provide the correct lubricity between work rolls and strip for reduced roll wear and control of strip temperature. This helps control strip flatness and avoids friction scratches.

Measurement	Routine Reporting of Data	Control Chart
Tools/Equipment – Standard chem. test set-up Frequency – Twice/turn By – Solution Attendant	Form No. Solution Attendant's Report By – Solution Attendant	Type – X & Moving Range By – Solution Att.

Corrective Action
- Solution concentration approaching limits - add rolling oil or water.
- Solution temp. approaching limits - adjust temperature control.
- SAP reading between 100 & 120, skim solution tank and add new oil. SAP reading below 100, retest immediately and contact Operating Supervisor. If retest below 100, shut down mill and switch to alternative solution tank.
- Iron fines approaching limit, skim tank for 2 hours and add 100 gallons oil. Retest after 30 minutes second time, repeat procedure if still near or above limit.

Operating Procedure
See attached sheet

Disposition of Non-Compliant Product
Identify coil(s) for special surface evaluation. Notify Metallurgical Supervisor.

Review Procedure
Once per turn the Operating Supervisor will:
- Check Solution Attendant's Report
- Visually check temperature of solution

Developed By: Robert Gordkin — IPC Coordinator; Richard H. Burg

Approved: D. T. Zid — Department Superintendent/Manager; Dale H. Dick Jr. — Manager–Quality Control; R. Vatch Jr. — General Superintendent/Print Manager

FIGURE 16.4
Process control standard procedure. (*From LTV Steel.*)

(variables that must be controlled to meet specification limits on the product). The procedure answers the following issues:

- What the process standards are.
- Why control is needed.
- Who is responsible for control.
- What and how to measure.
- When to measure.
- How to report routine data.
- Who is responsible for data reporting.
- How to audit.
- Who is responsible for audit.
- What to do with product that is out of compliance.
- Who developed the standard.

Often, detailed process instructions are not known until workers have become experienced with the process. Updating of process instructions based on job experience can be conveniently accomplished by posting a cause-and-effect diagram in the production department and attaching index cards to the diagram. Each card states additional process instructions based on recent experience.

A checklist must be created

The above discussion covers the first criterion of self-control: People must have the means of knowing what they are supposed to do. To evaluate adherence to this criterion, a checklist of questions can be created, including the following:

Adequate and complete work procedures

1. Are there written product specifications, process specifications, and work instructions? If written down in more than one place, do they all agree? Are they legible? Are they conveniently accessible to the worker?
2. Does the worker receive specification changes automatically and promptly?
3. Does the worker know what to do with defective raw material?
4. Have responsibilities in terms of decisions and actions been clearly defined?

Adequate and complete performance standards

5. Do workers consider the standards attainable?
6. Does the specification define the relative importance of different quality characteristics? If control charts or other control techniques are to be used, is their relationship to product specifications clear?
7. Are standards for visual defects displayed in the work area?
8. Are the written specifications given to the worker the same as the criteria used by inspectors? Are deviations from the specification often allowed?
9. Does the worker know how the product is used?
10. Does the worker know the effect on future operations and product performance if the specification is not met?

Adequate selection and training

11. Does the personnel selection process adequately match worker skills with job requirements?

12. Has the worker been adequately trained to understand the specification and perform the steps needed to meet the specification?
13. Has the worker been evaluated by testing or other means to see whether he or she is qualified?

Criterion 2: Knowledge of Performance

For self-control, people must have the means of knowing whether their performance conforms to standard. This conformance applies to

- The product in the form of specifications on product characteristics.
- The process in the form of specifications on process variables.

This knowledge is secured from three primary sources: measurements inherent in the process, measurements by production workers, and measurements by inspectors.

Criteria for good feedback to workers

The needs of production workers (as distinguished from supervisors or technical specialists) require that the data feedback can be read at a glance, deal only with the few important defects, deal only with worker-controllable defects, provide prompt information about symptom and cause, and provide enough information to guide corrective actions. Good feedback should

- *Be readable at a glance.* The pace of events on the factory floor is swift. Workers should be able to review the feedback while in motion. Where a worker needs information about process performance over time, charts can provide an excellent form of feedback, provided they are designed to be consistent with the assigned responsibility of the worker (Table 16.3). It is useful to use visual displays to highlight recurrent problems. A problem described as "outer hopper switch installed backwards" displayed on a wall chart in large block letters has much more impact than the same message buried away as a marginal note in a work folder.
- *Deal only with the few important defects.* Overwhelming workers with data on all defects will result in diverting attention from the vital few.

TABLE 16.3
Worker responsibility versus chart design

Responsibility of the worker is to	Chart should be designed to show
Make individual units of product meet a product specification	Measurements of individual units of product compared to product specification limits
Hold process conditions to the requirements of a process specification	Measurements of the process conditions compared to the process specification limits
Hold averages and ranges to specified statistical control limits	Averages and ranges compared to the statistical control limits
Hold percentage nonconforming below some prescribed level	Actual percentage nonconforming compared to the limiting level

- *Deal only with worker-controllable defects.* Any other course provides a basis for argument that will be unfruitful.
- *Provide prompt information about symptoms and causes.* Timeliness is a basic test of good feedback; the closer the system is to "real time" signaling, the better.
- *Provide enough information to guide corrective action.* The signal should be in terms that make it easy to decide on remedial action.

Feedback related to worker action

The worker needs to know what kind of *process* change to make to respond to a *product* deviation. Sources of this knowledge are

- The process specifications (see Figure 16.4 under "Corrective Action").
- Cut-and-try experience by the worker.
- The fact that the units of measure for both product and process are identical.

If lacking all of these, workers can only cut and try further or stop the process and sound the alarm.

Sometimes the data feedback can be converted into a form that makes easier the worker's decision about what action to take on the process.

For example, a copper cap had six critical dimensions. It was easy to measure the dimensions and to discover the nature of product deviation. However, it was difficult to translate the product data into process changes. To simplify this translation, use was made of a position-dimensions (P–D) diagram. The six measurements were first "corrected" (i.e., coded) by subtracting the thinnest from all the others. These corrected data were then plotted on a P–D diagram as shown in Figure 16.5. Such diagrams provided a way of analyzing the tool setup.

Feedback to supervisors

Beyond the need for feedback at workstations, there is a need to provide supervisors with short-term summaries. These take several forms.

Matrix summary. A common form of matrix is workers versus defects; i.e., the vertical columns are headed by worker names and the horizontal rows by the names of defect types. The matrix makes clear which defect types predominate, which workers have the most defects, and what the interaction is. Other matrixes include machine number versus defect type and defect type versus calendar week. When the summary is published, it is usual to circle matrix cells to highlight the vital few situations that call for attention.

An elaboration of the matrix is to split the cell diagonally, thus permitting the entry of two numbers, e.g., number defective and number produced.

Pareto analysis. Some companies prefer to minimize detail and provide information on the total defects for each day plus a list of the top three (or so) defects encountered and how many there were of each. In some industries a "chart room" displays performance against goals by product and by department.

Computer data analysis and reporting. Production volume and complexity are important factors in determining the role of the computer. Chapter 22, "Quality Infor-

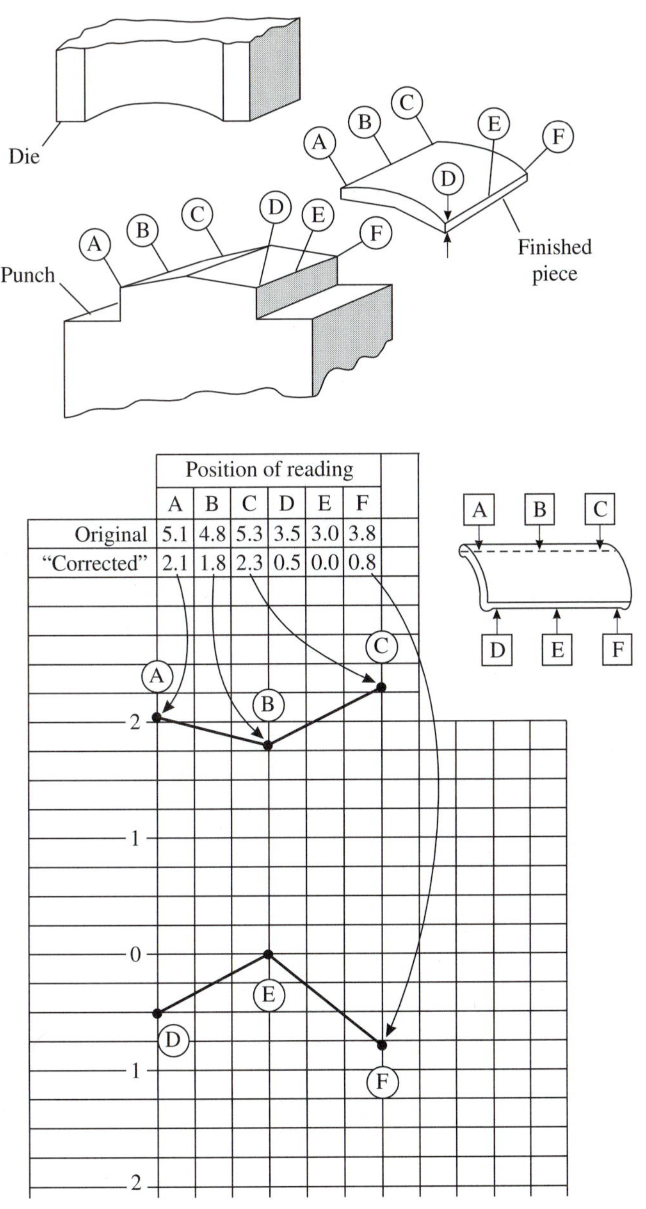

	Position of reading					
	A	B	C	D	E	F
Original	5.1	4.8	5.3	3.5	3.0	3.8
"Corrected"	2.1	1.8	2.3	0.5	0.0	0.8

FIGURE 16.5
Method of drawing P–D diagram.

mation Systems," explains the role of computers in analyzing and reporting data during production and other phases of the product life cycle.

Automated quality information. Some situations justify the mechanization of both the recording and analysis of data. Entry of data into computer terminals on production floors is now common. Many varieties of software are now available for analyzing, processing, and presenting quality information collected on the production floor.

The term *quality information equipment* (QIE) designates the physical apparatus that measures products and processes, summarizes the information, and feeds the information back for decision making. Sometimes such equipment has its own product development cycle to meet various product effectiveness parameters for the QIE.

Checklist

A checklist to evaluate the second criterion of self-control includes questions such as these:

Adequate review of work

1. Are gages provided to the worker? Do they provide numerical measurements rather than simply sorting good from bad? Are they precise enough? Are they regularly checked for accuracy?
2. Is the worker told how often to sample the work? Is sufficient time allowed?
3. Is the worker told how to evaluate measurements to decide when to adjust the process and when to leave it alone?
4. Is a checking procedure in place to ensure that the worker follows instructions on sampling work and making process adjustments?

Adequate feedback

5. Are inspection results provided to the worker, and are these results reviewed by the supervisor with the worker?
6. Is the feedback timely and in enough detail to correct problem areas? Have personnel been asked what detail is needed in the feedback?
7. Do personnel receive a detailed report of errors by specific type of error?
8. Does feedback include positive comments in addition to negative?
9. Is negative feedback given in private?
10. Are certain types of errors tracked with feedback from external customers? Could some of these be tracked with an internal early indicator?

Criterion 3: Ability and Desire to Regulate

Ability and desire to regulate is the third criterion for self-control. Regulating the process depends on various management-controllable factors, including these:

The process must be capable of meeting the tolerances. This factor is of paramount importance. In some organizations the credibility of specifications is a serious problem. Typically, a manufacturing process is created after release of the product design, a few trials are run, and full production commences. In cases where quality problems arise during full production, diagnosis sometimes reveals that the process is not capable of consistently meeting the design specifications. Costly delays to production then occur while the problem is resolved by changing the process or changing the specification. The capability of the manufacturing process should be verified during the product development cycle *before the product design is released for full production.* See Section 18.10 for a full discussion.

The process must be responsive to regulatory action in a predictable cause-and-effect relationship to minimize variation around a target value.

EXAMPLE 16.1. In a process for making polyethylene film, workers were required to meet multiple product parameters. The equipment had various regulatory devices, each of which could vary performance with respect to one or more parameters. However, the workers could not "dial in" a predetermined list of settings that would meet all parameters. Instead, it was necessary to cut and try in order to meet all parameters simultaneously. During the period of cut and try, the machine produced nonconforming product to an extent that interfered with meeting standards for productivity and delivery. The workers were unable to predict how long the cut-and-try process would go on before full conformance was achieved. Consequently, it became the practice to stop cut and try after a reasonable amount of time and to let the process run, whether in conformance or not.

The worker must be trained in how to use the regulating mechanisms and procedures. This training should cover the entire spectrum of action—under what conditions to act, what kind and extent of changes to make, how to use the regulating devices, and why these actions need to be taken.

EXAMPLE 16.2. Of three qualified workers on a food process, only one operated the process every week and became proficient. The other two workers were used only when the primary worker was on vacation or was ill, and thus they never became proficient. Continuous training of the relief people was considered uneconomical, and agreements with the union prohibited their use except under the situations cited above. This problem is management controllable, i.e., additional training or a change in union agreements is necessary.

The act of adjustment should not be personally distasteful to the worker, e.g., should not require undue physical exertion.

EXAMPLE 16.3. In a plant making glass bottles, one adjustment mechanism was located next to a furnace area. During the summer months, this area was so hot that workers tended to keep out of it as much as possible. When the regulation consists of varying the human component of the operation, the question of process capability arises in a new form: Does the worker have the capability to regulate? This important question is discussed in Section 3.9, which includes some examples of discovering worker "knack."

The process must be maintained sufficiently to retain its inherent capability. Without adequate maintenance, equipment breaks down and requires frequent adjustments—often with an increase in both defects and variability around a nominal value. Clearly, such maintenance must be both preventive and corrective. The importance of maintenance has given rise to the concept of total productive maintenance (TPM). Under this approach teams are formed to identify, analyze, and solve maintenance problems for the purpose of maximizing the uptime of process equipment. These teams consist of production line workers, maintenance personnel, process engineers, and others as needed. Problems are kept narrow in scope to encourage a steady stream of small improvements. Examples of improvements include a reduction in the number of tools lost and simplification of process adjustments.

Control systems and the concept of dominance

Specific systems for controlling characteristics can be related to the underlying factors that dominate a process. The main categories of dominance include the following:

- *Setup dominant.* Such processes have high reproducibility and stability for the entire length of the batch to be made. Hence the control system emphasizes verification of the setup before production proceeds. Examples of such processes are drilling, labeling, heat sealing, printing, and presswork.
- *Time dominant.* Such a process is subject to progressive change with time (wearing of tools, depletion of reagent, heating up of machine). The associated control system will feature a schedule of process checks with feedback to enable the worker to make compensatory changes. Screw machining, volume filling, wool carding, and papermaking are examples of time-dominant processes.
- *Component dominant.* Here the quality of the input materials and components is the most influential. The control system is strongly oriented toward supplier relations along with incoming inspection and sorting of inferior lots. Many assembly operations and food formulation processes are component dominant.
- *Worker dominant.* For such processes quality depends mainly on the skill and knack possessed by the production worker. The control system emphasizes such features as training courses and certification for workers, error-proofing, and rating of workers and quality. Workers are dominant in processes such as welding, painting, and order filling.
- *Information dominant.* In these processes the job information usually undergoes frequent change. Hence the control system places emphasis on the accuracy and up-to-dateness of the information provided to the worker (and everyone else). Examples include order editing and "travelers" used in job shops.

The various types of dominance differ also in the tools used for process control. Table 16.4 lists the forms of process dominance along with the usual tools used for process control.

Checklist

A checklist for evaluating the third criterion of self-control typically includes questions such as these:

Process capable

1. Has the quality capability of the process been measured to include both inherent variability and variability due to time? Is the capability periodically checked?
2. Has the design of the job made use of the principles of error-proofing?
3. Has equipment, including any software, been designed to be compatible with the abilities and limitations of workers?

Process adjustments

4. Has the worker been told how often to reset the process or how to evaluate measurements to decide when the process should be reset?

TABLE 16.4
Control tools for forms of process dominance

Setup dominant	Time dominant	Component dominant	Worker dominant	Information dominant
Inspection of process conditions	Periodic inspection	Supplier rating	Acceptance inspection	Computer-generated information
First-piece inspection	$\overline{X}$ chart	Incoming inspection	p chart	"Active" checking of documentation
Lot plot	Median chart	Prior operation control	c chart	Bar codes and electronic entry
Precontrol	$\overline{X}$ and R chart	Acceptance inspection	Operator scoring	Process audits
Narrow-limit gaging	Precontrol	Mockup evaluation	Recertification of workers	
Attribute visual inspection	Narrow-limit gaging		Process audits	
	p chart			
	Process variables check			
	Automatic recording			
	Process audits			

5. Can the worker make a process adjustment to eliminate defects? Under what conditions should the worker adjust the process? When should the worker shut down the machine and seek more help? Whose help?
6. Have the worker actions that cause defects, and the necessary preventive action, been communicated to the worker, preferably in written form?
7. Can workers institute job changes that they show will provide benefits? Are workers encouraged to suggest changes?

Worker training in adjustments

8. Do some workers possess a hidden knack that needs to be discovered and transmitted to all workers?
9. Have workers been provided with the time and training to identify problems, analyze problems, and develop solutions? Does the training include diagnostic training to look for patterns of errors and determine sources and causes?

Process maintenance

10. Is there an adequate preventive maintenance program on the process?

Strong quality culture/environment

11. Is there sufficient effort to create and maintain awareness of quality?
12. Is there evidence of management leadership?
13. Have provisions been made for self-development and empowerment of personnel?
14. Have provisions been made for participation of personnel as a means of inspiring action?
15. Have provisions been made for recognition and rewards for personnel?

For a comparison of Russian and American practices with respect to worker control, see Pooley and Welsh (1994) in the Supplementary Reading.

Use of Checklists on Self-Control

These checklists can help operations in the design (and redesign) of jobs to prevent errors, diagnosis of quality problems on individual jobs, and identification of common weaknesses in many jobs and assist supervisors to function as a coach with personnel, prepare for process audits, and conduct training classes on quality. For elaboration, see *JQH5,* pages 22.56–22.57.

16.4 AUTOMATED MANUFACTURING

The march to automation proceeds unabated. Several terms have become important:

- *Computer-integrated manufacturing (CIM).* CIM is the process of applying a computer in a planned fashion from design through manufacturing and shipping of the product.

- *Computer-aided manufacturing (CAM).* CAM is the process in which a computer is used to plan and control the work of specific equipment.
- *Computer-aided design (CAD).* CAD is the process by which a computer assists in the creation or modification of a design.

This trio of concepts is producing huge increases in factory productivity. But automation, with proper planning, can also benefit product quality in several other ways:

- Automation can eliminate some of the monotony or fatigue tasks that result in human errors. For example, when a manual seam-welding operation was turned over to a robot, the scrap rate plunged from 15 percent to zero.
- Process variation can be reduced by the automatic monitoring and continuous adjustment of process variables.
- An important source of process troubles, i.e., the number of machine setups, can be reduced.
- Machines can not only automatically measure product but also record, summarize, and display the data for line production operators and staff personnel. Feedback to the worker can be immediate, thus providing an early warning of impending troubles.
- With CAD, the quality engineer can pride inputs early in the design stage. When a design is placed into the computer, the quality engineer can review that design over and over again and thus keep abreast of design changes.

Collectively, these and other computer-based technologies for manufacturing are sometimes called "advanced manufacturing systems" (AMS).

With the emergence of an electronic information network provided by the Internet, a group of companies can operate as one virtual factory. This environment enables companies to exchange and act on information concerning inventory levels, delivery schedules, supplier lists, product specifications, and test data. It also means that CAD/CAM information and other manufacturing process information can be exchanged, data can be transferred to machines in a supplier's plant, and supplier software can be used to analyze producibility and to begin actual manufacturing.

Achieving the benefits of automated manufacturing requires a spectrum of concepts and techniques. Three of these are discussed below: the key functions of CIM, group technology, and flexible manufacturing systems.

Key Functions of Computer-Integrated Manufacturing

CIM integrates engineering and production with suppliers and customers (even globally) to interactively design, plan, and conduct the manufacturing activities. CIM activities include providing design tools to support innovation in remote sites, manufacturing planning, computer simulation models to evaluate and predict process performance, sensing and control tools to monitor processes, intelligence tools to gather and organize process data to share with other sites, information processing and transferring among geographically dispersed participants, and automated translation of text between different languages. For elaboration, see Lee (1995).

Group Technology

Group technology is the process of examining all items manufactured by a company to identify those with sufficient similarity that common design or manufacturing plans can be used. The aim is to reduce the number of new designs or new manufacturing plans. In addition to the savings in resources, group technology can improve both the quality of design and the quality of conformance by using proven designs and manufacturing plans. In many companies only 20 percent of the parts initially thought to require a new design actually need it; of the remaining new parts, 40 percent could be built from an existing design, and the other 40 percent could be created by modifying an existing design.

Location of production machines can also benefit from the group technology concept. Machines are grouped according to the parts they make and can be sorted into cells of machines, each cell producing one or several part families (thus "cellular manufacture"). For a discussion of shifting from traditional manufacturing organization to cellular manufacture and the impact on first-level supervisors, see Wright (1998) in the Supplementary Reading.

Flexible Manufacturing System

A flexible manufacturing system (FMS) is a group of several computer-controlled machine tools, linked by a materials handling system and a computer, to accommodate varying production requirements. The system can be reprogrammed to accommodate design changes or new parts. In contrast, in a fixed automation system, machinery, materials handling equipment, and controllers are organized and programmed for production of a single part or limited range of parts.

Typically, the individual machines are robots or other types of numerically controlled machine tools, each of which is run by a microcomputer. Several such tools are linked by a minicomputer, and then several of these minicomputers are tied into the mainframe computer.

Quality planning for automated processes requires some special precautions. These are identified and discussed in *JQH5,* pages 22.36–22.37.

From the one extreme of the (typically) mass production in automated industries, we can shift to the other extreme of the (typically) low-volume production in job shop industries. Applying quality concepts in job shop industries is covered in *JQH5,* Section 24.

Even traditional craft industries now benefit from technology. For example, the manufacture of fine pianos uses programmable logic controllers to make process adjustments; in wine making, aerial images of vineyards, taken by using digital sensors to collect data on the chlorophyll content of the vines, provides a strong indicator of taste (the aircraft will soon be replaced by satellites).

The potential benefits of the automated factory require significant time and resources for planning. Automation, however, will not be total—there will never be robot plumbers in the factory.

16.5 OVERALL REVIEW OF MANUFACTURING PLANNING

We incur great risk in going directly from a proposed manufacturing process plan into regular production. The time delays and extra costs involved in quality failures require a review of the proposed process, including software used with the process. Such a review is most effectively accomplished through preproduction trials and runs.

Ideally, product lots should be put through the entire system, with deficiencies found and corrected before full-scale production begins. In practice, companies usually make some compromises with this ideal approach. "Preproduction" may be merely the first runs of regular production but with special provision for prompt feedback and correction of errors as found. Alternatively, the preproduction run may be limited to features of product and process design that are so new that prior experience cannot reliably provide a basis for good risk taking. Although some companies do adhere to a strict rule of proving in the product and process through preproduction lots, the more usual approach is one of flexibility, in which the use of preproduction lots depends on various factors:

- The extent to which the product embodies new or untested quality features
- The extent to which the design of the manufacturing process embodies new or untried machines, tools, etc.
- The amount and value of product that will be out in the field before the extent of process, product, and use difficulties is fully known.

These trials sometimes include "production validation tests" to ensure that the full-scale process can meet the design intent. See *JQH5,* page 22.30, for an example.

The terms *process qualification* and *process certification* are also used to describe the review of manufacturing processes. Black (1993) describes how Caterpillar developed an internal certification program similar to its supplier certification program. The certification program is built around a 12-step manufacturing procedure to achieve process control and includes the identification of critical product characteristics and critical process parameters, the determination of process capability indexes, and actions for continuous quality improvement.

Preproduction trials and runs provide the ultimate evaluation—by manufacturing real product. Other techniques provide an even earlier warning before any product is made. For example, the failure mode, effect, and criticality analysis is useful in analyzing a proposed *product* design (see Section 13.5). The same technique can dissect potential failure modes and their effects on a proposed *process* design. Another technique makes use of highly detailed checklists for the review of proposed processes.

16.6 ORGANIZING FOR QUALITY IN MANUFACTURING OPERATIONS

The organization of the future will be influenced by the interaction of two systems: the technical system (design, equipment, procedures) and the social system (people, roles)—thus the name "sociotechnical systems" (STSs).

New ways of organizing work, particularly at the workforce level, are emerging. For example, supervisors are becoming "coaches"; they teach and empower rather than assign and direct. Operators are becoming "technicians"; they perform a multi-skilled job with broad decision making rather than a narrow job with limited decision making. Team concepts play an important role in these new approaches. In some organizations 40 percent of the people participate on a team; some organizations have a goal of 80 percent. Permanent teams (e.g., process teams, self-managing teams) are responsible for all output parameters, including quality; ad hoc teams (e.g., a quality project team) are typically responsible for improvement in quality. A summary of the most common types of quality is given in Section 8.7.

Although these various types of quality teams are showing significant results, the reality is that, for most organizations, daily work in a department is managed by a supervisor who has a complement of workers performing various tasks. This configuration is the "natural work team" in operations. But team concepts can certainly be applied to daily work. One framework for a team in daily operations work is the control process from the trilogy of quality processes. As applied to daily work, the steps are choose control subjects, establish measurement, establish standards of performance, measure actual performance, compare to standards, and take action on the difference. For a detailed discussion of the steps, see Chapter 5, "Quality Control." When the natural work team of the department is trained in these concepts, the work team gains greater control over the key work processes so that it can meet customer needs while increasing employee involvement and empowerment.

16.7 PLANNING FOR EVALUATION OF PRODUCT

The planning must recognize the need for formal evaluation of product to determine its suitability for the marketplace. Three activities are involved:

1. Measuring the product for conformance to specifications.
2. Taking action on the nonconforming product
3. Communicating information on the disposition of nonconforming product.

These activities are discussed in Section 19.3. But the activities impinge on the manufacturing planning process. For example, several alternatives are possible for determining conformance, i.e., have the activity done by production workers, by an independent inspection force, or by a combination of both. Increasingly, the combination approach is being employed.

What has evolved is the concept of self-inspection combined with product audit. Under this concept all inspection and all conformance decisions, both on the process and on the product, are made by the production worker. (Decisions on the action to be taken on a nonconforming product are *not,* however, delegated to the worker.) However, an independent audit of these decisions is made. The quality department inspects a random sample periodically to ensure that the decision-making process used by workers to accept or reject a product is still valid. The audit verifies the decision process. Note that, under a pure audit concept, inspectors are not transferred to do inspection work in the production department. Except for those necessary to do audits, inspection positions are eliminated.

If an audit reveals that wrong decisions have been made by the workers, the product evaluated since the last audit is reinspected—often by the workers themselves.

Self-inspection has decided advantages over the traditional delegation of inspection to a separate department:

- Production workers are made to feel more responsible for the quality of their work.
- Feedback on performance is immediate, thereby facilitating process adjustments. Traditional inspection also has the psychological disadvantage of an "outsider" reporting the defects to a worker.
- The costs of a separate inspection department can be reduced.
- The job enlargement that takes place by adding inspection to the production activity of the worker helps to reduce the monotony and boredom that are inherent in many jobs.
- Elimination of a specific station for inspecting all products reduces the total manufacturing cycle time.

EXAMPLE 16.4. In a coning operation of textile yarn, the traditional method of inspection often resulted in finished cones sitting in the inspection department for several days, thereby delaying any feedback to production. Under self-inspection, workers received immediate feedback and could get machines repaired and setups improved more promptly. Overall, the program reduced nonconformities from 8 percent to 3 percent. An audit inspection of the products that were classified by the workers as "good" showed that virtually all of them were correctly classified. In this company workers can also classify product as "doubtful." In one analysis worker inspections classified 3 percent of the product as doubtful, after which an independent inspector reviewed the doubtful product and classified 2 percent as acceptable and 1 percent as nonconforming.

EXAMPLE 16.5. A pharmaceutical manufacturer employed a variety of tests and inspections before a capsule product was released for sale. These checks included chemical tests, weight checks, and visual inspections of the capsules. A 100 percent visual inspection had traditionally been conducted by an inspection department. Defects ranged from "critical" (e.g., an empty capsule) to "minor" (e.g., faulty print). This inspection was time-consuming and frequently caused delays in production flow. A trial experiment of self-inspection by machine operators was instituted. Operators performed a visual inspection on a sample of 500 capsules. If the sample was acceptable, the operator shipped the full container to the warehouse; if the sample was not acceptable, the full container was sent to the inspection department for 100 percent inspection. During the experiment both the samples and the full containers were sent to the inspection department for 100 percent inspection with reinspection of the sample recorded separately. The experiment reached two conclusions: (1) the sample inspection by the operators gave consistent results with the sample inspection by the inspectors, and (2) the sample of 500 gave consistent results with the results of 100 percent inspection.

The experiment convinced all parties to switch to the sample inspection by operators. Under the new system, good product was released to the warehouse sooner, and marginal product received a highly focused 100 percent inspection. In addition, the level of defects *decreased.* The improved quality level was attributed to the stronger sense of responsibility by operators (they themselves decided if product was ready for sale) and the immediate feedback received by operators from self-inspection. But there was another benefit—the inspection force was reduced by 50 people. These 50 people were shifted to other types of work, including experimentation and analysis activities on the various types of defects.

Criteria for Self-Inspection

For self-inspection, some criteria must be met:

- Quality must be the number 1 priority within an organization. If this requirement is not clear, a worker may succumb to schedule and cost pressures and classify products as acceptable that should be rejected.
- Mutual confidence is necessary. Managers must have sufficient confidence in the workforce to be willing to give workers the responsibility of deciding whether the product conforms to specification. In turn, workers must have enough confidence in management to be willing to accept this responsibility.
- The criteria for self-control must be met. Failure to eliminate the management-controllable causes of defects suggests that management does not view quality as a high priority, and this environment may bias the workers during inspections. Workers must be trained to understand the specifications and perform the inspection.
- Specifications must be unequivocally clear. Workers should understand the use that will be made of their products (internally and externally) to grasp the importance of a conformance decision.
- The process must permit assignment of clear responsibility for decision making. An easy case for application is a worker running one machine, since there is clear responsibility for making both the product and the product-conformance decision. In contrast, a long assembly line or the numerous steps taken in a chemical process make it difficult to assign clear responsibility. Application of self-inspection to such multistep processes is best deferred until experience is gained with some simple processes.

Self-inspection should apply only to products and processes that are stabilized and meet specifications and only to personnel who have demonstrated their competence.

Worker response to such delegation of authority is generally favorable, the concept of job enlargement being a significant factor. However, workers who do qualify for self-inspection commonly demand some form of compensation for this achievement, e.g., a higher grade, more pay. Companies invariably make a constructive response to these demands, since the economics of making the delegation are favorable. In addition, the resulting differential tends to act as a stimulus to the nonqualified workers to qualify themselves.

An adjunct to self-inspection is the use of *poka-yoke* devices as part of the inspection. These devices are installed in the machine to inspect the process conditions and product results and provide immediate feedback to the operator. Devices such as limit switches and interference pins, used to ensure proper positioning of materials on machines, are *poka-yoke* devices for inspecting process conditions. Go/no go gages are examples of *poka-yoke* devices for inspecting the product. For elaboration, see The Productivity Press Development Team (1997). Grout and Downs (1998) compare the use of *poka-yoke* devices for controlling processes with statistical process control charts.

16.8
PROCESS QUALITY AUDITS

A quality audit is an independent review conducted to compare some aspect of quality performance with a standard for that performance. Application to manufacturing has been extensive and includes both audit of activities (process audits) and audit of product (product audit). A full discussion of quality audits is given in Chapter 23, "Quality Assurance."

Process quality audit includes any activity that can affect final product quality. This on-site audit is usually made on a specific process by one or more persons and makes use of the process operating procedures. Adherence to existing procedures is emphasized, but audits often uncover situations of inadequate or nonexistent procedures. The checklists presented earlier in this chapter on the three criteria for self-control can suggest useful specific subjects for process audits. Audits must be based on a foundation of hard facts that are presented in the audit report in a way that will help those responsible to determine and execute the required corrective action.

Peña (1990) explains an audit approach for processes. Two types of audits are employed: engineering and monitor. The engineering process audit is conducted by a quality assurance engineer and entails an intense review of all process steps including process parameters, handling techniques, and statistical process control. Table 16.5 shows the audit checklist.

TABLE 16.5
Audit checklist

1. Is the specification accessible to production staff?
2. Is the current revision on file?
3. Is the copy on file in good condition with all pages accounted for?
4. If referenced documents are posted on equipment, do they match the specification?
5. If the log sheet is referenced in specifications, is a sample included in the specification?
6. Is the operator completing the log sheet according to specifications?
7. Are lots with out-of-specification readings authorized and taken care of in writing by the engineering department or the proper supervisor?
8. Are corrections to paperwork made according to specification?
9. Are equipment time settings according to specification?
10. Are equipment temperature settings according to specification?
11. Is the calibration sticker on equipment current?
12. Do chemicals or gases listed in the specification match actual usage?
13. Do quantities listed in the specification match the line setup?
14. Are changes of chemicals or gases made according to specification?
15. Is the production operator certified? If not, is this person authorized by the supervisor?
16. Is the production operating procedure according to specification?
17. Is the operator performing the written cleaning procedure according to specification?
18. If safety requirements are listed in the specification, are they being followed?
19. If process control procedures are written in the specification, are the actions performed by process control verifiable?
20. If equipment maintenance procedures are written in the specification, are the actions performed verifiable? according to specification?

Source: Peña (1990). Reprinted with permission by the ASQ.

The monitor process audit covers a broad range of issues, e.g., whether specifications are correct and whether logs are filled in and maintained. Discrepancies (critical, major, or minor) are documented, and corrective action is required in writing. Critical defects must be corrected immediately; majors and minors must be resolved within five working days.

Product audit involves the reinspection of product to verify the adequacy of acceptance and rejection decisions. In theory such product audits should not be needed. In practice they can often be justified by field complaints. Such audits can take place at each inspection station for the product or after final assembly and packing. Sometimes an audit is required before a product may be moved to the next operation. For elaboration, see Section 23.12, "Product Audit."

16.9 QUALITY MEASUREMENT IN MANUFACTURING OPERATIONS

The management of key work processes must include provision for measurement. In developing units of measure, the reader should review the basics of quality measurement discussed in Section 5.2.

Table 16.6 shows examples for manufacturing activities. Note that many of the control subjects are forms of work output. In reviewing current units in use, a fruitful starting point is the measure of productivity. *Productivity* is usually defined as the amount of output related to input resources. Surprisingly, some organizations still mistakenly calculate only one measure of output, i.e., the total (acceptable *and* nonacceptable). Clearly, the pertinent output measure is that which is usable by customers (i.e., acceptable output).

The units in Table 16.6 become candidates for data analysis using statistical techniques such as control charts, discussed in Chapter 18. But there is a more basic point—the selection of the unit of measure and the periodic collection and reporting of data demonstrate to operating personnel that management regards the subject as having priority importance. This atmosphere sets the stage for improvement!

16.10 MAINTAINING A FOCUS ON CONTINUOUS IMPROVEMENT

Historically, the operations function has always been involved in troubleshooting sporadic problems. As chronic problems were identified, these were addressed using various approaches, such as quality improvement teams (see Chapter 3, "Quality Improvement and Cost Reduction"). Often the remedies for improvement involve quality planning or replanning (see Chapter 4, "Operational Quality Planning and Sales Income"). These three types of action are summarized in Table 16.7.

TABLE 16.6
Examples of quality measurement in manufacturing

Subject	Unit of measure
Quality of manufacturing output	Percentage of output meeting specifications at inspection ("first-time yield")
	Percentage of output meeting specifications at intermediate and final inspection
	Amount of scrap (quantity, cost, percentage, etc.), amount of rework (quantity, cost, percentage, etc.)
	Percentage of output shipped under waiver of specifications
	Number of defects found in product audit (after inspection)
	Warranty costs due to manufacturing defects
	Overall measure of product quality (defects in parts per million, weighted defects per unit, variability for critical characteristics, etc.)
	Amount of downgraded output
Quality of input to manufacturing	Percentage of critical operations with certified workers
	Amount of downtime of manufacturing equipment
	Percentage of product input meeting specifications
	Percentage of instruments meeting calibration schedules
	Percentage of specifications requiring changes after release

Kannan et al. (1999) in the Supplementary Reading present the results of a survey of the application of 38 quality management practices (e.g., use of benchmark data to improve quality practices), 39 tools and techniques (e.g., statistical process control), 29 areas of documentation (e.g., quality assurance manual), and 12 quality measurements (e.g., customer complaints) at the operations level in manufacturing industries. The analysis includes the degree of usage and the impact on five organization performance measures.

Maintaining the focus on improvement clearly requires a positive quality culture in the organization. Therefore, we must first determine the present quality culture (see Section 2.7) and then take the steps to change the culture to one that will foster continuous improvement (see Chapter 9, "Developing a Quality Culture"). In addition, the operations function must be provided with the support to maintain the focus on improvement. A key source of that support should be the quality department. Thus the quality department should view operations as its key internal customer and provide the training, technical quality expertise, and other forms of support to enable operations to maintain the focus on improvement. Also, a quality department can urge upper management to set up cross-functional teams to address operations problems that may be

TABLE 16.7
Three types of action

Type of action to take	When to take action	Basic steps
Troubleshooting (part of quality control)	Performance indicator outside control limits Performance indicator in clear trend toward control limits	Identify problem Diagnose problem Take remedial action
Quality improvement	The control limits are so wide that it is possible for the process to be in control and still miss the targets Performance indicator frequently misses its target	Identify project Establish project Diagnose cause Remedy the cause Hold the gains
Quality planning	Many performance indicators for this process miss their targets frequently Customers have significant needs that the product does not meet	Establish project Identify customers Discover customer needs Develop product Develop process Design controls

Source: Adapted from Juran Institute, Inc. (1995, pp. 5–7).

caused by other functional departments such as engineering, purchasing, and information technology.

SUMMARY

- Activities to integrate quality in manufacturing planning have two objectives: to prevent defects and to minimize variability.
- By creating a flow diagram, we can dissect a manufacturing process and plan for quality at each work station.
- To prevent defects and to minimize variability, we must discover the relationships between process variables and product results.
- Error-proofing a process is an important element of prevention.
- For human beings to be in a state of self-control, they must have knowledge of what they are supposed to do, knowledge of what they are actually doing, and a process that is capable of meeting specifications and can be regulated.
- Failure to meet all three of these criteria means that the quality problem is management controllable. About 80 percent of quality problems are management controllable.

- The key factors of dominance in manufacturing are setup, time, component, worker, and information.
- Under the self-inspection concept, conformance decisions are made by the production worker, with an independent audit by the quality department.
- Computer-integrated manufacturing is the process of applying a computer in a planned fashion from design through manufacturing and shipping; computer-aided manufacturing is the process in which a computer is used to plan and control the work of specific equipment; computer-aided design is the process by which a computer assists in the creation or modification of a design.
- In the group technology concept, all items are examined to identify those with sufficient similarity that a common design or manufacturing plan can be used.
- In a flexible manufacturing system, a group of several computer-controlled machine tools is linked by a materials handling system and a mainframe computer to process a group of parts completely.
- A process quality audit is an independent evaluation of any activity that can affect final product quality.

PROBLEMS

16.1. Visit a local manufacturing company and identify the departments that have the principal and collateral responsibilities for carrying out the planning activities for quality.

16.2. For a specific manufacturing operation with which you are familiar, describe how you would conduct a controllability study.

16.3. Your plant has just conducted a controllability study by analyzing the causes of a large sample of defective parts. The results showed that 45 percent of the causes were worker controllable and 55 percent management controllable. One program (diagnosis of specific causes, determination of remedies, etc.) is planned for the management-controllable problems. For the worker-controllable problems, a motivation program for the workers producing poor work is planned. You are asked to comment on this approach.

16.4. Study the system of performance feedback available to any of the following categories of people and report your conclusions on the adequacy of this feedback for controlling quality of performance.
(a) A motorist in city traffic.
(b) A student at school.
(c) A supermarket cashier.

16.5. For any process to which you can gain access, determine the major form of "dominance" affecting attainment of quality.

16.6. A steel manufacturing plant has had a chronic problem of scrap and rework in the wire mill. The costs involved have reached a level where they are a major factor in the profits of the division. All levels of personnel in the wire mill are aware of the problem, and there is agreement on the vital few product lines that account for most of the problem. However, no reduction in scrap and rework costs has been achieved. What do you propose as a next step?

16.7. Select a manufacturing task regularly performed by a worker. Use the checklists on self-control to determine whether the worker can be held responsible for the quality of the output.

16.8. Why are the three criteria of self-control a prerequisite for a successful motivation program?

REFERENCES

Black, S. P. (1993). "Internal Certification: The Key to Continuous Quality Success," *Quality Progress,* January, pp. 67–68.

Eibl, S., U. Kess, and F. Pukelsheim (1992). "Achieving a Target Value for a Manufacturing Process," *Journal of Quality Technology,* January, pp. 22–26.

Goldman, S. L., R. N. Nagel, and K. Preiss (1995). *Agile Competitors and Virtual Organizations,* Van Nostrand Reinhold, New York.

Grout, J. R. and B. T. Downs (1998). "Mistake-Proofing and Measurement Control Charts," *Quality Management Journal,* vol. 5, no. 2, pp. 67–75.

Juran, J. M. (1992). *Juran on Quality by Design,* Free Press, New York.

Lee, J. (1995). "Perspective and Overview of Manufacturing Initiatives in the United States," *International Journal of Reliability, Quality and Safety Engineering,* vol. 2, no. 3, pp. 227–233.

Nakajo, T. and H. Kume (1985). "The Principles of Foolproofing and Their Application in Manufacturing," *Reports of Statistical Application Research,* Union of Japanese Scientists and Engineers, Tokyo, vol. 32, no. 2, June, pp. 10–29.

Peña, E. (1990). "Motorola's Secret to Total Quality Control," *Quality Progress,* October, pp. 43–45.

The Productivity Press Development Team (1996). *5S for Operators—5 Pillars of the Visual Workplace,* Productivity Press, Portland, OR.

The Productivity Press Development Team (1997). *Mistake Proofing for Operators: The ZQC System,* Productivity Press, Portland, OR.

Shuker, T. J. (2000). "The Leap to Lean," *Annual Quality Congress Proceedings,* ASQ, Milwaukee, pp. 105–112.

Siff, W. C. (1984). "The Strategic Plan of Control—A Tool for Participative Management," *ASQC Quality Congress Transactions,* Milwaukee, pp. 384–390.

Snee, R. D. (1993). "Creating Robust Work Processes," *Quality Progress,* February, pp. 37–41.

Somerton, D. G. and S. E. Mlinar (1996). "What's Key? Tool Approaches for Determining Key Characteristics," *Proceedings of the Annual Quality Congress,* ASQ, Milwaukee, pp. 364–369.

Spear, S. and H. K. Bowen (1999). "Decoding the DNA of the Toyota Production System," *Harvard Business Review,* September–October, pp. 96–106.

Wright, J. R. (1998). "Cellular Manufacturing Leadership," *Proceedings of the Annual Quality Congress,* ASQ, pp. 646–655.

SUPPLEMENTARY READING

Manufacturing and quality: *JQH5,* Sections 22, 24, 27, and 28.

Skrabec, Q. R. Jr. (1997). "A Model for Customer Driven Manufacturing," *Quality Engineering,* vol. 9, no. 4, pp. 711–720.

Self-control: Dean, P. J., M. R. Dean, and R. M. Rebalsky (1996). "Employee Perceptions of Workplace Factors That Will Most Improve Their Performance," *Performance Improvement Quarterly,* vol. 9, no. 2, pp. 75–89.

Pooley, J. and D. H. B. Welsh (1994). "A Comparison of Russian and American Factory Quality Practices," *Quality Management Journal,* vol. 1, no. 2, pp. 57–70.

Quality management practices and tools: Kannan, V. R., K-C. Tan, R. B. Handfield, and S. Ghosh (1999). "Tools and Techniques of Quality Management: An Empirical Investigation of Their Impact on Performance," *Quality Management Journal,* vol. 6, no. 3, pp. 34–49.

WEBSITES

Best manufacturing practices: www.bmpcoe.org/search1/bestpractices.html
Lean manufacturing: www.nwlean.net
Poka-yoke: www.campbell.berry.edu/pokayoke

17

OPERATIONS—SERVICE SECTOR

17.1 THE SERVICE SECTOR

In many service organizations the cost of poor quality ranges from 25 to 40 percent of operating expenses. For example, in one large bank the cost of poor quality is 37 percent for automatic teller machines (ATMs), 26 percent for customer inquiry centers, and 50 percent for commercial loan operations (*JQH5,* page 33.11). Such numbers prove the need for addressing quality in the service sector.

The service sector of an economy consists of a wide spectrum of industries. The major categories of service industries are listed in Table 17.1.

The Supplementary Reading directs the reader to literature on quality in several service industries.

The variety of service industries gives rise to distinguishing characteristics:

- A physical product may be involved (a restaurant meal).
- An interaction with a service person may or may not take place—different delivery channels (an ATM machine or a bank teller).
- Contact with a service person may be verbal only or may be in person (a call center or a salesperson).
- Contact with a service person may have degrees of knowledge content (fast-food service or advice from a financial planner).
- The service may be brief or have an extended duration (retail transaction or energy services).
- Spatial proximity for personal encounter (an open counter at a fast-food restaurant or a remote lane at a drive-in bank).
- Backroom operations may or may not be involved (haircut or an insurance company).

TABLE 17.1
Categories of service industries

- Transportation (railroads, airlines, bus lines, subways, common carrier trucking, pipelines)
- Public utilities (telephone communication, energy services, sanitation services)
- Marketing (retail food, apparel, automotive, wholesale trade, department stores)
- Finance (banks, insurance, sales finance, investment)
- Real estate
- Restaurants, hotels and motels
- News media
- Business services (advertising, credit services, computer services)
- Health services (nursing, hospitals, medical laboratories)
- Personal services (amusements, laundry and cleaning, barber and beauty shops)
- Professional services (lawyers, doctors)
- Repair services (garages, television and home repairs)
- Government (defense, health, education, welfare, municipal services)

But there are also similarities among some service industries: The service usually must be provided when the customer requests it, the service output is created as it is delivered, the service usually cannot be stored in an inventory, and completion time is critical. The differences among service industries make it difficult to generalize on how to approach quality—even starting with a definition of the word *quality.*

As in the case of manufactured product, quality means customer satisfaction and loyalty (see Section 1.3). Satisfaction and loyalty are achieved through two components: product features and freedom from deficiencies. One approach to further defining quality for the service sector is the SERVQUAL model (Zeithaml et al., 1990). This model identifies five dimensions of quality:

- *Tangibles:* appearance of facilities, equipment, personnel, materials.
- *Reliability:* ability to perform dependably and accurately.
- *Responsiveness:* provide timely service.
- *Assurance:* trust and confidence in employees.
- *Empathy:* individualized attention to customers.

To go to the next level of detail requires customizing a definition of quality for the specific service industry to reflect the distinguishing characteristics of that industry. What evolves is a service design that incorporates the product features and a process design free of deficiencies needed to satisfy and make customers loyal.

17.2
INITIAL PLANNING FOR QUALITY

Section 4.11 presents a road map for planning quality in a new product. These steps are: Establish the project, identify the customers, discover customer needs, develop the product, develop the process, and establish process controls/transfer to operations.

In the service industries the *service design* defines the features of the output provided to customers to meet their needs. Thus the concepts presented in Chapter 12, "Understanding Customer Needs," and Chapter 13, "Designing for Quality," apply to both the manufacturing and service sectors. The service design is turned into a reality by the *service process*, i.e., the process features such as the work activities, people, equipment, and physical environment to meet customer needs. The concept of quality function deployment (see Section 13.4 under "Quality Function Deployment") is helpful in designing both service products and service processes.

The service process can be reviewed for quality by several means: analyze the process flow diagram, reduce the cycle time, error-proof the process, plan for a neat and clean workplace, provide for supplier quality, qualify the process by validating the process and measurement capability, and plan for personnel self-control. These topics are discussed below and in the next chapter.

Analysis of the Process Flow Diagram

Process flow diagrams (alias process maps or process blueprints) are also discussed in Chapters 3, 6, and 16. Figure 17.1 shows a flow diagram for handling a request for adjustments to customer bills (AT&T). The "line of interaction" is the boundary of activities where the customer and frontline employees ("on-line group") have discussions. The "line of invisibility" separates activities that are seen or not seen by customers. The "organization boundary" shows which activities occur in the three departments involved in the process. Note how this example illustrates both frontline direct customer contact and back-room or back-office operations. Also note that the time required is shown for some activities.

The symbol P denotes process points at which problems could occur that would cause customer dissatisfaction. To *prevent* problems, we must identify potential problems, usually based on past data or an analysis of the flow diagram. In the service sector, problems arise for recurring reasons (AT&T):

- Promises are not kept.
- Customers must contact several people to achieve problem resolution.
- Only partial service is provided, the service is performed incorrectly, the wrong service is performed, the wrong information is provided.
- Customers do not understand the service provided.
- Service is not provided when needed or takes too long.
- The customer is inconvenienced with paperwork or other matters.

These known or potential problems should be identified on the flow diagram and preventive actions put in place. If necessary, the problems can be regarded as process failures and analysis tracked using the approach of a failure mode, effect, and criticality analysis (see Section 13.5 under "Failure Mode, Effect, and Criticality Analysis").

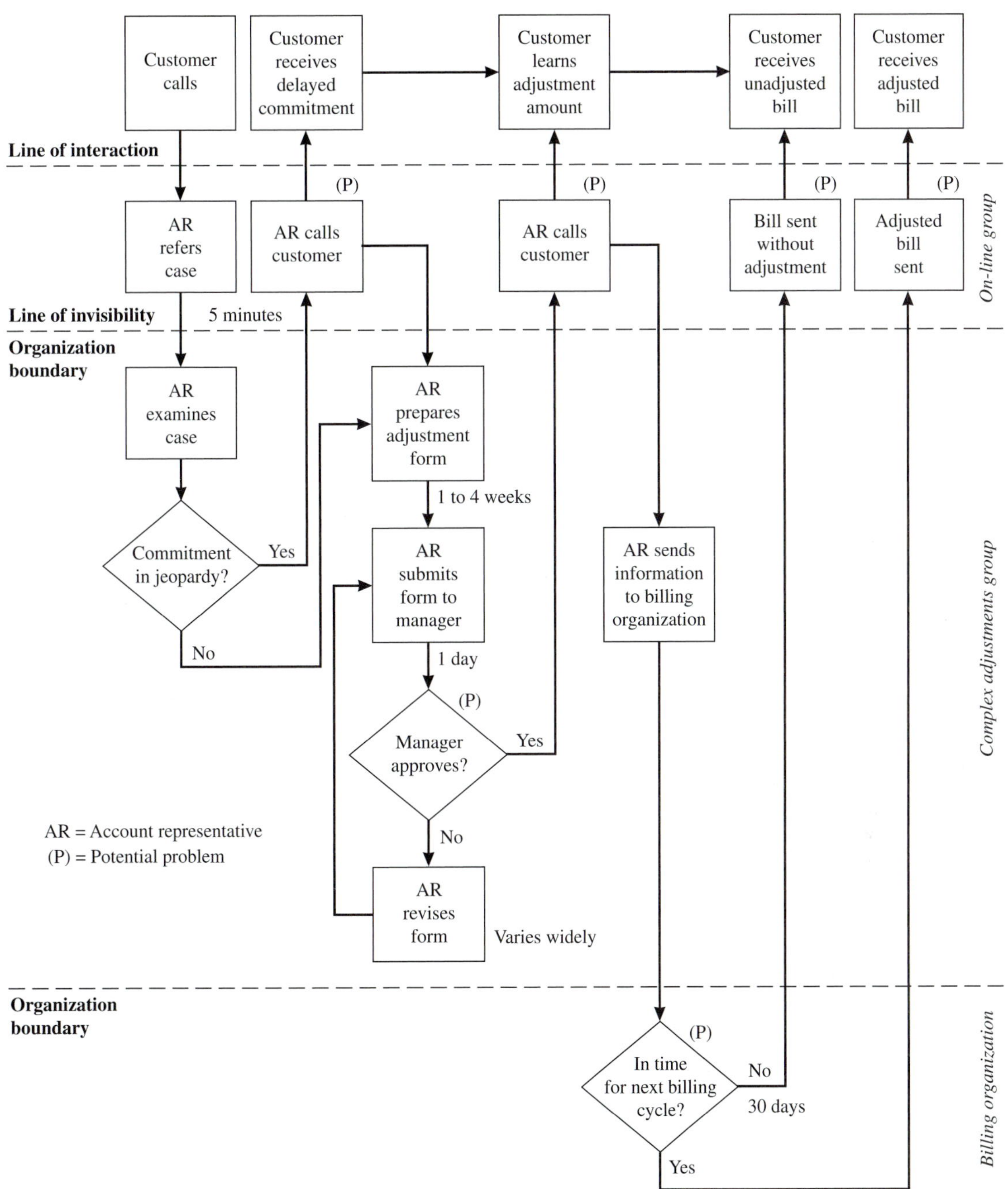

FIGURE 17.1
Service blueprint diagram. (*Reproduced with permission of AT&T.*)

Reduction of Process Cycle Time

Customers are demanding shorter delivery times for services provided, and thus time becomes a quality parameter. For existing processes the quality improvement process (see Chapter 3) is a structured approach for diagnosing the causes and taking corrective action on excessive cycle time. Among the common actions that can be taken are

- Eliminate rework loops to correct process errors.
- Eliminate or simplify steps of marginal value to the customer.
- Eliminate redundant steps such as inspections or reviews (but not until the causes of errors have been determined and eliminated).
- Combine steps and have them done by one worker, by several workers in a work cell, or by a multifunctional team.
- Transfer approval steps to lower levels.
- Change the sequence of activities from consecutive steps to simultaneous steps.
- Perform a step after serving a customer rather than before.
- Use technology to perform both routine and complex steps.

Marriott Hotels analyzed the elapsed time of the guest registration process. The analysis led to the integration of two subprocesses—the front-desk function and the bellperson function (to escort new guests to their room and carry their luggage). Under a revised process the front desk now preassigns a suitable guest room and assembles registration information and keys for the bellperson who greets and escorts the guests directly to their rooms (Hadley, 1995).

The flow diagram serves as a basic tool for analyzing cycle time.

Error-Proofing the Process

An important element of prevention of errors is error-proofing the process. Five important principles of error-proofing a process are presented in Section 16.2 under "Error-proofing the Process":

- Eliminate the error-prone activity.
- Substitute a more reliable process.
- Make the work easier for the worker.
- Detect errors earlier.
- Minimize the effect of errors.

In the service sector, examples of error-proofing include eliminating non-value-added activities, automating routine or unpleasant human tasks, using software to detect missing or grossly erroneous information, using bar codes at the supermarket checkout, using countdowns and checklists during surgical operations, entering medical prescriptions into a computer to replace handwritten prescriptions, and performing automatic built-in inspections. Computers and other forms of technology are an important source of error-proofing aids.

A useful adjunct to error-proofing is the neat and clean workplace. One framework is the 5S approach (sort, set in order, shine, standardize, and sustain). For elaboration, see Section 16.2 under "Plan for Neat and Clean Workplace."

Input from Information Technology as an Internal Supplier

Service industries need to plan, control, and improve the quality of inputs from external suppliers. The basic approach is provided in Chapter 15, "Supply Chain Management." The operations function within a service industry also has internal suppliers, and one of these deserves particular note: information technology (IT).

Many of the significant advances in providing new service product features, with amazing speed, are due to the unending contributions of computers and related IT software and equipment. A primary example is the integrated database that supports call centers by enabling personnel to handle a wide variety of customer problems. Without IT concepts these advances would never have taken place. Further, with the increasing role of the Internet and electronic commerce, we can look forward to additional advances that will benefit all areas of society.

These contributions have, unfortunately, been accompanied by significant problems in the services provided to operations people by the IT activity. Operations people report that these problems are sporadic and chronic quality-related problems that result in customer dissatisfaction and higher costs.

Research provides some understanding of the nature of these IT-related problems (Kittner et al., 1999). Thirteen operations managers from the financial services sector (but not from the IT or quality functions) were brought together in an electronic meeting room where GroupSystems software was used to enable anonymous and simultaneous responses from participants. The organizations included banking, investment banking, credit card processing, and insurance. The "operations" are activities that process customer transactions, including both direct contact with external customers and back-room activities that do not involve direct contact. Research determined the perceived impact of the quality of IT services on the quality of the output of the operations function.

First, the participants identified the outputs the IT department provided to operations. The question posed was, What services does the IT department provide to your area? One hundred and two responses were recorded, and these are summarized in 11 categories in Table 17.2.

Next, participants were asked to identify problems involved in these services. The managers identified 115 issues—some were general, such as "downtime," and others were quite specific, such as "can't hire C++ programmers fast enough to keep up with user demand." These issues were further analyzed to identify 57 issues related to customer satisfaction. Participants were then asked to score each item on a scale of 1 to 10, where 10 was most important and 1 was least important. The mean scores of the top 10 issues are shown in Table 17.3.

In summary, Table 17.3 shows the perceptions of operations managers on problems with input received from IT and the relative importance of these problems to output of the operations function. Conclusion: Significant sporadic and chronic problems

TABLE 17.2
IT services provided to operations

Response category	Examples
Provide reports	Summary of sales activities, list of "at risk" accounts
Provide hardware and software support	Personal computer support services, guidance on hardware and purchasing
Record information	Posting to accounts, current rate information
Provide data communications support	Local area network connections and software access, data security
Provide for system availability and maintenance	On-line availability of all systems, ensure e-mail systems are up and running
Provide on-line information	Automatic teller machine on-line, customer account information
Provide support for data processing	Download all accounts for customer information; download all workplace banking customers and their profiles
Develop new systems	Application development for new systems; systems development, testing, and design
Implement new systems	Assist in the development of testing plans and rollout, implementation of new systems
Provide programming support	Provide programming support
Provide training	Software training

TABLE 17.3
Top 10 quality issues

Issue	Mean
Accuracy of information	9.33
System downtime	8.92
System response time	8.58
Network reliability	8.42
System performance	8.33
Level of expertise	8.17
Hardware performance	8.09
Thorough testing—both IT and users need to be involved	8.08
Hardware problems	7.92
Lack of training	7.75

exist that cannot be solved by the traditional IT help desk. Chronic problems require the structured approach to quality improvement described in Chapter 3; sporadic problems require the troubleshooting approach described in Section 5.9 under "Troubleshooting."

The demands continually made on the IT function both for information services using current software and the development of new software make it almost impossible to effectively address chronic problems. But these problems often have a serious impact on external customer satisfaction. A quality department can help by taking a strong initiative to collect data to prove that action is necessary. Such proof can help to convince upper management to provide the IT and operations functions with the resources and a framework for improvement based on the project-by-project approach described in Chapter 3.

Input from Customers as an External Supplier

In service organizations the customer is often a supplier. Customers may provide the detailed data for a service transaction, e.g., buy 100 shares of American Express stock at a price not to exceed $Y; order a dress of size 16 in the rose color; reserve a hotel room for March 12 to March 18. The customer input can be in error, resulting in delays to the customer and extra time and costs to the organization that must rectify the customer's error. *JQH5,* page 33.17, describes three approaches financial service providers use to prevent customer errors: customer education, error-proofing, and monitoring and measuring customer input.

Measuring Process Capability

We need to assure that the process is able to meet the customer needs and goals under normal operating conditions. When customer needs can be quantified in particular parameters, then process capability indexes can be used to evaluate the process. In the six-sigma approach, the process capability can be described in units of sigma; e.g., a process might be at a level of 4.8 sigma out of the ideal of 6 sigma. These matters are discussed in Section 18.15. In the service sector an important quantitative parameter is the time to complete the service transaction.

A preliminary measure of capability can be obtained by simply collecting a sample of data and comparing it to the specifications for a process. For example, consider the process for repairing ATM machines. An analysis concluded that the service organization should be able to respond in 10 minutes, and customers indicated that a machine should not be out of service for more than 90 minutes. A team looked at historical data measuring the repair time each day over a six-month period. The results are shown in the histogram of Figure 17.2. Note that some observations exceeded the maximum time of 90 minutes. Thus the process is not capable. The process must be changed and tested again to verify that it is capable. This preliminary capability study is based on available data. A full study under controlled conditions should be conducted following the methods described in Section 18.10.

A broader approach to evaluating a service process measures four parameters: effectiveness (of output), efficiency, adaptability, and cycle time. For elaboration, see Section 6.6.

IBM defines five levels of process maturity. The highest level, level 1, designates a business process that operates at maximum effectiveness and efficiency and serves as a benchmark or leader; the lowest level, level 5, suggests a process that is ineffective and may have major deficiencies. Melan (1993) defines the specific criteria for

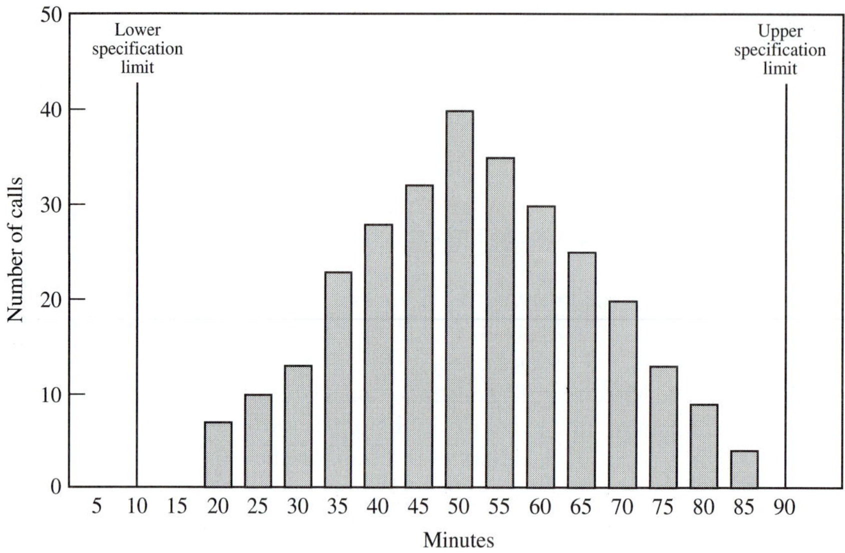

FIGURE 17.2
Time to repair machines. (*From Juran Institute, Inc.*)

each level. Criteria include organizational matters (e.g., a process owner) and technical matters (e.g., measurements for effectiveness and efficiency).

17.3 PLANNING FOR SELF-CONTROL

The concept of self-control was introduced in Section 5.3. Then, in Chapter 16, we applied the concept to manufacturing industries and provided checklists to evaluate manufacturing planning for quality. This chapter applies the concept to service industries and provides checklists to evaluate planning for quality in services.

For self-control, we must provide people with

1. Knowledge of what they are supposed to do.
   - Clear and complete work procedures.
   - Clear and complete performance standards.
   - Adequate selection and training of personnel.
2. Knowledge of what they are actually doing (performance).
   - Adequate review of work.
   - Feedback of review results.
3. Ability and desire to regulate the process for minimum variation.
   - A process and job design capable of meeting quality objectives.
   - Process adjustments that will minimize variation.

- Adequate worker training in adjusting the process.
- Process maintenance to maintain the inherent process capability.
- A strong quality culture and environment.

In designing for self-control, technology is of increasing help—for "supposed to do," information can be placed on line and kept up-to-date instantly; for "knowledge on how we are doing," we can now often provide instant feedback; for "regulating a process," mechanisms can be integrated into a process to provide adjustments when necessary.

Most service organizations do not adhere to the three elements and 10 subelements of self-control. Collins and Collins (1993) provide six examples from the manufacturing and service sectors illustrating problems that were originally blamed on people but that really were management controllable. Berry, Parasuraman, and Zeithaml (1994) identify 10 lessons learned in improving service quality. Three of these are service design (real culprit is poor service system design), employee research (ask employees why service problems occur and what they need to do their jobs), and servant leadership (managers must serve by coaching, teaching, and listening to employees). Note that these three lessons are directly related to the concept of self-control.

The concept of self-control applies to the service sector—both to frontline direct customer contact operations and to back-room operations. Based on research with personnel in the financial services industry, Shirley and Gryna (1998) developed checklists for self-control, and these are presented below.

Criterion 1: Knowledge of "Supposed to Do"

The manufacturing sector makes extensive use of product and process specifications and work procedures, but the use of such documents is still evolving in the service sector. Nevertheless, providing personnel in the service sector with the knowledge of what they are supposed to do is essential for self-control. The following checklist can help to evaluate this criterion.

Work procedures

1. Are job descriptions published, available, and up-to-date?
2. Do personnel know who their customers are? Have personnel and customers ever met?
3. Do personnel who perform the job have any impact on the formulation of the job procedure?
4. Are job techniques and terminology consistent with the background and training of personnel?
5. Do guides and aids (e.g., computer prompts) lead personnel to the next step in a job?
6. Do provisions exist to audit procedures periodically and make changes? Are changes communicated to all affected personnel?
7. Do provisions exist for deviations from corporate directives to meet local conditions?
8. Are procedures "reader friendly"?

9. Does supervision have a thorough knowledge of operations and provide assistance when problems arise?
10. Do procedures given to personnel fully apply to the job they do in practice?
11. Have personnel responsibilities been clearly defined in terms of decisions and actions?
12. Do personnel know what happens to their output in the next stage of operations and understand the consequences of not doing the job correctly?
13. If appropriate, is job rotation used?

Performance standards

14. Are formal job standards on quality and quantity needed? If yes, do they exist? Are they in written form?
15. Have personnel been told about the relative priority of quality versus quantity of output? Do personnel really understand the explanation?
16. Are job standards reviewed and changed when more tasks are added to a job?
17. Do personnel feel accountable for their output, or do they believe that shortcomings are not under their control?
18. Does information from a supervisor about how to do a job always agree with information received from a higher-level manager?

Training

19. Are personnel given an overview of the entire organization?
20. Is there regularly scheduled training to provide personnel with current information on customer needs and new technology?
21. Do personnel and their managers provide input to their training needs?
22. Does training include the why, not just the what?
23. Does the design of the training program consider the background of those to be trained?
24. Do the people doing the training provide enough detail? Do they know how to do the job?
25. Where appropriate, are personnel who are new to a job provided with mentors?

Criterion 2: Knowledge of Performance

For self-control, people must have the means of knowing whether their performance conforms to standard on product and process characteristics. The checklist below can help to evaluate this criterion.

Review of work

1. Are personnel provided with the time and instructions for performing self-review of their work?
2. Can errors be detected easily?
3. Are independent checks on quality needed? Are they performed? Are these checks performed by peer personnel or others?
4. Is a review of work performed at various checkpoints in process, not just when work is complete? Is the sample size sufficient?

5. Does an independent audit of an entire process ensure that individual work assignments are integrated to achieve process objectives?
6. Where appropriate, are detailed logs kept on customer contacts?

Feedback

7. Do upper management and supervision provide the same message and actions on the importance of quality versus quantity?
8. If needed, do standards exist on making corrections to output?
9. Where appropriate, is feedback provided to both individuals and a group of personnel? Is time provided for discussion with the supervisor, and does the discussion occur?
10. Is feedback provided to those who need it? Is it timely? Is it personnel specific?
11. Does feedback provide the level of detail needed particularly to correct problem areas? Have personnel been asked what detail is needed in the feedback?
12. Is feedback provided from customers (external or internal) to show the importance of the output and its quality?
13. Does feedback include information on both quality and quantity?
14. Are both positive and negative (corrective) feedback provided?
15. Is negative feedback given in private?
16. Do personnel receive a detailed report of errors by specific type of error?
17. Where appropriate, are reports prepared to describe trends in quality (in terms of specific errors)? Are such reports prepared for individual personnel and for an entire process performed by a group of people?
18. Are certain types of errors tracked with feedback from external customers? Could some of these be tracked with an internal early indicator?

A credit card provider has identified 18 key processes such as credit screening and payment processing. For the total of 18 processes, more than 100 internal and supplier process measures were identified. Daily and monthly performance results are available through video monitors and are also posted. Each morning the head of operations meets with senior managers to discuss the latest results, identify problems, and propose solutions. Employees can access a summary of this meeting via telephone or electronic mail. The measurement system is linked to compensation by a daily bonus system that provides up to 12 percent of base salary for nonmanagers and 8 to 12 percent for managers (Davis et al., 1995).

Criterion 3: Ability to Regulate

The process given to personnel must be capable of meeting requirements, and the job design must include the necessary steps and authority for personnel to regulate the process. Here is a checklist to evaluate the ability to regulate.

Job design

1. Is the process (including procedures, equipment, software, etc.) given to personnel capable of meeting standards on quality and quantity of output? Has this capability been verified by trial under normal operating conditions?

2. Has the design of the job made use of the principles of error-proofing?
3. Does the job design minimize monotonous or unpleasant tasks?
4. Does the job design anticipate and minimize errors due to normal interruptions in the work cycle?
5. Can special checks be created (e.g., balancing of accounts) to detect errors?
6. Can steps be incorporated in data entry processes to reject incorrect entries?
7. Does the job design include provisions for action when wrong information is submitted or information is missing as an input to a job?
8. Is paperwork periodically examined, and are obsolete records destroyed to simplify working condition?
9. When the volume of work changes significantly, do provisions exist to adjust individual responsibilities or add resources?
10. Do external factors (e.g., no account number on a check, cash received instead of a check) hinder the ability to perform a task?
11. Are enough personnel cross trained to provide an adequate supply of experienced personnel for filling in when needed?
12. If appropriate, is a "productive hour" scheduled each day in which phone calls and other interruptions are not allowed, thus providing time to be away from the work location to attend to other tasks?
13. Has equipment, including any software, been designed to be compatible with the abilities and limitations of personnel?
14. Does an adequate preventive maintenance program exist for computers and other equipment used by personnel?
15. Do some personnel possess a hidden knack that needs to be discovered and explained to all personnel?
16. For a job requiring special skills, have personnel been selected to ensure the best match of their skills and job requirements?

Changes in job design

17. Are proposed changes limited by technology (e.g., address fields on forms)?
18. Can personnel institute changes in a job when they show that the change will provide benefits? Are personnel encouraged to suggest changes?
19. What levels of management approval are required for proposed changes to be instituted? Could certain types of changes be identified as not needing any level of management approval?
20. Do management actions confirm that they are open to recommendations from all personnel?

Handling problems

21. Have personnel been provided with the time and training to identify problems, analyze problems, and develop solutions? Does this training include diagnostic training to look for patterns of errors and determine sources and causes?
22. Are personnel permitted to exceed process limits (e.g., maximum time on a customer phone call) if they believe it is necessary?
23. When personnel encounter an obstacle on a job, do they know where to seek assistance? Is the assistance conveniently available?

Participants in the research rated all three criteria of self-control as highly important. On a scale of 1 (not important at all) to 10 (most important), criterion 1 had a rating of 9.64, criterion 2 a rating of 9.45, and criterion 3 a rating of 9.36.

These checklists can help operations in the design (and redesign) of jobs to prevent errors, the diagnosis of quality problems on individual jobs, and the identification of common weaknesses in many jobs. The checklists can also help supervisors coach personnel, prepare for process audits, and develop training classes on quality. For elaboration, see *JQH5,* pp. 22.56–22.57. We proceed now from planning for quality in service operations to the control of quality in operations.

17.4 CONTROL OF QUALITY IN SERVICE OPERATIONS

The manufacturing sector has a long history of detailed procedures for controlling quality during operations. In the service sector the formalization of such procedures is still evolving. The basic steps for controlling quality are described in Chapter 5, "Quality Control." These steps are choose control subjects, establish measurement, establish standards of performance, measure actual performance, compare performance to standards, and take action on the difference. Applying these steps to service operations is discussed below.

Choose Control Subjects

To choose control subjects, we must first identify a major work process, identify the process objective, describe the process, identify customers, and discover customer needs. These steps, discussed in Chapter 5, lead to the selection of control subjects.

Examples of control subjects in several service industries are given in Table 17.4.

Establish Measurement

In this step we develop a unit of measure and the means of measurement (the sensor)—see Chapter 5. The unit of measure must be understandable to all, specific enough for decision making, and customer focused.

Control subjects can be a mixture of features of the product, features of the process, and side effects of the process. Quantification of control subjects involves two kinds of indicators that must be made explicit for those running the process:

1. *Performance indicators.* These indicators measure the output of the process and its conformance to customer needs as defined by the unit of measure for the control subject.
2. *Process indicators.* These indicators measure activities or variation within the process that affect the performance indicators.

TABLE 17.4
Control subjects

Major work product	Major work process	Control subjects
Medical insurance	Claims processing	Accuracy of claim form
		Completeness of supporting documentation
Catering services	Food preparation	Freshness of ingredients
		Oven temperature
24-hour banking services	Maintenance of ATM machines	Availability of cash
		Number of service people available
Photo developing	Film processing	Maintenance of chemicals
		Accuracy of placement of film on spool

Source: Juran Institute, Inc.

TABLE 17.5
Federal Express service quality indicators

Indicator	Weight
Abandoned calls	1
Complaints reopened	5
Damaged packages	10
International	1
Invoice adjustments requested	1
Lost packages	10
Missed pickups	10
Missing proofs of delivery	1
Overgoods (lost and found)	5
Right-day late deliveries	1
Traces	1
Wrong-day late deliveries	5

Source: American Management Association (1992).

Table 17.5 shows 12 quality indicators at Federal Express along with the importance weights assigned to each indicator.

These measures are tracked every day, both individually and in total.

Figure 17.3 shows how a division of AT&T related internal metrics to business processes and customer needs.

Specific measurements can be related to the underlying factors that dominate a process: setup, time, component, worker, and information. For elaboration, see Section 16.3 under "Control systems and the concept of dominance."

Deyong and Case (1998) provide a methodology for linking customer satisfaction attributes with process metrics in service industries.

	Business process	Customer need	Internal metric
Overall quality	Product (30%)	Reliability (40%)	% repair call
		Easy to use (20%)	% calls for help
		Features/functions (40%)	Function performance test
	Sales (30%)	Knowledge (30%)	Supervisor observations
		Response (25%)	% proposal made on time
		Follow-up (10%)	% follow-up made
	Installation (10%)	Delivery interval (30%)	Average order interval
		Does not break (25%)	% repair reports
		Installed when promised (10%)	% installed on due date
	Repair (15%)	No repeat trouble (30%)	% repeat reports
		Fixed fast (25%)	Average speed of repair
		Kept informed (10%)	% customers informed
	Billing (15%)	Accuracy, no surprises (45%)	% billing inquiries
		Resolve on first call (35%)	% resolved first call
		Easy to understand (10%)	% billing inquiries

FIGURE 17.3
Relating internal metrics to business processes and customer needs.
(*From Kordupleski et al., 1993. By permission of The Regents of the University of California. Reprinted from the California Management Review.*)

Establish Standards of Performance

This step involves setting not only target values on each performance indicator and each process indicator but also maximum and minimum limits.

Measure Actual Performance

Sometimes the measurement process can be automated. For example, the number of calls waiting to be answered in an insurance service center is clearly displayed; the elapsed time in filling a customer order at a fast-food franchise is clearly visible (the goal is 45 seconds).

In other cases the measuring process is more complex. For example, to measure the quality of customer contact activities, banks and other organizations use "mystery shoppers"—researchers posing as customers to assess key dimensions of quality delivery.

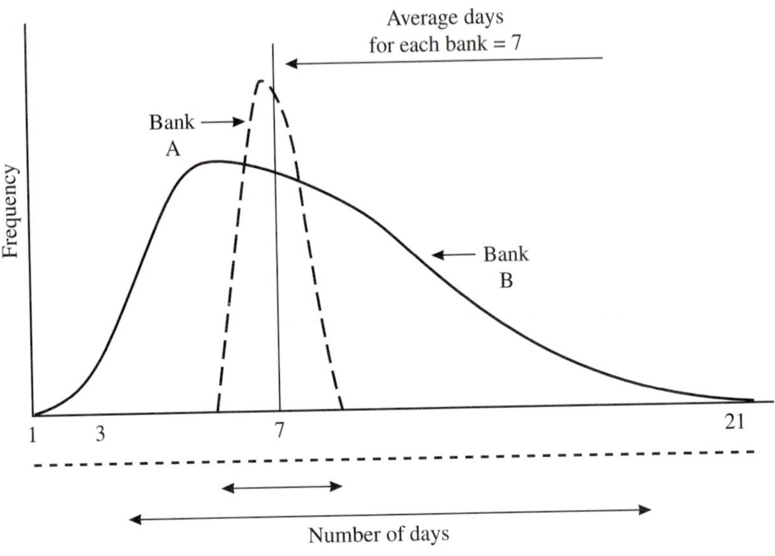

FIGURE 17.4
Time to process a loan. (*From Juran Institute, Inc.*)

Compare Performance to Standards

Sometimes the comparison must go beyond a simple analysis of average values. For example, in Figure 17.4 two banks are compared on the time to process a loan. Both banks average about seven days. For bank A almost all the loans are processed in a 5- to 8-day window; bank B has an average of 7 days, but some loans take about 21 days to process.

The comparison of actual to a target may make use of statistical process control techniques to distinguish between common and special causes. This approach is explained in Chapter 18, "Statistical Process Control."

Take Action on the Difference

Failure to meet standards may require one of three actions: troubleshooting, quality improvement, or quality planning. The steps for troubleshooting are similar to those for diagnosis in quality improvement. This process includes developing theories on possible causes and testing the theories (using data) to find the cause of the problem. The process for troubleshooting and diagnosing sporadic problems is usually simpler than the process for achieving improvement on chronic problems. For further discussion on the three types of action, see Section 16.10.

17.5
PROCESS QUALITY AUDITS

A *quality audit* is an independent review conducted to compare some aspect of quality performance with a standard for that performance. A discussion of various types of quality audits is presented in Chapter 23, "Quality Assurance."

A *process quality audit* includes any activity that can affect final product quality. This on-site audit is usually made on a specific process and makes use of process operating procedures. Adherence to existing procedures is emphasized, but audits often uncover situations of inadequate or nonexisting procedures. The self-control checklists presented earlier in this chapter can suggest useful specific subjects for process audits.

A major airline employs process audits to evaluate service in three areas: airport arrivals and departures, aircraft interior and exterior, and airport facilities. Forty-seven specific activities are audited periodically, and then performance measurements are made and compared with numerical goals. Two examples on the aircraft are the condition (appearance) of carpets inside the planes and the adhesion of paint on the planes.

McDonald's Corporation audits restaurants through a series of announced and unannounced visits. The audit includes quality, service, cleanliness, and sanitation. Highly detailed audit items include numerical standards on food-processing variables.

17.6
FRONTLINE CUSTOMER CONTACT

A basic activity in the service industries is the service encounter, i.e., the contact made with the client when meeting a customer's need. Typically examples involve a bank teller processing deposits or withdrawals of money, a flight attendant providing services on an airplane, or a hotel clerk registering a guest. In all cases the quality of the transaction involves both the technical adequacy of the result and the social skills of the "frontline" person who conducts the transaction. Three factors emerge as important: selection of the frontline employee, training of that employee, and "empowerment" of the employee to act to meet customer needs. (The reader should compare these factors to the three elements of self-control; see Section 5.3.)

In person-to-person type transactions, employee *selection* often has an immediate, direct, and lasting impact on customer perception. Some people have the necessary personal characteristics for frontline personnel; some people do *not* have these characteristics, even with training. Proper selection requires that we identify the personal characteristics required for a position (how about asking experienced employees?), use multiple interviews, train managers in interview procedures, identify nominees from among present employees, and ask for recommendations from present employees. As one example, Federal Express selects people using scientifically prepared profiles of successful performers.

Training, of course, is essential. Superior-quality service organizations devote from 1 to 5 percent of employee working hours to training. The content of the training depends on the job requirements, but often it stresses product knowledge. In addition, training involves activities such as role playing to handle situations when a transaction goes wrong and handling irate customers who have a complaint. Such training takes time and special effort, e.g., at Lands' End each mail-order telephone representative spends time in a warehouse viewing the products. Isn't it impressive when such a person is able to describe the color "medium gray" to a customer? Basically, however, the training must enable the employee to provide customers with dependable and regular service. Customers are impressed when they know that they can depend on uniformly good service from an organization. When an extraordinary situation arises and it is handled well, we have a delighted customer.

Empowerment, a key step beyond training, involves giving a new degree of authority to frontline employees. The term usually means encouraging employees to handle unusual situations that standard procedures don't cover. In the past the employee would check with a superior—while the customer waited—and waited. The concept minimizes the use of the rule book and maximizes the use of the frontline employee's knowledge, initiative, and judgment to take the action necessary to meet the need of a customer standing at a service counter. For example, the policy manual at Nordstrom's department store states: "Use your own best judgment at all times." Risky? Some people would think so, but the opportunities for customer satisfaction and the employee attitude toward "ownership" of his or her job are strongly convincing. Section 9.7, under "Empowerment," provides further discussion.

JQH5, page 25.25, discusses six human resource factors (including the three above) in terms of goals, metrics, and improvement approach.

Bank One Corp. uses an extensive shopper survey to measure the performance of bank tellers. The survey queries customers on specific contact items such as a friendly greeting, employee identification, eye contact, a smile, use of the customer's name during the transaction, undivided employee attention, accurate processing of the transaction, ability to give clear explanations, a professional appearance, and a neatly organized work area.

The investment in selection, training, and empowerment of employees can lead to the fulfillment of a company's dream—delighted customers. Table 17.6 provides some categories of action and examples of remarkable exploits to generate delighted customers.

The service encounter goes beyond "speed, content, and attitude." A research study of five service organizations and nine manufacturers identified 15 competencies that exemplify superior frontline service (Jeffrey, 1995). The top six competencies were considered the most important by the organizations and their customers. These competencies are

1. *Build customer loyalty and confidence.* Meet customer needs and do what is sensible to maintain customer goodwill.
2. *Empathize with customers.* Be sensitive to customer feelings and show genuine concern and respect.
3. *Communicate effectively.* Be articulate and diplomatic.
4. *Handle stress.* Stay organized and calm and show patience.
5. *Listen actively.* Interpret the meaning of the customer's words.
6. *Demonstrate mental alertness.* Process information quickly.

TABLE 17.6
Some action taken to achieve delighted customers

Action	Example
Providing a service far beyond the scope of the company service	An airline flight attendant accompanied a sick passenger and her daughter to a hospital.
Providing a service beyond the call of a normal effort	To eliminate the sound of tinkling glass in a chandelier due to air conditioning, hotel employees removed every second piece of glass 30 minutes before the start of the meeting.
Providing extraordinary recognition of customer inconvenience	An automobile manufacturer paid a customer $985 for "lost time" incurred by the customer on excessive and improperly made repairs.
Providing a recognition of personal customer loss	A customer reported that she lost a pen, having sentimental value, in a grocery store. A clerk searched for the pen but with no success. The clerk presented the lady with three $20 gift certificates.

These competencies are not easy to achieve. Organizations must make the investment of careful selection of employees, training, and empowerment (see above) and then develop the employees into professionals to assure the retention of these key people.

Frontline personnel can serve as "listening posts" for an organization. When the listening posts are well designed to probe and ask specific questions, the information generated can help to develop new products and to sell existing products of which the customer is not aware. Some frontline personnel are uncomfortable asking certain questions that may lead to answers that the customer and the bank teller often prefer not to discuss, e.g., Why are you transferring your account to another bank? For a discussion of designing appropriate questions for listening posts, see *JQH5,* page 33.12. Another way to gather customer information for future product development is to use customer management software to track every encounter with a customer.

Call centers and help desks are now a main mechanism for a service encounter. When the call concerns a customer problem, typically resolution of the problem is not achieved with the first phone call reporting the problem. At one health maintenance organization, 36 percent of the calls were related to a lack of complete, accurate, and timely information between suppliers and customers. Cross (2000) discusses some of the issues involved.

17.7 ORGANIZING FOR QUALITY IN SERVICE OPERATIONS

Teams are becoming a common means of organizing in the service industries. For a summary of various types of teams, see Section 8.7.

Teams may be ad hoc to address a specific problem or may be permanent to be responsible for a specific activity. For example, the American Express Consumer Card

Group uses semi-autonomous work teams. A team consists of 10 to 12 employees in the natural work group. Team members do customer service work, manage quality, inventory, and attendance; prepare work schedules; and prepare production reports and forecasts. Individual roles are defined to handle these team responsibilities. The team leader focuses on coaching, feedback, and special human resource issues.

One fast-food firm creates teams of crew members (workers at one location) who are trained to manage the site without a full-time manager (Harvard Business School, 1994). This approach means installing on-line sensor technology such as the time to prepare an order, and providing crew members with the same operating and financial information provided to a restaurant general manager to run the site. Crew members make operating decisions such as ordering food materials. Thus knowledge that long separated "brain workers" from "hand workers" now resides in a computer on the operations floor.

The Ritz-Carlton Hotel company uses self-directed teams. These process teams are aligned with the way customers come in contact with the hotel: (1) prearrival team; (2) arrival, stayover, and departure team; (3) dining services team; (4) banquet services team; and (5) engineering and security team. In a self-directed work team, members may have specific individual roles, but the team shares accountability for meeting performance objectives.

Kaiser Permanente uses quality-in-daily-work teams. These frontline work teams focus on both work process improvements and the definition of a service guarantee to communicate service performance levels. The projects span across departments and include both clinical and support services. Table 17.7 shows a few examples (Centano et al., 1995).

TABLE 17.7
Examples of Quality in Daily Work Projects

Department	Customer	Essential service	Service guarantee
Nursing unit 7-west	Ambulance emergency department	Pick up patients within specified time frame	The critical care unit staff will pick up the patient within 15 minutes of emergency room's call
Admitting	Elective surgery patients	Admission process including explanation of forms	Provide service within five minutes of entering department
Education	Diabetic patients	Out-patient diabetic education program	95% of all diabetic patients will be scheduled for diabetic education
Pediatrics	Member	Telephone access	98% of all calls will be handled within two minutes

TABLE 17.8
Examples of six-sigma projects at a financial services organization

Reduce cycle time of credit issuance
Reduce trading-call length
Reduce card issuance cycle time
Reduce encoding errors
Improve business travel contract-to-billing process
Eliminate nonreceived renewals
Eliminate incorrect fee adjustments
Improve XXX payment accuracy
Reduce YYY writeoffs

For a discussion of research conducted on teams, see Katzenbach and Smith (1993). Mann (1994) explains how managers need to develop skills as coaches, developers, and managers of activities that reside in different departments ("boundary managers").

17.8 SIX-SIGMA PROJECTS IN SERVICE INDUSTRIES

The six-sigma approach to improvement includes the phases of define, measure, analyze, improve, and control. These phases are explained in Section 3.5. Six sigma is increasingly being applied in the service sector. Table 17.8 shows examples of six-sigma projects at the American Express organization.

Note the wide variety of projects in Table 17.8.

Hahn et al. (2000) describe how GE Capital applied six sigma (including a modified full-factorial design of experiment) to reduce losses due to delinquent credit card customers. Bott et al. (2000) explain the six-sigma steps for two projects at American Express.

17.9 QUALITY MEASUREMENT IN SERVICE OPERATIONS

The management of key work processes must include provisions not only for measuring process control but also for monitoring overall operations. Readers should review the basics of quality measurement in Section 5.2.

The quality measures used in a service organization are unique to the type of service organization. Table 17.9 shows a few examples of quality measurements for overall operations in a credit card company.

Note the large number of measures in Table 17.9. This organization can easily calculate the measures from basic raw data that are routinely collected in daily operations.

King and Dickinson (1996) present a framework (Figure 17.5) of three process effectiveness measures that are particularly suited for service processes.

TABLE 17.9
Quality measures in a credit card company

Measure number	Quality measure
148	Number of abandoned calls/total calls
454	Average time to call pickup, seconds
458	Number of statement insertion errors
460	Number of applications not processed within standard
466	Number of hours credit card system down
467	Number of payments not posted

Source: JQH5, page 33.21.

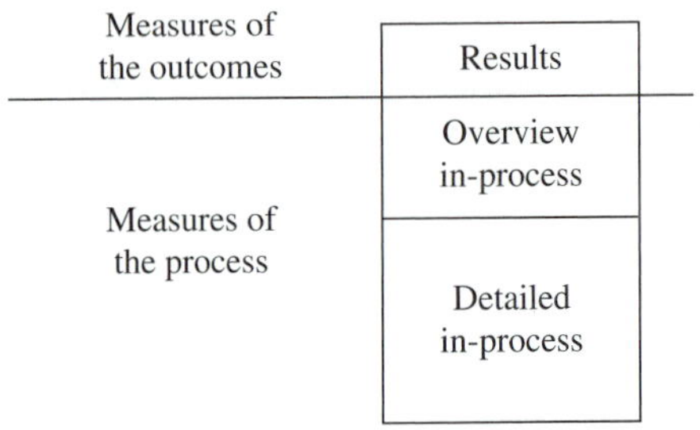

Measures of the outcomes	Results
Measures of the process	Overview in-process
	Detailed in-process

FIGURE 17.5
Framework of process effectiveness measures. (*Reprinted from King and Dickinson, 1996, by courtesy of Marcel Dekker, Inc.*)

Results measures are primarily customer perceptions of outcomes. These measures drive priorities for improvement and monitor performance improvement. Overview in-process measures are predictive of results measures and are lead indicators of process outcomes. Changes in an overview measure will thus lead to changes in the related results measure(s). Overview measures help to initiate the search for root causes of poor performance. Detailed in-process measures are predictive of overview measures and are lead indicators of subprocess outcomes. Detailed in-process measures control the day-to-day operation of the process and provide early warnings and diagnostic information for improvement. Thus the approach forms a hierarchy of process control measures—exterior and interior to the organization and at different levels of a process.

As an example, consider the process of having a new phone installed at an arranged time. The results measure is the proportion of customers who report that the technician arrived on time. The overview in-process measure is also the proportion of appointments met, but this measure is logged by the technicians. Thus the results and the overview measures in this case are the same measure but collected both within and outside the process. The detailed in-process measures are many and might include proportion of customers who could not be given appointments at the time they requested or incidences and reasons for staff shortages.

Quality measurements are candidates for data analysis using statistical techniques such as the control charts discussed in Chapter 18. But the more basic point is that the reporting of data demonstrates to operating personnel that management regards quality as having a high priority.

Fuchs (*JQH5,* page 25.23) presents a flowchart for defining, collecting, and analyzing metrics in a service organization.

17.10 MAINTAINING A FOCUS ON CONTINUOUS IMPROVEMENT

Operations personnel in the service sector (and the manufacturing sector) are involved in addressing not only sporadic problems (the fire drills) but also chronic problems. The actions required include troubleshooting, quality improvement, and quality planning. Maintaining the focus on improvement clearly requires a positive quality culture in an organization.

The broad approach to the three actions and the key elements of a positive quality culture are summarized in Section 16.10. A detailed discussion of each subject is provided in other chapters of this book.

For a summary of the results of more than 1000 quality improvement teams at a large bank, see Section 8.7.

Finally, the operations function must be provided with the support to maintain the focus on improvement. The quality department should regard operations as a key internal customer and provide training and technical quality expertise to support operations. In addition, operations managers must be guided in how to review the mountain of reports they receive to identify and prioritize quality problems and set up the teams and other mechanisms to address those problems. Also, the quality department can urge upper management to set up cross-functional teams to address operations problems that may be caused by other functional departments such as IT (see Section 17.2), marketing, and purchasing.

SUMMARY

- Activities to integrate quality in service planning have two objectives: to incorporate product features and to prevent defects (minimize variability).
- By creating a flow diagram, we can dissect a service process and plan for quality at each workstation.
- Error-proofing a process is an important element of prevention.
- Information technology is an important internal supplier for the service operations process.
- Process capability can be measured for a service process.

- For self-control, we must provide personnel with the knowledge of what they are supposed to do, knowledge of what they are actually doing, and a process that is capable of meeting specifications and can be regulated.
- Failure to meet all three of these criteria means that the quality problem is management controllable. About 80 percent of quality problems are management controllable.
- The basic steps for controlling quality (Chapter 5) can be applied to service operations.
- Process quality audits apply to any activity that can affect final service quality.
- Three factors are important in frontline customer contact: selection, training, and empowerment of personnel.
- Various types of quality teams play a key role in service processes.
- Six sigma applies to service processes.
- Quality measures must be designed for each type of service organization.

PROBLEMS

17.1. For a specific service activity with which you are familiar, create a flow process diagram and indicate current or potential problem points.

17.2. For a specific service activity with which you are familiar, describe how you would conduct a controllability study to determine whether the quality problems are primarily management controllable or employee controllable.

17.3. Select a service task regularly performed by an employee. Use the checklists on self-control to determine whether the employee can be held responsible for the quality of the output.

17.4. For a service process with which you are familiar, define two control subjects and at least one measurement for each control subject.

17.5. Your work team performs blood tests that physicians request for their patients. Identify one control subject that will help manage the work process while meeting customer needs. Then define a unit of measure and a sensor for the control subject.

17.6. For a service process with which you are familiar, determine the major form of "dominance" affecting attainment of quality.

17.7. For each scenario presented below, determine whether the action needed is troubleshooting, quality improvement, or quality planning (Juran Institute, Inc.):

(a) The mail services team at a large complex discovered that internal mail is taking five to eight days to be delivered, whereas U.S. mail is delivered within three days.

(b) Your team is responsible for housekeeping at a hotel. The housekeepers sometimes do not have enough towels to supply the rooms, although a relatively large stack of towels is usually available.

(*c*) The scheduling team at a hospital reports an increasing number of double bookings, mismatches of staff for surgery, and customer complaints. The team decides that the installation of a new centralized scheduling system will greatly reduce the problem.

REFERENCES

American Management Association (1992). *Blueprints for Service Quality,* American Management Association, New York, pp. 51–64.

AT&T Quality Steering Committee (1990). *Achieving Customer Satisfaction,* AT&T Customer Information Center, Indianapolis, IN.

Berry, L. L., A. Parasuraman, and V. A. Zeithmal (1994). "Improving Service Quality in America: Lessons Learned," *Academy of Management Executive,* vol. 8, no. 2, pp. 32–52.

Bott, C., E. Keim, S. Kim, and L. Palser (2000). "Service Quality Six Sigma Case Studies," *Annual Quality Congress Proceedings,* ASQ, Milwaukee, pp. 225–231.

Centeno, A. M., K. Ahn, and R. Tawell (1995). "Operationalizing Quality in Daily Work Concepts," *Impro Conference Proceedings,* Juran Institute, Inc., pp. 4C.2-1 to 4C.2-8.

Collins, W. H. and C. B. Collins (1993). "Differentiating System and Execution Problems," *Quality Progress,* February, pp. 59–62.

Cross, K. F. (2000). "Call Resolution: The Wrong Focus for Service Quality," *Quality Progress,* February, pp. 64–67.

Davis, R., S. Rosegrant, and M. Watkins (1995). "Managing the Link between Measurement and Compensation," *Quality Progress,* February, pp. 101–106.

Deyong, C. F. and K. E. Case (1998). "Linking Customer Satisfaction Attributes with Progress Metrics in Service Industries," *Quality Management Journal*, vol. 5, no. 2, pp. 76–90.

Hadley, H. (1995). Private communication to Patrick Mene, Marriott Hotels and Resorts, Washington, DC.

Hahn, G. J., N. Doganaksoy, and R. Hoerl (2000). "The Evolution of Six Sigma," *Quality Engineering,* vol. 12, no. 3, pp. 317–326.

Harvard Business School (1994). *Case 9-694-076, Taco Bell,* Boston.

Jeffrey, J. R. (1995). "Preparing the Front Line," *Quality Progress,* February, pp. 79–82.

Katzenbach, J. R. and D. K. Smith (1993). *Wisdom of Teams: Creating the High Performance Organization,* Harvard Business School Press, Boston.

King, M. N. and T. Dickinson (1996). "A Framework for Process Effectiveness Measures," *Quality Engineering,* vol. 9, no. 1, pp. 45–50.

Kittner, M., M. Jeffries, and F. M. Gryna (1999). "Operational Quality Issues in the Financial Sector: An Exploratory Study on Perception and Prescription for Information Technology," *Journal of Information Technology Management,* vol. X, nos. 1–2, pp. 29–39.

Kordupleski, R. E., R. T. Rust, and A. J. Zahorik (1993). "Why Improving Quality Doesn't Improve Quality (Or Whatever Happened to Marketing?)," *California Management Review,* Spring, vol. 35, no. 3, pp. 82–95.

Mann, D. W. (1994). "Re-engineering the Manager's Role," ASQC Quality Congress Transactions, Milwaukee, pp. 155–159.

Melan, E. H. (1993). *Process Management,* McGraw-Hill, New York.

Shirley, B. M. and F. M. Gryna (1998). "Work Design for Self-Control in Financial Services," *Quality Progress,* May, pp. 67–71.

Shostack, G. L. (1984). "Designing Services That Deliver," *Harvard Business Review,* January–February, pp. 133–139.

Zeithaml, V. A., A. Parasuraman, and Leonard L. Berry (1990). *Delivering Service Quality,* Free Press, New York.

SUPPLEMENTARY READING

Service industries quality: *JQH5,* Sections 25, 30, 31, 32, 33.

WEBSITES

Society of Consumer Affairs Professionals (SOCAP): www.socap.org
The Right Answer (frontline customer service): www.therightanswer.com
Best practices in one-stop customer service:
www.npr.gov/library/papers/benchmark/1stpcus.html

18

STATISTICAL PROCESS CONTROL

18.1 DEFINITION AND IMPORTANCE OF STATISTICAL PROCESS CONTROL

Dateline 1950: I was installing control charts as part of a statistical process control system to control the variation in the weight of packages of dry soup mix at the Lipton Company. We set up the system and quietly tried it out on one filling line by having the operator weigh samples of packages and plot the average and range of the samples. Within one day operators on the other lines enthusiastically asked us to show them how to plot and use the charts. Their message to us was clear: They wanted to be better able to control their process (self-control), and they wanted to have fun plotting the data. Moral: Statistical process control is not all numbers.

We define *statistical process control* as the application of statistical methods to the measurement and analysis of variation in a process. This technique applies to both in-process parameters and end-of-process (product) parameters. Let the reader be aware, however, that the term *statistical process control* has also assumed other definitions, even including some that involve little or no use of statistical analysis!

A process is a collection of activities that converts inputs into outputs or results. More specifically, a process is a unique combination of machine, tools, methods, materials, and people that attain an output in goods, software, or services—a circuit chip, a computer program, or answers provided on a consumer hot line.

Methods of collecting, summarizing, and analyzing data are discussed in Chapter 10, "Basic Concepts of Statistics and Probability," and Chapter 11, "Statistical Tools for Analyzing Data." In the current chapter we examine the significance of variation, use of control charts in analyzing and minimizing variation, quantification of process capability, the statistical basis of the six-sigma approach, and the relation of these concepts to other techniques for process improvement.

The evidence is clear that these fascinating techniques can make an important contribution to achieving quality objectives. For most organizations statistical process control techniques are essential. To help assure successful and continued application of these concepts in the reality of lean operating budgets, statistical process control techniques must not become an end in themselves.

Pragmatic operating managers correctly demand that each potential application show a tangible opportunity for significant benefits—a situation that quality professionals should never forget.

18.2 ADVANTAGES OF DECREASING PROCESS VARIABILITY

Reducing the variation in a process leads to some great benefits:

- Lower variability may result in improved product performance that is discernible by the customer. Sullivan (1984) describes a case of two Sony plants making the same television set (Figure 18.1). The San Diego plant had no product outside specifications, but the distribution was virtually rectangular, with a large percentage of product close to the specification limits. In contrast, the plant in Japan did have some product outside of specification limits, but the distribution was normal and was concentrated around the target value. Field experience revealed that product near the specification limits generated complaints from customers. This and other reasons led to a higher loss per unit at San Diego even though that plant was superior in meeting the specification. The higher internal loss due to complaints would, of course, likely result in lower future sales.
- Lower variability on a component characteristic may be the only way to compensate for high variability in other components and thereby meet performance require-

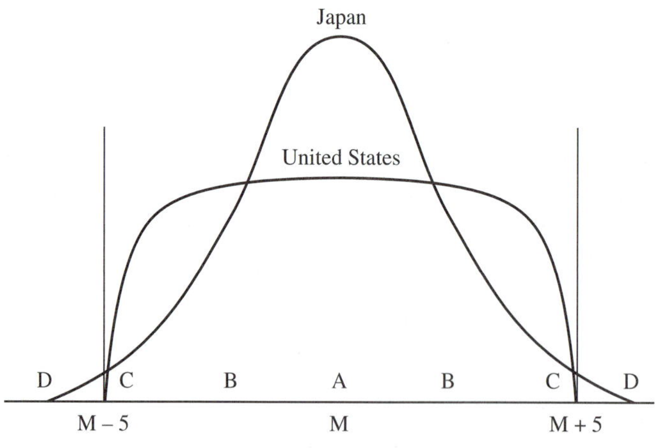

FIGURE 18.1
Uniformity and production quality of television sets produced in Japan and the United States. (*From Sullivan, 1984.*)

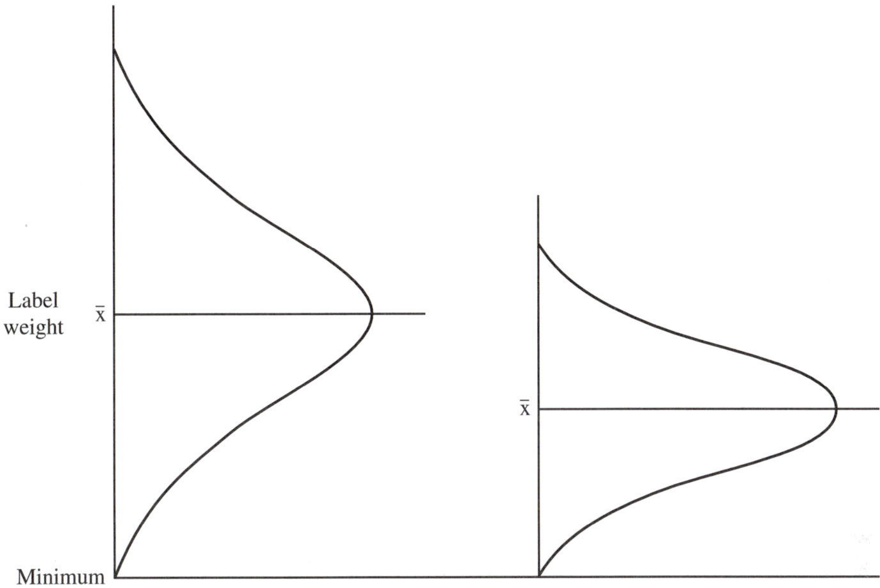

FIGURE 18.2
Reducing average overfill by reducing the standard deviation.

ments on an assembly or system. To meet these requirements may also require strict control of the average values of each component, as was the case in the design and manufacture of undersea cable.

- On some characteristics such as weight, lower variability may provide the opportunity to change the process average. Thus reducing the standard deviation of fill content in a food package permits a reduction in the *average* fill, thereby resulting in cost reduction (Figure 18.2). Imagine the cumulative cost reduction over millions of packages!
- Lower variability results in less need for inspection. In the extreme case, if there were *no* variability, inspection of only one unit of product would tell the whole story.
- Lower variability may command a premium price on a product. Some electronic components have traditionally been priced as a function of the amount of variability.
- Lower variability may be a competitive factor in determining market share. Increasingly, meeting specification limits is no longer sufficient. Industrial customers in particular realize that high variability of purchased material and components often requires frequent (and costly) adjustment to their own processes to compensate for the variability of purchased products. The result is that these customers compare suppliers on variability of important product characteristics.

EXAMPLE 18.1. The marketing manager of a commodity chemicals manufacturer describes two scenarios—old and new—between a customer and a salesperson.

Old scenario:
Customer: "Your product quality is no good."
Salesperson: "I'll lower the price."

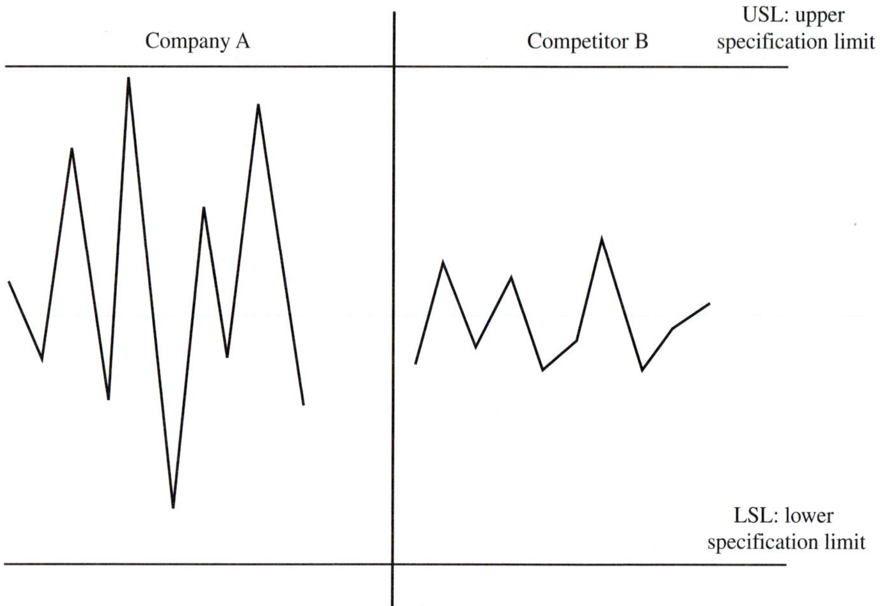

FIGURE 18.3
Variation and competition.

Customer: "Good, that's what I wanted to hear."

New scenario:

Customer: "Your product quality is no good."
Salesperson: "I'll lower the price."
Customer: "The price is acceptable; I said your quality is no good."
Salesperson: "Was some of the product out of specification?"
Customer: "No."
Salesperson: "I don't understand."
Customer: "Look at these data." (See Figure 18.3.)
Customer: "It's not enough to meet the specifications. Your competitor meets the same specification with less variability."

With a zeal for reducing variability, the Hughes Company has published a list of 219 "variability reduction specialists" who offer advice throughout the organization.

18.3 STATISTICAL CONTROL CHARTS—GENERAL

A *statistical control chart* compares process performance data to computed "statistical control limits," drawn as limit lines on the chart. The process performance data

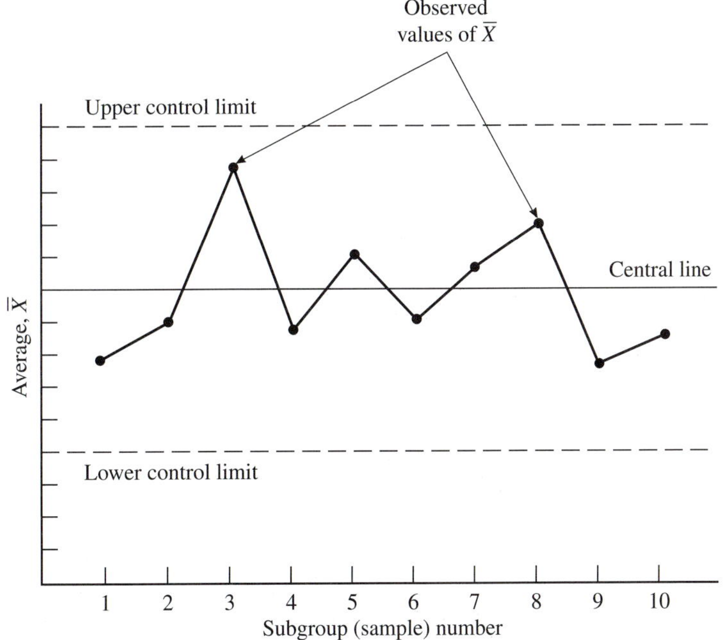

FIGURE 18.4
Generalized control chart for averages.

usually consist of groups of measurements (rational subgroups) that come from the regular sequence of production while preserving the order of the data.

A prime objective of a control chart is detecting *special* (or assignable) causes of variation in a process—by analyzing data from both the past and the future. Knowing the meaning of "special causes" is essential to understanding the control chart concept (see Table 5.5).

Process variations have two kinds of causes: (1) common (or random or chance), which are inherent in the process, and (2) special (or assignable), which cause excessive variation. Ideally, only common causes should be present in a process because these represent a stable and predictable process that leads to minimum variation. A process that is operating without special causes of variation is said to be "in a state of statistical control." The control chart for such a process has all of the data points within the statistical control limits. The object of a control chart is not to achieve a state of statistical control as an end in itself but to reduce variation.

The control chart distinguishes between common and special causes of variation through the choice of control limits (Figure 18.4). These are calculated by using the laws of probability in such a way that highly improbable causes of variation are presumed to be due not to random causes but to special causes. When the variation *exceeds* the statistical control limits, it is a signal that special causes have entered the process and the process should be investigated to identify these causes of excessive variation. Random variation *within* the control limits means that only common (random) causes are present; the amount of variation has stabilized, and minor process adjustments should be avoided. Note that a control chart detects the presence of a special cause but does not *find* the cause—that task must be handled by a subsequent investigation of the process.

18.4 ADVANTAGES OF STATISTICAL CONTROL

A state of statistical control exists when only common causes of variation exist in the process. This condition provides several important advantages:

- The process has stability, which makes it possible to predict its behavior, at least in the near term.
- The process has an identity in terms of a given set of conditions that are necessary for making predictions. An analogy to baseball is useful. If we say that a player has a batting capability of about .250, we mean than on average he will get one hit in four times at bat. Of course, this statement assumes that the player is in a normal state of health, i.e., free of any assignable or special causes that would detract from the batting capability. Predictions of any kind must be related to an assumed set of conditions. For manufacturing and business processes, these conditions are represented by a state of statistical control. Ideally, drawing conclusions from any set of data should be preceded by an analysis to see whether the data come from a process in statistical control.
- A process in statistical control operates with less variability than a process having special causes. Lower variability has become an important tool of competition (see Section 4.10, "Planning for Product Quality to Generate Sales Income").
- A process having special causes is unstable, and the excessive variation may hide the effect of changes being introduced to achieve improvement. Also, the removal of some special causes and subsequent replotting of the control chart may reveal that additional special causes exist that were masked earlier.
- Knowing that a process is in statistical control is helpful to the workers running a process. It says that when data fall within the statistical control limits, adjustments should *not* be made. Making such adjustments will add to the variability, not decrease it. Conversely, a control chart helps to avoid underadjustment as out-of-control points will signal the presence of special causes.
- Knowing that a process is in statistical control provides direction to those who are trying to make a long-term reduction in process variability. To reduce process variability, the process system must be analyzed and changed rather than management expecting the workers who run the process to reduce the variability by themselves.
- An analysis for statistical control, which includes the plotting of data in order of production, will easily identify trends over time that are hidden by other summarizations of data such as histograms.

- A stable process (as verified by statistical control) that also meets product specifications provides evidence that the process has conditions that, if maintained, will result in an acceptable product. Such evidence is needed *before* a process is transferred from the planning stage to full production.

These advantages of statistical control feed into the main objectives of statistical process control, which are to reduce process variation and prevent problems. Real prevention is achieved by having capable processes that are maintained in a state of statistical control. Process variation is much like the swing of a pendulum. Wouldn't it be nice if we could reduce the sweep of the arc?

18.5 STEPS IN SETTING UP A CONTROL CHART

Setting up a control chart requires the following steps:

1. Choosing the characteristic to be charted.
    - Giving high priority to characteristics that are currently running with a high defective rate. A Pareto analysis can establish priorities.
    - Identifying the process variables and conditions that contribute to the end-product characteristics, so as to define potential charting applications from raw materials through processing steps to final characteristics. For example, pH, salt concentration, and temperature of plating solution are process variables contributing to plating smoothness.
    - Verifying that the measurement process has sufficient accuracy and precision (see Section 19.8) to provide data that does not obscure variation in the manufacturing or service process. The observed variation in a process reflects not only the variation of the manufacturing process but also the *combined* variation of the manufacturing and measurement processes. Anthis et al. (1991) describe how the measurement process was a roadblock to improvement by hiding important clues to the sources of variation in a manufacturing process. Dechert et al. (2000) explain how large measurement variation can be controlled and result in effective statistical process control methods.
    - Determining the earliest point in the production process at which testing can be done to get information on assignable causes so that the chart serves as an effective early-warning device to prevent defectives.
2. Choosing the type of control chart. Table 18.1 compares three basic control charts. Schilling (1990) provides additional guidance in choosing the type of control chart to use.
3. Deciding the central line to be used and the basis of calculating the limits. The central line may be the average of past data, or it may be a desired average (i.e., a standard value). The limits are usually set at $\pm 3\sigma$, but other multiples may be chosen for different statistical risks.
4. Choosing the "rational subgroup." Each point on a control chart represents a subgroup (or sample) consisting of several units of product. For process control purposes,

TABLE 18.1
Comparison of some control charts

Statistical measure plotted	Average $\bar{X}$ and range R	Percentage nonconforming (p)	Number of nonconformities (c)
Type of data required	Variable data (measured values of a characteristic)	Attribute data (number of defective units of product)	Attribute data (number of defects per unit of product)
General field of application	Control of individual characteristics	Control of overall fraction defective of a process	Control of overall number of defects per unit
Significant advantages	Provides maximum utilization of information available from data	Data required are often already available from inspection records	Same advantages as p chart but also provides a measure of defectiveness
	Provides detailed information on process average and variation for control of individual dimensions	Easily understood by all personnel Provides an overall picture of quality	
Significant disadvantages	Not understood unless training is provided; can cause confusion between control limits and tolerance limits	Does not provide detailed information for control of individual characteristics	Does not provide detailed information for control of individual characteristics
	Cannot be used with go/no go type of data	Does not recognize different degrees of defectiveness in units of product	
Sample size	Usually four or five	Use given inspection results or samples of 25, 50, or 100	Any convenient unit of product such as 100 feet of wire or one television set

subgroups should be chosen so that the units *within* a subgroup have the greatest chance of being alike and the units *between* subgroups have the greatest chance of being different.

5. Providing a system for collecting the data. If the control chart is to serve as a day-to-day shop tool, it must be made simple and convenient for use. Measurement must be simplified and kept free of error. Indicating instruments must be designed to give prompt, reliable readings. Better yet, instruments should be designed that can record as well as indicate. Recording of data can be simplified by skillful design of data or tally sheets. Working conditions are also a factor.
6. Calculating the control limits and providing specific instructions on the interpretation of the results and the actions that various production personnel are to take (see

TABLE 18.2
Control chart limits—attaining a state of control

Chart for	Central line	Lower limit	Upper limit
Averages $\bar{X}$	$\bar{\bar{X}}$	$\bar{\bar{X}} - A_2\bar{R}$	$\bar{\bar{X}} + A_2\bar{R}$
Ranges R	$\bar{R}$	$D_3\bar{R}$	$D_4\bar{R}$
Percentage nonconforming p	$\bar{p}$	$\bar{p} - 3\sqrt{\frac{\bar{p}(1-\bar{p})}{n}}$	$\bar{p} + 3\sqrt{\frac{\bar{p}(1-\bar{p})}{n}}$
Number of nonconformities c	$\bar{c}$	$\bar{c} - 3\sqrt{\bar{c}}$	$\bar{c} + 3\sqrt{\bar{c}}$

below). Control limit formulas for the three basic types of control charts are given in Table 18.2. These formulas are based on $\pm 3\sigma$ and use a central line equal to the average of the data used in calculating the control limits. Values of the A_2, D_3, and D_4 factors used in the formulas are given in Table I in Appendix II. Each year, *Quality Progress* magazine publishes a directory that includes software for calculating sample parameters and control limits and for plotting the data.

7. Plotting the data and interpreting the results.

The control chart is a powerful statistical concept, but its use should be kept in perspective. The ultimate purpose of an operations process is to make product that is fit for use—not to make product that simply meets statistical control limits. Once the charts have served their purpose, many should be taken down and the effort shifted to other characteristics needing improvement. Schilling (1990) traces the life cycle of control chart applications (Table 18.3). A given application might employ several types of control charts. Note that, in the "phase out" stage, statistical control has been achieved, and some of the charts are replaced with spot checks.

18.6 CONTROL CHART FOR VARIABLES DATA

For variables data the control chart for sample averages and sample ranges provides a powerful technique for analyzing process data.

A small sample (e.g., five units) is periodically taken from the process, and the average ($\bar{X}$) and range (R) are calculated for each sample. A total of at least 50 individual measurements (e.g., 10 samples of five each) should be collected before the control limits are calculated. The control limits are set at $\pm 3\sigma$ for sample averages and sample ranges. The $\bar{X}$ and R values are plotted on separate charts against their $\pm 3\sigma$ limits.

Standard deviations are readily computed by modern calculators, but calculations can be avoided by using shortcuts.

TABLE 18.3
Life cycle of control chart applications

Stage	Step	Method
Preparatory	State purpose of investigation	Relate to quality system
	Determine state of control	Attributes chart
	Determine critical variables	Fishbone
	Determine candidates for control	Pareto
	Choose appropriate type of chart	Depends on data and purpose
	Decide how to sample	Rational subgroups
	Choose subgroup size and frequency	Sensitivity desired
Initiation	Ensure cooperation	Team approach
	Train user	Log actions
	Analyze results	Look for patterns
Operational	Assess effectiveness	Periodically check usage and relevance
	Keep up interest	Change chart, involve users
	Modify chart	Keep frequency and nature of chart current with results
Phase-out	Eliminate chart after purpose is accomplished	Go to spot checks, periodic sample inspection, overall *p, c* charts

Source: Schilling (1990).

The shortcut formulas for the upper control limit (UCL) and lower control limit (LCL) on sample averages are

$$\text{UCL} = \bar{\bar{X}} + A_2\bar{R}$$

$$\text{LCL} = \bar{\bar{X}} - A_2\bar{R}$$

where $\bar{\bar{X}}$ = grand average = average of the sample averages
$\bar{R}$ = average of the sample ranges
A_2 = constant found from Table I in Appendix II

The shortcut consists of (1) computing, for each sample, the range (difference between largest and smallest) of the individuals; (2) averaging the ranges thus obtained; and then (3) multiplying the average range by a conversion factor to get the distance from the expected average to the limit line. The central line is merely the average of all the individual observations.

The shortcut formulas for control limits on sample ranges are

$$\text{UCL} = D_4\bar{R}$$

$$\text{LCL} = D_3\bar{R}$$

where D_3 and D_4 are constants found in Table I in Appendix II.

TABLE 18.4
Constants for $\bar{X}$ and R chart

n	A_2	D_3	D_4	d_2
2	1.880	0	3.268	1.128
3	1.023	0	2.574	1.693
4	0.729	0	2.282	2.059
5	0.577	0	2.114	2.326
6	0.483	0	2.004	2.534
7	0.419	0.076	1.924	2.704
8	0.373	0.136	1.864	2.847
9	0.337	0.184	1.816	2.970
10	0.308	0.223	1.777	3.078

A partial tabulation of the A_2, D_3, and D_4 factors is reproduced in Table 18.4 for the convenience of the reader in following the text.

Consider the data for machines N-5 and N-7 in Figure 18.5. For each machine the data consist of 10 samples (with six units each) plotted in time order of production (sample number). Figure 18.5 shows the $\bar{X}$ and R charts for each machine. The upper part of the figure displays the individual observations.

For machine N-5 the UCL and LCL are calculated as

AVERAGES:

$$\text{UCL} = \bar{\bar{X}} + A_2\bar{R} = 9.59 + 0.483(6.6) = 12.77$$

$$\text{LCL} = \bar{\bar{X}} - A_2\bar{R} = 9.59 - 0.483(6.6) = 6.40$$

RANGES:

$$\text{UCL} = D_4\bar{R} = 2.004(6.6) = 13.23$$

$$\text{LCL} = D_3\bar{R} = 0(6.6) = 0$$

As all points fall within the control limits, it is concluded that the process is free of assignable causes of variation.

Control limits for a chart for averages represent three standard deviations of sample *averages* (not individual values). As specification limits usually apply to *individual* values, the control limits *cannot* be compared to specification limits, because averages inherently vary less than the individual measurements going into the averages (see Figure 11.1). Therefore, specification limits should *not* be placed on a control chart for averages. Sample averages, rather than individual values, are plotted because averages are more sensitive to detecting process changes than individual values are.

Another example is presented in Figure 18.5 for machine N-7. This machine has both within-sample variation shown by the range chart and between-sample variation as illustrated by the chart for sample averages. The $\bar{X}$ chart indicates that some factor such as tool wear is present that results in larger values of the characteristic with the passing of time (note the importance of preserving the order of the measurements). In

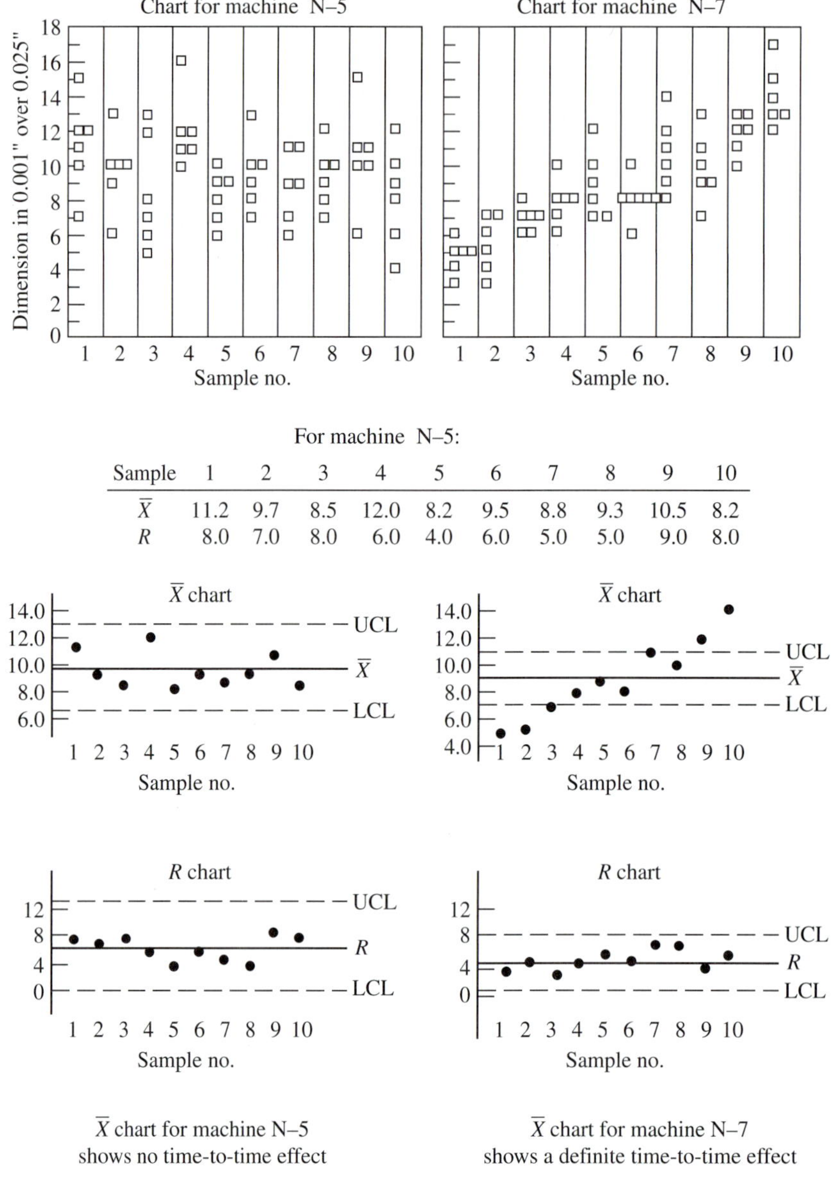

For machine N–5:

Sample	1	2	3	4	5	6	7	8	9	10
$\bar{X}$	11.2	9.7	8.5	12.0	8.2	9.5	8.8	9.3	10.5	8.2
R	8.0	7.0	8.0	6.0	4.0	6.0	5.0	5.0	9.0	8.0

FIGURE 18.5
$\bar{X}$ and R charts confirm suggested machine differences.

such cases measures of process capability should reflect both sources of variation. The inherent capability can be estimated in the usual way as 6σ (where $\sigma = \bar{R}/d_2$), and this will depict the variation about a given process aim. In addition, the time-to-time variation can be expressed separately as the difference in process aim over the time period covered by the averages plotted on the control chart.

Interpretation of Charts

Place the charts for $\bar{X}$ and R (or s) one above the other so the average and range for any one subgroup are on the same vertical line. Observe whether either or both indicate lack of control for that subgroup.

$\bar{X}$'s outside the control limits are evidence of a general change affecting all pieces after the first out-of-limits subgroup. The log kept during data collection, the operation of the process, and the worker's experience should be studied to discover a variable that could have caused the out-of-control subgroups. Typical causes are a change in material, personnel, machine setting, tool wear, temperature, or vibration.

R's outside control limits are evidence that the uniformity of the process has changed. Typical causes are a change in personnel, increased variability of material, or excessive wear in the process machinery. In one case a sudden increase in R warned of an impending machine accident.

A single out-of-control R can be caused by a shift in the process that occurred while the subgroup was being taken.

Look for unusual patterns and nonrandomness. Nelson (1984, 1985) provides eight tests to detect such patterns on control charts using 3σ control limits (Figure 18.6). Each of the zones shown is 1σ wide. (Note that test 2 in Figure 18.6 requires nine points in a row; other authors suggest seven or eight points in a row; see Nelson [1985] for elaboration.)

AT&T (1990) provides an example from the service industry:

EXAMPLE 18.2. A manager of a personnel database at AT&T collected data on the average time to implement an employee change notification.

The average and range control chart is shown in Figure 18.7. The manager learned from the chart that the average time to implement began in the 2- to 3-day range and, over the course of 30 days, continuously increased to the 10- to 12-day range. The increase began gradually and then accelerated, suggesting that the process had changed. Investigation revealed that more and more change notification requests arrived with incomplete information, requiring workers to call the originating organization and increasing the time to implement.

Ott and Schilling (1990) provide a definitive text on analysis after the initial control charts by presenting an extensive collection of cases with innovative statistical analysis clearly described.

Introducing Control Charts

To quality specialists, control charts serve as sensitive devices for detecting process changes; to operating forces, the charts represent a major change from the traditional "law of the shop," i.e., the specification limits. In introducing control charts, it is essential to prevent confusion about the role of control limits versus specification limits. Workers react to nonconforming product because specification limits have been the law of the shop; they do not react to control limits in the same way because the legitimacy of control limits may not have been fully established and clarified. For example, what is a worker to do if a control chart is frequently out of control but the product is well within specification limits? See Section 18.16.

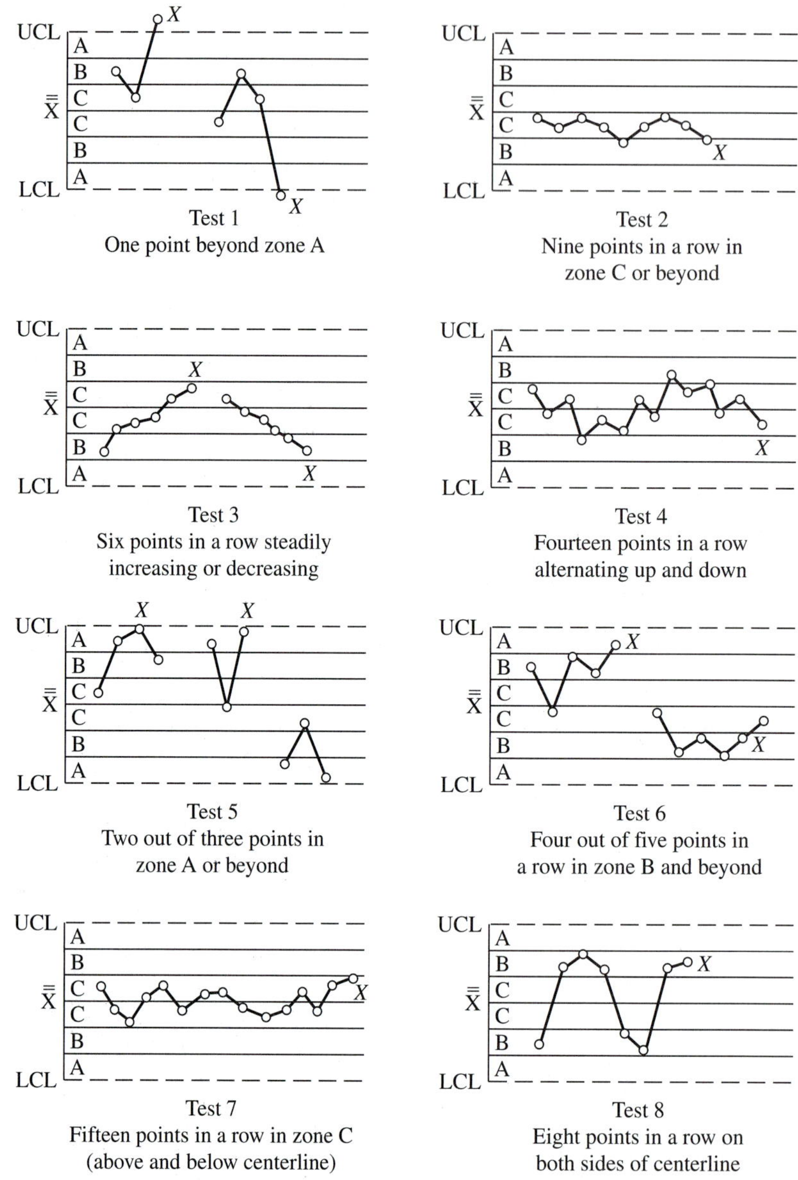

FIGURE 18.6
Illustrations of tests for special causes applied to $\bar{X}$ control charts. (*From Nelson, 1984.*)

Chart for Individuals

An alternative to the $\bar{X}$ and R chart is the chart for individual X values. This chart, often called a run chart, is a plot of individual observations against time. In the simplest case, specification limits are added to the chart; in other cases, $\pm 3\sigma$ limits of

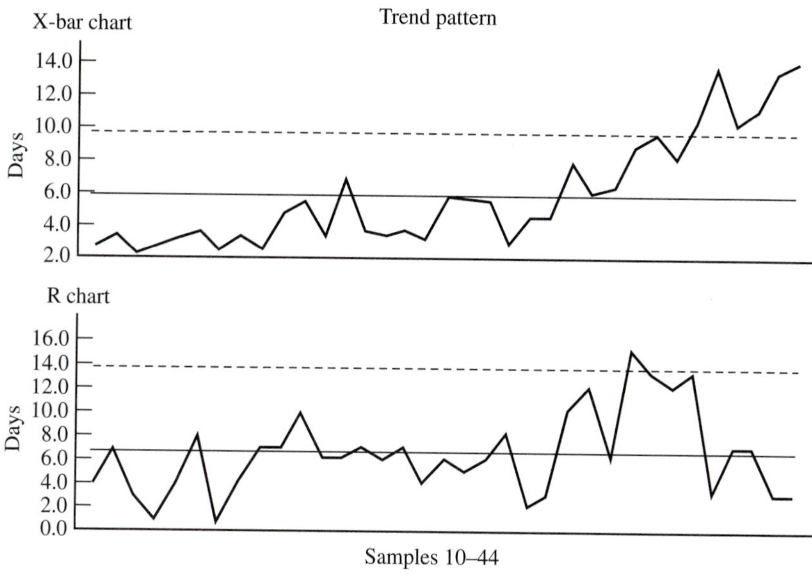

FIGURE 18.7
Average and range control charts.

individual values are added. A chart for individuals is not as sensitive as the $\bar{X}$ chart. See Section 10.6 for further elaboration.

EXAMPLE 18.3. Tom Pohlen, a quality engineering specialist, provides a personal example of a chart for individual values (Pohlen, 1999). His wife is diabetic, and she needed to try to control the variability of her blood glucose level. For nondiabetics the normal glucose range is 70 to 120 milligrams/deciliter; the level for Pohlen's wife sometimes varied from under 70 to over 300 within a 24-hour period—that's variability. High levels can be caused by a variety of factors including different foods, exercise, illness, infections, and emotional stress.

Figure 18.8 shows the control chart for several months of glucose readings. With the help of their physician and the chart, the couple learned about the causes of high blood glucose and successfully developed a "control strategy" for reducing the variability. This case is not an avocation of an engineer—any diabetic person will verify the seriousness of this matter. The reference makes fascinating reading and gives us a testimonial to the power of understanding variation—and the determination of Pohlen's wife.

A chart for individual measurements can be useful when the normal process measurements are spaced some time apart, e.g., one measurement per day from a chemical process or a single weekly measurement from accounting data.

18.7 PRE-CONTROL

PRE-Control is a statistical technique for detecting process conditions and changes that may cause defects (rather than changes that are statistically significant). PRE-Control

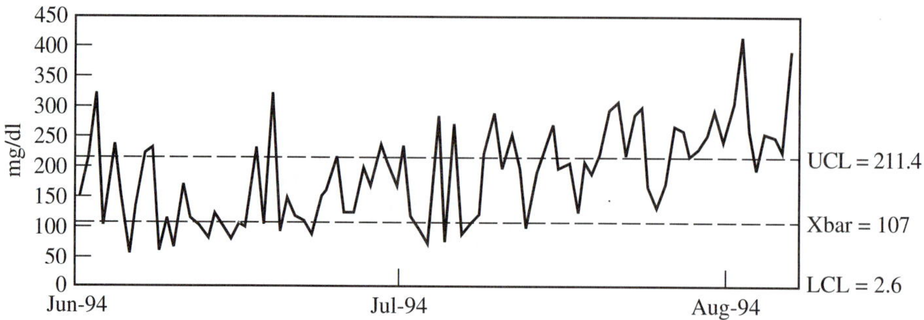

FIGURE 18.8
Blood glucose—summer 1994. (*Pohlen, 1999. Reprinted with permission by the ASQ.*)

focuses on controlling conformance to specifications, rather than statistical control. PRE-Control starts a process centered between specification limits and detects shifts that might result in making some of the parts outside a specification limit. PRE-Control requires no plotting and no computations, and it needs only three measurements to give control information. The technique utilizes the normal distribution curve in the determination of significant changes in either the aim or the spread of a production process that could result in increased production of defective work.

The principle of PRE-Control is demonstrated by assuming the worst condition that can be accepted from a process capable of quality production, i.e., when the natural tolerance is the same as the specification allows and when the process is precisely centered and any shift would result in some defective work.

If we draw two PRE-Control (PC) lines, each one-fourth of the way in from each specification limit (Figure 18.9), it can be shown that 86 percent of the parts will be inside the PC lines, with 7 percent in each of the outer sections. In other words, 7 percent, or 1 part in 14, will occur outside a PC line under normal circumstances.

The chance that two measurements in a row will fall outside a PC line is 1/14 times 1/14, or 1/196. Therefore, only once in about every 200 measurements should we expect to get two in a row in a given outer band. When two in a row do occur, the chance that the process has shifted is much greater (195/196). It is therefore advisable to reset the process to the center. It is equally unlikely to get a measurement beyond one given PC line and the next outside the other PC line. In this case the indication is not that the process has shifted, but that some factor has been introduced that has widened the pattern to an extent that defective pieces are inevitable. An immediate remedy of the cause of the trouble must be made before the process can safely continue.

The zone within the PC lines is the green zone; between the PC lines and the specification limits is the yellow zone; outside the specification limits is the red zone.

To qualify a process for PRE-Control:

1. Take consecutive individual measurements on a characteristic until five consecutive measurements fall within the green zone.
2. If one yellow occurs, restart the count.

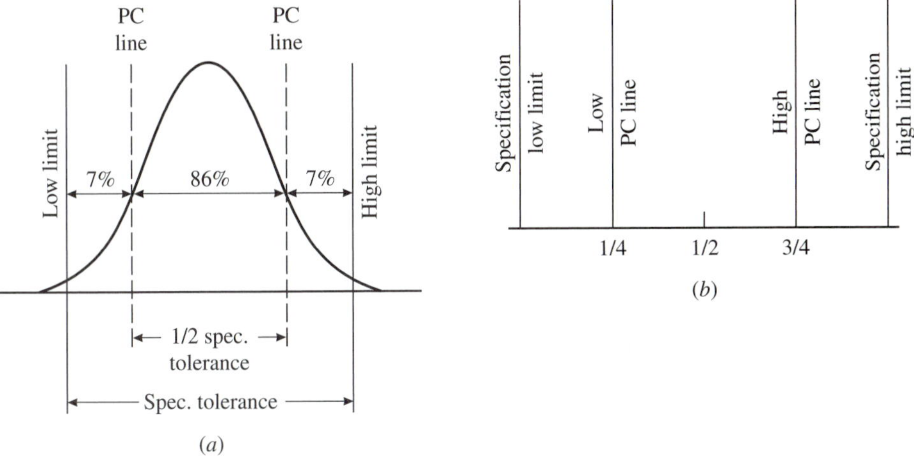

FIGURE 18.9
(a) Assumptions underlying PRE-Control. *(b)* Location of PRE-Control lines.

3. If two consecutive yellows occur, adjust the process.
4. Anytime an adjustment or other process change occurs, requalify the process.

When the process is qualified, the following PRE-Control rules are applied to running the process:

1. Use a sample of two consecutive measurements, A and B. If A is green, continue running the process. If A is yellow, take a second measurement, B.
2. If A and B are both yellow, stop the process and investigate.

During either the qualification or running stages, if a red occurs, stop the process and investigate.

Most processes require periodic adjustments to remain within specifications. Six A, B pairs of measurements between adjustments are viewed as sufficient to provide virtually no out-of-specification product. Thus if a process typically requires an adjustment about every two hours, then an A, B pair of measurements should be taken every 20 minutes.

PRE-Control is an example of a concept known as "narrow-limit gaging." The broader concept provides sampling procedures (sample size, location of the narrow limits, and allowable number of units outside the narrow limits) to meet predefined risks of accepting bad product. Narrow-limit gaging is discussed by Ott and Schilling (1990, Chapter 7).

The relative simplicity of PRE-Control versus statistical control charts can have important advantages in many applications. The concept, however, has generated some controversy. For a comparison of PRE-Control versus other approaches and the most appropriate applications of PRE-Control, see Ledolter and Swersey (1997) and Steiner (1997) in the Supplementary Reading. For a complete story, also see the references in both of these papers.

18.8
ATTRIBUTES CONTROL CHARTS

Control charts for $\bar{X}$, R, and X require that actual numerical measurements be made, e.g., line width from a photoresist process. Control charts for attributes data require only a count of observations on a characteristic, e.g., the number of nonconforming items in a sample.

Examples of Attributes Charts

The fraction nonconforming (p) chart can be illustrated with data on magnets used in electrical relays. For each of 19 weeks, the number of magnets inspected and the number of nonconforming magnets were recorded. The total number of magnets tested was 14,091. The total number found to be nonconforming was 1030. The average sample size was

$$\bar{n} = \frac{14{,}091}{19} = 741.6$$

The average fraction nonconforming was

$$\bar{p} = \frac{1030}{14{,}091} = 0.073$$

Control limits for the chart were placed at

$$\bar{p} \pm 3\sigma_p = \bar{p} \pm 3\sqrt{\frac{\bar{p}(1-\bar{p})}{\bar{n}}} = 0.073 \pm 3\sqrt{\frac{0.073(1-0.073)}{741.6}}$$

$$= 0.073 \pm 0.0287 = 0.102 \quad \text{and} \quad 0.044$$

Note that these control limits are based on the *average* sample size.

The resulting control chart is shown in Figure 18.10. Note that the last sample is below the LCL, indicating a significantly low fraction nonconforming. Although this sample might mean that some assignable cause is resulting in better quality, such points can also be due to (1) an inspector's accepting some nonconforming units in error or (2) the sample size being quite different from the average used to calculate the limits. Note that three points are beyond the control limits even though the data were included in calculating the control limits. A fascinating and powerful feature of control limits is their ability to detect the presence of (at least some) special causes, even though the control limits were influenced by those causes.

Leonard (1986) reports on the application of a p chart to recruiting new employees at the Rogers Corporation. Figure 18.11 shows a plot of the percentage of open job requisitions filled during calendar quarters. All of the points fall within the control limits, indicating that the variation is due to a system of common causes. Reducing that variation requires action "across the board" (as the team called it) rather than analyzing the cause of a low value such as 3 percent in the fourth quarter of 1981. Note

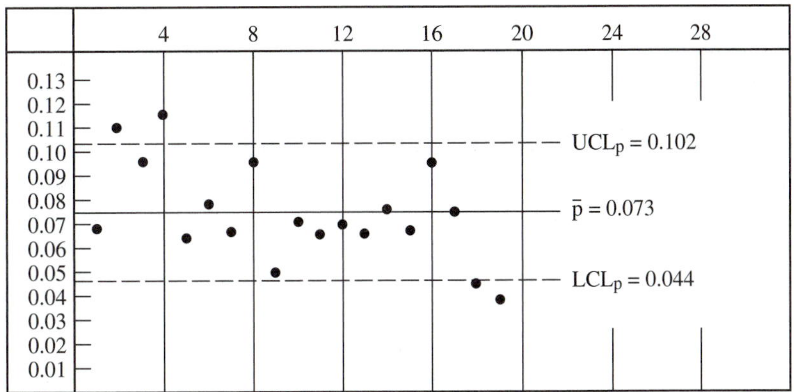

FIGURE 18.10
p chart for permanent magnets.

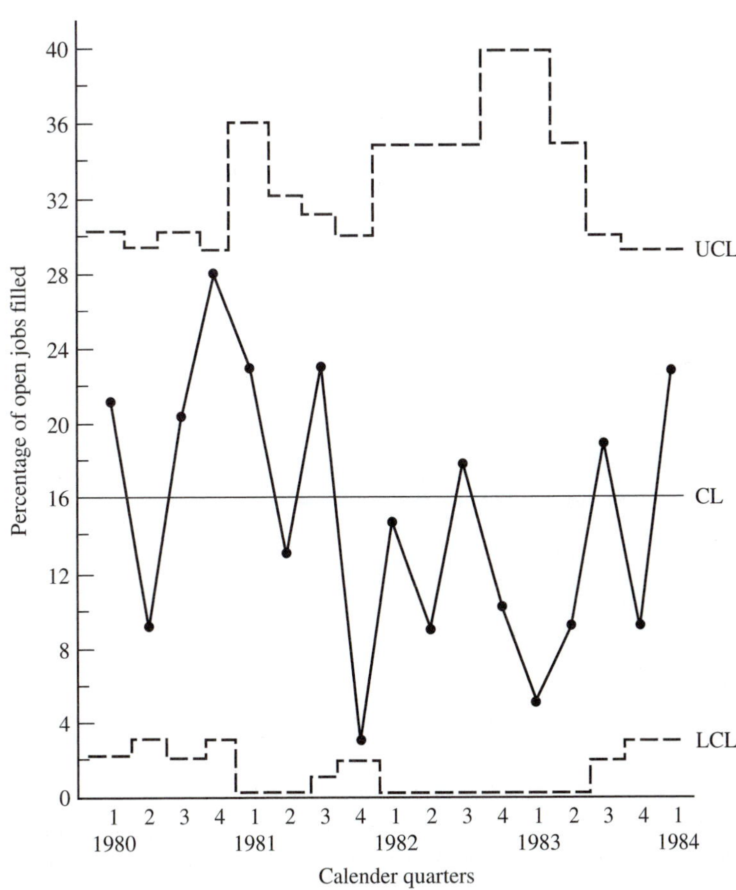

FIGURE 18.11
p chart of jobs filled as a percentage of jobs open by quarter (January 1, 1980–April 2, 1984). (*From Leonard, 1986.*)

that the control limits vary for each quarter. Instead of using an *average* sample size to calculate one set of control limits, the exact sample size is substituted in the formula to obtain the precise limits for each quarter. The price we pay for this precision is the difficulty in explaining why the control limits vary.

Heyes (1988) in the Supplementary Reading presents a vivid discussion of a p chart and a regression study. The analysis graphically convinced people that machines were aging, and the regression study quantified the age-performance relationship.

The c chart will be illustrated for information entered in 78 fields on a form for admitting patients to a hospital (Gitlow et al., 1995). The completed forms present many opportunities for clerical errors. To study the reasons for errors, 10 forms were selected each week for 25 weeks, and a total of 368 errors were found.

The central line of the chart is located at $\bar{c} = 368/25 = 14.72$. Control limits were calculated as $14.72 \pm 3\sqrt{14.72}$, or 3.21 and 26.24. The c chart is shown in Figure 18.12. Because the points all fall within statistical control limits, we conclude that there are no special causes of variation in the process and therefore the process is stable and predictable. If no changes are made, the process will continue to produce an average of 14.72 errors for every 10 forms processed.

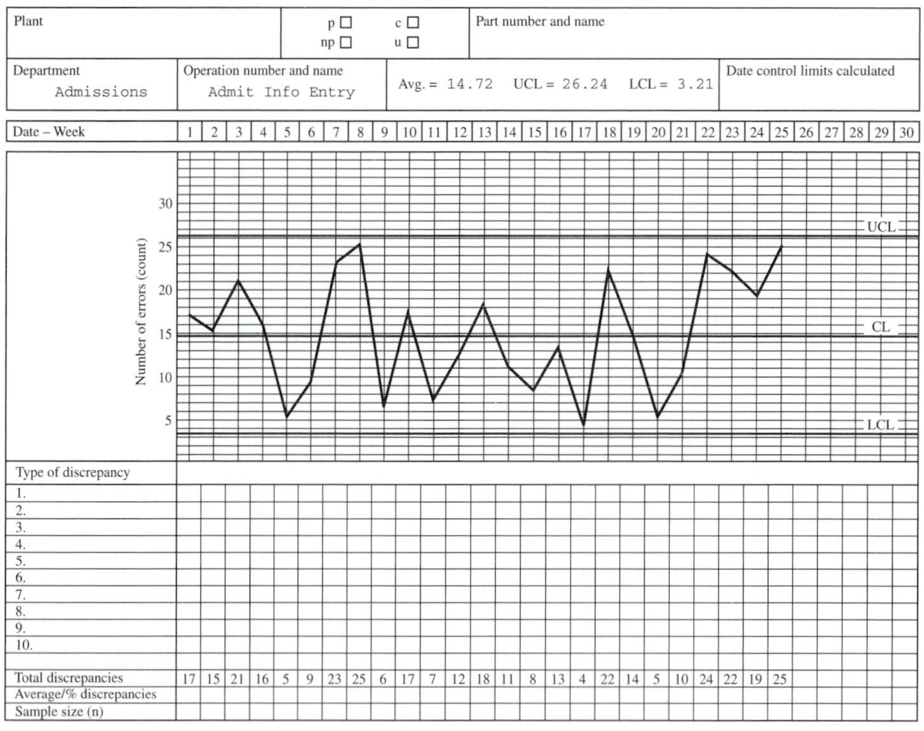

FIGURE 18.12
Control chart for errors on new patient admission forms.

18.9
SPECIAL CONTROL CHARTS

The previous paragraphs presented the basic types of control charts for variables and attributes data—the ones needed for most applications. Sometimes other types of control charts are employed to address special needs. Such special control charts have ingenious aspects, and several types are mentioned below to encourage the reader to explore further.

The *cumulative sum* (CUMSUM) *control chart* is a chronological plot of the cumulative sum of deviations of a sample statistic (e.g., $\bar{X}$, *R,* number of nonconformities) from a reference value (e.g., the nominal or target specification). By definition the CUMSUM chart focuses on a target value rather than the actual average of process data. The control limits are neither parallel nor fixed; the limits are typically displayed in a V-shaped mask (Figure 18.13) that is based on process data and is placed onto the chart and moved as a new point is plotted. Each point plotted contains information from all observations (i.e., a *cumulative* sum). CUMSUM charts are particularly useful in detecting small shifts in the process average (say 0.5σ to 2.0σ). Calculations for constructing the chart shown in Figure 18.13 are given in *JQH5,* pages 45.17–45.20.

Another special chart is the *moving average chart.* This chart is a chronological plot of the moving average, which is calculated as the average value updated by dropping the oldest individual measurement and adding the newest individual measurement. Thus a new average is calculated with each individual measurement. A further refinement is the *exponentially weighted moving average* (EWMA) *chart.* In the EWMA chart, the observations are weighted with the highest weight given to the most recent data. Moving average charts are effective in detecting small shifts, highlighting trends, and making use of data in processes in which it takes a long time to produce a single item.

Still another chart is the Box-Jenkins manual adjustment chart. The average and range, CUMSUM, and EWMA charts for variables focus on *monitoring* a process and

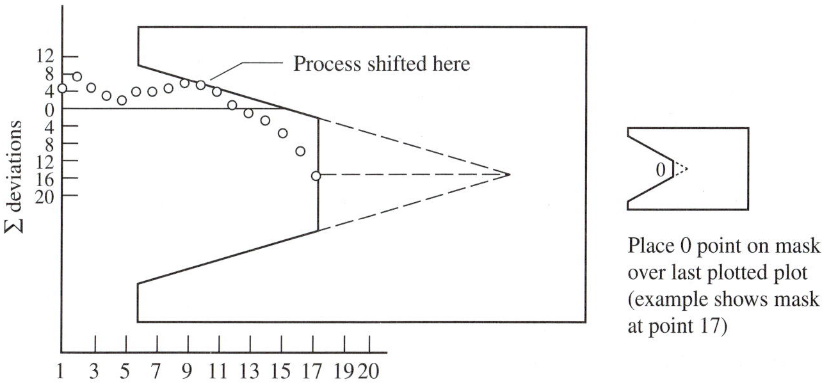

FIGURE 18.13
Cumulative sum control chart. (*From JQH5, page 45.19.*)

reducing variability due to special causes of variation identified by the charts. Box-Jenkins charts have a different objective: to analyze process data in order to *regulate* the process after each observation and thereby minimize the process variation. For elaboration on this advanced technique, see Box and Luceno (1997).

Finally, we consider the concept of multivariate control charts. When there are two or more quality characteristics on a unit of product, these could be monitored independently with separate control charts. Then the probability that a sample average on either control chart exceeds three sigma limits is 0.0027. But the joint probability that both variables exceed their control limits simultaneously when they are both in control is (0.0027)(0.0027) or 0.00000729, which is much smaller than 0.0027. The situation becomes more distorted as the number of characteristics increases. For this and other reasons, monitoring several characteristics independently can be misleading. Multivariate control charts address this issue. See Montgomery (1997, Section 8.4) for a highly useful discussion.

18.10 PROCESS CAPABILITY

In planning the quality aspects of operations, nothing is more important than advance assurance that the processes will be able to meet the specifications. In recent decades a concept of *process capability* has emerged to provide a quantified prediction of process adequacy. This ability to predict quantitatively has resulted in widespread adoption of the concept as a major element of quality planning.

Process capability is the measured, inherent variation of the product turned out by a process.

Basic Definitions

Each key word in this definition must itself be clearly defined, since the concept of capability has an enormous extent of application and since nonscientific terms are inadequate for communication within the industrial community.

- *Process* refers to some unique combination of machine, tools, methods, materials, *and people* engaged in production. It is often feasible and illuminating to separate and quantify the effect of the variables entering this combination.
- *Capability* refers to an ability, based on tested performance, to achieve measurable results.
- *Measured capability* refers to the fact that process capability is quantified from data that, in turn, are the results of measurement of work performed by the process.
- *Inherent capability* refers to the product uniformity resulting from a process that is in a state of statistical control, i.e., in the absence of time-to-time "drift" or other assignable causes of variation. "Instantaneous reproducibility" is a synonym for inherent capability.
- The *product* is being measured because product variation is the end result.

Uses of Process Capability Information

Process capability information serves multiple purposes:

1. Predicting the extent of variability that processes will exhibit. Such capability information, when provided to designers, provides important information in setting realistic specification limits.
2. Choosing from among competing processes that are most appropriate for the tolerances to be met.
3. Planning the interrelationship of sequential processes. For example, one process may distort the precision achieved by a predecessor process, as in hardening of gear teeth. Quantifying the respective process capabilities often points the way to a solution.
4. Providing a quantified basis for establishing a schedule of periodic process control checks and readjustments.
5. Assigning machines to classes of work for which they are best suited.
6. Testing theories of causes of defects during quality improvement programs.
7. Serving as a basis for specifying the quality performance requirements for purchased machines.

These purposes account for the growing use of the process capability concept.

Standardized Formula

The most widely adopted formula for process capability is

$$\text{Process capability} = \pm 3\sigma \text{ (a total of } 6\sigma)$$

where σ = the standard deviation of the process under a state of statistical control, i.e., under no drift and no sudden changes. See Section 18.15, however, for the six-sigma concept of process capability.

If the process is centered at the nominal specification and follows a normal probability distribution, 99.73 percent of production will fall within $\pm 3\sigma$ of the nominal specification.

Some industrial processes operate under a state of statistical control. For such processes the computed process capability of 6σ can be compared directly to specification limits, and judgments of adequacy can be made. However, most industrial processes exhibit both drift and sudden changes. These departures from the ideal are a fact of life, and the practitioner must deal with them.

Nevertheless, there is great value in standardizing on a formula for process capability based on a state of statistical control. Under this state, product variations are the result of numerous small variables (rather than being the effect of a single large variable) and, hence, have the character of random variation. It is most helpful for planners to have such limits in quantified form.

Relationship to Product Specifications

A major reason for quantifying process capability is to be able to compute the ability of the process to hold product specifications. For processes that are in a state of statistical control, a comparison of the variation of 6σ to the specification limits permits ready calculation of percentage defective by conventional statistical theory.

Planners try to select processes with the 6σ process capability well within the specification width. A measure of this relationship is the capability ratio:

$$C_p = \text{Capability ratio} = \frac{\text{Specification range}}{\text{Process capability}} = \frac{\text{USL} - \text{LSL}}{6s}$$

where USL = upper specification limit
LSL = lower specification limit

Note that $6s$ is used as an estimate of 6σ.

Some companies define the ratio as the reciprocal. Some industries now express defect rates in terms of parts per million. A defect rate of one part per million requires a capability ratio (specification range over process capability) of about 1.63.

Figure 18.14 shows four of many possible relations between process variability and specification limits and the likely courses of action for each. Note that, in all of these cases, the average of the process is at the midpoint between the specification limits.

Table 18.5 shows selected capability ratios and the corresponding level of defects assuming the process average is midway between the specification limits. A process that is just meeting specification limits (specification range $= \pm 3\sigma$) has a C_p of 1.0. The criticality of many applications and the reality that the process average will not remain at the midpoint of the specification range suggest that C_p should be at least 1.33.

Note that the C_p index measures whether the process variability can fit within the specification range. It does not indicate whether the process is actually running within the specification, because the index does not include a measure of the process average (this issue is addressed below).

Three capability indexes commonly in use are shown in Table 18.6. Of these, the simplest is *Cp*. The higher the value of any indexes, the lower the amount of product that is outside the specification limits.

Pignatiello and Ramberg (1993) provide an excellent discussion of various capability indexes. Bothe (1997) provides a comprehensive reference book that includes extensive discussion of the mathematical aspects. These references explain how to calculate confidence bounds for various process capability indexes.

The C_{pk} Capability Index

Process capability, as measured by C_p, refers to the variation in a process about the average value. This concept is illustrated in Figure 18.15. The two processes have equal capabilities (C_p) because 6σ is the same for each distribution, as indicated by the widths of the distribution curves. The process aimed at μ_2 is producing defectives because the aim is off center, not because of the inherent variation about the aim (i.e., the capability).

Thus the C_p index measures *potential* capability, assuming that the process average is equal to the midpoint of the specification limits and the process is operating in statistical control; as the average is often not at the midpoint, it is useful to have a

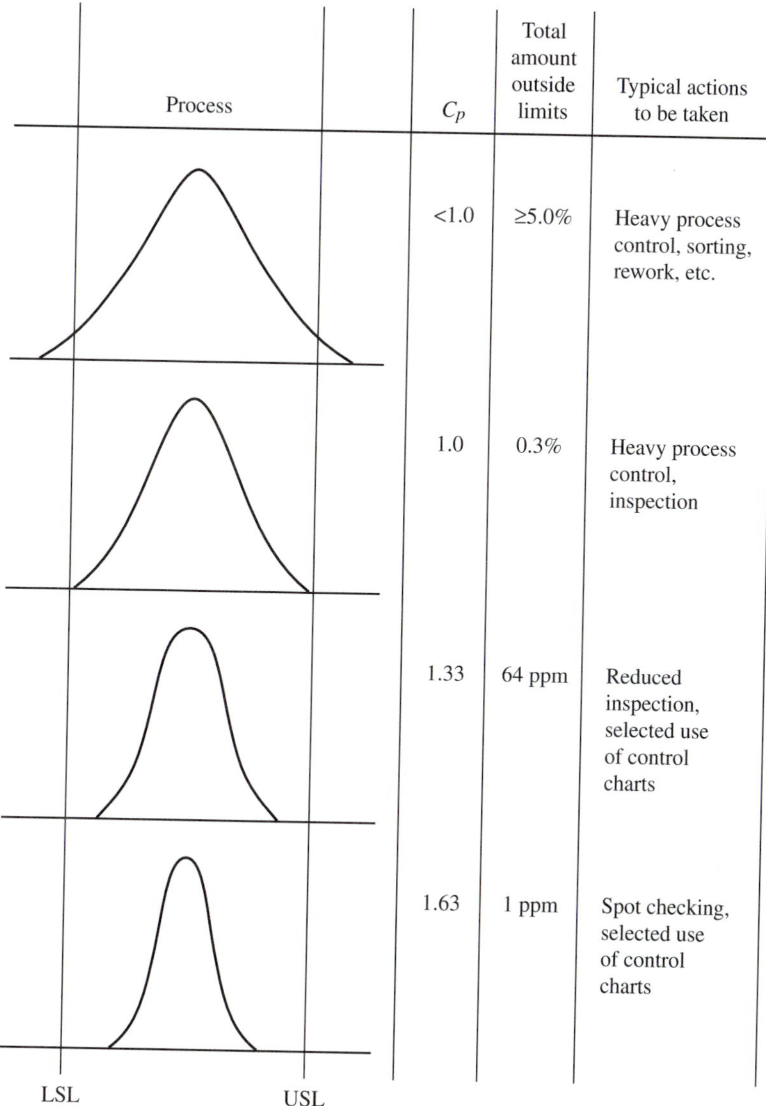

FIGURE 18.14
Four examples of process variability.

capability index that reflects both variation and the location of the process average. Such an index is C_{pk}.

C_{pk} reflects the current process mean's proximity to either the USL or LSL. C_{pk} is estimated by

$$\hat{C}_{pk} = \min\left[\frac{\bar{X} - \text{LSL}}{3s}, \frac{\text{USL} - \bar{X}}{3s}\right]$$

TABLE 18.5
Process capability index (C_p) and product outside specification limits

Process capability index (C_p)	Total product outside two-sided specification limits*
0.5	13.36%
0.67	4.55%
1.00	0.3%
1.33	64 ppm
1.63	1 ppm
2.00	0

*Assuming the process is centered midway between the specification limits.

TABLE 18.6
Process capability and process performance indexes

Process capability	Process performance
$C_p = \dfrac{\text{USL} - \text{LSL}}{6\sigma}$	$P_p = \dfrac{\text{USL} - \text{LSL}}{6s}$
$C_{pk} = \min\left[\dfrac{\text{USL} - \mu}{3\sigma}, \dfrac{\mu - \text{LSL}}{3\sigma}\right]$	$P_{pk} = \min\left[\dfrac{\text{USL} - \bar{X}}{3s}, \dfrac{\bar{X} - \text{LSL}}{3s}\right]$
$C_{pm} = \dfrac{\text{USL} - \text{LSL}}{6\sqrt{\sigma^2 + (\mu - T)^2}}$	$P_{pm} = \dfrac{\text{USL} - \text{LSL}}{6\sqrt{s^2 + (\bar{X} - T)^2}}$

In an example from Kane (1986):

$$\text{USL} = 20 \qquad \bar{X} = 16$$
$$\text{LSL} = 8 \qquad s = 2$$

The standard capability ratio is estimated as

$$\frac{\text{USL} - \text{LSL}}{6\sigma} = \frac{20 - 8}{12} = 1.0$$

which implies that *if* the process were centered between the specification limits (at 14), then only a small proportion (about 0.27 percent) of product would be defective.

However, when we calculate C_{pk}, we obtain

$$\hat{C}_{pk} = \min\left[\frac{16 - 8}{6}, \frac{20 - 16}{6}\right] = 0.67$$

which alerts us that the process mean is *currently* nearer the USL. (Note that, if the process were centered at 14, the value of C_{pk} would be 1.0.) An acceptable process will require reducing the standard deviation and/or centering the mean.

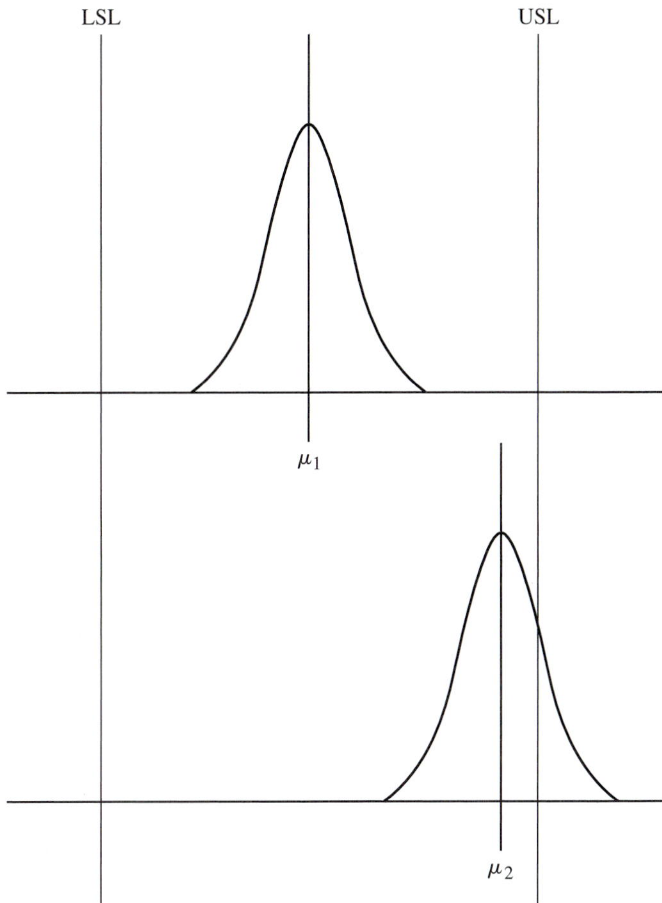

FIGURE 18.15
Process with equal process capability but different aim.

Note that, if the *actual* average is equal to the midpoint of the specification range, then $C_{pk} = C_p$.

The higher the value of C_p, the lower the amount of product outside specification limits. In certifying suppliers, some organizations use C_{pk} as one element of certification criteria. In these applications the value of C_{pk} desired from suppliers can be a function of the type of commodity being purchased.

A capability index can also be calculated around a target value rather than the actual average. This index, called C_{pm} or the Taguchi index, focuses on reduction of variation from a target value rather than reduction of variability to meet specifications.

Most capability indexes assume that the quality characteristic is normally distributed. Krishnamoorthi and Khatwani (2000) propose a capability index for handling normal and nonnormal characteristics by first fitting the data to a Weibull distribution.

Two types of process capability studies are as follows:

1. *Study of process potential.* In this study an estimate is obtained of what the process *can* do under certain conditions, i.e., variability under short-run defined conditions for a process in a state of statistical control. The C_p index estimates the process capability.
2. *Study of process performance.* In this study an estimate of capability provides a picture of what the process *is* doing over an extended period of time. A state of statistical control is also assumed. The C_{pk} index estimates the capability.

18.11 ESTIMATING INHERENT OR POTENTIAL CAPABILITY FROM A CONTROL CHART ANALYSIS

In a process potential study, data are collected from a process operating without changes in material batches, workers, tools, or process settings. This short-term evaluation uses consecutive production over one time period. Such an analysis should be preceded by a control chart analysis in which any assignable causes have been detected and eliminated from the process.

As specification limits usually apply to individual values, control limits for sample averages cannot be compared to specification limits. To make a comparison, we must first convert $\bar{R}$ to the standard deviation for individual values, calculate the $\pm 3\sigma$ limits, and compare them to the specification limits. This process is explained below.

If a process is in statistical control, it is operating with the minimum amount of variation possible (the variation due to chance causes). If, and only if, a process is in statistical control, the following relationship holds for using s as an estimate of σ:

$$s = \frac{\bar{R}}{d_2}$$

Table I in Appendix II and Table 18.4 provide values of d_2. If the standard deviation is known, process capability limits can be set at $\pm 3s$ and this value used as an estimate of 3σ.

For the data of Figure 18.5 (machine N-5):

$$s = \frac{\bar{R}}{d_2} = \frac{6.0}{2.534} = 2.37$$

and

$$\pm 3s = \pm 3(2.37) = 7.11$$

or

$$6s = 14.22 \text{ (or 0.0142 in the original data units)}$$

The specification was 0.258 ± 0.005.

Thus

$$\text{USL} = 0.263$$
$$\text{LSL} = 0.253$$

Then

$$C_p = \frac{\text{USL} - \text{LSL}}{6s} = \frac{0.263 - 0.253}{0.0142} = 0.72$$

Even if the process is perfectly centered at 0.258 (and it was not), it is not capable.

The Assumption of Statistical Control and Its Effect on Process Capability

All statistical predictions assume a stable population. In a statistical sense a stable population is one that is repeatable, i.e., a population that is in a state of statistical control. The statistician rightfully insists that this be the case before predictions can be made. The manufacturing engineer also insists that the process conditions (feeds, speeds, etc.) be fully defined.

In practice, the original control chart analysis will often show the process to be out of statistical control. (It may or may not be meeting product specifications.) However, an investigation may show that the causes cannot be economically eliminated from the process. In theory a process capability prediction should not be made until the process is in statistical control. However, in practice some kind of comparison of capability to specifications is needed. The danger in delaying the comparison is that the assignable causes may never be eliminated from the process. The resulting indecision will thereby prolong interdepartmental bickering on whether "the specification is too tight" or "manufacturing is too careless."

A good way to start is by plotting individual measurements against specification limits. This step may show that the process can meet the product specifications even with assignable causes present. If a process has assignable causes of variation but is able to meet the specifications, usually no economic problem exists. The statistician can properly point out that a process with assignable variation is unpredictable. This point is well taken, but in establishing priorities of quality improvement efforts, processes that are meeting specifications are seldom given high priority.

If a process is out of control and the causes cannot be economically eliminated, the standard deviation and process capability limits can nevertheless be computed (with the out-of-control points included). These limits will be inflated because the process will not be operating at its best. In addition, the instability of the process means that the prediction is approximate.

It is important to distinguish between a process that is in a state of statistical control and a process that is meeting specifications. A state of statistical control does not necessarily mean that the product from the process conforms to specifications. Statistical control limits on sample averages *cannot* be compared to specification limits because specification limits refer to individual units. For some processes that are not in control, the specifications are being met and no action is required; other processes are in control but the specifications are not being met, and action is needed (see Section 18.16).

In summary, we need processes that are both stable (in statistical control) and capable (meeting product specifications).

The increasing use of capability indexes has also led to the failure to understand and verify some important assumptions that are essential for statistical validity of the results. Five key assumptions are

1. *Process stability.* Statistical validity requires a state of statistical control with no drift or oscillation.
2. *Normality of the characteristic being measured.* Normality is needed to draw statistical inferences about the population.
3. *Sufficient data.* Sufficient data is necessary to minimize the sampling error for the capability indexes. Lewis (1991) provides tables of 95 percent lower confidence limits for values of C_p and C_{pk}.
4. *Representativeness of samples.* Samples must include random samples.
5. *Independence of measurements.* Consecutive measurements cannot be correlated.

These assumptions are not theoretical refinements—they are important conditions for properly applying capability indexes. Before applying capability indexes, readers are urged to read the paper by Pignatiello and Ramberg (1993). It is always best to compare the indexes with the full data versus specifications as depicted in a histogram.

18.12 MEASURING PROCESS PERFORMANCE

A process performance study collects data from a process that is operating under typical conditions but includes normal changes in material batches, workers, tools, or process settings. This study, which spans a longer term than the process potential study, also requires that the process be in statistical control.

The capability index for a process performance study is

$$C_{pk} = \min\left[\frac{\bar{X} - \text{LSL}}{3s}, \frac{\text{USL} - \bar{X}}{3s}\right]$$

EXAMPLE 18.4. Consider a pump cassette used to deliver intravenous solutions (Baxter Travenol Laboratories, 1986). A key quality characteristic is the volume of solution delivered in a predefined time. The specification limits are

$$\text{USL} = 103.5 \qquad \text{LSL} = 94.5$$

A control chart was run for one month, and no out-of-control points were encountered. From the control chart data, we know that

$$\bar{X} = 98.2 \qquad \text{and} \qquad s = 0.98$$

Figure 18.16 shows the process data and the specification limits.

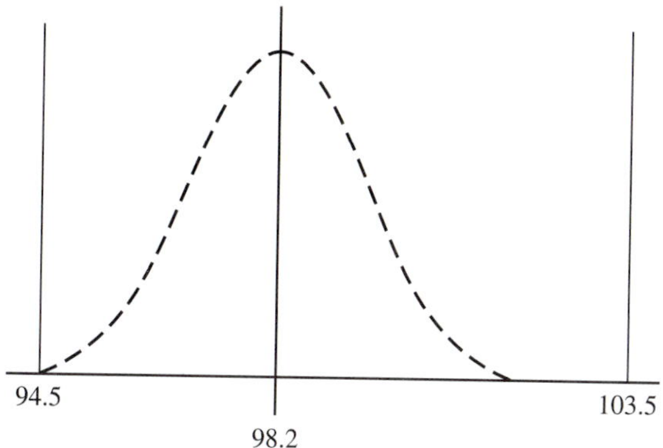

FIGURE 18.16
Delivered volume of solution. (*From Baxter Travenol Laboratories, 1986.*)

The capability index is

$$C_{pk} = \min\left[\frac{98.2 - 94.5}{3(0.98)}, \frac{103.5 - 98.2}{3(0.98)}\right]$$

$$C_{pk} = 1.26$$

For many applications, 1.26 is an acceptable value of C_{pk}.

Interpretation of C_{pk}

In using C_{pk} to evaluate a process, we must recognize that C_{pk} is an abbreviation of two parameters—the average and the standard deviation. Such an abbreviation can inadvertently mask important detail on these parameters; e.g., Figure 18.17 shows that three extremely different processes can all have the same C_{pk} (in this case $C_{pk} = 1$).

Increasing the value of C_{pk} may require a change in the process average, the process standard deviation, or both. For some processes, increasing the value of C_{pk} by changing the average value (perhaps by a simple adjustment of the process aim) may be easier than doing so by reducing the standard deviation (by investigating the many causes of variability). The histogram of the process should always be reviewed to highlight both the average and the spread of the process.

Note that Table 18.6 also includes the capability index C_{pm}. This index measures the capability around a target value T rather than the mean value. When the target value equals the mean value, the C_{pm} index is identical to the C_p index.

Attributes Data Analysis

The methods discussed above assume that numerical measurements are available from the process. Sometimes, however, the only data available are in attribute form, i.e., the number of nonconforming units and the number acceptable.

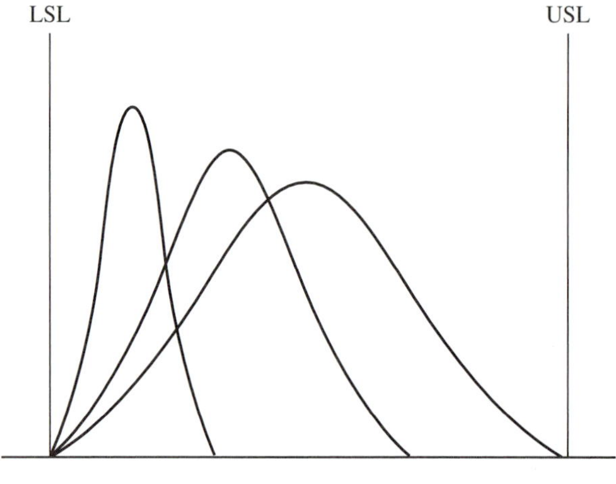

FIGURE 18.17
Three processes with $C_{pk} = 1$.

The data in Table 3.5 on errors in preparing insurance policies also can be used to illustrate process capability for attributes data. The data reported 80 errors from six policy writers or 13.3 errors per writer—the current *performance.* The process *capability* can be calculated by excluding the abnormal performance identified in the study—type 3 errors by worker B, type 5 errors, and errors of worker E. The error data for the remaining five writers becomes 4, 3, 5, 2, and 5, with an average of 3.8 errors per writer. The process capability estimate of 3.8 compares with the original performance estimate of 13.3.

This example calculates process capability in terms of errors or mistakes rather than the variability of a process parameter. Hinckley and Barkan (1995) point out that in many processes, nonconforming product can be caused by excessive variability or by mistakes (e.g., missing parts, wrong parts, wrong information, or other processing errors). For some processes, mistakes can be a major cause of failing to meet customer quality goals. The actions required to reduce mistakes are different from those required to reduce variability on a parameter.

Process Capability in Service Industries

Although the concept of process capability grew up in manufacturing, the concept also applies to service industry processes.

For certain parameters in service processes, process capability can be measured by the various capability indexes. For example, in a loan association the cycle time to complete the loan approval process is critical, and time data is readily available in quantitative form to calculate a capability index.

Other service processes may not have variables data available. For example, a firm provides a service of guaranteeing checks written by customers at retail stores. The decision to guarantee is based on a process that employs an on-line evaluation of six factors. Unfortunately, a percentage of checks default ("bounce"), and this percentage could be viewed as a measure of the capability of the process for making the decision to guarantee. This approach uses attributes data as was done in the example above on errors in insurance policies.

18.13 PROCESS CAPABILITY ANALYSIS USING PROBABILITY PAPER

Probability paper, introduced in Chapter 10, can also be used to determine process capability—without any calculation of the standard deviation.

For example, a sample of 100 measurements on total indicator reading (TIR) was grouped into a frequency distribution of 10 cells. The frequencies were then plotted on normal probability paper (Figure 18.18). The measurements are plotted and are observed to follow a straight line (indicating that the population is normally distributed). The vertical scale on the left shows the cumulative percentage of the population under a value of TIR; the vertical scale on the right shows the cumulative percentage over a value of TIR and also the number of standard deviations from the mean. The upper and lower horizontal grid lines represent $\pm 3\sigma$, respectively. When the plotted line is extended to intersect the grid lines for $\pm 3\sigma$, the values of 0.001 and 0.059 are read, indicating that 99.73 percent of the population will fall between 0.001 and 0.059. The $\pm 3\sigma$ capability is then calculated as $0.059 - 0.001$, or 0.058. As the C_{pk} specification range is 0.060, the value is $0.060/0.058 = 1.03$. As before, we are assuming that the process is in statistical control.

By extending the plotted line to intersect the specification limits (shown as vertical lines), the percentage of nonconforming product can be predicted. Essentially, zero percent will fall below the lower specification limit or above the upper specification limit.

The mean value can also be quickly estimated from the plot. Enter the vertical axis on the 50 percent line. This line intersects the diagonal at 0.030 on the horizontal scale.

The probability-paper approach has a few advantages. The term "standard deviation" (still a mystery to most people even after it is explained) is avoided. The plot provides an approximate test of normality. When data are limited, the probability-paper plot may be more valuable than comparing a histogram to the theoretical "bell shape" because small samples frequently yield histograms having many peaks and valleys, making it difficult to judge the underlying shape. Finally, Weibull probability paper could be used to allow for skewed or other unusual distributions.

Evaluating process capability by using probability paper (or histograms) is usually performed without evaluating the process for statistical control. This approach does not determine the inherent capability of the process. The data may include measurements from several populations, and time-to-time or other process changes may be present. Such conditions result in observed dispersions that are wider than the inherent capability of the process. Evaluating the inherent capability requires the use of a control chart.

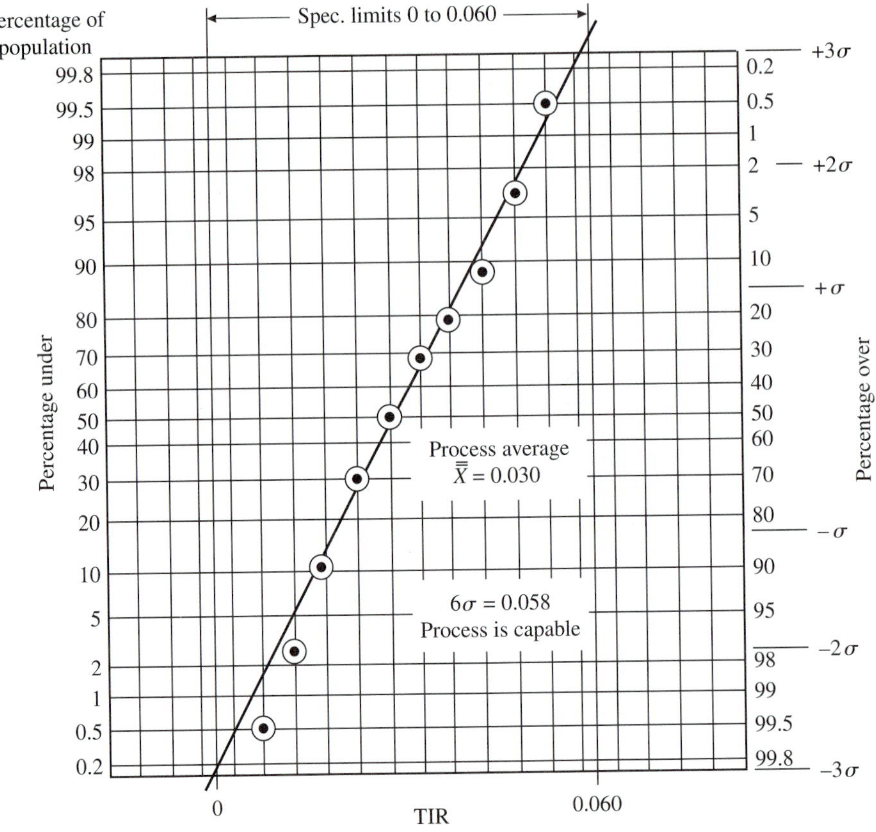

FIGURE 18.18

Capability analysis on probability paper. (*Adapted from Amsden et al., 1986.*)

18.14 PLANNING FOR THE PROCESS CAPABILITY STUDY

Capability studies are conducted for various reasons, e.g., to respond to a customer request for a capability index number or to evaluate and improve product quality. Prior to data collection, clarify the purpose for making the study and the steps needed to assure that it is achieved.

In some cases the capability study will focus on determining a histogram and capability index for a relatively simple process. Here the planning should assure that process conditions (e.g., temperature, pressure) are completely defined and recorded. All other inputs must clearly be representative, i.e., specific equipment; material; and of course, personnel.

For more complex processes or when defect levels of 1 to 10 parts per million are desired, the following steps are recommended:

1. Develop a process description including inputs, process steps, and output quality characteristics. This description can range from simply identifying the equipment

to developing a mathematical equation that shows the effect of each process variable on the quality characteristics.

2. Define the process conditions for each process variable. In a simple case this step involves stating the settings for temperature and pressure. But for some processes it means determining the optimum value or aim for each process variable. The statistical design of experiments provides the methodology. Also, determine the operating ranges of the process variables around the optimum because the range will affect the variability of the product results.
3. Make sure that each quality characteristic has at least one process variable that can be used to adjust it.
4. Decide whether measurement error is significant. This can be determined from a separate error of measurement study (see Section 19.8). In some cases the error of measurement can be evaluated as part of the overall study.
5. Decide whether the capability study will focus on variability only or will also include mistakes or errors that cause quality problems.
6. Plan for the use of control charts to evaluate stability of the process.
7. Prepare a data collection plan including adequate sample size that documents results on quality characteristics along with the process conditions (e.g., values of all process variables) and preserves information on the order of measurements so that trends can be evaluated.
8. Plan which methods will be used to analyze data from the study to assure, before starting the study, that all necessary data for the analysis will be available. The analyses should include not only process capability calculations on variability but also analysis of attribute data on mistakes and analysis of data from statistically designed experiments built into the study.
9. Be prepared to spend time investigating interim results before process capability calculations can be made. These investigations can include analysis of optimum values and ranges of process variables, out-of-control points on control charts, or other unusual results. The investigations then lead to the ultimate objective, i.e., improvement of the process.

Note that these steps focus on improvement rather than just on determination of a capability index.

18.15 SIX-SIGMA CONCEPT OF PROCESS CAPABILITY

The six-sigma approach is really a strategy for improvement. Chapter 3 explains the five steps in the approach along with the specific tools involved. In the current chapter, we present only the statistical basis of the concept.

For some processes, shifts in the process average are so common that they should be recognized in setting acceptable values of C_p. In some industries, shifts in the process average of ± 1.5 standard deviations (of individual values) are not unusual. To allow for such shifts, high values of C_p are needed. For example, if specification limits are at $\pm 6\sigma$ (Figure 18.19) and if the mean shifts $\pm 1.5\sigma$, then 3.4 parts per million will be beyond specification limits. The Motorola Company's six-sigma approach recognizes the likelihood of these shifts in the process average and uses a variety of quality engineering techniques to change the product, the process, or both to achieve a C_p of at least 2.0.

Table 18.7 shows capability indexes and defect levels in terms of parts per million (ppm) for a centered process and for a process with the mean shifted 1.5σ.

Note that at the 3σ level with a centered process, the defect level is 2700 parts per million (or 0.27 percent); with a shift of 1.5σ, the defect level is 66,803 parts per million (or 6.68 percent). These calculations apply to a single product or process characteristic. In reality, many products or processes have numerous characteristics, and thus the overall yield decreases with this complexity. McFadden (1993) discusses this matter. For example, assuming a process with a 1.5σ shift, a process with 100 steps each operating at the 3 sigma level would have a yield of only 0.10 percent; at the 5 sigma level the yield would still only be about 54 percent. This sobering outcome demonstrates the importance of low variability (i.e., a high sigma level), particularly for complex products or processes.

We must emphasize that the six-sigma approach to improvement is a strategy for improvement that assumes process shifts of 1.5σ (in practice a realistic assumption for many processes). This use of the term *six sigma* is quite different from the classical standardized formula for process capability as $\pm 3\sigma$ or a total of 6σ.

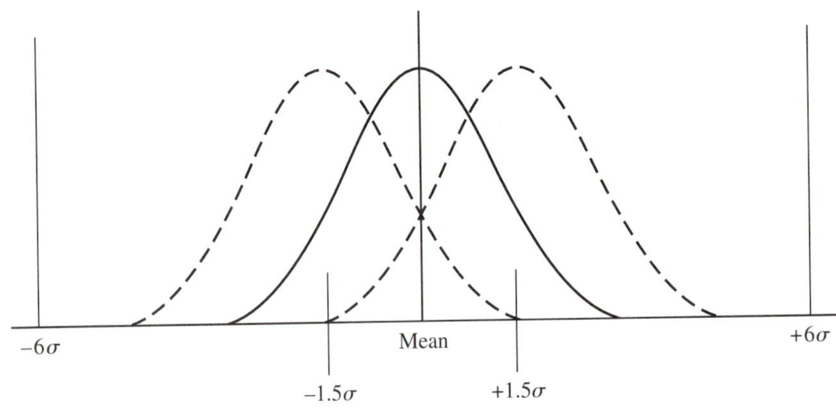

FIGURE 18.19
Six-sigma concept of process capability.

TABLE 18.7
Sigma levels and defect levels

	Centered process		Shifted ($\pm 1.5\sigma$) process	
Sigma level	C_p	**ppm***	C_{pk}	**ppm**
3	1	2,700	0.5	66,803
4	1.33	63	0.833	6,200
5	1.67	0.57	1.167	233
6	2	0.002	1.5	3.4

Source: McFadden, 1993. Reprinted with permission by the ASQ.
*ppm = parts per million.

18.16 STATISTICAL PROCESS CONTROL AND QUALITY IMPROVEMENT

In Chapter 1 we distinguish between control and improvement. The control process detects and takes action on *sporadic* quality problems; the improvement process identifies and takes action on *chronic* quality problems.

In the control process, statistical control charts detect the existence of special causes of variation that result in sporadic problems. The charts show sample data falling beyond statistical control limits; i.e., the process is "out of statistical control." Conversely, when a chart shows a process to be "in statistical control" the process is in a state of stability with variation due to a set of common causes inherent in the process. Statistical control means stability, but stability does not always mean customer satisfaction with the result. Unfortunately, a process in statistical control can have serious quality problems. Because the process is stable, the problems will continue (become chronic) unless a basic change in the system of common causes is made. Such a change, which typically affects the average or variation, is the job of improvement. "Removal of a special cause of variation, to move toward statistical control, important though it may be, is not improvement of the process" (Deming, 1986, page 338).

Process improvement is directed at several problems:

1. *The process average is misdirected.* Table 18.8 shows possible corrective action.
2. *The inherent variability of the process is too large.* Table 18.8 provides some of the arsenal of approaches available to reduce the variability.
3. *The instrumentation is inadequate.* See Section 19.8, "Errors of Measurement."
4. *A process drift exists.* Here the need is to quantify the amount of drift in a given period of time and to provide a means of resetting the process to compensate for this drift.
5. *Cyclical changes in the process exist.* The need is to identify the underlying cause and either remove it or reduce the effect on the process.
6. *The process is erratic.* Sudden changes can take place in processes. As the capability studies quantify the size of these changes and help to discover the reasons for them, appropriate planning action can be taken:
   - *Temporary phenomena* (e.g., a cold machine coming up to operating temperature) can be dealt with by scheduling warming periods plus checks at the predicted time of stability.
   - *More enduring phenomena* (e.g., changes due to new materials) can be dealt with by specifying setup reverification at the time of introducing such changes.

The statistical design of experiments is an essential analytical tool for improvement that goes far beyond the investigation of out-of-control points on a statistical control chart. This tool, when combined with the knowledge of those who plan and run the processes, replaces intuitive decision making with a scientific basis. See Section 11.11, "The Design of Experiments," and Section 13.4 under "Parameter Design and Robust Design."

Now, how do these matters relate to customer needs?

Clearly, in using statistical process control and taking subsequent actions of any sort, the focus must be on meeting customer needs. One definition—which is far from perfect—is given by the specification limits. Limits on statistical control charts

TABLE 18.8
Approaches to process improvement

Changing the average	Reducing variability
Adjust settings on the process equipment.	Investigate work methods and equipment factors. This includes identifying the process variables that affect the product result.
Change selected product design parameters so that the design is more robust to manufacturing conditions.	Change selected product design parameters so that the design is more robust to manufacturing conditions.
Identify the process variables that affect the product results, determine the optimum values for the variables, and set the process to these values.	Identify and reduce the causes of variability due to human inputs. The concept of system versus worker-controllable input and the concept of self-control are useful guides.
Employ automated process controls to continually measure, analyze, and adjust process variables that affect the average.	Reduce the variability of inputs to the process through an improvement program with internal and external suppliers.
	Employ automated process controls to continually measure, analyze, and adjust process variables that affect variability.

are different from specification limits. In some situations a process is not in statistical control but may not require action because the product specifications are easily being met. In other situations a process is in statistical control, but the product specifications are not being met.

If a product *does not* meet specifications, then some type of action is needed—changing the average value, reducing the variability, doing both, changing the specifications, sorting the product, etc. If a product *does* meet the specifications, the alternatives are different—taking no action, using a less precise process, or reducing the variability further (see below for reasons). Table 18.9 shows the more usual permutations encountered and provides suggestions on the type of action to be taken.

Our goal for processes is clear: Be in statistical control and be capable of meeting product specifications.

Note that this chapter focuses on the variability of product characteristics and its relationship to specification limits. A broader issue of improvement is the ability of existing product features to meet customer needs (see Chapter 12).

18.17 SOFTWARE FOR STATISTICAL PROCESS CONTROL

The availability of software has contributed substantially to the use of statistical process control techniques. Software has made it practical to collect large quantities of data and perform complex analyses.

Statistical process control software will calculate sample statistics, calculate control limits, and plot a control chart. Additional summaries and analyses can be pro-

TABLE 18.9
Action to be taken

	Product meets specifications	
	Process variation small relative to specifications*	**Process variation large relative to specifications***
Process is in control	Consider value in marketplace of tighter specifications. Reduce inspection.	Continue tight controls on process average.
Process is out of control	Process is erratic and unpredictable and may be heading for trouble. Investigate causes of lack of control.	
	Product does not meet specifications	
Process is in control	Process is "misdirected" to wrong average. Generally easy to correct permanently.	Process may be misdirected and also too scattered. Correct misdirection. Consider economics of more precise process versus wider specifications versus sorting the product.
Process is out of control	Process is misdirected or erratic or both. Correct misdirection. Discover cause of lack of control. Consider economics of more precise process versus wider specifications versus sorting the product.	

*As a rule of thumb, a process variation (sometimes called natural tolerance = 6σ) less than one-third of the specification range is small; more than two-thirds is large.

vided such as a listing of raw data, out-of-specification values, histograms, checks for runs and other patterns on a control chart, Pareto charts, and process capability calculations.

Excel® is an example of software for statistical process control. An example of a spreadsheet for data and control charts for averages and ranges is shown in Figure 18.20. The Chart Wizard feature allows you to display charts on the spreadsheet along with data on separate sheets. If changes are made to the data, the values in the chart are updated and displayed. Zimmerman and Icenogle (1999) provide step-by-step instructions on using Excel® in statistical quality control, including basic statistics, various types of control charts, and acceptance sampling. MINITAB® has similar capabilities for statistical process control and includes software for designing and analyzing experiments.

Quality Progress (ASQ) and *Quality Digest* publish annual directories of software for process control and other statistical uses. The *Journal of Quality Technology* (ASQ) and *Quality* (Hitchcock Publishing Company) have computer columns describing programs for process control.

SUMMARY

- Statistical process control is the application of statistical methods to the measurement and analysis of any process.
- Decreased process variability has important advantages.

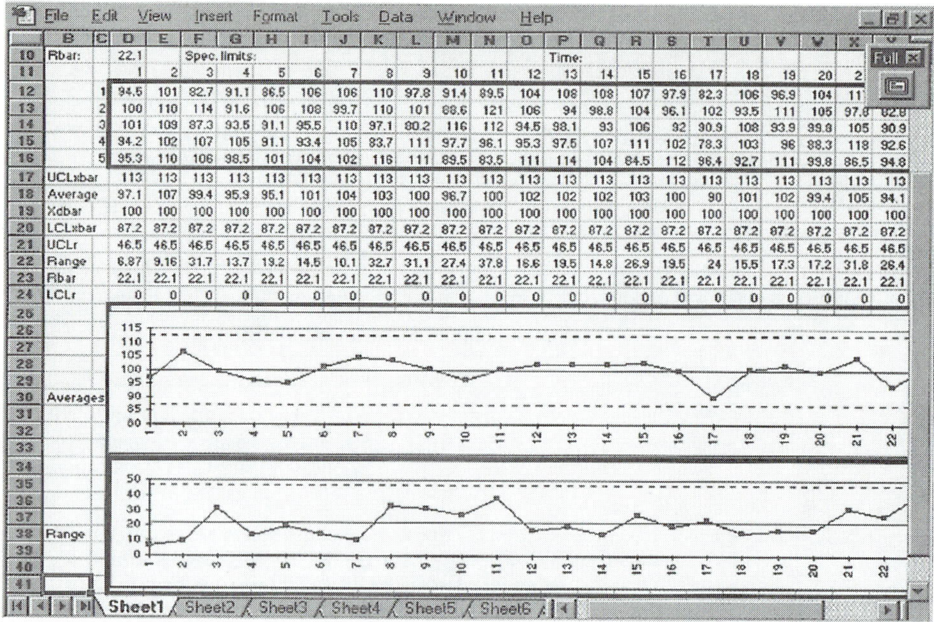

FIGURE 18.20
Excel $\overline{X}$ and range charts.

- A statistical control chart is a graphic comparison of process performance data to computed statistical control limits drawn as limit lines on the chart.
- A control chart distinguishes between common causes and special causes of variation.
- A state of statistical control has important advantages for any process.
- Control charts come in many types for both variables data and attributes data.
- PRE-Control is a statistical technique for detecting process conditions and changes that may cause defects.
- Process capability is the measured, inherent reproducibility of a product turned out by a process.
- Capability ratios help to quantify process capability.
- We want processes that are in statistical control and capable of meeting specification limits.
- The six-sigma approach is a strategy for improvement. The goal is for process variation to be so small that the specification limits are 6σ above and below the process average.

PROBLEMS

18.1. A large manufacturer of watches makes some of its own parts and buys other parts from a supplier. The supplier submits lots of parts that meet the specifications of the horologist. The supplier thus wishes to keep a continuous check on its production of watch parts. One gear has been a special problem. A check of 25 samples of five pieces gave the following data on a key dimension:

$$\bar{\bar{X}} = 0.3175 \text{ cm} \qquad \bar{R} = 0.00508 \text{ cm}$$

What criterion should be set up to determine when the process is out of control? How should this criterion compare with the specification? What are some alternatives if the criterion is not compatible with the specification?

18.2. A manufacturer of dustless chalk is concerned with the density of the product. Previous analysis has shown that the chalk has the required characteristics only if the density is between 4.4 g/cm^2 and 5.04 g/cm^2. If a sample of 100 pieces gives an average of 4.8 and a standard deviation of 0.2, is the process aimed at the proper density? If not, what should the aim be? Is the process capable of meeting the density requirements? Calculate C_p and C_{pk}.

18.3. The head of an automobile engine must be machined so that both the surface that meets the engine block and the surface that meets the valve covers are flat. These surfaces must also be 4.875 ± 0.001 in. apart. Presuming that the valve-cover side of the head is finished correctly, compare the capability of two processes for performing the finishing of the engine-block side of the head. A broach set up to do the job gave an average thickness of 4.877 in. with an average range of 0.0005 in. for 25 samples of four each. A milling machine gave an average of 4.875 in. and average range of 0.001 in. for 20 samples of four each. For each machine, calculate C_p and C_{pk}.

Answer: Broach $\pm 3s = \pm 0.00072$, milling machine $\pm 3s = 0.00144$.

18.4. A critical dimension on a double-armed armature has been causing trouble, and the designer has decided to change the specification from 0.033 ± 0.005 in. to 0.033 ± 0.001 in. To evaluate the proposed change, the manufacturing planning department has obtained the following coded data from the process:

Time	Left arm			Right arm			Comments
8:00	331	330	331	329	330	328	
8:30	332	331	329	327	331	329	
9:00	330	329	329	330	329	327	
9:30	332	330	331	331	332	328	
10:00	333	332	333	326	331	326	
10:30	332	331	332	329	330	331	
11:00	333	331	331	330	326	327	
11:30	332	332	333	327	326	329	
12:00	331	332	334	337	328	337	Adjustment
12:30	335	334	336	326	325	325	
1:00	333	332	332	329	332	330	
1:30	336	331	330	331	328	329	
2:00	332	334	329	332	330	329	
2:30	336	336	330	329	329	327	
3:00	329	335	338	333	330	331	Adjustment
3:30	341	333	330	329	331	332	

Comment on the proposal to change the specification.

18.5. A company manufactures an expensive chemical. The net package weight has a minimum specification value of 25.0 lb. Data from a control chart analysis show (based on 20 samples of five each):

$$\bar{\bar{X}} = 26.0 \qquad \bar{R} = 1.4$$

The points on both the average and range chart are all in control.
(a) Draw conclusions about the ability of the process to meet the specification.
(b) What action, if any, would you suggest on the process? If any action is suggested, are there any disadvantages to the action?

18.6. Samples of four units were taken from a manufacturing process at regular intervals. The width of a slot on a part was measured, and the average and range were computed for each sample. After 25 samples of four, the following coded results were obtained:

	Averages	**Ranges**
Upper control limit	626	37.5
Average value	$\bar{\bar{X}} = 614$	$\bar{R} = 16.5$
Lower control limit	602	0

All points on the $\bar{X}$ and R charts fell within control limits. The specification requirements are 610 $\pm$15. If the width is normally distributed and the distribution centered at $\bar{\bar{X}}$, what percentage of product would you expect to fall outside the specification limits?
Answer: 9.4 percent.

18.7. *(a)* What kind of conclusion about a process can be made from an average and range control chart that cannot be made from a histogram?
(b) What kind of conclusion can be made from a histogram that cannot be made from an average and range control chart?

18.8. The percentage of water absorption is an important characteristic of common building brick. A certain company occasionally measured this characteristic of its product, but never kept records. Management decided to analyze the process with a control chart. Twenty-five samples of four bricks each yielded these results:

Sample number	$\bar{X}$	R	**Sample number**	$\bar{X}$	R
1	15.1	9.1	14	9.8	17.5
2	12.3	9.9	15	8.8	10.5
3	7.4	9.7	16	8.1	4.4
4	8.7	6.7	17	6.3	4.1
5	8.8	7.1	18	10.5	5.7
6	11.7	9.1	19	9.7	6.4
7	10.2	12.1	20	11.7	4.6
8	11.5	10.8	21	13.2	7.2
9	11.2	13.5	22	12.5	8.3
10	10.2	6.9	23	7.5	6.4
11	9.6	5.0	24	8.8	6.9
12	7.6	8.2	25	8.0	6.4
13	7.6	5.4			

Plot the data on an average and range control chart with control limits. Comment.
Answer: 15.8 and 3.97; 18.4 and 0.

18.9. The specification on a certain dimension is 3.000 ± 0.004 in. A large sample from the process indicates an average of 2.998 in. and a standard deviation of 0.002 in. Suppose that controls are instituted to shift the process average to the nominal specification of 3.000. Each part outside specification limits results in a loss of $5. In a lot of 1000 parts, how much money would be saved by shifting the average versus keeping it at 2.998? *Answer:* $573.

18.10. A statistical control chart for averages and ranges has been used to help control a manufacturing process. The sample data are consistently within control limits, and the control limits are inside the engineering tolerance limits. The supervisor is confused because a high percentage of product is outside the tolerance limits, even though the process is within control limits. What is your explanation?

18.11. The following data represent the number of defects found on each sewing machine cabinet inspected:

Sample number	Number of defects	Sample number	Number of defects
1	8	14	6
2	10	15	4
3	7	16	7
4	7	17	5
5	8	18	8
6	6	19	6
7	9	20	4
8	8	21	5
9	4	22	7
10	7	23	4
11	9	24	5
12	6	25	5
13	5		

Plot a control chart with control limits. Comment on the chart.

18.12. A sample of 100 electrical connectors was inspected each shift. Three characteristics were inspected on each connector but each connector was classified simply as defective or acceptable. The results follow:

Sample number	Defective, %	Sample number	Defective, %
1	4	14	4
2	3	15	4
3	5	16	5
4	6	17	3
5	7	18	0
6	5	19	3
7	4	20	2
8	2	21	1
9	5	22	3
10	6	23	4
11	4	24	2
12	3	25	2
13	3		

(*a*) Plot a control chart with control limits. Comment on the chart.
(*b*) If the inspection results had been recorded in sufficient detail, what other type of chart could also have been plotted?

Answer: (*a*) 9.2 percent, 0.

18.13. An average and range chart based on a sample size of five has been run with the following results:

	Averages	Ranges
Upper control limit	78.0	8.0
Average value	75.8	3.8
Lower control limit	73.6	0

How large an increase in the overall process average would have to occur to have a 30 percent chance that a sample average will exceed the upper control limit?

Answer: 1.83.

18.14. As part of a quality improvement program, a textile manufacturer decides to use a control chart to monitor the number of imperfections in bolts of cloth. The data from the last 25 inspections are recorded in the following table. From these data, compute control limits for an appropriate type of control chart. Plot the chart.

Bolt of cloth	Number of imperfections	Bolt of cloth	Number of imperfections
1	14	14	22
2	5	15	1
3	10	16	6
4	19	17	14
5	0	18	8
6	6	19	6
7	2	20	9
8	9	21	7
9	8	22	1
10	7	23	5
11	3	24	12
12	12	25	4
13	1	Total	191

Answer: Limits are 15.93 and 0.

18.15. Control chart data were collected for the softening point (in degrees) in a polymerization process. Based on 25 samples of four each, the following control limits were calculated:

	Averages	Ranges
Upper control limit	12.9	13.4
Average value	9.4	5.9
Lower control limit	5.9	0

Suppose that the population average shifts to 12.4. How large a sample is necessary to have a 25 percent probability that the control chart for averages will signal out of control?

18.16. A p chart is to be used to analyze the September record for 100 percent inspection of certain radio components. The total number inspected during the month was 2196, and the total number of defectives was 158. Compute $\bar{p}$. Compute control limits for the following three days and state whether the percentage defective falls within control limits for each day.

Date	Number inspected	Number of defectives
Sept. 14	54	8
Sept. 15	162	24
Sept. 16	213	3

18.17. A control chart with upper and lower control limits can detect when a process is out of control. It cannot, however, identify the causes. Select a process from a service industry and describe an approach to identify the variable(s) that caused an out-of-control process.

18.18. What is the main objective of statistical control for the process of evaluating applications for mortgage loans at a financial institution?

REFERENCES

Amsden, R. T., H. E. Butler, and D. M. Amsden (1986). *SPC Simplified: Practical Steps to Quality,* UNIPUB, White Plains, NY.

Anthis, D. L., R. F. Hart, and R. J. Stanula (1991). "The Measurement Process: Roadblock to Product Improvement," *Quality Engineering,* vol. 3, no. 4, pp. 461–470.

AT&T (1990). *Analyzing Business Process Data: The Looking Glass,* AT&T Customer Information Center, Indianapolis, IN.

Baxter Travenol Laboratories (1986). *Special Process Control Guideline,* Deerfield, IL, p. 17.

Bothe, D. R. (1997). *Measuring Process Capability,* McGraw-Hill, New York.

Box, G. E. P. and A. Luceno (1997). *Statistical Control by Monitoring and Adjustment,* 3rd ed., John Wiley & Sons, New York.

Dechert, J., K. E. Case, and T. L. Kautiainen (2000). "Statistical Process Control in the Presence of Large Measurement Variation," *Quality Engineering,* vol. 12, no. 3, pp. 417–423.

Deming, W. E. (1986). *Out of the Crisis,* Massachusetts Institute of Technology, Center for Advanced Engineering Study, Cambridge, MA.

Gitlow, H., A. Oppenheim, and R. Oppenheim (1995). *Quality Management: Tools and Methods for Improvement,* 2nd ed., Irwin, Burr Ridge, IL.

Hinckley, C. M. and P. Barkan (1995). "The Role of Variation, Mistakes, and Complexity in Producing Nonconformities," *Journal of Quality Technology,* vol. 27, no. 3, pp. 242–249.

Kane, V. E. (1986). "Process Capability Indices," *Journal of Quality Technology,* vol. 18, no. 1, pp. 41–52.

Krishnamoorthi, K. S. and S. Khatwani (2000). "A Capability Index for All Occasions," *Annual Quality Congress Proceedings,* ASQ, Milwaukee, pp. 77–81.

Leonard, J. F. (1986). "Quality Improvement in Recruiting and Employment," *Juran Report Number Six,* Winter, Juran Institute, Inc., Wilton, CT, pp. 111–118.

Lewis, S. S. (1991). "Process Capability Estimates from Small Samples," *Quality Engineering,* vol. 3, no. 3, pp. 381–394.

McFadden, F. R. (1993). "Six Sigma Quality Programs," *Quality Progress,* June, pp. 37–42.

Montgomery, D. C. (1997). *Introduction to Statistical Quality Control,* 3rd ed., John Wiley & Sons, New York.

Nelson, L. S. (1985). "Interpreting Shewhart Charts," *Journal of Quality Technology,* vol. 17, no. 2, pp. 114–116.

Nelson, L. S. (1984). "The Shewhart Control Chart-Tests for Special Causes," *Journal of Quality Technology,* vol. 16, no. 4, October, pp. 237–239.

Ott, E. R. and E. G. Schilling (1990). *Process Quality Control,* McGraw-Hill, New York.

Pignatiello, J. H. Jr. and J. S. Ramberg (1993). "Process Capability Indicies: Just Say No," *Annual Quality Congress Transactions,* ASQC, Milwaukee, pp. 92–104.

Pohlen, T. (1999). "Statistical Thinking: A Personal Application," *Proceedings of the Annual Quality Congress Proceedings,* ASQ, Milwaukee, pp. 230–235.

Schilling, E. G. (1990). "Elements of Process Control," *Quality Engineering,* vol. 2, no. 2, p. 132. Reprinted by courtesy of Marcel Dekker, Inc.

Sullivan, L. P. (1984). "Reducing Variability: A New Approach to Quality," *Quality Progress,* July, pp. 15–21.

Zimmerman, S. M. and M. L. Icenogle (1999). *Statistical Quality Control Using EXCEL,* Quality Press, ASQ, Milwaukee.

SUPPLEMENTARY READING

Statistical process control: *JQH5,* Section 45.

Grant, E. L. and R. S. Leavenworth (1996). *Statistical Quality Control,* 7th ed., McGraw-Hill, New York.

Automotive Industry Action Group (1995). *Statistical Process Control,* AIAG, Southfield, MI.

PRE-Control: Ledolter, J. and A. Swersey (1997). "An Evaluation of PRE-Control," *Journal of Quality Technology,* vol. 29, no. 2, pp. 163–171.

Steiner, S. H. (1997). PRE-Control and Some Simple Alternatives," *Quality Engineering,* vol. 10, no. 1, pp. 65–74.

p charts: Heyes, G. B. (1988) "Do We Need New Machines? A p-Chart and Regression Study," *Quality Engineering,* vol. 1, no. 1, pp. 13–28.

WEBSITE

ASQ statistics division: www.asq.org/about/divisions/stats

19

INSPECTION, TEST, AND MEASUREMENT

19.1 THE TERMINOLOGY OF INSPECTION

Inspection and test typically include measurement of an output and comparison to specified requirements to determine conformity. Inspection is performed for a wide variety of purposes, e.g., distinguishing between good and bad product, determining whether a process is changing, measuring process capability, rating product quality, securing product design information, rating the inspectors' accuracy, and determining the precision of measuring instruments. Each of these purposes has its special influence on the nature of the inspection and on the manner of doing it.

The distinction between "inspection" and "test" has become blurred. Inspection, typically performed under static conditions on items such as components, can vary from simple visual examination to a series of complex measurements. The emphasis in inspection is to determine conformance to a standard. Test, on the other hand, is performed under either static or dynamic conditions and is typically performed on more complex items such as subassemblies or systems. Test results not only determine conformance but can also be used as input for other analyses such as evaluating a new design, diagnosing problems, or making physical adjustments on products. Some industries have their own terms for inspection or testing; e.g., assay is used in the mining and pharmaceutical industries.

Although the terms *inspection* and *test* usually refer to manufacturing industries, the concepts also apply to other industries. In service industries, different terms are used, e.g., *review, checking, reconciliation, examination.* The evaluation made of the correctness of an income tax return, the cleanliness of a hotel room, or the accuracy of a bank teller's closing balance are really all forms of inspection—a measurement, a comparison to a standard, and a decision.

19.2
CONFORMANCE TO SPECIFICATION AND FITNESS FOR USE

Of all the purposes of inspection, the most ancient and the most extensively used is product acceptance, i.e., determining whether product conforms to standard and therefore be accepted. *Product* can mean a discrete unit, a collection of discrete units (a "lot"), a bulk product (a tank car of chemicals), or a complex system.

Product can also mean a service, such as a transaction at a bank; an inquiry to an agency about tax regulations; or the performance of personnel before, during, and after an airline flight. In these examples inspection characteristics can be identified, standards set, and a judgment made on conformance.

Product acceptance involves the disposition of product based on its quality. This disposition involves several important decisions:

1. *Conformance.* Judging whether the product conforms to specification.
2. *Fitness for use.* Deciding whether nonconforming product is fit for use.
3. *Communication.* Deciding what to communicate to insiders and outsiders.

The Conformance Decision

Except in small companies, the number of conformance decisions made each year is huge. There is no possibility that the supervisory body can become involved in the details of so many decisions. Hence the work is organized so that inspectors or production workers can make these decisions themselves. To this end, they are trained to understand the products, the standards, and the instruments. Once trained, they are given the jobs of making inspections and judging conformance. (In many cases the delegation is to automated instruments.)

Associated with the conformance decision is the disposition of conforming product. The inspector is authorized to identify the product ("stamp it up") as acceptable product. This identification then serves to inform packers, shippers, etc., that the product should proceed to its next destination (further processing, storeroom, customer). Strictly speaking, this decision to "ship" is made not by inspectors but by management. With some exceptions, product that conforms to specification is also fit for use. Hence company procedures (which are established by the managers) provide that conforming products should be shipped as a regular practice.

The Fitness-for-Use Decision

In the case of nonconforming products, a new question arises: Is this nonconforming product fit or unfit for use? In some cases the answer is obvious—the nonconformance is so severe that the product is clearly unfit. Hence it is scrapped or, if economically repairable, brought to a state of conformance. However, in many cases the answer as to fitness for use is not obvious. In such cases, if enough is at stake, a study is made to determine fitness for use. This study involves securing inputs such as the following:

- *Who will the user be?* A technologically sophisticated user may be able to deal successfully with the nonconformance; a consumer may not. A nearby user may have easy access to field service; a distant or foreign user may lack such easy access.
- *How will the product be used?* For many materials and standard products, the specifications are broad enough to cover a variety of possible uses, and the actual use to which the product will be put is not known at the time of manufacture. For example, sheet steel may be cut up to serve as decorative plates or as structural members; a television receiver may be stationed at a comfortable range or at an extreme range; chemical intermediates may be employed in numerous formulas.
- *Are there risks to human safety or to structural integrity?* Where such risks are significant, all else is academic.
- *What is the urgency?* For some applications, the client cannot wait because the product in question is critical to putting some broader system into operation. Hence the product may demand delivery now and cause repairs in the field.
- *What are the company's and the users' economics?* For some nonconformances, the economics of repair are so forbidding that the product must be used as is, although at a price discount. In some industries, e.g., textiles, the price structure formalizes this concept by use of a separate grade—"seconds."
- *What are the users' measures of fitness for use?* These may differ significantly from those available to the manufacturer. For example, a manufacturer of abrasive cloth used a laboratory test to judge the *ability* of the cloth to polish metal; a major client evaluated the *cost* per 1000 pieces polished.

These and other inputs may be needed at several levels of fitness for use, i.e., the effects on the economics of subsequent processors, the marketability requirements of the merchants, the qualities that determine fitness for the ultimate user, and the qualities that influence field maintenance.

The job of securing such inputs is often assigned to a staff specialist, e.g., a quality engineer who "makes the rounds," contacting the various departments that are able to provide pertinent information. There may be a need to contact the customer and even to conduct an actual tryout. A typical list of sources is shown in Table 19.1.

Once all the information has been collected and analyzed, the fitness-for-use decision can be made. If the amount of money at stake is small, this decision will be

TABLE 19.1
Sources of information

Input	Usual sources
Who will the user be?	Marketing
How will this product be used?	Marketing; client
Are there risks to human safety or to structural integrity?	Product research and design
What is the urgency?	Marketing; client
What are the company's and users' economics?	All departments; client
What are the users' measures of fitness for use?	Market research; marketing; client

delegated to a staff specialist, to the quality manager, or to some continuing decision-making committee such as a material review board. If the amount at stake is large, the decision will usually be made by a team of upper managers.

Deliberations on the fitness-for-use decision are often a dramatic blend of voices—some balanced and judicious, other bowing to the pressures of delivery deadlines even if it means tossing up gems of earnest nonsense.

The Communication Decision

The conformance and fitness-for-use decisions are a source of essential information, although some of the data is not well communicated.

Data on nonconforming products are usually communicated to the producing departments to aid them in preventing a recurrence. More elaborate data collection systems may require periodic summaries to identify "repeaters," which then become the subject of special studies.

When nonconforming products are sent out as fit for use, the need for two additional categories of communication arises:

1. *Communication to "outsiders"* (usually customers) who have a right and a need to know. All too often manufacturing companies neglect or avoid informing their customers when shipping nonconforming products. Such avoidance can be the result of bad experience; i.e., some customers will seize on such nonconformances to secure a price discount despite the fact that use of the product will not add to their own costs. Neglect is more usually a failure even to face the question of what to communicate. A major factor here is the design of the forms used to record the decisions. With rare exceptions these forms lack provisions that force those involved to make recommendations and decisions on (a) whether to inform the outsiders and (b) what to communicate to them.
2. *Communication to insiders.* When nonconforming goods are shipped as fit for use, the reasons for doing so are not always communicated to inspectors and especially not to production workers. The resulting vacuum of knowledge has been known to breed some bad practices. When the same type of nonconformance has been shipped several times, an inspector may conclude (in the absence of knowing why) that it is a waste of time to report such nonconformances in the first place. Yet in some future case, the special reasons that were the basis of the decision to ship the nonconforming goods may not be present. In like manner, a production worker may conclude that it is a waste of time to exert all that effort to avoid some nonconformance that will be shipped anyway. Such reactions by well-meaning employees can be minimized if the company squarely faces the question, What shall we communicate to insiders?

19.3 DISPOSITION OF NONCONFORMING PRODUCT

Once an inspector finds a lot of product to be nonconforming, he or she prepares a report to that effect. Copies of this report are sent to the implicated departments. This action sets a planned sequence of events into motion. The lot is marked "hold" and is often sent to a special holding area to reduce the risk of mixups. The product is put into quarantine. Schedulers look into the possibility of shortages and the need for replacement. An investigator is assigned to collect the type of information needed as inputs for the fitness-for-use decision as discussed above.

Decision Not to Ship

The investigation may conclude that the lot should not be shipped as is. In that event the economics are studied to find the best disposition: sorting, repairing, downgrading, scrapping, etc. Supplemental accounting efforts may charge the costs to the responsible source, especially if supplier responsibility is involved. Some degree of action to prevent a recurrence also occurs (see below).

Decision to Ship

This decision may come about in several ways:

- *Waiver by the designer.* Such a waiver is a change in specification as to the lot in question that thereby puts the lot into a state of conformance.
- *Waiver by the customer* or by the marketing department on behalf of the customer. Such a waiver in effect supersedes the specification. (The waiver may have been "bought" by a change in warranty or by a discount in price.)
- *Waiver by the quality department* under its delegation to make fitness-for-use decisions on noncritical matters. The criteria for "noncritical" may be based on prior seriousness classification of characteristics, on the low cost of the product involved, or on still other bases. For minor categories of seriousness, the delegation may even be made by the quality engineers or by inspection supervisors. However, as to major and critical defects, the delegation is typically by the technical manager, the quality manager, or some team of managers.
- *Waiver by a formal material review board.* This board concept was originally evolved by military buyers of defense products as a means of expediting decisions on nonconforming lots. Membership on the board includes the military representative plus the designer and the quality specialist. A unanimous decision is required to ship nonconforming product. The board procedures provide for formal documentation of the facts and conclusions, thereby creating a data source of great potential value.
- *Waiver by upper managers.* This part of the procedure is restricted to cases of a critical nature involving risks to human safety, marketability of the product, or risk of loss of large sums of money. For such cases, the stakes are too high to warrant decision making by a single department. Hence the managerial team takes over. Waivers, however, have an insidious way of becoming part of a culture. It is valuable to continuously track the amount of product shipped under the waiver of specifications, e.g., the percentage of lots shipped each month under waiver.

The service sector also has a variety of actions for disposition of a nonconforming service. Peach (1997, page 435) provides some examples:

- *For a cable television distribution failure,* the disposition is to repair by reinitializing the connection.
- *For an incorrect crediting in an account by a bank,* the disposition is to rework by crediting the correct account and debiting the incorrect amount.
- *For tainted food in a restaurant,* the disposition is to scrap by returning the food to the supplier, disposing in the trash, and notifying local health authorities.

Corrective Action

Aside from a need to dispose of the nonconforming lot, there is a need to prevent a recurrence. This prevention process is of two types, depending on the origin of the nonconformance.

1. Some nonconformances originate in some isolated, sporadic change that took place in an otherwise well-behaved process. Examples are a mixup in the materials used, an instrument that is out of calibration, or a human mistake in turning a valve too soon. For such cases the local supervision is often able to identify what went wrong and to restore the process to its normal good behavior. Sometimes this troubleshooting may require the assistance of a staff specialist. In any case, no changes of a fundamental nature are involved, since manufacturing planning has already established an adequate process.
2. Other nonconformances are "repeaters." They arise over and over again, as evidenced from their recurring need for disposition by the material review board or other such agency. Such recurrences point to a chronic condition that must be diagnosed and remedied if the problem is to be solved. The local supervision is seldom able to find the cause of these chronic nonconformances, mainly because the responsibility for diagnosis is vague. Lacking agreement on the cause, the problem goes on and on amid earnest debates about who or what is to blame—unrealistic design, incapable process, poor motivation, etc. The need is not for troubleshooting to restore the normal good behavior, since the normal behavior is bad. Instead, the need is to organize for an improvement project, as discussed in Chapter 3.

19.4 INSPECTION PLANNING

Inspection planning is the activity of (1) designating the "stations" at which inspection should take place and (2) providing those stations with the means for knowing what to do plus the facilities for doing it. For simple, routine quality characteristics, the planning is often done by the inspector. For complex products made in large multidepartmental companies, the planning is usually done by specialists such as quality engineers.

Locating the Inspection Stations

The basic tool for choosing the location of inspection stations is the flowchart (see, e.g., Figure 16.1). The most usual locations are

- At receipt of goods from suppliers, usually called "incoming inspection" or "supplier inspection."
- Following the setup of a production process to provide added assurance against producing a defective batch. In some cases this "setup approval" also becomes approval of the batch.
- During the running of critical or costly operations, usually called "process inspection."
- Prior to delivery of goods from one processing department to another, usually called "lot approval" or "tollgate inspection."
- Prior to shipping completed products to storage or to customers, usually called "finished-goods inspection."
- Before performing a costly, irreversible operation, e.g., pouring a melt of steel.
- At natural "peepholes" in the process.

The concept of inspection stations also applies to the service sector. For example, in arranging travel for customers, "receiving inspection" often includes verifying information on a customer credit card; "process inspection" includes verifying that the customer has a passport, visa, and driver's license; "final inspection" includes confirming that tickets match agreements (Peach, 1997, page 432).

The inspection station is not necessarily a fixed zone where the work comes to the inspector. In some cases the inspector goes to the work by patrolling a large area and performing inspections at numerous locations including the shipping area, at the supplier's plant, or on the customer's premises.

Process inspection is often within the responsibility of production operators. An adjunct to this approach is the use of "poka-yoke" devices as part of the inspection. These devices are installed in the machine to inspect process conditions and product results and provide immediate feedback to the operator (see Section 16.7). See also Section 5.3 for a discussion of the self-adjusting concept where operators employ simple, direct methods (including inspection) continuously. Increasingly, inspection is built into the process rather than being placed at the end of the process (see Durkee and Gookins, 1999).

Choosing and Interpreting Quality Characteristics

The planner prepares a list of which quality characteristics are to be checked at which inspection stations. For some of these characteristics, the planner may find it necessary to provide information that supplements the specifications. Product specifications are prepared by comparatively few people, each generally aware of the needs of fitness for use. In contrast, these specifications must be used by numerous inspectors and operators, most of whom lack such awareness. The planner can help bridge this gap in a number of ways:

- By providing inspection and test environments that simulate the conditions of use. This principle is widely used, for example, in testing electrical appliances. It is also extended to applications such as the type of lighting used for inspecting textiles.
- By providing supplementary information that goes beyond the specifications as prepared by the product designers and process engineers. Some of this information is

available in published standards—company, industry, and national. Other information is specifically prepared to meet the specific needs of the product under consideration. For example, in an optical goods factory, the generic term *beauty defects* was used to describe several conditions that differed widely as to their effect on fitness for use. A scratch on a lens surface in the focal plane of a microscope made the lens unfit for use. A scratch on the large lens of a pair of binoculars, although not functionally serious, was visible to the user and hence was not acceptable. Two other species of scratches were neither adverse to fitness for use nor visible to the user and hence were unimportant. Through planning analysis these distinctions were clarified and woven into the procedures.

- By helping to train inspectors and supervisors to understand the conditions of use and the "why" of the specification requirements.
- By providing seriousness classification (see Section 19.5).

Detailed Inspection Planning

For each quality characteristic, the planner determines the detailed work to be done. This determination covers matters such as:

- The type of test to be done. This area may require a detailed description of the testing environment, testing equipment, testing procedure, and associated tolerances for accuracy.
- The number of units to be tested (sample size).
- The method of selecting the samples to be tested.
- The type of measurement to be made (attributes, variables, other).
- Conformance criteria for the units, usually the specified product tolerance limits.

Beyond this detailed planning for the characteristics and units is further detailed planning applicable to the product, the process, and the data system:

- Conformance criteria for the lot, usually consisting of the allowable number of nonconforming units in the sample.
- The physical disposition to be made of the product—the conforming lots, the nonconforming lots, and the units tested.
- Criteria for decisions on the process—should it run or stop?
- Data to be recorded, forms to be used, reports to be prepared.

This planning is usually included in a formal document that must be approved by the planner and the inspection supervisor. For an example from Baxter Travenol, see Figure 19.1.

Sensory Characteristics

Sensory characteristics are those for which we lack measuring instruments and for which the senses of human beings must be used as measuring instruments. Sensory qualities may involve technological performance of a product (e.g., adhesion of a protective coating), aesthetic characteristics (e.g., odor of a perfume), taste (e.g., food), or human services characteristics (e.g., the spectrum of hotel services).

Part Number: XXXX **Part Name: YYYY**

Process	Characteristics	C_p[1] Index	C_{pk}[1] Index	Frequency[2]	Sample size[2]	Analysis methods	Out-of-control conditions are encountered[4]
Incoming inspection	Stock thickness	1.6	1.0	Every shipment	—	Review control charts provided with each lot	Impound lot—contact supplier for resolution
In-process inspection	Thickness	1.9	1.1	Every 1000 parts	2 pieces	Micrometers/$\overline{X}$ and s chart	Correct process
	Width	1.5	1.4	Every 10,000 parts	5 pieces	Micrometer/median chart	Correct process
	Length	1.6	1.2	Every 4 hours	75 pieces	Tapered ring gage/p chart	Correct process
Assembly area	Thickness	2.0	1.8	Hourly	30 pieces	Special gage/p chart	Correct process
	Width	2.2	1.9	Chart hourly	100%	Automatic tester/u chart	Repair by responsible operator
Outgoing[3]	Complete assembly	2.8	1.9	Hourly	20 pieces	Automatic tester/$\overline{X}$ and s chart	Correct process
	Complete assembly	NA	1500 DPM	Each lot	50 pieces	Complete visual inspection plus gage and test stand/ c chart	Reject lot and sort for identified nonconformance

[1]Explanations and formulas are contained in the SPC Guideline.
[2]The frequencies and sample size are determined from the performance study of the stability of each process. They are periodically reviewed and updated as required.
[3]After 6 months production experience, the process control and inspection records will be reviewed to determine if outgoing inspection can be reduced.
[4]If any nonconforming products are found in the process samples, then there will be performed 100% inspection of all products produced since the last in control point.

FIGURE 19.1
Control plan. (*From Baxter Travenol Laboratories, 1986.*)

An important category of sensory characteristics is the visual quality characteristic. Typically, written specifications are not clear because of their inability to quantify the characteristics. Among the approaches employed to describe the limits for the characteristics are the following:

1. Providing photographs to define the limits of acceptability of the product.

   EXAMPLE 19.1. A fast-food enterprise has the problem of defining quality standards for suppliers of hamburger buns. The solution is photographs showing the ideal and the maximum and minimum acceptable limits for "golden brown" color, symmetry of bun, and distribution of sesame seeds.

2. Providing physical standards to define the limits of acceptability.

   EXAMPLE 19.2. A government agency needed to define the lightest and darkest acceptable shades of khaki for suppliers of uniforms. Color swatches of cloth were prepared for the limiting shades and issued to inspectors. Imagine the follow-up required to periodically replace the swatches when fading was imminent!

3. Specifying the *conditions* of inspection instead of trying to explicitly define the limits of acceptability.

   EXAMPLE 19.3. Riley (1979) describes a special inspection procedure for cosmetic (appearance) defects of electronic calculators. Part drawings indicate the relative importance of different surfaces, using a system of category numbers and class letters. Three categories identify the surface being inspected:

   I. Plastic window (critical areas only)
   II. External
   III. Internal

   Three classes indicate the frequency with which the surface will be viewed by the user:

   A. Usually seen by the user.
   B. Seldom seen by the user.
   C. Never seen by user (except during maintenance).

   For example, a sheet-metal part that will seldom be seen carries a grade of Coating IIB.

   The conditions of inspection are stated in terms of viewing distance, viewing time, and lighting conditions. The distance and time are specified for each combination of surface being inspected and the frequency of viewing by the user. Lighting conditions are required to be between 75 and 150 foot-candles from a nondirectional source.

The guidelines help to establish cosmetic gradings on parts drawings. However, a judgment must still be made by the inspector as to whether or not the end user would consider the flaw(s) objectionable, using the specified time and distance.

Elaboration on sensory characteristics is provided in *JQH5,* pages 23.25–23.29.

19.5 SERIOUSNESS CLASSIFICATION

Quality characteristics are decidedly unequal in their effect on fitness for use. A relative few are "serious," i.e., of critical importance; many are of minor importance.

Clearly, the more important the characteristic, the greater the attention it should receive in matters such as extent of quality planning; precision of processes, tooling, and instruments; sizes of samples; strictness of criteria for conformance; etc. However, making such discrimination requires that the relative importance of the characteristics be made known to the various decision makers involved: process engineers, quality planners, inspection supervisors, etc. To this end, many companies utilize formal systems of seriousness classification. The resulting classification is used not only in inspection and quality planning but also in specification writing, supplier relations, product audits, executive reports on quality, etc. This multiple use of seriousness classification dictates that the system be prepared by an interdepartmental committee which then

1. Decides how many classes or strata of seriousness to create (usually three or four).
2. Defines each class.
3. Classifies each characteristic into its proper class of seriousness.

Characteristics and Defects

There are actually two lists that need to be classified. One is the list of quality characteristics derived from the specifications. The other is the list of "defects," i.e., symptoms of nonconformance during manufacture and of field failure during use. There is a good deal of commonality between these two lists, but there are differences as well. (For example, the list of defects found on glass bottles has little resemblance to the list of characteristics.) In addition, the two lists do not behave alike. The design characteristic "diameter," for example, gives rise to two defects—oversize and undersize. The amount by which the diameter is oversize may be decisive as to seriousness classification.

Normally, it is feasible to make one system of classification applicable to both lists. However, the uses to which the resulting classifications are put are sufficiently varied to make it convenient to publish the lists separately.

Definitions for the Classes

Most sets of definitions show the influence of the pioneering work of the Bell System in the 1920s. Study of numerous such systems reveals an inner pattern that is a useful guide to any committee faced with applying the concept to its own company. Table 19.2 shows the nature of this inner pattern as applied to a company in the food industry.

Classification

Classification is a long and tedious but essential task. However, it yields some welcome by-products through pointing up misconceptions and confusion among departments and thereby opening the way to clear up vagueness and misunderstandings. Then, when the final seriousness classification is applied to several different purposes, it is subjected to several new challenges that provide still further clarification of vagueness.

TABLE 19.2
Composite definitions for seriousness classification in the food industry

Defect	Effect on consumer safety	Effect on usage	Consumer relations	Loss to company	Effect on conformance to government regulations
Critical	Will surely cause personal injury or illness	Will render the product totally unfit for use	Will offend consumers' sensibilities due to odor, appearance, etc.	Will lose customers and will result in losses greater than value of product	Fails to conform to regulations for purity, toxicity, identification
Major A	Very unlikely to cause personal injury or illness	May render the product unfit for use and may cause rejection by the user	Will likely be noticed by consumers and will likely reduce product salability	May lose customers and may result in losses greater than the value of the product; will substantially reduce production yields	Fails to conform to regulations on weight, volume, or batch control
Major B	Will not cause injury or illness	Will make the product more difficult to use, e.g., removal from package, or will require improvisation by the user; affects appearance, neatness	May be noticed by some consumers and may be an annoyance if noticed	Unlikely to lose customers; may require product replacement; may result in loss equal to product value	Minor nonconformance to regulations on weight, volume, or batch control, e.g., completeness of documentation
Minor	Will not cause injury or illness	Will not affect usability of the product; may affect appearance, neatness	Unlikely to be noticed by consumers and of little concern if noticed	Unlikely to result in loss	Conforms fully to regulations

A problem often encountered is the reluctance of the designers to become involved in seriousness classification of characteristics. They may offer plausible reasons: All characteristics are critical, the tightness of the tolerance is an index of seriousness, etc. Yet the real reasons may be unawareness of the benefits, a feeling that other matters have higher departmental priority, etc. In such cases it may be worthwhile to demonstrate the benefits of classification by working out a small-scale example. In one company the classification-of-characteristics program reduced the number of dimensions that had to be checked from 682 to 279, the effect being to reduce inspection time from 215 minutes to 120 minutes.

19.6 AUTOMATED INSPECTION

Automated inspection and testing are widely used to reduce inspection costs, reduce error rates, alleviate personnel shortages, shorten inspection time, avoid inspector monotony, and provide still other advantages. Applications of automation have successfully been made to mechanical gaging, electronic testing (for high volumes of components as well as system circuitry), nondestructive tests of many kinds, chemical analyses, color discrimination, visual inspection (e.g., of large-scale integrated circuits), etc. In addition, automated testing is extensively used as a part of scheduled maintenance programs for equipment in the field.

Examples in nonmanufacturing activities range from the spelling check provided within word processors to the checking of bank transactions for errors.

A company contemplating the use of automated inspection first identifies the few tests that dominate the inspection budgets and use of personnel. The economics of automation are computed, and trials are made on some likely candidates for a good return on investment. As experience is gained, the concept is extended further and further.

With the emphasis on defect levels in the parts-per-million range, many industries are increasingly accepting on-machine automated 100 percent inspection and testing. Orkin and Olivier (1999) provide an extensive table that identifies seven categories of potential applications of automated inspection ranging from dimensional gaging to nondestructive testing.

A dramatic example of automated inspection is the concept of "machine vision" where electronic eyes inspect and guide an array of industrial processes. The applications include steering robots to place doors on cars, finding blemishes on vegetables in a frozen food processing line, examining wood for knots in veneer panels, and checking that the right color drug capsule goes into correctly labeled packages before being shipped to pharmacies (*Fortune,* February 16, 1998, page 104B). High-speed visual inspection devices are either integrated or are slightly off-line from manufacturing operations. Products can automatically undergo multicharacteristic checks. At the end of the inspection cycle, the computer monitor tells the operator whether the product is acceptable and then updates quality process statistics. Human inspection methods usually detect only 80 to 90 percent of the defects (see below); with machine visual inspection, essentially all defects are detected. For elaboration, see *JQH5,* page 23.15.

A critical requirement for all automated test equipment is precision measurement; i.e., repeated measurements on the same unit of product should yield the "same" test results within some acceptable range of variation. This repeatability is inherent in the design of the equipment and can be quantified by the methods discussed in Section 19.8. In addition, means must be provided to keep the equipment "accurate," i.e., in calibration with respect to standards for the units of measure involved.

Still another aspect of automated test equipment is the problem of processing the data that are generated by the tests. Modern systems of electronic data processing allow these test data to be entered directly from the test equipment into the computer without the need for intermediate documents. Such direct entry supports the prompt preparation of data summaries, conformance calculations, comparisons with prior lots, etc. In turn, it is feasible to program the computer to issue instructions to the test equipment with respect to frequency of test, disposition of units tested, alarm signals relative to improbable results, etc.

Cooper (1997) discusses how to design, test, implement, and maintain a paperless inspection system.

19.7 INSPECTION ACCURACY

Inspection accuracy depends on (1) the completeness of the inspection planning (see above), (2) the bias and precision of the instruments (see later in this section), and (3) the level of human error.

High error rates are particularly prevalent in inspection tasks having a high degree of monotony, e.g., viewing jars of a food product for foreign particles, screening luggage at an airport security gate. Surprisingly, monotony that causes an inspector to miss defects can build up in a short time. With monotonous inspection, inspectors detect about 80 to 90 percent of the defects and miss the remainder. Thus 100 percent inspection that is monotonous is *not* 100 percent effective in detecting defects. One of the advantages of automated inspection is the elimination of human error.

Human errors in inspection arise from multiple causes, of which four are most important: technique errors, inadvertent errors, conscious errors, and communication errors. The nature of these errors is similar to the same categories for personnel in other activities (see Section 3.9 under "Test of Theories Involving Human Error"). For specific elaboration on inspection errors, see *JQH5,* pages 23.42–23.53.

Measure of Inspector Accuracy

Some companies carry out regular evaluations of inspector accuracy as part of the overall evaluation of inspection performance. The plans employ a check inspector who periodically reviews random samples of work previously inspected by the various inspectors. The check inspection findings are then summarized, weighted, and converted into some index of inspector performance. *JQH5,* pages 23.51–23.53, explains this procedure.

Harris and Chaney (1969) provided some early research on inspector accuracy. Among their findings were inspection accuracy decreases with reductions in defect rates; inspection accuracy increases with repeated inspections (up to a total of six); inspection accuracy decreases with additional product complexity, and the effect cannot be overcome by increasing the allowable inspection time. These conclusions are sobering.

19.8 ERRORS OF MEASUREMENT

Variation in a process has two sources: variation of the process making the product and variation of the measurement process (see Figure 16.3). Particularly with the low defect levels demanded under the six-sigma approach, we must understand not only the capability of the manufacturing process but also the capability of the measurement process.

Even when correctly used, a measuring instrument may not give a true reading of a characteristic. The difference between the true value and the measured value can be due to one or more of five sources of variation (Figure 19.2).

There is much confusion as to terminology. *Bias* is sometimes referred to as "accuracy." Because *accuracy* has several meanings in the literature (especially in measuring instrument catalogs), its use as an alternative for "bias" is not recommended.

The distinction between repeatability and bias is illustrated in Figure 19.3. *Repeatability* is often referred to as "precision."

Any statement of bias and repeatability (precision) must be preceded by three conditions:

1. *Definition of the test method.* This definition includes the step-by-step procedure, equipment to be used, preparation of test specimens, test conditions, etc.
2. *Definition of the system of causes of variability,* such as material, analysts, apparatus, laboratories, days, etc. ASTM recommends that modifiers of the word *precision* be used to clarify the scope of the precision measure. Examples of such modifiers are single operator, single analyst, single-laboratory-operator-material-day, and multilaboratory.
3. *Existence of a statistically controlled measurement process.* The measurement process must have stability for the statements on bias and precision to be valid. This stability can be verified by a control chart (see Section 18.6).

Effect of Measurement Error on Acceptance Decisions

Error of measurement can cause incorrect decisions on (1) individual units of product and (2) lots submitted to sampling plans.

In one example of measuring the softening point of a material, the standard deviation of the test precision is 2 degrees, yielding two standard deviations of ±4 degrees. The specification limits on the material are ±3 degrees. Imagine the incorrect decisions that are made under these conditions.

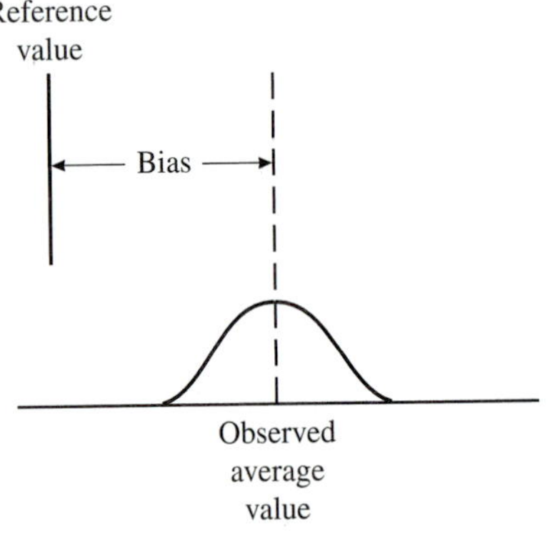

Bias

Bias is the difference between the observed average of measurements and the reference value. The reference value, also known as the accepted reference value or master value, is a value that serves as an agreed-upon reference for the measured values.[1] A reference value can be determined by averaging several measurements with a higher level (e.g., metrology lab or layout equipment) of measuring equipment.

[1]ASTM D 3980-88.

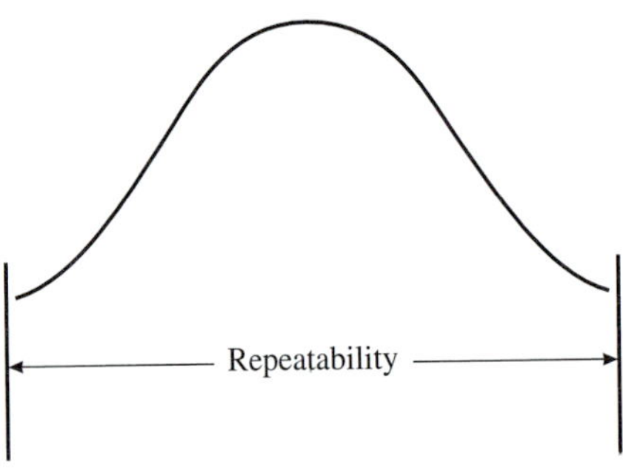

Repeatability

Repeatability is the variation in measurements obtained with one measurement instrument when used several times by an appraiser while measuring the identical characteristic on the same part.

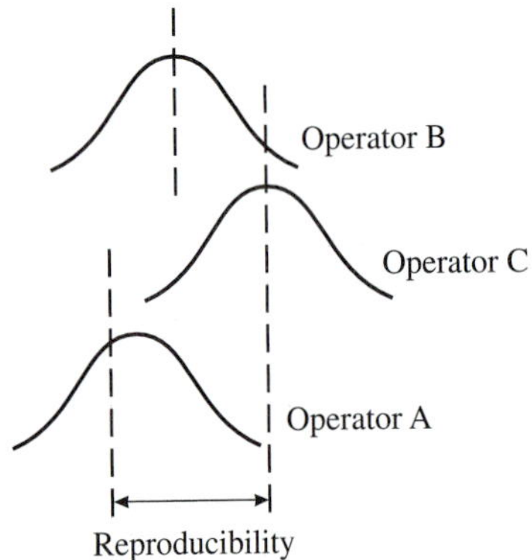

Reproducibility

Reproducibility is the variation in the average of the measurements made by different appraisers using the same measuring instrument when measuring the identical characteristic on the same part.

FIGURE 19.2

Five sources of measurement variation. (*Reprinted with permission from the MSA Manual. DaimlerChrysler, Ford, General Motors Supplier Quality Requirements Task Force.*)

Two types of errors can occur in the classification of a product: (1) a nonconforming unit can be accepted (the consumer's risk) and (2) a conforming unit can be rejected (the producer's risk). In a classic paper, Eagle (1954) showed the effect of precision on each of these errors.

The probability of accepting a nonconforming unit as a function of measurement error (called test error, σ_{TE}, by Eagle) is shown in Figure 19.4. The abscissa expresses

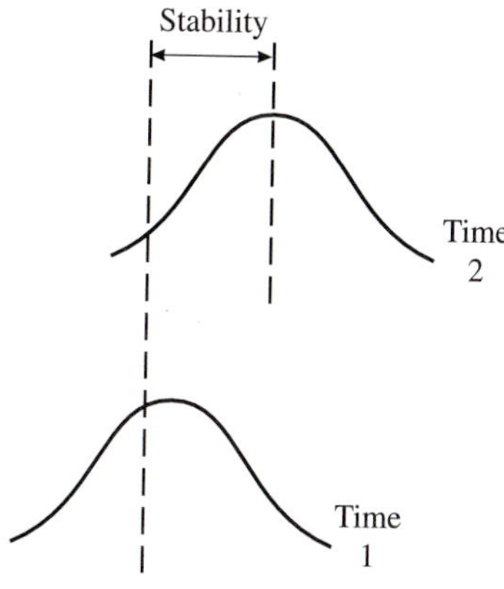

Stability
Stability (or drift) is the total variation in the measurements obtained with a measurement system on the same master or parts when measuring a single characteristic over an extended time period.

Linearity
Linearity is the difference in the bias values through the expected operating range of the gage.

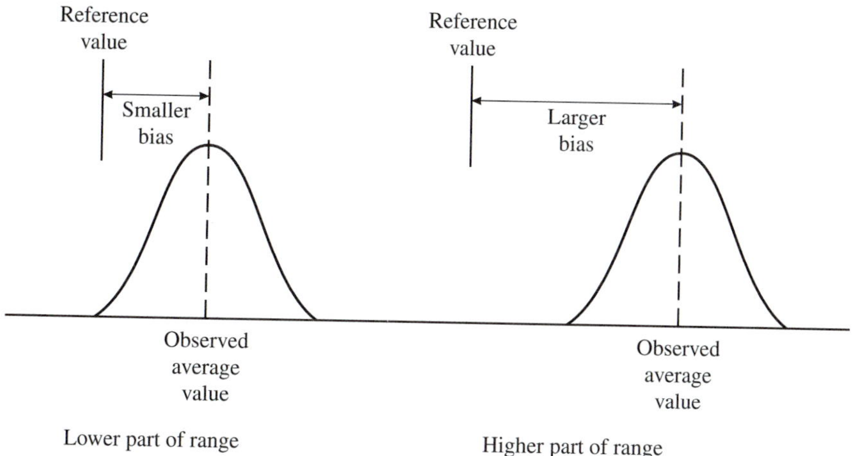

FIGURE 19.2 *(Continued)*
Five sources of measurement variation.

the test error as the standard deviation divided by the plus-or-minus value of the specification range (assumed equal to two standard deviations of the product).

For example, if the measurement error is one half of the tolerance range, the probability is about 1.65 percent that a nonconforming unit will be read as conforming (due to the measurement error) and therefore will be accepted.

Figure 19.5 shows the percentage of *conforming* units that will be *rejected* as a function of the measurement error. For example, if the measurement error is one half of the plus-or-minus tolerance range, about 14 percent of the units that are really within specifications will be rejected because the measurement error will show these conforming units as being outside specification.

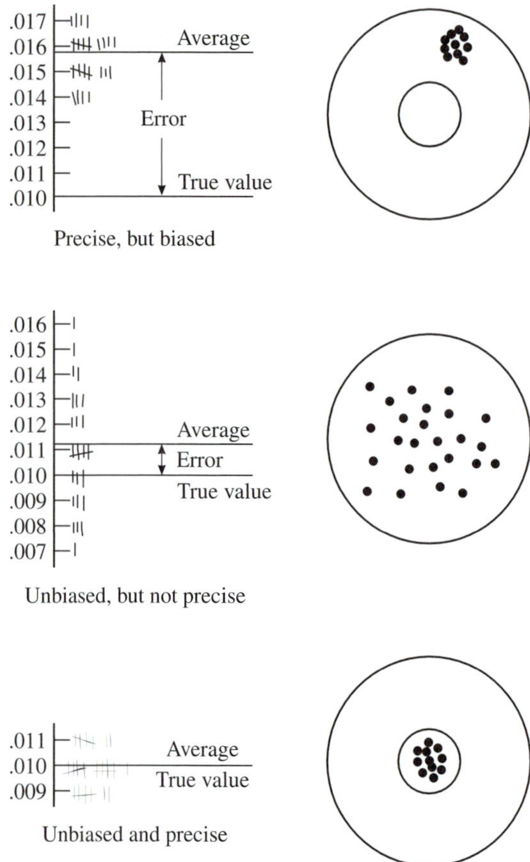

FIGURE 19.3
Distinction between bias and repeatability (precision).

The test specification can be adjusted with respect to the performance specification (see Figures 19.4 and 19.5). Moving the test specification inside the performance specification reduces the probability of accepting a nonconforming product but increases the probability of rejecting a conforming product. The reverse occurs if the test specification is moved outside the performance specification. Both risks can be reduced by increasing the precision of the test, i.e., by reducing the value of σ_E (see below under "Reducing and Controlling Errors of Measurement").

Hoag et al. (1975) studied the effect of inspector errors on the type I (α) and type II (β) risks of sampling plans. For a single sampling plan and an 80 percent probability of the inspector detecting a defect, the real value of β is two to three times that specified, and the real value of α is about one fourth to one half of that specified.

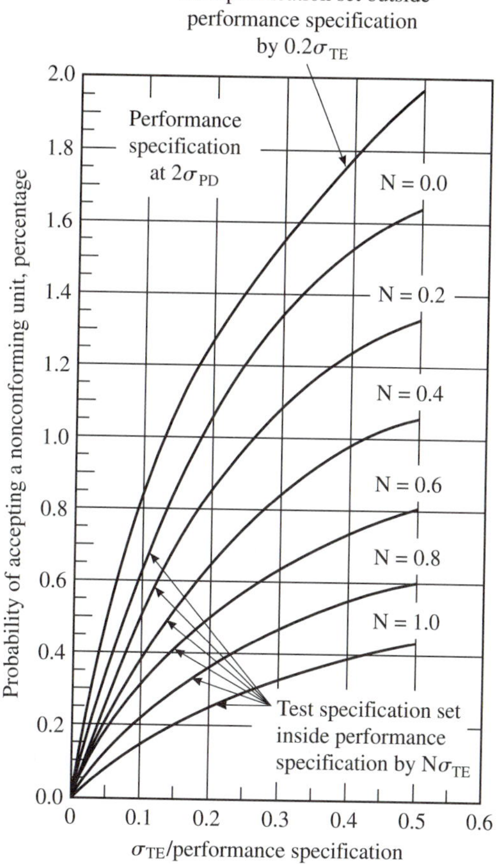

FIGURE 19.4
Probability of accepting a nonconforming unit.
(*From Eagle, 1954.*)

Case et al. (1975) investigated the effect of inspection error on the *average outgoing quality* (AOQ) of an attributes sampling procedure. They concluded that not only do the AOQ values change, but significant changes can occur in the shape of the AOQ curve.

Automotive Industry Action Group (1995, page 77) presents the concept of a gage performance curve to determine the probability of accepting or rejecting a part when the gage repeatability and reproducibility (R & R) is unknown.

All these investigations concluded that measurement error can be a serious problem.

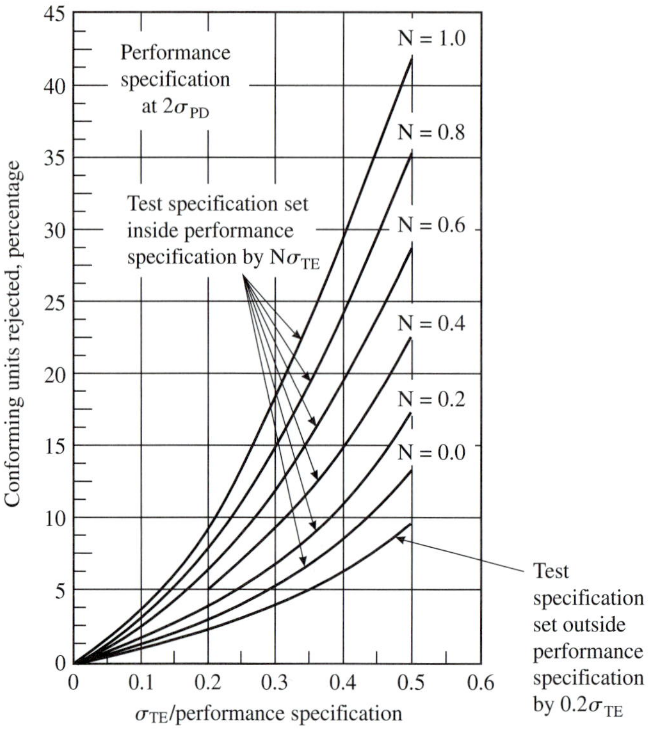

FIGURE 19.5
Conforming units rejected, percentage. (*From Eagle, 1954.*)

Components of Variation

In drawing conclusions about measurement error, it is worthwhile to study the causes of variation of observed values. The relationship is

$$\sigma_{\text{observed}} = \sqrt{\sigma^2_{\text{cause } A} + \sigma^2_{\text{cause } B} + \cdots + \sigma^2_{\text{cause } N}}$$

The formula assumes that the causes act independently.

It is valuable to find the numerical values of the components of observed variation because the knowledge may suggest where effort should be concentrated to reduce the variation in the product. A separation of the observed variation into the product variation plus the other causes of variation may indicate important factors other than the manufacturing process. Thus, if the *measurement* error is found to be a large percentage of the total variation, this finding must be analyzed before preceding with a quality improvement program. Finding the components (e.g., instrument, operator) of this error may help to reduce the measurement error, which in turn may completely eliminate a problem.

Observations from an instrument used to measure a series of different units of product can be viewed as a composite of (1) the variation due to the measuring method and (2) the variation in the product itself. This value can be expressed as

$$\sigma_O = \sqrt{\sigma_P^2 + \sigma_E^2}$$

where $\sigma_O = \sigma$ of the observed data
$\sigma_P = \sigma$ of the product
$\sigma_E = \sigma$ of the measuring method

Solving for σ_P yields

$$\sigma_P = \sqrt{\sigma_O^2 + \sigma_E^2}$$

The components of measurement error often focus on repeatability and reproducibility (R & R). *Repeatability* concerns variation due to measurement gages and equipment; *reproducibility* concerns variation due to human "appraisers" who use the gages and equipment. Studies to estimate these components are often called "gage R & R" studies.

A gage R & R study can provide separate numerical estimates of repeatability and reproducibility. Two methods are usually used to analyze the measurement data. Each method requires a number of appraisers, a number of parts, and repeat trials of appraisers measuring different parts. For example, an R & R study might use three appraisers, 10 parts, and two trials.

One method analyzes averages and ranges of the measurement study data. This method requires minimum statistical background and does not require a computer. The second method is the analysis of variance, ANOVA (see Section 11.14). Compared to the first method, ANOVA requires a higher level of statistical knowledge for interpretation of the results but can evaluate the data for a possible interaction between appraisers and parts. ANOVA is best done on a computer using MINITAB or other software. Overall, the ANOVA method is preferred over analyzing the averages and ranges. Detailed illustrations of each method are provided in the Automotive Industry Action Group booklet *Measurement Systems Analysis* (1995). Also, see Tsai (1988) for an example using ANOVA and considering both no interaction and interaction of operators and parts. Burdick and Larsen (1997) provide methods for constructing confidence intervals on measures of variability in R & R studies.

When the total standard deviation of repeatability and reproducibility is determined from the ANOVA, a judgment must then be made on the adequacy of the measurement process. A common practice is to calculate 5.15σ ($\pm 2.575\sigma$) as the total spread of the measurements that will include 99 percent of the measurements. If 5.15σ is equal to or less than 10 percent of the specification range on the quality characteristic, the measurement process is viewed as acceptable for that characteristic; if the result is greater than 10 percent, the measurement process is viewed as unacceptable. Engel and DeVries (1997) examine how the practice of comparing measurement error with the specification interval relates to making correct decisions at product testing.

Reducing and Controlling Errors of Measurement

Steps can be taken to reduce and control errors for all sources of measurement variation. The systematic errors that contribute to bias can sometimes be handled by applying a numerical correction to the measured data. If an instrument has a bias of −0.001, then, on the average, it reads 0.001 too low. The data can be adjusted by adding 0.001 to each value of the data. Of course, it is preferable to adjust the instrument as part of a calibration program.

In a calibration program the measurements made by an instrument are compared to a reference standard of known accuracy. If the instrument is found to be out of calibration, an adjustment is made.

A calibration program can become complex for these reasons:

1. The large number of measuring instruments.
2. The need for periodic calibration of many instruments.
3. The need for many reference standards.
4. The increased technological complexity of new instruments.
5. The variety of types of instruments, i.e., mechanical, electronic, chemical, etc.

A calibration program should include provisions for periodic audits. These follow the general approach for quality audits (see Chapter 23).

Precision of measurement can be improved through either or both of the following procedures:

- *Discovery of the causes of variation and remedy of these causes.* A useful step is to resolve the observed values into components of variation (see above). This process can lead to the discovery of inadequate training, perishable reagents, lack of sufficient detail in procedures, and other such problems. This fundamental approach also points to other causes for which the remedy is unknown or uneconomic, i.e., basic redesign of the test procedure.
- *Multiple measurements and statistical methodology to control the error of measurement.* The use of multiple measurements is based on the following relationship (see Section 11.3):

$$\sigma_{\bar{x}} = \frac{\sigma}{\sqrt{n}}$$

The formula states that halving the error of measurement requires quadrupling (not doubling) the number of measurements.

As the number of tests grows larger and larger, a significant reduction in the error of measurement can be achieved only by taking a still *larger* number of additional tests. Thus the cost of the additional tests versus the value of the slight improvement in measurement error becomes an issue. The alternatives of reducing the causes of variation (by control charts or other techniques) must also be considered.

For reducing other forms of measurement error, see Automotive Industry Action Group (1995).

19.9 HOW MUCH INSPECTION IS NECESSARY?

The amount of inspection to decide the acceptability of a lot can vary from no inspection to a sample to 100 percent inspection. The decision is governed mainly by the amount of prior knowledge available as to quality, the *homogeneity* of the lot, and the allowable degree of risk.

Prior knowledge that is helpful in deciding on the amount of inspection includes

- Previous quality history on the product item and the supplier (internal or external).
- Criticality of the item on overall system performance.
- Criticality on later manufacturing or service operations.
- Warranty or use history.
- Process capability information. A process that is in statistical control with good uniformity around a target value (e.g., a 6σ process, see Section 18.15) will require minimum inspection.
- Measurement capability information, e.g., the availability of accurate and precise instruments.
- The nature of the manufacturing process. For example, some operations primarily depend on the adequacy of the setup.
- Inspection of the first few and the last few items in a production run. This is usually sufficient.
- Product homogeneity. For example, fluid product is homogeneous and reduces the need for large sample sizes.
- Data on process variables and process conditions, e.g., as provided by automatic recording charts.
- Degree of adherence to the three elements of self-control for the personnel operating the process (see Section 5.3).

Competition to reduce costs has resulted in pressures to reduce the amount of inspection. The concept of inspection by the producers (self-inspection) has added to the focus of reducing inspection. Indeed, opportunities do exist for cost reduction in inspection activities. First, however, the causes of the high failure costs must be diagnosed and removed and the prerequisites for self-inspection must be met.

EXAMPLE 19.4. The Datapoint Corporation manufactures office and computer products (Adams, 1987). Part of the operation was 100 percent in-line inspection of visual characteristics by the quality staff. A dramatic shift was planned—production personnel would do their own visual inspection; the quality staff would perform an audit inspection and do diagnostic work on the causes of nonconformities. But a number of steps were required: a quality education process for first-line management, supervisors, and line personnel; special training in workmanship standards to help people recognize nonconformances; an 18-month implementation plan to phase in the new approach; use of data from functional acceptance tests for process yield reports; analysis of data; and a process audit system to review documentation, tools, materials, and people.

The results were dramatic: The staff of 35 in-line inspectors was reduced to five process auditors, while scrap and rework plunged from 15 percent to 2 percent.

Economics of Inspection

We have several alternatives for evaluating lots:

1. *No inspection.* This approach is appropriate if prior inspections on the same lot have already been made by qualified laboratories, e.g., in other divisions of the same company or in supplier companies. Prior inspections by qualified production workers have the same effect.
2. *Small samples.* Small samples can be adequate if the process is inherently uniform and the order of production can be preserved. For example, in some punch press operations, the stamping dies are made to a high degree of stability. As a result, the successive pieces stamped out by such dies exhibit a high degree of uniformity for certain dimensional characteristics. For such characteristics, if the first and last pieces are correct, the remaining pieces are also correct, even for lot sizes running to many thousands of pieces. In its generalized form, the press example is one of a high degree of process capability combined with "stratified" sampling—sampling based on knowledge of the order of production.

   Small samples can also be used when the product is homogeneous due to its fluidity (gases, liquids) or to prior mixing operations. This homogeneity need not be assumed—it can be verified by sampling. Even solid materials may be homogeneous due to *prior* fluidity. Once the fact of homogeneity has been established, the sampling needed is minimal.

   For product with a continuing history of good quality, sampling can be periodic, e.g., "skip-lot" or "chain sampling" (see *JQH5,* Section 46).
3. *Large samples.* In the absence of prior knowledge, the information about lot quality must be derived solely from sampling, which means random sampling and hence relatively large samples. The actual sample sizes depend on two main variables: (a) the tolerable percentage of defects and (b) the risks that can be accepted. Once values have been assigned to these variables, the sample sizes can be determined scientifically in accordance with the laws of probability (see Section 10.15). However, the choice of defect levels and risks is largely based on empirical judgments.

   Random sampling is clearly needed in cases where there is no ready access to prior knowledge, e.g., purchases from certain suppliers. However, there remain many, many cases in which random sampling is used despite the availability of inputs such as process capability, order of manufacture, fluidity, etc. A major obstacle is a lack of publications which show how to design sampling plans in ways which make use of these inputs. In the absence of such publications, quality planners are faced with creating their own designs. This means added work amid the absence of protection derived from the use of recognized, authoritative published materials.

   See *JQH5,* pages 46.15–46.17, for a discussion of the formation of inspection lots and the selection of samples.
4. *One hundred percent inspection.* This technique is used when the results of sampling show that the level of defects present is too high for the product to go on to the users. In critical cases, added provisions may be needed to guard against inspector fallibility, e.g., automated inspection or redundant 200 percent inspection.

An economic evaluation of these alternatives requires a comparison of *total* costs under each one.

Let N = number of items in lot
n = number of items in sample
p = proportion defective in lot
A = damage cost incurred if a defective slips through inspection
I = inspection cost per item
P_a = probability that lot will be accepted by sampling plan

A and I are sometimes denoted as k_1, and k_2, respectively.

Consider the comparison of sampling inspection versus 100 percent inspection. Suppose it is assumed that no inspection errors occur and the cost to replace a defective found in inspection is borne by the producer or is small compared to the damage or inconvenience caused by a defective. The total costs are summarized in Table 19.3. These costs reflect both inspection costs and damage costs and recognize the probability of accepting or rejecting a lot under sampling inspection. The expressions can be equated to determine a break-even point. If the sample size is assumed to be small compared to the lot size, the break-even point, p_b, is

$$p_b = \frac{I}{A}$$

If the lot quality (p) is less than p_b, the total cost will be lowest with sampling inspection or no inspection. If p is greater than p_b, 100 percent inspection is best. This principle is often called the Deming *kp* rule.

For example, a microcomputer device costs \$.50 per unit to inspect. A damage cost of \$10.00 is incurred if a defective device is installed in the larger system. Therefore

$$p_b = \frac{.50}{10.00} = .05 = 5.0\%$$

If the percentage defective is expected to be greater than 5 percent, then 100 percent inspection should be used. Otherwise, use sampling or no inspection.

The variability in quality from lot to lot is important. If past history shows that the quality level is much better than the break-even point and is stable from lot to lot, little if any inspection may be needed. If the level is much worse than the break-even point, and consistently so, it will usually be cheaper to use 100 percent inspection

TABLE 19.3
Economic comparison of inspection alternatives

Alternative	Total cost
No inspection	NpA
Sampling	$nI + (N - n)pAP_a + (N - n)(1 - P_a)I$
100% inspection	NI

rather than sampling. If the quality is at neither of these extremes, a detailed economic comparison of no inspection, sampling, and 100 percent inspection should be made. Sampling is usually best when the product is a mixture of high-quality lots and low-quality lots, or when the producer's process is not in a state of statistical control.

The high costs associated with component failures in complex electronic equipment coupled with the development of automatic testing equipment for components has resulted in the economic justification of 100 percent inspection for some electronic components. The cost of finding and correcting a defective can increase by a ratio of 10 for each major stage that the product moves to from production to the customer; i.e., if it costs $1 at incoming inspection, the cost increases to $10 at the printed circuit board stage, $100 at the system level, and $1,000 in the field.

19.10 THE CONCEPT OF ACCEPTANCE SAMPLING

Acceptance sampling is the process of evaluating a portion of the product in a lot for the purpose of accepting or rejecting the entire lot.

The main advantage of sampling is economy. Despite some added costs for designing and administering the sampling plans, the lower costs of inspecting only part of the lot result in an overall cost reduction.

In addition to this main advantage, there are others:

- The smaller inspection staff is less complex and less costly to administer.
- There is less damage to the product, i.e., handling incidental to inspection is itself a source of defects.
- The lot is disposed of in shorter (calendar) time so that scheduling and delivery are improved.
- The problem of monotony and inspector error induced by 100 percent inspection is minimized.
- Rejection (rather than sorting) of nonconforming lots tends to dramatize the quality deficiencies and to urge the organization to look for preventive measures.
- Proper design of the sampling plan commonly requires study of the actual level of quality required by the user. The resulting knowledge is a useful input to the overall quality planning.

The disadvantages are sampling risks, greater administrative costs, and less information about the product than is provided by 100 percent inspection.

Acceptance sampling is used when (1) the cost of inspection is high in relation to the damage cost resulting from passing a defective product, (2) 100 percent inspection is monotonous and causes inspection errors, or (3) the inspection is destructive. Acceptance sampling is most effective when it is preceded by a prevention program that achieves an acceptable level of quality of conformance.

We must also emphasize what acceptance sampling does not do. It does not provide refined estimates of lot quality. (It does determine, with specified risks, an acceptance or rejection decision on each lot.) Also, acceptance sampling does not provide judgments on whether or not rejected product is fit for use. (It does give a decision on a lot with respect to the defined quality specification.)

In recent years the emphasis on statistical process control has led some practitioners to conclude that acceptance sampling is no longer a valid concept. Their belief, stated here in oversimplified terms, is that only two levels of inspection are valid—no inspection or 100 percent inspection. This text takes the viewpoint that the concept of prevention (using statistical process control and other statistical and managerial techniques) is the foundation for meeting product requirements. Acceptance sampling procedures are, however, important in a program of *acceptance control.* Under this latter approach, described at the end of this chapter, sampling procedures are continually matched to process history and quality results. This step ultimately leads to phasing out acceptance sampling in favor of supplier certification and process control.

This chapter presents examples of specific acceptance sampling plans.

For some perceptive discussions of the modern role of acceptance sampling, see Schilling (1994) and Taylor (1994). *JQH5* provides specifics on the various types of sampling plans.

19.11 SAMPLING RISKS: THE OPERATING CHARACTERISTIC CURVE

Neither sampling nor 100 percent inspection can guarantee that every defective item in a lot will be found. Sampling involves a risk that the sample will not adequately reflect the conditions in the lot; 100 percent inspection has the risk that monotony and other factors will result in inspectors missing some of the defectives (see Section 19.7). Both of these risks can be quantified.

Sampling risks are of two kinds:

1. Good lots can be rejected (the producer's risk). This risk corresponds to the α risk.
2. Bad lots can be accepted (the consumer's risk). This risk corresponds to the β risk.

The α and β risks are discussed in Section 11.7.

The operating characteristic (OC) *curve* for a sampling plan quantifies these risks. The OC curve for an attributes plan is a graph of the percentage defective in a lot versus the probability that the sampling plan will accept a lot. As p is unknown, the probability must be stated for all possible values of p. It is assumed that an infinite number of lots are produced. Figure 19.6 shows an "ideal" OC curve where it is desired to accept all lots 1.5 percent defective or less and reject all lots having a quality level greater than 1.5 percent defective. All lots less than 1.5 percent defective have a probability of acceptance of 1.0 (certainty); all lots greater than 1.5 percent defective have a probability of acceptance zero. Actually, however, no sampling plan exists that can discriminate perfectly; there always remains some risk that a "good" lot will be rejected or that a "bad" lot will be accepted. The best that can be achieved is to make the acceptance of good lots more likely than the acceptance of bad lots.

An acceptance sampling plan basically consists of a sample size (n) and an acceptance criterion (c). For example, a sample of 125 units is to be randomly selected

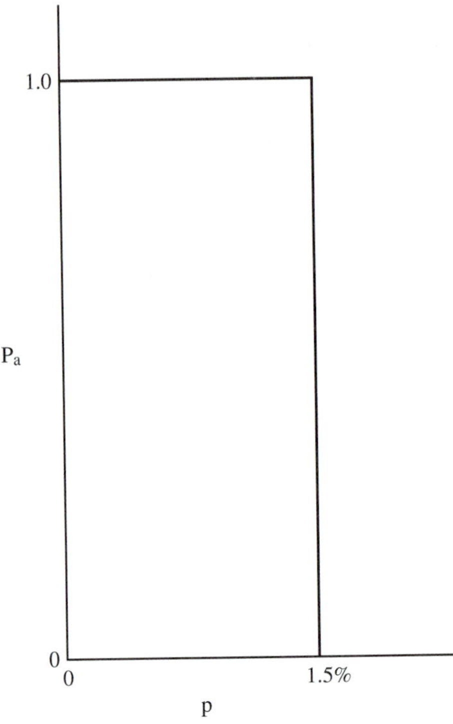

FIGURE 19.6
Ideal OC curve.

from the lot. If five or fewer defectives are found, the lot is accepted. If six or more defectives are found, the lot is rejected.

The sample of 125 could, by the laws of chance, contain 0, 1, 2, 3, even up to 125 defectives. It is this *sampling variation* that causes some good lots to be rejected and some bad lots to be accepted. The OC curve for $n = 125$ and $c = 5$ is curve A, Figure 19.7. (The other curves will be discussed later.) A 1.5 percent defective lot has about a 98 percent chance of being accepted. A much worse lot, say 6 percent defective, has a 23 percent chance of being accepted. With the risks stated in quantitative form, a judgment can be made on the adequacy of the sampling plan.

The OC curve for a specific plan states *only* the chance that a lot having p percent defective will be accepted by the sampling plan. The OC curve does *not*

- Predict the quality of lots submitted for inspection. For example (Figure 19.7), it is incorrect to say that there is a 36 percent chance that the lot quality is 5 percent defective.
- State a "confidence level" in connection with a specific percentage defective.
- Predict the final quality achieved after all inspections are completed.

These and other myths about the OC curve require a careful explanation of the concept to those using it. (Acceptable quality level is explained in Section 19.12.)

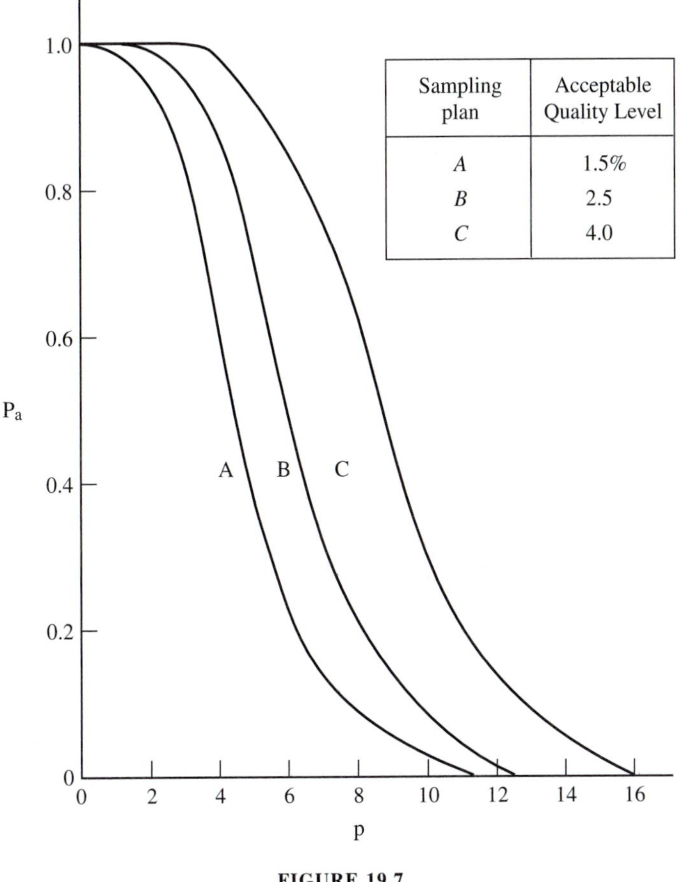

FIGURE 19.7
OC curves.

Constructing the Operating Characteristic Curve

An OC curve can be developed by determining the probability of acceptance for several values of incoming quality, p. The probability of acceptance is the probability that the number of defectives in the sample is equal to or less than the acceptance number for the sampling plan. Three distributions can be used to find the probability of acceptance: the hypergeometric, binomial, and Poisson distributions. When its assumptions can be met, the Poisson distribution is preferable because of the ease of calculation.

Grant and Leavenworth (1996, pages 183–193) describe the use of the hypergeometric and binomial distributions.

The Poisson distribution yields a good approximation for acceptance sampling when the sample size is at least 16, the lot size is at least 10 times the sample size, and p is less than 0.1. The Poisson distribution function as applied to acceptance sampling is

$$P\begin{pmatrix}\text{exactly} \\ r \text{ defectives} \\ \text{in sample of } n\end{pmatrix} = \frac{e^{-np}(np)^r}{r!}$$

The equation can be solved using a calculator or by using Table C in Appendix II. This table gives the probability of *r or fewer* defectives in a sample of n from a lot having a fraction defective of p. To illustrate Table C, consider the sampling plan previously cited; i.e., $n = 125$ and $c = 5$. To find the probability of accepting a 4 percent defective lot, calculate np as $125(0.04) = 5.00$. Table C then gives the probability of five or fewer defectives as 0.616. Figure 19.7 (curve A) shows this probability as the value of P_a for 4 percent defective lot quality.

The preceding discussion of sampling risks assumes that the proportion defective of incoming lots is reasonably constant. This assumption is often made in practice. Chun and Rinks (1998) derive modified producer's and consumer's risks when the incoming quality is not constant.

19.12 QUALITY INDEXES FOR ACCEPTANCE SAMPLING PLANS

Many of the published plans can be categorized in terms of one of several quality indexes:

1. *Acceptable quality level (AQL).* The units of quality level can be selected to meet the particular needs of a product. Thus ANSI/ASQC Z1.4 (1993) defines AQL as "the maximum percent nonconforming (or the maximum number of nonconformities per hundred units) that, for purposes of sampling inspection, can be considered satisfactory as a process average." If a unit of product can have a number of different defects of varying seriousness, then demerits can be assigned to each type and product quality measured in terms of demerits. As an AQL is an *acceptable* level, the probability of acceptance for an AQL lot should be high (see Figure 19.8).
2. *Limiting quality level (LQL).* LQL defines *unsatisfactory* quality. Different titles are sometimes used to denote an LQL; for example, Dodge-Romig plans use the term *lot tolerance percentage defective (LTPD).* As an LQL is an *unacceptable* level, the probability of acceptance for an LQL lot should be low (see Figure 19.8). In some tables this probability is known as the consumer's risk, is designated as P_c, and has been standardized at 0.1. The consumer's risk is not the probability that the consumer will actually receive product at the LQL. The consumer will, in fact, not receive 1 lot in 10 at LQL fraction defective. What the consumer actually gets depends on the actual quality in the lots *before* inspection and on the probability of acceptance.
3. *Indifference quality level (IQL).* IQL is a quality level somewhere between the AQL and LQL. It is frequently defined as the quality level having a probability of acceptance of 0.5 for a given sampling plan (see Figure 19.8).

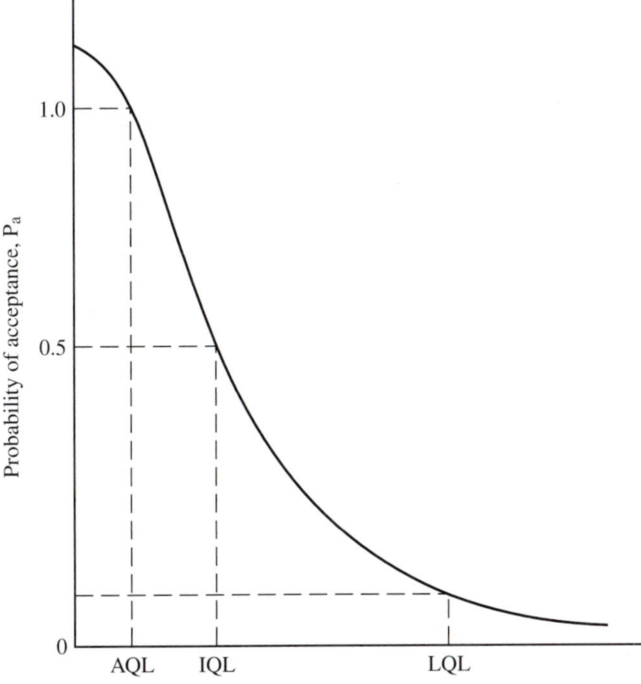

FIGURE 19.8
Quality indexes for sampling plans.

It should be emphasized to both internal and external suppliers that all product submitted for inspection is expected to meet specifications. An acceptable quality level does not mean the submission of a certain amount of nonconforming product is approved. The AQL simply recognizes that, under sampling, some nonconforming product will pass through the sampling scheme.

4. *Average outgoing quality limit (AOQL).* An approximate relationship exists between the fraction defective in the material before inspection (incoming quality p) and the fraction defective remaining after inspection (outgoing quality AOQ): AOQ $= pP_a$. When incoming quality is perfect, outgoing quality must also be perfect. However, when incoming quality is bad, outgoing quality will also be perfect (assuming no inspection errors) because the sampling plan will cause all lots to be rejected and inspected in detail. Thus at either extreme—incoming quality excellent or terrible—the outgoing quality will tend to be good. Between these extremes is the point at which the percentage of defectives in the outgoing material will reach its maximum. This point is the average outgoing quality limit (AOQL). For a sample calculation, see *JQH5,* page 46.11.

These indexes apply primarily when the production occurs in a continuing series of lots. For isolated lots the LQL concept is recommended. The indexes were originally developed by statisticians to help describe the characteristics of sampling plans.

Misinterpretations (particularly of the AQL) are common and are similar to those mentioned in Section 15.3. For example, a sampling plan based on AQL *will* accept some lots having a quality level worse than the AQL.

19.13 TYPES OF SAMPLING PLANS

Sampling plans are of two types:

1. *Attributes plans.* A random sample is taken from the lot, and each unit is classified as acceptable or defective. The number defective is then compared with the allowable number stated in the plan, and a decision is made to accept or reject the lot. This chapter illustrates attributes plans based on AQL.
2. *Variables plans.* A sample is taken, and a *measurement* of a specified quality characteristic is made on each unit. These measurements are then summarized into a sample statistic (e.g., sample average), and the observed value is compared with an allowable value defined in the plan. A decision is then made to accept or reject the lot.

The key advantage of a variables sampling plan is the additional information provided in each sample that, in turn, results in smaller sample sizes as compared with an attributes plan having the same risks. However, if a product has several important quality characteristics, each must be evaluated against a separate variables acceptance criterion (e.g., numerical values must be obtained and the average and standard deviation for each characteristic calculated). In a corresponding attributes plan, the sample size required may be higher, but the several characteristics can be treated as a group and evaluated against one set of acceptance criteria. *JQH5,* Section 46, provides examples of variables plans.

Single Sampling, Double Sampling, and Multiple Sampling

Many published sampling tables give a choice among single, double, and multiple sampling. In single-sampling plans a random sample of n items is drawn from the lot. If the number of defectives is less than or equal to the acceptance number (c), the lot is accepted. Otherwise, the lot is rejected. In double-sampling plans (Figure 19.9), a smaller initial sample is usually drawn, and a decision to accept or reject is reached on the basis of this smaller first sample if the number of defectives is either quite large or quite small. A second sample is taken if the results of the first are not decisive. Since it is necessary to draw and inspect the second sample only in borderline cases, the average number of pieces inspected per lot is generally smaller with double sampling. In multiple-sampling plans one, two, or several still smaller samples are taken, usually continuing as needed until a decision to accept or reject is obtained. Thus, double- and multiple-sampling plans may mean less inspection but are more complicated to administer.

In general, it is possible to derive single-, double-, or multiple-sampling schemes with essentially identical OC curves (see *JQH5,* page 46.17 and Table 46.6).

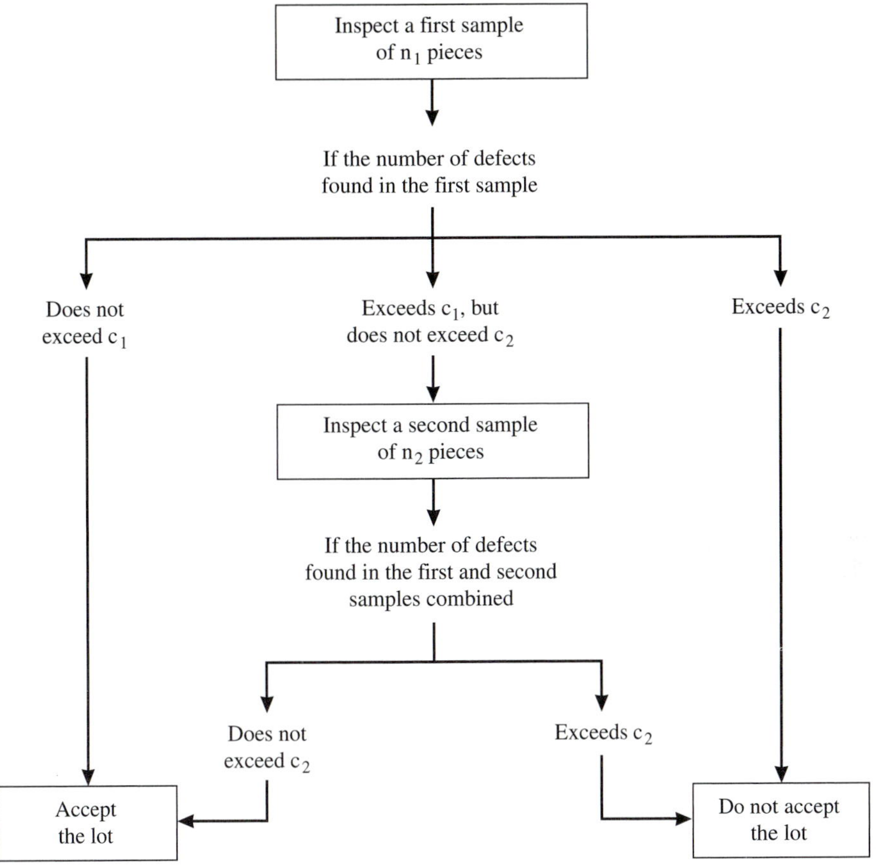

FIGURE 19.9
Schematic operation of double sampling.

19.14 CHARACTERISTICS OF A GOOD ACCEPTANCE PLAN

An acceptance sampling plan should have these characteristics:

- The index (AQL, AOQL, etc.) used to define "quality" should reflect the needs of the consumer and the producer and not be chosen primarily for statistical convenience.
- The sampling risks should be known in quantitative terms (the OC curve). The producer should have adequate protection against the rejection of good lots; the consumer should be protected against the acceptance of bad lots.
- The plan should minimize the *total* cost of inspection of all products. This requires careful evaluation of the pros and cons of attributes and variables plans, as well as single, double, and multiple sampling. It should also reflect product priorities, particularly from the fitness-for-use viewpoint.

- The plan should make use of other knowledge, such as process capability, supplier data, and other information.
- The plan should have built-in flexibility to reflect changes in lot sizes, quality of product submitted, and any other pertinent factors.
- The measurements required by the plan should provide information useful in estimating individual lot quality and long-run quality.
- The plan should be simple to explain and administer.

See *JQH5,* Section 46, for elaboration of these characteristics. Fortunately, published tables are available that meet many of these characteristics. We now proceed to a discussion of AQL plans.

19.15 ANSI/ASQC Z1.4

ANSI/ASQC Z1.4 (1993) is an attributes sampling system. Its quality index is the acceptable quality level (AQL). The AQL is the maximum percentage nonconforming (or the maximum number of nonconformities per 100 units) that, for purposes of sampling inspection, can be considered satisfactory as a process average. (The standard uses the term *nonconformity* rather than *defective unit.*) The probability of accepting material of AQL quality is always high but not exactly the same for all plans. For lot quality just equal to the AQL, the "percentage of lots expected to be accepted" ranges from about 89 to 99. The choice may be made from 26 available AQL values ranging from 0.010 to 1000.0. (AQL values of 10.0 or less may be interpreted as percentage nonconforming or nonconformities per 100 units; values above 10.0 are interpreted as nonconformities per 100 units.)

The tables specify the relative amount of inspection to be used as "inspection level" I, II, or III; level II is regarded as normal. The inspection-level concept permits the user to balance the cost of inspection against the amount of protection required. The three levels involve inspection in amounts roughly in the ratio 0.4:1.0:1.6. (Four additional inspection levels are provided for situations requiring "small-sample inspection.")

A plan is chosen from the tables as follows:

1. The following information must be known:
   - AQL.
   - Lot size.
   - Type of sampling (single, double, or multiple).
   - Inspection level (usually level II).
2. Knowing the lot size and inspection level, a code letter is obtained from Table 19.4.
3. Knowing the code letter, AQL, and type of sampling, the sampling plan is read from Table 19.5. (Table 19.5 is for single sampling; the standard also provides tables for double and multiple sampling.)

For example, suppose a purchasing agency has contracted for a 1.5 percent AQL. Suppose also that the parts are bought in lots of 1500 pieces. The table of sample-size code letters (Table 19.4) shows that K plans are required for inspection level II. Table

TABLE 19.4
Sample-size code letters

Lot or batch size	Special inspection levels S-1	S-2	S-3	S-4	General inspection levels I	II	III
2–8	A	A	A	A	A	A	B
9–15	A	A	A	A	A	B	C
16–25	A	A	B	B	B	C	D
26–50	A	B	B	C	C	D	E
51–90	B	B	C	C	C	E	F
91–150	B	B	C	D	D	F	G
151–280	B	C	D	E	E	G	H
281–500	B	C	D	E	F	H	J
501–1200	C	C	E	F	G	J	K
1201–3200	C	D	E	G	H	K	L
3201–10,000	C	D	F	G	J	L	M
10,001–35,000	C	D	F	H	K	M	N
35,001–150,000	D	E	G	J	L	N	P
150,001–500,000	D	E	G	J	M	P	Q
500,001 and above	D	E	H	K	N	Q	R

19.5 states the sample size is 125. For AQL = 1.5, the acceptance number is given as five and the rejection number as six. Therefore, the entire lot of 1500 articles may be accepted if five or fewer nonconforming articles are found but must not be accepted (rejected) if six or more are found.

Sampling risks are defined by the OC curve published in the standard. The curve for this plan is shown in Figure 19.7 as curve A.

The standard provides single, double, and multiple plans for each code letter (i.e., lot-size category). The plans for code letter K are shown in Table 19.6. Thus, the three plans can be found under the AQL column of 1.5. For example, the double-sampling plan calls for a first sample of 80 units. If two or fewer nonconforming are found, the lot is accepted. If five or more nonconforming are found, the lot is not accepted. If three or four nonconforming are found in the sample of 80, then a second sample of 80 is taken, giving a cumulative sample size of 160. If the total number of nonconforming in both samples is six or less, the lot is accepted; seven or more nonconforming means lot rejection.

Switching Procedures in ANSI/ASQC Z1.4

ANSI/ASQC Z1.4 includes provision for tightened inspection if quality deteriorates. If two out of five consecutive lots are not acceptable (rejected) on original inspection, a tightened inspection plan is imposed. The sample size is the same as usual, but the acceptance number is reduced. (The tightened plans do require larger sample sizes if the probability of acceptance for an AQL lot is less than 0.75.) For the example

TABLE 19.5

Master table for normal inspection (single sampling)

Sample-size code letter	Sample size	0.010 Ac Re	0.015 Ac Re	0.025 Ac Re	0.040 Ac Re	0.065 Ac Re	0.10 Ac Re	0.15 Ac Re	0.25 Ac Re	0.40 Ac Re	0.65 Ac Re	1.0 Ac Re	1.5 Ac Re
A	2												
B	3												
C	5												↓
D	8											↓	0 1
E	13										↓	0 1	↑
F	20									↓	0 1	↑	↓
G	32								↓	0 1	↑	↓	1 2
H	50							↓	0 1	↑	↓	1 2	2 3
J	80						↓	0 1	↑	↓	1 2	2 3	3 4
K	125					↓	0 1	↑	↓	1 2	2 3	3 4	5 6
L	200				↓	0 1	↑	↓	1 2	2 3	3 4	5 6	7 8
M	315			↓	0 1	↑	↓	1 2	2 3	3 4	5 6	7 8	10 11
N	500		↓	0 1	↑	↓	1 2	2 3	3 4	5 6	7 8	10 11	14 15
P	800	↓	0 1	↑	↓	1 2	2 3	3 4	5 6	7 8	10 11	14 15	21 22
Q	1250	0 1	↑	↓	1 2	2 3	3 4	5 6	7 8	10 11	14 15	21 22	↑
R	2000	↑		1 2	2 3	3 4	5 6	7 8	10 11	14 15	21 22	↑	

Notes: ↓, use first sampling plan below arrow. If sample size equals, or exceeds, lot of batch size, do 100 percent inspection.
↑, use first sampling plan above arrow.
Ac, acceptance number.
Re, rejection number.

previously cited, the tightened plan can be read from Table 19.6 as a sample size of 125 and an acceptance number of three.

ANSI/ASQC Z1.4 also provides for reduced inspection where the supplier's record has been good. The preceding 10 lots must have had a normal inspection with all lots accepted. A table of lower limits for the process average is provided to help decide whether the supplier's record has been good enough to switch to reduced inspection. The plan does, however, provide an option of switching to reduced inspection without using the table of lower limits. Under reduced sampling, the sample size is usually 40 percent of the normal sample size.

These switching rules apply when production is submitted at a steady rate. The plan provides other rules concerning the use of normal, tightened, and reduced inspection.

Other Provisions of ANSI/ASQC Z1.4

The standard provides OC curves for most of the individual plans along with "limiting quality" values for a probability of acceptance of 10 percent and 5 percent. Average sample-size curves for double and multiple sampling are also included. The latter curves show the average sample sizes expected as a function of the product quality submitted. Although the OC curves are roughly the same for single, double, and mul-

TABLE 19.5

Master table for normal inspection (single sampling) (*cont.*)

Acceptable quality levels (normal inspection)																											
2.5		4.0		6.5		10		15		25		40		65		100		150		250		400		650		1000	
Ac	Re	Ac	Re	Ac	Re	Ac	Re	Ac	Re	Ac	Re	Ac	Re	Ac	Re	Ac	Re	Ac	Re	Ac	Re	Ac	Re	Ac	Re	Ac	Re
↓		↓		0	1	↓		↓		1	2	2	3	3	4	5	6	7	8	10	11	14	15	21	22	30	31
↓		0	1	↑		↓		1	2	2	3	3	4	5	6	7	8	10	11	14	15	21	22	30	31	44	45
0	1	↑		↓		1	2	2	3	3	4	5	6	7	8	10	11	14	15	21	22	30	31	44	45	↑	
↑		↓		1	2	2	3	3	4	5	6	7	8	10	11	14	15	21	22	30	31	44	45	↑			
↓		1	2	2	3	3	4	5	6	7	8	10	11	14	15	21	22	30	31	44	45	↑					
1	2	2	3	3	4	5	6	7	8	10	11	14	15	21	22	↑		↑		↑							
2	3	3	4	5	6	7	8	10	11	14	15	21	22	↑													
3	4	5	6	7	8	10	11	14	15	21	22	↑															
5	6	7	8	10	11	14	15	21	22	↑																	
7	8	10	11	14	15	21	22	↑																			
10	11	14	15	21	22	↑																					
14	15	21	22	↑																							
21	22	↑																									
↑																											

tiple sampling, the average sample-size curves vary considerably because of the inherent differences among the three types of sampling. The standard also states the AOQL that would result if all rejected lots were screened for nonconforming units.

In ANSI/ASQC Z1.4, *a sampling scheme* is defined as "a combination of sampling plans with switching rules and possibly a provision for discontinuance of inspection." For the sampling schemes associated with the individual plans, the standard provides OC curves and information on AOQL, limiting quality, and average sample sizes—all for single sampling.

Dodge-Romig Sampling Tables

Dodge and Romig (1959) provide four sets of attributes plans emphasizing either lot-by-lot quality (LTPD) or long-run quality (AOQL);

Lot tolerance percentage defective (LTPD): single sampling
double sampling

Average outgoing quality limit (AOQL): single sampling
double sampling

These plans differ from those in ANSI/ASQC Z1.4 in that the Dodge-Romig plans assume that all rejected lots are 100 percent inspected and the defectives replaced with acceptable items. Plans with this feature are called *rectifying inspection plans.* The tables provide protection against poor quality on either a lot-by-lot basis or average

TABLE 19.6

Sampling plan for sample-size code letter K

Type of sampling plan	Cumulative sample size	Acceptable quality levels (normal inspection)													
		Less than 0.10		0.10		0.15		×		...	1.0		1.5		...
		Ac	Re	Ac	Re	Ac	Re	Ac	Re		Ac	Re	Ac	Re	
Single	125	∇		0	1	0					3	4	5	6	
Double	80	∇		*		Use letter J		Use letter M			1	4	2	5	
	160										4	5	6	7	
Multiple	32	∇		*							#	3	#	4	
	64										0	3	1	5	
	96										1	4	2	6	
	128										2	5	3	7	
	160										3	6	5	8	
	192										4	6	7	9	
	224										6	7	9	10	
		Less than 0.15		0.15		×		0.25			1.5		2.5		
		Acceptable quality levels (tightened inspection)													

Notes: Δ, use next preceding sample-size code letter for which acceptance and rejection numbers are available.
∇, use next subsequent sample-size code letter for which acceptance and rejection numbers are available.
Ac, acceptance number.
Re, rejection number.
*, use single-sampling plan above (or alternatively use letter N).
#, acceptance not permitted at this sample size.

long-run quality. The LTPD plans assure that a lot having poor quality will have a low probability of acceptance, i.e., the probability of acceptance (or consumer's risk) is 0.1 for a lot with LTPD quality. The LTPD values range from 0.5 to 10.0 percent defective. The AOQL plans assure that, after all sampling and 100 percent inspection of rejected lots, the *average* quality over many lots will not exceed the AOQL. The AOQL values range from 0.1 to 10.0 percent. Each LTPD plan lists the corresponding AOQL, and each AOQL plan lists the LTPD.

19.16 SELECTION OF A NUMERICAL VALUE OF THE QUALITY INDEX

The problem of selecting a value of the quality index (e.g., AQL, AOQL, or lot tolerance percentage defective) is one of balancing the cost of finding and correcting a defective against the loss incurred if a defective slips through an inspection procedure.

TABLE 19.6

Sampling plan for sample-size code letter K (*cont.*)

Acceptable quality levels (normal inspection)												
×		4.0		×	...	×		10		Higher than 10		Cumulative sample size
Ac	Re	Ac	Re	Ac	Re	Ac	Re	Ac	Re	Ac	Re	
8	9	10	11	12	13	18	19	21	22	Δ	10	125
3	7	5	9	6	10	9	14	11	16	Δ		80
11	12	12	13	15	16	23	24	26	27			160
0	4	0	5	0	6	1	8	2	9	Δ		32
2	7	3	8	3	9	6	12	7	14			64
4	9	6	10	7	12	11	17	13	19			96
6	11	8	13	10	15	16	22	19	25			128
9	12	11	15	14	17	22	25	25	29			160
12	14	14	17	18	20	27	29	31	33			192
14	15	18	19	21	22	32	33	37	38			224
4.0		×		6.5		10		×		Higher than 10		
Acceptable quality levels (tightened inspection)												

Enell (1954), in a classic paper, has suggested that the break-even point (see Section 19.9) be used in the selection of an AQL. The *break-even point* for inspection is defined as the cost to inspect one piece divided by the damage done by one defective. For the example cited, the break-even point was 5 percent defective.

As a 5 percent defective quality level is the break-even point between sorting and sampling, the appropriate sampling plan should provide for a lot to have a 50 percent probability of being sorted or sampled; i.e., the probability of acceptance for the plan should be 0.50 at a 5 percent defective quality level. The OC curves in a set of sampling tables such as ANSI/ASQC Z1.4 can now be examined to determine an AQL. For example, suppose that the device is inspected in lots of 3000 pieces. The OC curves for this case (code letter K) are shown in ANSI/ASQC Z1.4 and Figure 19.7. The plan closest to having a P_a of 0.50 for a 5 percent level is the plan for an AQL of 1.5 percent. Therefore, this is the plan to adopt.

Some plans include a classification of defects to help determine the numerical value of the AQL. Defects are first classified as critical, major, or minor according to definitions provided in the standard. Different AQLs may be designated for groups of defects considered collectively or for individual defects. Critical defects may have a 0 percent AQL, whereas major defects may be assigned a low AQL, say 1 percent, and minor defects a higher AQL, say 4 percent. Some manufacturers of complex products specify quality in terms of number of defects per million parts.

In practice, the quantification of the quality index is a matter of judgment based on the following factors: past performance on quality, effect of nonconforming product on later production steps, effect of nonconforming product on fitness for use, urgency of delivery requirements, and cost of achieving the specified quality level.

A thorough discussion of this difficult issue is provided by Schilling (1982, pages 571–586) in the classic, comprehensive book on acceptance sampling.

19.17
HOW TO SELECT THE PROPER SAMPLING PROCEDURES

Sampling procedures can serve different purposes. As itemized by Schilling 1982, these include:

- Guaranteeing quality levels at stated risks.
- Maintaining quality at AQL level or better.
- Guaranteeing an AOQL.
- Reducing inspection after good history.
- Checking inspection.
- Ensuring compliance to mandatory standards.
- Reliability sampling.
- Checking inspection accuracy.

For each purpose, Schilling recommends specific attributes or variable sampling plans. Selection of a plan depends on the purpose, the quality history, and the extent of knowledge of the process.

The steps involved in the selection and application of a sampling procedure are shown in Figure 19.10. Emphasis is on the feedback of information necessary for the proper application, modification, and evolution of sampling in a manner that encourages continuous improvement and reduced inspection costs. This can be achieved by moving from a system of acceptance sampling to acceptance control.

Moving from Acceptance Sampling to Acceptance Control

Acceptance sampling is the process of evaluating a portion of the product in a lot for the purpose of accepting or rejecting the entire lot as either conforming or not conforming to a quality specification. *Acceptance control* is a "continuing strategy of selection, application, and modification of acceptance sampling procedures to a changing inspection environment" (Schilling, 1982, page 564). This evaluation of a sampling plan application is shown in the life cycle of acceptance control (Table 19.7). The cycle is applied over a lifetime of a product to achieve (1) quality improvement (using process control and process capability concepts) and (2) reduction and elimination of inspection (using acceptance sampling).

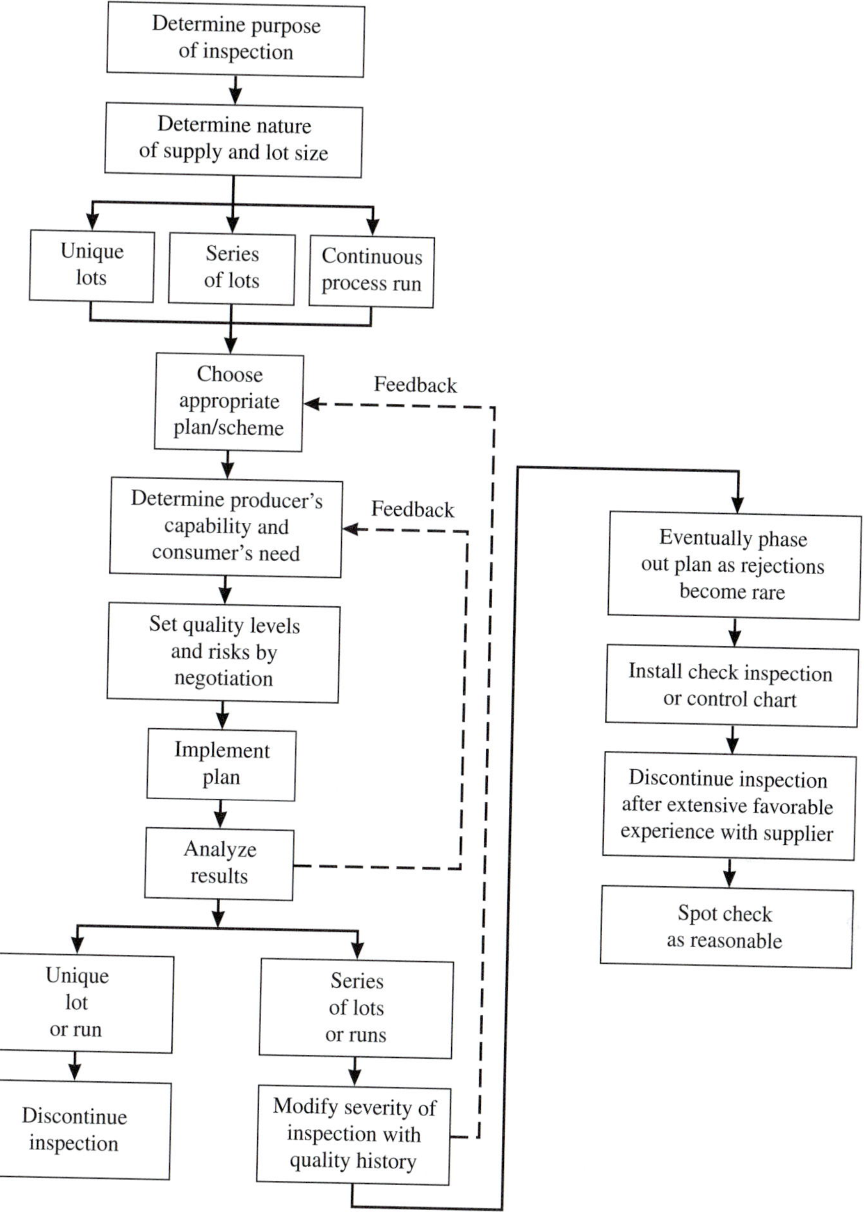

FIGURE 19.10
Check sequence for implementation of sampling procedure. (*From Schilling, 1982. Reprinted by courtesy of Marcel Dekker, Inc.*)

TABLE 19.7
Life cycle of acceptance control application

Stage	Step	Method
Preparatory	Choose plan appropriate to purpose	Analysis of quality system to define the exact need for the procedure
	Determine producer capability	Process performance evaluation using control charts
	Determine consumer needs	Process capability study using control charts
	Set quality levels and risks	Economic analysis and negotiation
	Determine plans	Standard procedures if possible
Initiation	Train inspector	Include plan, procedure, records, and action
	Apply plan properly	Ensure random sampling
	Analyze results	Keep records and control charts
Operational	Assess protection	Periodically check quality history and OC curves
	Adjust plan	When possible, change severity to reflect quality history and cost
	Decrease sample size if warranted	Modify to use appropriate sampling plans taking advantage of credibility of supplier with cumulative results
Phase out	Eliminate inspection effort where possible	Use demerit rating or check inspection procedures when quality is consistently good
		Keep control charts
Elimination	Spot check only	Remove all inspection when warranted by extensive favorable history

Source: Schilling (1982, page 566). Reprinted by courtesy of Marcel Dekker, Inc.

SUMMARY

- Product acceptance involves three decisions: conformance, fitness for use, and communication.
- In deciding whether nonconforming product is fit for use or not, inputs must be secured from several sources.
- The communication decision involves both outsiders (customers) and insiders.
- Inspection planning includes the designation of inspection stations and spelling out the instructions and facilities required.
- A classification of characteristics is a list of quality characteristics derived from the specifications; a classification of defects is a list of symptoms of nonconformance during manufacture and field use.
- The amount of inspection necessary depends mainly on the amount of prior knowledge about product quality, homogeneity of the lot, and the allowable risk.

- Human errors in inspection arise from technique errors, inadvertent errors, conscious errors, and communication errors.
- Errors of measurement have two parts: precision and bias. Both of these parts can be quantified.
- Acceptance sampling is the process of evaluating a portion of the product in a lot for the purpose of accepting or rejecting the entire lot.
- We have four alternatives for evaluating lots: no inspection, small samples, large samples, or 100 percent inspection.
- Sampling risks are of two kinds: good lots can be rejected and bad lots can be accepted. An operating characteristic (OC) curve quantifies these risks.
- Published sampling plans categorize quality in terms of acceptable quality level (AQL), limiting quality level (LQL), indifference quality level (IQL), or average outgoing quality limit (AOQL).
- Attributes sampling plans evaluate each unit of product as simply acceptable or defective: ANSI/ASQC Z1.4 (1993) and the Dodge-Romig sampling tables are examples of attributes plans.
- Variables sampling plans involve measurements on product units.

PROBLEMS

19.1. Discuss the inspection of new homes with the appropriate municipal department. Determine the purpose of its inspection, the specification used, and how the inspection is conducted. Comment on this inspection from the viewpoint of the purchaser of a home.

19.2. In one large company making consumer durable products, the chief inspector operated on the principle that disposition of nonconforming lots of components must be done in only three ways: (1) by scrapping them, (2) by repairing them to bring them into conformance, or (3) by securing a waiver from the design department. What she did not permit was a tryout to see whether the components were usable despite the nonconformance. Her stated reason was that if she authorized such tryouts, the production people would devote their energies to tryouts rather than to making the components right in the first place. What do you think of this philosophy?

19.3. Certain plates stamped out in a stamping press include holes for which there are close tolerances for diameter and for the distance between holes. In the discussion of how many pieces to gage for these dimensions, one proposal is to measure the first and the last piece for each lot and to accept the lot if both pieces conform to specification. A statistician objects to this proposal on the grounds that the sample is only two pieces and that, if the lot were 50 percent defective, it could easily be accepted due to statistical variation. What is your opinion?

19.4. A large manufacturer of minicomputers is incurring high costs due to the need for assembling and testing the computers to discover and eliminate defects before the computers are installed in clients' computer rooms. After this assembly, testing, and repair, the computers are disassembled, shipped to the clients' facilities, reassembled, and checked out. How would you go about reducing the cost of preassembly and pretest?

19.5. You are a quality manager engaged in a seminar to discuss common problems with other quality managers. A lively discussion has developed over some problems associated with pressures applied to quality managers. These pressures concern the shipment of nonconforming, unfit, or even unsafe products. In addition, the pressures concern the matter of the quality manager signing a test certificate or other document that puts him on record as having approved something when he was really against it.

Here are several of the problem categories identified by the group:

- A lot of product has been made with a nonconformance to specification. All company managers (including the quality manager) are convinced that the nonconforming product is fit for use. They are not agreed on whether to tell the client about the nonconformance. The marketing manager is against informing the client on the ground that some clients may use such information to wring a price concession out of the manufacturer.
- A product lot contains a small percentage of units that are clearly unfit for use. There is a debate on whether to sort the lot to remove the defective units or to ship the lot as is and pay any claims as they arise. The production manager (who wants to ship the product without sorting) contends that the problem is purely economic and that the quality considerations are secondary.
- A large electromechanical system has been made under a contract that includes a penalty clause for late delivery. The system has already met its test requirements and is being crated for shipment. At this point it is discovered that the test equipment used to test one of the subsystems was out of calibration at the time of making the test. Under the accepted practice in the industry, such a discovery throws suspicion on the quality of the subsystem and hence on the quality of the system. Unfortunately, the subsystem is not easily accessible. It is buried deep within the system so that it would involve a serious delay as well as a large expense to take the system apart, test the subsystem, and then put it all back together again. The manufacturing people take the position that the subsystem is okay despite the condition of the test equipment. They note that another subsystem built by the same people using the sample process has just tested okay. They urge that the system be shipped based on this evidence of a reliable process and work force.
- A product with a good safety record has resulted in a serious injury to a user. The injury involved a most unusual combination of unlikely events plus an obvious misuse by the user. The design manager defends the design on the record—the only known serious injury was associated with misuse.

What are your conclusions as to the position to be taken by the quality manager in the foregoing cases with respect to (*a*) shipping the product and (*b*) signing the documents?

19.6 An instrument has been used to measure the length of a part. The result was 6.70052 cm. An error-of-measurement study was made on the instrument with the following results:

Bias: +0.00254 cm (on the average, the instrument reads 0.00254 cm high).

Repeatability (precision): 0.001016 cm (1σ).

Make a statement concerning the true value of the part just measured. State all assumptions needed.

19.7. The precision of a certain mechanical gage is indicated by a standard deviation (of individual repeat measurements) of 0.00254 cm. Investigate the effect on precision of making multiple measurements. Consider 2, 3, 4, 5, 10, 20, and 30 as multiples. Graph the results.

19.8 One ball bearing was measured by one inspector 13 times with each of two vernier micrometers. The results are shown below:

Measurement number	Model A	Model B
1	0.6557	0.6559
2	0.6556	0.6559
3	0.6556	0.6559
4	0.6555	0.6559
5	0.6556	0.6559
6	0.6557	0.6559
7	0.6556	0.6559
8	0.6558	0.6559
9	0.6557	0.6559
10	0.6557	0.6559
11	0.6556	0.6559
12	0.6557	0.6560
13	0.6557	0.6560

Suppose that the true diameter is 0.65600. Calculate measures of bias and precision for each micrometer. What restrictions must be placed on the applicability of the numbers you determined?
Answer: A: bias $= -0.00035$, precision $= 1\sigma$ of 0.000075; *B:* bias $= -0.000085$, precision $= 1\sigma$ of 0.000036.

19.9. A large sample of product has been measured. The mean was 2.506 in, and the standard deviation was 0.002 in. A separate error-of-measurement study yielded the following results:

Bias $= +0.001$ in.

Precision: 0.0005 in. (1σ). The product has only a minimum specification limit. What should this limit be to take into account bias and precision and reject only 5 percent of the product?
Answer: 2.502.

19.10. A sample of measurements shows a mean and standard deviation of 2.000 in and 0.004 in, respectively. These results are for the *observed* values. A separate error-of-measurement study indicates a precision of 0.002 in (1σ). There is no bias error. The dimension has a specification of 2.000 $\pm$ 0.006 in. What percentage of the population has *true* dimensions outside the specification?

19.11. A large gray-iron foundry casts the base for precision grinders. These bases are produced at the rate of 18 per day and are 100 percent inspected at the foundry for flaws in the metal. The castings are then stored and subsequently shipped in lots of 300 to the grinder manufacturer. The grinder manufacturer has found these lots to be 10 percent defective. Upon receiving a lot, the manufacturer inspects 12 of them and rejects the lot if two or more defectives are found. What is the chance that the manufacturer will reject a given lot?

19.12. The following double-sampling plan has been proposed for evaluating a lot of 50 pieces:

Sample	Sample size	Acceptance number	Rejection number
1	3	0	3
2	3	2	3

Using only the addition and multiplication rules for probability, calculate the probability of accepting a lot that is 10 percent defective.
Answer: 0.984.

19.13. Prepare the operating characteristic curves for a single-sampling plan with an acceptance number of zero. Use sample sizes of 2, 5, 10, and 20.

19.14. You are quality manager for a company receiving large quantities of materials from a supplier in lots of 1000. The cost of inspecting the lots is \$0.76/unit. The cost that is incurred if bad material is introduced into your product is \$15.20/unit. A sampling plan of 75 with acceptance number equal to two has been submitted to you by one of your engineers. In the past, lots submitted by the supplier have averaged 3.4 percent defective.
(*a*) Is a sampling plan economically justified?
(*b*) Prepare an operating characteristic curve.
(*c*) If you want to accept only lots of 4 percent defective or better, what do you think of the sampling plan submitted by the engineer?
(*d*) Suppose that rejected lots are 100 percent inspected. If a supplier submits many 4 percent defective lots, what will be the average outgoing quality of these lots?

19.15. Refer to ANSI/ASQC Z1.4 with the following conditions:
- Lot size = 10,000.
- Inspection level II.
- Acceptable quality level = 4 percent.

(*a*) Find a single-sampling plan for normal inspection.
(*b*) Suppose that a lot is sampled and accepted. Someone makes the statement, "This means the lot has 4 percent or less defective." Comment on this statement. (Assume that the sample was randomly selected and no inspection errors were made.)
(*c*) Calculate the probability of accepting a 4 percent defective lot under normal inspection.
Answer: (*c*) 0.98.

19.16. A manufacturer sells his product in large lots to a customer who uses a sampling plan at incoming inspection. The plan calls for a sample of 200 units and an acceptance number of two. Rejected lots are returned to the manufacturer. If a lot is rejected and returned, the manufacturer has decided to gamble and send it right back to the customer without screening it (and without telling the customer it was a rejected lot). The manufacturer hopes that another random sample will lead to an acceptance of the lot. What

is the probability that a 2 percent defective lot will be accepted on either one or two submissions to the customer?
Answer: 0.42.

19.17. A customer is furnished resistors under ANSI/ASQC Z1.4. Inspection level II has been specified with an AQL of 1.0 percent. Lot sizes vary from 900 to 1200.
(*a*) What single-sampling plan will be used?
(*b*) Calculate the quality (in terms of percentage defective) that has an equal chance of being accepted or rejected.
(*c*) What is the chance that a 1 percent defective lot will be accepted?
Answer: (b) 3.4 percent; *(c)* 0.95.

19.18. A rule-of-thumb sampling plan states that, for any lot size, the sample size should be 10 percent of the lot and the acceptance number should be zero. It is believed that this procedure will hold the sampling risks constant. Prepare the operating characteristic curves using this plan for lot sizes of 100, 200, and 1000. Calculate points for quality levels of 0, 2, 4, 6, 10, and 14 percent defective. Compare the three curves and draw conclusions about the sampling risks.

REFERENCES

Adams, R. (1987). "Moving from Inspection to Audit," *Quality Progress,* January, pp. 30–31.

ANSI/ASQC Z1.4 (1993). "Sampling Procedures and Tables for Inspection by Attributes," American Society for Quality Control, Milwaukee.

Automotive Industry Action Group (1995). "Measurement Systems Analysis," Southfield, MI.

Baxter Travenol Laboratories (1986). "Statistical Process Control Guideline," Baxter Travenol Laboratories, Deerfield, IL, p. 23.

Burdick, R. K. and G. A. Larsen (1997). "Confidence Intervals on Measures of Variability in R&R Studies," *Journal of Quality Technology,* vol. 29, no. 3, pp. 261–273.

Case, K. E., G. K. Bennett, and J. W. Schmidt (1975). "The Effect of Inspection Error on Average Outgoing Quality," *Journal of Quality Technology,* vol. 7, no. 1, pp. 1–12.

Chun, Y. H. and D. B. Rinks (1998). "Three Types of Producer's and Consumer's Risks in the Single Sampling Plan," *Journal of Quality Technology,* vol. 30, no. 3, pp. 254–268.

Cooper, J. (1997). "Implementing a Paperless Inspection System," *Annual Quality Congress Proceedings,* ASQ, Milwaukee, pp. 231–235.

Dodge, H. F. and H. G. Romig (1959). *Sampling Inspection Tables,* 2nd ed., John Wiley & Sons, New York.

Durkee, D. and B. Gookins (1999). "The Inspector's Role in Organizational Quality," *Annual Quality Congress Proceedings,* ASQ, Milwaukee, pp. 117–121.

Eagle, A. R. (1954). "A Method for Handling Errors in Testing and Measuring," *Industrial Quality Control,* March, pp. 10–14.

Enell, J. W. (1954). "What Sampling Plan Shall I Choose?" *Industrial Quality Control,* vol. 10, no. 6, pp. 96–100.

Engel, J. and B. DeVries (1997). "Evaluating a Well-Known Criterion for Measurement Precision," *Journal of Quality Technology,* vol. 29, no. 4, pp. 469–476.

Grant, E. L. and R. S. Leavenworth (1996). *Statistical Quality Control,* 7th ed., McGraw-Hill, New York.

Harris, D. H. and F. B. Chaney (1969). *Human Factors in Quality Assurance,* John Wiley & Sons, New York, pp. 77–85.

Hoag, L. L., B. L. Foote, and C. Mount-Campbell (1975). "The Effect of Inspector Accuracy on Type I and II Errors of Common Sampling Techniques," *Journal of Quality Technology,* vol. 7, no. 4, pp. 157–164.

Orkin, F. I. and D. P. Olivier (1999). In *JQH5,* Table 10.2.

Peach, R. W., ed. (1997). *The ISO 9000 Handbook,* 3d ed., McGraw-Hill, New York.

Quality Engineering (1990). "Letters to the Editor," vol. 3, no. 2, pp. vii–xii.

Riley, F. D. (1979). "Visual Inspection—Time and Distance Method," *ASQC Annual Technical Conference Transactions,* Milwaukee, p. 483.

Schilling, E. G. (1982). *Acceptance Sampling in Quality Control,* Marcel Dekker, New York.

Schilling, E. G. (1994). "The Importance of Sampling in Inspection," *Annual Quality Congress Proceedings,* ASQ, Milwaukee, pp. 809–812.

Taylor, W. A. (1994). "Acceptance Sampling in the 90's," *Annual Quality Congress Proceedings,* ASQ, Milwaukee, pp. 591–598.

Tsai, P. (1988). "Variable Gauge Repeatability and Reproducibility Study Using the Analysis of Variance Method," *Quality Engineering,* vol. 1, no. 1, pp. 107–115.

SUPPLEMENTARY READING

Inspection and testing, general: *JQH5,* Section 23.

Pareto concept and testing errors: Gambino, R., P. Mallon, and G. Woodrow (1990). "Managing for Total Quality in a Large Laboratory," *Archives of Pathology and Laboratory Medicine,* November, pp. 1145–1148.

Measurement concepts in the automotive industry: Automotive Industry Action Group (AIAG) (1990). "Measurement Systems Analysis Reference Manual," AIAG, Southfield, MI.

Measurement error in batch processes: Basnet, C. and K. E. Case (1993). "The Effect of Measurement Error on Accept/Reject Probabilities for Homogeneous Products," *Quality Engineering,* vol. 4, no. 3, pp. 383–397.

Acceptance sampling, general: *JQH5,* Section 46.

Shilling, E. G. (1982). *Acceptance in Sampling in Quality Control,* Marcel Dekker, New York.

WEBSITES

ASQ inspection division: www.asq.org/display_web.cgi?13

American Society for Testing and Materials (ASTM): www.astm.org

National Institute of Standards and Technology (NIST): icdb.nist.gov or www.bipm.fr

20

MARKETING, FIELD PERFORMANCE, AND CUSTOMER SERVICE

20.1 QUALITY CONCEPTS IN THE MARKETING FUNCTION

All functions in an organization, including the marketing function, can benefit from the application of quality concepts. Table 20.1 shows some marketing activities and their related quality activities. Further elaboration is provided in *JQH5,* Section 18.

The marketing function has opportunities to apply the trilogy of quality processes: quality improvement (Chapter 3), quality planning (Chapter 4), and quality control (Chapter 5). To cite a dramatic illustration, Nickell (1985) describes some pioneering efforts at IBM concerning the amount of product shipped that was canceled before installation. Surprisingly, 13 percent of signed and shipped orders were canceled. This problem became a quality improvement project, and the team effort reduced the cancellations to 3 percent.

Hurley (1994) in the Supplementary Reading describes how seven organizations (manufacturing and service) transformed their marketing function from traditional to quality focused. Five of these organizations were Baldrige Award winners. The research revealed that the companies (1) learn, in detail, the customer view of quality; (2) know how the company core processes relate to customer satisfaction; and (3) emphasize relationship marketing (see below) rather than competing for sales transactions.

Several activities in Table 20.1 concern the traditional act of selling the product to customers. An emerging concept that replaces the traditional short-range selling process of closing the sale with a longer-range process of achieving customer loyalty is called customer relationship management (CRM).

TABLE 20.1
Marketing and quality-related activities

Marketing activity	Quality-related activity
Launching new products	Conducting a "test market" to identify product weaknesses and weaknesses in the marketing plan
Labeling	Ensuring that products conform to label claims
	Ensuring that label information is accurate and complete
Advertising	Identifying the product features that will persuade customers to purchase a product
	Verifying the accuracy of quality claims included in advertising
Assistance to customers in product selection	Presenting quality-related data to help customers evaluate alternative products
Assistance to merchants	Providing merchants with quality-related information for use by salespeople
	Providing merchants with technical advice on product storage and handling and sales demonstrations
Preparation of sales contract	Defining product requirements on performance, other technical requirements, and level of quality (e.g., defects per million)
	Defining requirements on execution of a contract, e.g., a quality plan, submission of specified documentation during contract
	Defining warranty provisions
	Defining incentive provisions on quality and reliability
Order entry and filling	Applying quality improvement concepts to reduce lead time or reduce errors

20.2 CUSTOMER RELATIONSHIP MANAGEMENT

CRM is a continuous and systematic effort to establish interactions with customers in order to determine individual customer needs and provide products that meet them. This concept certainly sounds old. Some of the elements of CRM are old, but like many concepts with a new title the collection of specifics of CRM do present a new viewpoint.

The basic elements of CRM are

- *Identify customers.* Here the emphasis is on groups of different types of customers or even individual customers. This approach is in contrast to identifying a broad customer base and then designing a product for that broad group. The focus on a broad customer base leads to mass production of goods and services, but eventually competitors enter the market with similar products.
- *Differentiate customers by their value to the company.* Clearly, some customers contribute more to the profit of a company than other customers do. These key cus-

tomers can be identified by analyzing data in the customer database that is now becoming an integral part of any business. Identifying these important customers leads to focusing on their needs.
- *Differentiate customers by their needs.* Determining those needs involves market research and other activities as described in Chapter 12, "Understanding Customer Needs." But it also involves making continuous use of information gathered during routine interactions with customers.
- *Interact with customers in a way that builds a lasting relationship.* These interactions include one-on-one, on-site sales discussions, websites, and call centers for providing information and resolving complaints. This aspect of CRM goes far beyond closing a sale or solving a product problem. It means solving the immediate issue in a way that convinces a customer that the company really wants to be helpful.
- *Customize some aspect of the goods or service that is provided to these key customers.* Current technology often makes it possible to provide a product uniquely suited to meet the needs of these customers.

Note how electronic commerce (e-commerce) will contribute to customer relationship management. E-commerce contacts a large number of customers; e-commerce presents and collects information in depth to customers; e-commerce helps customers in making decisions. For elaboration, see Section 4.7 under "Electronic Commerce."

Peppers and Rogers (1999) and Corcoran et al. (1995) provide further discussion on CRM. Anton (1996) describes the use of qualitative and quantitative techniques in CRM, including the application of SPSS software to analyze and display data. Selden (2000) in the Supplementary Reading contrasts the "sale as discrete" (SAD) event view of sales with a broader and systematic approach of understanding and meeting customer needs that he calls "sales quality engineering."

Customer satisfaction is the result of the comparison between customer expectations and the performance delivered. One element of customer expectations is the warranty provided with a product.

20.3 WARRANTY OF QUALITY

A warranty is a form of assurance that a product is fit for use or, failing this, that the user will receive some kind of compensation. In this sense, a warranty constitutes a system for reducing user costs of poor quality.

In most jurisdictions, a seller, by the mere act of sale, makes two *implied* warranties:

1. A *general* warranty of "merchantability," i.e., fitness for the customary use of such products.
2. An added *special* warranty of fitness for the specific use to which the product will be put. This warranty is implied only if the seller knows what that specific use is.

Beyond the warranties (which are implied by law) are added *express* warranties made by the seller or negotiated between the parties. For consumer products most such warranties are made unilaterally by the seller through oral representations about the product, display of samples, descriptions in catalogs, claims made in advertising, etc. In addition, specific statements of warranty are published in documents that are headed up by the word *warranty*. In the case of sales made to industrial companies, warranties are sometimes specially negotiated, as are other aspects of the sales contract.

Traditionally, warranties as to quality were limited to replacement or repair of the product during the warranty period. With the proliferation of long-life products, warranties have been extended into parameters such as reliability, downtime, and maintenance costs. This extension is squarely in line with the need to optimize users' costs. This trend can be expected to continue.

In the case of warranties to consumers, the complications associated with long-life products have led to much confusion as to their meaning, especially as to where the responsibility for action lies. This situation has given rise to national legislation that requires warranties to be clear and that sets out some criteria for clarity.

Consumer product warranties are either "full" or "limited." The term *full warranty* refers to the consumer's rights, not to the portion of the physical product that is covered by the warranty; i.e., it does not have to cover the entire product. A full warranty means that

1. The manufacturer will fix or replace any defective product free of charge.
2. The warranty is not limited in time.
3. The warranty does not exclude or limit payment for consequential damages (see below).
4. If the manufacturer is unable to make an adequate repair, the consumer may choose between a refund and a replacement.
5. The manufacturer cannot impose unreasonable duties on the consumer. For example, the warranty cannot require the consumer to ship a piano to the factory (one manufacturer listed such a condition).
6. The manufacturer is not responsible if the damage to the product was caused by unreasonable use.

The full warranty also provides that not only the original purchaser but also any subsequent owners of the product during the warranty period are entitled to make claims.

A limited warranty is a warranty that does not meet the requirements for a full warranty. Typically, a limited warranty may exclude labor costs, require the purchaser to pay for transportation charges, and be limited to the original purchaser of the product. As a practical matter, most warranties on consumer products are limited warranties and must be so labeled.

The duration and coverage of a warranty, for both industrial and consumer products, vary greatly. For example, for one industrial commodity product, the warranty may call only for compliance to specifications at the time of formal "acceptance"; another industrial product may have a warranty for a period of time and even provide for payment of consequential damage costs if the product fails. For a consumer product, the warranty covers normal performance during a limited period (usually about 10 percent of the useful life); in a few cases (e.g., a pen), a lifetime warranty is provided for specified conditions.

Because the warranty period covers only a portion of the product life, users must contend with any service needs that arise during the remainder of the item's life. Recognizing this condition as a business opportunity, manufacturers have responded in several ways, including product redesign, competition in warranties, extended warranties, and reliability improvement warranties.

20.4 FIELD PERFORMANCE

The act of final product acceptance may be regarded as terminating the manufacturing phase. Following this act and before customer use of the product, a number of preuse phases take place: packing, shipping, receiving, and storage. Finally, there are the use phases: installation, checkout, operation, and maintenance.

As product complexity grows, the extent of field problems increases. Field factors cause 20 to 30 percent of problems concerning fitness for use on long-life products of moderate to high complexity.

Those who do quality planning for service activities can learn much from the experience of formalizing quality planning in design and manufacturing. Those who perform service activities that affect quality often believe that their activities are sufficient to provide a quality service. In cases where service is in fact deficient, there is often a need to revise the service plan as it relates to the topics discussed below. Such planning must show—explicitly—what to do differently. Generalities are unacceptable.

Packaging, Transportation, and Storage

One major department store found that more customer complaints are caused by activities in packaging, storing, and delivering the product than were caused by the original manufacturing.

The most critical aspect of the quality of packaging, transportation, and storage is the design and acquisition of effective packaging materials. Packaging requires a sequence of activities similar to those used to achieve fitness for use for the product itself. These activities include package design and the purchasing, manufacturing, and testing of packaging materials.

Package Design

Designing the package and the product simultaneously is the ideal approach. Sometimes a minor alteration in product design can strengthen the product and thereby eliminate the need for extensive packaging materials.

This simultaneous approach requires that packaging design start during the product design stage rather than after manufacture begins. If little or no product information is available to the packaging engineer, the result may be a package that either underprotects or overprotects the product. Design of packaging calls for laboratory testing (e.g., impact, vibration, compression, and drop tests) and field testing of materials.

Especially for products that undergo numerous or critical handlings, it becomes important to look at handling and packaging from a systems viewpoint as opposed to numerous department viewpoints. The extent of handling during production, packaging, and transportation can be surprising. Handling costs run about 20 percent of manufacturing costs, with some estimates as high as 60 percent. A system review can identify opportunities to improve the overall handling by methods such as:

- Modifying supplier packaging. Some electronic components are blister packaged to allow testing without unpacking.
- Starting unitized and other containerization at early operations or even at the supplier's location. For example, the drug industry has evolved unitized packaging of dosages in which the identity of the dose (and even its own environment) is designed into the unit package and carries through to the patient.
- Designing racks, trays, bins, tote boxes, etc., to provide optimal service to all companies and departments involved rather than to require added handling and repacking for some.

Purchasing, Manufacturing, and Testing of Packaging Materials

The techniques described in other sections of this book for manufacturing and testing of products apply equally to packaging materials. In many cases packaging materials are obtained from a supplier, and thus the elements of a supplier quality program are applicable.

Transportation

Handling and transport introduce many perils to the product. Some of these are fully predictable: climatic temperature changes, humidity, vibration, and shock (during automated handling). Others are the result of ignorance, carelessness, blunder, and even sabotage. For some of these perils, the product is in greater danger from handling and transport than from usage.

Protecting a product during transportation involves determining the ability of the product to withstand the transportation hazards. Tests can be run to simulate shock, vibration, and other transport damage. These stresses are measured in terms of cycles per second, "g levels" of deceleration, pulse shapes, and still other quantified measures. The results are then used to develop specifications for packaging materials and vehicle loading.

A useful innovation is the placement of small sensors into transport containers. These sensors record—during the actual transportation of materials—data on shock, vibration, temperature, humidity, acceleration, container's drop height, etc. This technology provides a reading of what goes on during transportation of the item. The information is stored in a microprocessor-based digital data system inside the sensor and later is transferred to computers for processing and analysis (Hicks, 1991). The experience gained finds its way into later specifications for packaging and vehicle loading.

Storage

Immense quantities of raw materials, components, and finished products are constantly in storage, awaiting further processing, sale, or use. To minimize deterioration and degradation, various actions can be taken: establishing the shelf life of the product based on laboratory and field data, establishing standards to place limits on time in storage, dating the product conspicuously to make it easy to identify the age of the product in stock, and designing the package and controlling the environment to minimize both expected and unexpected degradation.

A common weakness of these programs is a failure to date the product conspicuously. Sometimes this failure is simply due to poor technique, e.g., iron in open storage rusts away because the color of the rust preservative is not changed annually. However, some failure to date conspicuously is the result of marketing decisions, e.g., the dates are put on the back of the product or on the front in tiny print because the advertising has priority; there is a fear of dating the product in a way that discourages effective stock rotation. Such reasons, if they were ever valid, now have become obsolete. Conspicuous dating on outer cartons as well as on the unit package aids in traceability, stock rotation, and establishing age of inventories.

Some products such as drugs are made in batches that may be released only after tests are completed on a sample. While the batches are being held, special inventory procedures are needed to assure that unreleased product is not inadvertently sent to a customer.

Installation and Use

Before the packaged product is put into use, it undergoes additional processing during distribution, assembly, installation and checkout, etc. These operations are quite as much a part of the progression of the product as the design and manufacture are, and they demand corresponding controls.

Processing during distribution

The distribution process carries out operations such as breaking bulk, readjusting, adding reagents, touching up finishes, and repackaging. The planning of these operations should be a part of the overall product planning. The results of this planning should then be embodied in specifications to be used by the distribution organizations. Compliance with these specifications should be audited independently.

In some cases the distribution process cannot be relied on to carry out these operations, e.g., there are international problems of language or culture or small retailers lack the technological skills to do so. In such cases the need is for a systems redesign that eliminates the necessity for technological skills or even for the operations. These problems can be at their worst when the operations are to be performed by the end user.

On-site installation by specialists

The assembly setup is conducted at the user's premises to put the product into a state of readiness to operate. Installation may require the services of specialists, or users may perform their own installation.

On-site installations may require

- *Special facilities to house the product* (plus its auxiliary equipment), including means of controlling the environment.
- *Special tools and instruments.* All too often these are not as completely engineered as the corresponding facilities in the factory are because many marketing and service departments have lagged behind the factories in the use of formal quality planning and quality specialists.
- *Instructions for installing and checking out the product.* Written instructions for complex products are forever subject to omissions and mistakes because of their sheer complexity. One failure in an early space shot was traced to the omission in checkout procedures of a certain adjustment on the missile. The adjustment was not specified and hence was not made. The launch was a complete failure. Such problems are not, however, restricted to complex products.

Some issues that must be addressed are

- The setup must be simple—capable of being accomplished by an amateur with a minimum of tools.
- The instructions must be error-proof—photographs are helpful.
- Each step of the instructions should contain only one task.
- Clarity of the instructions should be verified by having a typical user follow the instructions without any assistance.
- Telephone hot lines for customer assistance should be provided.

Use

Some field problems can be traced to improper operation of the product. As products become increasingly complex, the problem of human error during operation of the equipment becomes more critical. Manufacturers have done much to prepare operating manuals or other instructions for proper use and maintenance. These manuals are rudimentary for simple products and grow into elaborate handbooks for complex products. An intermediate example is the "owner's manual" for owners of automobiles.

While installation and use by specialists tend to be rather professional, use by consumers is characterized by much ignorance in several ways:

- *Failure to allow sufficient time for training and learning the operation of new and complex products.* For example, the amazing capabilities of word processing software can be realized only when personnel are thoroughly trained and given sufficient time to practice. The capabilities of the equipment have tended to overshadow the importance of learning time and practice.

- *Failure to use available information.* For example, a vacuum cleaner rotary brush encounters an obstruction and stops rotating. The owner's manual states clearly what to do: Remove the obstruction and reset the little red button. The user does not know this because the owner's manual is lost or it is simpler to have the unit serviced.
- *Use under conditions that were never contemplated.* For example, an automobile door lock freezes on a subzero day. The car's owner uses a portable hair dryer in an attempt to thaw out the lock. The dryer fails because it was not designed to operate in subzero temperatures.
- *Application of stresses that were never contemplated.* For example, a home owner stands on a washing machine to paint the ceiling overhead.
- *Failure to maintain.* Consumers are notoriously lax in following prescribed schedules for lubrication, cleaning, replacement of expendables, etc.

Of paramount importance is the need for the manufacturer to find out the actual use that takes place. As this knowledge becomes available—through field observations, complaint analysis, etc.—the manufacturer has a wide variety of options for improving use: consumer education, product redesign, systems redesign, etc. In many cases the prevention of human errors during use requires changes in the product design.

Interaction of the worker and the product is important. Factors involved can be the anatomical dimensions of the worker; placement on products of push buttons and switches; physical loads imposed on the worker by the product; monotony or other psychological effects of product operation; environmental effects produced by the product, such as noise and vibration; and workplace design, such as lighting and space.

Collectively, such matters are called "ergonomics" and need to be considered during the product design.

Maintenance

One of the elements of fitness for use is availability. Availability is a function of reliability and maintainability (see Section 13.6, "Availability"). An adequate field maintenance program requires fixing product problems correctly and in a timely manner. For some consumer products complaints from customers on repair service are more numerous than complaints on the original product quality. These complaints include both faulty repairs and long waits for the return of the product, resulting in lower availability of the product to the customer.

For simple products a product malfunction can often be corrected by replacing the defective element. As products become more complex, the process of finding and correcting malfunctions becomes more difficult. The usual steps are shown in Figure 20.1. These steps often require more resources than originally anticipated. Trained personnel, technical data, and test equipment must be planned for far in advance to be

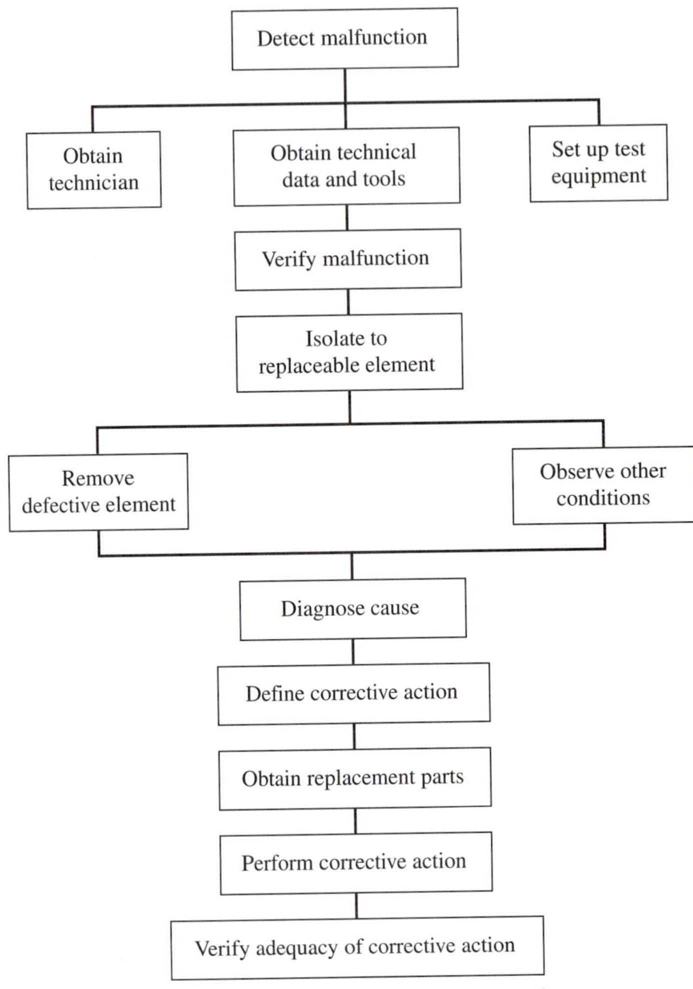

FIGURE 20.1
Steps in correcting a malfunction.

completely ready when a product enters the field. In addition, spare parts must be planned for in sufficient quantity and quality. A failure mode, effect, and criticality analysis or a fault tree analysis (see Sections 13.5 and 13.7) can be helpful in planning both preventive and corrective maintenance programs.

Quality control of repair work is troublesome for a service center. Most shops are small, and the tradition has been self-inspection by the mechanic, with only sporadic review by the service manager. Some organizations establish standards of performance for service centers and provide audit and certification to assure that these standards are met. The Chevrolet division of General Motors identifies 22 areas that are essential to total service satisfaction (e.g., quality, facilities, personnel, etc.). For each of these 22 areas, specific "evaluators" help to determine whether a dealer conforms to the standards. The quality parameter has 15 evaluators, e.g., "inspection procedures are available." Chevrolet provides technical assistance to enable dealers to achieve certification. Certification involves both a review of the dealer's procedures and facilities and formal feedback from customers on the performance of the dealer. The program also provides for periodic recertification of dealers.

20.5 OBTAINING FEEDBACK ON FIELD PERFORMANCE

Collecting, analyzing, and responding to customer complaints on a product are essential to minimize customer *dissatisfaction.* But activities on product complaints are not sufficient to increase customer *satisfaction.* Satisfaction involves a much broader scope of factors that relate to the need for an in-depth understanding of customer needs (see Chapter 12, "Understanding Customer Needs").

Customer needs can be more effectively addressed when field data are available. Such data have spawned the concept of customer-based (rather than company-based) measurement—obviously logical but *not* a reality for many companies.

An example of a customer-based measurement system comes from a manufacturer (Xerox) of reproduction equipment (Ekings, 1986). A monthly customer survey covers four categories: equipment, service, sales, and administrative support. The survey includes 25 questions covering equipment, service support, administrative support, and sales support (see Table 20.2).

TABLE 20.2
Customer satisfaction survey questions

Equipment	Service support	Administrative support	Sales support
Frequent equipment failure	Poor telephone service	Frequent billing errors	Sales do not meet needs
Frequent paper jams	Slow delivery service	Difficulty in correcting errors	Reps do not return calls
Copy quality not consistent	Failure to repair	Invoices difficult to understand	Reps are incompetent
	Too much downtime	Mishandling of multiple invoices	Reps are not interested
	No response in emergency	Problems are crediting delays	Difficulty in ordering supplies
	Service reps unprofessional	Problems with collections	Supply deliveries are late
	Poor key-op training	Mishandling of phone messages	Frequent supply failures
		Unprofessional administrators	

Each month, 40,000 survey forms are mailed, and the response rate is 45 percent. The results are used to take immediate action on customer problems, identify problems requiring generic corrective action, and provide a quantitative measurement of customer satisfaction. Similar information is obtained from users of competitor products, thus providing competitive benchmarking data.

Mail surveys are one component of an array of methods for collecting field information. See Section 2.6, "Standing in the Marketplace," for a discussion of some of these methods.

Most feedback of field performance emanates from company personnel. Historically, obtaining adequate feedback from personnel has been a continuing problem for many companies. Improving the quality and promptness of feedback requires a variety of actions to make the process as convenient and effective as possible. These include

- Providing personnel with well-designed data sheets.
- Providing incentives to encourage adequate feedback.
- Providing a glossary of terms to improve communication and a mnemonic code number to simplify the data entry and analysis.
- Providing training in the how and why.
- Conducting audits of the data feedback process.
- Using modern technology to collect the field information.
- Using modern methods of analysis to provide managers with valid summaries for decision making.
- Minimizing the number of data relay stations.
- Using the sample concept.
- Obtaining an operations log.
- Buying the data.
- Using the concept of controlled usage.

20.6 ANALYZING FIELD DATA

For Problem Identification

The analysis of field data can have several objectives, including problem identification, measurement of actual performance, and prediction of future performance. Among the techniques for problem identification are

1. *Defect matrices.* An example of this form is a table in which the horizontal lines list the principal defect types and the vertical columns list the product types. The cognizant managers and specialists are able to recognize significance in the patterns that emerge, e.g., defects that are restricted to certain product types or that affect all product types.
2. *Listing in order of importance.* This technique involves sorting and listing the basic data so that the resulting lists show the items in their order of importance. Typically, the listing appears in such forms as

- Failure rate by defect type.
- Unit repair cost by product type.
- Total repair cost by product type.
- Complaint rate by customer.

Often these listings include columns for cumulative totals of all the elements under study. A principal purpose of these listings is to permit focus on the vital few.

3. *Cost analyses.* "External failures" is a standard category of quality costs (see Chapter 2). Most data systems evaluate these costs by using conventional subcategories such as repair labor, parts and supplies, and travel costs. The quality department uses cost analyses and quality cost figures to justify quality improvement programs.
4. *Spare-parts usage.* Two methods of analysis are available:
   - The record of actual usage of products as derived from service reports. These records reflect the real consumption of these parts, but only if they are accurate and complete.
   - The sale of spare parts to the distribution chain. This information is also potentially unreliable because of the use of spare parts made by competitors and the fact that a sale to the distribution chain is not use—it is a transfer of inventory from one part of the chain to another.

Computer software can, of course, be extremely helpful in these data analyses.

For Problem Analysis

The basic analytical techniques mentioned in earlier chapters also apply to field data. Two such techniques are the Pareto analysis and a simple plot of frequency of failures over time. For example, Figure 20.2 shows a Pareto diagram for circuit pack failures reported at a telephone company field office. Circuit pack codes on which reliability improvements should be focused can be identified readily from this diagram. Pareto analysis also is performed on other parameters such as diagnostic failures and component failures, to further define areas with the greatest potential for improvement.

An example of failure-versus-time plot is given in Figure 20.3. This figure shows the cumulative infant mortality percentage for a particular circuit pack code compared to that code's infant mortality theoretical cumulative failure rate. The theoretical rate is based on worst-case component reliability modeling. Comparisons of performance among different offices are also conducted.

Because of the cyclical nature of many service industries, trend analysis is performed to isolate "peak times," i.e., error-prone periods, in poor service performance. When measurements are recorded in sufficient detail, trend analysis can also reveal peak times of poor service on a daily or hourly basis.

- For example, First Tennessee Bank studied tellers and waiting lines and found regular, predictable patterns of bank traffic. These were accommodated by rearranging staffing patterns. Some part-time assignments were added, some full-time assignments were discontinued, and cross training was given to provide more flexibility for handling peaks. Without any layoffs, more than $1 million a year was saved, and the maximum waiting time was reduced by 80 percent.

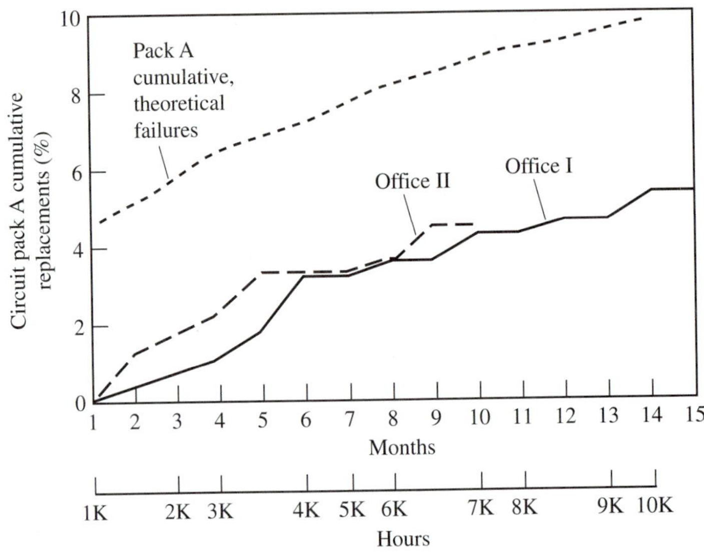

FIGURE 20.2
Circuit pack failure vs. time. (*From Fitch et al., 1990. Reprinted with permission by the ASQ.*)

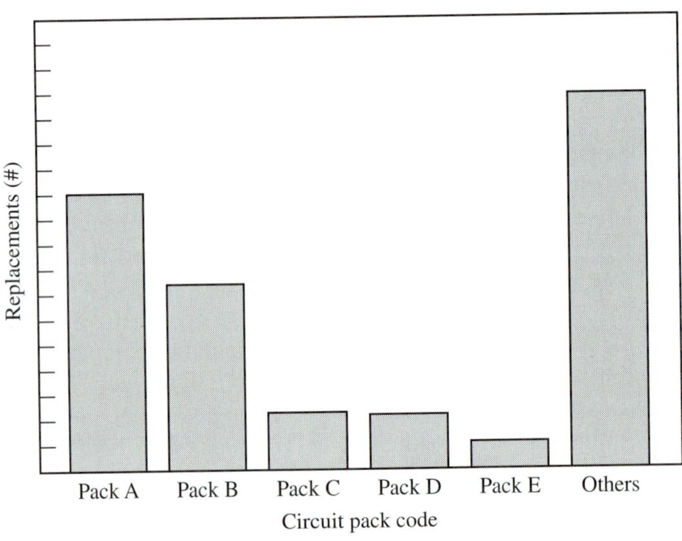

FIGURE 20.3
Circuit pack failures (Pareto diagram example). (*From Fitch et al., 1990. Reprinted with permission by the ASQ.*)

- In another example, data on aircraft engine maintenance were analyzed. One important technique was plotting the probability of failure versus operating age. It had been assumed that the probability of failure would increase with operating age. This simple plot of probability of failure versus time revealed a surprise—for 89 percent of the items, the probability did *not* increase with age. This study eventually led to changing the traditional maintenance approach (replacing an item after it accumulated X hours of operation) to a different concept of maintenance.

For Measurement and Prediction

In the measurement and prediction of field performance, several techniques are useful. One is the cumulative complaint analysis, which requires the product to be dated to show when it was made, sold, or installed.

The various dates are useful not only in disposing of complaints and claims on product of a slow-perishable character (candy, photographic film, etc.), but also in predicting the failure rate of various product designs.

EXAMPLE 20.1. Certain articles of women's clothing were tearing in service, and some were being returned. When the products made during one specific month were code dated, the cumulative returns reached 2 percent within two years after the date of manufacture. The resulting cumulative curve is shown in Figure 20.4. This 2 percent was regarded as tolerable. (At the unit price level of the product, it was likely that about 20 percent of the product was actually tearing in service.)

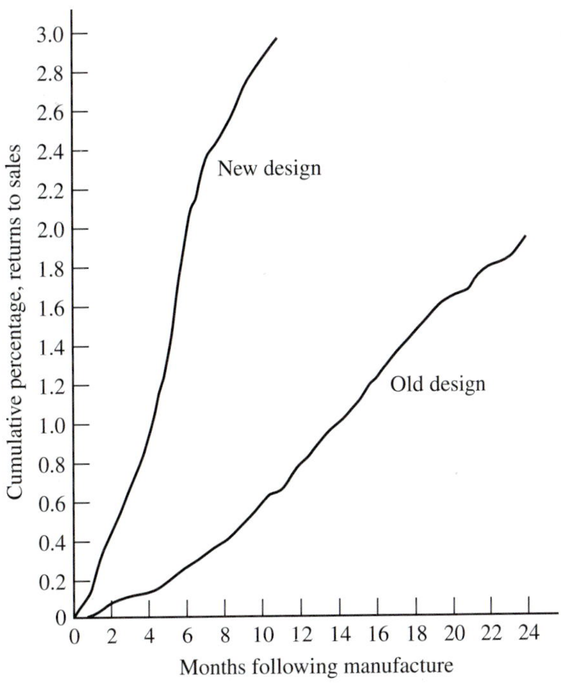

FIGURE 20.4
Comparison of cumulative returns based on number of months since month of manufacture.

Meanwhile, the research department evolved a new product that went into full-scale production 14 months after the "specific month" noted above. The cumulative returns for the new product are also charted on Figure 20.4. Note that, by use of a horizontal time scale based on "months following manufacture," both curves start at the origin and hence can readily be compared with each other. By such analysis the company could be informed *within the first few months* of the life of the new design that with regard to the problem of tearing, the new design had made the situation worse rather than better.

The concept of using the cumulative graphing approach to analyze data has broad application.

Use of Probability Paper for Predicting Complaint Level

Weibull probability paper (see Section 10.12) can be used to analyze early field data on warranty claims. Through such analysis, it is possible to predict what the cumulative number of claims will be at the end of the warranty period.

Consider the repair information on an electrical subassembly listed in Table 20.3. The cumulative repairs per 100 units are interpreted as a cumulative failure rate in percentage. These data are summarized from a large number of warranty reports, and thus the data can be plotted directly without using the mean rank approach discussed in Chapter 10.

The Weibull plot is shown in Figure 20.5. The eight months of data have been plotted, and a line drawn through the points. This line has been extended, and it predicts the repair rate at the end of the 12-month warranty period to be 2.6 repairs per 100 units. Extrapolation of the line beyond the plotted points is valid only on the assumption that the failure pattern does not change.

Note that this Weibull paper includes a scale for estimating the Weibull slope or shape parameter (see Section 10.12). The slope can be found by drawing a line parallel to the line of best fit and through the point circled on the vertical scale. Here the slope is read as 0.7. This defines the shape of the failure distribution and can aid in problem definition. Table J in Appendix II shows a blank sheet of this same paper.

TABLE 20.3
Repair information on electrical subassembly

Time in service, months	Repairs per 100 units (*R*/100)	Cumulative *R*/100
1	0.49	0.49
2	0.32	0.81
3	0.24	1.05
4	0.24	1.29
5	0.21	1.50
6	0.19	1.69
7	0.19	1.88
8	0.23	2.11

Source: Ford (1972).

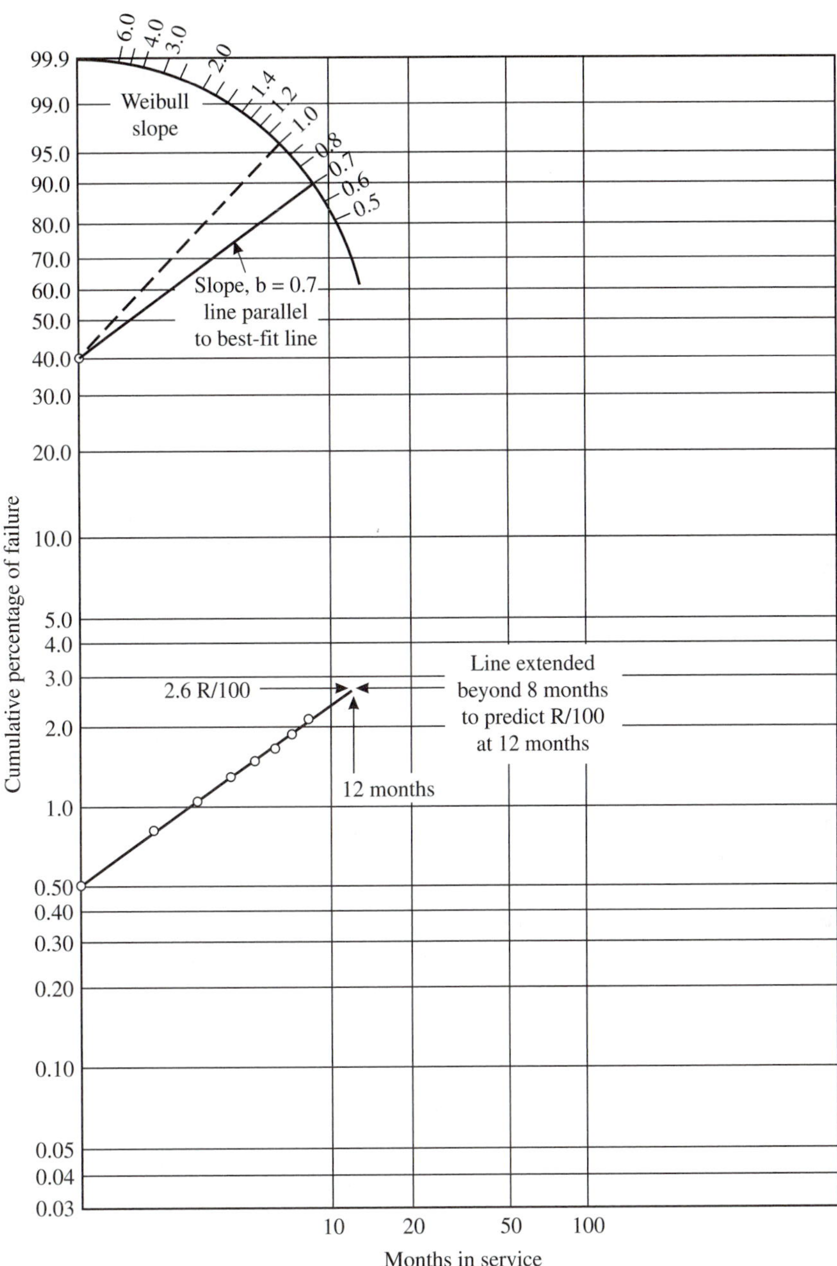

FIGURE 20.5
Warranty data plot. (*From Ford, 1972.*)

For slopes <1, the distribution is exponential in shape and generally means that the failures are early failures due to manufacturing or assembly deficiencies. When the slope is 1, the Weibull reduces to the exponential distribution (see Section 10.11).

For slopes >1, the distribution is skewed and generally means that the failures are due to wear-out or fatigue, especially for higher slopes. When the slope is about 3.5 (and $\alpha = 1$ and $\gamma = 0$), the Weibull approximates the normal distribution.

Extrapolation beyond the limits of actual data is always questionable. However, when extrapolated performance has been confirmed by actual performance on similar product lines, the approach may be justified. It should be stressed that decisions on the vital-few warranty problems deserving major attention *must* and *will* be made by someone. Although the approach suggested here is not rigorous, it is a major step beyond intuitive decision making.

20.7 SAFETY AND PRODUCT LIABILITY

The responsibility of manufacturers to provide safe products is clear. "Product liability" refers to the legal obligation of a manufacturer or seller to compensate for injury or damage caused by a product.

The growth in both the number of lawsuits and the size of awards to victims has been dramatic. Awards in excess of a million dollars are no longer rare. The reasons are clear: More manufactured products are in the hands of amateurs, creating more opportunities for injury; some traditional legal defenses of manufacturers have become eroded, making it easier for a victim to sue a manufacturer successfully; the consumer movement has encouraged consumer action generally; the publicity given to large awards has stimulated more lawsuits.

The basic defense against liability is prevention—by eliminating the causes of the injuries or damage. All company functions contribute to this prevention activity, as shown in Table 20.4.

Most product liability lawsuits—civil lawsuits—are aimed at manufacturing companies because of their ability to pay, not at individuals such as design specialists.

But in a criminal liability case, the state is the plaintiff. In recent decades public prosecutors have pursued individuals such as the head of a company or a manager of product development or quality. Criminal liability requires proof that the manager knowingly performed illegal actions or was grossly negligent in executing duties. As it is difficult to establish such proof, the likelihood that a manager will be found guilty of criminal liability is small.

Responsibility for investigating injuries varies but is typically divided among the insurance company, the legal department, and the appropriate technical departments. Irrespective of the allocation of responsibility, certain precautions should be taken:

- Notifying the insurance company promptly.
- Assigning only qualified experts to make the investigation.
- Securing and retaining the product unit asserted to have caused the injury.

TABLE 20.4
Activities for product safety

Upper management	Product design	Manufacturing inspection	Marketing	Advertising	Customer service
Creating a policy on product safety	Making product safety a formal design parameter	Emphasizing error-proofing	Providing product labels with warnings, dangers, antidotes	Having advertising reviewed for technical and legal aspects	Providing service that minimizes likelihood of claims
Creating a companywide safety effort	Adopting fail-safe design practices	Providing documentation and traceability	Supplying safety information to distributors and dealers	Emphasizing safety through education	Providing feedback on how product is actually used
Directing that procedures be created for traceability and product recalls	Using design evaluation techniques for safety	Identifying safety-critical characteristics, operations, and tests	Training sales force in safety aspects of contracts		Ensuring that product repairs leave the product in safe condition
Establishing a safety scoreboard	Organizing formal design review for safety	Providing training on safety-related items	Training users in safety matters		
	Conforming to all safety standards	Assuring proper disposition of nonconforming products			
	Analyzing and providing feedback on injury data to designers				

- Analyzing the injury environment and product thoroughly to identify the precise failure mode.
- Conducting laboratory tests with a view of reproducing the same failure mode. (If needed, verification from independent laboratories should be secured.)
- Using modern analysis techniques: high-speed cameras, scanning electron microscope, etc.

See *JQH5,* pages 35.125–35.18 for further discussion.

20.8 CUSTOMER SERVICE

As organizations strive to differentiate their products from competition, customer service has emerged as a key success factor. Fuchs in *JQH5* provides a list of typical customer service elements (Table 20.5).

Note that customer service includes transactions and relationships with customers that occur before and after the sale. Also note that the activities include both the traditional, such as sales ordering, and the innovative, such as value-added information access, e.g., that provided with the purchase of a computer software package.

An exemplary organization for customer service is the American Society for Quality (ASQ). Rarely is the caller transferred to another party at ASQ. The people answering the phone have been selected and trained to handle a wide variety of customer requests. ASQ practices what it preaches.

The emphasis on building relationships with customers has led to a changing role for customer service representatives (CSRs). Traditionally, CSRs had one primary responsibility: answer customer questions and resolve customer problems. This task involves careful listening about a problem, but this listening and discussion can create a relationship with a customer. Relationships can go beyond problem solving and lead to increased sales of products. Financial service organizations find that immediately after a problem has been fixed, customers can be candidates to be cross-sold a service—about 10 to 20 percent of the customers will buy another service (*JQH5,* page 33.20). Brennan

TABLE 20.5
Traditional and innovative elements of customer service

Traditional presale elements	Innovative presale elements	Traditional postsale elements	Innovative postsale elements
Sales information Ordering	Market research for key customer satisfiers	Packaging Transport Delivery Installation Maintenance Complaint handling Problem solving	Report cards Customer care Value-added information access

(1997) proposes four phases for transforming a service call to a sales order: (1) establishing a relationship with the customer; (2) building relationships through questions, information gathering, listening, and getting the customer to agree on the need for change; (3) presenting the company's product as a solution to the customer's needs; and (4) gaining a customer commitment to the next step in the sales process such as a sales order or a review of additional information. Brennan also notes that some CSRs dislike selling because they view it as being aggressive. Thus when the CSR responsibility is to include selling, careful CSR selection and training is important.

20.9 SIGNIFICANCE OF FIELD COMPLAINTS

Field complaints are a poor measure of product performance. Some users complain despite the fact that a product is fit for use; others do not complain despite the fact that a product is not fit for use. Based on research studies, Goodman (2000) estimates that of people who encounter a problem, 50 percent do not complain; for business customers, 25 percent do not complain. When complaints are made, how well they are acted upon has a decided effect on repeat sales (see Section 20.11 below and also 4.7).

Whether or not a complaint is made depends on several factors:

- *Economic climate.* The number of complaints falls in a sellers' market and rises in a buyers' market, even for the same product.
- *Age, affluence, technological skills, etc., of users.* Individuals react differently to defects—the same product generates complaints from some customers but not from others.
- *Seriousness of the defect as seen by the user.* This factor is often influenced by the temperament of the user. For example, to a child a toy missing from a cereal box is a serious problem. In one company, complaint rates on cereals are about one per million packages compared to four per million packages on missing premiums.
- *Unit price of the product.* When the unit price of a product is low, the complaint rate can greatly understate field difficulties; the complaint rate may need to be inflated from 20 to more than 100 times to arrive at the actual defect rate.

Thus a low complaint rate is not proof of customer satisfaction. However, a high complaint rate is proof of dissatisfaction, and therefore the rate should be measured and watched closely.

20.10 PROCESSING AND RESOLUTION OF CUSTOMER COMPLAINTS

The process of fixing a customer problem is called "recovery." Recovery should include the following elements (*JQH5,* page 33.20):

- Making an apology.
- Replacing the product quickly.
- Conveying a sense of understanding and empathy to the customer.

- Providing a substitute or compensation for any functional value lost by the customer.
- Following up to be sure that the problem has been solved to the satisfaction of the customer.

Beyond the recovery for individual complaints, steps must be taken for the collection of complaints. This process includes:

- Preventing a recurrence of isolated complaints. A common practice is to bring isolated complaints to the attention of those who are suspected of having caused them and to ask how those individuals plan to prevent a recurrence.
- Identifying the vital few serious complaints that demand in-depth studies to discover the basic causes and to remedy them. Usually, the decisive issues here are (1) whether the complaint is isolated or widespread and (2) whether the complaint is on critical matters (e.g., safety, government regulations).
- Performing in-depth analysis to discover the basic causes of the complaint. This action is oriented to the product and is normally needed only in the vital few cases that are responsible for the bulk of the failures.
- Performing further analysis to discover and apply remedies for the basic causes.

An approach to handling the vital few problems is shown in Figure 20.6. Because the causes and remedies can involve many departments, a "corrective action group" is often formed with representatives from design, manufacturing, purchasing, quality, and field service. The group jointly analyzes the data to select the vital few problems, conducts an initial investigation to prepare a thorough problem statement, and assigns responsibility for determining a remedy.

The group meets regularly to review all new complaints and to review progress on current problems. A problem agenda is sent out by the chairperson several days ahead of the meeting date. Minutes of the meetings, including a problem-status log, are formally recorded, and actions to be taken are documented and distributed to all concerned. This log summarizes each problem before the committee. It also shows the scheduled start and completion of activity, assigns responsibility, and lists action taken. It gives project management an indication of major problems and the status of corrective efforts. Additional effort may then be placed on troublesome areas as deemed necessary.

An example of a complaint/inquiry management system in the financial services industry is shown in Figure 20.7. Note the similarity to Figure 20.6 (manufacturing industry) in terms of data collection, problem description, problem selection and priority, response, and follow-up.

The American Productivity and Quality Center (1999) conducted a benchmarking study on complaint management. The nine organizations that participated in the study received an average of 500,000 telephone calls each month—33 percent were complaints and 42 percent were inquiries.

Five of the nine organizations are considered best-practice companies. The success factors for the top organizations are

- Complaint management visions are linked to the corporate mission.
- Customer complaints are viewed as opportunities for improvement, and companies proactively solicit complaints from customers and employees.

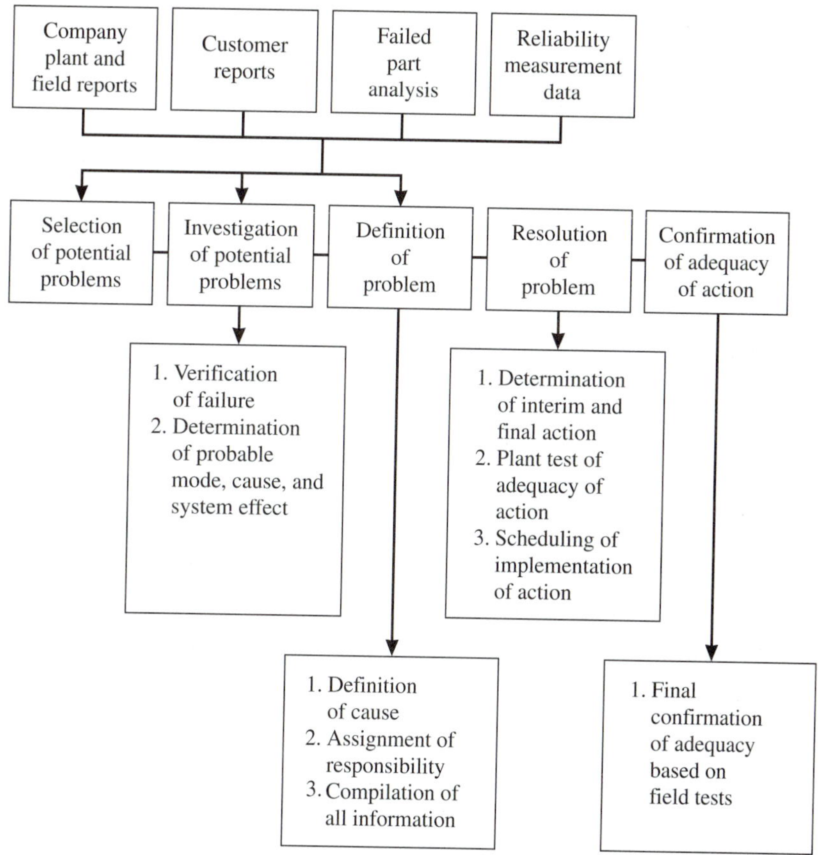

FIGURE 20.6
Failure reporting, analysis, and corrective action system (FRACAS).

- Complaints and other customer feedback are used to make process improvements, increase customer satisfaction and loyalty, and increase profits.
- Complaint information is shared across the organization so that other employees and business units can benefit from the information.
- Four of the five best-practice organizations believe that their centralized complaint process is a key factor for their success.
- Best-practice organizations use a combination of individual and team-based compensation, rewards, and recognition in customer contact and complaint activities.
- Complaint management measures are included in overall customer satisfaction measures.
- Companies measure the reasons for complaints and perform root-cause analysis for each complaint.
- Time standards are set for resolving complaints—60 percent of the companies take longer than three days to resolve, 20 percent take one or two days, and 20 percent resolve complaints during the initial contact.

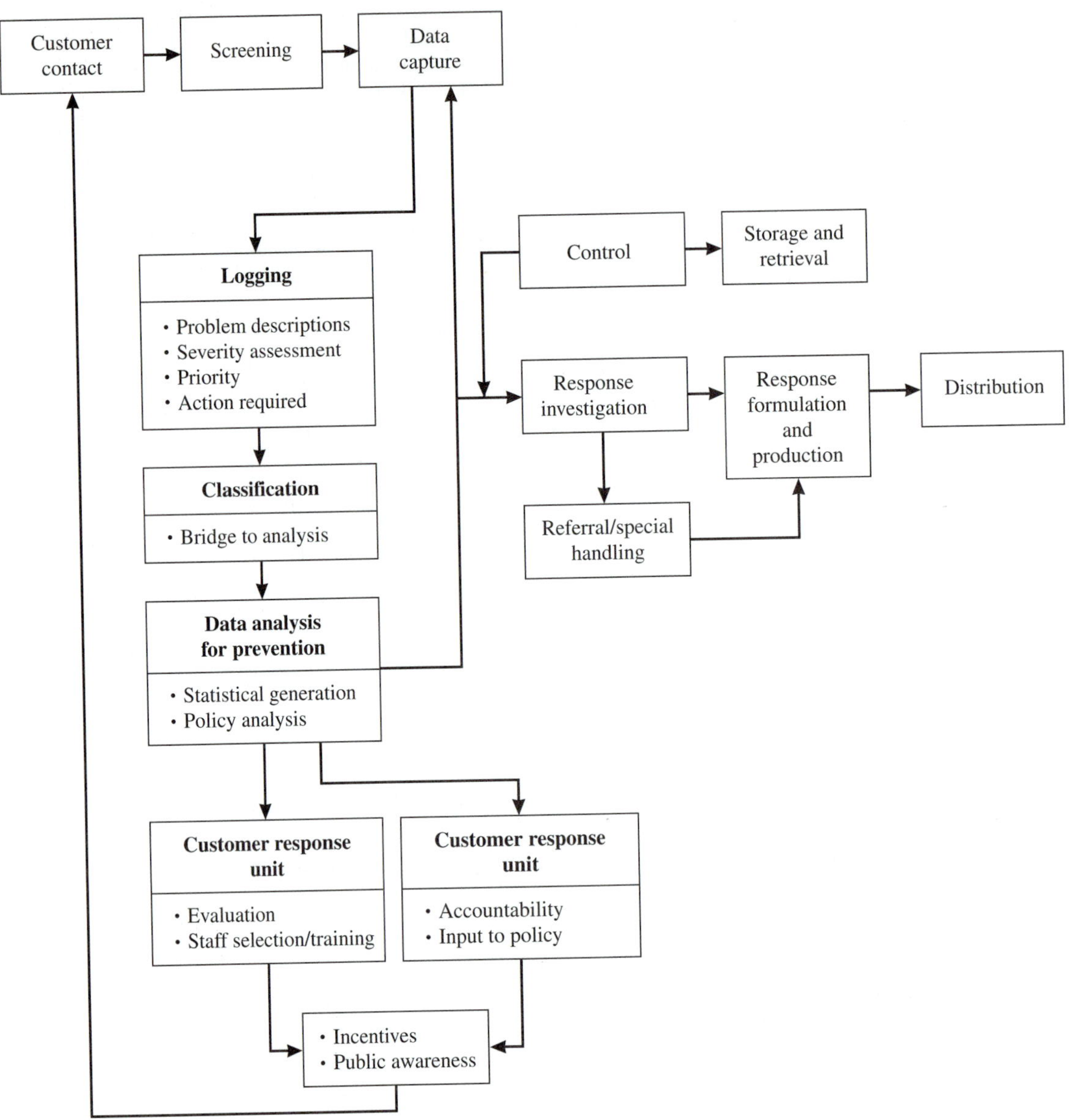

FIGURE 20.7
Customer-response framework: complaint/inquiry management. (*Source: Banc One, Columbus, OH.*)

- Follow-up mechanisms such as phone surveys are used to ensure that problems are resolved to the customer's satisfaction.
- The overall cost of handling complaints is measured in 60 percent of the companies.

These practices can serve as a guide to improving the complaint-handling process.

The speed of response to a complaint clearly has a significant impact on customer satisfaction. The ideal is to solve the problem on the first call and preferably on the same day. Research shows that each time it is necessary to recontact the customer or transfer the inquiry, customer satisfaction drops by 8 to 10 percent (*JQH5,* page 33.19). Mitchell (2000) describes how control charts and Pareto diagrams are used to identify special causes of variation related to engineering complaints in a telecommunication company.

The process of handling customer complaints has emerged as an important tool of competition in achieving sales. Data are now available showing how the level of customer satisfaction in handling complaints is related to lost sales (see Section 4.7, "Level of Satisfaction to Retain Present Customers").

The concepts of quality improvement can help to improve the process of handling complaints. In an application at the Mobil Chemical Company Films Division, the average response time on complaints was reduced from 92 to 30 days (Gust, 1985). The steps in the breakthrough sequence (see Chapter 3 generally), starting with proof of the need and ending with holding the gains, were the road to success.

20.11 ESTIMATING LOST PROFIT DUE TO PRODUCT PROBLEMS

A model has been developed that estimates the profit lost annually because of problems that customers experience and the way those problems are handled by the company. An example (Goodman, 2000) is used with permission by the ASQ. The model uses company data and customer survey data (sample values are shown in parentheses).

1. Company-provided data:
   - Average period of loyalty of the customer (five years).
   - Average number of purchases made by the customer over the period of loyalty (10).
   - Size of the customer base (500,000).
   - Average profit per purchase ($20).
2. Data obtained from a survey of customers:
   - Percentage of customers who experience problems (70).
   - Percentage of customers who experience problems and request assistance (50).
   - Percentage of customers who don't experience problems and will repurchase the product/service (88).
   - Percentage of customers who experience problems, do not request assistance, and will repurchase (55).
   - Percentage of customers who are satisfied, mollified, and dissatisfied by the contact-handling system (40, 35, 25).

- Percentages of customers who are satisfied, mollified, and dissatisfied by the contact-handling system and who will repurchase (95, 75, 35).
- Word of mouth generated by customers who experience problems and don't request assistance (two).
- Word of mouth generated by customers who are satisfied, mollified, and dissatisfied by the contact-handling system (one, three, six).

Sales lost over the period of customer loyalty that are attributable to customers' problem experiences are estimated as sales lost due to problems less sales lost if no problems had been experienced.

The sales lost due to customers' problems have two components:

1. *Sales lost due to the way the company handles customer problems.* We start by calculating the number of sales lost over the period of loyalty as a result of *satisfied* contacters (customers) who will not recommend the company. The calculation is

$$(500{,}000 \times 0.70)(0.50)(0.40)(0.05)(10) = 35{,}000 \text{ sales.}$$

   Similar calculations are made for customers who have problems but where the company's action has resulted in a mollified or dissatisfied customer rather than the "satisfied" customer assumed in the above calculation. Table 20.6 shows a summary.

   But these calculations reflect only the 50 percent of customers who do contact the company about the problem; the other 50 percent who have a problem do not contact the company. The number of sales lost from this latter 50 percent are calculated as 350,000(0.50)(0.45)(10) = 787,500 sales (see Table 20.6).
2. *Sales lost due to decreased brand loyalty and negative word of mouth.* We start by calculating the number of sales lost over the period of loyalty as a result of negative word of mouth from satisfied customers (with problems): 35,000(1.0)(0.02) = 700 sales.

   Similar calculations are made for the mollified and dissatisfied customers. Table 20.6 shows a summary.

TABLE 20.6
Sales lost due to problems customers experience

Response of complainant	Sales lost due to company's handling of problems	Sales lost due to negative word of mouth	
Satisfied	35,000	700	
Mollified	153,130	9,190	
Dissatisfied	306,250	36,750	
Total	494,380	46,640	541,020
Sales lost if customer does not complain	787,500	31,500	819,000
Sales lost due to problems customers experience			1,360,020

But these calculations reflect only the 50 percent of customers who complain. For the 50 percent who do not complain, the number of sales lost is calculated as 350,000(0.50)(0.45)(2)(10)(0.02) = 31,500 (see Table 20.6).

Thus 1,360,020 is the estimated number of sales lost due to problems that customers experience. But some sales will be lost even if no problems are experienced. In this case assume that 12 percent of sales will fall into this category. The sales lost are 350,000(10)(0.12) = 420,000 sales.

Thus the total number of sales lost over the loyalty period that are attributable to customers' problem experience is 1,360,020 − 420,000 = 940,020.

The *annual* number of sales lost is 940,020/5, or 188,004. If profit is $20 per unit, the annual profit lost due to customer problems is estimated as $3,760,080.

Of course, the credibility of this estimate depends on assumptions that directly relate to consumer behavior. This issue can be addressed by running a sensitivity analysis, i.e., changing the numbers in the assumptions and repeating the calculations to estimate the lost profit. See Goodman (2000) for the results of such a sensitivity analysis.

The approach described above is called the "projection method"; i.e., financial data is combined with market research data to project the financial impact of addressing customer needs. Brandt (2000) contrasts this method with another method that links customer satisfaction indicators with actual market or financial performance (the "direct linkage" method). He also provides guidelines for choosing a method for a given organization, market, or industry. The key point is this: The profit lost due to product problems and the actions to satisfy the complaints can be quantified. Such quantification can be instrumental in gaining the support of financial and other executives to lead the actions for customer loyalty and retention—our next subject.

20.12 CUSTOMER LOYALTY AND RETENTION

Customer satisfaction and customer loyalty are distinct, although related, concepts. In brief, a satisfied customer will buy from our company but also from our competitors; a loyal customer will buy primarily (or exclusively) from our company (see Section 4.4). Also, customer satisfaction concerns what customers *say*, their opinions about a product; customer loyalty relates to what customers *do*, their buying decisions.

Loyal customers not only provide continuing sales revenue but also contribute other benefits:

- Adding new sales by referring other potential customers.
- Paying (often) a price premium.
- Buying other products from the company.
- Cooperating in the development of new products.
- Reducing company internal costs such as selling costs.

Thus the benefits of having a high percentage of loyal (not just satisfied) customers warrants specific action steps to achieve high customer loyalty and minimize defections. Some of these steps are discussed in earlier chapters of this book; some of them are presented below for the first time. To provide an overall perspective, the nine actions below are collectively a road map for achieving high customer loyalty.

Continually Assess Customer Needs and Translate These Needs into Product Improvements

Product development must be based on a thorough understanding of customer needs. But those needs are continually changing. Thus marketing research to define customer needs must be ongoing (see Chapter 12). The results of the research are then translated into new or modified products (see Chapter 4 and Chapter 13).

Periodically Assess Market Standing Relative to Competition

This process typically means conducting multiattribute market research studies on quality (see Section 2.6). These studies not only provide status in the marketplace but also identify differences in satisfaction that are likely to result in customer defections. This research should incorporate one or more questions on the likelihood that the customer will repurchase or recommend the product.

Track Retention and Loyalty Information

As a simple customer retention measure, an insurance company measures the percentage of customers who do not allow a policy to lapse due to nonpayment of the annual premium. But as a loyalty measure the company estimates what percentage of insurance products being purchased by its customers are purchased from the company versus competitor insurance companies, i.e., the share of spending a firm earns from its customers.

Fidelity Investments (Nash, 1998) has determined that highly loyal customers have almost twice the assets and share of wallet with Fidelity than low loyal customers do (see Figure 20.8). Reichheld (1996, Chapter 8) discusses the share-of-wallet concept and other measures of customer loyalty.

Sun Microsystems (Lynch, 1998) calculates a customer loyalty index based on four components of loyalty questions: customer satisfaction, likelihood to repurchase, likelihood to recommend, and customer delight.

It is useful to set goals on customer retention and customer loyalty. For example, Lexus set an owner retention goal of 75 percent (Waltz, 1996).

Determine the Drivers of Customer Loyalty

These drivers are the specific elements that have a significant impact on customer loyalty.

Fidelity Investments (Nash, 1998) analyzed data obtained from a 12-page questionnaire mailed to a random sample of customers. Loyalty data were matched against marketing data to determine actual behavior. Regression analysis was used to deter-

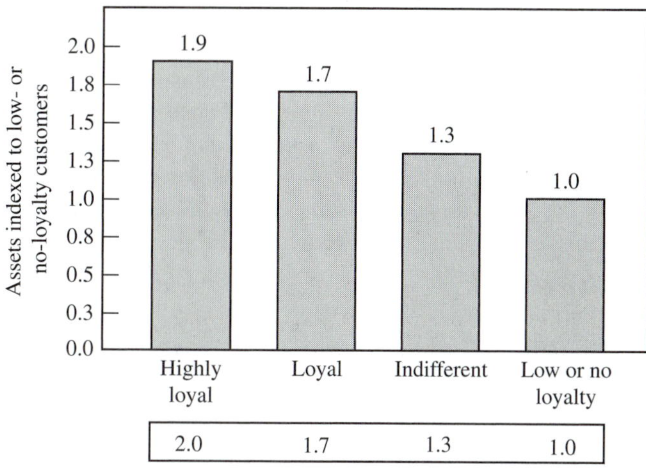

FIGURE 20.8
Share of wallet indexed to low loyalty

mine 10 drivers of loyalty. For example, the top driver (satisfaction with customer service) had a 17 percent impact on loyalty; a driver called satisfaction with investment performance had a 14 percent impact on loyalty.

Sun Microsystems (Lynch, 1998) identified 20 loyalty drivers and classified them based on relative importance and Sun performance relative to competition.

Determine the Impact on Profit of Reducing Customer Defections

Customer problems must be converted to cost and revenue implications. The sales revenue from a loyal customer measured over the period of potential repeat purchases can be dramatic. Research concluded that a five percentage point decrease in defection rate can increase profit by 35 to 95 percent, depending on the service industry involved (see Section 4.5).

Understand the Impact of Handling Complaints on the Likelihood of Repurchasing

Customer satisfaction with the handling of their complaints has a significant impact on repurchasing and their intention to recommend purchase to others. The numbers can be dramatic (see Section 4.7) and should be determined and made known throughout the organization.

Analyze Complaints

Complaints are an early indicator of potential customer defections. The frequency and nature of complaints must be analyzed—see Section 20.9.

Determine the Reasons for Customer Defections

This step means conducting research to discover the reasons for defections by asking customers why they left. But the reasons stated by customers are often not the real reasons; e.g., price is often mentioned as a key reason but probing usually reveals other reasons. Reichheld (1996, Chapter 7) discusses how a "failure analysis" (a *failure* is a customer defection) can help to discover the causes of defections. The analysis includes applying the classical improvement technique of five whys—i.e., asking why an event occurred at least five times to get to the *root* cause of a defection.

Present the Results of Loyalty Analyses for Action

Market research and loyalty analysis results must be acted upon if they are to contribute to customer retention and loyalty. To assure action, one bank employs a guide to help each branch manager interpret and act on the research results (for elaboration, see *JQH5,* page 18.19).

A related issue is designing customer surveys to collect the right information in sufficient detail so that line managers can take action on the results. This activity highlights the importance of obtaining input from line managers in designing customer surveys.

Graphing the research results can be an effective stimulus for action. Section 2.7 under "Field Studies" shows an example of mapping customer satisfaction and importance ratings to relate customer views and potential action. The four quadrants in the map show the priorities for action and the types of actions required. For additional examples of maps for customer retention modeling, see Lowenstein (1995, Chapter 9) in the Supplementary Reading.

Finally, it is useful to relate satisfaction research results to operational processes or to specific activities. Figure 17.3 provides an example from AT&T that dissects overall quality into business processes, customer needs, and internal metrics. This approach helps to ensure that improvement efforts have a strong customer emphasis.

The nine actions discussed above involve many units in an organization. Therefore, some mechanism is necessary to plan, coordinate, and ensure that the activities are assigned sufficient resources and executed effectively. A quality department could perform this role, perhaps by using the project management concept (see Section 8.12).

20.13
QUALITY MEASUREMENT IN MARKETING, FIELD PERFORMANCE, AND CUSTOMER SERVICE

The management of quality-related activities in marketing, field performance, and customer service must include provision for measurement.

TABLE 20.7
Examples of quality measurement in marketing, field performance, and customer service

Subject	Unit of measure
Marketing	Percentage of orders canceled Percentage of bids that become contracts Order entry errors Number of misprocessed orders
Complaints	Total number of complaints Number of complaints per \$1 million of sales Number of complaints per million units of product Value of material under complaint per \$100 of sales for such products
Returns	Values of material returned per \$100 of sales
Claims	Cost of claims paid Cost of claims per \$1 million of sales
Failures	Mean time between failures (MTBF) Mean usage between failures, e.g., cycles, miles Mean time between repair calls Failure per 1000 units under warranty
Shipments	Amount of product shipped that did not meet specifications
Maintainability	Mean time to repair (MTTR) Mean downtime Number of repeat service calls for same complaint
Service cost	Ratio of maintenance hours to operating hours Repair cost per unit under warranty Cost per service call

The reader is urged to review the ten basic principles of quality measurement in Section 5.2.

Table 20.7 shows control subjects for various subject areas of marketing, field performance, and customer service.

SUMMARY

- Customers form perceptions of quality during three phases—before purchase, at point of purchase, and after purchase.
- A warranty is a form of assurance that a product is fit for use. Consumer product warranties are either full or limited.
- Preuse phases of field performance include packaging, shipping, receiving, and storage; use phases consist of installation, checkout, operation, and maintenance. Field factors cause about 20 to 30 percent of problems concerning fitness for use of complex products.

- Product liability refers to the legal obligation of a manufacturer or seller to compensate for injury or damage caused by a product.
- The process of handling customer complaints has become an important tool of competition in achieving sales.
- Feedback on field performance must use a customer-based measurement system.
- A customer defection occurs when a customer switches to another brand. When customers defect, the profit-making potential over many years goes with them.
- Measurements for marketing, field performance, and customer service should be based on input from customers; provide for both evaluation and feedback; and include early, concurrent, and lagging indicators of performance.
- The incidence of field complaints often underestimates field problems because many users do not complain (they simply go elsewhere to make their next purchase).
- Profit lost due to product problems can be estimated.
- Simple forms of data analysis serve three purposes: problem identification, problem analysis, and measurement and prediction.
- Plotting data on Weibull probability paper can provide predictions on the level of complaints.

PROBLEMS

20.1. Study the methods used for advertising quality for any of the following categories of products or services: beverages, cigarettes, automobiles, household appliances, airline travel, movies, electronic components, e.g., resistors, and mechanical components, e.g., bearings. Report on the extent to which the advertising: (*a*) identifies specific qualities, (*b*) quantifies the extent to which the product possesses these qualities, (*c*) engages in exaggeration, and (*d*) appeals to various human traits, e.g., vanity.

20.2. Visit a transportation service involving loading, storage, crating, unloading, transport, etc. Some possible examples: a trucking terminal, a ship pier, an air freight terminal, a commercial warehouse, the shipping area of a factory, the parcel room of the post office, and a rail freight yard. Study the processes in use and their likely effect on the qualities of the products undergoing these processes. Report your findings and conclusions.

20.3. Visit a manager who is exposed to problems of installation and operation of consumers' hardware, e.g., the buyer in the home appliance department of a department store or the service manager of a factory making home appliances. Secure information on major troubles encountered by users in installation and operation and also on what steps are being taken to minimize these troubles. Report your findings and conclusions.

20.4. Study the operating and/or maintenance instructions for a product. Report your findings and conclusions.

20.5. Select one of the field activities discussed in this chapter. Apply the concept of self-control (see Section 5.3) to evaluate whether the activity has been properly planned with respect to the effect on product quality. For each of the three criteria of self-control, list specific questions that can be asked to evaluate the planning.

20.6. Select a consumer product and make an ergonomics study to evaluate the adequacy of the design with respect to the operation of the product by a human being.

20.7. Prepare a plan of market research to secure data on the opinion of owners concerning repairs made on a specific product by a dealer or manufacturer.

20.8. Visit a repair facility and observe typical repairs made. Trace the steps taken and make a comparison to Figure 20.1.

20.9. Propose one or more complaint indexes for one of the following products: (*a*) checking and savings account services at a bank, (*b*) household refrigerators, (*c*) passenger automobiles, (*d*) corn processing by-products sold primarily to beer brewers and to pharmaceutical companies (for making medicine capsules), (*e*) the totality of products sold by a large department store, (*f*) jet engines for passenger planes.

20.10. Select a product that you or a friend owns and that has been tested and reported on in a consumer magazine.
(*a*) Comment on the adequacy of the tests made to evaluate competing brands.
(*b*) Compare the opinions of the user to the evaluation published in the magazine.

20.11. Visit a local organization and report on the procedures used to process complaints and to summarize complaint information for executive action.

20.12. Warranty data on an ignition switch show that 0.15 percent have failed by 3000 miles, 0.25 percent by 6000 miles, and 0.40 percent by 12,000 miles. Predict the percentage of failure at 24,000 miles and 50,000 miles. State the assumptions necessary.

Answer: 0.7 percent, 1.2 percent.

20.13. Past data on a certain type of windshield wiper motor indicate a Weibull slope of about 0.5. A goal of no more than 0.5 percent failures by 24,000 miles has been set. What percentage of failure observed at 6000 miles would indicate that the goal would probably not be met? State the assumptions necessary.

Answer: 0.22 percent.

20.14. Plants A and B are part of the same company and manufacture the same product. The following data show the percent of returned product, by month, for each plant:

Month	A	B	Month	A	B
Jan.	0.4	0.2	July	0.5	0.2
Feb.	0.3	0.1	Aug.	0.2	0.1
Mar.	0.2	0.2	Sept.	0.3	0.2
Apr.	0.4	0.5	Oct.	0.3	0.2
May	0.3	0.3	Nov.	0.5	0.3
June	0.2	0.4	Dec.	0.4	0.1

Construct two plots using ordinary graph paper. One plot should use noncumulative data to compare the two plants. The other plot should use cumulative data. Comment on the two methods of plotting.

20.15. Data on returned products are summarized weekly for each of three products (A, B, and C). For each return, the primary reason for the return is noted: V for visual defects, E for poor electrical performance, M for poor mechanical performance. Data are available for three weeks. For the first week, 26 units of product A were returned with a distribution of the reason as 5 V, 8 E, and 13 M. Eight units of product B were returned with 0 V, 3 E, and 5 M. Product C showed a tally of 34 units with 22 V, 10 E, and 2 M. For the second week, product A had 37 returns with 6 V, 11 E, and 20 M. Product B had 15 returns with 2 V, 9 E, and 4 M. Product C had 24 returns with 11 V, 8 E, and 5 M. Assuming that the data are representative of weekly returns, prepare a summary that indicates priorities for an improvement effort.

REFERENCES

American Productivity and Quality Center (1999). *APQC White Paper Based on Findings from APQC's Complaint Management & Problem Resolution Consortium Benchmarking Study I,* Houston.

Anton, J. (1996). *Customer Relationship Management,* Prentice Hall, Upper Saddle River, NJ.

Brandt, D. R. (2000). "Linking Measures of Customer Satisfaction, Value, and Loyalty to Market and Financial Performance: Basic Methods and Key Considerations," *Annual Quality Congress Proceedings,* ASQ, Milwaukee, pp. 113–122.

Brennan, C. D. Jr. (1997). *Proactive Customer Service,* AMACOM, New York.

Corcoran, K. J., L. K. Petersen, D. B. Baitch, and M. F. Barrett (1995). *High Performance Sales Organizations*, Irwin, Chicago.

Ekings, J. D. (1986). "A Nine Step Quality Improvement Program to Increase Customer Satisfaction," *Proceedings of the 30th Annual Conference,* European Organization for Quality Control, Berne, Switzerland, pp. 399–408.

Fitch, D. J., F. S. Beltrano, and A. A. Frigo (1990). "Use of Field Data for Basis of Design Improvements," *Annual Quality Congress Transactions,* ASQ, Milwaukee, pp. 875–880.

Ford (1972). *Reliability Methods, Module No. XII*, Ford Motor Company, pp. 11–13.

Goodman, J. (2000). "Modeling Customer Satisfaction and Loyalty in a Manner Which Evokes Action," *Proceedings of the 12th Annual Customer Satisfaction and Quality Measurement Conference,* American Marketing Association and ASQ, San Antonio, TX.

Goodman, J., P. O'Brien, and E. Segal (2000). "Turning CFOs into Quality Champions," *Quality Progress,* March, pp. 47–54.

Gust, L. J. (1985). "Non-Manufacturing Quality Improvement," *Juran Report Number Four,* Winter, Juran Institute, Inc., Wilton, CT, pp. 112–120.

Hicks, J. P. (1991). "Sensors That Tell Just How Good the Packaging Is," *New York Times,* September 15, p. F-7.

Lynch, J. (1998). "Measuring the Pursuit of Customer Loyalty at Sun," *Proceedings of the 10th Annual Customer Satisfaction and Quality Measurement Conference,* ASQ and American Marketing Association, Atlanta.

Mitchell, G. L. (2000). "A Quality Progress Approach to Internal Complaint System Analysis," *Annual Quality Congress Proceedings,* ASQ, Milwaukee, pp. 791–798.

Nash, M. (1998). "The Business Impact of Customer Loyalty," *Proceedings of the 10th Annual Customer Satisfaction and Quality Measurement Conference,* ASQ and American Marketing Association, Atlanta.

Nickell, W. L. (1985). "Quality Improvement in a Marketing Organization," *Quality Progress,* June, pp. 46–51.

Peppers, D. and M. Rogers (1999). *The One to One Manager: Real-World Lessons in Customer Relationship Management,* Currency/Doubleday, New York.

Reichheld, F. F. (1996). *The Loyalty Effect,* Harvard Business School Press, Boston.

Waltz, B. (1996). "Managing Customer Loyalty," *Proceedings of the Impro Conference,* Juran Institute, Inc., Wilton, CT, p. 7A-9.

SUPPLEMENTARY READING

Quality in the marketing function: *JQH5,* Section 18.

Hurley, R. F. (1994). "TQM and Marketing: How Marketing Operates in Quality Companies," *Quality Management Journal,* July, pp. 42–51.

Quality and sales process: Selden, P. (2000). "The Power of Sales Quality Engineering," *Annual Quality Congress Proceedings,* ASQ, Milwaukee, pp. 143–149.

Acceptance sampling: *JQH5,* Section 46.

Customer retention: Lowenstein, M. W. (1995). *Customer Retention*, Quality Press, ASQ, Milwaukee.

WEBSITES

American Productivity and Quality Center (APQC): www.apqc.org
Society of Consumer Affairs Professionals (SOCAP): www.socap.org
CRM forum: www.crm-forum.com

21

ADMINISTRATIVE AND SUPPORT OPERATIONS

21.1 DEFINITION AND SCOPE

Staff organizations are constantly under fire to demonstrate their value. Quality concepts can help them pack more punch.

Administrative operations are required for the organization to complete its mission. Examples include finance and accounting, human resources, training, information technology, safety, security, facilities management, legal, and office services. *Support operations* have some effect on the product itself, e.g., shipping, receiving, storage, traffic, product publications, and order filling. The terms *shared services, internal support services,* and *business services* are sometimes used to describe these activities. Administrative and support activities are prevalent throughout manufacturing and service industries.

Traditionally, quality systems focus on activities that have a direct connection with the product or service provided to external customers, i.e., production activities in the manufacturing industries and customer-related activities in the service industries. But it is now clear that quality concepts apply equally to administrative and support activities.

Customer orientation is the most important concept in quality systems. The concept has always been clear for activities such as design and operations that focus on external customers. That concept has not been clear for administrative and support (A&S) departments because many A&S customers are internal departments that are "captive customers." But the current emphasis on lean operations and outsourcing requires that internal departments have a strong focus on their internal and external customers.

TABLE 21.1
Quality concepts for potential application in administrative and support operations

Concept	Reference in this book
Definition of quality	Chapter 1
Triple role	Chapter 1
Assessment	
Cost of poor quality	Chapter 2
Marketing research study	Chapter 2
Quality culture study	Chapter 2
Quality improvement	Chapter 3
Quality planning	Chapter 4
Quality control	Chapter 5
Understanding customer needs	Chapter 12

Internal and external customers insist on reductions in errors, costs, and cycle time. In addition, these customers want new services in order to adapt to the changing needs of their external customers. The manager of a staff department must always be alert to the possibility that the department will be outsourced, i.e., the work shifted to an outside supplier. Also, "service agreements" (similar to a product specification) between support departments and internal customers and "charge backs" for activities provided are changing the operational role of staff departments. Quality concepts can improve the effectiveness and efficiency of staff departments and thereby help to preserve their role.

This chapter describes applications of quality concepts, presented in earlier chapters, to administrative and support activities of manufacturing and service industries. Table 21.1 lists quality concepts that can be applied within administrative and support departments.

Managers of A&S departments will need to embrace fully the customer orientation if they are to provide services to internal and external customers that have a value greater than the budget for the department. The first step is to do internal marketing research to obtain the perceptions of their internal customers on the services provided and then to benchmark the department's performance against A&S departments in other companies. This input then leads to identifying and pursuing specific quality projects, using the concepts and techniques in this book, to satisfy customers. Readers are urged to review earlier chapters and then identify applications, e.g., defining quality for employee benefit activities, explaining the triple roles in a shipping function, assessing quality performance in an information technology department, applying the three quality processes in a finance department.

We proceed to describe how quality concepts can be applied to three A&S departments: finance, human resources, and information technology. Additional examples are also discussed for facilities management, order entry, and legal services.

21.2 FINANCE

The finance function typically includes three segments (Werner and Stoner, 1995):

1. Controller—accounting, reporting, taxes.
2. Financial analysis—budgeting, planning, project analysis.
3. Treasury—bank relationships, investment management, credit management.

All three areas have opportunities for applying quality concepts.

One example involves American Express customers who failed to receive their renewal credit card (Bott et al., 2000). The following scenario uses the six-sigma approach of the define, measure, analyze, improve, and control phases.

Define and Measure

American Express received 1000 returned renewal cards each month. In 65 percent of the returns, the card member failed to notify American Express of an address change.

Analyze

Data were collected and analyzed to search for the root causes. Some conclusions:

- *By type of card.* One type of card, Optima, had the highest incidence of defects but was not significantly different, in the percentage of defects, from the other card types.
- *Issuance reason.* Renewal cards had a far higher defect rate than cards issued for other reasons, e.g., replacement and new accounts.
- *Validated reason for return.* Of five reasons for returns, returns with forwardable addresses had the highest percentage and quantity of returns.

Improve

The use of the National Change of Address service was instituted, and an experiment run on renewal files issued. The defect rate dropped from 13,552 to 6036 defects per million opportunities (DPMO); see Table 21.2.

TABLE 21.2
Combined test results

	Baseline	Test results
Defect rate	1.35%	0.6%
DPMO	13,552	6,036
Cost of poor quality		$3,360
Total annual savings		$1,228
Sigma level	3.71	4.01

The experiment validated the solution, which was then adopted to achieve larger scale savings and customer satisfaction.

Control

To ensure that the process continues to perform within acceptable limits, several steps were taken. A *p* control chart (see Section 18.8) tracks proportions of returned cards over time, cooperation with a supplier enables American Express to monitor the defect rate on a monthly basis, and American Express is alerted to any returned cards that used the National Change of Address database.

For elaboration, see Bott et al. (2000).

Another application in the finance function concerns the value-added concept. This concept questions the need for doing a task at all (What will the impact be if this task is not done?). Over time, activities to satisfy a supposed customer need or to compensate for some process inadequacy can creep into processes and become embedded. Sometimes these added activities are justified, temporarily or permanently; sometimes they are justified temporarily but not permanently; sometimes they are not justified at all (no value added). For example, a department of 33 people was created to check the validity and accuracy of "transfer charges" to one division from other divisions of a utility. Experience had shown that the number of errors and the amount of money involved justified the checking activity. No action was taken on the causes of the errors ("Our job is to find the errors"). The annual budget of the department—well over a million dollars—thus represented a cost of poor quality, year after year.

Another example involves the application of self-control (Section 5.3) to finance. Self-control requires that the process-planning stage provide personnel with the means of knowing what they are supposed to do, the means of knowing their performance, and the means of regulating their work.

The basic concept is described in Chapter 5; application to product development and operations is presented in Chapters 13, 16, and 17. Typically, planning is weak in one or more of the three criteria.

> EXAMPLE 21.1. At a bank, procedures for processing loans focus on financial criteria but exclude time standards for making a decision on a loan application (criterion 1); many measurements are taken, but feedback is primarily to management and not to the line personnel (criterion 2); a "proof operator" is not provided with any means of correcting a process in which a microencoding machine is unable to read a check because of handling of the check during previous process steps (criterion 3).

Flow diagrams and other industrial engineering tools are useful in planning and replanning both for activities within a department and for complete processes going across departments. For example, IBM uses a department activity analysis to examine activities within a department. Three steps are involved:

1. Listing all major activities. For example, a finance department might have 12 activities such as payroll processing and supplier payment.
2. For *each* activity:
   - Listing the inputs: What are they? What is the source?

TABLE 21.3
Application of quality management to financial processes

Financial process	Company
Monthly closing of books	Honeywell, IBM, Motorola
Payroll	Paradyne, Motorola
Expense reports	Honeywell
Accounts payable	Paradyne
Cash management	Corning
Audit	Gulf-Canada, Motorola, Baxter
Capital decisions	Alcoa
Financial forecasting	Solectron

- Analyzing the work: Why do it? What is its value? Suppose it is not done?
- Listing the outputs: What are they? Who receives them?

3. For *each* activity:
   - Meeting with the supplier and agreeing on requirements.
   - Meeting with the customer and agreeing on requirements.
   - Defining the measurements that will evaluate output against requirements.

   Suppliers and customers may be internal or external to the company.

Jankowski and Gryna (1996) show applications of quality management concepts to financial processes in various companies (see Table 21.3).

Finance is an example of a department that may have a changing set of responsibilities in an era of total quality management (TQM). Jankowski and Gryna (1996) identified six issues that challenge the traditional roles of a finance department. The six issues are:

- Should the finance department continue in a controlling and policing role or shift to focus on a service role for its internal and external customers and other indirect stakeholders? A change would mean a shift from conformance to requirements and procedures to a focus on internal and external customer satisfaction. Gulf-Canada, Motorola, and Baxter Healthcare are leaders in this arena. This issue has significant ramifications to a finance department.
- What opportunities exist for using quality management concepts in specific processes within the finance function? See Table 21.3 for some examples. The two components of quality (see Section 1.3) apply to financial processes: (1) output features, e.g., the scope of technical content in a financial report, and (2) output deficiencies, e.g., errors and omissions in the financial report. Addressing these two components in a formal way means applying the road maps of quality planning, quality control, and quality improvement.
- Should we evaluate capital budgeting proposals primarily on hard, quantitative financial measures, or should we include soft criteria such as maximizing customer satisfaction? Often we can quantify the benefits of improving quality: reduction in waste, reduction in complaints and warranty costs, reduction in cycle time, increase in selling price due to quality enhancements, increase in market share due to quality enhancements, and reduction in lost sales income due to customer defections. Previous chapters have developed these points. The finance department has a pivotal

role in the overall financial process and can therefore exert an important influence by presenting or supporting quality improvement proposals to upper management.

- Are reductions possible in working capital investment through the application of quality management concepts? The answer is yes. Quality improvement projects throughout an organization are reducing many forms of waste that are related to capital expenditures. Suppliers are providing high-quality just-in-time materials. In high-quality financial processes, paperwork is almost eliminated; the transfer of documents between departments is automated; and invoice verification, billing, collections, or inventory management become automated. General Electric has reduced its working capital investment by $400 million through a quality working capital approach (Tully, 1994).
- How can competitive benchmarking help to improve the effectiveness of the finance function? Benchmarking concentrates on efforts to improve both the effectiveness (e.g., cycle time) and efficiency (e.g., cost per transaction processed) of financial processes. Also, the scope of benchmarking can include three broad areas of finance: (1) transaction processing (e.g., accounts payable, payroll); (2) reporting and control (e.g., general accounting, internal audit); (3) business decision making (e.g. financial analysis, risk management). Benchmarking is only a first step on a journey of improvement in the finance function but can be a catalyst to show the urgency and opportunities for improvement. Sometimes we need a spark to ignite the status quo.
- What types of skills will be needed in the corporate finance department in the future? The traditional skills for finance are essential. These include analytical skills, a control orientation, and conformance to financial goals. But the team-oriented structure present today in many organizations means that finance personnel will require supplementary skills. These include assisting line departments, acting as a catalyst and facilitator, training line personnel to perform financial activities, and serving on cross-functional teams. The new process-oriented, decentralized organization requires a wider range of expertise than a centralized audit and control-type finance department offer. This area presents some wonderful opportunities for finance professionals.

In summary, finance departments can benefit from learning and applying quality management concepts. One organization, Honeywell, has even developed self-assessment criteria for the finance function based on the Malcolm Baldrige National Quality Award criteria. Finance activities within Honeywell employ these criteria to determine strengths, weaknesses, and improvement opportunities within the finance function.

21.3 HUMAN RESOURCES

The human resources function usually includes several activities within its scope: planning for human resource needs; recruiting and selecting employees; providing socialization to help new employees fit smoothly into an organization; arranging for training and development programs, performing performance appraisals; and handling promotions, transfers, demotions, and separations (Stoner et al., 1995). Applications of quality management concepts to the human resource function are presented below.

One example from DuPont involves the improvement of cycle time for an employee's application for long-term disability benefits. This project followed the six-sigma approach (Bott et al., 2000; Lisa Palser reported for DuPont). Reprinted with permission by the ASQ.

Introduction (Define)

The application process is initiated in the fifth month of the employee's six-month short-term disability leave of absence. The company tries to make a decision prior to the end of the short-term-benefits period; otherwise, the short-term benefits are extended, resulting in an extra cost to DuPont. The cycle time objective has an upper limit of 45 days.

Measure

A project team consisting of a process owner, subject matter experts, and the supplier's account managers was formed. The team mapped and documented the process, developed a cause-effect diagram, and surveyed a sample of employees at different employment sites. To assess where significant breakdowns in the process occurred, baseline data were collected, including overall cycle time and data at various milestone dates between the initial application and the decision.

Analyze

Data showed that the mean cycle time was 30 days, but the variability was high (one more time—variability—the demon of quality problems). The sigma value for the cycle time was 0.83, but in the six-sigma approach a sigma value of 6 is desired, and the lower the sigma value, the more the variability relative to the specification (see Section 18.15). Thus variability was the crux of the problem. Various analysis tools were applied, and the most significant factors affecting cycle time included incomplete medical and other information, delays in receiving medical information from physicians, processing of the application, substantiation of the employee's medical condition, and mail cycle time.

Improve

Process improvements within DuPont included electronic distribution of materials, use of fax-server technology to process applications, and improvement of the application packages for the employee and for the physician.

Process improvements at the benefit supplier included staffing changes, tighter standards with financial penalties for poor performance by independent medical exam suppliers, and extensive status reporting on each case.

Control

Controls included a new database to track each application and to monitor the key subcycles and periodic training for human resources personnel on benefits and benefit processing particularly when changes are implemented.

For elaboration, see Bott et al. (2000).

Another example comes from the Bureau of Labor Statistics.

EXAMPLE 21.2. Galvin (1991) describes how a team used a simple flow diagram to replan an employee separations process. Key transactions in the process included disposition of the employee's unused vacation time, disposition of retirement contributions, and issuance of the final paycheck. The primary goal was to improve process timeliness. Figure 21.1 is a simplified flow diagram developed by the team. At the start, boundaries on the study were set around the personnel and payroll departments because management believed that these were the primary sources of process variation. Team discussions of the flowchart concluded that client offices (where the separating employee worked) were the

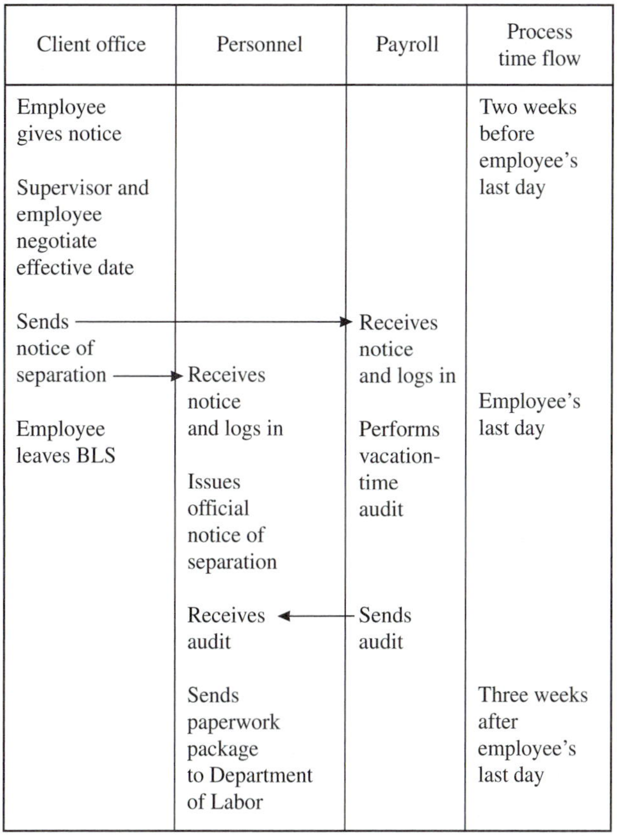

Client office	Personnel	Payroll	Process time flow
Employee gives notice			Two weeks before employee's last day
Supervisor and employee negotiate effective date			
Sends notice of separation → (to Payroll)		Receives notice and logs in	
Sends notice of separation → (to Personnel)	Receives notice and logs in		
Employee leaves BLS		Performs vacation-time audit	Employee's last day
	Issues official notice of separation		
	Receives audit ←	Sends audit	
	Sends paperwork package to Department of Labor		Three weeks after employee's last day

FIGURE 21.1
An employee separation process. (*From Galvin, 1991.*)

TABLE 21.4
Application of quality management to human resources processes

Human resources process	Companies
Recruiting	Rogers Corp., James River, Sykes Enterprises
Responding to submitted résumés	Universal Card
Compiling payrolls	Universal Card
Individual and team compensation	GTE

main suppliers and that measurement and analysis were needed to pinpoint the sources of process variation. As a result, project team boundaries were expanded to include client office functions. This boundary redefinition was instrumental in leading to process changes that increased on-time performance for the separation process from 62 percent to 88 percent.

Table 21.4 shows other applications of quality management to human resources processes.

Human resources is another example of a department that may have a changing role in the era of TQM. Stoner et al. (*JQH5,* pp. 26.19–26.22) summarize modern human resources practices that are consistent with quality management concepts. These include

- *Staffing.* Recruiting for skills in teamwork, problem solving, and quality improvement as well as for technical skills.
- *Performance and evaluation.* Emphasizing developmental feedback rather than judgments on current performance and emphasizing team and organizational goals rather than individual goals.
- *Compensation.* Providing a variety of rewards rather than only base pay and benefits.
- *Training and development.* Setting aggressive goals for the amount of training provided to all employees every year.
- *Management development.* Preparing managers for roles as facilitators and coaches and providing training in leadership.
- *Relationship to organized labor.* Seeking collaborative rather than adversarial relations with unions.
- *Role of the human resources function in management.* Emphasizing a strategic role in acquiring and developing human resources rather than preoccupation with routine personnel activities.

Ulrich (1998) proposes a new role for human resources: focus on business results that enrich the company's value to customers, investors, and employees. To achieve this role requires human resources to become (1) a partner in strategy execution, (2) an expert in the way work is organized and executed, (3) a champion for employees, and (4) an agent for continual change. The details behind these broad tasks would require a revolution in human resources departments.

Wilkinson et al. (1993) studied human resources functions and quality management in the United Kingdom and identified five phases of an organization's transformation to quality and the role that the human resources function can play in each. The phases are formulation or developmental, introductory, maintenance and reinforcement, companywide review, and review of the human resources activities. The researchers concluded that human resources can play a change-agent role at the strategic decision-making level in the organization. Again, this responsibility goes far beyond the traditional operational activities of the human resources function.

21.4 INFORMATION TECHNOLOGY

All activities in all organizations are subject to errors, delays, and other forms of poor quality. The information technology (IT) function is no exception. Pinter (1992) cites some sobering evidence:

- 15 percent of all IT projects never deliver anything.
- Cost overruns of 100 to 200 percent are not uncommon.
- Staffs to perform rework are generally more than double the size of those that build computer information systems.

Section 17.2 under "Input from Information Technology as an Internal Supplier" discusses the problems that internal customers have with the output of IT departments. Table 17.3 shows that the top 10 quality issues with IT output reported by operations managers customers were (in order of importance) accuracy of information, system downtime, system response time, network reliability, system performance, level of expertise, hardware performance, thorough testing, hardware problems, and lack of training.

IT personnel work under strong pressure from line operations to deliver new information systems and fix those systems and hardware that are not operating properly. Thus IT performs a complex task in an ever-changing technological environment. The road maps for quality improvement (Chapter 3), quality planning (Chapter 4), and quality control (Chapter 5) provide IT personnel and others with a useful framework for IT output; see Figure 21.2. This figure provides the trilogy of quality processes—quality planning for developing new IT products, quality control for sporadic problems, and quality improvement for chronic problems. Note that critical inputs to these processes are continuous feedback, such as daily measurements, help desks, and problem tracking, as well as special studies, such as cost of poor quality, customer surveys, and IT personnel surveys (see below).

A sporadic problem is a sudden, unexpected adverse change in a process. Examples of sporadic problems in the IT area are unscheduled system downtime and storage limitations. Often this type of problem is handled by a help desk in the IT department. The problem may be logged in a problem-tracking system, given a priority based on severity of the problem, and followed until resolved. The road map for handling sporadic problems is the quality control process—also called "troubleshooting" (see Chapter 5).

A chronic problem is a long-standing, adverse situation. IT examples include inadequate system response time and poor documentation. Initially, this type of problem is

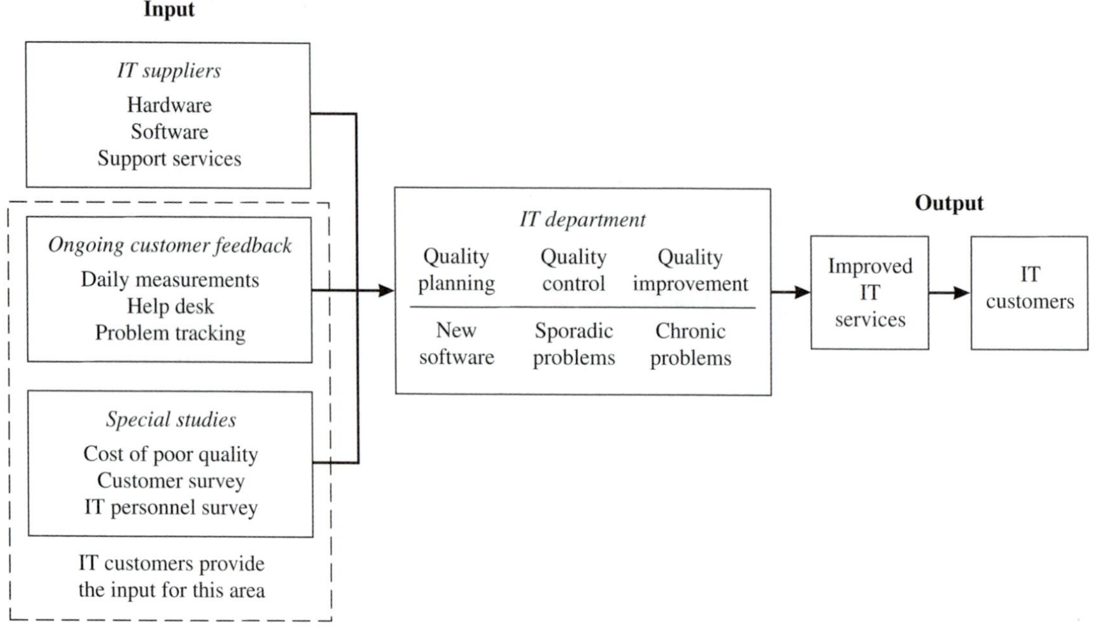

FIGURE 21.2
Framework for IT quality. (*From Kittner et al., 1999.*)

referred to a help desk, but as the problem becomes chronic, it may be escalated in the organization. Chronic problems are usually cross functional in nature and require a team of individuals for achieving resolution. The road map for handling chronic problems is the quality improvement process (see Chapter 3).

A pioneer in using the structured quality improvement road map in IT is Tektronix. Examples of early projects included computer program execution errors (systems failures, program failures, etc.), errors in introducing computer programs in the production environment, timeliness of reports, interdepartmental communications, system design (development time, implementation errors, cost of operations), and unusable reports (Gray et al., 1986; Green and Lamper, 1987).

Just as the cost-of-poor-quality concept has been so useful in justifying improvement efforts on physical products, so it can be used to justify efforts on software quality. Section 2.3 explains one categorization as prevention, appraisal, internal failures, and external failures. Kaner (1998) uses the same categories to show examples of quality cost elements for software products. He also reminds us that these are the internal costs to the software developer and do not include the costs of poor quality to the user of the software.

In addition to addressing on-going problems, we also need an instrument to tell us current user perceptions about the quality of services provided by the IT function (see Figure 21.2). The tools of market research can be beneficial. For example, a survey of 292 companies describes practices to learn about user satisfaction with IT services.

The survey revealed five methods of obtaining user feedback: (1) written user surveys, (2) focus groups with participants from several departments, (3) meetings with individual departments, (4) steering committee meetings, and (5) user-IT contact through a subunit of IT dedicated to user service and liaison.

Formal user satisfaction surveys were employed by 43 percent of the companies. Typically, surveys were conducted on an annual basis, but the range ran from monthly to biannually. Companies reported that surveys were used to make IT more responsive to users, identify dissatisfied users, pinpoint problem areas in specific IT units, and monitor progress over time. For an explanation of the survey results and examples of survey instruments, see Newman (1989).

Sometimes the results of a user satisfaction survey can be sobering. Rustogi and Bajawa (1996) report the results of a survey at a hospital in which users from 25 departments were asked to rate the quality of IT services in terms of several dimensions. The overall result: User perception of the IT department *failed* to meet expectations in *any* of the dimensions. An analysis of the detailed responses would be fertile ground for using the structured quality improvement road map of Chapter 3.

Finally, readers are referred to Section 22.5 for a discussion of preferred practices in developing a quality information system.

21.5 OTHER ADMINISTRATIVE AND SUPPORT AREAS

Other applications of quality concepts span all administrative and support areas. To illustrate, we present examples from three diverse areas: facilities management, order entry, and legal services.

The example from facilities management illustrates a simple diagnostic tool of quality improvement.

EXAMPLE 21.3. In an example of histograms, data were plotted to analyze turnover time in rooms in a laboratory at Brigham and Women's Hospital (Laffel and Plsek, 1989). Turnover time was defined as the time between the moment all catheters and sheaths are removed from one patient and the time local anesthetic is injected into the next patient. Simply collecting data yielded some surprises: The mean time was 78 minutes (45 minutes had been the usual estimate); the variation ranged from 20 to 150 minutes. In one part of the analysis, data were stratified by room, and histograms on turnover time were plotted by room (see Figure 21.3). When the total data were stratified by room, we learned that room 1 had a shorter mean time and much less variation than room 2. Further analysis showed that when a nurse called for the next patient before the previous case was complete, turnover time was relatively short. No one had been aware, until the data were recorded and the analysis made, that the timing of the call was a critical determinant of turnover time.

The examples from order entry illustrate two tools of diagnosis—a flow diagram and a cause-and-effect diagram.

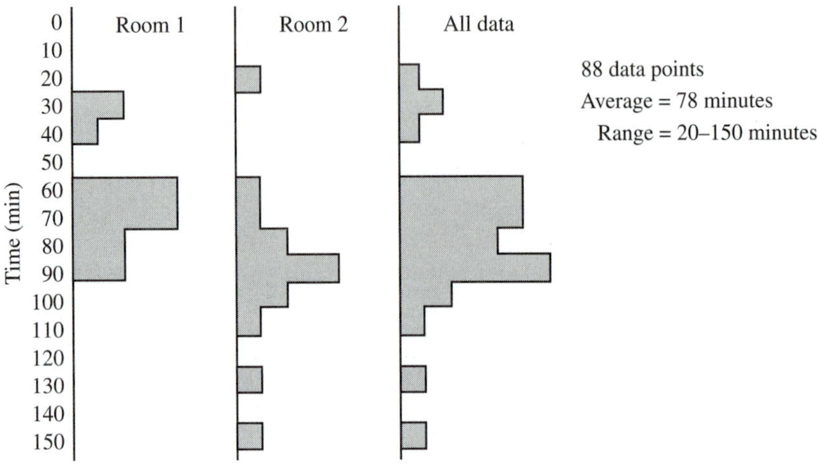

FIGURE 21.3
Room turnover times. (*From Laffel and Plsek, 1989.*)

Figure 21.4 shows a flow diagram for a segment of an order receipt and entry process. Note the use of some standard symbols: A rounded rectangle denotes the beginning or end of a process; a square or rectangle, a step in the process; a square with an irregular bottom, a document; a diamond, a decision or branch point; and a circle, a continuation of the diagram.

Another useful tool is the cause-and-effect diagram (see Section 3.10, "Diagnose the Causes").

EXAMPLE 21.4. An order entry department was running a 50 percent error rate on sales order documentation (yes, 50 percent). A project team created a cause-and-effect diagram (Figure 21.5). Analysis led to changes that, in a short time, cut the error rate in half.

A few organizations are pursuing the application of quality concepts to their internal legal services department. Motorola and AT&T are two pioneers.

- At Motorola a first effort involved the preparation of patent applications (Rauner, 1990). The department mapped the activity from the receipt of an invention disclosure to searching for prior art (optional), drafting a patent, obtaining patent drawings, obtaining information from the inventor, finalizing the documents, getting inventor signatures, and filing the documents in the U.S. Patent and Trademark Office. Data were collected on the time for each step (some were "unnecessarily and harmfully long"), and internal and external customers of the patent process were identified. Analysis of the time data and potential defects in the patent process resulted in significant improvement in both quality and cycle time.
- At AT&T, an Intellectual Property Process Quality Improvement Team was formed. The original team established seven process quality improvement teams to deal with processes such as identifying patent users, enforcing AT&T patent rights, licensing

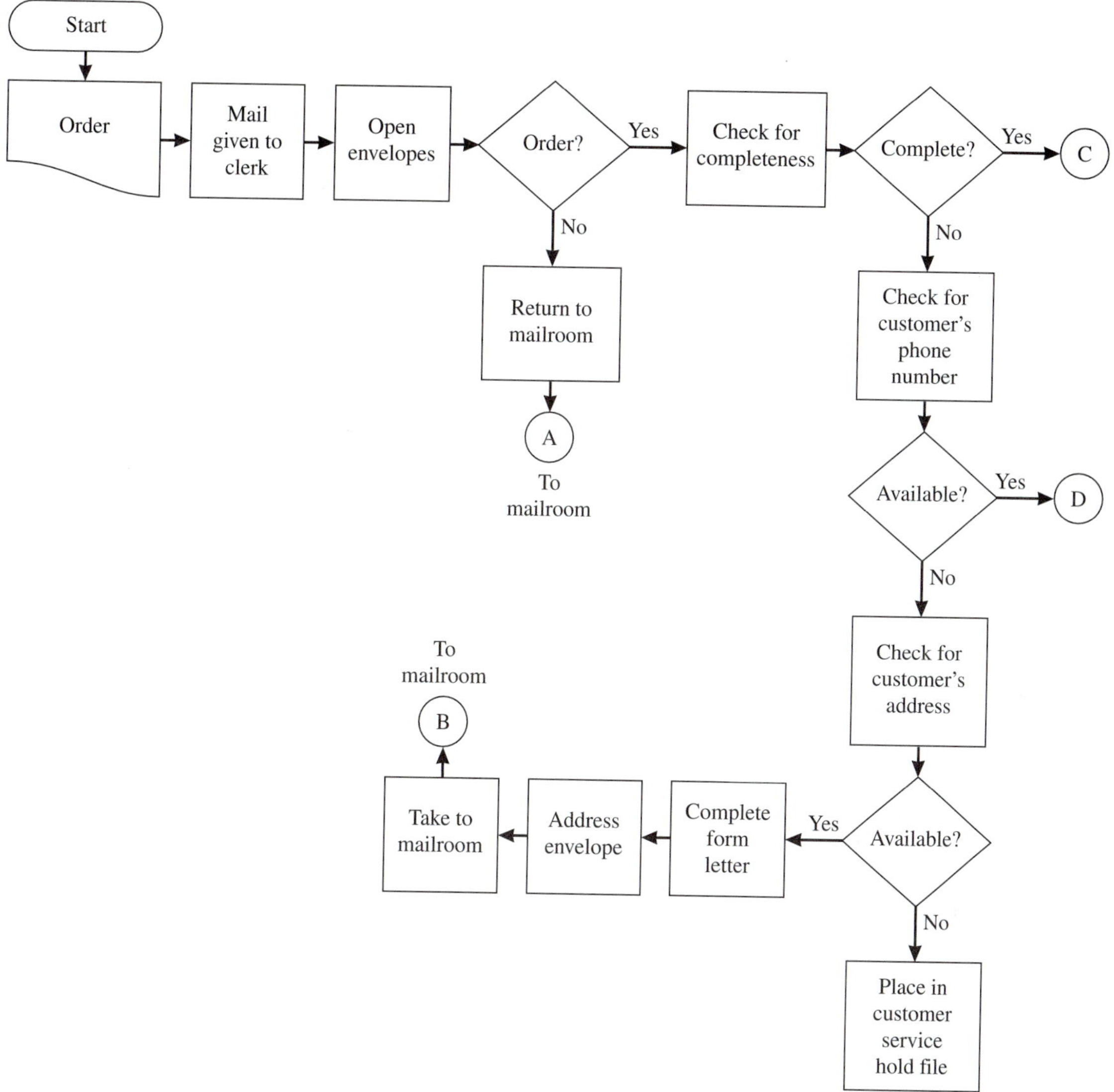

FIGURE 21.4
Segment of flow diagram for order receipt and entry. (*From Juran Institute, Inc., 1989.*)

agreements, and monitoring and ensuring compliance with intellectual property agreements. One conclusion: "AT&T could earn more than five times the economic return on its intellectual property assets than was being achieved" (Greene, 1995).

Stoner et al. in *JQH5,* Section 26, summarize these and other applications of quality concepts to legal services.

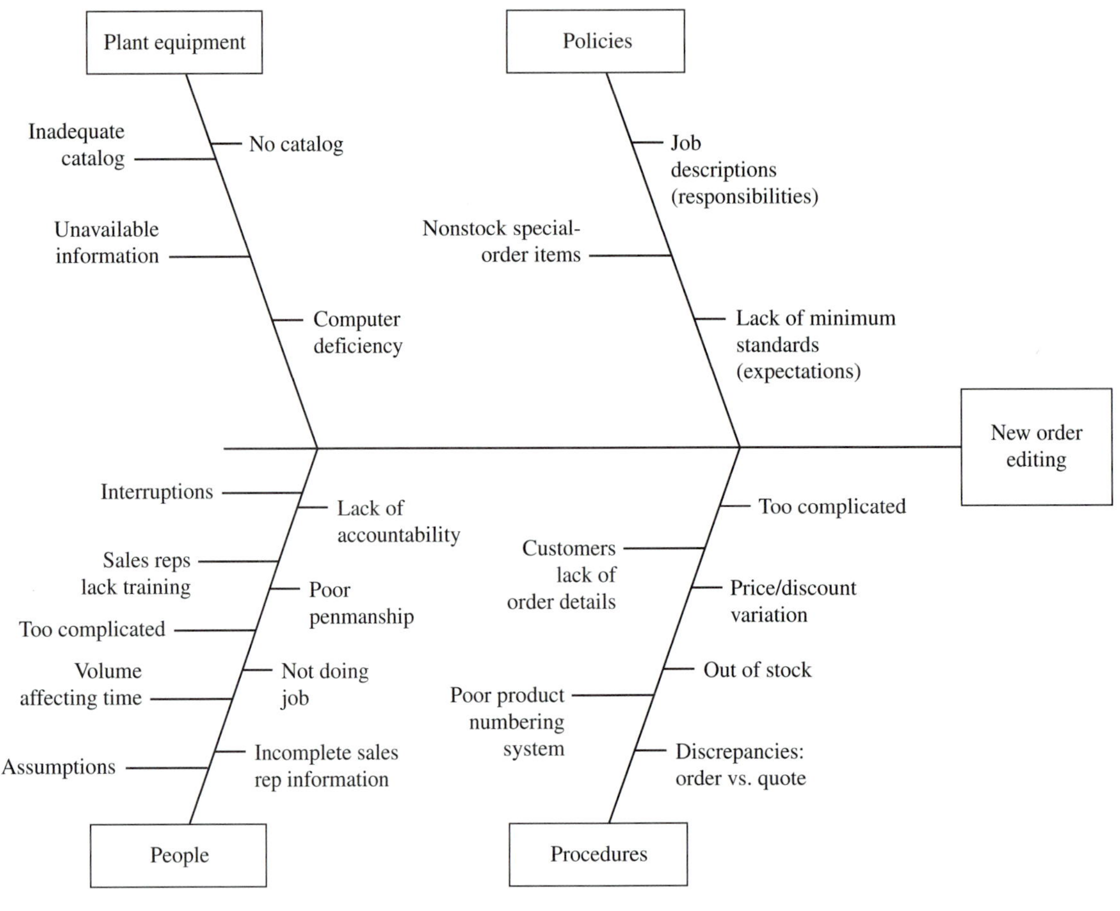

FIGURE 21.5
Cause-and-effect diagram. (*From Nader, 1989.*)

21.6 QUALITY MEASUREMENT IN ADMINISTRATIVE AND SUPPORT OPERATIONS

Basic to the quality control of any activity is quality measurement. (Readers should review the basics of quality measurement discussed in Section 5.2.)

Table 21.5 shows units of measure for some activities (functions) and control subjects (features).

SUMMARY

- Administrative and support activities have both internal and external customers and can therefore benefit from the application of quality concepts.

TABLE 21.5
Quality measurement in administrative and support activities

Activity/control subject	Units of measure
Finance	
Payment of invoices	Monetary value of invoices paid late
Issuance of invoices	Average number of days to issue
Errors in invoices	Percentage of invoices returned due to errors
Accounts receivable	Monetary value of unrecoverable accounts receivable
Personnel	
Quality of candidates' résumés	Percentage of résumés received that result in interviews
Yield of recruitment process	Number of candidates interviewed before an offer is made and accepted
Time required	Average number of days from request for personnel to initial employment date
Qualification of skilled workers	Percentage of certified workers in critical activities
Information technology	
System availability	Percentage of system uptime
Reliability of equipment	Mean time between failures
Downtime of equipment	Mean time to restore service
Response time	Turnaround time for reports
Software errors	Number of errors per thousand lines of code—at initial review, at final review

- The quality planning road map (and associated techniques such as flow diagrams) is beneficial in planning or revising processes.
- The quality control road map (including techniques such as statistical quality control) helps to measure and regulate processes.
- The quality improvement road map (and techniques such as diagnosis) provides a means of addressing chronic problems in administrative and support processes.
- Finance, human resources, and information technology are examples of administrative and support activities that use quality management concepts to improve their effectiveness and efficiency.

PROBLEMS

21.1. Select one administrative or support department in an organization such as a business organization, a local government, a college, or a volunteer organization. List all major activities. For one major activity, list the inputs, analyze the work, and list the outputs. For the same activity, meet with the supplier and customer and agree on requirements and define measurements to evaluate the output.

21.2. Select one administrative or support department in an organization with which you are familiar. For that department, identify at least three chronic quality-related problems. For one of these problems, write a brief problem and mission statement for a project team. Follow the guidelines for these statements given in Section 3.7, "Define Phase."

21.3. Prepare a flow diagram for one activity that takes place primarily within one administrative or support department.

21.4. For an organization with which you are familiar, identify a quality planning (or replanning) project in each of two administrative or support departments.

22.5. For an organization with which you are familiar, define two quality units of measure for each of two administrative or support departments.

REFERENCES

Bott, C., E. Keim, S. Kim, and L. Palser (2000). "Service Quality Six Sigma Studies," *Annual Quality Congress Proceedings,* AQC, Milwaukee, pp. 225–231.

Galvin, J. M. (1991). "Structural Problem-Solving for Administrative Processes: A Case Study," *ASQC Quality Congress Transactions,* Milwaukee, pp. 692–697.

Gray, S., Y. Green, and M. Lamper (1986). "Quality Improvement in Information Systems," *Juran Institute Impro Conference Proceedings,* Wilton, CT, pp. 86–89.

Green, Y. and M. Lamper (1987). "Maintaining Continuity and Momentum for Quality Improvement in Information Systems," *Impro Conference Proceedings,* Juran Institute, Inc., Wilton, CT, pp. 4B-13 to 4B-19.

Greene, R. M. (1995). "Intellectual Property Management in a Dynamic Organizational Context," 1995 *Management for Quality Research and Development Symposium,* Juran Institute, Inc., Wilton, CT, pp. 4-29 to 4-38.

Jankowski, J. R. and F. M. Gryna (1996). "Changing Responsibilities for Financial Management in an Era of TQM," *International Journal of Quality Science,* vol. 1, no. 1, pp. 9–18.

Juran Institute, Inc. (1989). "Quality Improvement Tools—Flow Diagrams," Wilton, CT.

Kaner, C. (1998). "Bad Software—Who Is Liable?" *Annual Quality Congress Proceedings,* ASQ, Milwaukee, pp. 607–626.

Kittner, M., M. Jeffries, and F. Gryna (1999). "Operational Quality Issues in the Financial Sector: An Exploratory Study on Perception and Prescription for Information Technology," *Journal of Information Technology Management,* vol. X, nos. 1–2, pp. 29–39.

Laffel, G. and P. E. Plsek (1989). "Preliminary Results from a Quality Improvement Demonstration Program at Brigham and Women's Hospital," *Impro Conference Proceedings,* Juran Institute, Inc., Wilton, CT, pp. 8A-21 to 8A-27.

Nader, G. J. (1989). "Applying Quality Methods: Non-Manufacturing Areas," *ASQC Quality Congress Transactions,* Milwaukee, pp. 14–21.

Newman, G. (1989). *Measuring User Satisfaction with Information Management.* The Conference Board, New York.

Pinter, L. F. (1992). "Computer Information Systems Quality: From Commitment to Reward," *Annual Quality Congress Transactions,* ASQ, Milwaukee, pp. 44–49.

Rauner, V. J. (1990). "Pursuing Quality in Patent Applications," *IMPRO 90 Conference Proceedings,* Juran Institute, Inc., Wilton, CT, pp. 2C-1 to 2C-9.

Rustogi, H. and D. S. Bajawa (1996). "Perceived Service Quality with the IS Function. An Exploratory Study," *Proceedings of the 25th Annual Northeast Decision Sciences Institute Conference,* St. Croix, VI.

Stoner, J. F., R. E. Freeman, and D. R. Gilbert Jr. (1995). *Management,* 6th ed., Prentice Hall, Englewood Cliffs, NJ.

Tully, S. (1994). "Raiding a Company's Hidden Cash," *Fortune,* August 22, p. 86.

Ulrich, D. (1998). "A New Mandate for Human Resources," *Harvard Business Review,* January–February, pp. 124–134.

Werner, F. M. and J. A. F. Stoner (1995). *Modern Financial Managing,* HarperCollins College Publishers, New York.

Wilkinson, A., M. Marchington, and B. Dale (1993). "Enhancing the Contribution of the Human Resource Function to Quality Improvement," *Quality Management Journal,* vol. 1, no. 1, pp. 35–46.

SUPPLEMENTARY READING

Administrative and support operations: *JQH5,* Section 26.

Software development: *JQH5,* Section 20.

Quality in information technology: Woodall, J., D. K. Rebuck, and F. Voehl (1997). *Total Quality in Information Systems and Technology,* St. Lucie Press, Delray Beach, FL.

McLeod, R. Jr. (1996). *Management Information Systems,* 6th ed., Chapter 4, Prentice Hall, Englewood Cliffs, NJ.

WEBSITE

Software Program Managers Network: www.spmn.com

22

QUALITY INFORMATION SYSTEMS

22.1 SCOPE AND OBJECTIVE OF A QUALITY INFORMATION SYSTEM

A *quality information system* (QIS) is an organized method of collecting, storing, analyzing, and reporting information on quality to assist decision makers at all levels. The scope may vary from a simple system covering in-process inspection data or data on customer complaints to a broad system covering all information on the overall effectiveness of both product and process quality.

The objective of a QIS is to align quality-related activities with the vision of the organization and measure performance to drive improvement. The QIS identifies problems and provides data for diagnosing causes and determining action for improvement. Note that a QIS goes far beyond an information system with a primary objective of reporting status. The viewpoint here is that a QIS that does not lead to improvement is a failure.

In the past information systems on quality were mainly concerned with data such as customer complaints and inspection data. But now quality spans the spectrum of all functional and staff departments, and emphasis is placed on fitness for use rather than conformance to specifications. These changing conditions have resulted in a broader viewpoint toward information on quality in both the manufacturing and service sectors.

This broader viewpoint requires inputs from many functional areas. It also recognizes that "information" consists not only of data but also of other knowledge needed for decision making. Input for a QIS includes the following elements:

- *Market research information on quality.* Examples are data on customer satisfaction, data on customer experiences that suggest opportunities for improving fitness for use, and data on quality of competitors.

- *Product design test data.* Examples are development test data, data on parts and components under consideration from various suppliers, and data on the environment that the product may encounter.
- *Information on design evaluation for quality.* Examples are minutes of design review meetings; reliability predictions; and failure mode, effect, and criticality analyses.
- *Information on purchased parts and materials.* Examples are inspection data, data on tests conducted by a supplier, data on tests conducted by an independent laboratory of a procured item, supplier survey information, and supplier rating data.
- *In-process data.* These data cover the entire manufacturing or service operation inspection system including all forms of waste such as nonconformances and rework. Also included are cycle time data and process control and process capability data.
- *Final inspection data.* These data are the routine data at final inspection.
- *Field performance data.* Examples are mean time between failures (MTBF) and other reliability data.
- *Improvement data.* These data can include a summary of recent improvement results along with a compilation of improvement projects—both completed and in progress.
- *Results of departmental quality measurements.* Examples include data from functional and administrative and support activities such as design, operations, supplier relations, marketing, finance, human resources, and information technology. Many examples are provided in tables of quality measurements in Chapters 13, 15, 16, 17, 20, and 21.
- *Audit results.* These data include audits of processes, specific activities, and product (see Chapter 23).
- *Complaint data.* These data include both warranty data and customer complaints along with data on the responsiveness to handling the complaint.
- *Management control data.* Cost-of-poor-quality data including sales revenue lost and market share illustrate this category.
- *Information on the quality system.* Examples include the quality policy manual, quality procedures, quality system documentation, and process control plans.

Note that these categories cover customer-related, financial, marketing, operational, and competitive data.

We proceed in this chapter to discuss reports on quality, the development of a QIS, and the issue of data quality.

22.2 REPORTS ON QUALITY

The bulk of quality information is derived from multiple sources of operational information as described above.

This information is used in the first instance for operational controls, e.g., day-to-day regulation of factory and office process and field performance. The same information, when summarized and converted into suitable form, becomes a major input to the quality instrument panel—a system of information that enables busy managers to become adequately informed as to quality performance and trends without becoming heavily involved in day-to-day operations. Table 22.1 shows the interrelationship between operational and executive controls.

Operational Reports

Operational control reports assist in conducting day-to-day operations, with particular emphasis on achieving improvement. Table 22.2 depicts an inspection reporting system for an electronics manufacturer. The system translates gross information into lower levels of detail so managers and engineers can isolate problems by product and by worker. Note that Pareto analysis is used extensively in this system.

Executive Reports

Early forms of executive reports tended to be limited to summaries of operational quality information (e.g., defects, errors, rework) plus summaries of customer complaints. The emphasis now is on showing how this basic data can be translated into extra costs and loss of sales revenue (see Section 20.11). Also, as the performance of business systems has grown in importance, reports have included additional control subjects such as promptness of service, competitiveness in quality, software quality, invoicing errors, and other quality measurements mentioned earlier in this book.

TABLE 22.1
Operational quality controls versus executive quality controls

Aspects	Application to operational quality controls	Application to executive quality controls
Control subjects	Physical, chemical, specification requirements	Summarized performance for product lines, departments, etc.
Units of measure	Natural physical, chemical (ohms, kilograms, etc.)	Various: often in money
Sensing devices	Physical instruments, human senses	Summaries of data
Who collects the sensed information?	Operators, inspectors, clerks, automated instruments	Various statistical departments
When is the sensing done?	During current operations	Days, weeks, or months after current operations
Standards used for comparison	Specification limits for materials, press, product	History, competitors, plan
Who acts on the information?	Servomechanisms, nonsupervisors, first-line supervisors	Managers
Action taken	Process regulation, repair, sorting	Replanning, quality improvement, motivation

The information needed by managers for executive control varies widely from company to company, depending on the nature of the product, the extent to which the control problems have been solved, etc.

EXAMPLE 22.1. General Dynamics Corporation, a defense contractor, undertook a form of corporate quality improvement. One element of the approach is a report battery. The reports deal with numerous parameters of which 12 are corporate, i.e., they are common to all divisions of the corporation. The corporate parameters are as follows:

- Avoidable engineering changes.
- Deviations/waivers.
- First-time yield.
- Scrap (labor-hour content).
- Scrap (material value content).
- Repair and/or rework (labor-hour content).
- On-time delivery by production.
- Purchased item acceptability.
- Service report response time.
- Material review actions.
- Inspection escapes.
- Overtime.

For elaboration, see Talley (1986).

TABLE 22.2
Summary of reports in an inspection information system

Name of summary	Description	Example
Part-operation accept/reject	Summarizes accept/reject information for each work area	In work area 5924, operation 85 on 40432 was highlighted as a poor performer; 46 inspections were made and 23 units were rejected.
Part-operation listing	Summarizes fault categories for each part operation	For part operation 40432-85, there were 17 instances of "terminal solder missing" or 31% of the total faults for operation 85.
Fault listing	Summarizes fault categories for each work area	For work area 5924, there were 52 instances of terminal solder missing or 28% of the total; of these 52, six were charged to operator 37157.
Operator listing	Summarizes faults charged to each operator	Operator 37157 was charged with 24 faults, six of which were terminal solder missing.
Monthly inspection report	Summarizes in a graphical and tabular format overall performance; also lists the major problem areas; a separate page is prepared for each work area	For work area 5924, a graph of percentage defective by month is plotted; for December, the major problem was on part operation 40432-85; 1693 units were inspected; 455 faults were found, of which 361 were due to operators; operator 37157 was a major contributor to defects, particularly missing and improper solder and wiring errors.

In many companies, these summarized executive reports are supplemented by independent audit reports. Such audits help to provide assurance that the report system correctly reflects what is actually going on with respect to quality.

Some organizations use phased indicators of quality performance.

EXAMPLE 22.2. Texas Instruments (TI) designed its quality reporting system around three types of indicators: leading indicators, concurrent indicators, and lagging indicators (Onnias, 1986).

Figure 22.1 depicts this concept. Leading indicators include data on raw material or piece parts purchased from suppliers (e.g., parts per million defective material, purity level). Concurrent indicators include internal manufacturing data (e.g., dust count average, reliability of manufacturing equipment, warehouse errors). Lagging indicators are of two types: material rejected and returned by customers and customer feedback data. The TI report battery comprises these three types of indicators plus the cost of quality.

Some executive control subjects are lagging indicators because the reports appear weeks or even months after the operations have been performed. Such reports are nevertheless of great value in showing trends, identifying substantial failures in meeting goals, measuring performance of managers, and so forth.

In contrast, some subjects are leading indicators. Market research on quality may lead product development by months. Design review is a leading indicator of failure rates. Product audit lags behind date of manufacture but is a leading indicator of quality as received by the user.

A well-balanced system of executive reports makes use of leading indicators ("early warning signals") as well as summaries that lag behind operations.

TI's report (called the *Quality Blue Book*) is issued monthly and is the basis of annual performance appraisal of all managers for their contribution to quality.

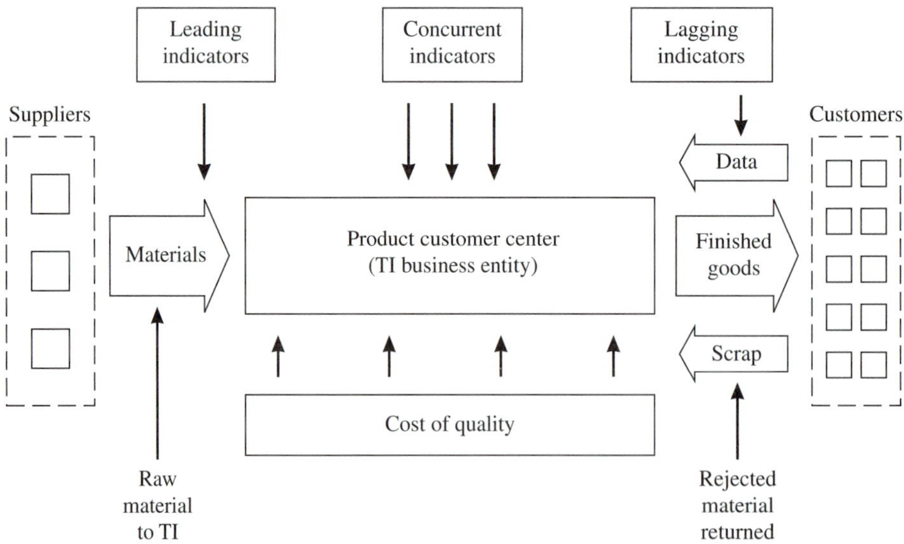

FIGURE 22.1
Content of the *Quality Blue Book*. (*From Onnias, 1986.*)

TABLE 22.3
Key performance indicator measurement system

1. External customer satisfaction—external measurements. Examples include customer rating, auditing, customer satisfaction team survey, net new customers, and percentage of market share.
2. External customer satisfaction—internal measurements. Examples include on-time delivery, customer complaints, and responsiveness to customers.
3. Cost of quality. This indicator includes prevention costs, appraisal costs, internal failure costs, and external failure costs.
4. Productivity. This indicator includes value added per colleague, yields, physical output per colleague, unit costs and/or cost reduction program implementation, and inventory turns.
5. Cycle time of key processes. This indicator includes new-product-development cycle time, order-entry-to-ship cycle time, material-procurement cycle time, and customer-complaint-resolution cycle time.
6. Innovation and learning. This indicator includes rate of new product introduction, percentage of sales from new products in the last three years, resources devoted to technological leadership, effectiveness of technological resources, and percentage of sales from new products from acquisitions made in the last three years.
7. Human resource excellence. This indicator includes development of managerial and leadership talent and general human resource excellence with measurements such as colleague development and colleague satisfaction.

Balanced Scorecard for Quality

The concepts of the balanced scorecard and key results indicators (see Section 7.8) provide measurements at the strategic level that integrate quality and business objectives. Table 22.3 shows a balanced scorecard for quality developed at AMETEK Inc. (Wain, 1996). Reprinted by permission of the ASQ.

Even though financial indicators such as return on assets or cash flow are not included in Table 22.3, the company believes that it has identified the drivers of its future financial performance. Each division or plant develops key performance indicators (KPIs) and presents the data along with supporting information (e.g., trend graphs, survey results, and program completion reports) as part of the annual corporate review of plant performance. KPIs showing improvement are highlighted at this annual review; KPIs showing no improvement are supplemented with an action plan for improvement.

22.3 INITIAL PLANNING FOR A QUALITY INFORMATION SYSTEM

The planning of a computer-based QIS can be complex. The road starts with an analysis of customer needs, creation of a design specification for the system, and preparation of a proposal indicating costs and time required. When management approves the proposal, the system is developed, tested, and implemented. Finally, provisions are made for review of system performance.

A system must be tailored to meet the needs of both internal and external customers of an organization. The following principles are generally applicable:

- Plan the system to receive information in almost any form imaginable. Although much of the information will be received electronically, the system should also receive and process information by means of special forms, a telephone call, letters, or other media.
- Provide flexibility for meeting new data needs. A cardinal example of this principle is the failure reporting form that must be revised periodically because someone suddenly discovers a critical need to record an additional item of information.
- Provide for collection of data on three time phases: (a) real time (continuous), (b) recent (minutes to hours), and (c) historical (extended time).
- Provide for eliminating collection of data that are no longer useful as well as reports that are no longer needed. Conduct a periodic audit of the use (or lack of use) of the data and reports.
- Issue reports that are readable, timely, and have sufficient useful detail on current problems to facilitate investigations and corrective action and also provide early warning of potential problems.
- Prepare summary reports covering long periods of time to highlight potential problem areas and show progress on known problems.
- Keep track of the *cost* of collecting, processing, and reporting information and compare this cost to the *value* of the information.

The QIS becomes a reality through *software.* Software is the collection of computer programs, procedures, and associated documentation necessary for the operation of the information system. The computer software program is found either in existing computer software packages or by creating new software.

22.4 SELECTION OF OFF-THE-SHELF SOFTWARE

Application software packages are available for many needs. The June 2000 issue of *Quality Progress* magazine listed packages for the following categories: auditing, Baldrige Award, calibration, capability studies, data acquisition, design of experiments, gage repeatability and reproducibility, inspection, ISO 9000 measurement, problem solving, quality assurance for software development, quality costs, quality function deployment, reliability, sampling, simulation, statistical methods, statistical process control, supplier quality assurance, Taguchi techniques, training, and other categories. The *Journal of Quality Technology,* published by the American Society for Quality, regularly runs a column providing the detailed programming steps for statistical techniques. Such programs can be used as is or incorporated as part of a larger program created to meet the needs of a specific user.

Table 22.4 shows the steps in acquiring a software application package. In examining software packages, it is useful to have a checklist of the attributes (often called "factors") of software along with a checklist of specific questions. Table 22.5 lists a time-tested set of software quality factors.

TABLE 22.4
Steps in acquiring an application package

1. Identify present and future requirements.
2. Survey available packages.
3. Examine documentation and manuals.
4. Determine data and communication interoperability with existing programs.
5. Survey existing users as to product acceptability. (Internet queries provide an optimal way to obtain this information.)
6. Request quality assurance and testing information from the developer.
7. Request a list of known bugs.
8. Review copyright and licensing requirements.
9. Obtain programs for internal trial execution.
10. Negotiate contract to include maintenance services and upgrades.

Source: Orkin and Olivier in *JQH5,* page 10.8.

TABLE 22.5
Software quality factors

Factor	Definition
Correctness	Extent to which a program satisfies its specifications and fulfills the user's mission objectives
Reliability	Extent to which a program can be expected to perform its intended function with required precision
Efficiency	Amount of computing resources and code required by a program to perform a function
Integrity	Extent to which access to software or data by unauthorized persons can be controlled
Usability	Effort required to learn, operate, prepare input of, and interpret output of a program
Maintainability	Effort required to locate and fix an error in an operational program
Testability	Effort required to test a program to ensure that it performs its intended function
Flexibility	Effort required to modify an operational program
Portability	Effort required to transfer a program from one hardware configuration and/or software system environment to another
Reusability	Extent to which a program can be used in other applications—related to the packaging and scope of the functions that programs perform
Interoperability	Effort required to couple one system with another

Source: Adapted from McCall et al. (1977).

22.5 PREFERRED PRACTICES IN DEVELOPING A QUALITY INFORMATION SYSTEM

Basic to developing an effective QIS is the concept of alignment. As applied to a QIS, *alignment* means understanding and supporting organizationwide goals and developing measures for the planning, tracking, analysis, and improvement at three levels: the organization level, the key process level, and the work unit level (see Section 5.2). Alignment includes how businesses allocate and integrate goals to achieve organizationwide goals and how senior management deploys measurement requirements to track work-group- and process-level performance.

Before we discuss preferred practices for developing a QIS, readers are urged to review the 10 principles of measurements presented in Section 5.2. These principles cover the purpose, emphasis on customer use of measurements, and attributes of the measurements and the collection process. Such principles must be reflected in the QIS development process. *JQH5,* Section 9, provides further elaboration on measurement.

Research has been conducted to discover preferred practices for developing a QIS. One example of such research is by Long and Gryna (1999). Four research sites were involved: a hospital, a financial services organization, an energy company, and a telecommunications organization. These sites represent profit and nonprofit, regulated and nonregulated, and large and small organizations. With the exception of one person at one site, everyone interviewed used a QIS. Thus this research emphasized user needs of a QIS.

None of the organizations had a comprehensive QIS, but each had developed an information system for a specific quality-related activity. These systems are summarized as follows:

- Recording and monitoring patient care, including billing of costs (hospital).
- Capturing and responding to customer comments (complaints) regarding credit cards (financial services organization).
- Recording and summarizing data on outages of failed equipment (energy company).
- Recording and summarizing data on incidents reported by customers in using software (telecommunications organization).

To discover preferred practices, researchers posed 15 questions about QIS development at each site. Table 22.6 summarizes the responses.

The research discussions led to the identification of the following preferred practices:

1. *Define the purpose and scope of the QIS.* The initial scope of the system should be limited, with provisions made for expansion. Start small, learn what works and what does not work, and do the development in digestible pieces. Clear and concise deliverables are essential.
2. *Focus on core business functions* to start the development. This approach gives the QIS credibility. Initial discussions should also address flexibility in the system to adapt to changes. Alignment of the QIS with business goals is a must.

TABLE 22.6
Summary of responses to questions about QIS development

Questions	Majority response
1. Who was the compelling force?	Top-level company officer.
2. Were purposes defined?	No.
3. How was information selected?	Issues selected by the user.
4. How was change built into the system?	Did not address change initially. Later incorporated change.
5. Was the system externally benchmarked?	Yes. Companies were compared to external standards.
6. What were the effective ways to analyze performance results	Comparative charts and training.
7. What were the effective ways to transmit performance results.	Meetings and the Internet with the use of peer pressure.
8. What is the quality of measurements?	All companies used filters to improve usefulness and believability.
9. What measurements are made?	Turnaround time; report length; usefulness.
10. How can the QIS be more effective?	Meetings to share information and research reports from their industry.
11. Was training provided?	Training was provided both during the development process and after.
12. What did not work in the development process?	Defining the scope of the QIS to be too expansive.
13. What are the three most important problems?	(1) Relying on the IT department; (2) lack of ownership of system by employees; (3) no developmental process flow model.
14. What are the three most important steps to take?	(1) Recognition of project importance; (2) accurate definition of inputs and outputs; (3) use of project management tools.
15. Can quality be improved in your current system?	Use more technology such as computers and networks to distribute information.

3. *Assemble a group of people to discuss a complete QIS* but then identify one or more core functions related to quality and develop a QIS for the core function(s).
4. *Define who has primary responsibility for development of the QIS.* Do not rely entirely on the IT department. The IT staff views information as a by-product, not as a product (see Section 22.7). In addition, IT department data were originally collected for other purposes and often are not what the QIS needs.
5. *Use project management tools to develop the QIS.* This step includes task definition, schedule, responsibilities, plus a person to direct the effort. Appoint someone from the quality department to serve as the manager of the QIS during both development and operational use.
6. *Define performance issues, not metrics.* Users should define issues to be investigated. The quality department and the users should collaborate to develop the metrics. This approach illustrates the importance of end-user involvement through the development process.
7. *Develop useful data from the user viewpoint.* The transmission media is not as important as the functionality of the results (see Section 22.7).

8. *Evaluate data* for usefulness, relevance, comprehensiveness, level of detail, readability, and interpretability (see Section 22.7).
9. *Use peer pressure to ensure action on problems uncovered by the QIS and the quality system.* Using comparative analysis tools in the context of positive competition seems to be the most effective way to promote problem resolution.
10. *Disseminate information* about development progress and system use as the system develops.
11. *Publish a list of reports or outputs from the QIS* and distribute it to all concerned parties. The list should include a brief description of the purpose of each report and some of the key metrics used. This step allows users to see what others are using and can reduce duplication.
12. *Filter data for accuracy.* Involve the IT department, the quality department, and the users.
13. *Minimize metrics that involve codes.* Expansive code sets are both underutilized and inappropriately used. Trade off simplicity for granularity.
14. *Explain the reasons for and the benefits of the QIS and provide thorough training.* The training should include how to use the QIS and how to do problem diagnosis to facilitate action.

Other useful research is available. Weimer and Munyan (1999) report on the results of a survey of 75 information services professionals and end users from business and government organizations. In order of importance and absence in development efforts the priorities were adequate training, managing change, adequate budget, thorough plan for conversion, end-user involvement, clear and concise deliverables, strategic objectives, technical and end-user documentation, project management, development methodology, rapid design tools, leadership, and executive sponsorship.

Now suppose we view a QIS as a product. The approach to develop a QIS can be related to the steps in operational quality planning (Chapter 4).

Quality Planning Steps as Applied to Developing a Quality Information System

The following list of operational quality planning steps relates each step to the 14 preferred practices (PP) given above.

1. Establish the project (PP 1, 2, 3, 4, 5).
   - Create a mission statement for developing the QIS (purpose, scope, goals).
   - Establish a team.
   - Plan the project (responsibilities, schedule, follow-up).
2. Identify the customers—users of the QIS (PP 3).
   - Internal to the organization.
   - External.
3. Discover customer needs (PP 6, 7).
   - Plan how to discover customer needs.
   - Collect information on customer needs (preferably quantify).
   - Analyze and prioritize customer needs.

4. Develop the QIS product (PP 7, 8, 9, 10, 11).
   - Group together related customer needs.
   - Identify alternative product features of the QIS.
   - Develop detailed QIS product features and goals.
   - Finalize product design of the QIS.
5. Develop process features to create the QIS product (PP 12, 13, 14).
   - Identify alternative process features for collection of data and analysis, presentation, and usage of results.
   - Develop detailed process features and goals.
   - Establish initial process capability.
   - Finalize process design.
6. Develop process controls and transfer to operations (PP 10, 11).
   - Identify controls needed.
   - Optimize self-control and self-inspection.
   - Establish audit of process.
   - Verify process capability in operations.
   - Transfer plans to operations.

From these preferred practices for developing a QIS, we next present an overview of the process of creating new software.

22.6 CREATING NEW SOFTWARE

The management dimensions of developing a QIS are addressed by the preferred practices discussed above. The technical dimensions of developing the software involve defining the software requirements, designing the software system, implementing the system, and maintaining the system. Table 22.7 lists eight steps and their "deliverables" (Gryna, 1988).

Software programming spans the spectrum of complexity, depending on the processing that is desired for the information. *JQH5,* Section 20, and Ireson et al. (1996), Section 22, are two of the many excellent references.

The Software Engineering Institute (SEI) is a source of good software practices. The SEI People Capability Maturity Model defines five levels of organizational maturity concerning software development practices:

- *Level 1.* Initial: ad hoc, chaotic.
- *Level 2.* Repeatable: processes depend on individuals repeating mastered tasks.
- *Level 3.* Defined: processes are codified and institutionalized.
- *Level 4.* Managed: software metrics beyond tracking costs and schedules are in place.
- *Level 5.* Optimized: continuous process improvement is in place.

Few organizations are at level 5.

The Software Program Managers Network (SPMN) is funded by the U. S. Congress to provide direct support to Department of Defense software-intensive programs. The SPMN identifies "16 best practices for large software-intensive system development and

TABLE 22.7
System life cycle phases

Step	Deliverable/activity
1. Requirements analysis	System requirements specification
	Disaster recovery plan*
	Risk assessment*
	Resource requirements analysis*
	Requirements analysis review
2. External design	External design specifications
	User's manual
	Maintenance manual
	Preliminary test plan
	External design review
	System requirements specifications
3. Internal design	Internal design specifications
	Conversion/implementation plan
	System test plan
	Internal design review
	System requirements specifications
4. Detailed development	Code review
	Disaster recovery plan
	Security risk assessment
	Detailed development review
5. System test	Test report
	System test review
6. Data management	Software development file (SDF)
	User's manual
	Maintenance manual
7. Production/implementation	Implementation plan
	Maintenance manual
8. Maintenance	Scheduled activities
	Documentation of changes
	Maintenance manual

*May be excluded from system requirements specification.
Source: Gryna, D. (1988).

sustainment." These best practices are management or technical practices that significantly improve one or more of the following dimensions: productivity, development and sustainment cost, schedule, quality, user satisfaction, and predictability of cost and schedule. The website is provided at the end of this chapter.

22.7 DATA QUALITY

Organizations recognize that data fed into any information system must be of "high quality." What does this mean and how do we achieve high-quality data? The definition of data quality concerns the intrinsic content of the data, but other matters such as accessibility are also important (see below). But a more basic question is: Should we consider information as a product or as a by-product?

Wang et al. (1998) recommend that, from the viewpoint of the *user* of information, we should view information as a product that will be used by a customer, not as a by-product of computer hardware and software (see Table 22.8). This perspective is basic to good data quality.

Accordingly, quality management methodologies should be applied to the product called "information." Thus the trilogy of quality processes applies: quality planning to plan for data quality, quality control to track and monitor data quality, and quality improvement to improve data quality. Wang et al. (1998) present four principles: (1) understand customer needs, (2) manage the information production process, (3) manage the life cycle of information products, and (4) appoint an information product manager. Note how these principles relate to the three quality processes.

With this background we next consider the definition of data quality. Wang et al. (1998) define data quality in terms of user needs. The result is 16 dimensions of data quality grouped into four categories (see Table 22.9).

Elaboration on these dimensions is provided by Wang and Strong (1996). Redman in *JQH5,* Section 34, also discusses dimensions of data quality. Thoroughly defining the dimensions of data quality for a given application is essential for developing a QIS that is useful to end users. Murine and Murine (1997) in the Supplementary Reading describe how a coroner's office uses quality improvement techniques in criminal investigations. Quality improvement can be everywhere—call it ubiquitous.

TABLE 22.8
Information as product or by-product

	Product	**By-product**
What is managed?	• Information • Information product life cycle	• Hardware and software • Systems life cycle
How is it managed?	• Integrated, cross-functional approach • Encompass information suppliers, manufacturers, and consumers	• Integrate stovepipe (individual) systems • Control of individual components • Cost controls
Why manage it?	Deliver quality information products to consumers	Implement quality hardware and software systems
What is success?	• Quality information product continuously delivered over the product life cycle • No garbage in, garbage out (GIGO)!	• The system works • No bugs • Short-term perspective
Who manages it?	• Chief information officer • Information product manager	• Chief information officer • IT director and database administrators

TABLE 22.9
Information quality categories and dimensions

Category	Dimensions
Intrinsic information quality	Accuracy Objectivity Believability Reputation
Accessibility information quality	Accessibility Ease of operations Security
Contextual information quality	Relevance Value added Timeliness Completeness Amount of information
Representational information quality	Interpretability Ease of understanding Concise representation Consistent representation

SUMMARY

- A quality information system (QIS) is an organized method of collecting, storing, analyzing, and reporting information on quality to assist decision makers at all levels.
- Achieving improvements in quality is a key purpose of a QIS. This system goes far beyond reporting quality status.
- A QIS must be aligned with organizationwide goals.
- Reports on quality focus on both operational and executive matters. A useful framework is the concept of leading, concurrent, and lagging indicators of performance.
- Application software packages are available for numerous needs. Be alert to the lessons learned about acquiring software packages.
- Preferred practices have been identified to help develop a QIS.
- Data quality has many dimensions and must be thoroughly defined for a QIS.

PROBLEMS

22.1. For any of the following, obtain the necessary information and draw up a flowchart for the data collection and analysis of *(a)* final inspection results at a plant; *(b)* goods returned to a plant; *(c)* traffic fines; *(d)* complaints at a department store; *(e)* automobile accident insurance claims; *(f)* performance deficiencies by an athletic team; *(g)* loss of utility service to homes. Make recommendations for changes or additions.

22.2. For any institution, evaluate the usefulness of at least two reports on product quality by speaking with those who receive the report. Set up a scale of frequency of use and obtain opinions on use and on shortcomings of the present report.

22.3. Speak with people engaged in writing computer programs. Develop a list of common programming errors. For a sample of 10 programs, tally the frequency of occurrence for each type of programming error.

22.4. Speak with people responsible for collecting information for a database on quality or other subjects. Determine the steps taken, if any, to verify the quality of the input data and the output results, i.e., accuracy, completeness, and other criteria.

22.5. Assemble a group of at least three people and discuss how the steps in the quality control road map (Chapter 5) could be applied to control the quality of data inputs to a QIS.

22.6. Assemble a group of at least three people and discuss how the steps in the quality improvement road map (Chapter 3) could be applied to improve the quality of data inputs to a QIS.

REFERENCES

Gryna, D. S. (1988). "Data Processing—A Software Quality Challenge," *ASQC Quality Congress Transactions,* Milwaukee, pp. 423–428.

Ireson, W. G., C. F. Coombs Jr., and R. Y. Moss (1996). *Handbook of Reliability Engineering and Management,* 2nd ed., McGraw-Hill, New York.

Long, C. S. and F. M. Gryna (1999). *Preferred Practices in Developing a Quality Information System.* Report No. 907, College of Business, University of Tampa, Tampa, FL.

McCall, J., P. Richards, and G. Walters (1977). "Factors in Software Quality," *Joint General Electric-U.S. Air Force Report No. RADC TR-77-369,* vol. 1, November, pp. 3–5.

Onnias, A. (1986). "The Quality Blue Book," *Juran Report Number Six,* Juran Institute, Inc., Wilton, CT, pp. 127–131.

Talley, D. J. (1986). "The Quest for Sustaining Quality Improvement," *Juran Report Number Six,* pp. 188–192.

Wain, H. W. (1996). "Developing a Balanced Scorecard: The Ametek Experience," *Proceedings of the Annual Quality Congress,* ASQ, Milwaukee, pp. 84–91.

Wang, R. Y., Y. W. Lee, L. L. Pipino, and D. M. Strong (1998). "Manage Your Information as a Product," *Sloan Management Review,* Summer, pp. 95–105.

Wang, R. Y. and D. M. Strong (1996). "Beyond Accuracy: What Data Quality Means to Data Consumers," *Journal of Management Information Systems,* vol. 12, Spring, pp. 5–34.

Weimer, A. L. and R. J. Munyan (1999). "Recipe for a Successful System," *Software Quality Professional,* September, pp. 22–30.

SUPPLEMENTARY READING

Measurement systems: *JQH5,* Section 9.
Data quality: *JQH5,* Section 34.

Murine, K. E. and G. E. Murine (1997). "Forensic Data Quality Improvement Techniques," *Proceedings of the Annual Quality Congress,* ASQ, Milwaukee, pp. 1049–1057.

Data collection, storage, and retrieval: Bersbach, P. L. (1992). "Quality Information Systems," in T. Pyzdek and R. W. Berger, eds., *Quality Engineering Handbook,* Marcel Dekker, New York, pp. 61–83.

WEBSITE

Software Program Managers Network: www.spmn.com

23

QUALITY ASSURANCE; QUALITY AUDIT

23.1
DEFINITION AND CONCEPT OF QUALITY ASSURANCE

In this book, *quality assurance* is the activity of providing evidence to establish confidence that quality requirements will be met. ISO 8402-1994 defines quality assurance as all the planned and systematic activities implemented within the quality system, and demonstrated as needed, to provide adequate confidence that an entity will fulfill requirements for quality. Readers are warned that other meanings are common; e.g., "quality assurance" is sometimes the name of a department concerned with many quality management activities such as quality planning, quality control, quality improvement, quality audit, and reliability.

Many quality assurance activities provide protection against quality problems through early warnings of trouble ahead. The assurance comes from evidence—a set of facts. For simple products, the evidence is usually some form of inspection or testing of the product. For complex products, the evidence is not only inspection and test data but also reviews of plans and audits of the execution of plans. A family of assurance techniques is available to cover a wide variety of needs.

Quality assurance is similar to the concept of the financial audit, which provides assurance of financial integrity by establishing, through "independent" audit, that the plan of accounting is (1) such that, if followed, it will correctly reflect the financial condition of the company and (2) actually being followed. Today independent financial auditors (certified public accountants) have become an influential force in the field of finance.

Many forms of assurance previously discussed in this book are performed within functional departments (Table 23.1). This chapter discusses three forms of company-wide quality assurance: quality audits, quality assessments, and product audit.

TABLE 23.1
Examples of departmental assurance activities

Department	Assurance activity
Marketing	Product evaluation by a test market Controlled use of product Product monitoring Captive service activity Special surveys Competitive evaluations
Product development	Design review Reliability analysis Maintainability analysis Safety analysis Human factors analysis Manufacturing, inspection, and transportation analysis Value engineering Self-control analysis
Supplier relations	Qualification of supplier design Qualification of supplier process Evaluation of initial samples Evaluation of first shipments
Production	Design review Process capability analysis Preproduction trials Preproduction runs Failure mode, effect, and criticality analysis for processes Review of manufacturing planning (checklist) Evaluation of proposed process control tools Self-control analysis Audit of production quality
Inspection and test	Interlaboratory tests Measuring inspector accuracy
Customer service	Audit of packaging, transportation, and storage Evaluation of maintenance services

Source: JQH4, p. 9.3.

23.2
CONCEPT OF QUALITY AUDITS AND QUALITY ASSESSMENTS

A *quality audit* is an independent review conducted to compare some aspect of quality performance with a standard for that performance. The term *independent* is critical and is used in the sense that the reviewer (called the "auditor") is neither the person responsible for the performance under review nor the immediate supervisor of that person. An independent audit provides an unbiased picture of performance. The terms *quality assessment* (or *quality evaluation*) and *quality audit* have similar meanings, but in common usage *assessment* refers to a total spectrum of quality activities often including managerial matters such as the cost of poor quality, standing in the market place, and quality culture (see Chapter 2).

The ISO 10011-2-1994 definition spells out some additional aspects: Quality audit is a systematic and independent examination to determine whether quality activities and related results comply with planned arrangements and whether these arrangements are implemented effectively and are suitable to achieve objectives. (Product audit, discussed later in this chapter, is a review of *physical product*; quality audit is a review of an *activity.)*

An internal audit (i.e., an audit conducted within an organization by an auditor employed by that organization) is called a first-party audit. External audits are either second party or third party. A second-party audit is conducted within a supplier's organization by the organization that is making a purchase from the supplier. A third-party audit is conducted by an auditing organization that is independent of the purchaser or supplier organization

Companies use quality audits to evaluate their own quality performance and the performance of their suppliers, licensees, agents, and others; regulatory agencies use quality audits to evaluate the performance of organizations they regulate.

The specific purpose of quality audits is to provide independent assurance that:

- Plans for attaining quality are such that, if followed, the intended quality will, in fact, be attained.
- Products are fit for use and safe for the user.
- Standards and regulations defined by government agencies, industry associations, and professional societies are being followed.
- There is conformance to specifications.
- Procedures are adequate and are being followed.
- The data system provides accurate and adequate information on quality to all concerned.
- Deficiencies are identified, and corrective action is taken.
- Opportunities for improvement are identified, and the appropriate personnel alerted.

A key question in establishing an audit program is whether the audits should be compliance oriented or effectiveness oriented or both. In practice, many quality audits are compliance oriented; i.e., the audits compare quality-related activities to some standard or requirement for those activities (e.g., do written work instructions exist for production and service operations?). The emphasis is on determining conformance to the requirement of written work instructions and maintenance of procedures for those instructions as evidence of that conformance. Effectiveness audits evaluate whether the requirement is achieving the desired result (for external and internal customers) and whether the activity is making efficient use of resources (for elaboration, see Russell and Regel, 1996).

At first glance, it seems that audits should be both compliance oriented and effectiveness oriented—and sometimes they can be both. When audits are conducted internally, they can and should be both compliance and effectiveness oriented. But when audits are conducted by external parties, the companies audited can have serious and reasonable issues if an audit concerns matters of effectiveness of operations including use of resources. Effectiveness evaluations of internal operations by external auditors open a broad range of considerations about customer satisfaction and requirements and internal management processes that make it difficult to conduct such audits in a fair and useful way.

23.3
PRINCIPLES OF A QUALITY AUDIT PROGRAM

Five principles are essential to a successful quality audit program:

1. An uncompromising emphasis on conclusions based on facts. Any conclusions lacking a factual base must be so labeled.
2. An attitude on the part of auditors that the audits provide not only assurance to management but also a useful *service* to line managers in managing their departments. Thus audit reports must provide sufficient detail on deficiencies to facilitate analysis and action by line managers.
3. An attitude on the part of auditors to identify opportunities for improvement. Such opportunities include highlighting good ideas used in practice that are not part of formal procedures. Sometimes an audit can help to overcome deficiencies by communicating through the hierarchy the reasons for deficiencies that have a source in another department.
4. Addressing the human relations issues discussed.
5. Competence of auditors. The basic education and experience of the auditors should be sufficient to enable them to learn in short order the technological aspects of the operations they are to audit. Lacking this background, they will be unable to earn the respect of the operations personnel. In addition, they should receive special training in the human relations aspects of auditing. The American Society for Quality provides a program for the certification of quality auditors.

These five essentials for a successful quality audit activity were responsible for a dramatic tribute to an audit activity within one company. Line managers voluntarily give up part of their own budget each year to provide funds for a quality audit group.

23.4
SUBJECT MATTER OF AUDITS

For simple products the range of audits is also simple and is dominated by product audits (see below). For complex products the audit is far more complex. In large companies even the division of the subject matter is a perplexing problem. For such companies the programs of audit use one or more of the following approaches for dividing up the subject matter:

- *Organizational units.* Large companies comprise several layers of organization, each with specific assigned missions: corporate office, operating divisions, plants, etc. Such companies commonly use multiple teams of quality auditors, each reviewing its specialized subject matter and reporting the results to its own "clientele."
- *Product lines.* Here the audits evaluate the quality aspects of specific product lines (e.g., printed circuit boards, hydraulic pumps) all the way from design through field performance.
- *Quality systems.* Here the audits are directed at the quality aspects of various segments of the overall systematic approach to quality such as design, manufacturing, supplier quality, and other processes. A system-oriented audit reviews any such sys-

tem over a whole range of products. Table 23.2 provides an example from Mallinckrodt Inc., a medical products manufacturer.

- *Specific activities.* Audits may also be designed to single out specific procedures that have special significance to the quality mission: disposition of nonconforming products, documentation, instrument calibration, software (see Table 23.3).

TABLE 23.2
Quality systems evaluation—components and elements

A. Organizational design
  1. Management responsibility
  2. Job descriptions
B. Customer management practices
  1. Corrective action
  2. Servicing
  3. Complaint and inquiry handling
  4. Recall and field correction
C. Organizational and individual development practices
  1. Training
  2. Personnel hygiene
D. Product development practices
  1. Device design control
  2. Concept generation
  3. Device development
  4. Transfer to operations
  5. Life cycle maintenance/postmarket surveillance
E. Product and process control practices
  1. Process control
  2. Special processes
  3. Process capability
  4. Facilities and equipment
  5. Control of contamination
  6. Recovered material
F. Procurement practices
  1. Purchasing
  2. Contract review
G. Warehousing and distribution practices
  1. Handling, storage, distribution, and installation
H. Quality assurance practices
  1. Product identification and traceability
  2. Acceptance activities
  3. Nonconforming goods
  4. Labeling
  5. Internal quality audits
  6. Electronic data processing
I. Information analysis practices
  1. Inspection, measuring, and test equipment
  2. Statistical techniques
  3. Analytical methods and laboratories
J. Document management practices
  1. Document control
  2. Quality records
  3. Product registration and approval dossiers

TABLE 23.3
Examples of audits of tasks

Scope or activity	Examples of specific tasks audited
Engineering documentation	Use of latest issue of specifications by operators; time required for design changes to reach shop
Job instructions	Existence and adequacy of written job instructions
Machines and tools	Use of specified machines and tools; adequacy of preventive maintenance
Calibration of measuring equipment	Existence of calibration procedures and degree to which calibration intervals are met
Production and inspection	Adequacy of certification program for critical skills; adequacy of training
Production facilities	General cleanliness and control of critical environmental conditions
Inspection instructions	Existence and adequacy of written instructions
Documentation of inspection results	Adequacy of detail; feedback and use by production personnel
Material status	Identification of inspection status and product configuration; segregation of defective product
Materials handling and storage	Procedure for handling critical materials; protection from damage during handling; control of in-process storage environments

Audits of quality systems as well as specific activities may take the form of (1) audit of the plans or (2) audit of the execution versus the plans. Further, the subject matter may include internal activities or external activities such as those conducted by suppliers.

Identifying Opportunities

Experienced auditors are often able to discover opportunities for improvement as a by-product of their search for discrepancies. These opportunities may even be known to the operations personnel so that the auditor is only making a rediscovery. However, these personnel may have been unable to act due to any of a variety of handicaps: preoccupation with day-to-day control, inability to communicate through the layers of the hierarchy, and lack of diagnostic support.

The auditor's relatively independent status and lack of preoccupation with day-to-day control may enable him or her to prevail over these handicaps. In addition, the auditor's reports go to multiple layers of the hierarchy and thereby have a greater likelihood of reaching the ear of someone who has the power to act on the opportunity. For example, the auditor may find that the quality cost reports are seriously delayed owing to backlogs of work in the accounting department. His or her recommendation to expedite the reports may reach the person who can act, whereas the same proposal made by the operations personnel may never reach that level.

ISO 9000-2000 will likely have a clear and strong emphasis on continuous improvement.

23.5 STRUCTURING THE AUDIT PROGRAM

Audits of individual tasks or systems of tasks are usually structured; e.g., they are designed to carry out agreed purposes and are conducted under agreed rules of conduct. Reaching agreement on these rules and purposes requires collaboration among three essential participating groups:

- The heads of the activities which are to be the subject of audit.
- The heads of the auditing department(s).
- The upper management, which presides over both.

Without such collective agreements, the audit program may fail. The usual failure modes are (1) an abrasive relationship between auditors and line managers or (2) a failure of line managers to heed the audit reports.

Table 23.4 depicts the typical flow of events through which audit programs are agreed on and audits are carried out. A published statement of purposes, policies, and methods becomes the charter that legitimizes the audits and provides continuing guidelines for all concerned.

Audits are often done by full-time auditors who are skilled in both technical and human relations aspects. Audit teams of upper managers, middle managers, and specialists can also be effective. See *JQH5*, pages 41.16–41.19, for a discussion.

TABLE 23.4
Steps in structuring an audit program

	Audit department	Line department	Upper management
Discussion of purposes to be achieved by audits and general approach for conducting audits	X	X	X
Draft of policies, procedures, and other rules to be followed	X	X	
Final approval			X
Scheduling of audits	X	X	
Conduct of audits	X		
Verification of factual findings		X	
Publication of report with facts and recommendations	X		
Discussion of reports	X	X	X
Decisions on action to be taken		X	
Subsequent follow-up	X		

23.6
PLANNING AUDITS OF ACTIVITIES

The main steps in performing an audit are planning, performing, reporting, follow-up on corrective action, and closure. The flowchart in Figure 23.1 describes these steps in some detail. An excellent reference for the auditing process is ASQ Quality Audit Division (2000). Malsbury (1999) provides specific guidance during various phases of an audit.

Behind the steps in Figure 23.1 are a number of important policy issues:

Legitimacy

The basic right to conduct audits is derived from the "charter" that has been approved by upper management, following participation by all concerned. Beyond this basic right are other questions of legitimacy: What is the scope and objective? What shall the subject matter for audit be? Should the auditor be accompanied during the tour? Whom may the auditor interview? etc. The bulk of auditing practice provides for legitimacy—the auditor acts within the provisions of the charter plus supplemental agreements reached after discussion with all concerned.

Scheduled versus unannounced

Most auditing is done on a scheduled basis. "No surprises, no secrets." This practice enables all concerned to organize workloads, assign personnel, etc., in an orderly manner. It also minimizes the irritations that are inevitable when audits are unannounced.

Customer

The *customer* of the audit is anyone who is affected by the audit (see Section 1.3). The key customer is the person responsible for the activity being audited. Other customers include upper management and functions affected by the activity. Each customer has needs that should be recognized during the planning of the audit (see Chapter 12). Note that this orientation of service to the activity audited means that the audit must go beyond compliance to a requirement. Such an orientation is *not* practiced (or even accepted) by all auditors, but the author believes that the concept is basic to useful audits.

Audit team

Audits are conducted by individuals or by a team. A team usually has a lead auditor who plans the audit, conducts the meetings, reviews the findings and comments of the auditors, prepares the audit report, evaluates corrective action, and presents the audit report.

Clearly, auditors must be open-minded and possess sound judgment, have the trust and respect of line management, and be knowledgeable in the area being audited. ANSI/ASQ Q10011-2-1994 recommends other qualifications for auditors including education, training, experience, personal attributes, and management capabilities.

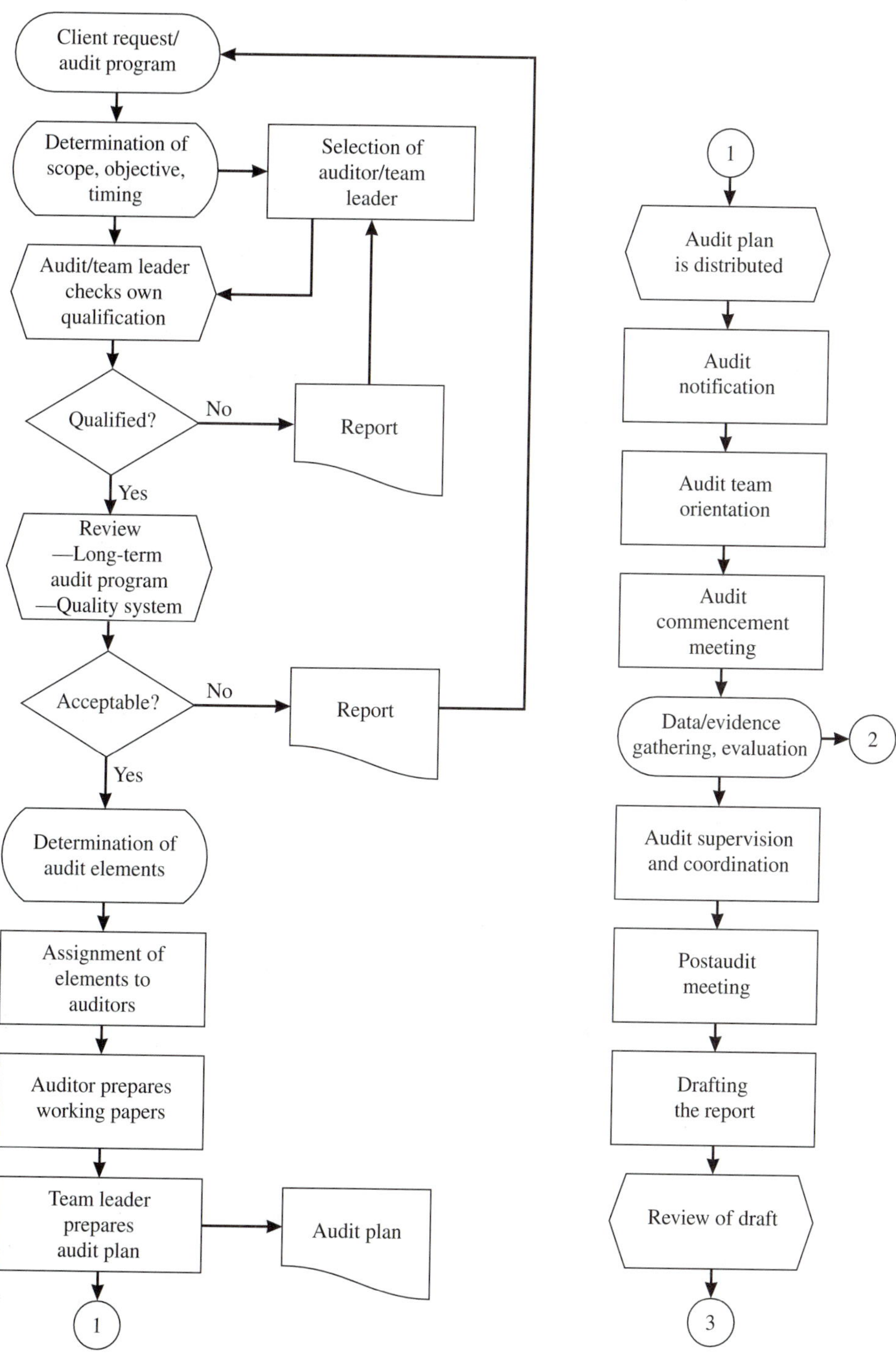

FIGURE 23.1
Flowchart for quality audit. (*Adapted from ANSI/ASQC, 1986, pp. 9–13.*)

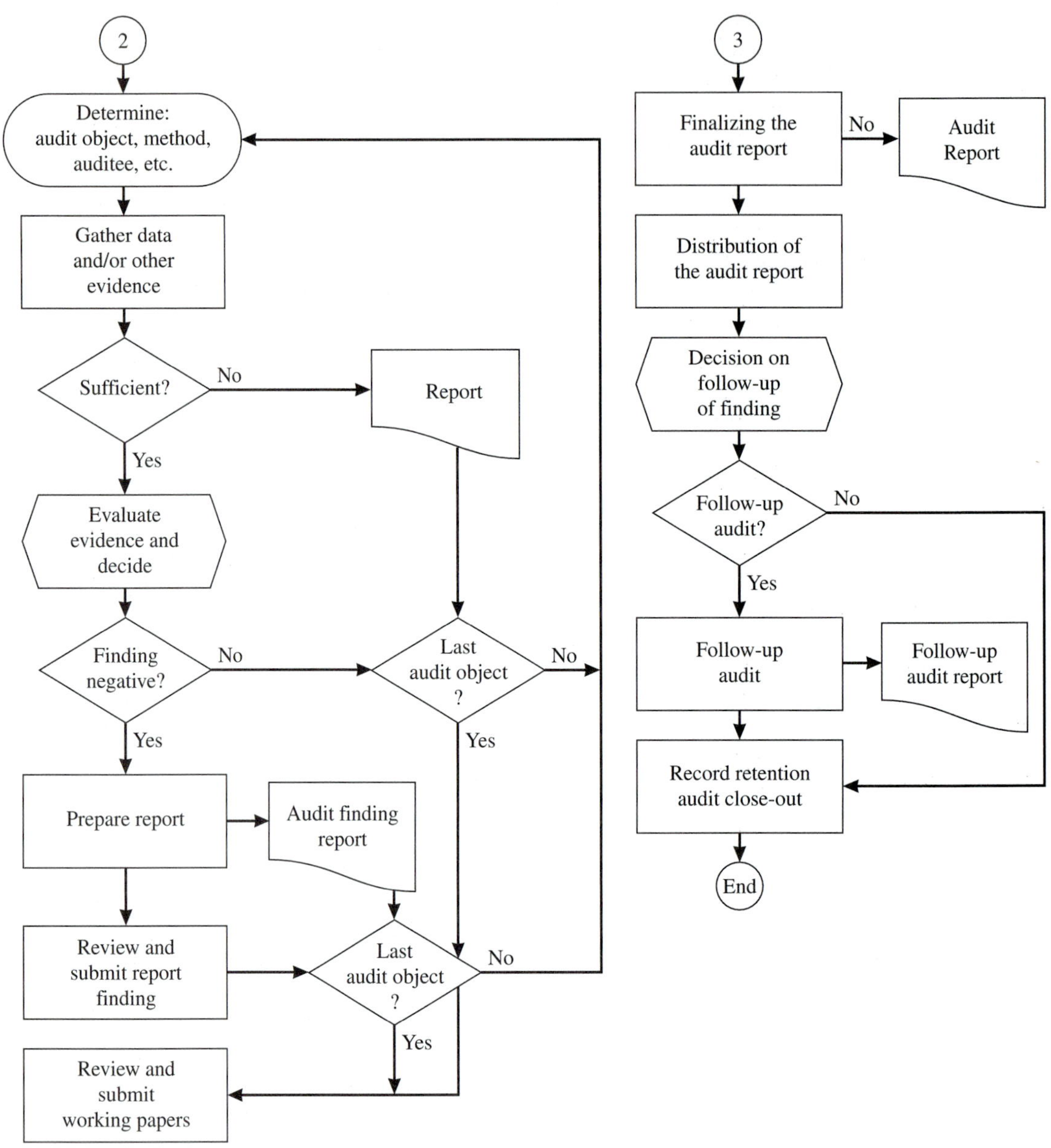

FIGURE 23.1 *(continued)*

Auditing is a sensitive task. A survey of auditors and auditees in the financial services industry investigated five attributes of the audit and auditor: professionalism, business knowledge, risk perspective, audit planning and conduct, and reporting audit results. Several surprises: Auditees viewed professionalism (objectivity, knowledge of the area being audited) as three times more important than auditors did; auditors thought risk perspective (coverage of key risk areas, audit in sufficient detail) as three times more important than auditees did. Notice how this disparity illustrates the importance of understanding the needs of audit customers.

Use of reference standards and checklists

As far as possible, the auditor is expected to compare activities as they are with some objective standard of what they should be. Where such standards are available, there is less need for the auditor to make a subjective judgment and thereby less opportunity for wide differences of opinion. However, provision should be made for challenge of the standard itself. The reference standards normally available include

- Written policies of the company as they apply to quality.
- Stated objectives in the budgets, programs, contracts, etc.
- Customer and company quality specifications.
- Pertinent government specifications and handbooks.
- Company, industry, and other pertinent quality standards on products, processes, and computer software.
- Published guides for conduct of quality audits.
- Pertinent quality departmental instructions.
- General literature on auditing.

One type of checklist identifies areas of subject matter that are to be checked, leaving it to the auditor to supply the detailed checklist. Typical examples of such areas are maintenance of machines and tools or control of engineering change orders. Having a standard for comparison in audits is an important matter. Steven Ehrhardt of Mallinckrodt Inc. uses a perceptive principle. When his audit group is asked to make an audit, he first determines whether the company has a standard or clear job specification for the activity. If not, then *before* making the audit, he asks those responsible for the activity to define a standard. Without a standard, Ehrhardt believes, people do not know what they are supposed to do (the first element of self-control, see Section 5.3), and thus an audit is not appropriate. Clarification of standards may be more important than making the audit itself. Imagine how this initiative to provide service by an auditor helps to build a trusting relationship with operations people.

Some checklists go into great detail, requiring the auditor to check numerous items of operational performance (and to record the fact that such items were checked). For example, an auditor checking a test performed by an inspector might be required to check the work of the inspector as to the correctness of the specification issue number used, the list of characteristics checked, the type of instruments employed, the sample size, data entries, etc. In a hospital an audit checklist could include questions such as: Are applicable intravenous solutions stored under refrigeration prior to delivery? Are all drugs, chemicals, and biologicals clearly, accurately, and appropriately labeled? For an example, see Table 16.5.

Section 16.8 discusses process quality audits in manufacturing operations; Section 17.5 explains applications of process quality audits in service operations.

23.7 AUDIT PERFORMANCE

Several policy issues affect audit performance.

Verification of facts

Auditors are universally expected to review with the line supervision the facts (outward symptoms) of any deficiencies discovered during the audit. The facts should be agreed on before the item enters a report that will go to higher management.

Discovery of causes

Many companies expect the auditor to investigate major deficiencies in an effort to determine their causes. This investigation then becomes the basis of the auditor's recommendation. Other companies expect the auditor to leave such investigations to the line people; audit recommendations will then include proposals that such investigations be made.

Recommendations and remedies

Auditors are invariably expected to make recommendations with a view to reducing deficiencies and improving performance. In contrast, auditors are commonly told to avoid becoming involved in designing remedies and making them effective. However, auditors are expected to follow up recommendations to assure that something specific is done, i.e., that the recommendation is accepted or else considered and rejected.

Policy issues are often incorporated into a "quality audit manual." Such a manual also includes details on the subject matter to be covered in audits; checklists of items to be checked and questions to be asked; classification of the seriousness of deficiencies observed; use of software for entry, processing, storage, and retrieval of audit data; and guidelines for audit reports.

Status of the audit

The key customer should be kept informed about progress of the audit—what has been covered and what remains to be done. The status of lengthy audits can be reported through debriefing meetings, informal discussions, and electronic mail. Status reporting includes explaining what deficiencies or problems have been detected—even before preparing a draft of the audit report. Status reports enable the company to check the accuracy of the auditors' observations and give the people responsible for the activity audited a chance to explain their plan to correct the deficiency.

23.8 AUDIT REPORTING

Audit results should be documented in a report, and a draft should be reviewed (preferably at the postaudit meeting) with the management of the activity that was audited.

Auditors and the activity audited should agree in advance on the distribution of the audit report. If desired, the report may be issued by the auditor and the auditee. All members of an audit team and the auditee should sign the report.

The report should include the following items:

- Executive summary.
- Purpose and scope of the audit.
- Details of the audit plan, including audit personnel, dates, the activity that was audited (personnel contacted, material reviewed, number of observations made, etc.). Details should be placed in an appendix.
- Standards, checklist, or other reference documents that were used during the audit.
- Audit observations, including supporting evidence, conclusions, and recommendations—using the audit customer's terminology.
- Recommendations for improvement opportunities.
- Recommendations for follow-up on the corrective action that is to be proposed and implemented by line management, along with subsequent audits if necessary.
- Distribution list for the audit report.

Bucella (1988) describes, for the Warner-Lambert pharmaceutical company, the audit approach used for "Ten Systems of Quality" (e.g., quality information). A histogram provides an overall picture of the effectiveness of each system. The histogram also shows the corrective action response time required for each system.

Summarizing audit data

In an audit most elements of performance are found to be adequate, while some are found to be in a state of discrepancy. Reporting of these findings requires two levels of communication:

1. *Reports of each discrepancy to secure corrective action.* These reports are made promptly to the responsible operating personnel, with copies to some of the managerial levels.
2. *A report of the overall status of the subject matter under review.* To meet these requirements, the report should
   - Evaluate overall quality performance in ways that provide answers to the major questions raised by upper managers, e.g., Is the product safe? Are we complying with legal requirements? Is the product fit for use? Is the product marketable? Is the performance of the department under review adequate?
   - Provide evaluations of the status of the major subdivisions of the overall performance—the quality systems and subsystems, the divisions, the plants, the procedures, etc.
   - Provide some estimate of the frequency of inadequacies in relation to the number of opportunities for inadequacies (see below under "Units of Measure").
   - Provide some estimate of the trend of this ratio (of inadequacies found to inadequacies possible) and of the effectiveness of programs to control the frequency of occurrence of inadequacies.

Seriousness classification

Some audit programs make use of seriousness classification of inadequacies. This approach is quite common in product audits, where defects found are classified in terms such as critical, major, and minor, each with some "weight" in the form of demerits. These systems of seriousness classification are highly standardized (see Section 19.5, "Seriousness Classification").

Some audit programs also apply seriousness classification to discrepancies found in planning, in procedures, in decision making, in data recording, and so on. The approach parallels that used for product audits. Definitions are established for such terms as "serious," "major," and "minor"; demerit values are assigned; and total demerits are computed.

Units of measure

For audits of plans, procedures, documentation, etc., it is desirable to compare the inadequacies found against some estimate of the opportunities for inadequacies. Some companies provide an actual count of the opportunities, such as the number of criteria or check points called out by the plans and procedures. Another form is to count the inadequacies per audit with a correction factor based on the length of time consumed by the audit. The obvious reason is that more time spent in auditing means more ground covered and more inadequacies found.

Distribution of audit report

Traditionally, copies of the audit report are sent to upper management for notification, review, and possible follow-up. Clearly, managers of audited activities are not happy with audit reports listing various deficiencies that are sent to their superiors. With a view to promoting harmony and a constructive viewpoint on audits, some organizations have adopted a different policy. The audit report is sent only to the manager whose activity is audited, and a follow-up audit is scheduled. If the deficiencies are corrected in time for the follow-up audit, the audit file is closed; otherwise, a copy of both audit reports is sent to upper management.

In the spirit of ongoing improvement, after the report is issued the auditees should be asked about the value they received from the audit and the report.

Regel (2000) reports on a survey of quality auditors. Of their seven primary concerns in achieving closure on their audits, the one ranked first in importance was "report issues," accounting for 25.3 percent of the total concerns. Examples of reasons included "insufficient detail," "not linked to business goals," "not stated clearly," and "not stated in management language."

23.9 CORRECTIVE ACTION FOLLOW-UP

The final phase of the audit is follow-up to confirm that corrective action has been taken by the audited activity and that the corrective action is effective. The steps are shown in Figure 23.2 (from Russell and Regal, 1996; reprinted by permission of the ASQ). ASQ Quality Audit Division (2000) provides details on this process.

It is important to remember a key purpose of an audit: achieve improvement. If corrective action is not implemented for some reason, the auditor should first verify that the conclusions in the report are correct and agreed upon by the audited area. If

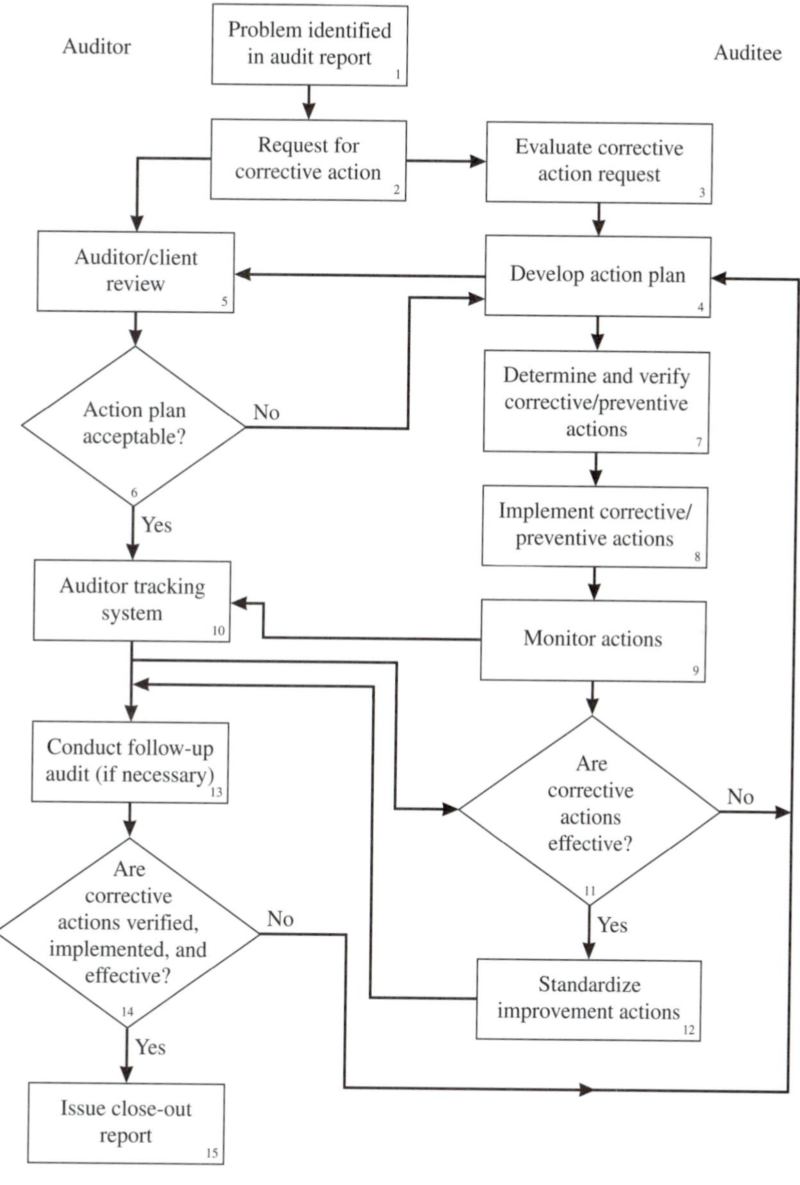

FIGURE 23.2
The audit function improvement process.

this is not the case, then the lack of agreement must be resolved. If there is agreement with the report but the audited area has not been able to implement corrective action because of lack of resources or other reasons, the auditor should determine whether he or she can somehow help the audited area. One possible approach is to review how the seriousness of the deficiency is presented to management. Restating the deficiency in monetary or other terms that will have an impact on management can help to obtain the necessary resources or remove obstacles to implementation of the corrective action. For an example, see Section 23.14.

23.10 HUMAN RELATIONS IN AUDITING

In theory, the audit is a sort of instrument plugged into operations to secure an independent source of information. Where it is a physical instrument, e.g., the propeller speed indicator on the bridge of a ship, there is no problem of clash of personalities. However, auditors are human beings, and in practice, their relationships with those whose work is being audited can become quite strained. Deficiencies turned up in the audit may be resented because of the implied criticism therein.

Recommendations in the audit may be resented as an invasion of responsibilities. In the reverse direction, auditors may regard slow responses to requests for information as a form of grudging cooperation. These and other human relations problems are sufficiently important to warrant extensive discussion plus indoctrination of both auditing personnel and operations personnel with respect to the following issues:

- *The reasons behind the audits.* These reasons may have been well discussed during the basic formulation of the audit program. However, that discussion was held among the managers. There is also a need to explain to both supervisors and nonsupervisors the "why" of the audits. (It is not enough to explain that upper management wants audits done.) Obviously, all employees are also customers, consumers, and concerned citizens, so it is easy to point out the benefits they derive from audits conducted in other companies. In addition, it can be made clear that the managers, customers, regulators, etc., of this company likewise require added assurance.
- *Avoiding an atmosphere of blame.* A sure way to cause a deterioration in human relations is to look for someone to blame rather than how to achieve improvement. Line managers as well as auditors can fall into this trap. An atmosphere of blame not only breeds resentment but also dries up the sources of information. Audit reports and recommendations should be problem oriented rather than person oriented.
- *Balance in reporting.* An audit that reports only deficiencies may be factual as far as it goes. Yet it will be resented because nothing is said about the far greater number of elements of performance that are done well. ("Even a broken clock is correct twice a day"—Anonymous.) Some companies require the auditors to start their reports with "commendable observations." Others have evolved overall summaries or ratings that consider not only deficiencies but also the opportunities for deficiencies (see below).

- *Depersonalizing the report.* In many companies auditors derive much influence from the fact that their reports are reviewed by upper management. Auditing departments should be careful to avoid misusing this influence. The ideal is to depersonalize the reports and recommendations. The real basis of the recommendations should be the facts rather than the opinion of the auditor. Where there is room for a difference of opinion, auditors have a right and a duty to give their opinion as an input to the decision-making process. However, any position of undue advocacy should be avoided, as this tends to reduce the auditor's credibility as an objective observer. (The ultimate responsibility for results rests on the line managers, not on the auditors.)
- *Postaudit meeting.* An important part of the implementation phase is the postaudit meeting that is held with the manager of the audited activity. At this meeting the audit observations are presented so that the manager can plan for corrective action. In addition, the manager can point out to the auditor any mistakes with respect to the facts that have been collected.

A self-audit and an independent audit can be combined to provide a two-tier audit. Each audit has an audit plan, execution, and report. The advantages include using the expertise of the person responsible for the function, assuring objectivity with an independent auditor, and minimizing some of the human relationship issues.

The aim of both the self-audit and the independent audit is to build an atmosphere of trust based on the prior reputation of the auditors, the approach used during the audit, and an emphasis on being helpful to the activity audited. Even such small matters as the title of the audit process should be carefully considered. Occasionally, people try to avoid the use of the term *audit* when, indeed, what will be done *is* an observation and evaluation. Also, audits may be hidden in a company education program. Such subterfuges detract from the trust that must be developed for audits to be effective and useful.

For a discussion of some unique human relations aspects in a research and development activity, see Frank and Voigt (1988).

Auditors must function within the culture of an organization. Anthropology is the study of the origin; behavior; and physical, social, and cultural development of human beings. For the astute comments of a quality manager (who is a certified quality auditor) with a formal background in anthropology, see Hunt (1997).

Quality auditors must, of course, adhere to the highest standards of ethics and professional conduct. For a discussion of the specifics including the American Society for Quality Code of Ethics and the Institute of Internal Auditors Code of Ethics, see ASQ Quality Audit Division (2000).

23.11 QUALITY ASSESSMENTS

Audits, as described above, are concerned almost exclusively with conformance of various sorts: conformance of plans to standards of good planning and conformance of execution to plan. Such audits provide answers to some vital questions and must be regarded as an essential element of quality assurance. These audits are, however, not sufficient to provide full assurance to upper management that all is well with respect to quality, since they commonly are not concerned with such matters as:

- Relative standing in the marketplace with regard to quality.
- Analysis of users' situations with respect to cost, convenience, etc., over the life of the product.
- Opportunities for reducing costs of poor quality.
- Challenges to product development, design engineering, and other "monopolistic" departments on quality adequacy, perfectionism, cost, etc.
- Challenges to top management with respect to policies, goals, premises, and axiomatic beliefs.
- Employee perceptions on quality.

Providing such missing elements of quality assurance requires a broader view than the structured audit. The broader review is often called a "quality assessment," a "quality survey," or a "companywide audit." In this section the word *audit* implies the existence of established criteria against which plans and their execution can be checked. In contrast, the word *survey* implies the inclusion of matters not covered by agreed-on criteria. (In a sense, the audit discovers discrepancies and alarm signals; the survey goes further and also discovers opportunities and unexpected threats.)

Such assessments can be accomplished in several ways:

1. Using an overall framework that includes an assessment of both quality results and the quality system. Chapter 2 presented such a framework consisting of four components: (a) cost of poor quality, (b) standing in the marketplace, (c) company culture on quality, and (d) assessment of current quality activities.
2. Assessing the quality system by using published criteria that emphasize quality *results*. One such criterion is the Malcolm Baldrige National Quality Award (see Section 2.8).
3. Assessing the quality system by using published criteria that emphasize defined elements of the quality system. One such criterion is the ISO 9000 specification.
4. Assessing the quality system by using criteria developed within a company for evaluating its own operations. For an example, see Table 23.2. Often such criteria draw upon those contained in the National Quality Award or the ISO 9000 specification.
5. Assessing the quality system by using criteria developed within a company for evaluating its suppliers (see Section 15.5). This approach may involve a third-party audit or survey. Two parties are involved in the purchase of products: the purchaser and the supplier. In the past the purchaser evaluated the quality from the supplier. Purchasers are now increasingly using the concept of a third party to evaluate supplier quality. The third party (a person or an organization) performs the evaluation service for the purchaser but is independent of the purchaser and the supplier. This practice relieves the purchaser of maintaining a staff with the necessary skills. A third party can inspect the product, evaluate the quality system, or both.
6. Assessing the quality system for a specific purpose. An example is the quality system inspection technique (QSIT) used by the Food and Drug Administration to

assess the quality system for medical device manufacturers. This technique focuses on management controls, design controls, corrective and preventive actions, and production and process controls. For elaboration, see Layloff (2000).

In a different type of survey, a consultant was asked to define specific task responsibilities in the quality program for all major departments of a health care company. The consultant used five questions to interview the department managers:

1. What tasks in your department affect quality?
2. Should any additional quality-related tasks be performed in your department?
3. Should any additional quality-related tasks be performed anywhere else in the company?
4. What quality-related tasks have unclear responsibility?
5. What quality-related tasks currently done in your department require more definitive written procedures?

The consultant summarized the findings as follows: (1) the quality program consisted of 178 tasks performed in 26 functional areas, (2) responsibility for 19 tasks was not clearly defined, and (3) for 27 quality-related tasks, the managers expressed concern about the clarity of the task.

The report included detailed comments on the overall scope of the quality program, key tasks requiring improvement, organization for quality, and the role of upper management.

Reference standards that provide criteria for quality systems are periodically revised. The "Standards Column" in *Quality Engineering* magazine provides the practitioner with updates on current standards and descriptions of new standards.

23.12 PRODUCT AUDIT

Product audit is an independent evaluation of product quality to determine its fitness for use and conformance to specification. Product auditing takes place after inspections have been completed. The purposes of product auditing include

1. Estimating the quality level as delivered to customers.
2. Evaluating the effectiveness of the inspection decisions in determining conformance to specifications.
3. Providing information useful in improving the outgoing product quality level and improving the effectiveness of inspection.
4. Providing additional assurance beyond routine inspection activities.

There is a good deal of logic behind creating such product audits. In many cases the inspection and testing department is subordinate to a manager who is also responsible for meeting other standards (schedules, costs, etc.). In addition, there is a value in reviewing the performance of the entire quality control function, which includes inspection and test planning as well as the conduct of the tests themselves. Finally, the more critical the product, the greater is the need for some redundancy as a form of assurance.

Stage of evaluation

Ideally, the product audit should compare actual service performance with users' service needs. This ideal is so difficult and costly to administer that most product auditing consists of an approximation (see Table 23.5).

For many simple, stable products, the approximation of test results versus specifications is a useful, economic way of conducting the product audit. Even for products not so simple, most quality characteristics identifiable by the user are also completely identifiable while the product is still at the factory. Thus product characteristics that are essential to use are properly evaluated at some appropriate stage, whether in the factory or in some more advanced stage of progression.

As products become increasingly complex, the product auditing is increasingly conducted at several of the stages shown in Table 23.5. The bulk of the characteristics may be evaluated at the most economical stage, i.e., shortly after factory inspection. However, the remaining (and usually more sophisticated) characteristics may be evaluated at other stages.

Scope of the product audit

The scope of some product audits completely misses the mark in measuring customer reaction.

As one example the plant manager of an electronics manufacturing firm received a rating of 98 percent on a product audit from the plant. For this rating the plant received an award for quality. When the mean time between failures of that same product was measured in the field, the value was only 200 hours. This problem was a known reason for customer complaints, but such matters had not been evaluated by the product audit.

TABLE 23.5
Potential stages of product auditing

Stage at which product auditing is conducted	Pros and cons of using this stage
After acceptance by inspectors	Most economical, but does not reflect effect of packing, shipping, storage, or usage
After packing but before shipment to field	Requires unpacking and repacking, but evaluates effect of original packing
Upon receipt by dealers	Difficult to administer at such multiple locations, but reflects effect of shipping, storage
Upon receipt by users	Even more difficult to administer, but evaluates the added effects of dealer handling and storage plus effects of shipment to user and unpacking
Performance in service	The ideal, but also the most difficult to administer because of the number and variety of usages; can be simplified through sampling

In another case a vehicle manufacturer had a system of taking a weekly product audit sample from production. A comparison of separate market research results with the internal product audit was devastating. Only 18 percent of the characteristics that customers claimed were important to them were being checked in the product audit.

For simple products a representative sample of finished goods may be bought on the open market. These samples are then checked for fitness for use and conformance to specification. Some companies conduct such audits annually as part of the broad annual planning for the product line. Such audits may include a review of competitive product as well.

For complex consumer products, e.g., household appliances, it is feasible to secure product audit data at multiple stages of the product progression shown in Table 23.5. The most extensive product audit takes place immediately following factory inspection and testing. Additional audit data are then secured from selected distributors and dealers under a special joint "open and test" audit. Similar arrangements are made to secure data from selected servicing dealers. In addition, use is made of the data from consumer "arrival cards." When properly arranged with due regard to time lags, all of these data sources can be charted in a way that shows trends as well as levels.

Audit plans must spell out, or give guidance on, the selection of detailed product dimensions or properties that are to be checked. Provision should be made for two types of audit—random and focused. The former is based on a random selection of product characteristics in order to yield an unbiased picture of the quality status. A focused audit, on the other hand, concentrates on a specific area of the product that experience suggests needs to be studied. Many companies use audit manuals to spell out the design of the audit for the auditor, almost to the last level of detail. For example, the manual may specify particular categories of dimensions to be audited (i.e., length) but may rely on the auditor to select which length dimension to audit.

23.13 SAMPLING FOR PRODUCT AUDIT

For products manufactured by mass production, sample sizes for product audit can often be determined by using conventional statistical methods. These methods determine the sample size required for stated degrees of risk (see Section 11.4). Sample sizes for product audit determined by these methods when applied to mass production still represent a small fraction of the product that needs to be sampled. In contrast, for products manufactured as large units or in small quantities, the conventional concepts of statistical sampling are prohibitively costly. In such cases sample sizes are often arbitrary, and they seem small from the viewpoint of probability considerations. For example, a vehicle manufacturer uses a product audit sample consisting of 2 percent of production per shift with a minimum of five vehicles—whichever number is larger. Even though the number of vehicles sampled may be small, the total number of characteristics that is sampled may be quite large. Auditors check 380 items on each vehicle, and the product audit test includes a 17-mile road test. In some cases of highly homogeneous production, a sample of one unit taken from batch production can be adequate for product audit.

23.14 REPORTING THE RESULTS OF PRODUCT AUDIT

The results of product audit appear in the form of the presence or absence of defects, failures, etc. A continuing score or "rating" of quality is then prepared based on the audit results.

Product audit programs often make use of seriousness classification of defects. Defects are classified in terms such as critical, major, minor A, minor B, each with some "weight" in the form of demerits. In product audits the usual unit of measure is *demerits per unit of product.*

EXAMPLE 23.1. A product audit system makes use of four classes of seriousness of defects. During one month the product auditors inspected 1200 finished units of product, with the following results:

Type of defects	Number found	Demerits per defect	Total demerits
Critical	1	100	100
Major	5	25	125
Minor A	21	5	105
Minor B	64	1	64
Total	91		394

Although the 91 defects found represented many defect types and four classes of seriousness, the total of 394 demerits, when divided by the 1200 units inspected, gives a single number, i.e., 0.33 demerit per unit.

The actual number of demerits per unit for the current month is often compared against historical data to observe trends. (Sometimes it is compared with competitors' products to judge the company's quality versus market quality.) A major value of a measure such as demerits per units is that it compares discrepancies found with the opportunity for discrepancies. Such an index appeals to operating personnel as being eminently fair.

The scoreboard in terms of demerits per unit is by no means universally accepted. Managers in some industries want ready access to the figures on critical and major defects. These managers believe that such figures represent the real problems regardless of demerits per unit.

It is often useful to summarize the product results in other languages. A manufacturer of consumer products classifies defects at a product audit as visual (V), electrical (E), and performance (P) and then predicts service costs on products in the field. This is done by first establishing classes for each type of defect in terms of the probability of receiving a field complaint (e.g., a class 2 visual defect has a 60 percent probability). Service call costs are then combined with the audit data. For example,

TABLE 23.6
Audit data of 50 units

Class of defect	Probability	Number of defects revealed by audit	Cost per service call, $	Expected costs, $	Expected number of service calls
V1	1.00	1	15.00	15.00	1.00
V2	0.60	3	15.00	27.00	1.80
E1	1.00	3	30.00	90.00	3.00
E2	0.60	4	30.00	72.00	2.40
P1	1.00	1	25.00	25.00	1.00
P2	0.60	2	25.00	30.00	1.20
P3	0.20	2	25.00	10.00	0.40
Totals				269.00	10.80

Table 23.6 shows the results of an audit of 50 units. The expected cost is the product of the probability, the number of defects, and the cost per service call. The expected service cost per unit is then estimated as \$269/50 = \$5.38. Alternatively, as indicated in Table 23.6, the expected number of service calls is the product of the probability and the number of defects. The expected number of service calls per unit is then estimated as 10.8/50 = 0.22, or about 22 out of every 100 products delivered to the field can be expected to have a service call.

In addition to summarizing the defects found (in both number and relative seriousness), the audit results can be tallied by functional responsibility (i.e., design, purchasing, production).

Audit results can also be summarized to show the effectiveness of the previous inspection activities. Typically, a simple ratio is used, such as the percentage of total defects detected by inspection. For example, if the previous inspection revealed a total of 45 defects in a sample of N pieces and if the product audit inspection revealed five additional defects, the inspection effectiveness would be (45/50)(100), or 90 percent.

Stravinskas (1989), Lane (1989), and Williams (1989) describe how AT&T Microelectronics changed from a traditional audit approach of reinspection of product (product audit) to an approach that uses system audits and process audits combined with a reduced amount of product audit. During one period, inspection costs were reduced by 12 percent, and the savings were then used to provide additional prevention activity, which resulted in a \$2 saving in failure costs for each \$1 added in prevention cost.

SUMMARY

- Quality assurance is the activity of providing evidence to establish confidence that quality requirements will be met.
- Quality audit is an independent review conducted to compare some aspect of quality performance with a standard for that performance. We conduct quality audits on activities that have an impact on product quality.

- Five ingredients are essential for successful audits: emphasis on facts, attitude of service on the part of auditors, identification of opportunities for improvement, addressing human relations issues, and the competence of auditors.
- Quality surveys provide a broader review of quality activities than audits of specific activities do.
- Product audit is an independent evaluation of product quality to determine fitness for use and conformance to specifications.

PROBLEMS

23.1. Visit a retail establishment that sells consumer electronics and observe the extent to which customers make use of their senses in securing quality assurance of the products they buy. Report your findings.

23.2. Visit an apartment building and discuss with the superintendent the various means used to obtain early warning of potential dangers, e.g., burglary, fire. Report your findings.

23.3. List the early-warning devices in use in a private home. Report your findings.

23.4. You are a quality manager. On one of your company's product lines, a report shows that power consumption has risen from the usual level of 35.5 W to a level of 35.9 W. This difference is, without a doubt, statistically significant. However, the line manager has taken no action to investigate the reason for the change on the grounds that (1) the product still conforms to the specification limit of maximum 36.4 W, and (2) he must give priority to several other products in which there are failures to comply with specification. What action do you take?

23.5. Visit a nearby facility that is a part of a chain of such facilities, e.g., food market, restaurant, motel, gasoline station. You will most likely find that it is subject to periodical quality audits from some headquarters. Obtain a copy of the auditor's checklist, study it, and report on its contents with respect to the various aspects of quality audits discussed in this chapter.

23.6. For any manufacturing company to which you have access, secure a copy of a quality audit. Study it and report on its contents with respect to the various aspects of quality audits discussed in this chapter.

REFERENCES

ANSI/ASQC Q10011-1,2,3, (1994). *Guidelines for Quality Auditing Quality Systems,* ASQ, Milwaukee.

ASQ Quality Audit Division (2000). *The Quality Audit Handbook,* 2nd ed., ASQ, Milwaukee.

Bucella, J. E. (1988). "Auditing—A New View," *ASQC Quality Congress Transactions,* Milwaukee, pp. 98–102.

Frank, N. C. and J. V. Voigt (1988). "Technical Auditors—A Positive Response for Auditing," *ASQC Quality Congress Transactions,* Milwaukee, pp. 94–97.

Hunt, J. R. (1997). "The Quality Auditor: Helping Beans Take Root," *Quality Progress,* December, pp. 27–33.

Lane, P. A. (1989). "Continuous Improvement—AT&T QA Audits," *ASQC Quality Congress Transactions,* Milwaukee, pp. 772–775.

Layloff, G. A. (2000). "The Quality System Inspection Technique," *Annual Quality Congress Proceedings,* ASQ, Milwaukee, pp. 266–267.

Malsbury, J. A. (1999). "Audits That Make a Difference," *Annual Quality Congress Proceedings,* ASQ, Milwaukee, pp. 559–563.

Regal, T. (2000). "Management Audit and Compliance Audit Compatibility," *Annual Quality Congress Proceedings,* ASQ, pp. 606–609.

Russell, J. P. and T. Regel (1996). *After the Quality Audit: Closing the Loop on the Audit Process,* Quality Press, ASQ, Milwaukee.

Stravinskas, J. M. (1989). "Manufacturing System and Process Audits," *ASQC Quality Congress Transactions,* Milwaukee, pp. 91–94.

Williams, C. A. (1989). "Improving Your Quality Auditing Systems," *ASQC Quality Congress Transactions,* Milwaukee, pp. 797–799.

SUPPLEMENTARY READING

ISO 9000: *JQH5,* Section 11.
Quality assurance: *JQH5,* pp. 2.13–2.15.
Quality audit: *JQH5,* pp. 22.60–22.62, pp. 27.35–27.36, pp. 29.18–29.19.

WEBSITE

American Society for Quality, quality audit division: www.asq.org/gad

Appendix I

STUDY GUIDE EXAMPLES OF CERTIFICATION EXAMINATION QUESTIONS AND ANSWERS

These examples were selected from the American Society for Quality certification brochures for Quality Engineer, Reliability Engineer, Quality Manager, Software Quality Engineer, and Quality Technician. Reprinted with permission by the ASQ.

Chapter 1

None.

Chapter 2

1. Which of the following is NOT an appropriate use of the Baldrige Award criteria? (a) self-assessment model; (b) quality system registration; (c) quality award application; (d) quality system model.

Chapter 3

1. A senior-level director is considering a \$10,000 investment to increase the quality rating of a piece of equipment from 85 percent to 95 percent and asks the quality manager for an opinion. The manager knows the equipment will require increased setup time that, in turn, will cause the overall availability of the equipment to decrease from 87 percent to 74 percent. In this situation the quality manager should respond in which of the following ways? (a) endorse the investment to improve quality; (b) discuss with the director the effectiveness measure that would result

from the investment; (c) consider the future value of the cost of this improvement in quality; (d) determine the costs of the downtime required to install the system before proceeding to implement the request.

2. Scatter diagrams are best described as (a) histograms; (b) correlation analysis; (c) Pareto analysis; (d) Ishikawa diagrams.
3. A form, in either diagram or table format, that is prepared in advance for recording data is known as a (a) cause-and-effect diagram; (b) Pareto chart; (c) flowchart; (d) check sheet.
4. Which of the following tools would be of the greatest use for finding the most efficient path and realistic schedule for the completion of a project? (a) interrelationship digraph; (b) activity network diagram; (c) tree diagram; (d) affinity diagram.
5. Which of the following best describes a diagnosis? (a) any state of unfitness for use or nonconformance to industry-specific specification; (b) an unproved assertion as to reasons for existence of defects and symptoms; (c) a proven reason for the existence of the symptom as illustrated by previous occurrences; (d) the process of studying symptoms, theorizing as to causes, testing theories, and discovering causes.
6. At what stage of the problem-solving process would a team most likely use a cause-effect diagram? (a) description of the process associated with the problem; (b) definition of the problem and its scope; (c) organization of possible problem causes; (d) collection of data to identify actual causes.
7. One of the best analytical methods to identify failure costs that represent the greatest return on the invested prevention dollar is to (a) perform an internal financial audit; (b) study the budget variance report; (c) apply the Gompertz curve technique; (d) apply the Pareto principle.

Chapter 4

1. The value of an exceptional guarantee is that it (a) builds long-term customer relationships while minimizing defects; (b) is possible for all customers to attain; (c) ensures that top-priority customers will pay the least amount of money; (d) allows a C_{pk} process to be in control.

Chapter 5

1. A company that supports the concept of an "internal supplier-customer relationship" should require (a) measurement of how well the supplier meets customer expectations; (b) internal feedback from the customer to the supplier through the quality department; (c) that each internal worker be responsible for satisfying the requirements of external customers; (d) that each individual worker be a customer to the previous operation and a supplier to the next operation.

Chapter 6

1. A process improvement team has studied the flow of product through the company's production system. To increase output, the most effective action would be to (a) shorten the critical path; (b) eliminate bottlenecks; (c) reduce quality check points; (d) change the sampling plan.

Chapter 7

1. Which of the following approaches to quality improvement planning connects quality and profits? (a) identifying, analyzing, and controlling all cost-of-quality costs for the business; (b) concentrating efforts on improving nonfinancial measures of quality; (c) developing a strategic quality plan that has financial and nonfinancial goals and that integrates business and financial planning processes; (d) focusing on reforms in management-employee relationships, worker training, new measurement schemes, and increased employee awareness of customer attitudes.

Chapter 8

1. To ensure success in implementing quality initiatives, the most important factor is (a) an empowered workforce; (b) a training program that explains and promotes the quality initiative; (c) upper-management support; (d) a reward and recognition program.
2. Rank order, from first to last, the steps listed below in the development of an employment requirements plan for a department or organization: (1) make an organization chart; (2) determine the amount of time and skills required to complete the activities; (3) list all activities required to produce the end product; (4) determine end products or output of the organization; (5) determine the number of people and skills needed. (a) 1,3,2,4,5; (b) 1,5,3,2,4; (c) 3,4,2,5,1; (d) 4,3,2,5,1.
3. One of the most effective means of implementing quality initiatives is for executive management to (a) establish quality goals tied to organizational performance; (b) conduct meetings on quality and demonstrate support for initiatives; (c) make public announcement explaining the company's quality goals; (d) hire a quality consultant to develop a total quality plan and lead its implementation.
4. Which of the following is the most effective way for a quality manager to lead the work activities of a quality department? (a) hold regular meetings to review performance against established goals and objectives; (b) review weekly written reports of activities submitted by staff; (c) conduct periodic meetings to flow down information about ongoing operations; (d) discuss the activities with the supervisors within the department.
5. Which of the following is the most critical role for a quality manager? (a) staffing the quality function to support the organization's stated quality objectives; (b) defining,

fully supporting, and providing leadership of the quality policy; (c) implementing changes in the cost-of-quality system; (d) examining the current quality level of all products and services.

Chapter 9

1. Training effectiveness can be measured most accurately by (a) trainers rating their own performance against a professionally developed grading system; (b) trainers analyzing trainees' job performance before training and after training; (c) trainees using a rating scale to assess their training session; (d) supervisors rating the improvement of employees' skills at the end of the training session.

Chapter 10

1. Weibull analysis is a way to quickly and easily analyze field data or interval test data. The limits of the use of this technique include having a good estimate for the: (a) MTBF; (b) expected life; (c) shape parameter; (d) average quality of the production lots.
2. A Weibull distribution has been found to describe the reliability distribution with characteristic life = 12,000 hours and shape parameter $\beta = 2.2$. If these are good parameters, at what time will reliability decrease to .85? (a) 2204 hours; (b) 3503 hours; (c) 4838 hours; (d) 5254 hours.
3. Which of the following tests may be used to determine whether a sample comes from a population with an exponential distribution? (a) t; (b) F; (c) Chi-square; (d) ANOVA.
4. A major drawback of using histograms in process control is that they (a) do not readily account for the factor of time; (b) are relatively difficult to construct and interpret; (c) require too many data points; (d) require too many intervals.
5. Statistical validity implies that (a) meaningful conclusions can be drawn from the data; (b) there is a mathematical derivation behind the technique used; (c) ANSI/ASQC Z1.4-1993 must be used; (d) the sample was obtained in a random manner.

Chapter 11

1. In an analysis of variance, which of the following distributions is the basis for determining whether the variance estimates are all from the same population? (a) Chi square; (b) Student's t; (c) Normal; (d) F.
2. A full factorial design of experiments has four factors. The first factor has two levels, the second factor has three levels, the third factor has two levels, and the final factor has four levels. How many runs are required for this analysis? (a) 16; (b) 48; (c) 192; (d) 256.

3. The correlation coefficient for the length and weight of units made by a process is 0.27. If the process were adjusted to reduce the weight of each unit by 0.5 ounce, the correlation coefficient of the length and weight of the units made by the new process will be equal to (a) 0.50; (b) 0.27; (c) 0.23; (d) -0.23.
4. A manufacturer of air conditioners wants to estimate the mean life (years from installation to replacement) of its units. The error level is set at 0.5 year, a desired probability $(1 - \alpha)$ of 95 percent is selected, and the standard deviation of unit life is given as 6.0 years. If unit life is normally distributed, then the required sample size for the desired estimate is equal to (a) 283; (b) 291; (c) 554; (d) 585.
5. The correct value for the expected frequency of cell I in the contingency table is (a) 28; (b) 42; (c) 52; (d) 78.

	Alternative		
Result	**A**	**B**	**Total**
X	I	II	80
Y	III	IV	120
Total	130	70	200

6. It is most appropriate to use a t-test for a hypothesis test to compare the (a) variances of two distributions; (b) standard deviations of two distributions; (c) means of two distributions; (d) medians of two distributions.

Chapter 12

None.

Chapter 13

1. Balancing a reliability requirement against other design parameters, such as performance, cost, or schedule, and then analyzing the consequences of placing special emphasis on one of these factors is called (a) reliability allocation; (b) reliability predictions; (c) trade-off decisions; (d) system modeling.
2. Software reliability planning includes all of the following EXCEPT (a) selecting models for data analysis and prediction; (b) modeling acquisition of computer software systems; (c) trade-offs of general purpose programs versus commercially available programs; (d) trade-offs involving cost, schedule, and failure intensity of software products.

Questions 3, 4, and 5 refer to the following situation: A high incidence of failures has developed during aircraft acceptance testing over the last several months. The identified failure is that an instrument panel light has malfunctioned on 6 of the last 10 aircraft tested. This problem needs to be investigated and a Failure Reporting and Corrective Action System (FRACAS) needs to be completed without stopping aircraft production.

3. The first step of the investigation should be to (a) collect additional data on similar events over the last two years; (b) conduct failure analysis to determine the failure mode and mechanism; (c) conduct surveillance testing on suspect components; (d) establish a cross-functional team to brainstorm on the cause and effect.
4. If the cause of the failure is determined to be a faulty subassembly manufactured only by a single supplier and this situation is threatening to shut down aircraft production, the next step should be to (a) visit the supplier to assist in determining the root cause of the problem; (b) initiate a supplier corrective action and return all of the unsorted inventory; (c) issue a Government and Industry Data Exchange Program (GIDEP) alert; (d) update the inspection instruction and retrain receiving inspection.
5. If a corrective action notice was sent to the supplier of a faulty subassembly and the supplier's response states that the root cause is simply an operation error, the next step should be to (a) accept the response and close the FRACAS; (b) visit the supplier to develop a better understanding of the root cause; (c) issue a Government and Industry Data Exchange Program (GIDEP) alert; (d) begin looking for a new supplier.
6. Which of the following is an appropriate use for experimental design? (a) establishing product requirements; (b) developing a fault tree analysis; (c) ensuring the robust design of a product; (d) analyzing customer complaint reports.
7. According to Taguchi, robustly designed experiments should employ all of the following techniques EXCEPT (a) inner and outer arrays; (b) signal-to-noise ratios; (c) linear graphs; (d) foldover capabilities.
8. All of the following are purposes of a production reliability assurance test (PRAT) EXCEPT (a) detect significant shifts between the as-built reliability requirements and the as-designed reliability requirements; (b) assess performance against reliability requirements; (c) assess actual product reliability against reliability requirements; (d) minimize the need for specific process controls.
9. Which of the following are important to elements of the concept of risk? (I) frequency, (II) schedule, (III) damage. (a) I and II only; (b) I and III only; (c) II and III only; (d) I, II, and III.
10. Which of the following tools is used to analyze the safety of a system? (a) fault tree analysis; (b) failure reporting and corrective action system; (c) reliability allocation; (d) environmental stress screening.
11. System-safety analytical techniques include all EXCEPT (a) hazards analyses; (b) fault tree analyses; (c) logic diagram analyses; (d) design readiness reviews.
12. The best way to set an overall reliability goal is to (a) write a specification calling for a product to have high reliability and incorporate it into a contract; (b) put down specific numerical requirements for reliability, statements of operating environments, and a definition of successful product performance; (c) insist that the goal be expressed in terms of mean-time-between-failures for all components and assemblies; (d) indicate who would be at fault if the desired reliability is not obtained during the warranty.
13. Compared to traditional engineering design, the time required to do concurrent engineering design is (a) greater during the requirement development phase;

(b) less during the requirement development phase; (c) greater during the implementation phase; (d) less during the equipment start-up phase.

14. The formal, documented, comprehensive, and systematic examination of a design that ensures requirements are met, identifies problems, and proposes solutions is known as a (a) quality review; (b) design review; (c) design examination; (d) failure mode, effect, and criticality analysis.
15. Which of the following analyses is best used to study the potential failures in a system? (a) failure analysis; (b) fault tree analysis; (c) reliability allocation analysis; (d) Pareto analysis.

Chapter 14

1. Which of the following is best defined as the practice of using parallel components and subsystems? (a) maintainability; (b) reliability; (c) optimization; (d) redundancy.
2. The lifetime of a mechanical lifter is normally distributed with a mean of 100 hours and a standard deviation of 3 hours. What is the reliability of the lifter at 106 hours? (a) 0.0228; (b) 0.0570; (c) 0.9430; (d) 0.9772.
3. In terms of reliability engineering, the failure rate is the reciprocal of the (a) mean time to repair; (b) mean-time-between-failures; (c) system effectiveness; (d) probability of failure.
4. Which of the following is NOT considered good practice in reliability design? (a) using proven parts; (b) using series design; (c) using failure mode and effects analysis (FMEA); (d) simplifying item configuration.
5. Which of the following measures can be used to find a quick approximation of the availability of a system? (a) mean time to failure (MTTF) and mean time to repair (MTTR); (b) failure rate and failure mode; (c) mission time and failure rate; (d) downtime and time to repair.
6. For a company operating multiple units of production equipment, the observed failure rate is 42×10^6 failures per operating hour, and the preventive maintenance rate is 320×10^6 actions per hour. What is the mean time between corrective and preventive maintenance (MTBM)? (a) 2688.2 hours; (b) 2762.4 hours; (c) 2840.9 hours; (d) 26,935.0 hours.
7. A go/no-go device is tested until it fails. If X is the number to tests to first failure with no wearout present and the probability of success on each test is .99, then the probability that X is greater than 5 is (a) 0.9310; (b) 0.9410; (c) 0.9510; (d) 0.9610.
8. A system consists of four parallel units each having a reliability of 0.80. The system can still complete its mission with only two units functioning. If the failure rate is constant and failures are independent then the system reliability will be (a) 0.4096; (b) 0.5376; (c) 0.8192; (d) 0.9728.
9. Given a reliability growth test in progress having accumulated four failures during 5000 test hours. Assume a growth rate of 0.3, what is the expected MTBF at 25,000 hours? (a) 1250 hours; (b) 1895 hours; (c) 2026 hours; (d) 3856 hours.

10. A data entry system consists of an input terminal and a host system. The eight-hour reliability of the terminal is .98, and the eight-hour reliability of the host is .95. What is the eight-hour reliability of the system? (a) 0.931; (b) 0.965; (c) 0.980; (d) 0.950.

Chapter 15

1. Vendor audits are usually arranged and coordinated by the (a) quality assurance department; (b) manufacturing department; (c) engineering department; (d) purchasing department.

Chapter 16

1. Which of the following is an example of the hub system of document control? (a) area document control coordinators are responsible for issuing and controlling documents from departments in their area; (b) a document control coordinator is responsible for establishing and implementing a companywide standard for document control; (c) satellite document stations are established throughout the facility; (d) each department is responsible for issuing and controlling its own documents.
2. A control plan is designed to do which of the following? (a) supplement information contained in operator instructions; (b) support the production scheduling system; (c) provide a documented system for controlling processes; (d) provide a method for tracking the design review process.
3. Using fuses for overloaded circuits in a computer is an example of which of the following error-proofing principles? (a) mitigation; (b) detection; (c) facilitation; (d) replacement.
4. Process control audits are used to (a) assist in determining process capability; (b) replace thorough systems and procedures audits; (c) replace thorough in-line product audits; (d) ensure that some minimal auditing is conducted on a product.

Chapter 17

None.

Chapter 18

1. Which of the following tools are appropriate for a quality engineer to use in qualifying a process that has variables data?

   I. An $\bar{X}$ and R control chart.
   II. Histogram.

III. A *c* chart
IV. A *p* chart.

(a) I and II only; (b) II and III only; (c) III and IV only; (d) I, II, and IV only.

2. To determine the average number of nonconforming parts over time, which of the following attribute control charts would be most appropriate? (a) c chart; (b) np chart; (c) p chart; (d) u chart.
3. Steel bars are cut to cylindrical shafts by means of a lathe. The diameter and allowable tolerance of the shaft is 2.000 ± 0.001 inch. A control chart is used to monitor the quality level of the process. Which of the following plots on the control chart might indicate a problem of wear on the lathe? (a) the diameter of a single shaft above 2.001 inch; (b) the diameter of a single shaft below 1.999 inch; (c) an apparent increasing trend in the shaft diameters; (d) erratic in-tolerance or out-of-tolerance diameter measurements.
4. Special process studies that involve investigation and tests are used to do which of the following:

I. Locate the causes of nonconforming products.
II. Determine the possibility of improving quality characteristics.
III. Ensure that improvement and corrective actions are permanent and complete.
IV. Increase production throughput.

(a) I and IV only; (b) II and III only; (c) I, II, and III only; (d) II, III, and IV only.

Chapter 19

1. A lot size of 450 pieces is being inspected for attributes, single sampling, normal inspection, and general inspection II with an AQL of 1.0 using ANSI/ASQC Z1.4-1993 (MIL-STD-105E). On completion of the inspection, three nonconformances were found. Which of the following actions should be taken? (a) reject the lot; (b) accept the lot; (c) take a second sample; (d) reinspect the sample.
2. Nonconformances were identified and segregated during a production shift. The manager of the next shift on reviewing the nonconformances, determined that they were slight, and requested that the material be released. What action should be taken next? (a) The material should be released; (b) all concerned groups should concur; (c) engineering approval should be obtained; (d) documented company procedures should be followed.
3. In deciding whether sampling inspection of a part would be more economical than conducting 100 percent inspection, a technician needs to determine all of the following EXCEPT the cost of (a) inspecting parts; (b) destructive testing; (c) finding no defective parts; (d) improving the production process.
4. A sample consists of one or more units of product drawn from a lot or batch on the basis of (a) defect of the product; (b) random selection; (c) size of the product; (d) when the inspection process was completed.
5. A classification of defects is a list of possible product defects that are classified according to (a) acceptability; (b) seriousness; (c) quantity; (d) size.

6. Attribute sampling should be used when (a) the population contains attributes; (b) a yes-or-no decision is to be made; (c) the population has variability; (d) multistage sampling plan is needed.
7. The investment in automated test equipment is often justified under which of the following circumstances? (a) Numerous tests must be performed; (b) repair times must be short; (c) conformance records are required; (d) traceable records are required.

Chapter 20

1. A quality engineer discovers that an alloy with a strength lower than specification was being used due to tremendous cost savings. There have been no instances of field failure. At this point, the quality engineer should (a) increase the number of field failure evaluations performed; (b) adjust the specifications based on the field performance history; (c) document the findings and take no further action; (d) report the discovery to appropriate supervision and request corrective action.
2. Customer complaint data can reveal which of the following?

   I. That a low complaint rate is not proof of customer satisfaction.
   II. That a high complaint rate is proof of customer dissatisfaction.
   III. Conclusive measures of product/service performance.

   (a) I and II only; (b) I and III only; (c) II and III only; (d) I, II, and III.

Chapter 21

None.

Chapter 22

None.

Chapter 23

1. When a company evaluates its own performance, it is conducting what type of audit? (a) first party; (b) second party; (c) third party; (d) extrinsic.
2. When an audit team concludes that a finding demonstrates a breakdown of the quality management system, the finding should be documented as (a) a minor nonconformance; (b) a major nonconformance; (c) a deficiency; (d) an observation.
3. There are several types of quality assurance audits. Which of the following is not a type of quality assurance audit? (a) product audit; (b) primary audit; (c) process audit; (d) software audit.

4. Which of the following is not a benefit of using checklists in conducting an audit? (a) The checklist is a helpful memory aid; (b) the checklist provides uniformity to the auditing process; (c) the checklist will help identify who has the responsibility for deficiencies; (d) the checklist is a useful training aid.
5. Consider the statement "Even if corrective action is taken immediately for a finding during the audit, all significant deficiencies should be detailed in the written report." This statement is (a) true, because the auditor is responsible for an accurate report of all findings; (b) false, because the audit report should include only unresolved deficiencies; (c) false, because the corrective action can be verified during the audit; (d) true, because a follow-up audit will need to be made on all findings.
6. Qualifications for a lead auditor must include (a) the ability to field antagonistic responses; (b) the ability to differentiate "the vital few from the trivial many"; (c) technical expertise in the engineering field that will be audited; (d) answers (a) and (b) only.
7. Follow-up and closeout of audit-initiated corrective actions should be performed by (a) the corrective action board (where one exists) with the audit function's participation; (b) the audit function; (c) quality engineering with the audit function's participation; (d) an authority defined by company policy.
8. Which of the following is potentially the most costly part of an audit? (a) the auditor's time spent on the audit; (b) the auditee's time spent being audited; (c) the cost of processing nonconformance reports; (d) costs generated by using untrained or unsuitable auditors.
9. The primary purpose of audit working papers is to provide (a) evidence of the analysis of internal control; (b) support for the audit report; (c) a basis for evaluation audit personnel; (d) a guide for subsequent audits of the same area.
10. Audits should be scheduled with regard to (a) the availability of objective evidence; (b) obtaining an unbiased sample of evidence; (c) the demand pattern for end items; (d) answers (a) and (b) only.
11. Which one of the following activities best reflects the fact that the systematic management of quality audits includes planning and control functions? (a) verifying the implementation of an inspection system; (b) establishing an audit schedule and preparing an appropriate checklist; (c) reviewing the audit procedures and instructions; (d) determining the traceability within a change control system.
12. The audit report is an important information source for the client of the auditor. Many auditors may be employed in auditing organizations or departments. The audit supervisor should (a) avoid interfering with the audit once the auditor has been assigned to conduct an audit and write the report; (b) review and approve the report before submitting it to the client; (c) decide on the distribution of the report; (d) send the entire report to the auditor in charge of the follow-up audit.
13. When can an audit be discontinued? (a) when the audit objectives appear to have become unattainable; (b) when the auditee disputes important observations; (c) as soon as it appears that everything is in order; (d) once started, it must be completed.
14. The auditor should always let the auditee know about significant deficiencies promptly so that the auditee has the chance to take corrective action before the

postaudit conference (if possible). Which of the following is not a rationale for doing this? (a) It gives the auditee the chance to show a genuine concern for quality improvement; (b) it saves reaudit or follow-up and therefore saves time and money; (c) it allows the auditor to omit the deficiencies during the postaudit conference; (d) it shows that the audit can get results.

15. After auditors report deficiencies that have not yet been corrected, their role should be to (a) ensure that corrective action is done right; (b) leave all further action up to the auditee; (c) evaluate corrective action after it is taken; (d) inform the auditee's top management that they should monitor the corrective action.

ANSWERS TO EXAMPLES OF EXAMINATION QUESTIONS PROVIDED IN ASQ CERTIFICATION BROCHURES

Chapter 1:	—						
Chapter 2:	1(b)						
Chapter 3:	1(b)	2(b)	3(d)	4(b)	5(d)	6(c)	7(d)
Chapter 4:	1(a)						
Chapter 5:	1(d)						
Chapter 6:	1(b)						
Chapter 7:	1(c)						
Chapter 8:	1(c)	2(d)	3(a)	4(a)	5(b)		
Chapter 9:	1(b)						
Chapter 10:	1(c)	2(d)	3(c)	4(a)	5(a)		
Chapter 11:	1(d)	2(b)	3(b)	4(c)	5(c)	6(c)	
Chapter 12:	—						
Chapter 13:	1(c)	2(b)	3(b)	4(a)	5(b)	6(c)	
	7(d)	8(d)	9(b)	10(a)	11(d)	12(b)	
	13(a)	14(b)	15(b)				
Chapter 14:	1(d)	2(a)	3(b)	4(b)	5(a)	6(b)	7(c)
	8(d)	9(c)	10(a)				
Chapter 15:	1(d)						
Chapter 16:	1(b)	2(c)	3(a)	4(a)			
Chapter 17:	—						
Chapter 18:	1(a)	2(b)	3(c)	4(c)			
Chapter 19:	1(a)	2(d)	3(b)	4(d)	5(b)	6(b)	7(a)
Chapter 20:	1(d)	2(a)					
Chapter 21:	—						
Chapter 22:	—						
Chapter 23:	1(a)	2(b)	3(b)	4(c)	5(a)	6(d)	7(b)
	8(d)	9(b)	10(d)	11(b)	12(b)	13(a)	
	14(c)	15(c)					

Appendix II

TABLES

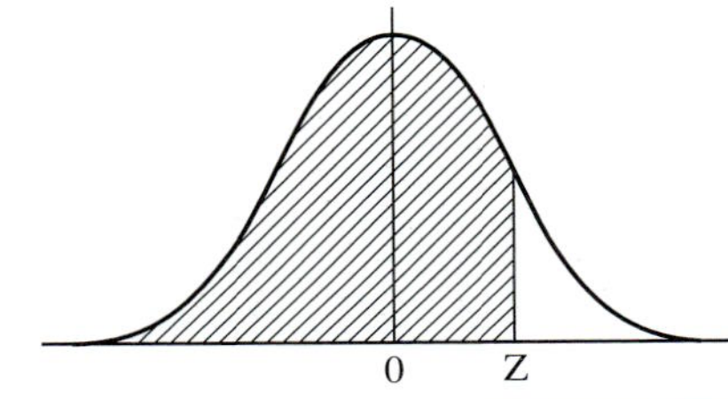

TABLE A

Normal distribution

Proportion of total area under the curve from $-\infty$ to $Z = \frac{X - \mu}{\sigma}$. To illustrate: when $Z = 2$, the probability is 0.9773 of obtaining a value equal to or less than X.

Z	0.09	0.08	0.07	0.06	0.05	0.04	0.03	0.02	0.01	0.00
−3.0	0.00100	0.00104	0.00107	0.00111	0.00114	0.00118	0.00122	0.00126	0.00131	0.00135
−2.9	0.0014	0.0014	0.0015	0.0015	0.0016	0.0016	0.0017	0.0017	0.0018	0.0019
−2.8	0.0019	0.0020	0.0021	0.0021	0.0022	0.0023	0.0023	0.0024	0.0025	0.0026
−2.7	0.0026	0.0027	0.0028	0.0029	0.0030	0.0031	0.0032	0.0033	0.0034	0.0035
−2.6	0.0036	0.0037	0.0038	0.0039	0.0040	0.0041	0.0043	0.0044	0.0045	0.0047
−2.5	0.0048	0.0049	0.0051	0.0052	0.0054	0.0055	0.0057	0.0059	0.0060	0.0062
−2.4	0.0064	0.0066	0.0068	0.0069	0.0071	0.0073	0.0075	0.0078	0.0080	0.0082
−2.3	0.0084	0.0087	0.0089	0.0091	0.0094	0.0096	0.0099	0.0102	0.0104	0.0107
−2.2	0.0110	0.0113	0.0116	0.0119	0.0122	0.0125	0.0129	0.0132	0.0136	0.0139
−2.1	0.0143	0.0146	0.0150	0.0154	0.0158	0.0162	0.0166	0.0170	0.0174	0.0179
−2.0	0.0183	0.0188	0.0192	0.0197	0.0202	0.0207	0.0212	0.0217	0.0222	0.0228
−1.9	0.0233	0.0239	0.0244	0.0250	0.0256	0.0262	0.0268	0.0274	0.0281	0.0287
−1.8	0.0294	0.0301	0.0307	0.0314	0.0322	0.0329	0.0336	0.0344	0.0351	0.0359
−1.7	0.0367	0.0375	0.0384	0.0392	0.0401	0.0409	0.0418	0.0427	0.0436	0.0446
−1.6	0.0455	0.0465	0.0475	0.0485	0.0495	0.0505	0.0516	0.0526	0.0537	0.0548
−1.5	0.0559	0.0571	0.0582	0.0594	0.0606	0.0618	0.0630	0.0643	0.0655	0.0668
−1.4	0.0681	0.0694	0.0708	0.0721	0.0735	0.0749	0.0764	0.0778	0.0793	0.0808
−1.3	0.0823	0.0838	0.0853	0.0869	0.0885	0.0901	0.0918	0.0934	0.0951	0.0968
−1.2	0.0985	0.1003	0.1020	0.1038	0.1057	0.1075	0.1093	0.1112	0.1131	0.1151
−1.1	0.1170	0.1190	0.1210	0.1230	0.1251	0.1271	0.1292	0.1314	0.1335	0.1357

Z	0.00	0.01	0.02	0.03	0.04	0.05	0.06	0.07	0.08	0.09
−1.0	0.1379	0.1401	0.1423	0.1446	0.1469	0.1492	0.1515	0.1539	0.1562	0.1587
−0.9	0.1611	0.1635	0.1660	0.1685	0.1711	0.1736	0.1762	0.1788	0.1814	0.1841
−0.8	0.1867	0.1894	0.1922	0.1949	0.1977	0.2005	0.2033	0.2061	0.2090	0.2119
−0.7	0.2148	0.2177	0.2207	0.2236	0.2266	0.2297	0.2327	0.2358	0.2389	0.2420
−0.6	0.2451	0.2483	0.2514	0.2546	0.2578	0.2611	0.2643	0.2676	0.2709	0.2743
−0.5	0.2776	0.2810	0.2843	0.2877	0.2912	0.2946	0.2981	0.3015	0.3050	0.3085
−0.4	0.3121	0.3156	0.3192	0.3228	0.3264	0.3300	0.3336	0.3372	0.3409	0.3446
−0.3	0.3483	0.3520	0.3557	0.3594	0.3632	0.3669	0.3707	0.3745	0.3783	0.3821
−0.2	0.3859	0.3897	0.3936	0.3974	0.4013	0.4052	0.4090	0.4129	0.4168	0.4207
−0.1	0.4247	0.4286	0.4325	0.4364	0.4404	0.4443	0.4483	0.4562	0.1562	0.4602
−0.0	0.4641	0.4681	0.4721	0.4761	0.4801	0.4840	0.4880	0.4920	0.4960	0.5000
+0.0	0.5000	0.5040	0.5080	0.5120	0.5160	0.5199	0.5239	0.5279	0.5319	0.5359
+0.1	0.5398	0.5438	0.5478	0.5517	0.5557	0.5596	0.5636	0.5675	0.5714	0.5753
+0.2	0.5793	0.5832	0.5871	0.5910	0.5948	0.5987	0.6026	0.6064	0.6103	0.6141
+0.3	0.6179	0.6217	0.6255	0.6293	0.6331	0.6368	0.6406	0.6443	0.6480	0.6517
+0.4	0.6554	0.6591	0.6628	0.6664	0.6700	0.6736	0.6772	0.6808	0.6844	0.6879
+0.5	0.6915	0.6950	0.6985	0.7019	0.7054	0.7088	0.7123	0.7157	0.7190	0.7224
+0.6	0.7257	0.7291	0.7324	0.7357	0.7389	0.7422	0.7454	0.7486	0.7517	0.7549
+0.7	0.7580	0.7611	0.7642	0.7673	0.7704	0.7734	0.7764	0.7794	0.7823	0.7852
+0.8	0.7881	0.7910	0.7939	0.7967	0.7995	0.8023	0.8051	0.8079	0.8106	0.8133
+0.9	0.8159	0.8186	0.8212	0.8238	0.8264	0.8289	0.8315	0.8340	0.8365	0.8389
+1.0	0.8413	0.8438	0.8461	0.8485	0.8508	0.8531	0.8554	0.8577	0.8599	0.8621
+1.1	0.8643	0.8665	0.8686	0.8708	0.8729	0.8749	0.8770	0.8790	0.8810	0.8830
+1.2	0.8849	0.8869	0.8888	0.8907	0.8925	0.8944	0.8962	0.8980	0.8997	0.9015
+1.3	0.9032	0.9049	0.9066	0.9082	0.9099	0.9115	0.9131	0.9147	0.9162	0.9177
+1.4	0.9192	0.9207	0.9222	0.9236	0.9251	0.9265	0.9279	0.9292	0.9306	0.9319
+1.5	0.9332	0.9345	0.9357	0.9370	0.9382	0.9394	0.9406	0.9418	0.9429	0.9441

(continued)

TABLE A *(continued)*

Z	0.00	0.01	0.02	0.03	0.04	0.05	0.06	0.07	0.08	0.09
+1.6	0.9452	0.9463	0.9474	0.9484	0.9495	0.9505	0.9515	0.9525	0.9535	0.9545
+1.7	0.9554	0.9564	0.9573	0.9582	0.9591	0.9599	0.9608	0.9616	0.9625	0.9633
+1.8	0.9641	0.9649	0.9656	0.9664	0.9671	0.9678	0.9686	0.9693	0.9699	0.9706
+1.9	0.9713	0.9719	0.9726	0.9732	0.9738	0.9744	0.9750	0.9756	0.9761	0.9767
+2.0	0.9773	0.9778	0.9783	0.9788	0.9793	0.9798	0.9803	0.9808	0.9812	0.9817
+2.1	0.9821	0.9826	0.9830	0.9834	0.9838	0.9842	0.9846	0.9850	0.9854	0.9857
+2.2	0.9861	0.9864	0.9868	0.9871	0.9875	0.9878	0.9881	0.9884	0.9887	0.9890
+2.3	0.9893	0.9896	0.9898	0.9901	0.9904	0.9906	0.9909	0.9911	0.9913	0.9916
+2.4	0.9918	0.9920	0.9922	0.9925	0.9927	0.9929	0.9931	0.9932	0.9934	0.9936
+2.5	0.9938	0.9940	0.9941	0.9943	0.9945	0.9946	0.9948	0.9949	0.9951	0.9952
+2.6	0.9953	0.9955	0.9956	0.9957	0.9959	0.9960	0.9961	0.9962	0.9963	0.9964
+2.7	0.9965	0.9966	0.9967	0.9968	0.9969	0.9970	0.9971	0.9972	0.9973	0.9974
+2.8	0.9974	0.9975	0.9976	0.9977	0.9977	0.9978	0.9979	0.9979	0.9980	0.9981
+2.9	0.9981	0.9982	0.9983	0.9983	0.9984	0.9984	0.9985	0.9985	0.9986	0.9986
+3.0	0.99865	0.99869	0.99874	0.99878	0.99882	0.99886	0.99889	0.99893	0.99896	0.99900

Source: Adapted with permission from Eugene L. Grant and Richard S. Leavenworth, *Statistical Quality Control,* 4th ed., McGraw-Hill, New York, 1972, pp. 642–643.

And we can keep going:

Z	0.00
+4.0	0.9999683
+5.0	0.9999997133
+6.0	0.9999999990

TABLE B

Exponential distribution values of $e^{-X/\mu}$ for various values

Fractional parts of the total area (1.000) under the exponential curve greater than X. To illustrate: if X/μ is 0.45, the probability of occurrence for a value greater than X is 0.6376.

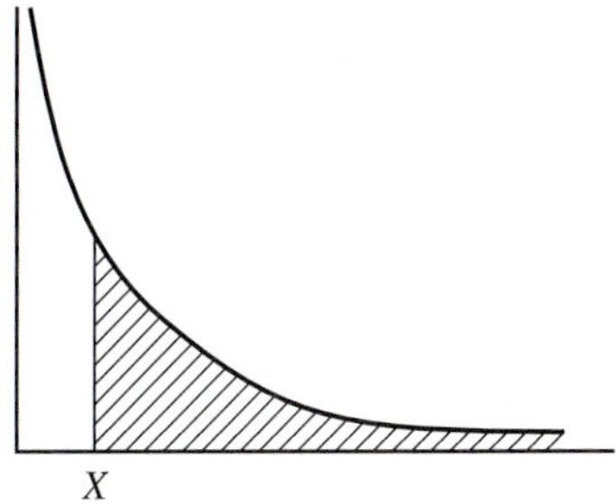

$\frac{X}{\mu}$	**0.00**	**0.01**	**0.02**	**0.03**	**0.04**	**0.05**	**0.06**	**0.07**	**0.08**	**0.09**
0.0	1.000	0.9900	0.9802	0.9704	0.9608	0.9512	0.9418	0.9324	0.9231	0.9139
0.1	0.9048	0.8958	0.8860	0.8781	0.8694	0.8607	0.8521	0.8437	0.8353	0.8270
0.2	0.8187	0.8106	0.8025	0.7945	0.7866	0.7788	0.7711	0.7634	0.7558	0.7483
0.3	0.7408	0.7334	0.7261	0.7189	0.7118	0.7047	0.6977	0.6907	0.6839	0.6771
0.4	0.6703	0.6637	0.6570	0.6505	0.6440	0.6376	0.6313	0.6250	0.6188	0.6126
0.5	0.6065	0.6005	0.5945	0.5886	0.5827	0.5769	0.5712	0.5655	0.5599	0.5543
0.6	0.5488	0.5434	0.5379	0.5326	0.5273	0.5220	0.5169	0.5117	0.5066	0.5016
0.7	0.4966	0.4916	0.4868	0.4819	0.4771	0.4724	0.4877	0.4630	0.4584	0.4538
0.8	0.4493	0.4449	0.4404	0.4360	0.4317	0.4274	0.4232	0.4190	0.4148	0.4107
0.9	0.4066	0.4025	0.3985	0.3946	0.3906	0.3867	0.3829	0.3791	0.3753	0.3716

$\frac{X}{\mu}$	**0.0**	**0.1**	**0.2**	**0.3**	**0.4**	**0.5**	**0.6**	**0.7**	**0.8**	**0.9**
1.0	0.3679	0.3329	0.3012	0.2725	0.2466	0.2231	0.2019	0.1827	0.1653	0.1496
2.0	0.1353	0.1225	0.1108	0.1003	0.0907	0.0821	0.0743	0,0672	0.0608	0.0550
3.0	0.0498	0.0450	0.0408	0.0369	0.0334	0.0302	0.0273	0.0247	0.0224	0.0202
4.0	0.0183	0.0166	0.0150	0.0130	0.0123	0.0111	0.0101	0.0091	0.0082	0.0074
5.0	0.0067	0.0061	0.0055	0.0050	0.0045	0.0041	0.0037	0.0033	0.0030	0.0027
6.0	0.0025	0.0022	0.0020	0.0018	0.0017	0.0015	0.0014	0.0012	0.0011	0.0010

Source: Adapted from S. M. Selby, ed., *CRC Standard Mathematical Tables,* 17th ed., CRC Press, Cleveland, OH, 1969, pp. 201–207.

TABLE C
Poisson distribution

1000 × probability of *r* or fewer occurrences of event that has average number of occurrences equal to *np*.

np \ *r*	0	1	2	3	4	5	6	7	8	9
0.02	980	1,000								
0.04	961	999	1,000							
0.06	942	998	1,000							
0.08	923	997	1,000							
0.10	905	995	1,000							
0.15	861	990	999	1,000						
0.20	819	982	999	1,000						
0.25	779	974	998	1,000						
0.30	741	963	996	1,000						
0.35	705	951	994	1,000						
0.40	670	938	992	999	1,000					
0.45	638	925	989	999	1,000					
0.50	607	910	986	998	1,000					
0.55	577	894	982	998	1,000					
0.60	549	878	977	997	1,000					
0.65	522	861	972	996	999	1,000				
0.70	497	844	966	994	999	1,000				
0.75	472	827	959	993	999	1,000				
0.80	449	809	953	991	999	1,000				
0.85	427	791	945	989	998	1,000				
0.90	407	772	937	987	998	1,000				
0.95	387	754	929	984	997	1,000				
1.00	368	736	920	981	996	999	1,000			
1.1	333	699	900	974	995	999	1,000			
1.2	301	663	879	966	992	998	1,000			
1.3	273	627	857	957	989	998	1,000			
1.4	247	592	833	946	986	997	999	1,000		
1.5	223	558	809	934	981	996	999	1,000		
1.6	202	525	783	921	976	994	999	1,000		
1.7	183	493	757	907	970	992	998	1,000		
1.8	165	463	731	891	964	990	997	999	1,000	
1.9	150	434	704	875	956	987	997	999	1,000	
2.0	135	406	677	857	947	983	995	999	1,000	
2.2	111	355	623	819	928	975	993	998	1,000	
2.4	091	308	570	779	904	964	988	997	999	1,000
2.6	014	267	518	736	877	951	983	995	999	1,000
2.8	061	231	469	692	848	935	976	992	998	999
3.0	050	199	423	647	815	916	966	988	996	999
3.2	041	171	380	603	781	895	955	983	994	998
3.4	033	147	340	558	744	871	942	977	992	997
3.6	027	126	303	515	706	844	927	969	988	996
3.8	022	107	269	473	668	816	909	960	984	994
4.0	018	092	238	433	629	785	889	949	979	992

TABLE C (*continued*)

np \ r	**0**	**1**	**2**	**3**	**4**	**5**	**6**	**7**	**8**	**9**
4.2	015	078	210	395	590	753	867	936	972	989
4.4	012	066	185	359	551	720	844	921	964	985
4.6	010	056	163	326	513	686	818	905	955	980
4.8	008	048	143	294	476	651	791	887	944	975
5.0	007	040	125	265	440	616	762	867	932	968
5.2	006	034	109	238	406	581	732	845	918	960
5.4	005	029	095	213	373	546	702	822	903	951
5.6	004	024	082	191	342	512	670	797	886	941
5.8	003	021	072	170	313	478	638	771	867	929
6.0	002	017	062	151	285	446	606	744	847	916
10	**11**	**12**	**13**	**14**	**15**	**16**				
2.8	1,000									
3.0	1,000									
3.2	1,000									
3.4	999	1,000								
3.6	999	1,000								
3.8	998	999	1,000							
4.0	997	999	1,000							
4.2	996	999	1,000							
4.4	994	998	999	1,000						
4.6	992	997	999	1,000						
4.8	990	996	999	1,000						
5.0	986	995	998	999	1,000					
5.2	982	993	997	999	1,000					
5.4	977	990	996	999	1,000					
5.6	972	988	995	998	999	1,000				
5.8	965	984	993	997	999	1,000				
6.0	957	980	991	996	999	999	1,000			
6.2	002	015	054	134	259	414	574	716	826	902
6.4	002	012	046	119	235	384	542	687	803	886
6.6	001	010	040	105	213	355	511	658	780	869
6.8	001	009	034	093	192	327	480	628	755	850
7.0	001	007	030	082	173	301	450	599	729	830
7.2	001	006	025	072	156	276	420	569	703	810
7.4	001	005	022	063	140	253	392	539	676	788
7.6	001	004	019	055	125	231	365	510	648	765
7.8	000	004	016	048	112	210	338	481	620	741
8.0	000	003	014	042	100	191	313	453	593	717
8.5	000	002	009	030	074	150	256	386	523	653
9.0	000	001	006	021	055	116	207	324	456	587
9.5	000	001	004	015	040	089	165	269	392	522
10.0	000	000	003	010	029	067	130	220	333	458

(*continued*)

TABLE C (*continued*)

np \ *r*	**10**	**11**	**12**	**13**	**14**	**15**	**16**	**17**	**18**	**19**
6.2	949	975	989	995	998	999	1,000			
6.4	939	969	986	994	997	999	1,000			
6.6	927	963	982	992	997	999	999	1,000		
6.8	915	955	978	990	996	998	999	1,000		
7.0	901	947	973	987	994	998	999	1,000		
7.2	887	937	967	984	993	997	999	999	1,000	
7.4	871	926	961	980	991	996	998	999	1,000	
7.6	854	915	954	976	989	995	998	999	1,000	
7.8	835	902	945	971	986	993	997	999	1,000	
8.0	816	888	936	966	983	992	996	998	999	1,000
8.5	763	849	909	949	973	986	993	997	999	999
9.0	706	803	876	926	959	978	989	995	998	999
9.5	645	752	836	898	940	967	982	991	996	998
10.0	583	697	792	864	917	951	973	986	993	997

	20	**21**	**22**
8.5	1,000		
9.0	1,000		
9.5	999	1,000	
10.0	998	999	1,000

Source: Adapted with permission from E. L. Grant and Richard S. Leavenworth, *Statistical Quality Control,* 4th ed., McGraw-Hill, New York, 1972.

TABLE D
Distribution of t

Value of t corresponding to certain selected probabilities (i.e., tail areas under the curve). To illustrate: the probability is 0.975 that a sample with 20 degrees of freedom would have $t = +2.086$ or smaller.

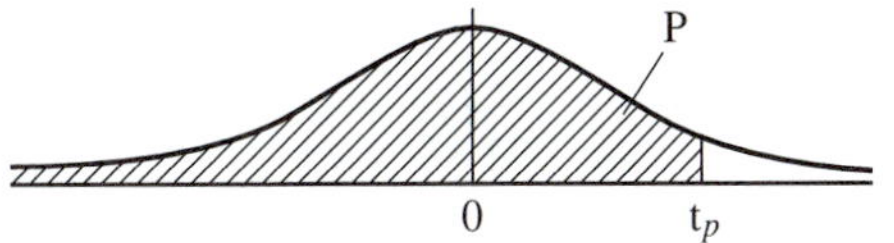

DF	$t_{.60}$	$t_{.70}$	$t_{.80}$	$t_{.90}$	$t_{.95}$	$t_{.975}$	$t_{.99}$	$t_{.995}$
1	0.325	0.727	1.376	3.078	6.314	12.706	31.821	63.657
2	0.289	0.617	1.061	1.886	2.920	4.303	6.965	9.925
3	0.277	0.584	0.978	1.638	2.353	3.182	4.541	5.841
4	0.271	0.569	0.941	1.533	2.132	2.776	3.747	4.604
5	0.267	0.559	0.920	1,476	2.015	2.571	3.365	4.032
6	0.265	0.553	0.906	1.440	1.943	2.447	3.143	3.707
7	0.263	0.549	0.896	1.415	1.895	2.365	2.998	3.499
8	0.262	0.546	0.889	1.397	1.860	2.306	2.896	3.355
9	0.261	0.543	0.883	1.383	1.833	2.262	2.821	3.250
10	0.260	0.542	0.879	1.372	1.812	2.228	2.764	3.169
11	0.260	0.540	0.876	1.363	1.796	2.201	2.718	3.106
12	0.259	0.539	0.873	1.356	1.782	2.179	2.681	3.055
13	0.259	0.538	0.870	1.350	1.771	2.160	2.650	3.012
14	0.258	0.537	0.868	1.345	1.761	2.145	2.624	2.977
15	0.258	0.536	0.866	1.341	1.753	2.131	2.602	2.947
16	0.258	0.535	0.865	1.337	1.746	2.120	2.583	2.921
17	0.257	0.534	0.863	1.333	1.740	2.110	2.567	2.898
18	0.257	0.534	0.862	1.330	1.734	2.101	2.552	2.878
19	0.257	0.533	0.861	1.328	1.729	2.093	2.539	2.861
20	0.257	0.5.33	0.860	1.325	1.725	2.086	2.528	2.845
21	0.257	0.532	0.859	1.323	1.721	2.080	2.518	2.831
22	0.256	0.532	0.858	1.321	1.717	2.074	2.508	2.819
23	0.256	0.532	0.858	1.319	1.714	2.069	2.500	2.807
24	0.256	0.531	0.857	1.318	1.711	2.064	2.492	2.797
25	0.256	0.531	0.856	1.316	1.708	2.060	2.485	2.787
26	0.256	0.531	0.856	1.315	1.706	2.056	2.479	2.779
27	0.256	0.531	0.855	1.314	1.703	2.052	2.473	2.771
28	0.256	0.530	0.855	1.313	1.701	2.048	2.467	2.763
29	0.256	0.530	0.854	1.311	1.699	2.045	2.462	2.756
30	0.256	0.530	0.854	1.310	1.697	2.042	2.457	2.750
40	0.255	0.529	0.851	1.303	1.684	2.021	2.423	2.704
60	0.254	0.527	0.848	1.296	1.671	2.000	2.390	2.660
120	0.254	0.526	0.845	1.289	1.658	1.980	2.358	2.617
∞	0.253	0.524	0.842	1.282	1.645	1.960	2.326	2.576

Source: Adapted with permission from W. J. Dixon and F. J. Massey Jr., *Introduction to Statistical Analysis,* 3rd ed., McGraw-Hill, New York, copyright © 1969. Entries originally from Table III of R. A. Fisher and F. Yates, *Statistical Tables,* Oliver & Boyd Ltd., London.

TABLE E
Distribution of χ^2

Values of χ^2 corresponding to certain selected probabilities (i.e., tail areas under the curve). To illustrate: the probability is 0.95 that a sample with 20 degrees of freedom, taken from a normal distribution, would have $\chi^2 = 31.41$ or smaller.

P

χ^2_P

Values of χ^2_P corresponding to P

DF	$\chi^2_{.005}$	$\chi^2_{.01}$	$\chi^2_{.025}$	$\chi^2_{.05}$	$\chi^2_{.10}$	$\chi^2_{.90}$	$\chi^2_{.95}$	$\chi^2_{.975}$	$\chi^2_{.99}$	$\chi^2_{.995}$
1	0.000039	0.00016	0.00098	0.0039	0.0158	2.71	3.84	5.02	6.63	7.88
2	0.0100	0.0201	0.0506	0.1026	0.2107	4.61	5.99	7.38	9.21	10.60
3	0.0717	0.115	0.216	0.352	0.584	6.25	7.81	9.35	11.34	12.84
4	0.207	0.297	0.484	0.711	1.064	7.78	9.49	11.14	13.28	14.86
5	0.412	0.554	0.831	1.15	1.61	9.24	11.07	12.83	15.09	16.75
6	0.676	0.872	1.24	1.64	2.20	10.64	12.59	14.45	16.81	18.55
7	0.989	1.24	1.69	2.17	2.83	12.02	14.07	16.01	18.48	20.28
8	1.34	1.65	2.18	2.73	3.49	13.36	15.51	17.53	20.09	21.96
9	1.73	2.09	2.70	3.33	4.17	14.68	16.92	19.02	21.67	23.59
10	2.16	2.56	3.25	3.94	4.87	15.99	18.31	20.48	23.21	25.19
11	2.60	3.05	3.82	4.57	5.58	17.28	19.68	21.92	24.73	26.76
12	3.07	3.57	4.40	5.23	6.30	18.55	21.03	23.34	26.22	28.30
13	3.57	4.11	5.01	5.89	7.04	19.81	22.36	24.74	27.69	29.82
14	4.07	4.66	5.63	6.57	7.79	21.06	23.68	26.12	29.14	31.32
15	4.60	5.23	6.26	7.26	8.55	22.31	25.00	27.49	30.58	32.80
16	5.14	5.81	6.91	7.96	9.31	23.54	26.30	28.85	32.00	34.27
18	6.26	7.01	8.23	9.39	10.86	25.99	28.87	31.53	34.81	37.16
20	7.43	8.26	9.59	10.85	12.44	28.41	31.41	34.17	37.57	40.00
24	9.89	10.86	12.40	13.85	15.66	33.20	36.42	39.36	42.98	45.56
30	13.79	14.95	16.79	18.49	20.60	40.26	43.77	46.98	50.89	53.67
40	20.71	22.16	24.43	26.51	29.05	51.81	55.76	59.34	63.69	66.77
60	35.53	37.48	40.48	43.19	46.46	74.40	79.08	83.30	88.38	91.95
120	83.85	86.92	91.58	95.70	100.62	140.23	146.57	152.21	158.95	163.64

Source: Adapted with permission from W. J. Dixon and F. J. Massey Jr., *Introduction to Statistical Analysis,* 3rd ed., McGraw-Hill, New York, copyright © 1969.

TABLE F
Ninety-five percent confidence belts for population proportion

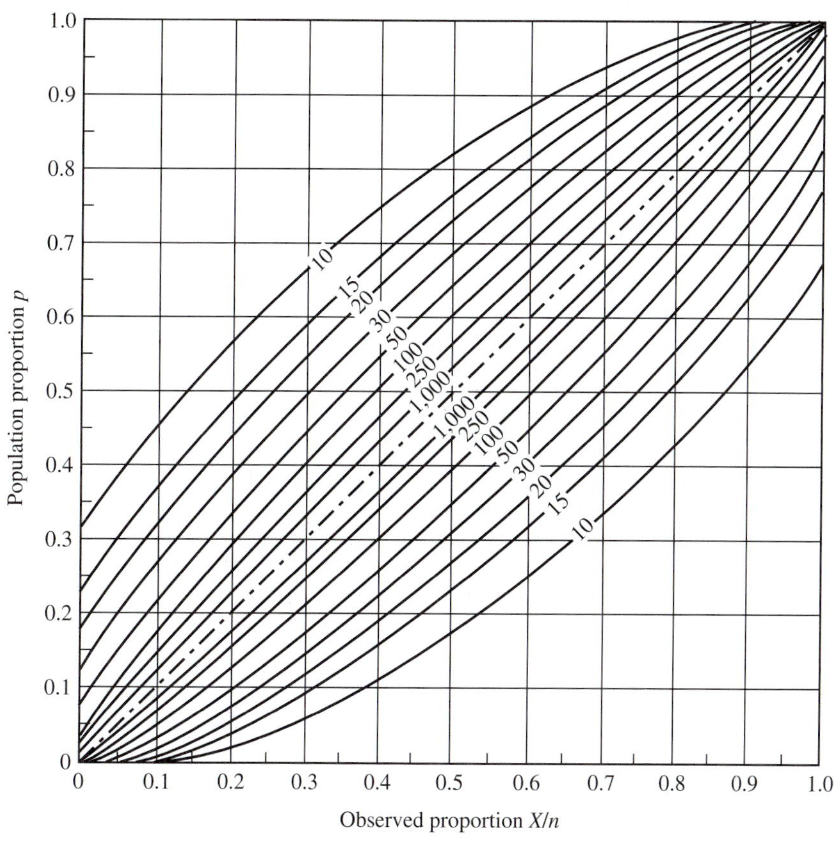

Example In a sample of 10 items, 8 were defective ($X/n = 8/10$). The 95 percent confidence limits on the population proportion defective are read from the two curves (for $n = 10$) as 0.43 and 0.98.

Source: C. Eisenhart, M. W Hastay, and W. A. Wallis, *Selected Techniques of Statistical Analysis—OSRD,* McGraw-Hill, New York, 1947.

TABLE G
Distribution of F

Values of F corresponding to certain selected probabilities (i.e., tail areas under the curve). To illustrate: the probability is 0.05 that the ratio of two sample variances obtained with 20 and 10 degrees of freedom in numerator and denominator, respectively, would have $F = 2.77$ or larger. For a two-sided test, a lower limit is found by taking the reciprocal of the tabulated F value for the degrees of freedom in reverse. For the above example, with 10 and 20 degrees of freedom in numerator and denominator respectively, F is 2.35 and $1/F$ is 1/2.35, or 0.43. The probability is 0.10 that F is 0.43 or smaller or 2.77 or larger.

n_2 \ n_1	1	2	3	4	5	6	7	8	9
				$F_{.95}(n_1, n_2)$					
1	161.4	199.5	215.7	224.6	230.2	234.0	236.8	238.9	240.5
2	18.51	19.00	19.16	19.25	19.30	19.33	19.35	19.37	19.38
3	10.13	9.55	9.28	9.12	9.01	8.94	8.89	8.85	8.81
4	7.71	6.94	6.59	6.39	6.26	6.16	6.09	6.04	6.00
5	6.61	5.79	5.41	5.19	5.05	4.95	4.88	4.82	4.77
6	5.99	5.14	4.76	4.53	4.39	4.28	4.21	4.15	4.10
7	5.59	4.74	4.35	4.12	3.97	3.87	3.79	3.73	3.68
8	5.32	4.46	4.07	3.84	3.69	3.58	3.50	3.44	3.39
9	5.12	4.26	3.86	3.63	3.48	3.37	3.29	3.23	3.18
10	4.96	4.10	3.71	3.48	3.33	3.22	3.14	3.07	3.02
11	4.84	3.98	3.59	3.36	3.20	3.09	3.01	2.95	2.90
12	4.75	3.89	3.49	3.26	3.11	3.00	2.91	2.85	2.80
13	4.67	3.81	3.41	3.18	3.03	2.92	2.83	2.77	2.71
14	4.60	3.74	3.34	3.11	2.96	2.85	2.76	2.70	2.65
15	4.54	3.68	3.29	3.06	2.90	2.79	2.71	2.64	2.59
16	4.49	3.63	3.24	3.01	2.85	2.74	2.66	2.59	2.54
17	4.45	3.59	3.20	2.96	2.81	2.70	2.61	2.55	2.49
18	4.41	3.55	3.16	2.93	2.77	2.66	2.58	2.51	2.46
19	4.38	3.52	3.13	2.90	2.74	2.63	2.54	2.48	2.42
20	4.35	3.49	3.10	2.87	2.71	2.60	2.51	2.45	2.39
21	4.32	3.47	3.07	2.84	2.68	2.57	2.49	2.42	2.37
22	4.30	3.44	3.05	2.82	2.66	2.55	2.46	2.40	2.34
23	4.28	3.42	3.03	2.80	2.64	2.53	2.44	2.37	2.32
24	4.26	3.40	3.01	2.78	2.62	2.51	2.42	2.36	2.30
25	4.24	3.39	2.99	2.76	2.60	2.49	2.40	2.34	2.28
26	4.23	3.37	2.98	2.74	2.59	2.47	2.39	2.32	2.27
27	4.21	3.35	2.96	2.73	2.57	2.46	2.37	2.31	2.25
28	4.20	3.34	2.95	2.71	2.56	2.45	2.36	2.29	2.24
29	4.18	3.33	2.93	2.70	2.55	2.43	2.35	2.28	2.22
30	4.17	3.32	2.92	2.69	2.53	2.42	2.33	2.27	2.21
40	4.08	3.23	2.84	2.61	2.45	2.34	2.25	2.18	2.12
60	4.00	3.15	2.76	2.53	2.37	2.25	2.17	2.10	2.04
120	3.92	3.07	2.68	2.45	2.29	2.17	2.09	2.02	1.96
∞	3.84	3.00	2.60	2.37	2.21	2.10	2.01	1.94	1.88

Note: n_1 = degrees of freedom for numerator; n_2 = degrees of freedom for denominator.

Source: Adapted with permission from E. S. Pearson and H. O. Hartley, eds., *Biometrika Tables for Statisticians,* 2nd ed., vol. I, Cambridge University Press, New York, 1958.

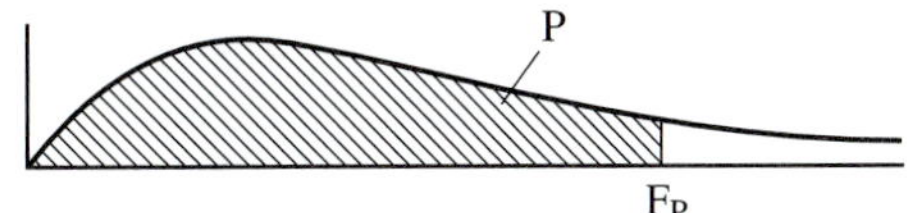

10	12	15	20	24	30	40	60	120	∞
				$F_{.95}(n_1, n_2)$					
241.9	243.9	245.9	248.0	249.1	250.1	251.1	252.2	253.3	254.3
19.40	19.41	19.43	19.45	19.45	19.46	19.47	19.48	19.49	19.50
8.79	8.74	8.70	8.66	8.64	8.62	8.59	8.57	8.55	8.53
5.96	5.91	5.86	5.80	5.77	5.75	5.72	5.69	5.66	5.63
4.74	4.68	4.62	4.56	4.53	4.50	4.46	4.43	4.40	4.36
4.06	4.00	3.94	3.87	3.84	3.81	3.77	3.74	3.70	3.67
3.64	3.57	3.51	3.44	3.41	3.38	3.34	3.30	3.27	3.23
3.35	3.28	3.22	3.15	3.12	3.08	3.04	3.01	2.97	2.93
3.14	3.07	3.01	2.94	2.90	2.86	2.83	2.79	2.75	2.71
2.98	2.91	2.85	2.77	2.74	2.70	2.66	2.62	2.58	2.54
2.85	2.79	2.72	2.65	2.61	2.57	2.53	2.49	2.45	2.40
2.75	2.69	2.62	2.54	2.51	2.47	2 43	2.38	2.34	2.30
2.67	2.60	2.53	2.46	2.42	2.38	2.34	2.30	2.25	2.21
2.60	2.53	2.46	2.39	2.35	2.31	2.27	2.22	2.18	2.13
2.54	2.48	2.40	2.33	2.29	2.25	2.20	2.16	2.11	2.07
2.49	2.42	2.35	2.28	2.24	2.19	2.15	2.11	2.06	2.01
2.45	2.38	2.31	2.23	2.19	2.15	2.10	2.06	2.01	1.96
2.41	2.34	2.27	2.19	2.15	2.11	2.06	2.02	1.97	1.92
2.38	2.31	2.23	2.16	2.11	2.07	2.03	1.98	1.93	1.88
2.35	2.28	2.20	2.12	2.08	2.04	1.99	1.95	1.90	1.84
2.32	2.25	2.18	2.10	2.05	2.01	1.96	1.92	1.87	1.81
2.30	2.23	2.15	2.07	2.03	1.98	1.94	1.89	1.84	1.78
2.27	2.20	2.13	2.05	2.01	1.96	1.91	1.86	1.81	1.76
2.25	2.18	2.11	2.03	1.98	1.94	1.89	1.84	1.79	1.73
2.24	2.16	2.09	2.01	1.96	1.92	1.87	1.82	1.77	1.71
2.22	2.15	2.07	1.99	1.95	1.90	1.85	1.80	1.75	1.69
2.20	2.13	2.06	1.97	1.93	1.88	1.84	1.79	1.73	1.67
2.19	2.12	2.04	1.96	1.91	1.87	1.82	1.77	1.71	1.65
2.18	2.10	2.03	1.94	1.90	1.85	1.81	1.75	1.70	1.64
2.16	2.09	2.01	1.93	1.89	1.84	1.79	1.74	1.68	1.62
2.08	2.00	1.92	1.84	1.79	1.74	1.69	1.64	1.58	1.51
1.99	1.92	1.84	1.75	1.70	1.65	1.59	1.53	1.47	1.39
1.91	1.83	1.75	1.66	1.61	1.55	1.50	1.43	1.35	1.25
1.83	1.75	1.67	1.57	1.52	1.46	1.39	1.32	1.22	1.00

TABLE H
Tolerance factors for normal distributions (two sided)

P	γ = 0.75					γ = 0.90				
N	**0.75**	**0.90**	**0.95**	**0.99**	**0.999**	**0.75**	**0.90**	**0.95**	**0.99**	**0.999**
2	4.498	6.301	7.414	9.531	11.920	11.407	15.978	18.800	24.167	30.227
3	2.501	3.538	4.187	5.431	6.844	4.132	5.847	6.919	8.974	11.309
4	2.035	2.892	3.431	4.471	5.657	2.932	4.166	4.943	6.440	8.149
5	1.825	2.599	3.088	4.033	5.117	2.4.54	3.494	4.152	5.423	6.879
6	1.704	2.429	2.889	3.779	4.802	2.196	3.131	3.723	4.870	6.188
7	1.624	2.318	2.757	3.611	4.593	2.034	2.902	3.452	4.521	5.750
8	1.568	2.238	2.663	3.491	4.444	1.921	2.743	3.264	4.278	5.446
9	1.525	2.178	2.593	3.400	4.330	1.839	2.626	3.125	4.098	5.220
10	1.492	2.131	2.537	3.328	4.241	1.775	2.535	3.018	3.959	5.046
11	1.465	2.093	2.493	3.271	4.169	1.724	2.463	2.933	3.849	4.906
12	1.443	2.062	2.456	3.223	4.110	1.683	2.404	2.863	3.758	4.792
13	1.425	2.036	2.424	3.183	4.059	1.648	2.355	2.805	3.682	4.697
14	1.409	2.013	2.398	3.148	4.016	1.619	2.314	2.756	3.618	4.615
15	1.395	1.994	2.375	3.118	3.979	1.594	2.278	2.713	3.562	4.545
16	1.383	1.977	2.355	3.092	3.946	1.572	2.246	2.676	3.514	4.484
17	1.372	1.962	2.337	3.069	3.917	1.552	2.219	2.643	3.471	4.430
18	1.363	1.948	2.321	3.048	3.891	1.535	2.194	2.614	3.433	4.382
19	1.355	1.936	2.307	3.030	3.867	1.520	2.172	2.588	3.399	4.339
20	1.347	1.925	2.294	3.013	3.846	1.506	2.152	2.564	3.368	4.300
21	1.340	1.915	2.282	2.998	3.827	1.493	2.135	2.543	3.340	4.264
22	1.334	1.906	2.271	2.984	3.809	1.482	2.118	2.524	3.315	4.232
23	1.328	1.898	2.261	2.971	3.793	1.471	2.103	2.506	3.292	4.203
24	1.322	1.891	2.252	2.950	3.778	1.462	2.089	2.480	3.270	4.176
25	1.317	1.883	2.244	2.948	3.764	1.453	2.077	2.474	3.251	4.151
26	1.313	1.877	2.236	2.938	3.751	1.444	2.065	2.460	3.232	4.127
27	1.309	1.871	2.229	2.929	3.740	1.437	2.054	2.447	3.215	4.106
30	1.297	1.855	2.210	2.904	3.708	1.417	2.025	2.413	3.170	4.049
35	1.283	1.834	2.185	2.871	3.667	1.390	1.988	2.368	3.112	3.974
40	1.271	1.818	2.166	2.846	3.635	1.370	1.959	2.334	3.066	3.917
100	1.218	1.742	2.075	2.727	3.484	1.275	1.822	2.172	2.854	3.646
500	1.177	1.683	2.006	2.636	3.368	1.201	1.717	2.046	2.689	3.434
1,000	1.169	1.671	1.992	2.617	3.344	1.185	1.695	2.019	2.654	3.390
∞	1.150	1.645	1.960	2.576	3.291	1.150	1.645	1.960	2.576	3.291

Source: From C. Eisenhart, M. W. Hastay, and W. A. Wallis, *Selected Techniques of Statistical Analysis,* McGraw-Hill, New York, 1947. Used by permission.

γ = confidence level
P = percentage of population within tolerance limits
N = number of values in sample

$\gamma = 0.95$					$\gamma = 0.99$				
0.75	**0.90**	**0.95**	**0.99**	**0.999**	**0.75**	**0.90**	**0.95**	**0.99**	**0.999**
22.858	32.019	37.674	48.430	60.573	114.363	160.193	188.491	242.300	303.054
5.922	8.380	9.916	12.861	16.208	13.378	18.930	22.401	29.055	36.616
3.779	5.369	6.370	8.299	10.502	6.614	9.398	11.150	14.527	18.383
3.002	4.275	5.079	6.634	8.415	4.643	6.612	7.855	10.260	13.015
2.604	3.712	4.414	5.775	7.337	3.743	5.337	6.345	8.301	10.548
2.361	3.369	4.007	5.248	6.676	3.233	4.613	5.488	7.187	9.142
2.197	3.136	3.732	4.891	6.226	2.905	4.147	4.936	6.468	8.234
2.078	2.967	3.532	4.631	5.899	2.677	3.822	4.550	5.966	7.600
1.987	2.839	3.379	4.433	5.649	2.508	3.582	4.265	5.594	7.129
1.916	2.737	3.259	4.277	5.452	2.378	3.397	4.045	5.308	6.766
1.858	2.655	3.162	4.150	5.291	2.274	3.250	3.870	5.079	6.477
1.810	2.587	3.081	4.044	5.158	2.190	3.130	3.727	4.893	6.240
1.770	2.529	3.012	3.955	5.045	2.120	3.029	3.608	4.737	6.043
1.735	2.480	2.954	3.878	4.949	2.060	2.945	3.507	4.605	5.876
1.705	2.437	2.903	3.812	4.865	2.009	2.872	3.421	4.492	5.732
1.679	2.400	2.858	3.754	4.791	1.965	2.808	3.345	4.393	5.607
1.655	2.366	2.819	3.702	4.725	1.926	2.753	3.279	4.307	5.497
1.635	2.337	2.784	3.656	4.667	1.891	2.703	3.221	4.230	5.399
1.616	2.310	2.752	3.615	4.614	1.860	2.659	3.168	4.161	5.312
1.599	2.286	2.723	3.577	4.567	1.833	2.620	3.121	4.100	5.234
1.584	2.264	2.697	3.543	4.523	1.808	2.584	3.078	4.044	5.163
1.570	2.244	2.673	3.512	4.484	1.795	2.551	3.040	3.993	5.098
1.557	2.225	2.651	3.483	4.447	1.764	2.522	3.004	3.947	5.039
1.545	2.208	2.631	3.457	4.413	1.745	2.494	2.972	3.904	4.985
1.534	2.193	2.612	3.432	4.382	1.727	2.460	2.941	3.865	4.935
1.523	2.178	2.595	3.409	4.353	1.711	2.446	2.914	3.828	4.888
1.497	2.140	2.549	3.350	4.278	1.668	2.385	2.841	3.733	4.768
1.462	2.090	2.490	3.272	4.179	1.613	2.306	2.748	3.611	4.611
1.435	2.052	2.445	3.213	4.104	1.571	2.247	2.677	3.518	4.493
1.311	1.874	2.233	2.934	3.748	1.383	1.977	2.355	3.096	3.954
1.215	1.737	2.070	2.721	3.475	1.243	1.777	2.117	2.783	3.555
1.195	1.709	2.036	2.676	3.418	1.214	1.736	2.068	2.718	3.472
1.150	1.645	1.960	2.576	3.291	1.150	1.645	1.960	2.576	3.291

TABLE I

Factors for $\bar{X}$ and R control charts;* factors for estimating s from R†

$$\begin{cases} \text{Upper control limit for } \bar{X} = UCL_{\bar{X}} = \bar{\bar{X}} + A_2\bar{R} \\ \text{Lower control limit for } \bar{X} = LCL_{\bar{X}} = \bar{\bar{X}} - A_2\bar{R} \end{cases}$$

$$\begin{cases} \text{Upper control limit for } R = UCL_R = D_4\bar{R} \\ \text{Lower control limit for } R = LCL_R = D_3\bar{R} \end{cases}$$

$$s = \bar{R}/d_2$$

Number of observations in sample	A_2	D_3	D_4	Factor for estimate from $\bar{R}$: $d_2 = \bar{R}/s$
2	1.880	0	3.268	1.128
3	1.023	0	2.574	1.693
4	0.729	0	2.282	2.059
5	0.577	0	2.114	2.326
6	0.483	0	2.004	2.534
7	0.419	0.076	1.924	2.704
8	0.373	0.136	1.864	2.847
9	0.337	0.184	1.816	2.970
10	0.308	0.223	1.777	3.078
11	0.285	0.256	1.744	3.173
12	0.266	0.284	1.717	3.258
13	0.249	0.308	1.692	3.336
14	0.235	0.329	1.671	3.407
15	0.223	0.348	1.652	3.472

*Factors reproduced from *1950 ASTM Manual on Quality Control of Materials* by permission of the American Society for Testing and Materials, Philadelphia. All factors in Table I are based on a normal distribution.

†Reproduced by permission from *ASTM Manual on Presentation of Data,* American Society for Testing and Materials, Philadelphia, 1945.

TABLE J
Weibull paper

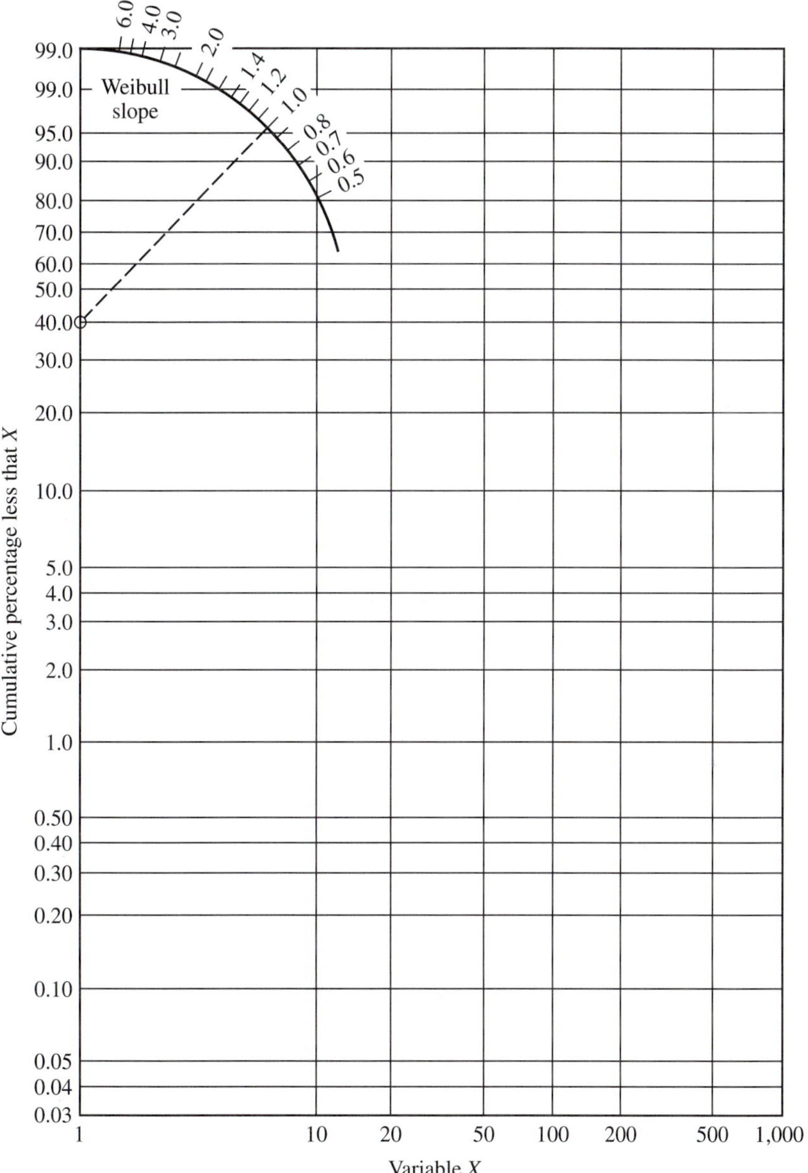

NAME INDEX

SUBJECT INDEX